Springer Series on Atomic, Optical, and Plasma Physics

Volume 113

The Springer Series on Atomic, Optical, and Plasma Physics covers in a comprehensive manner theory and experiment in the entire field of atoms and molecules and their interaction with electromagnetic radiation. Books in the series provide a rich source of new ideas and techniques with wide applications in fields such as chemistry, materials science, astrophysics, surface science, plasma technology, advanced optics, aeronomy, and engineering. Laser physics is a particular connecting theme that has provided much of the continuing impetus for new developments in the field, such as quantum computation and Bose-Einstein condensation. The purpose of the series is to cover the gap between standard undergraduate textbooks and the research literature with emphasis on the fundamental ideas, methods, techniques, and results in the field.

More information about this series at http://www.springer.com/series/411

Isak Beilis

Plasma and Spot Phenomena in Electrical Arcs

Volume 1

Isak Beilis
Department of Electrical Engineering
Tel Aviv University
Tel Aviv, Israel

ISSN 1615-5653 ISSN 2197-6791 (electronic)
Springer Series on Atomic, Optical, and Plasma Physics
ISBN 978-3-030-44746-5 ISBN 978-3-030-44747-2 (eBook)
https://doi.org/10.1007/978-3-030-44747-2

This Springer imprint is published by the registered company Springer Nature Switzerland AG
The registered company address is: Gewerbestrasse 11, 6330 Cham, Switzerland

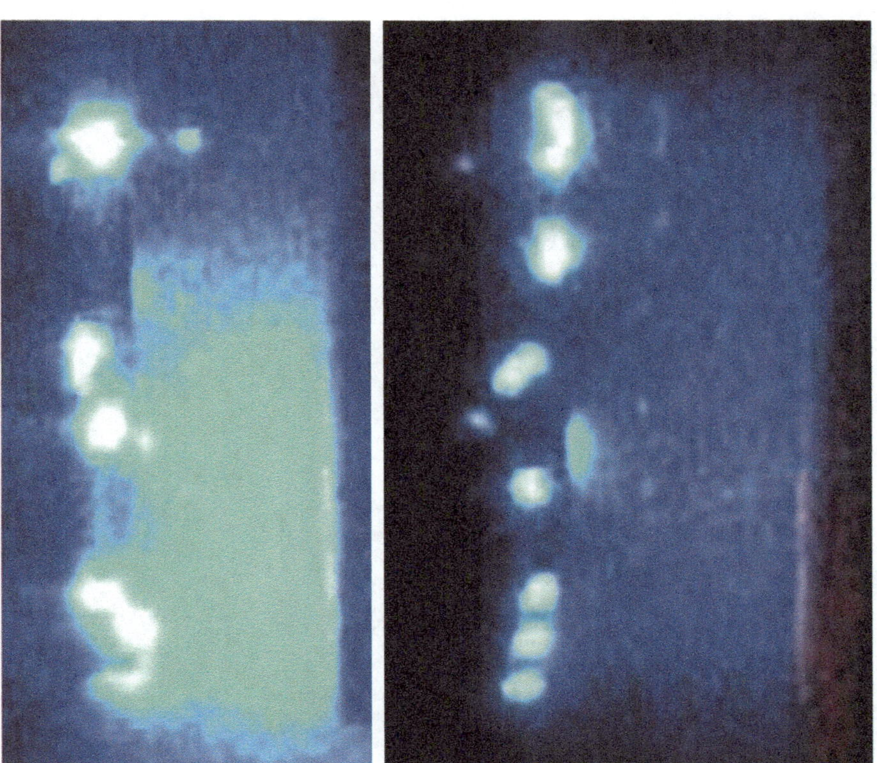

Photographs illustrating momentary distribution of the cathode spots and presence of an inter-electrode (gap 10 mm) plasma, which is detected after 45 s (left) and 70 s (right) of 200-A vacuum arc ignition with Zn cathode and Mo anode. In the left image, the glass of the observation window is weakly deposited and the plasma is observed better, whereas in the right image the window is strongly deposited and the plasma is more shielded

My special dedication of this book is to the blessed memory of my parents, Ioil and Sima, and my brother, Gregory

Preface

...and you will hear it, and investigate thoroughly,
and behold, the matter will be found true

(Deuteronomy—Chap. 17)

The electrical arc is commonly defined as an electrical discharge with relatively high current density and low gap voltage. Electrical arcs occur in vacuum, as well as in low- and high-pressure gases. An important issue is the origin of a conductive medium in the electrode gap. In vacuum, the conductive medium is generated by vaporization of electrode material during arcing. This arc may also be called an electrode vapor discharge. Even in most cases where a gas is present, electrode vaporization is the plasma source process in the near-electrode region. Many phenomena in these arcs are similar to those in vacuum arcs.

The electrical arc was first reported in the nineteenth century after the observation of a continuous arc between carbon electrodes while measuring carbon resistance by Petrov in 1802 in Petersburg and by Humphry Davy in 1808 in London, while experimenting with pulsed discharges. After some applications, preliminary investigations of the arc were conducted even in nineteenth century and systematic research was conducted from the beginning of twentieth century, firstly with graphite and then with metallic electrodes.

Different phenomena were observed in the arcs. Strongly constricted and highly mobile luminous forms appeared near the electrodes, later called cathode and anode "spots." In these regions, the electrical conductivity transits from very high in the metal to low in the plasma. Curious phenomena were observed, including fragmentation of the plasma in at the arc roots, plasma jet formation, and retrograde arc spot motion in a transverse magnetic field. Retrograde cathode spot motion in magnetic fields, supersonic plasma jet flow, extremely high electron emission density at the cathode, and the high current density in the arc root plasma could not be explained using earlier existing physical models. This was a consequence of the then limited knowledge of the plasma processes, leading to inadequate assumptions for the calculated models. Therefore, the behavior of the arc cathode region was considered mysterious for the long even though it was intensively studied. With time, experimental and theoretical investigations contributed knowledge that improved our understanding of some of these phenomena.

Improved diagnostic techniques were applied to the near-electrode region provided new information, but also required further explanations. Numerous investigators contributed experimental and theoretical knowledge. In the second half of nineteenth century, electrode repulsion, plasma fluxes, and the arc force on the electrodes were detected and studied by Dewar, Schuster, Hemsalech, Duffield Tyndall, and Sellerio. Langmuir, Tonks, and Compton developed the theory of the space charge sheath, and volt–current and plasma characteristics in the arcs in the first decades of twentieth century. Stark, and then Tanberg, Cobine, Gallagher, Slepian, Finkelnburg, Maecker, Gunterschulze, Holm, Smith, Froom, Reece, Ecker, Lee, Greenwood, Robson, Engel, Kesaev, Plyutto, Hermoch, Davies, Miller, Rakhovsky, Kimblin, Daalder, Juttner, Hantsche, Anders, Goldsmith, Boxman, the Mesyats group, the Shenli Jia group and many others significantly contributed to different aspects of the electrode and plasma characteristics of electrical arcs. The understanding of observed arc behavior, its characteristics in different regions of the electrode gap expanded with the progress of plasma science, especially after 1960 in low-temperature plasma, plasma fusion, and astrophysics.

This book consists of four main parts that describe progress in study of cathodic arc phenomena. It began with the early separate hypotheses of charged particle generation in the cathode region up to the present time including our own investigations extending over more than 50 years. The first introductory part contains five chapters and characterizes the basic plasma and electrode processes needed to study cathodic arc. Plasma concepts and particle interactions in the plasma volume and at the surface are presented in Chap. 1. Chapter 2 presents the kinetics of cathode vaporization and mechanisms of electron emissions. The mathematical formulation of heat conduction for different cathode geometries subjected to local moving heat sources are considered in Chap. 3. Chapter 4 presents the transport equations and mathematical formulation of particle diffusion, and mass and heat transfer phenomena in a multi-component plasma. The plasma–wall transition including cathode sheath formation is described in Chap. 5.

Part II presents a review of experimental studies of electrode phenomena in vacuum arcs. Chapter 6 describes the electrical breakdown of a vacuum gap and the methods of vacuum arc ignition. Chapter 7 reviews cathode spot dynamics, different spot types, and the problem of current density in the cathode spot. The spot types are described by their physical characteristics as determined experimentally in contrast to the previous confused numerical classification. The total eroded cathode mass loss is considered in Chap. 8, and cathode erosion in the form of macroparticles is considered in Chap. 9. Chapter 10 presents data concerning electrode energy losses and the effective voltage measured during the arc operation. The repulsive effect in the arc gap discovered in the second half of nineteenth century as a plasma flow reaction and the further measurements of the electrode force phenomena in vacuum arcs is reviewed in Chap. 11. Chapter 12 describes experimental investigation of the cathode jets, including jet velocity and ion current. Chapter 13 presents the state of the art of observations of spot motion in transverse and oblique magnetic fields. Current continuity in the anode region and the conditions for anode spot formation are described in Chap. 14.

Theory of the cathodic arc is presented in Part III. The evolution of cathode spot theory starting from the beginning of the twentieth century up to present time is described in Chap. 15. The progress and weakness of the developed theories are analyzed. The reason why the cathode spot was considered mysterious is discussed. The key issue is electrical current continuity from the metallic cathode to the plasma. In order to explain continuity, charge-exchange ion–atom interactions were examined, and it was shown that the ion current to the cathode is determined by the collisions in the dense near-cathode plasma. Consequently, a gasdynamic theory was advanced (Chap. 16), which consistently considers the cathode and multi-component plasma processes and determines the cathode spot parameters using a mathematically closed formulation. This approach describes the parameters and current continuity at cathode plasma region as functions of the cathode potential drop using measured cathode erosion rate and current per spot.

The cathode spot parameters for materials with a wide range of thermophysical properties were investigated. This study includes the cathode spot characteristics for three groups of materials. (1) Cathode materials with intermediate thermal properties (e.g., Cu, Al, Ni, and Ti). In cathodic arcs with these materials, the electron emission flux is comparable to the evaporated atom flux for a wide region of cathode temperatures. (2) Cathodes with refractory materials are characterized by larger flux of emitted electrons than the flux of evaporated atoms over a wide range of cathode temperatures. A modified model using a "virtual plasma cathode" allows obtaining an electron emission current density lower than the saturated density. (3) Volatile cathode material (e.g., Hg) forms an opposite case, which is characterized by an electron emission flux much lower than the vapor flux. The modified model considers an additional plasma layer adjacent to the cathode sheath, which serves as a "plasma cathode" supplying electrons for self-sustained arc operation.

A general physically closed approach is advanced considering the kinetics of cathode vaporization and gasdynamic of plasma expansion, which described as a kinetic theory in Chap. 17. The kinetic theory simultaneously considered atom and electron evaporation, taking into account conservation laws for particle density, flux, momentum, and energy in a non-equilibrium Knudsen layer. In the rarefied collisional Knudsen layer, the heavy particle return flux is formed at a length of few mean free paths from the cathode surface. A relatively high atom ionization rate is due to large energy dissipation which produced the high ion fraction in the back flux. The difference between direct evaporated atom flux and the returned flux of atom and ions determines the cathode mass loss, i.e., the cathode erosion rate G_k. It was obtained that the plasma velocity at external boundary of Knudsen layer is significantly lower than the sound speed. This very important result indicated that cathode spots operate self-sufficiently only when plasma velocity is lower than the sound speed, i.e., when the plasma flow near the cathode is impeded.

The cathode potential drop u_c was determined by considering the kinetics for emitted and plasma electrons. Electron emission was considered as electron evaporation. The electron beam relaxation zone (analogous to the Knudsen layer for atoms) depends on the electron beam energy, which is increased in the cathode sheath by eu_c. The returned electron flux depends on the plasma density and by the potential barrier in the sheath (u_c). Conservation of electron momentum and electron energy and the quasineutrality condition on the boundary of Knudsen layer allow determine u_c directly. *Two principle rules* based on necessity of impeded plasma flow near the cathode and satisfaction to a positive cathode energy balance at low spot current, which determine conditions of the cathode spot ignition and spot development. Chapter 18 describes formation of the cathode plasma jet. The jet begins as a relatively slow plasma flow, initially produced due to difference of plasma parameters (plasma velocity formation) in Knudsen layer. The plasma is further accelerated up to supersonic speed, determined by the cathode dense plasma, and the energy dissipated by electron beam relaxation due to gasdynamic mechanisms to the expansion region. The calculated plasma velocity agreed well with the measurements. The potential distribution is not monotonic, but rather has a hump with a height of the electron temperature.

Transient cathode spots were studied on the protrusions and bulk cathodes. The nanolife spot parameters, the mechanism of spot operation on film cathodes, specific spot parameters at large rate of current rise, and the nature of high voltage arc initiation were also studied.

Cathode spot motion in magnetic fields is studied in Chap. 19. Three important experimental phenomena are described: (1) cathode spot grouping and (2) cathode spot retrograde motion, i.e., in the anti-Amperian direction, when the arc runs in a transverse magnetic field, and (3) cathode spot motion in oblique magnetic fields. The retrograde spot motion model takes in account the kinetic principle rules of impeded plasma flow in the non-symmetric total pressure (gas−kinetic+magnetic) in the near-cathode plasma. The current per group spot was calculated taking into account that the plasma kinetic pressure is comparable to the self-magnetic pressure in the cathode plasma jet. The theory showed that the current per group spot and spot velocity increase linearly with the magnetic field and arc current, and these dependencies agree with experimental observations. Retrograde cathode spot motion in short electrode gaps and in atmospheric pressure as well as reversed motion in strong magnetic fields (>1 T) observed by Robson and Engel is described. The observed acute angle effect and spot splitting in oblique magnetic fields are explained. Robson's experimentally obtained dependencies of the drift angle on magnetic field and on the acute angle are described by this theory by using kinetic principle rules.

Chapter 20 presents a theoretical study of anode spots. An anode spot theory was developed using a kinetic treatment of anode evaporation and modeling the plasma flow in the plasma acceleration region. The plasma energy balance was described considering the Joule energy in the anode plasma, the energy dissipation caused by ionization of atoms, energy convection by the electric current, and the energy required for anode plasma acceleration. The anode surface temperature, current density, plasma density, plasma temperature, and plasma velocity were calculated self-consistently. The main result is a significantly lower degree of anode plasma ionization (10^{-3} to 5×10^{-2}) than in the cathode plasma, due to mobility of electrons higher than the ions, which both supports current continuity in the anode and cathode regions, respectively.

The applications of cathode spot theory are discussed in Part IV. Chapter 21 describes unipolar arcs on the metal elements in fusion devices. The results of experimental and theoretical investigation (including details of the primary work of Robson and Thonenmann) are reviewed, beginning from the first reported data. The specifics of unipolar arcs on nanostructured surfaces in tokamaks are analyzed. Our own experimental results for a tokamak wall with a thin tungsten coating on carbon demonstrate the spot motion mechanism which is similar to spots on film cathodes.

Chapter 22 presents the application of vacuum arc plasma sources for thin film deposition. Novel arc plasma sources were developed in the last decades at Tel Aviv University. Low macroparticle density in the films was reached by converting cathode droplets to plasma through their re-evaporations on a hot refractory anode surface. Arc characteristics (plasma density, temperature) and deposition rate as functions of the distances to the substrate from the arc axis, gap distance, electrode configuration, and electrode materials are discussed. A theory was developed to describe self-consistently transient anode heating and anode plasma generation. The advantages of the method are compared with traditional deposition approaches.

Another vacuum arc application is microthrusters for spacecraft. Chapter 23 presents the general problems of vacuum arc plasma thrusters, their main characteristics, and their efficiency. Phenomena in arcs with small electrode gaps are considered, and processes in very short cathodic arcs relevant to thrusters are modeled. Simulation produced thrust efficiency results. Microthruster devices operating in space in the last decade are analyzed.

Chapter 24 applies cathode spot theory to laser–metal interaction and laser plasma generation. Experiments and previously developed theories of laser–target interaction are analyzed. It is shown that the plasma generated near the target by laser and the plasma generated in the cathode region of a vacuum arc have similar characteristics. Both plasmas are very dense and appear in a minute region, known as the laser spot and the cathode spot, respectively. The plasma interactions with matter in both cases are also similar. Taking into account a detailed analysis of the features of the laser spot, a physical model based on a previously presented kinetic approach of the cathode spot is described by modifying it for the particular features of laser-generated plasma. The model considers that the net electrical current is zero in the laser spot, in contrast to the current carried in a vacuum arc cathode spot. The specifics of target vaporization, breakdown of neutral vapor, near-target electrical sheath, electron emission from the hot area of the target, plasma heating, and plasma acceleration mechanism are detailed. The mathematical description and simulation for different target materials show an unusual new result. The laser power absorbed in the plasma is converted into kinetic and potential energy of the plasma particles and is returned to the target and the target surface temperature calculated, taking into account the plasma energy flux is significantly larger than that without this flux.

Finally, Chap. 25 reports about application of cathode spot theory to arcs formed in different technical devices: low-temperature plasma generators, plasma accelerators, devices using plasma of combustion products, and rail guns that accelerate the solid bodies. Cathode spots on the electrodes of these devices are explained using spot theory for film cathodes and considering a plasma flow consisting of lightly ionized dopants. Transition from a diffuse to an arc mode was modeled, considering electrode overheating instability with an exponential dependence of the electrical conductivity on temperature. Plasma accelerator experiments are discussed: A significantly larger electron emission current density than that described by the Richardson law was detected, and the models are analyzed. The contribution of ion current is demonstrated using the gasdynamic model of the cathode spot, taking into account the experimentally observed plasma expansion from the spot. A mathematical approach described the long-time problem of plasma column constriction in atmospheric pressure arcs, without recourse to Steinbeck's minimal principle. The arc mechanism and the observed arc parameters in rail gun are explained using cathode and anode spot models. Experimental reduction of the rate of solid body acceleration can be explained by the calculated arc velocity dependence on applied electrical power, which tends to saturate when the power increases, due to increasing arc plasma mass needed to support increasing arc current.

The text includes not only the cathode current continuity mechanism usually published in the spot theories, but also various ideas to understand many different spot aspects. The studies presented here cannot address all observed arc discharge phenomena—they cannot be completely explained by the developed theories. Nevertheless, I hope that the developed models systematically lead to further research, increase our understanding of different arc spot mechanisms, and reduce the mystery surrounding them. I hope this book will be useful for researchers in the physics of low temperature plasma and gas discharges, practicing engineers and students learning electrical engineering, plasma physics, and gas discharge physics.

I gratefully acknowledge colleagues for useful discussion, help, and support at various stages of this work. First, thanks to Vadim Rakhovsky who introduced me to this subject, worked with me during my postgraduate studies, and helped me to understand the early experiments on near-electrode phenomena conducted in widely known Veniamin Granovsky's laboratory. Thanks to laboratory colleagues N. Zykova and V. Kantsel for helpful direct discussion for the experiment at first stage of my spot study. I am grateful to Gregory Lyubimov for discussions and nice joint theoretical work on the electrode phenomena based on the hydrodynamics of plasma flow. Great thanks to Igor Kesaev, who explain the specifics of his world known experiments and the difficulties in my way by choosing this confused problem.

Many thanks to the supervisors of a number of All-Union Seminars: Academician Nikolay Rykalin (heat conduction in solids caused by a moving concentrated heat source), Academician Vitaly Ginsburg, Anri Rukhadze, Yuri Raizer (general physics and physics of gas discharges), and Sergey Anisimov (kinetics of solid vaporization). I appreciate Oleg Firsov, which noticed about the charge-exchange process during the discussions and to Boris Smirnov (elementary collisions of plasma particles), Alexey Morosov and Askold Zharinov (mechanisms of plasma acceleration).

I am grateful to Academician Vladimir Fortov, which initiated and supported the doctoral work in the department and Academician Gennady Mesyats for support the presentation of a thesis of Doctor of Science. As a result, I acquired much scientific experience and broad scientific knowledge, which played an essential role in my understanding of the many aspects of cathode spot science.

Since January 1992, my investigations were continued in the Electrical Discharges and Plasma laboratory at the Faculty of Engineering of Tel Aviv University. My great thanks to Samuel Goldsmith and Ray Boxman for very important support and efficient joint work at the first stage and for the excellent lab atmosphere as my work progressed. I gratefully acknowledge N. Parkansky, V. Zitomirsky, H. Rosenthal, V. Paperny A. Shashurin, D. Arbilly, A. Nemirovsky, A. Snaiderman, D. Grach, Y. Koulik, Ben Sagi, and Y. Yankelevich for their contributions at different stages of the investigation in the laboratory. A special thanks is addressed to Ray Boxman for helpful discussion of my cathode spot research and English help at the publication stage. I am grateful to Michael Keidar for our joint work that began at Tel Aviv University, continues presently and whose discussions give me great pleasure. Many thanks also for his help during processing of the manuscript. I appreciate Burhard Juttner, Erhard Hantzsche, Michael Laux, Boyan Djakov, Heinz Pursch, Peter Siemroth, Lorenzo Torrisi, Joachim Heberlein, and Emil Pfender for their help during my visits to their respective institutions.

Finally, I would like to thank my family, my loved children for being always patient with me. Greatly I appreciate my wife Galina, supporting my way of life in the science, for her understanding and taking care of all my needs.

Tel Aviv, Israel
2020

Isak Beilis

Contents

Part I
Plasma, Particle Interactions, Mass and Heat Phenomena

Chapter 1
Base Plasma Particle Phenomena at the Surface and in Plasma Volume

The plasma phenomena are widely described in the literature focused on general plasma physics and on specific applications [1–3]. More recent book [4] described the plasma and particle collision effects that is relevant to some emerging applications in aerospace, nanotechnology, and medical applications. This Chapter presents some details of plasma characteristics and determines the elementary particle processes in plasma volume and at the plasma–surface interface, which mainly related to understand the specifics of metallic plasmas formed in low pressure or vacuum arcs.

1.1 Plasma

In order to understand the fundamentals, the main plasma characteristics will be defined briefly, and state of the art of the experimental and theoretical studies will be described.

1.1.1 Quasineutrality

The ionized gas consists of charge particles having density n_e and neutral atoms having density n_a. Such gas can be low ionized when $n_e/n_a \leq 10^{-3}$–10^{-2} or highly ionized when $n_e/n_a \geq 10^{-2}$. Each charge induces an electric field that influences the neighboring particle field, and therefore, the positive charge is always surrounded by the negative charges. Therefore, an electric field can be localized at some radius r_D supporting the charges shielding and gas neutrality. Let us define r_D using Poisson equation for potential φ distribution

$$\Delta\varphi = -4\pi e(n_i - n_e) \tag{1.1}$$

I. Beilis, *Plasma and Spot Phenomena in Electrical Arcs*, Springer Series on Atomic, Optical, and Plasma Physics 113, https://doi.org/10.1007/978-3-030-44747-2_1

Boltzmann distribution can be used for electrons density n_e distribution and ion density n_i with, respectively, temperatures T_e and T_i (e is the electron charge)

$$n_e = n\exp\left(\frac{e\varphi}{kT_e}\right) \quad \text{and} \quad n_i = n\exp\left(\frac{e\varphi}{kT_i}\right) \tag{1.2}$$

Assuming $T_e = T_i = T$ and $n_e = n_i = n$ and $e\varphi \ll kT$, (1.1) can be reduced to:

$$n_e - n_i = \frac{2en}{kT}\varphi \quad \Delta\varphi = -\frac{2\pi e^2 n}{kT}\varphi \tag{1.3}$$

Equation (1.3) in spherical coordinates is

$$\frac{\mathrm{d}^2\varphi}{\mathrm{d}r^2} + \frac{2}{r}\frac{\mathrm{d}\varphi}{\mathrm{d}r} - \frac{\varphi}{r_D} = 0 \tag{1.4}$$

Using the condition $\varphi = 0$ for $r \to \propto$, the solution of (1.4) is

$$\varphi = \frac{e}{r}\exp\left(-\frac{r}{r_D}\right) \tag{1.5}$$

where $r_D = \sqrt{\frac{kT}{4\pi e^2 n}}$ was named as Debye radius and (1.5) indicated that the potential decreases by exponentially at radius $r = r_D$. This characteristic length was introduced by Langmuir [5] as analogy to the phenomena described in Debye and Huckel [6] work in which the behavior of the charge particles in an electrolytes was studied. When $T_e > T_i$, Debye radius is

$$r_D = \sqrt{\frac{kT_i}{4\pi e^2 n}} = 6.9\sqrt{\frac{T_i K}{n(\mathrm{cm}^{-3})}}, \ \mathrm{cm}$$

The ionized gas with characteristic size $L > r_D$ is quasineutral and has been named as “plasma” by Langmuir [5]. The conductivity media in a vacuum arc gap were produced from the vapor of electrode material, and therefore, the plasma consists of heavy particles that are mostly the ions and neutrals of the same electrode atoms as well as electrons.

1.1.2 Plasma Oscillations

Due to some local fluctuation of the charge particle density or by particle separation in the plasma, a local volume charge can be formed. In such case, the force appears that acts in the direction leading to recover the previous quasineutral state. According to the conservation law, the charge changes in time with induced current density j is

$$\frac{dq}{dt} = -\mathrm{div}j; \ j = \mathrm{env} \tag{1.6}$$

The momentum equation of charge with mass m and velocity v in the induced electric field E is

$$m\frac{dv}{dt} = -eE \tag{1.7}$$

By differentiation of (1.6) and using (1.7), the following can be obtained:

$$\frac{d^2q}{dt^2} = -\frac{ne^2}{m}\mathrm{div}E \tag{1.8}$$

Taking into account the Poisson equation in form $\mathrm{div}E = -4\pi q$, (1.8) can be written as

$$\frac{d^2q}{dt^2} = -\omega_p^2 q \tag{1.9}$$

On the other hand, the charge separation can be considered as a capacitor formation that in the planar configuration has capacity $C = S/4\pi x$ for cross section areas S and distance x between planes. As the charge $q = S\mathrm{nex}$ and the voltage is $U = q/C = 4\pi enx$, then the electric field will be $E = U/x$, and the equation of charge motion is

$$\frac{d^2x}{dt^2} = -\frac{eE}{m} = \omega_p^2 x \tag{1.10}$$

The solutions of (1.9) and (1.10) indicate harmonic oscillations of the charge and charge separation in the plasma with ω_p frequency. The expression $\omega_p^2 = \frac{4\pi ne^2}{m}$ was introduced and described by Langmuir and named as frequency of *plasma* or *Langmuir oscillation* [5]. For electrons, $\omega_p = 5.6 \times 10^4 (n_e)^{0.5}$, where ω_p is in s^{-1} and electron density n_e is in cm^{-3}. The time charge separation is $\tau = 1/\omega_p$, and the characteristic scale is $d = v/\omega_p$. For thermal velocity, $v = (kT/m)^{0.5}$ is $d = r_D$. The plasma oscillations are important phenomena because of its influence on the plasma wave formation and wave expansion as well as on the relaxation processes of some plasma perturbations caused by the electromagnetic or particle beam injection [7].

1.1.3 Interaction of Electron Beam with Plasma

When an energetic electron beam is injected into the plasma volume, a plasma instability appears due to energy transfer from the beam to the plasma [7–9]. As result, the experiments showed strong increase of the electric field in the vicinity of plasma boundary in the discharges (including arcs) [4]. In case of electron beam, electron velocity v_b is significantly larger than the average thermal velocity v_{th} of the plasma electrons. As a result, the energy exchange occurs between beam electrons and resonance wave of plasma oscillation. Thus, a particular wave from the entire spectrum of plasma oscillation with phase velocity in vicinity of v_b will be pumped by the beam electron energy. When the velocity associated with plasma oscillation transport, denoted as v_g, is much lower than v_b, the increment of the plasma oscillation amplitude is increased. A development of the mono-energetic electron beam instability can occur in two stages [8]. At initial linear stage, the oscillation amplitude increases exponentially with an increment γ_{lin} and can be described in the framework of hydrodynamic approximation. At the next stage (named as kinetic stage), nonlinear effects appear limiting the amplitude increase of the excited waves with an increment $\gamma_{n-\text{lin}}$ and the plasma beam system relaxed to a new state. The total relaxation time consists of $\tau_{\text{rel}} = \tau_{\text{lin}} + \tau_{n-\text{lin}}$ and can be described as [7, 8]:

$$\tau_{\text{lin}} = \frac{1}{\gamma_{\text{lin}}} = \frac{\sqrt[3]{n/n_b}}{\omega_p} \tag{1.11}$$

According to (1.11), the beam energy loss is relatively low $(n/n_b)^{1/3}$ at the linear stage, where the mono-kinetic beam was characterized by spread distribution of the velocities.

$$\tau_{n-\text{lin}} = \frac{1}{\gamma_{n-\text{lin}}} = \frac{n}{\omega_p n_b} \tag{1.12}$$

At the kinetic stage, the beam energy was significantly transferred to the plasma electrons, and the velocity dispersion in the beam is approximately equal to the plasma electron velocity. The characteristic length of the beam relaxation will be determined as $L_b = v_b \tau_{\text{rel}}$. The beam relaxation in dense cathode spot plasma can be also occurred due to collisions with plasma particles (see Chapter 16).

1.1.4 Plasma State

Plasma can be considered as ideal gas or non-ideal gas dependent on the plasma state. The plasma state is determined by the relation between a kinetic energy E_k of particles dependent on the particle temperature T and the potential energy E_p that depends on the mutual charge particle interaction via Coulomb force. When the distance between charge particles defined by $L = (n)^{1/3}$, the condition for plasma state as ideal was determined by parameter [10]

$$\gamma_{\text{id}} = \frac{E_p}{E_k} < \frac{e^2}{LkT} = \frac{e^2 \sqrt[3]{n}}{kT} \tag{1.13}$$

The distance between charge particles can be also defined as $L = r_D$, and then, the condition for plasma state as ideal gas was determined as

$$\frac{e^2}{r_D kT} = e^3\sqrt{\frac{4\pi n}{kT^3}} < 1, \text{ or } \gamma_{id} = \frac{4\pi e^6 n}{kT^3} < 1 \tag{1.14}$$

Another condition of plasma treatment as ideal gas indicates that the total number of charge particle N_D in a Debye sphere should relatively large (>1):

$$N_D = \frac{4\pi}{3} n r_D = \sqrt{\frac{4\pi n}{9kTe^2}} > 1 \tag{1.15}$$

1.2 Particle Collisions with Surfaces

Different effects arise during the particle collisions with the surface. The ion and atoms might transfer their kinetic energy during the collisions. The fraction of the transferred energy is characterized by coefficient of energy accommodation a_n. The particles can be absorbed (or condensed) or reflected from the surface, and the fraction of particles remained on the surface is characterized by coefficient of condensation a_c. The ion charge can be neutralized during ion collision with surface, and the probability of this effect is described by a coefficient of ion neutralization a_i. Ion spattering coefficient a_s illustrates how much ejected atoms from the surface per one ion bombarded the cathode surface. The ratio of de-excited atom flux from the surface to the excited atom flux toward the electrode surface is characterized by coefficient of atom de-excitation a_{*s}.

Analysis of the experimental data [11–14] shows that the kinetic and potential energy transfer and the ion neutralization have high probability by ion incident with low ion energy (10–20 eV) on the surface of the same material. The mechanism of ion neutralization consists of two steps via Auger effect. Hagstrum investigated this effect in general cases of different ions and body materials [15, 16]. When the low energy ion approaches to the surface, first the high-level energy electron from solid body neutralizes the ion producing an excited atom. At the next step, the excitation energy transferred to the solid body electrons causing or their heating or Auger electron emission (see below).

The ratio of the probability p_i of reflection of the incident ion as ion to probability p_a of reflection the incident ion as neutral atom is determined by

$$\frac{p_i}{p_a} = \exp\left(\frac{\varphi - u_i}{kT_s}\right) \tag{1.16}$$

where φ is the surface work function, u_i is the ionization potential, k is the Botzmann constant and T_s is the surface temperature. As example, for Cu $\varphi = 4.5$ eV, $u_i = 7.7$ eV, $kT_s = 0.5$ eV, and therefore $p_i \ll p_a$. The theory predicts a critical distance from the metal surface outside which resonance neutralization was possible [17]. The high probability of ions neutralization is due to very small distance from the surface (about few angstroms), at which the process occurred.

In general, when distribution function of the incident particles reaches a thermal equilibrium with the wall temperature, the coefficient of the energy accommodation is $a_n = 1$. For individual ion-wall collision with relatively high ion energy with respect to the wall temperature, the large a_n also indicate that the ions have probability to be condensed on the surface. In this case, the probability is [12]:

$$a_n = \frac{4M_1M_2}{M_1^2 + M_2^2}\exp\left(\frac{T_{com}}{T_s}\right) \tag{1.17}$$

where M_1 and M_2 are the mass of incident and wall atoms, T_{com} is the characteristic Compton temperature that determined by the solid body heat capacity and is about 150 K. According to (1.17) for $M_1 \ll M_2$, the coefficient a_n is significantly smaller than unit. However, for $M_1 = M_2$ atoms and for sufficiently hot arc electrodes mostly all energy of the incident ion will be transferred to the wall and $a_n = 1$. This coefficient will be also larger in case of ion collisions with rough surface as compare with smooth mirror-like surfaces.

The values of ion sputtering coefficient a_s vary in a wide range depending on material combination of ion-wall systems. The threshold energy was defined as energy of incident ions when the sputtered atoms were observed in the experiment [12]. The threshold energy of incident ions is determined by energy of lattice atom shift which is about 20–25 eV. This coefficient is significantly larger for ions of noble gas than for metal ions striking the metallic targets. For example, $a_n = 2 \times 10^{-3}$ for system Cr–Ar^+, while $a_n = 6 \times 10^{-5}$ for system Cr–Hg^+ and $a_n \sim 10^{-5}$ for system Ni–Hg^+ [12, 13]. Hayward and Wolter [18] measured the self-sputtering yields versus ion energy for Ag, Au, Cu, Al, and Cr. The self-sputtering yields of all five metals increased with energy over the used energy range (10–500 eV) and the yields varied almost linearly with energy above 50 eV. At 50 eV, a_s is about 0.05 for Al and Cr, while a_s is in range 0.1–0.25 for Cu, Au, and Ag. At 100 eV, a_s is about 0.2 for Al and Cr, while a_s is in range 0.8–0.5 for Cu, Au, and Ag. The spatter coefficient approaches to zero with ion energy decreasing below 50 eV.

Several models were developed to describe the atom sputtering. The models can be classified as thermal and collisional dominated. The thermal model [19] assumes that the energy of the incident ion was dissipated in a small region on the metal surface near the impact point. As a result, the atoms were evaporated due to increase of the temperature of this region. This approach describes the atom sputtering by energetic ions from thin foils, but it is not suitable for understanding the phenomena when the size of local heating is smaller than the characteristic mean free path and the heating time is comparable with the time of lattice atom oscillations.

The collisional model involves momentum change between surface atom and incident ion using the shielding Coulomb potential. Mainly, the theory describes the sputtering caused by relatively high energy ions (>1 keV). Harrison developed a semi-quantitative model indicating that the sputtering coefficient a_s can be obtained by determine four atomic parameters included ratio of mass (incident ion to the target atom), trapping parameter (dependent on penetrate of incident ion in lattice of the target), ion mean free path and sputtering threshold energy [20]. Rol et al. [20] and then Almen and Bruce [21] consider the collisional model assuming that the energy dissipated by the impinging particle in the first atomic layers and a_s is proportional to the mass M_1 and M_2 in form:

$$a_s = K\frac{E}{\lambda(E)\cos\varphi}\frac{M_1M_2}{M_1^2 + M_2^2} \tag{1.18}$$

where K is a constant depending on the target material, E is the energy of impinging particle, φ is the angle between the normal on the target surface and the direction of incidence, λ is the mean free path, which was determined by the collision radius R and by the number of lattice atoms per unit volume. R was calculated by assuming that the slowing down of an ion takes place in a screened coulomb field of a type proposed by Bohr. Almen and Bruce [21] measured the sputtering yields for different metals in the noble ion energy >1 keV, and showed good correlation of their experimental data with the momentum-exchange model [20]. Hayward and Wolter [18] studied this model in order to compare the theory with their measurements in the range of low energy ions. It was shown that the momentum-exchange model, which was verified primarily for higher ion energies, might also apply to very low ion energies and the dependence implied by the model on the mass of the impinging ion can be used to approximate some self-sputtering results.

The sputtering coefficient weakly depends on the target temperature when it is smaller than about 1000 K [12, 21]. The small changes in a_s at higher temperature were due to phase transition (Fe) or with target vaporization (Ag). Miller [22] calculated the effective condensation coefficient for the ion flux coming from the Cu cathode spot region of a vacuum arc and impinging upon the anode. It was used the multi-charged ion velocity distributions produced in plasma jet. Miller assumed that (i) sputtering data for argon ions on a copper surface can be applied to the sputtering of copper by copper ions, (ii) the charge state of the incoming copper ion does not affect its sputtering ability, only its kinetic energy (momentum) being important (E of Cu^{2+} ion was treated as a $2E$ of Cu^{+} ion), and (iii) normal incidence of the ion flux upon a smooth surface. Using experimental values for ion flux and sputtering coefficients, a sputtered flux of 6% the incident ion flux was found for the anode in a copper vacuum arc. The sputtered flux is significantly greater, 20%, for a surface at cathode potential. Miller shows that for a copper vacuum arc, the sputtering alone will limit the effective ion condensation coefficient to, at most, 0.94 or 0.80 for copper surfaces at anode or cathode potential, respectively.

In conclusion, the sputtering effect produced by ion impinging the anode of a vacuum arc is not zero but relatively weak for ion energy equal or smaller than 50 eV for typical ions as Cu, Al, and Cr. In condition of plasma cathode interface in the arc spot, this effect can be neglected due to low ion energy (~20) and high cathode temperature (~4000 K).

1.3 Plasma Particle Collisions

The near-electrode plasma in a vacuum arc is produced by ionization of the electrode vapor and therefore consists of atom and ions of electrode material and electrons. The collisions that occur in this region are between charge particles as ion-ion, electron-electron, and electron–ion that are elastic resulting in change of the particle momentum or charge. The collision between electron and neutral atom can be elastic or non-elastic resulting in atom excitation or ionization. The process of particle collisions can be occurred in the direct and opposite directions, i.e., as electron-excited atom (de-excitation) or electron–ion (recombination). The principle of detailed equilibrium was characterized by the case when the frequency of direct and opposite collisions is equal. All type of collisions was characterized by a cross section that indicated a probability of the collision in plasma with density n.

1.3.1 Charged Particle Collisions

A Coulomb force that decreases with distance as r^{-2} characterizes the interaction between the charged particles. The differential cross section indicated the particle scattering in an angle θ is described by Rutherford's formula (see in [11]).

$$Q(E,\theta) = \left(\frac{Z_1 Z_2 e^2}{4E^2}\right)^2 \sin^{-4}\theta \tag{1.19}$$

As the velocity v changes in course of each interaction of the particles with charge number Z_1 and Z_2, the velocity component relative to the initial direction changed as $v(1-\cos\theta)$. The transport cross section can be obtained by the integral of the cross sections varied in range of different θ.

$$Q(E,\theta) = \int (1-\cos\theta)\mathrm{d}Q(\theta) \tag{1.20}$$

In the case of Coulomb interaction, the mentioned integral diverges due to $Q(\theta) = \rightarrow \propto$ when $\theta \rightarrow 0$. Therefore, the expression (1.20) integrated taking into account that the scattering is determined by the Coulomb electric field at distance lower than Debye radius r_D, while at the large distance, the scattering is due to a field decreased exponentially. As a result, the scattering interaction determined by two impact parameters: maximal $b_{max} = r_D$ and minimal $b_{min} = e^2/E$, and the electron–ion cross section Q_{ei} can be presented in form:

$$Q_{ei} = \frac{Ze^4 Ln(b_{max}/b_{min})}{16\varepsilon_0 \pi E^2} = \frac{2.07 \times 10^{-14} Z\Lambda}{E^2(\text{eV})}(\text{cm}^2) \quad (1.21)$$

where $\Lambda = \text{Ln}(b_{max}/b_{min})$ is the Coulomb logarithm. According to [23]:

$$\Lambda = \ln(1.55 \times 10^{10} \frac{T_e^{1.5}(\text{eV})}{\sqrt{n_e(\text{cm}^{-3})}}) \quad (1.22)$$

For relatively large plasma density >10^{17} cm^{-3} and electron temperature T_e ~ 1–5 eV, is Λ ~ 5 [23, 24].

Another approach [11] is based on the assumption that the maximal impact parameter is determined by the distance between the charge particles, i.e., $b_{max} = n^{-1/3}$, and in this case, the transport cross section was obtained in form

$$Q_{ei} = 4\pi\left(\frac{e^2}{2E}\right)^2 \text{Ln}\left(\frac{2E}{\sqrt[3]{n}e^2}\right) = \frac{6.5 \times 10^{-14}}{E^2}\text{Ln}\left(\frac{1.4 \times 10^7 E}{\sqrt[3]{n}}\right) \quad (1.23)$$

For electron–electron collisions, E should replace $2E$ in Formula (1.23).

1.3.2 Electron Scattering on Atoms

The electron energy loss by the collision is proportional to $2m_e/m$, with electron m_e and atom m mass, respectively. This type of collisions was mostly studied for hydrogen, nitrogen, oxygen, and noble gases at low electron energy <10 eV where a non-monotonic dependence (with a minimum) of scatter cross section on electron energy (0.1–1 eV) was detected [11, 13, 25]. The non-monotonic dependence was explained in frame of quantum mechanics as a change between impact and inside atom electrons depended on electron de Broglie wave. The electron–atom scatter cross section Q_{ea} increase with electron energy (>1 eV) passing Q_{ea} through a maximum and then weakly decreased with the energy. The literature review showed that Q_{ea} changed mostly between 10^{-15} and 10^{-16} cm^{-2} [11, 13, 23]. The scattering was anisotropy with small angles preferable [11, 26].

The electron impact cross section Q_{ea} with molecule H_2O was calculated and measured in range (5–3) × 10^{-15} cm^{-2} when the temperature increases from 2500 to 4000 K [27]. Using kinetic Boltzmann equation method and Monte Carlo simulation, the cross section Q_{ea} ~ 10^{-15} cm^{-2} was calculated for electron scattering on SF_6 molecule [28]. While the experiments with noble gases were widely studied, a limited data are present for electron sputtering on metallic atoms. McDaniel [29] reported that $Q_{ea} = 1.8 \times 10^{-15}$ cm^{-2} (Hg), 2.3×10^{-15} cm^{-2} (Zn), and 3.5×10^{-15} cm^{-2} (Cd) when the electron energy was in range 10–60 eV. The relatively larger scattering cross section was measured for alkaline atoms [30, 31]. Perel et al. [32] measured this cross section for alkaline atoms at electron energy in region from 0.25 to 13 eV, and for 13 V, the results show that $Q_{ea} = (1.5–2) \times 10^{-14}$ cm^{-2} (K), 1.3×10^{-14} cm^{-2} (Na), and 2.5×10^{-14} cm^{-2} (Cs). Romanyuk et al. [33] reported the scattering cross section for alkaline-earth metals Ca, Sr, and Ba. According to this measurement for all materials, Q_{ea} decreased from about 3×10^{-14} cm^{-2} to about 7×10^{-15} cm^{-2} with electron energy in region of 0.1–10 eV.

1.3.3 Charge-Exchange Collisions

This type of collisions is related to a non-elastic heavy particle interaction because a neutral atom can be ionized by electron exchange with an ion. In the vacuum arc cathode plasma, the ions and neutral atoms are from the same cathode material. Therefore, there the ion–atom interaction is determined by a resonance charge-exchange collision with a cross section of one order of magnitude larger than that for different ion and atom materials [29, 31]. The main transfer mechanism of electron from atom to the ion is due to potential barrier decreasing tunneling effect when the ion approaches the atom. The larger particle velocity is determined by a condition that the time of ion residence at distance r of order of atom radius will be much larger than the electron rotation time in the atom. As potential barrier for the electron in the atom (ionization potential u_i) changes with distance as $W_{ia} = eu_i-(4e^2/r)$, then in the first approximation of $W_{ia} = 0$, the cross section is [34]:

$$Q_{ia} = \frac{8\pi e^2}{u_i^2}(\text{cm}^2) \tag{1.24}$$

Equation (1.24) was improved taking into account the ion velocity and the barrier transparence for a case of $W_{ia} \neq 0$. The result showed that Q_{ia} approaches to that given by (1.24) with ion velocity increase. The charge-exchange cross section estimation of multi-charge ions was proposed in [35]

$$Q_{ia} = \pi a_0^2 Z \frac{Ry^2}{e^2 u_i^2}(\text{cm}^2) \tag{1.25}$$

where $\pi a_0^2 = 0.88 \times 10^{16}$ cm^2, $Ry = 13.6$ eV, and Z-ion charge number. The resonance charge-exchange cross section weekly depends on ion velocity v_i. A simple expression for resonance charge-exchange cross section depended on v_i was obtained from Firsov's work [36]

$$Q_{ia}(v_i) = \frac{h^2}{16\pi m_e u_i} A_{\text{ch}} \text{Ln}\left(\frac{2 \times 10^{10}}{v_i}\right) \quad (1.26)$$

where h is the Planck constant, m_e is the electron mass. The dependence on v_i was also indicated in work [37]:

$$Q_{ia}(v_i) = A_{ch} Ln^2 \frac{v_{ch}}{v_i} \quad (1.27)$$

where A_{ch} and v_{ch} are some constants. The calculated [37] data of resonance Q_{ia} for different materials and ion energy were presented in Table 1.1. The results show that Q_{ia} moderately decreases when ion energy (i.e., velocity) grows from 0.1 to 10 eV, and it is in range (1–2) $\times$ 10^{-14} cm^{-2} for metallic atoms.

1.3.4 Excitation and Ionization Collisions

This type of collisions named as *non-elastic* particle interactions, in which kinetic energy is redistributed between the particles in bounded and free states. The atom or ion (A) will be *excited* when the inside electron shifted from a lower to a higher energy level (A*):

A + e → A* + e

or *ionized* when the inside electron passed from the ground state to a free state due to interaction with an external electron.

A + e → A^+ + 2e

Ionization can arise by electron impact with previously excited atom (*stepwise ionization*).

Excitation: A + e → A* + e; and then ionization: A* + e → A^+ + 2e

Or by photon impact (*photoionization*):

A* + hν → A^+ + e

Or by collisions energetic heavy particle B_{ener} in which its kinetic energy was transferred leading to slow particle B_{slow}.

A + B_{ener} → A^++ B_{slow} + e

The energy which is necessary to move atom to excited or ionized state named as *potential* of excitation u^* or ionization u_i. The excitation or ionization probability was characterized by the heavy particle density and a cross section [29, 35]. The cross section σ was determined by reducing a particle beam (e.g., electron flux N_e) interacted with a background assembly of atoms with density density n_a. The change of the particle flux dN_e in the beam along *x-axis* will be proportional by σ in form:

$$dN_e(x) = -\sigma N_e n_a dx$$

Table 1.1 Resonance charge-exchange cross section (in 10^{-15}, cm^2) of positive ion in fixed atoms [37]

Material	Ion energy, eV		
	0.1	1	10
H	6.2	5.0	3.8
He	3.5	2.8	2.1
Li	26	22	18
Be	13	11	19
B	9.6	7.4	5.8
C	5.3	4.3	3.2
N	5.0	3.8	3.0
O	4.8	3.5	2.8
F	3.1	2.5	1.9
Ne	3.2	2.5	1.9
Na	31	26	22
Mg	19	16	13
Al	16.1	12.9	10
Si	8.7	6.5	4.9
P	8.1	6.5	5.0
S	8.5	6.8	5.3
Cl	4.9	3.9	3.0
Ar	5.5	4.5	3.6
K	41	35	29
Ca	19	16	13
Ti	22	19	15
V	23	19	16
Cr	21	18	14
Mn	20	16	13
Fe	21	18	15
Co	22	18	15
Ni	19	16	13
Cu	19	16	13
Zn	16	13	11
Ga	17	14	11
Ge	9.4	7.6	6.1
As	9.8	8.0	6.3
Se	9.3	7.3	5.7
Br	5.9	4.6	3.7
Kr	7.3	5.9	4.6
Rb	45	39	32

(continued)

Table 1.1 (continued)

Material	Ion energy, eV		
	0.1	1	10
Sr	30	35	21
Zr	24	20	16
Nb	22	19	15
Mo	21	17	14
Ag	20	17	14
Cd	17	14	12
In	19.5	16	13
Sn	10.7	8.7	6.9
Sb	11.4	9.1	7.2
Te	10.6	8.6	6.8
I	7.0	5.6	4.4
Xe	9.1	7.5	6.0
Cs	5.3	45	38
Ba	35	30	25
Ta	19	16	13
W	18	15	13
Re	21	17	14
Pt	17	16	13
Au	15	14	11
Tl	15	12	10
Hg	18.6	15.1	12.1
Pb	11.0	9.2	7.3
Bi	15.4	12.7	10.3

The reduction of the particle flux can be obtained integrating the expression:

$$\frac{\mathrm{d}N_e}{N_e} = -n_a \sigma \,\mathrm{d}x$$

and the flux dependence on distance x will be obtained as:

$$N_e(x) = N_{e0} \exp(-n_a \sigma x) \tag{1.28}$$

The effective cross section σ can be calculated from (1.28) by given the measured $N_e(x)$ for known initial N_{e0} at distance $x = 0$ and n_a of the background gas. The characteristic length $\lambda = (n_a \sigma)^{-1}$ named *mean free path* between two collisions. The cross section depends on the impacted particle energy E_p. Theoretically, the excitation or ionization cross section function on E_p was obtained using classical or quantum mechanics approaches.

1.3.4.1 Classical Approach

Thomson [38] first developed a method of cross section calculation using the classical description of the electron–atom interaction. It was assumed pair interaction of impact electron with valence electron in the atom, which was considered as free. It was taken in account that the transferred energy ε by interaction of two particles with mass m_1 and m_2 and impact particle energy E_p is follows for sputtered angle θ [30, 39]:

$$\varepsilon = \frac{4m_1 m_2}{(m_1 + m_2)^2} E_p \sin^2\theta \quad (1.29)$$

For two electrons, the following relation was obtained [38]:

$$\sin^2\theta = \left(1 + \frac{p^2}{e^4} E_p\right)^{-1}$$

where e is the electron charge, p is an interaction parameter indicating a length of perpendicular to a direct line between scattered and moved particles. In this case, the cross section was expressed as $\sigma = \pi p^2$. Assuming then that the neutral atom was ionized when the transferred energy was equal to ionization potential u_i, the respective cross section for atoms with N valence electrons was obtained in form:

$$\sigma_i = \frac{N\pi e^4}{E_p}\left(\frac{1}{u_i} - \frac{1}{E_p}\right) \quad (1.30)$$

Although that Formula (1.30) qualitatively well describes the experiment, in the case of low energy of the impact electron, the calculated value of σ significantly exceeds the one predicted by Formula (1.30). When the energy of impact electrons is large, the cross section predicted by Formula (1.30) is lower than that measured experimentally. A very simplest model was proposed by Devis [40] by considering the impact of hard elastic spheres. One of these was the cathode ray, the other one something like the atom were equivalent to taking the mass of the atomic sphere as that of an electron. The atom ionization would occur according to classical mechanics when transfer of energy was as great as the ionization energy, and the cross section with arbitrary *const* is:

$$\sigma_i = \mathrm{const}\left(1 - \frac{u_i}{E_p}\right) \quad (1.31)$$

Rosseland [41] approach was similar to Davis's except that the spheres were electrons with the classical inverse-square repulsion. He predicted:

$$\sigma_i = 2\frac{\pi e^2}{u_i^2}\left(1 - \frac{u_i}{E_p}\right)\frac{u_i}{E_p} \quad (1.32)$$

Similar simple formula was introduced by Killian [42] to describe the initial part of ionization function in a mercury vapor of positive column of an electrical discharge when E_p is lower than $3u_i$:

$$\sigma_i = c(E_p - u_i)$$

The value of c was defined as being proportional to the ratio of vapor pressure to the temperature. In order to consider the orbital atom electron motion, the next development was conducted by Thomas [43, 44], and by Vebster et al. [45] taking into account that electron has a speed greater than it had before entering the atom, because it has been attracted by the nucleus. Vebster et al. developed Thomas result and was presented the following expression

$$\sigma_i = 2\frac{\pi e^2}{u_i^2}\frac{1 - U^{-1} + \frac{2}{3}T(1 - U^{-2})}{U + T + 1} \tag{1.33}$$

where $U = E_p/u_i$ and T are the ratio of the orbital kinetic energy to the ionization energy, which as example for silver is 1.278. They improve expression (1.33) by introducing innovation, which take into account the deflection of the impact electron by nuclear attraction within an atom, before it reaches the valence electron ionize as follows.

$$\sigma_i = 2\frac{\pi e^2}{u_i^2}\frac{1 - U^{-1} + \frac{2}{3}T(1 - U^{-2})}{U} \tag{1.34}$$

A wide range of experimental data for cross section of atom ionization (number of gases, Ag, Ni, etc.) as dependences on impact electron energy was analyzed by Drawin [46]. As a result, an empirical analytical expression for the ionization cross sections is given in the following form:

$$\sigma_i = 2.66\pi a_0^2\left(\frac{u_i^H}{u_i}\right)^2 f_1\frac{U - 1}{U^2}\mathrm{Ln}(1.25 f_2 U) \tag{1.35}$$

It was indicated that using f_1 in range of 0.7–1.3 and f_2 in range 0.8–3 for different atoms, Formula (1.35) very good approximates the measured cross sections and which is able to give a large part of tile ionization curves as a function of the energy of the impacting electrons. Usually, Formula (1.35) was used with $f_1 = f_2 = 1$.

The most extended development of the classical theory was suggested by Gryzinski [47–50] considering the pair interaction of impact electron of velocity v_1, energy E_1 with orbital electron of velocity v_2, energy E_2. According to Gryzinski theory, the transferred energy ε depends on few parameters: (i) the angle θ between the two vectors v_1 and v_2, (ii) the azimuthal angle φ referred to a system of coordinates, in which the z-axis coincides with the direction of v_2, (iii) the impact parameter p, and (iv) the angle Θ between the orbital plane and the fundamental plane containing the vectors v_1 and v_2. In this case, the energy transferred to the atom electron is

$$\varepsilon = \left(1 + \frac{p^2 V^2 m^2}{4e^4}\right)^{-1} \left(E_2 - E_1 + \frac{p V^2 v^1 v^2 m^2}{2e^2} \sin\theta \cos\Theta\right) \tag{1.36}$$

where $V^2 = v_1^2 + v_2^2$. Integrating (1.36) over the mentioned parameters, the ionization cross section for a collision in which particle 2 undergoes a change in energy equal to ionization potential is obtained as follows:

$$\sigma = \frac{\sigma_0}{u_1^2}\left(\frac{U-1}{U+1}\right)^{3/2} \times \begin{Bmatrix} \frac{2T}{3U} + \frac{1}{U}\left(1 - \frac{T}{U}\right) - \frac{1}{U^2} - - - - - - - - - - - - - - \rightarrow U > 1 + T \\ \frac{2}{3}\left[\frac{1}{U} + \frac{1}{U}\left(1 - \frac{T}{U}\right) - \frac{1}{U^2}\right]\sqrt{\left(1 - \frac{1}{U}\right)\left(1 + \frac{1}{T}\right)} - - - \rightarrow U \le 1 + T \end{Bmatrix} \tag{1.37}$$

$$\sigma_0 = \pi e^4 = 6.56 \times 10^{-14}\,\mathrm{eV\,cm^2}$$

The later improving of the classical theory by comparing with a number of experimental data Gryzinski [51] obtains a simpler expression:

$$\sigma = \frac{\sigma_0}{u_1^2}\frac{1}{U}\left(\frac{U-1}{U+1}\right)^{3/2}\left\{1 + \frac{2}{3}\left(1 - \frac{1}{2U}\right)\mathrm{Ln}(2.72 + (U-1)^{0.5})\right\} \tag{1.38}$$

Ochkur and Petrun'kin [52] developed the classical theory more precise and still within the framework of the two-body problem by taking into account the velocity of the atomic electron but assuming it to be free as before. The suggestions were applied to cross section calculations of atom excitation and ionization. In case of excitation, the expression is:

$$\sigma_n = \begin{Bmatrix} N\frac{\pi e^4}{E_p}\frac{1}{U}\left(\frac{1}{U_n} - \frac{T}{E_p}\right) - - - \rightarrow U_n \le E_p \le U_{n+1} \\ N\frac{\pi e^4}{E_p}\left(\frac{1}{U_n} - \frac{T}{U_{n+1}}\right) - - - - - - - \rightarrow U_n \le E_p \end{Bmatrix} \tag{1.39}$$

where U_n and U_{n+1} are the excitation energies of the n-th and $(n + 1)$th levels, respectively, and N is the number of valence electrons. In contrast to Gryzinski, the atom ionization considered by Ochkur and Petrun'kin more accurately take into account the relative velocity as $V^2 = v_1^2 + v_2^2 - 2v_1 v_2 \cos\vartheta$. The expression for ionization cross section, presented in [52], is more complicated and can be solved only

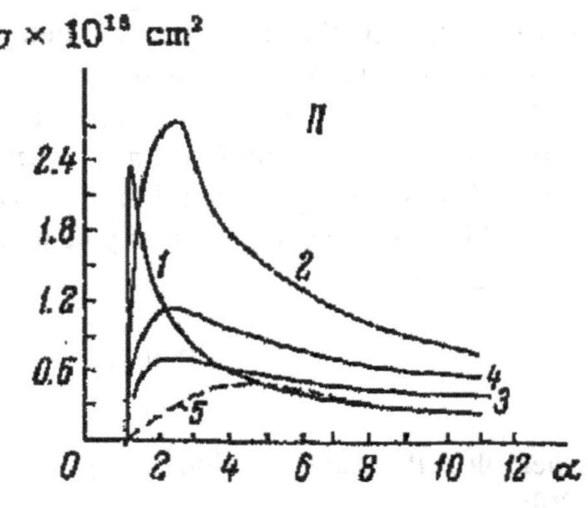

Fig. 1.1 Excitation of He atom. 1—Calculating according to Formula (1.39), 2—numerical calculation, 3—calculating according to Drawin's Formula (1.35), 4—calculating according to Gryzinski's Formula (1.38). 5—experiment [52] Permission 4776030484766

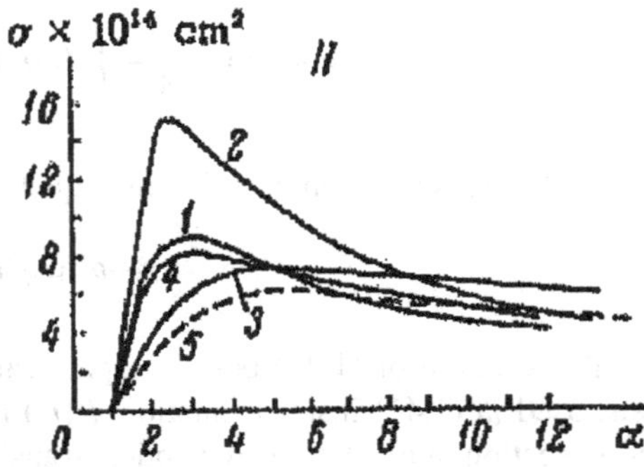

Fig. 1.2 Ionization of Hg atom. 1—Calculating according to Tomson Formula (1.30), 2—numerical calculation, 3—calculating according to Drawin's Formula (1.35), 4—calculating according to Gryzinski's Formula (1.38). 5—experiment [52] Permission 4776030484766

numerically. The solution for He atom excitation (Fig. 1.1) and Hg atom ionization (Fig. 1.2) showed that, although the more accurately of this model, the result was worse agreed with the experiment than that agreement proposed by Gyzinski. This result can be attributed to limited applicability of the classical approach of free electrons in general.

1.3.4.2 Quantum Mechanical Approach

According to the quantum approach, the problem can be solved when the energy and the angular distributions of scattering particles can be determined after the electron–atom interaction. The problem of atom ionization is similar to that for the atom excitation. However, it is necessary to use continuous spectrum functions. Also, in case of atom ionization, a three-body problem should be analyzed considering Schrödinger equation. As the exact solution is impossible, different approximations were used. For this reason, the Born theory [53, 54] is most often utilized. There two main assumption were made (i) the kinetic energy of impact electron is significantly larger than the bounded energy of atom's electron, (ii) the distortion of impact electron wave function by the atom electrical field is weak. In this case, the solution can be obtained using the perturbation theory.

Also, in quantum theory, the impact particle beam was assumed as a plane wave in the form of Ce^{ikx} with $k = 2\pi/\lambda = 2\pi mv/h$, and λ is the wavelength. Constant C was determined by particle flux A flowing per unit area and per second as $C = A/v$. Taking into account an atom potential field $U(r)$, Schrödinger equation can be formulated as

$$\nabla^2 \Psi + (k^2 - U(r))\Psi = 0 \tag{1.40}$$

where Ψ is the wave function. Using Green function, (1.40) will have a solution as [30]:

$$\Psi(r) = \Psi_0(r) - \frac{1}{4\pi} \int U(r')\Psi(r') + \frac{e^{ik|r-r'|}}{|r-r'|} dr' = 0 \tag{1.41}$$

where $\Psi_0(r)$ is solution of following equation:

$$\nabla^2 \Psi + k^2 \Psi = 0 \tag{1.42}$$

The solution of (1.42) before impact can be presented as $\Psi_0 e^{ikz}$. This solution was used in (1.41) instead function $\Psi(r')$ in Born approximation. As a result, the cross section of excitations for optical transition from level E_0 to E_n including atom ionization was obtained in form [30]

$$\sigma \approx \frac{4\pi m_e^2 e^4}{k^2 h^4} |z_{0n}|^2 \mathrm{Ln} \frac{2m_e v_e^2}{E_n - E_0} \tag{1.43}$$

where z_{0n} is the matrix element. For threshold energy, an electron exchange phenomena (due to indecomposability of impact and atom electrons) could influence the probability of the excitation and to change of multiplicity. The effect of multiplicity change was considered by Oppenheimer [55], which corrected an expression for wave amplitude of the scattering electron. However, the cross section calculated in framework of Born–Oppenheimer theory indicates larger result in comparison with that obtained by Born approximation. Later, this problem was considered by Peterkop [56] and Ochkur [57]. Ochkur indicates that Born–Oppenheimer theory described the scattered electron by plane wave, and it is correct when $k \gg 1$ and $k' \gg 1$. Only the first term of the series was remained in this case that not available at low energy by exchange scattering. Ochkur obtained an improved formula for cross section of the transition from the state m to state n based on quantum mechanical analysis in form:

$$\sigma \approx \frac{24\pi}{k^2} \int\limits_{q_{\min}}^{q_{\max}} \left| \langle n | e^{iqr} | \alpha \rangle \right|^2 q \mathrm{d}q \tag{1.44}$$

where $q = k - k'$. Ochkur showed also an important point indicating an analogy between expressions for cross section of exchange scattering obtained using quantum and classical mechanics. The following classical expression for cross section of the transition from the state m to state n was derived [57]:

$$\sigma \approx \frac{\pi}{E_p + \Delta} \left\{ \begin{array}{c} (E_p + U_{n+1})^{-1} - (E_p + U_n)^{-1} \rightarrow\rightarrow\rightarrow E_p \geq U_m - U_{n+1} \\ U_m^{-1} - (E_p + U_n)^{-1} \rightarrow\rightarrow\rightarrow\rightarrow\rightarrow U_m - U_{n+1} \geq E_p \geq U_m - U_n \end{array} \right\} \tag{1.45}$$

where Δ is the certain constant, which is at least equal to excitation or ionization potential.

According to Fisher et al. [58], the electron impact excitation cross sections for allowed transitions in atoms are

$$Q_{ea}^{*} = \frac{8\pi}{\sqrt{3}} \pi a_0^2 f_{ij} \frac{Ry^2}{U_{ij}^2} \frac{G_{ij}(\varepsilon)}{\varepsilon} (\mathrm{cm}^2) \tag{1.46}$$

Here, Q_{ea}^{*} is the electron–atom excitation cross section for transition from the lower state i into the upper state j; $\varepsilon = U/E_{ij}$ is the kinetic energy of relative motion between electron and atom (or ion), E_{ij} is the transition energy, a_0 is the Bohr radius, $Ry =$ 13.6 eV is the Rydberg constant, f_{ij} is the absorption oscillator strength, and $G_{ij}(\varepsilon)$ is the Gaunt factor which may be treated as a fitting function of order unity. In case of direct ionization, most of the data for the cross sections were retrieved via the Atomic Database of the National Institute of Fusion Science (NIFS), Nagoya, Japan, and for the cross section of the direct transition, an analysis resulted in the expression [59]:

$$Q_{zqq}^{ion} = C_l \left(\frac{Ry}{I_{zqq'}} \right)^{2-\delta l} \xi_{zqq'} b_{zqq'} \frac{\ln(I_{zqq'})}{\varepsilon / I_{zqq'}} (\mathrm{cm}^2) \tag{1.47}$$

where $I_{zqq'}$ is the minimal energy required for ionization from state q into state q', b_{zqq} is the branching ratio [58], i.e., the probability that the removal of an electron from a proper nl-subshell generates the q' state out of a few possible ones (say, removal of a 1 s electron from the $1s^2 2s$ configuration may result in $1s^2 s\ ^1S$ and 1s2s ^3S terms with branching ratios 1/4 and 3/4, respectively). Coefficients C_l and δl were chosen from the best fitting of expression (1.47) to the known *direct* ionization cross sections explained in work of [58].

Recently, Gupta et al. [60] calculated electron impact total and ionization cross sections for Sr, Y, Ru, Pd, and Ag atoms. The cross sections were computed in the energy range from ionization threshold to 2000 eV. The results obtained are compared with other theories and measurements wherever available and were found to be quite consistent and uniform (Table 1.2). The total cross section Q_T was defined as the sum of elastic Q_{el} and inelastic Q_{inel} cross sections. The inelastic interaction included excitation and total ionization cross section Q_{ion}.

1.3.4.3 Experimental Data

Most of experimental data of energy-dependent cross sections were obtained for different gas atoms. However, the important issue is the knowledge of the cross sections dependence on electron energy for metallic atoms in order to describe the phenomena in a vacuum arc operated in an electrode vapor. Let us consider such data from limited number of publications. Most extensive data were presented for ionizations of Hg, Ni, and alkali atoms [26, 29, 30, 61].

The electron impact ionization function for lead atoms was measured over an energy range extending from the ionization threshold to 150 eV [62, 63]. The result (Fig. 1.3) showed relatively small initial section of the ionization function that has a slope 0.7×10^{-16} cm^2/V.

At the maximum of the ionization function, which corresponds to energy of 55 eV, the cross section is 8×10^{-16} cm^2, and at 150 eV, this value was 6.5×10^{-16} cm^2. Figure 1.3 also presents the cross sections of the ionization function calculated in according to Gryzinski and Drawin by classical theory and to Born's quantum mechanics approximations (using Vainshtein's model [35]). The calculated results show that the better agreement was reached by Drawin's formula, while the cross sections calculated by Gryzinski and Born approximations exceed the measured maximum value by about 30–60%. For electron energy larger than 60–80 eV, the calculations are lower than measured data, and at 150 eV, the calculated values were about 60–70% of that measured. It was indicated that the mentioned discrepancy was due the multiple atom ionizations.

The electron impact multiple ionization function for lead atoms (from Pb^+ to Pb^{5+}) was measured over an energy range extending from the ionization threshold to 400 eV [64]. Figure 1.4 shows each cross sections in scale according to $n\sigma^{n+}$, where $n = 1, 2, 3, 4, 5$. The maximal values of σ^{n+} were 1.4, 0.32, 0.056, 0.011 x 10^{-16} cm^2 at electron energy 100, 215, 260, and 400 eV for σ^{2+},σ^{3+} σ^{4+}, σ^{5+}, respectively. The total cross section was determined as $s_{\mathrm{tot}} = \sum_{n=1}^{n=5} n\sigma_n$.

The electron impact ionization function for magnesium, calcium, strontium, and barium atoms was measured over an energy range extending from the ionization threshold to 200 eV (Fig. 1.5) [65, 66].

The ratio of cross sections for electron impact ionization of Fe, Co, Ni to the cross section for Ag has been measured by Cooper et al. [67]. These data obtained for ionizing electron voltage of 60 eV that was chosen to be approximately at the maximum ionization probability for all four species and can be considered in the cited work. The measured ratios for each species were about unit. The electron impact ionization cross sections for Ag, Ge, Sn, and Pb have been measured relative to Ag by Lin and Stafford [68]. For 60-eV impacting electrons, the relative single

Table 1.2 Q_{ion} and Q_T for all the atoms studied in 10^{-16} cm^2 [60]

E_i(eV)	Sr		Y		Ru		Pd		Ag	
	Q_{ion}	Q_T	Q_{ion}	Q_T	Q_{ion}	Q_T	Q_{ion}	Q_T	Q_{ion}	Q_T
6	0.004	93.47	–	–	–	–	–	–	–	–
7	0.323	91.94	0.042	90.31	–	–	–	–	–	–
8	1.237	90.00	0.365	85.02	0.008	56.94	–	–	0.003	50.35
9	2.517	87.25	0.990	80.36	0.107	51.73	0.004	39.58	0.078	43.45
10	3.905	84.372	1.793	76.903	0.331	46.87	0.055	35.13	0.284	38.47
11	5.238	81.621	2.659	73.90	0.647	42.75	0.175	31.98	0.592	34.69
12	6.430	78.69	3.509	71.33	1.011	40.00	0.353	31.80	0.955	31.76
13	7.451	75.832	4.310	68.78	1.392	38.35	0.569	30.53	1.338	29.39
14	8.297	73.25	5.011	66.48	1.765	37.04	0.806	29.04	1.715	27.56
15	8.982	70.63	5.635	64.20	2.119	35.88	1.050	27.64	2.073	26.68
18	10.268	63.76	7.016	58.17	3.01	33.08	1.750	24.32	2.974	25.03
20	10.673	59.89	7.603	54.71	3.461	31.57	2.155	22.72	3.428	24.14
25	10.871	52.19	8.337	47.57	4.205	28.5	2.913	19.87	4.163	22.30
30	10.581	46.55	8.513	42.33	4.598	26.10	3.382	18.03	4.533	20.81
40	9.737	39.04	8.302	35.24	4.898	22.48	3.819	15.73	4.768	18.49
50	8.976	34.34	7.919	30.77	4.938	19.84	3.941	14.30	4.746	16.75
60	8.354	31.10	7.539	27.75	4.887	17.82	3.939	13.23	4.646	15.35
70	7.844	28.61	7.193	25.54	4.802	16.28	3.887	12.37	4.526	14.13
80	7.419	26.58	6.884	23.75	4.705	15.05	3.817	11.57	4.405	13.16
90	7.056	24.86	6.607	22.28	4.606	14.05	3.741	10.85	4.29	12.33
100	6.741	23.40	6.358	21.02	4.508	13.31	3.665	10.20	4.182	11.66
120	6.218	21.00	5.926	18.96	4.321	12.11	3.521	9.34	3.989	10.59
200	4.885	15.31	4.745	13.95	3.724	9.05	3.079	7.11	3.430	8.07
250	4.363	13.28	4.262	12.16	3.444	7.98	2.878	6.29	3.183	7.13
300	3.966	11.82	3.886	10.85	3.214	7.21	2.711	5.72	2.984	6.44
400	3.389	9.83	3.335	9.08	2.854	6.16	2.448	4.92	2.672	5.53
500	2.984	8.49	2.944	7.89	2.585	5.47	2.246	4.40	2.437	4.92
600	2.584	7.52	2.604	7.02	2.375	4.96	2.082	4.02	2.249	4.47
700	2.302	6.78	2.295	6.36	2.204	4.56	1.946	3.73	2.096	4.13
800	2.106	6.19	2.096	5.82	2.061	4.24	1.832	3.48	1.969	3.85
900	1.946	5.71	1.934	5.39	1.937	3.97	1.734	3.28	1.860	3.62
1000	1.81	5.312	1.798	5.02	1.835	3.74	1.650	3.11	1.769	3.42
2000	1.21	3.24	1.096	3.118	1.234	2.51	1.142	2.15	1.218	2.33

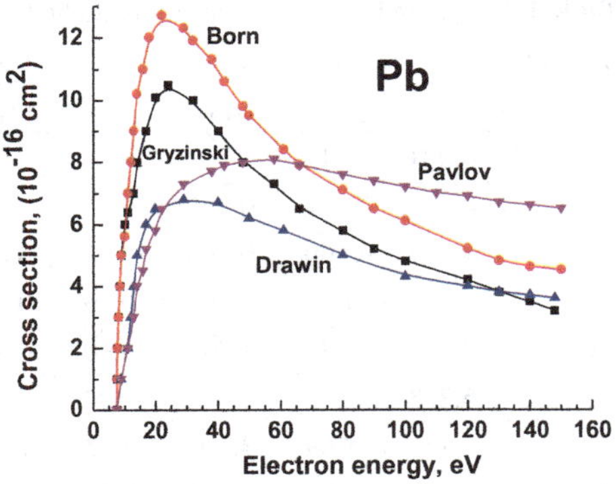

Fig. 1.3 Cross sections of Pb atom measured by Pavlov et al. [62, 63] compared with that calculated by Gryzinski [48], Drawin [46], and Born approximations [35]

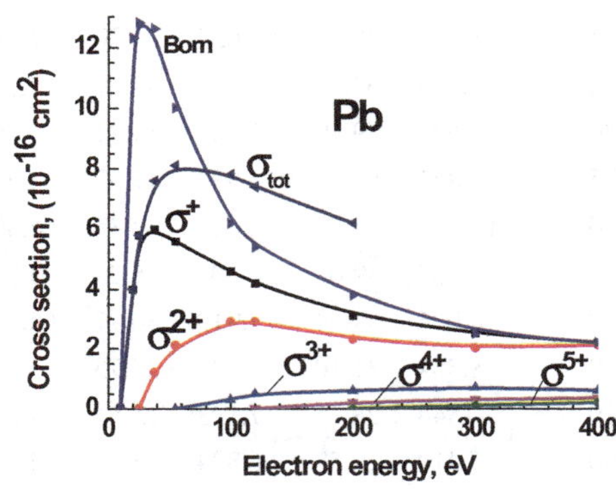

Fig. 1.4 Measured multiple ionization cross sections for Pb atom from Pb^{+} to Pb^{5+}, including σ_{tot} [64]

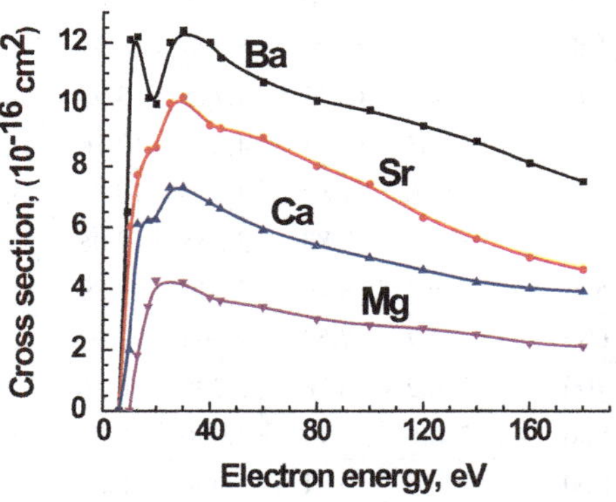

Fig. 1.5 Measured cross sections of Ba, Sr, Cs, and Mg atoms [65, 66]

ionization cross sections are found to be (1.00), 1.46 ± 0.3, 1.46 ± 0.16, and 1.43 ± 0.1, respectively. The absolute cross section for silver Ag^{+} there indicated as 5.6×10^{-16} cm^2.

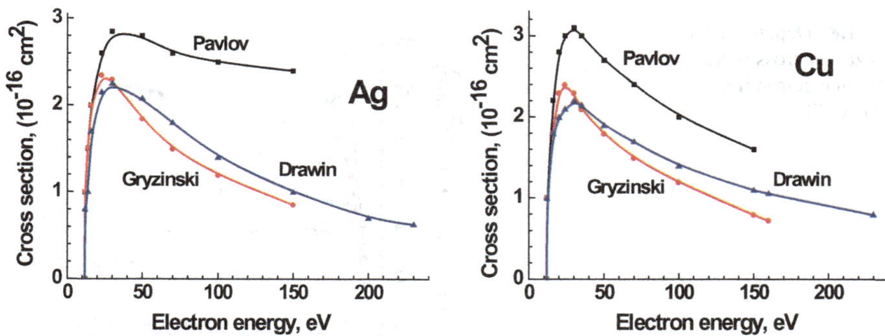

Fig. 1.6 Measured cross sections of Ag (left) and Cu (right) atoms and compared with that calculated by approximations developed by Gryzinski and Drawin [63]

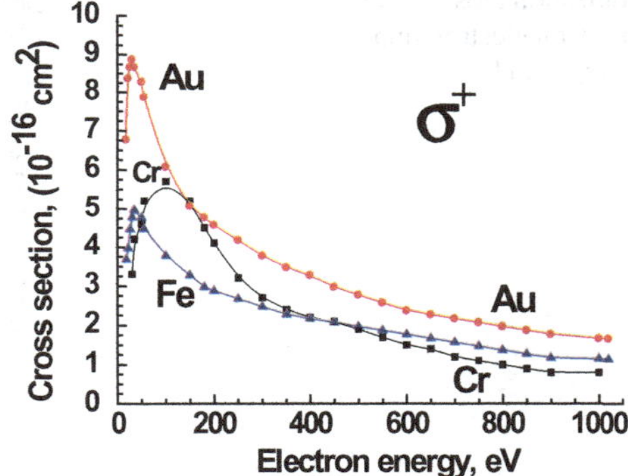

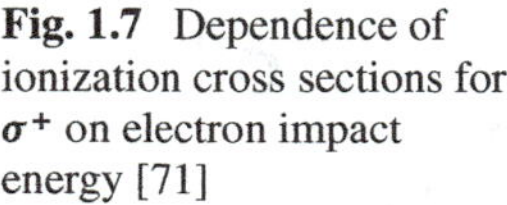
Fig. 1.7 Dependence of ionization cross sections for σ^+ on electron impact energy [71]

The absolute value of cross sections of the electron impact ionization has been measured for silver by Crawford [69]. The measured values at 75 eV are (4.65 $\pm$ 1.0) $\times$ 10^{-16} cm^2 for the production of Ag$^+$ (agree with result [68]) and (4.75 $\pm$ 3.0) $\times$ 10^{-17} cm^2 for the Ag^{2+}. Pavlov et al. [63] have measured the cross sections for electron impact ionization of Ag and Cu. The results illustrated in Fig. 1.6 compared with calculations by Gryzinski [48] classical approximation and Drawin's semi-empirical approach [46]. The maximal ionization cross section obtained by Gryzinski approximation is lower by 25% for Cu and 15% for Ag than measured data. The calculation according to Drawin's formula is lower by about 30% for both Cu and Ar than measured data in [63].

The cross section for production of Cs$^+$ has been determined from threshold to 100 eV [70]. At threshold, the slope of the cross section versus energy curve measured to be 2.7 $\times$ 10^{-16} cm^2/eV. Two maximums near 1 $\times$ 10^{-15} cm^2 were observed at about 12 and 30 eV, respectively.

Relative and absolute electron impact multiple ionization cross sections were measured for Au, Cr, and Fe by Nelson [71]. The results of ionization cross sections for σ^+, σ^{2+}, σ^{3+} are presented in Figs. 1.7, 1.8, 1.9 and 1.10. Total cross section results on the elements are also presented as $s_{\text{tot}} = \sum_{n=1}^{n=3} n\sigma_n$, where $n = 1, 2, 3$.

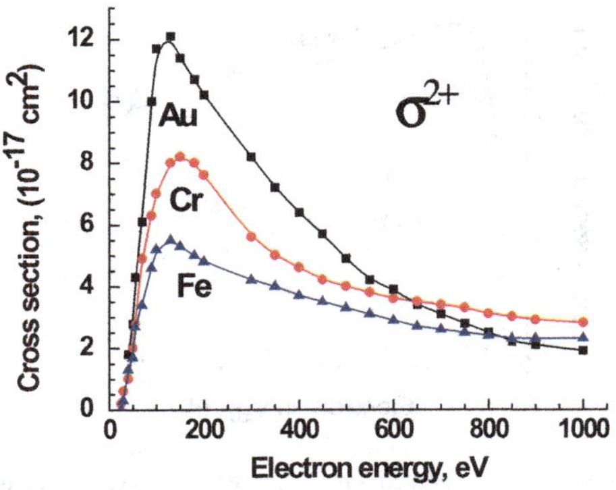

Fig. 1.8 Dependence of ionization cross sections for σ^{++} on electron impact energy [71]

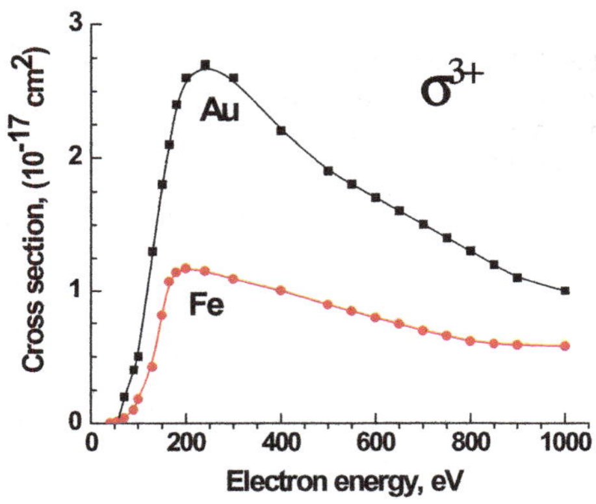

Fig. 1.9 Dependence of ionization cross sections for σ^{+++} on electron impact energy [71]

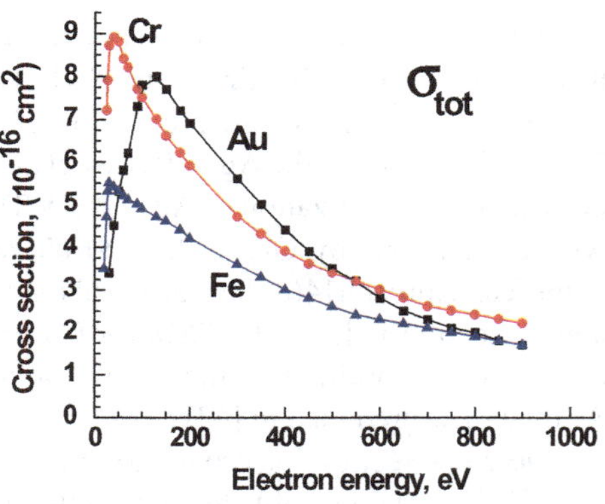

Fig. 1.10 Dependence of total ionization cross sections ($s_{\text{tot}} = \sum_{n=1}^{n=3} n\sigma_n$) on electron impact energy [71]

Boivin and Srivastava [72] measured the cross sections of single, double, and triple ionization of magnesium over the 0–700 eV range of electron energy using a crossed beam technique. They indicated that the measurements are in good agreement with theory of McGuire [73] and Peach [74] using Born approximation.

Their results are presented in Table 1.3. These measurements of the single ionization cross section as dependence on electron energy Boivin and Srivastava [72] compared with the other experiments showed in Fig. 1.11 indicating agreements and differences with their data depended on electron energy. The details discussed in [72].

1.3.5 Electron–Ion Recombination

This is a process, which is inverse to the ionization. As result, a neutral atoms or molecules are formed during collisions of free electrons with positive ions, which in general can be also in an excited state. The charge density decreases during the recombination according to equation

$$\frac{dn}{dt} = -R \tag{1.48}$$

where R is the recombination rate in cm^{-3}/s. Two main types of electron–ion recombination were determined the expression for R. In relatively low dense plasma mainly by a radiative recombination, when one electron was captured by the ion and released energy is radiated in form of a quantum, while the atom remains in neutral or exited state:

$A^+ + e \leftrightarrow A$ (or A^*) $+ h\nu$

The radiative recombination is a process reverse to the photo ionization. When an addition electron captured the released energy, the ion recombination occurs as three-body process

$A^+ + e \leftrightarrow A^+ + 2e$ or

$A^+ + e \leftrightarrow A^+ + e + e(h\nu)$

where A^+ is a singly charged ion, e is an electron, h is Planck's constant, ν is the frequency of the radiation. For radiative case,

$R = \beta n^+ n$

and for in three-body recombination

$R = \beta n^+ n^2$

Coefficient β characterized the recombination rate, i.e., decrease rate of the charge particle density. Recombination coefficient can be obtained from (1.48) when the electron density decaying known from an experiment [13, 29, 30]. Probe methods and optical techniques used to determine β from observations of the (optically thin) plasma. In the dense plasma of high current discharges, mainly three-body recombination was occurred. In general, β was determined by a recombination cross section σ_r and electron velocity v function distribution $f(v)$ as:

Table 1.3 Cross sections for single σ_1, double σ_2, and triple σ_3 ionization of Mg by electron impact [72]

Electron energy (eV)	σ_1 10^{-16} cm^2	σ_2 10^{-17} cm^2	σ_3 10^{-18} cm^2
5.0	0.000	–	–
10.0	2.100	–	–
15.0	4.566	–	–
20.0	5.080	0.000	–
25	4.926	0.049	–
30.0	4.731	0.124	–
35.0	4.499	0.236	–
40.0	4.232	0.324	–
45.0	3.972	0.364	–
50.0	3.733	0.385	–
55.0	3.513	0.436	–
60.0	3.312	0.538	–
65.0	3.128	0.643	–
70.0	2.959	0.745	–
75.0	2.804	0.846	–
80.0	2.663	0.944	–
85.0	2.534	1.039	–
90.0	2.416	1.131	–
95.0	2.308	1.220	–
100.0	2.209	1.306	0.000
110.0	2.037	1.467	0.011
120.0	1.893	1.614	0.030
130.0	1.771	1.747	0.057
140.0	1.669	1.865	0.089
150.0	1.582	1.969	0.126
160.0	1.507	2.059	0.169
170.0	1.441	2.136	0.216
180.0	1.383	2.202	0.267
190.0	1.330	2.256	0.321
200.0	1.282	2.301	0.376
225.0	1.175	2.374	0.519
250.0	1.080	2.406	0.657
275.0	0.994	2.411	0.781
300.0	0.916	2.400	0.885
325.0	0.848	2.382	0.966
350.0	0.792	2.360	1.022

(continued)

Table 1.3 (continued)

Electron energy (eV)	σ_1 10^{-16} cm^2	σ_2 10^{-17} cm^2	σ_3 10^{-18} cm^2
375.0	0.748	2.337	1.055
400.0	0.716	2.313	1.070
425.0	0.693	2.286	1.071
450.0	0.677	2.255	1.064
475.0	0.664	2.219	1.052
500.0	0.650	2.178	1.038
525.0	0.633	2.132	1.023
550.0	0.612	2.085	1.008
575.0	0.587	2.039	0.990
600.0	0.560	1.997	0.967
625.0	0.535	1.959	0.937
650.0	0.518	1.919	0.904
675.0	0.513	1.862	0.877

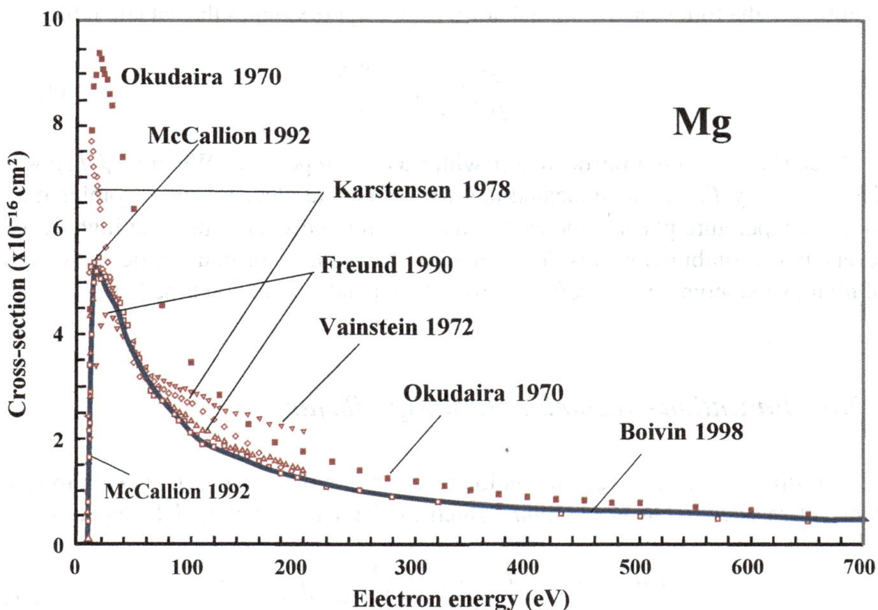

Fig. 1.11 Single ionization cross section of Mg by electron impact measured by Boivin and Srivastava (▬) [72] and comparison with other experiments: (◇) Karstensen and Schneider 1978 [75]; (□) McCallion et al. 1992 [76]; (Δ) Freund et al. 1990 [77]; (∇) Vainshtein et al. 1972 [66]; (■) Okudaira et al. 1970 [78]

$$\beta = \int_0^{\infty} v\sigma_r(v)f(v)\mathrm{d}v \tag{1.49}$$

As example, the function distribution in energy space $f(\varepsilon)$ was obtained using Fokker–Planck equation, and assuming that the atom temperature is much lower than the ionization potential, the recombination coefficient was [31]

$$\beta \sim \frac{\sigma_{el}N_a}{T^{2.5}}$$

More extensive theory of three-body recombination was considered in collisional–radiative approximation when the transitions between exited states in the atom were determined by collisions with electrons and also by the radiations [29, 30]. Using Gryzinski cross sections for the transitions between exited states and assumption that the atoms in ground state the coefficient of collisional–radiative recombination was calculated by Bates et al. [79, 80]. The data of β were presented in Table 1.4.

A simple approach assuming collisions in an ideal plasma was developed in [31] resulting in the following formula that very good approximates the data of Table 1.4.

$$\beta = \frac{\mathrm{e}^4}{3\hbar^2}\left(\frac{m}{T}\right)^{0.5}\frac{\mathrm{e}^{10}N_i}{T^5} \tag{1.50}$$

As can be seen, β sharply decreased with plasma temperature. When $T_e \neq T$, β was determined by T_e. It was indicated that for collisional electron–ion recombination in low-temperature plasma, the captured electron mainly takes place at high exited levels and recombination was determined by frequencies of non-elastic transitions of high exited atom, and therefore, β weakly depends on the ion type [31].

1.3.6 Ionization–Recombination Equilibrium

For the single ionization–recombination reaction of atomic gas (A), the equilibrium constant was defined as $K = n_e n_i/n_a$ which can be determined by Saha equation

$$\frac{n_e n_i}{n_a} = \frac{G_e G_i}{G_a}\left(\frac{2\pi m_e kT}{h}\right)^{3/2}\exp\left(-\frac{E_i}{kT}\right) \tag{1.51}$$

whereas T is the equilibrium plasma temperature, m_e, k, and h are the electron mass, the Boltzmann constant, and the Planck constant, respectively. E_i is the ionization energy, G_a and G_i are the internal partition functions of atoms and ions, respectively, expressed as

Table 1.4 Coefficient of collisional–radiative recombination β (cm^3/s) [79]

N_e, cm^{-3}	Temperature, K								
	250	500	1000	2000	4000	8000	16,000	32,000	64,000
$N_e \to 0$	4.8×10^{-12}	3.1×10^{-12}	2.0×10^{-12}	1.3×10^{-12}	7.9×10^{-13}	4.8×10^{-13}	2.9×10^{-13}	1.7×10^{-13}	1.0×10^{-13}
10^8	8.8×10^{-11}	1.4×10^{-11}	4.1×10^{-12}	1.8×10^{-12}	9.2×10^{-13}	5.1×10^{-13}	3.0×10^{-13}	1.8×10^{-13}	1.0×10^{-13}
10^9	4.0×10^{-10}	3.8×10^{-11}	7.5×10^{-12}	2.5×10^{-12}	1.0×10^{-12}	5.3×10^{-13}	3.0×10^{-13}	1.8×10^{-13}	1.0×10^{-13}
10^{10}	2.8×10^{-9}	1.6×10^{-10}	1.9×10^{-11}	4.1×10^{-12}	1.4×10^{-12}	6.1×10^{-13}	3.2×10^{-13}	1.8×10^{-13}	1.0×10^{-13}
10^{11}	2.7×10^{-8}	1.0×10^{-9}	6.9×10^{-11}	9.1×10^{-12}	2.2×10^{-12}	8.1×10^{-13}	3.4×10^{-13}	1.8×10^{-13}	1.0×10^{-13}
10^{12}	2.6×10^{-7}	9.0×10^{-9}	3.9×10^{-10}	2.9×10^{-12}	4.4×10^{-12}	1.2×10^{-12}	4.3×10^{-13}	2.0×10^{-13}	1.0×10^{-13}
10^{13}	2.6×10^{-6}	8.9×10^{-8}	3.1×10^{-9}	1.4×10^{-10}	1.2×10^{-11}	2.1×10^{-12}	6.2×10^{-13}	2.4×10^{-13}	1.1×10^{-13}
10^{14}	2.6×10^{-5}	8.8×10^{-7}	2.9×10^{-8}	9.8×10^{-10}	5.1×10^{-11}	5.1×10^{-12}	1.0×10^{-12}	3.1×10^{-13}	1.2×10^{-13}
10^{15}	–	8.8×10^{-6}	2.9×10^{-7}	8.7×10^{-9}	2.7×10^{-10}	1.7×10^{-11}	2.3×10^{-12}	4.9×10^{-13}	1.6×10^{-13}
10^{16}	–	–	2.9×10^{-6}	8.5×10^{-8}	2.3×10^{-9}	8.4×10^{-11}	5.0×10^{-12}	7.3×10^{-13}	1.9×10^{-13}
10^{17}	–	–	–	8.4×10^{-7}	2.1×10^{-8}	3.4×10^{-10}	1.4×10^{-11}	1.8×10^{-12}	4.4×10^{-13}
10^{18}	–	–	–	–	2.0×10^{-7}	2.5×10^{-9}	9.6×10^{-11}	1.2×10^{-11}	2.8×10^{-12}
$N_e \to \infty$	1.9×10^{-25} N_e	1.9×10^{-25} N_e	1.9×10^{-25} N_e	1.9×10^{-25} N_e	1.9×10^{-25} N_e	2.4×10^{-27} N_e	9.1×10^{-29} N_e	1.1×10^{-29} N_e	2.7×10^{-30} N_e

$$G_a = \sum_k g_{a,k} \exp\left(-\frac{E_{a,k}}{kT}\right)$$

$$G_i = \sum_k g_{i,k} \exp\left(-\frac{E_{i,k}}{kT}\right)$$

whereas $g_{a,k}$ or $g_{i,k}$ is the degree of degeneracy, $E_{a,k}$ or $E_{i,k}$ is the energy difference between the kth energy state and the ground state of the atoms or ions, respectively. Usually, for G_a calculation, the first two terms and for G_i only the first term were used and $G_e = 2$. The degree of degeneracy was defined as $g = (2S + 1)(2L + 1)$, where S is the spin a L orbital quantum numbers that determined the spin and orbital moment of the state. When the plasma temperature is high relatively to the multiple energy the partition functions were determined mainly by degree of degeneracy g .

In many cases, the state of discharge plasmas was not in a total equilibrium because the energy loss fraction of the electrons is relatively small due to the elastic collisions between them and the heavy particles and transferred kinetic energy is proportional to m_e/m. Therefore, the local electron temperature T_e of a plasma is higher than the heavy particle temperature T [81]. A number of improved approaches were developed taking into account that $T_e \neq T$.

Chen and Han [82] reviewed these approaches, and they proposed an original thermodynamic derivation of the two-temperature Saha equation. The model considered for small plasma region in which all parameters were assumed uniform. The plasma system was divided into two parts, i.e., subsystem 1 and subsystem 2. Subsystem 1 includes all the free electrons and all the internal excited energy states of the atoms, the ions, while all the translational degrees of freedom of the atoms and the ions are grouped into subsystem 2. Subsystem 1 and subsystem 2 were assumed to be adiabatic to each other due to the extreme weakness of the translational kinetic energy coupling between the electrons and heavy particles, as mentioned above, but work or particle transfer may take place between the two subsystems due to the ionization and recombination (by three-body collision) reactions. Subsystem 1 is characterized by the electron temperature. The average translational kinetic energy of atoms and ions in subsystem 2 was correspondent the heavy particle temperature T. It is assumed that the free electrons and the heavy particles satisfy the Maxwellian velocity distributions at T_e and T, respectively, while the population densities of the different excited states of atoms or ions satisfy the Boltzmann distribution with the characteristic excitation temperature $T_{ex} = T_e$. Equilibrium condition was used as $\Sigma_j \nu_j \mu_j / T_j = 0$ (ν_j and μ_j are the stoichiometric coefficient and the chemical potential of the j-th species in the reaction). With these assumptions and using the basic thermodynamic principle, expressed that the entropy of an isolated system will reach its maximum at the equilibrium state, and the modified Saha equation was derived in form [82]

$$\frac{n_e n_i}{n_a} = \frac{2G_i(T_e)}{G_a(T_a)} \left(\frac{2\pi m_e k T_e}{h}\right)^{3/2} \exp\left(-\frac{E_i}{kT_e}\right) \tag{1.52}$$

Equation (1.52) is identical to (1.51) using instead T the electron temperature T_e and to obtained that previously derived by using the kinetic theory. The expression for multiple ionization can be written as ($j = 1, 2, 3\ldots$)

$$\frac{n_e n_{ij}}{n_{ij-1}} = \frac{2G_{ij}(T_e)}{G_{ij-1}(T_e)}\left(\frac{2\pi m_e k T_e}{h}\right)^{3/2} \exp\left(-\frac{E_{ij}}{kT_e}\right) \quad . \tag{1.53}$$

References

1. Lieberman, M. A., & Lichtenberg, A. J. (2005). *Principles of plasma discharges and materials processing*. John Wiley and Sons.
2. Popov, O.A. (Ed.). (1995). *High density plasma sources. Design, physics and performance.* Park Ridge, New Jersey: Noyes Publications.
3. Raizer, Yu P. (1997). *Gas discharge physics*. Berlin: Springer-Verlag.
4. Keidar, M., & Beilis, I. I. (2016). *Plasma engineering* (2nd ed.). London-NY: Acad Press, Elsevier.
5. Langmuir, I. (1928). Oscillations in ionized gases. *Proceedings of National Academy of Sciences, 14,* 627.
6. Debye, P., & Huckel, E. (1923). Zur Theorie der Elektrolyte. *Physicalische Zetschrift, 24*(9), 185.
7. Artsimovich, L. A., & Sagdeev, R. Z. (1983). *Plasma physics for physicists*. Berlin: Springer-Verlag.
8. Alexandrov, A. F., Bogdankevich, L. S., & Rukhadze, A. A. (1984). *Principles of plasma electrodynamics*. Heidelberg: Springer Verlag.
9. Neslin, M. V. (1991). *Dynamics of beams in plasmas*. Heidelberg: Springer Verlag.
10. Fortov, V. E., & Yakubov, I. T. (1994). *Non-ideal plasma* (p. 367). Moscow: Energoatomizdat Publ (in Russian).
11. Granovsky, V. L. (1952). *Electrical current in gases. General questions of gas electrodynamics.* Moscow: Gostekhisdat. (in Russian).
12. Kaminsky, M. (1965). *Atomic and ionic impact phenomena on metal surfaces*. Berlin: Springer-Verlag.
13. Massey, H. S. W., & Burhop, E. H. S. (1952). *Electronic and ionic impact phenomena*. Oxford: Clarendon Press.
14. Mashkova, E., & Molchanov, V. (1980). *The average ion energy sputtering by surfaces of solid body*. Atomisdat (in Russian).
15. Hagstrum, H. D. (1954). Auger ejection of electrons from Tungsten by noble gas ions. *Physical Review, 96*(2), 325–335.
16. Hagstrum, H. D. (1956). Auger ejection of electrons from molybdenum by noble gas ions. *Physical Review, 104*(3), 672–683.
17. Hagstrum, H. D. (1954). Theory of auger ejection of electrons from metals by ions. *Physical Review, 96*(2), 336–365.
18. Hayward, W. H., & Wolter, A. R. (1969). Sputtering yield measurements with low energy metal ion beams. *Journal of Applied Physics, 40*(7), 2911–2916.
19. Townes, C. H. (1944). Theory of cathode sputtering in low voltage gaseous discharges. *Physical Review, 65*(11–12), 319–327.
20. Harrison, D. E. (1960). Theory of the sputtering process. *Physics Review, 102*(6), 1473–1480. Rol, P. K., Fluit, J. M., & Kistemaker, J. (1960). Theoretical cathode sputtering energy 5–25 keV, *Physica, 26*, 1009–1011.

21. Almen, O., & Bruce, G. (1961). Collection sputtering noble gas ions. *Nuclear Instruments and Methods, 11,* 257–278.
22. Miller, H. C. (1979). On sputtering in vacuum arcs. *Journal of Physics. D. Applied Physics, 12*(8), 1293–1298.
23. Biberman, L. M., Vorob'ev, V. S., & Yakubov, I. T. (1987). *Kinetics of nonequilibrium low-temperature plasmas.* New York: Consultants Bureau.
24. Braginsky, S. I. (1965). Phenomena of transferring in a plasma. In M. A. Leontovich (Ed.). *Reviews of the plasma physics* (Vol. 1). New York: Consultants Bureau.
25. Mott, N. F., & Massey, H. S. W. (1965). *The theory of atomic collisions.* Oxford: The Clarendon Press.
26. Francis, G. (1960). *Ionization phenomena in gases.* London: Butterworths Scientific Publication.
27. Bagenova, T. V., Kotlyarov, A. D., & Uvarov, V. M. (1980). Determination of effective cross-section molecule H_2O with electrons in a plasma. *High Temperature, 18*(5), 906 (in Russian).
28. Itoh, H., Musumara, T., Satch, K., Date, H., Nakao, H., & Tagashira, H. (1993). Electron transport coefficient in SF_6. *Journal of Physics. D. Applied Physics, 26,* 1975.
29. McDaniel, E. W. (1964). *Collision phenomena in ionized gases.* N.Y.-London: John Wiley.
30. Hasted, J. B. (1964). *Physics of atomic collisions.* London: Butterworths Scientific Publication.
31. Smirnov, B. M. (1968). *Atomic collisions and elementary processes in plasma.* Moscow: Atomizdat.
32. Perel, J., Englander, P., & Bederson, B. (1962). Measurement of total cross sections for the scattering of low-energy electrons by lithium, sodium, and potassium. *Physical Review, 128*(3), 1148–1154.
33. Romanyuk, N. I., Shpenin, O. B., & Zapesochny, I. P. (1980). The cross-section of electron scattering on atoms of Ca, Sr, and Ba. *Pis'ma JETP, 32*(7), 472–475.
34. Sena, L. A. (1948). *Electron and ion collisions with gas atoms.* L-M: Gostechisdat. (in Russian).
35. Vainshtein, L. A., Sobelman, I. I., & Yukov, E. A. (1978). *Atom excitation and broadening of spectral lines.* Moscow: Nauka. (in Russian).
36. Firsov, O. B. (1951). Resonance charge-exchange of ions for slow collisions. *Soviet Physics JETP, 21,* 1001.
37. Smirnov, B. M. (1975). *Introduction in plasma physics.* Moscow: Nauka. (in Russian).
38. Thomson, J. J. (1912). Ionization by moving electrified particles. *Philosophical Magazine and Journal of Science. Series, 6*(23), 449–457.
39. Landau, L. D., & Lifshitz, E. M. (1965). *Mechanics.* Moscow: Nauka.
40. Devis, B. (1918). Characteristic X-ray emission as a function of the applied voltage. *Physical Review, 11*(6), 433–444.
41. Rosseland, G. (1923). On the theory of ionization by swiftly moving electrified particles and the production of characteristic X-rays. *Philosophical Magazine and Journal of Science. Series, 7*(45), 65.
42. Killian, T. J. (1930). The uniform positive column of an electric discharge in Hg vapor. *Physics Review, 35,* 1238–1252.
43. Tomas, L. H. (1927). The effect of the orbital velocity of the electrons in heavy atoms on their stopping of α-particles. *Proceedings of the Cambridge Philological Society, 23,* 713–716.
44. Tomas, L. H. (1927). The production of characteristic X rays by electronic impact. *Proceedings of the Cambridge Philological Society, 23,* 829–831.
45. Webster, D. L., Hansen, W. W., & Duveneck, F. B. (1933). Probabilities of K-electron ionization of silver by cathode rays. *Physical Review, 43*(N11), 839–858.
46. Drawin, H. W. (1961). Zur formelm ~ ifligen DarsteHung der Ionisierungsquerschnitte gegenuber Elektronenstoss. *Zeitschrift fiir Physik, 164,* 513–521.
47. Gryzinski, M. (1957). Stopping power of a medium for heavy, charged particles, *Physics Review Series 2, 107*(6), 1471–1475.
48. Gryzinski, M. (1959). Classical theory of electronic and ionic inelastic collisions. *Physical Review, 115*(2), 374–383.

49. Gryzinski, M. (1965). Two-particle collisions I. General relations for collisions in the laboratory system. *Physics Review, 138*(2), A305–A321.
50. Gryzinski, M. (1965). Two-particle collisions II. Coulomb collisions in the laboratory system of coordinates. *Physics Review, 138*(2), A322–A335.
51. Gryzinski, M. (1965). Classical theory of atomic collisions. I. Theory of inelastic collisions. *Physics Review, 138*(2), A336–A358.
52. Ochkur, V. I., & Petrun'kin, A. M. (1963). The classical calculation of the probabilities of excitation & ionization of atoms by electron impact. *Optics and Spectroscopy, 14*(4), 457–464.
53. Born, M. (1926). Zur Quantenmechanik der Stoßvorgänge. *Zeitschrift für Physik, 37*(12), 863–867.
54. Born, M. (1926). Quantenmechanik der Stoßvorgänge. *Zeitschrift für Physik, 38*(11–12), 803–827.
55. Oppenheimer, J. R. (1928). On the quantum theory of electronic impacts. *Physics Review, 32*, 361.
56. Peterkop, R. (1961). Consideration of exchange in ionization. *Proceedings of the Physical Society, 77*(6), 1220–1222.
57. Ochkur, V. I. (1964). Born-Oppenheimer method in the theory of collisions. *Soviet Physics JETP, 18*(2), 503–508.
58. Fisher, V., Bernshtam, V., Golten, H., & Maron, Y. (1996). Electron-impact excitation cross sections for allowed transitions in atoms. *Physical Review, 53*(4), 2425–2432.
59. Bernshtam, V. A., Ralchenko, Yu V, & Maron, Y. (2000). Empirical formula for cross section of direct electron-impact ionization of ions. *Journal of Physics B: Atomic, Molecular and Optical Physics, 33*, 5025–5032.
60. Gupta, D., Naghma, R., & Antony, B. (2013). Electron impact total ionization cross sections for Sr, Y, Ru, Pd, and Ag. *Canadian Journal of Physics, 91*(9), 744–750.
61. Engel, A. (1955). *Ionized gases*. Oxford: Clarendon Press.
62. Beilina, G. M., Pavlov, S. I., Rakhovskii, V. I., & Sorokaletov, O. D. (1965). Measurement of electron impact ionization function for metal atoms. *Journal of Applied Mechanics and Technical Physics, 6*(2), 86–88.
63. Pavlov, S. I., Rakhovskii, V. I., & Fedorova, G.M. (1967). Measurement of the cross section for ionization by electron impact in substances with low vapor pressure. *Soviet Physics JETP, 52*(1), 21.
64. Pavlov, S. I., & Stotsky, G. I. (1970). Single and multiple ionization of lead atoms by electrons. *Soviet Physics JETP, 58*(1), 108.
65. Rakhovsky, V. I., & Stepanov, A. M. (1969). Measurement of the cross section for ionization by electron impact of Ca atoms. *High Temperature, 7*(6), 1071.
66. Vainshtein, L. A., Ochkur, V. I., Rakhovsky, V. I., & Stepanov, A. M. (1971). Absolute values of the electron impact ionization cross section for magnesium, calcium, strontium and barium. *JETP, 61*(8), 511.
67. Cooper, J. L., Pressley, G. A., & Stafford, F. E. (1066). Cooper-electron impact ionization cross section for atoms. *The Journal of Chemical Physics, 44*(10), 3946–3949.
68. Lin, S.-S., & Stafford, F. E. (1067). Electron impact ionization cross section IV. Group IVb atoms. *The Journal of Chemical Physics, 47*(11), 4664–4666.
69. Crawford, C. K., & Wang, K. I. (1967). Electron impact ionization cross section for silver. *The Journal of Chemical Physics, 47*(11), 4667–4669.
70. Nygaard, K. J. (1968). Electron impact ionization cross section in cesium. *The Journal of Chemical Physics, 49*(5), 1995–2002.
71. Nelson, A. V. (1976). Electron impact ionization cross sections of gold, chromium and iron. *Technical Report* AFML-TR-75–198 (pp. 1–88). Massachusetts: MIT, Cambridge.
72. Boivin, R. F., & Srivastava, S. K. (1998). Electron-impact ionization of Mg. *Journal of Physics B, 31*(10), 2381–2394.
73. McGuire, E. J. (1977). Electron ionization cross sections in the Born approximation. *Physical Review A, 16*(1), 62–72.

74. Peach, G. (1970). Ionization of neutral atoms with outer 2p, 3s and 3p electrons by electron and proton impact. *Journal of Physics B: Atomic and Molecular Physics, 3,* 328–349.
75. Karstensen, F., & Schneider, M. (1978). Absolute cross sections for single and double ionization of Mg atoms by electron impact. *Journal of Physics B: Atomic and Molecular Physics, 11*(1), 167.
76. McCallion, P., Shah, M. B., & Gilbody, H. B. (1992). Multiply ionization of magnesium by electron impact. *Journal of Physics B: Atomic, Molecular and Optical Physics, 25,* 1051–1060.
77. Freund, R. S., Wetzel, R. C., Shul, R. J., & Hayes, T. R. (1990). Cross-section measurements for electron-impact ionization of atoms. *Physics Review, 41*(7), 3575–3595.
78. Okudaira, S., Kaneko, Y., & Kanomata, I. (1970). Multiple ionization of Ar and Mg by electron impact. *Journal of the Physical Society of Japan, 28*(6), 1536–1541.
79. Bates, D. R., Kingston, A. E., & McWhirter, R. W. P. (1962). Recombination between electrons and atomic ions. I. Optically thick plasmas. *Proceedings of the Royal Society, 267*, 297–312.
80. Bates, D. B., Kingston, A. E., & McWhirter, R. W. P. (1962). Recombination between electrons and atomic ions. II. Optically thick plasmas. *Proceedings of the Royal Society, 270,* 155–167.
81. Boulos, M. I., Fauchais, P., & Pfender, E. (1994). *Thermal plasma: Fundamentals and applications* (Vol. 1). New York: Plenum).
82. Chen, X., & Han, P. (1999). On the thermodynamic derivation of the Saha equation modified to a two-temperature plasma. *Journal of Physics D: Applied Physics, 32,* 1711–1718.

Chapter 2
Atom Vaporization and Electron Emission from a Metal Surface

Two important phenomena at electrode surfaces occurred during the plasma formation in the arcs. These processes will be described in this chapter. When the electrode is heated, the neutral atoms evaporate from the hot surface, a vapor plume is generated, and due to its expansion, it has a pressure lower than saturated pressure. The rarefied collisions determine the kinetics of the plume flow immediately near the surface. Vaporization of the heated electrode material will be analyzed using the kinetic approach. Simultaneously, the electrons are also emitted from the hot electrode surface. The current of electron emission is enhanced by the electric field produced at the surface in case of cathode plasma generation. In this Chapter, we will describe the electron emission caused by different mechanisms due to high body temperature and electric field at the surface as well as the influence of an individual field and explosive phenomena.

2.1 Kinetics of Material Vaporization

Two regimes could describe the material vaporization and its expansion: (i) kinetic-rarefied flow and (ii) continue hydrodynamic flow. Below the first region will be considered.

2.1.1 Non-equilibrium (Kinetic) Region

In 1982, Hertz [1] studied the rate of mercury evaporation by its condensation on a cooled substrate and measuring the vapor pressure. As a result, Hertz concluded that the maximal rate of material evaporation determined by the following flux density:

I. Beilis, *Plasma and Spot Phenomena in Electrical Arcs*, Springer Series on Atomic, Optical, and Plasma Physics 113, https://doi.org/10.1007/978-3-030-44747-2_2

$$W_H = \frac{1}{4} n v, \tag{2.1.1}$$

where n is the equilibrium vapor density at the surface and v is the average atom velocity. Langmuir [2] considered a metal surface in equilibrium with its saturated vapor when the rate of evaporation and condensation is balanced. In general, this balance depends on surface phenomena—condensation, evaporation, and atom reflection [3]. Assuming that the rates of evaporation and condensate are equal, Langmuir proposed that the rate relation between material evaporation W_L (atom/cm^2/s) into vacuum and the equilibrium vapor pressure p_0 can be described as [4]:

$$W_L = \frac{p_0}{\sqrt{2\pi m k T_0}} \tag{2.1.2}$$

where m is the atom mass, k is Boltzmann's constant, and T_0 is the surface temperature. Equation (2.1.2) is applicable for relatively low temperatures when $p_0 \leq 1$ torr when the evaporated material flows into surrounding space without collisions. With the surrounding pressure increase, the rate of evaporation is reduced. Langmuir hypothesized that in some gas pressure, the evaporated atom diffuses through the gas before they leave the surface.

The character of the mass flow depends on relation between atom mean free path λ and the chamber size L, and it is determined by the Knudsen number $Kn = \lambda/L$ [5, 6]. The molecular flow is produced when $Kn > 1$, while the flow as continue media occurred when $Kn \ll 1$. Knudsen [7] showed that for strong evaporation, the returned atom flux is required to be considered due to rarefied collisions in a near wall non-equilibrium layer ($Kn < 1$). It was proposed that the returned flux was determined by the vapor temperature T and pressure p, and these values were different from the surface temperature T_0 and saturated pressure p_0. The atom evaporation rate was described by so-called Hertz–Knudsen formula [1, 7] that can be expressed using relation (2.1.2) in form:

$$W_{HK} = \frac{p_0}{\sqrt{2\pi m k T_0}} - \frac{p}{\sqrt{2\pi m k T}} \tag{2.1.3}$$

Fonda [8, 9] measured the evaporation rate of tungsten filament. The measurement showed that the evaporation rate was reduced at a gas pressure presence in comparison with that in vacuum. The results were described in the framework of Langmuir atom diffusion hypothesis by the following model. The filament is surrounded by tungsten vapor at same pressure. The atoms of this vapor, however, instead of being projected directly from the filament, as in a vacuum, were pictured as diffusing through the stationary layer of gas. Once an atom reached the outer boundary of the layer, it would be carried away in the convection current of gas and would be lost to the filament. The path within the film was so irregular that an atom might return to the filament and be deposited on it. As result, the rate of evaporation was reduced as compared with that in a vacuum.

To determine the returned flux, a kinetic treatment of atoms in the vapor is requested. Thus, with increasing T_0, some atoms flow back to the surface due to collisions, and therefore, the net rate of material evaporation is less than that given by (2.1.1). While a dense vapor flow can be treated hydrodynamically for relatively large distances from the surface, within an atom mean free path of the surface collisions are rare, and a non-equilibrium layer of several mean free path lengths forms, through which atoms are returned back to the surface. The particle velocity distribution function approaches equilibrium in this "Knudsen layer" or atom relaxation zone. Collisions in the vapor relax the velocity distribution function (VDF) toward a Maxwellian form. Thus, two regions appear near the surface: (i) a non-equilibrium (kinetic) region with rare collisions and (ii) a collision dominated region with hydrodynamic vapor flow.

2.1.2 Kinetic Approaches. Atom Evaporations

Granovsky [10] studied in 1951 these two regions in order to describe the steady material vaporization in form close to (2.1.3). The atom flux density at the external Knudsen boundary was calculated as a difference between evaporated $m\Gamma_{ev}$ as (2.1.3) and the returned (condensed) mass fluxes density in form:

$$\rho_k v_k = m\Gamma_{ev} - \rho v k_m / 4 \tag{2.1.4}$$

where ρ_k is the mass density and v_k is the vapor velocity at the external Knudsen boundary, ρ is the mass density, v is the atom thermal velocity at the surface and k_m is the coefficient that takes in account a non-Maxwellian velocity DF due to the presence of the evaporated direct flux. To describe the hydrodynamic region, the Bernoulli and mass continuity equations were used. Granovsky considered two cases for (a) constant temperature and (b) adiabatic process, which the relation between expanding velocity v and density ρ obtained in form.

(a) Isothermal process:

$$v - v_k = \frac{2RT}{m} \ln \frac{\rho_k}{\rho} \tag{2.1.5}$$

(b) Adiabatic process:

$$v - v_k = \frac{\gamma}{1-\gamma} \frac{2RT_k}{m} \left[1 - \left(\frac{\rho}{\rho_k} \right)^{\gamma - 1} \right] \tag{2.1.6}$$

The above simple approach showed that the vapor velocity at the Knudsen layer is significantly lower than the sound speed and that the vapor expansion was close for spherical and plane flows.

Afanas'ev and Krokhin [11] considered the solid body vaporization by the laser power irradiation. The vapor expansion was studied in framework of one-dimension gasdynamic system of equations. The fact of strong laser power absorption mainly in a thin layer on the surface of the condensed body was used. The conservation laws of mass flow, momentum, and energy within this layer with sharp variation of all gasdynamic quantities were used to describe the boundary condition for the gasdynamic problem. Analytical solutions of the gasdynamic equations were derived to describe the process of vaporization and heating of matter for broad range of the radiation flux density and the parameters of condensed matter. The equilibrium between solid and vapor phase temperatures was assumed, and in a general case, the exact relation between the temperatures of the vapor and the condensed phase should be established by analyzing the kinetics of vapor flow at the vaporized surface.

In general, the kinetics of metal vaporization studied by solving the Boltzmann equation [12]. This study determines the flow parameters (density n, temperature T, and velocity v) in the non-equilibrium (Knudsen) layer. These parameters at the external boundary of the Knudsen layer served as boundary condition for the hydrodynamic region, for flow along a path from the surface. The Boltzmann equation in general case of atom flow is

$$\frac{\partial f}{\partial t} + \mathbf{v}\frac{\partial f}{\partial \mathrm{r}} = \left(\frac{\mathrm{d}f}{\mathrm{d}t}\right)_{\mathrm{col}} \tag{2.1.7}$$

where v is the velocity vector, r is a spatial coordinate, and f is the distribution function. Bhatnagar et al. (BGK) [13] modeled the collision term, and their approach is the simplest for the present case of sharp variation of the vapor parameters near the wall:

$$\left(\frac{\mathrm{d}f}{\mathrm{d}t}\right)_{\mathrm{col}} = \frac{f_0 - f}{\tau} \tag{2.1.8}$$

where f_0 is the local equilibrium distribution function, and τ is a constant collision time. The approach using (2.1.8) is called the relaxation model. In most practical cases, the characteristic time of the hydrodynamic parameter set is smaller than the characteristic time of the heat flux change to the surface [12]. This allows formulation of a steady-state one-dimension equation:

$$v\frac{\partial f}{\partial x} = \frac{f_0 - f}{\tau} \tag{2.1.9}$$

the following boundary conditions must be imposed to (2.1.9):

$$f_0 = n_0\left(\frac{m}{\pi 2kT_0}\right)^{3/2} \exp\left[-\frac{mv^2}{2kT_0}\right] \quad v_x > 0 \text{ at } x = 0 \tag{2.1.10}$$

$$f_\infty = n_\infty \left(\frac{m}{\pi 2kT_\infty} \right)^{3/2} \exp\left[-\frac{m}{2kT_\infty} \left((v_x - u)^2 + v_y^2 + v_z^2 \right) \right] \text{ on the equilibrium side.}$$

The evaporated atoms have a half-Maxwellian DF (f_0) characterized by T_0 and a given atom density n_0 previously shown in [14, 15]. Later, the measurements [16] by time of flight (TOF) showed that when low density particles were vaporized due to a pulsed heat source, the flow was characterized by collisionless regime. The particle emission was truly thermal and a half-Maxwellian DF, i.e., a Maxwellian with only positive velocities normal to the target (f_0), will then describe the velocities. For intense emitted particle flux, the near-surface collisions occurred, and a Knudsen layer was formed, i.e., there is a layer within a few mean free paths of the target surface in which the velocity distribution function evolves to a full Maxwellian ($f\infty$) at the external boundary.

Equation (2.1.10) for f_0 served as the boundary condition at the evaporated surface, i.e., at $x = 0$. The conservation laws in the non-equilibrium layer should be fulfilled taking into account the velocity distribution function in the following form [12, 17]:

$$\int v_x f(x, \mathbf{v}) d\mathbf{v} = C_1 \tag{2.1.11}$$

$$\int v_x^2 f(x, \mathbf{v}) d\mathbf{v} = C_2 \tag{2.1.12}$$

$$\int v_x v^2 f(x, \mathbf{v}) d\mathbf{v} = C_3 \tag{2.1.13}$$

Equations (2.1.9), (2.1.11)–(2.1.13) with boundary conditions (2.1.9) were solved using different approaches. Let us consider the different attempts to study the kinetics of the vaporizations.

Kucherov and Rikenglaz [18] developed a simple model of temperature and density jump at the boundary with a solid surface. Method of moments with a test distribution function close to the locally equilibrium function was used. The approach corresponded to small variation of the gas parameters in the layer, and the result is satisfactory for a case of low flow velocity in comparison with sonic velocity [12]. Anisimov considered more appropriate method consisting in the solution of the kinetic equation [12, 17]. He studied (see [17]) the Knudsen layer as a discontinuity surface in the hydrodynamic treatment of the vapor expanding from laser irradiation of a solid, using bimodal approximation of DF similar to earlier treatments of the shock wave problem [19]. Three equations of continuity, momentum, and energy conservations included four unknown vapor parameters with velocity at the external boundary of the Knudsen layer. Assuming that this velocity is the sonic velocity $u = u_{sn}$ (i.e., Mach number $M = 1$, evaporation in vacuum), Anisimov [17] obtained that the evaporated atom flux fraction is about 0.82 W_L. The dependence between the rate of propagation of the evaporation front and the heat flux was also studied showing that the evaporation front velocity is directly proportional to the specific heat flux [20]. Anisimov and Rakhmatulina [21] studied the transition of material vaporization into vacuum from the free molecular in the initial stage to a regime of continuity vapor expansion. Later Fisher [22] solved the kinetic problem using the BGK approach and obtained that the rate of evaporation into an infinite half-space is about 0.85 W_L. The system of (2.1.7)–(2.1.8) was solved numerically by iteration method and the regime when the bulk vapor flow that can be described by gasdynamic approach was obtained. It was indicated that the returned flux was finally formed at the time of ~10 times of mean free path.

Kogan and Makashev studied the Knudsen layer structure [23], boundary conditions for the vapor flow [24], and recondensation phenomena with arbitrary Knudsen number [25] considering material evaporation in the presence of chemical reactions at the surface. The Knudsen layer was investigated using the Hamel's model [26] of kinetic Boltzmann equation where the collision integral in right side of (2.1.7) replaced by collisional transfer of the momentum and energy for Maxwell molecules in a binary mixture.

Knight [27] also considers evaporation into vacuum from cylindrical surface extending Bhatnagar et al. approach [13] in assuming ellipsoidal approximation for the function distribution with different temperatures in radial, polar, and axial directions and using an expression for entropy production. The numerical solution was obtained for different Reynolds number, which reflected the vapor velocity indicating that the flow starts as a subsonic in non-equilibrium layer and becomes supersonic in future hydrodynamic expansion as condition of an entropy production. The asymptotic solution for large Reynolds number is in agreement with previous results. Knight [28] studied the vaporization into ambient gas using Anisimow's approach. The ambient pressure was considered in range from air 1 atm down to vacuum for laser interaction with an aluminum surface. It was concluded that the flow is subsonic and Mach number $M < 1$ at the layer edge. The temperature profile between parallel evaporated and condensate plates was discussed [29]. It was showed that the temperature jump is mostly appeared in a small boundary layer, which can be much lower or comparable with the distance between the plates. Baker [30] performed analysis of the Knudsen layer in order to predict the jump conditions across the layer for polyatomic gas as a function of the specific heat ratio γ and the Mach number M at the layer edge as parameters. It was shown that the jump of particle temperature and density increases with γ and M.

The kinetic aspects of arbitrary vaporization in boundary Knudsen layer were studied by Ytrehus and Ostmo [31] solving the stationary Boltzmann equation with the collision integral using the moments method for a single component system. The solution [17] was extended to arbitrary Mach number below 1 and calculated the structure of the Knudsen layer for Maxwell distribution function for molecules. The ratios of the gasdynamic parameters at the surface and at the external edge of the Knudsen layer were obtained as function external Mach number. The simulation showed that for $M = 1$, the returned net flow is about 18% of the emitted flux W_L in agreement with [17]. The evaporated flux rate dependence on M strongly deviated from that obtained according classical Hertz–Knudsen formula which neglects the mass velocity at the external boundary.

Mach number at the Knudsen external boundary is a parameter of the kinetic evaporation problem and depends on the flow conditions. In addition to the conservation laws, a condition for kinetics of the evaporation can be obtained using Boltzmann H-theorem [32]. The thermodynamic aspects of the vaporization characterized the entropy production was considered in [31, 33]. Previously, Muratova and Labuntsov [34] developed a kinetic analysis to study the solids vaporization using momentum method solving the Boltzmann equation, and then, Labuntsov and Kryukov [35] used the Boltzmann H-theorem as additional information to check the validity of approximate methods for the study of Knudsen problem. The authors of this work solved the Euler flow in hydrodynamic region together with equations for Knudsen layer obtained using approximation on the distribution function and conservation laws in a way similarly to that used in [17]. The parameters of Knudsen layer edge were obtained using the atom temperature as parameter, and the returned atom flux is 18% of the flux W_L at M close to 1.

Later, Rebhan considered the principle of entropy production to equilibrium system (Navier–Stokes flow) [36] and non-equilibrium system (shock wave, boundary layer) [37]. The entropy production was evaluated as parameter ignoring one of three conservation laws. The important point indicated that the maximal entropy change is appeared for strong shock wave as that obtained in the simulation approach used all three conservation laws. This result showed that the principle of maximum entropy production could be used to characterize the flow regime as one of conservation laws. It was concluded that the far from equilibrium flow regime was characterized by maximum entropy production, while the flow close to equilibrium was characterized by minimum entropy production.

It was shown above that Mach number is small ($M < 1$) in the case of material vaporization in an ambient media with arbitrary pressure. In this case, the velocity at the external boundary of the Knudsen layer varied as parameter solving three equations of continuity, momentum, and energy conservations included four unknown vapor parameters. Ford and Lee [38] extended the model [17] closing the system of equations by fourth condition and determined the range of Mach numbers at the boundary for the flow using Boltzmann H-theorem. Different types of velocity distribution function [17, 31] for particles incident onto the evaporating surface were assumed calculating the rate of entropy production as dependence on Mach number. It was concluded that the condition of $M = 1$ describing the material vaporization in vacuum can be justified by principle of maximum entropy production. And at the lower rate, the condition $M < 1$ was fulfilled. The range of gasdynamic parameters at the external boundary layer of Knudsen layer for arbitrary evaporation and condensation of one or multi-component particles was described considering the boundary problem and Boltzmann H-theorem [39].

Sibold and Urbassek [40] studied the kinetics of strong evaporation using Monte Carlo simulations in a one-dimensional approach of particle flow between a desorbing and a condensing surfaces. The numerical study used the Boltzmann equation solution taking into account a collision term. The boundary condition at the desorbing surface takes in account the half-Maxwellian distribution with the surface body temperature. The particle velocity distribution and Knudsen layer formation for one or binary light and a heavy species were studied as dependence on distance from the surface. It was indicated that at the end of the Knudsen layer, the calculated velocity distributions reached the equilibration process, while at the surface, however, the velocity cannot be fitted by a Maxwell distribution. In this case, a kinetic treatment should be used, and the present result deviates from that obtained by Ytrehus [31], which employed an ansatz distribution function. The distance at which the overall equilibrium is established is around 10–20λ.

The non-equilibrium evaporation and condensation phenomena in a Knudsen layer were considered by using gas kinetic moments method and molecular dynamics method (MD) in [41, 42] to study the layer structure as dependence on distance from the evaporating surface for different M at the external boundary. The velocity distribution functions for evaporation and reflection modes at various edges of Mach numbers as well as comparison of DSMC, MD, and BGK solutions were presented. It was shown that for strong evaporation, the main change of gas parameters (velocity, temperature components) occurred at few mean free path λ and then saturated after about 10λ.

2.1.3 Kinetic Approaches. Evaporations Into Plasma

According to the above brief survey, it can be emphasized that the intensive solid body vaporization has been widely studied considering mainly the kinetics of neutral atom flow in the Knudsen layer and consequent gasdynamic expansion. Such phenomena occur in a relatively low power density of the laser irradiation (<10^8 W/cm^2). In the case of a larger power density, the electron emission, laser plasma, and electric field near the target are formed. The electron emission can be supported by two mechanisms: multi-photon photoelectric effect and thermionic emission due to intensive target heating. The experimental investigation pointed out that the mechanism of electron emission is determined by incident energy fluence and laser pulse time [43]. Rhim and Hwang [44] studied Knudsen gas flow under an external potential field but without wall ablation. The Boltzmann equation in the relaxation form was solved for flow through parallel slit in which the particle collisions were occurred under given specific potential function. The diffusive reflection of gas molecule at the wall was assumed. The calculated result indicated larger gas flow velocity near the walls than that in the bulk flow. This result was explained by "surface diffusion" effect.

The vaporization phenomena become more complex when the intense local heat flux leads to the near-surface plasma generation. Such cases occur in the vacuum arc cathode spot in which the extremely high current density (10^6–10^8 A/cm^2) was formed in a minute area or by intense laser power density larger 10^7–10^8 W/cm^2. The kinetic model of complicated electrode vaporization in a vacuum arc spot plasma was developed by Beilis [45–47] taking into account the near-cathode electrical sheath and cathode electron emission. It was found that the plasma flow was strongly impeded and it is characterized by $M < 1$. The impeded mass flow was due to electrical current and energy dissipation in the dense cathodic vapor plasma, and it was found that the vapor flux into current carrying plasma differs from the flux W_L calculated with Langmuir's formula (2.1.2). The spot kinetic model, system of equation, and the results of simulation will be described below (Chap. 17). The condition of impeded plasma flow was also studied in the case of Teflon propellant in thrusters [48, 49].

A number of researchers studied the plasma production by high intense laser–target interaction theoretically [50]. A numerical model was developed by Bogaerts et al. [51] and Chen and Bogaerts [52] for Cu target. The target heating and the plasma formation was considered near the target surface. The ions and electrons emission from the heated surface was described by the Langmuir–Saha model, and in the vapor volume, the Cu atom ionization by the Saha–Eggert equations was calculated, assuming common temperature for the electrons, ions, and neutrals. This temperature was determined by the laser energy beam absorption in plasma, which was accelerated during its expansion. The plasma expansion was calculated using an equilibrium condition for the pressure and vapor density, and thermal atom velocity and W_L for surface evaporation rate as boundary condition [51]. The range of laser power density 10^8–10^{10} W/cm^2 was considered. The stage of heating, melting, and vaporization of the target material was studied. The expansion of the evaporated material plume in vacuum, the plasma formation, generating electrons, Cu^+ and Cu^{2+} ions, beside the Cu atoms was described, and the plasma shielding of the incoming laser light was indicated.

A 3D combined model of the laser induced plasma plume expansion in a vacuum or into a background gas was proposed by Itina et al. [53]. The model takes into account the mass diffusion and energy exchange between the ablated and background species, as well as the collective motion of the ablated species, and the background gas particles. Anisimov's result [17] was used as a boundary condition, including the Mach number $M = 1$ at the Knudsen layer for one-atomic vapor. Only singly charged ions were assumed. It was found that the temperature of electrons deviates from that of ions and neutrals. However, no information about the data of electron temperature as a boundary condition and in the expanding plasma was presented.

Recently, Gusarov and Aoki [54] considered the thermionic electron emission from Cu and Al targets, kinetics of ions and neutrals, and an electrostatic sheath that was formed at the surface. The temperature equilibrium for the plasma particles and weakly ionized vapor were assumed. The problem was solved for given plasma pressure at the external boundary of the Knudsen layer, as a parameter. The potential drop in the electrostatic sheath was determined taking into account the electron fluxes from the target and plasma.

The common weakness can be emphasis while considering the previously developed different approaches of modeling the laser plasma generation. As a rule, it was assumed an equilibrium boundary condition for plasma expansion and thermal equilibrium in the plasma. The weakly ionized vapor is described, neglecting the collisions between charged particles. The local vapor flow at the external boundary of the Knudsen layer was assumed at the sound speed. At the same time, the previous calculation indicated a relatively large particle temperature and a large degree of ionization of the vapor. Therefore, a space charge region with high potential drop and large electrical field at the surface, and as a result, large ion and electron emission currents at the target can be expected.

An extensive physical model and a mathematical formulation describing the laser generated plasma interaction with a target was developed by Beilis [55] using previously validated approaches developed for cathode spot description [45, 46, 56]. The kinetics of target ablation due to laser irradiation and the plasma energy flux of ions and electrons, the space charge sheath, as well as the adjacent plasma heating by emitted electron beam and plasma expansion into a vacuum were considered. The review of experimental study of laser plasmas and plasma acceleration was presented [57] (see details in Chap. 24).

2.2 Electron Emission

2.2.1 Work Function. Electron Function Distribution

Similarly, to free electrons that must acquire energy at the level of atom ionization to overcome the force of attraction, the electrons in solids must also acquire energy to leave the solid bulk due to presence on its border of a potential barrier, i.e., retarding forces. These forces have different origins. Firstly, when the electron is removed from a metal, some positive charge was induced in the body near its boundary (Fig. 2.1). The force of the interaction between the electron and the induced charge was called as the force of the mirror image and is constant at distance a of lattice constant $F = e^2/4a^2$ while depends on distance x as $F = e^2/4x^2$ for $x > a$, e-electron charge.

In addition, the electrons coming out from the metal form the electron cloud near the surface. This leads to formation at the boundary of a negative electrical layer retarding the output of electrons from the metal and contributing to formation of the potential barrier (Fig. 2.1). Thus, the transition of electrons from the metal in a vacuum requires certain energy to overcome such barrier. This energy is called the electron work function φ that is the difference between the energy states of the metal before and after electron transition per electron $\varphi = -(\partial \varepsilon/\partial N)_T$ at constant temperature T.

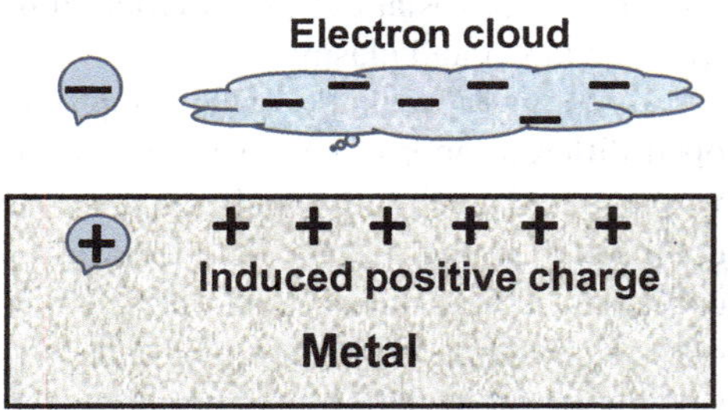

Fig. 2.1 Schematics of the contributors to the potential barrier formation

Electron energy ε_e in metals (with a temperature T) is distributed according to quantum statistics, which take into account Pauli principle, and described by Fermi–Dirac function distribution f:

$$f(\varepsilon) = \frac{1}{\exp\left(\frac{\varepsilon_e - \varepsilon_0}{kT}\right) + 1} \tag{2.2.1}$$

2.2.2 *Thermionic or T-Emission*

For the case of $T \rightarrow 0$ and for Fermi energy level $\varepsilon_0 > \varepsilon_e$, the function $f = f_0 = 1$. With increasing temperature ($T > 0$), the edge of the distribution function becomes blurred, and electrons are distributed by classical statistics with Maxwellian distribution. Thus, the electron with energy $\varepsilon_e > \varepsilon_0$ from the tail of the Maxwellian distribution can be able to escape the metal surface. Let us consider the electron flux in direction N_x (normal to the surface) through the metal edge per unit area with any velocities v_y and v_z but having velocity component v_x lying within dv_z using (2.1.1) [58]

$$\mathrm{d}N_x = \frac{2mv_x}{h^3} \int_{-\infty}^{\infty} \int_{-\infty}^{\infty} \frac{\mathrm{d}v_y\, \mathrm{d}v_z}{\mathrm{e}^{\frac{1}{kT}\left(\frac{m(v_x^2 + v_y^2 + v_z^2)}{2} - \varepsilon_0\right)}} \tag{2.2.2}$$

Integration of (2.2.2) leads to the following expression

$$\mathrm{d}N_x(\varepsilon_x, T) = \frac{4\pi mkT}{h^3} \ln\left(1 + \mathrm{e}^{-\left(\frac{\varepsilon_x - \varepsilon_0}{kT}\right)}\right) \mathrm{d}\varepsilon_x \tag{2.2.3}$$

When a metal is hot, the electron cloud at the surface is in thermodynamically equilibrium with the electron inside the metal. This means that the electron flux from the metal is equal to the electron back flux to the metal. Taking into account this specific condition, the electron current density j was obtained in accordance with the kinetic theory integrating (2.2.3) in form

$$j = AT^2(1 - R)\mathrm{e}^{-\frac{\varphi}{kT}};\ A = \frac{4\pi emk^2}{\hbar^3} = 120 \frac{\mathrm{A}}{\mathrm{cm}^2\, \mathrm{grad}^2} \tag{2.2.4}$$

where A is the universal constant and R is the coefficient of electron reflection from the edge. The mathematical form of emission law (2.2.4) based on kinetic theory of electron emission from a metal was developed by Richardson in 1911–1915 that discussed in [59, 60] and also by Dushman [61] and Nordheim [62]. Equation (2.2.4) describes early experimentally deduced exponential dependence of electron emission current from hot platinum on the temperature in 1901 and known as Richardson effect [63].

2.2.3 Schottky Effect. Field or F-Emission

Considering the classical statistics, electrons having energy larger than $\geq\varphi$ able to leave the metal, while according to the quantum theory, such electrons can be reflected from the metal edge.

Therefore, the edge metal, where a potential barrier is formed, can be characterized by a transmission coefficient D of this barrier. To understand this phenomenon, let us consider behavior of the electron potential energy at the edge of the metal (Fig. 2.2). When passing through the metal border, the potential increases from $-W$ (bottom of conduction band) to zero in the absence of external forces. However, in the presence of an electric field E, and by taking into account the image forces, the work function $\varphi(x)$ passes through a maximum and then decreases linearly. Integrating the force dependence, it can be obtained that

$$\varphi(x) = \varphi(0) - \int_{x_0}^{\infty} \left(eEx - \frac{e^2}{4x^2} \right) \mathrm{d}x \tag{2.2.5}$$

At the distance $x = x_0$ is $e^3E = e^2/4x_0$ (Fig. 2.2) and then $x_0 = (e/E)^{0.5}$. Taking into account this expression for x_0, a reduction of the potential barrier can be obtained as $\varphi(x_0) = \varphi(0) - \Delta\varphi = \varphi(0) - (e^3E)^{0.5}$ that is known as Schottky effect [64, 65]. In this case, the thermal emission in presence of an electric field is

$$j = AT^2 \mathrm{e}^{-\frac{\varphi - \sqrt{e^3 E}}{kT}} \tag{2.2.6}$$

According to this effect, E should be about 10^8 V/cm to reach $\Delta\varphi \sim \varphi$. However, significant electron emission was observed by $E = 10^6$ V/cm [65, 66]. This experimental evidence was studied within the framework of wave mechanics for which the electron can be emitted even for jump barrier by tunneling mechanism. Let us determine the probability of an electron passing through the barrier characterized by the wave function Ψ satisfies to Schrodinger equation [67]

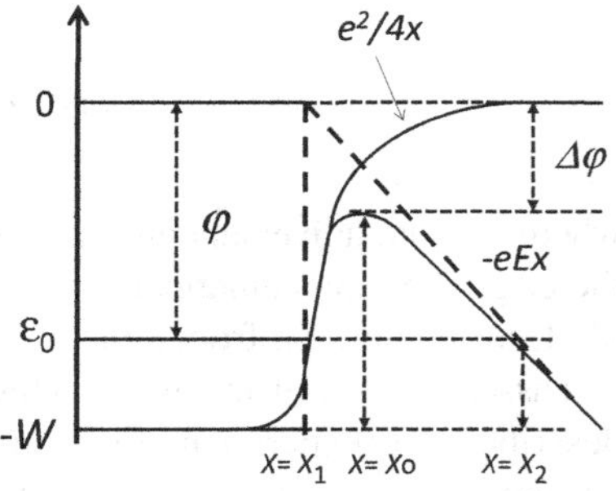

Fig. 2.2 Schematics of electron energy distribution near the metal surface with electric field and energy due to image force

$$\varepsilon\Psi + \frac{8\pi^2 m}{\hbar}\left(\varepsilon + eEx + \frac{e^2}{4x}\right)\Psi = 0 \tag{2.2.7}$$

The transmission coefficient D was defined as $D = \frac{|a_f|^2 - |a_r|^2}{|a_f|^2}$, where a_f, a_r are the amplitudes of incoming and reflected waves, respectively. The solution is searched for $x_1 < x < x_2$ outside the metal and $x < 0$ inside the metal. So, the transmission coefficient was obtained in form [67]:

$$D = 1 - R = 1 - \frac{\left(1 - \mathrm{e}^{-2K}/4\right)^2}{\left(1 + \mathrm{e}^{-2K}/4\right)^2} = \frac{\mathrm{e}^{-2K}}{\left(1 + \mathrm{e}^{-2K}/4\right)^2} \approx \mathrm{e}^{-2K} \tag{2.2.8}$$

$$K = \frac{\sqrt{2m}}{\hbar}\int\limits_{x_1}^{x_2}\sqrt{\left(|\varepsilon| + eEx + \frac{e^2}{4x}\right)}\mathrm{d}x \tag{2.2.9}$$

Expression (2.2.9) is an elliptical integral that was calculated by Nordheim as

$$K = \frac{4\pi\sqrt{2m}}{3\hbar}\frac{|\varepsilon|^{3/2}}{eE}\theta(y) \quad y = \frac{\sqrt{e^3 E}}{|\varepsilon|} \tag{2.2.10}$$

where y is the ratio of Schottky work function reduction to the work function with given electron energy. $\theta(y)$ is the Nordheim function indicating that $\theta(y) = 1$ for $y = 0$ and $\theta(y) = 0$ for $y = 1$. So, the transmission coefficient is

$$D(\varepsilon_x, E) = \mathrm{Exp}\left\{-\frac{8\pi\sqrt{2m}}{3\hbar}\frac{|\varepsilon_x|^{3/2}}{eE}\theta(y)\right\} \tag{2.2.11}$$

The electron emission current density j_e can be obtained taking into account (2.2.3) and (2.2.11) by

$$j_e = \frac{4\pi mkT}{h^3}\int\limits_0^{\infty} Ln\left(1 + \mathrm{e}^{-\frac{\varepsilon_x - \varepsilon_0}{kT}}\right)D(\varepsilon_x, E)\mathrm{d}\varepsilon_x \tag{2.2.12}$$

When electron emission was only under electric field, a condition $\boldsymbol{T \to 0}$ was considered. In this case, $\ln\left(1 + \mathrm{e}^{-\left(\frac{\varepsilon_x - \varepsilon_0}{kT}\right)}\right) \to -\frac{\varepsilon_x - \varepsilon_0}{kT}$ for $\varepsilon_x - \varepsilon_0 < 0$ and is equal to 0 for $\varepsilon_x - \varepsilon_0 > 0$. Taking into account that the electrons emitted from Fermi level and from levels close to ε_0, the emission is [68]

$$j = \frac{e^3 E^2}{8\pi h\varphi}\mathrm{Exp}\left\{-\frac{8\pi\sqrt{2m}}{3\hbar}\frac{\varphi^{3/2}}{eE}\theta\left(\frac{\sqrt{e^3 E}}{\varphi}\right)\right\} \tag{2.2.13}$$

Expression (2.2.13) describes the current density by field emission mechanism (F-emission) that also known as Fowler–Nordheim formula in form

$$j = 1.55 \times 10^{-6} \frac{E^2}{\varphi} \mathrm{Exp}\left\{ -\frac{6.85 \times 10^6 \varphi^{3/2}}{E} \theta\left(\frac{3.72 \times 10^{-4} \sqrt{E}}{\varphi} \right) \right\} \quad (2.2.14)$$

Function $\theta(y)$ was calculated and tabulated by Nordheim. The table of $\theta(y)$ was approximated by the following expression [69]

$$\theta(y) = 1 - y^2 \left\{ 1 + 0.85 \sin\left(\frac{1 - y}{2} \right) \right\} \quad (2.2.15)$$

Expression (2.2.15) indicates that $\theta(y) = 1$ for $y = 0$ and $\theta(y) = 0$ for $y = 1$.

2.2.4 Combination of Thermionic and Field Emission Named TF-Emission

When electron emission is under both electric field and temperature, conditions $\boldsymbol{T \neq 0}$ **and** $\boldsymbol{E \neq 0}$ must be considered. In this case, the electron emission is caused by solid heating and by electric field influence. As a result, the electrons can be emitted by tunneling through the potential barrier as well as by energetic electrons with energy larger than Schottky barrier. This type of emission called thermal field, ***T-F electron emission***. Different analytical approaches were developed assuming that the electrons leaved the metal from energetic levels in zones closed to the Fermi level, or to the Schottky level or to zone with levels between these [70–72]. The obtained analytical approximations were limited by the need of their use for determining the current of electrons emission from a specific energy bands, which correspond to given T and E. However, in practice, the necessity for calculating T-F electron emission occurs in a much broader range of temperatures and electrical fields than those that can be investigated by the limited analytical expressions. A review of these early published works was conducted in [73].

A more accurate calculation of the emission current j can be carried out by taking into account that electrons can be emitted from any energy level in the conducted band. Such general equation has the form

$$j = e \int_{-\infty}^{\infty} N_x(\varepsilon_e, T) D(\varepsilon_e, E) \mathrm{d}\varepsilon_e \quad (2.2.16)$$

Fermi–Dirac statistic for the electron population can be employed, and thus, the main issue is to obtain an accurate expression for transmission coefficient D. The one-dimensional model of the electron transmission gives several different approximations for D depending on additional assumptions used in the derivations [62, 67].

As noted by Elinson Vasil'ev [74] and Itskovich [75], the results of calculating by these approximations differ by a factor of about 2.

Based on generalized by WKB (Wentzel [76]–Kramers [77]–Brillouin) method, Kemble [78] obtains an expression for D, which differs from these mentioned above, in that, it describes the emission by one relation for both tunneling electrons and for electrons above the barrier. However, it is precisely near the summit of the barrier, the features of which were taken into account by Kemble, that the potential distribution used by Kemble is invalid [75]. As noted by Itskovich [75], any refinement in the calculation of D within framework of the free-electron model, which does not allow account an arbitrary dispersion law, cannot increase the accuracy of results deriving by this model itself. Itskovich [75] constructed a theory of electron emission for the case of an arbitrary dispersion law. The obtained results show that allowing for the dispersion law in theory of electron emission leads to the familiar relations obtained early but with a slightly increase of the work function φ. Thus, allowing for a complex of dispersion law is equivalent to making a certain correction to the value of φ, which is for most materials is relatively small.

Thus, Sommerfield theory [67] holds for the one-dimensional case and is based on the free-electron model. This approach was used for the explanation of different phenomena of emission electronics obtained experimentally [74]. The advancement of the free-electron theory was so successful that it was used as example for approving the base of quantum mechanics. Taking into account the above-mentioned arguments, let us consider the electron emission from the cathode of a vacuum arc caused by high temperatures and high electrical fields appearing in the spots. Below the electron emission will be analyzed utilizing the simplest and most frequently employed expression for transmission coefficient in form proposed by Sommerfeld [67]

$$D(\varepsilon, E) = \mathrm{Exp}\left\{-\frac{6.85 \times 10^6(\varphi-\varepsilon)^{3/2}}{E}\theta\left(\frac{3.72 \times 10^{-4}\sqrt{E}}{|\varphi-\varepsilon|}\right)\right\}$$

$$\varepsilon = \varepsilon_x - \varepsilon_0 \tag{2.2.17}$$

Utilizing (2.2.3), (2.2.16), and (2.2.17) different mechanisms the electron emission can be described which were determined by corresponding distribution in energy of the emitted electrons. To study of the effect of varying T and E, it is important to analyze how the energy distribution of the emitted electrons varies as a function of two aforementioned parameters, i.e., T and E. Such distribution was determined for relatively low temperatures (<3000 K) by Dolan and Dyke [79]. This approach [79] was used to describe the emitted electron distribution for wide range of variables T and E characterized for the vacuum arc at the cathode region [80]. An expression for the emission current density $j_e(\varepsilon)$, in A/(cm^2 eV), as a function of electron energy distribution using the product $N_e(\varepsilon, T)\, D(\varepsilon, E)$, can be expressed as

$$j_e(\varepsilon) = \frac{4\pi emkT}{h^3}\frac{Ln\left(1+\exp\left(-\frac{\varepsilon}{kT}\right)\right)}{\mathrm{Exp}\left\{-\frac{6.85\times10^6(\varphi-\varepsilon)^{3/2}}{E}\theta\left(\frac{3.72\times10^{-4}\sqrt{E}}{|\varphi-\varepsilon|}\right)\right\}} \tag{2.2.18}$$

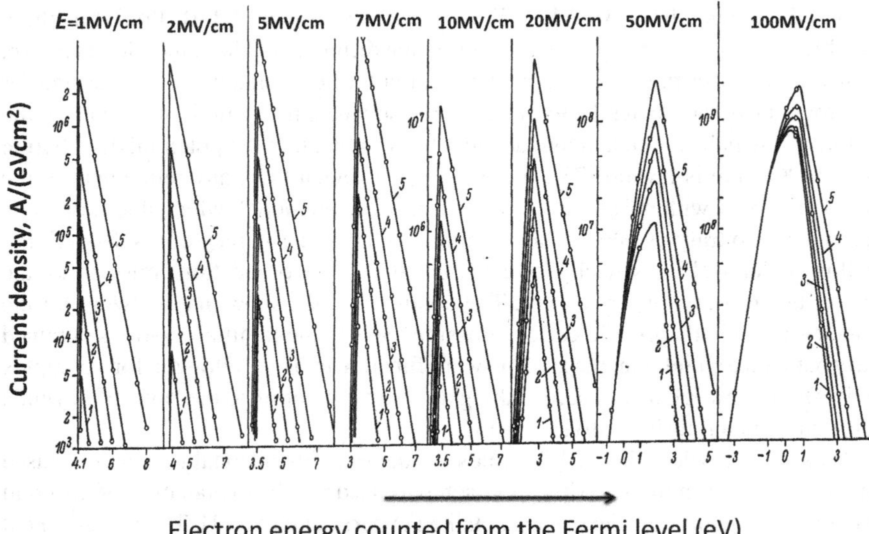

Fig. 2.3 Emission current density as function on energy of emitted electrons relatively Fermi level ($\varepsilon = 0$) for various cathode temperatures and electrical fields

The calculated results very obtained for ranges of $T = 3500 - 6000$ K and $E = 10^6 - 10^8$ V/cm, $\varphi = 4.5$ eV and presented in Fig. 2.3 as series of dependences. For each section, the captions cite the respective electric field and reduced by Schottky effect the work function $\varphi_s = \varphi - (e^3/E)^{0.5}$.

It can be seen that increase of the electric field shifts the curves into a region of negative values of ε, indicating the increasing role of field emission. With increasing temperature, the number of electrons excited to higher levels increases leading to more clear indication of the T-F character of the emission. These features quite clearly demonstrate the transition from one type of emission to other. A characteristic of the dependences is that the position of the maximum of each of temperature curves corresponds to electron energy equal to φ_s.

An exception of this rule occurs at $E = 10^8$ V/cm and $T < 5000$ K in electron energy distribution displaced toward smaller electron energy. For fields $E < 10^7$ V/cm, the number of emitted electrons with energy less than φ_s is negligible for any temperature in the investigated range of E and T. When $E = 5 \times 10^6$ V/cm and $T =$ 4000 K, the current density j_e can reach a value of 10^5 A/cm^2 per eV. An interesting and very evident fact that follows from the calculations of Fig. 2.3 is that the presence of Nottingham effect, i.e., heating of the cathode through the emission of electrons with energy less than the energy at the Fermi level. This effect begins when $E = 5 \times 10^6$ V/cm (small contribution) and becomes significant when $E = 10^8$ V/cm. It turns out that the Nottingham effect shows up in the investigated range of T only when the height of the potential barrier is small ($\varphi_s \sim 0.7$ eV; Fig. 2.3). The emission current j_e can exceed a value of 10^9 A/cm^2 in such case.

In general, the above calculation allows to conclude that the role of Nottingham effect is weak at the arc stage ($dI/dt \leq 10^7$ A/s) and therefore can be negligible in the thermal balance of the cathodes. Nevertheless, this effect manifests itself mainly under conditions, characterized the prebreakdown and spark stages with large rate of current I rise ($E \geq 10^8$ V/cm, $dI/dt \geq 10^8$ A/s). At the same time, in calculating the effect of cooling of the cathode through the emission of excited electrons, one can assume that the electrons carry away from the cathode an energy of about φ_s when $E \leq 2\times10^7$ V/cm. The contribution from tunneling electrons to the total emission flux, as follows from Fig. 2.3, is possibly not large.

2.2.5 *Threshold Approximation*

An original attempt to develop the theory of electron emission free from the essentially classical approach (e.g., existence of a potential barrier at metal–vacuum boundary, free-electron model, [62, 67, 74]) was performed by Brodskii and Gurevich [81, 82]. They use a threshold approximation that depends on existence of small parameter determined by a relatively small range of energies of emitted electrons. The developed concepts are based on the theory of potential scattering. A clear explanation was given on the electron emission phenomena at large transmission coefficient ($D \sim 1$) based on a comparison of the wavelength of emitted electrons with the characteristic thickness of the potential barrier. It turns out that the emission process in this case does not depend on the shape of the potential barrier and that the expression for D was specified solely by classical model. Accordingly, varying the specific form of this expression cannot have a decisive effect on the results of emission calculations. The most advance of the threshold theory [82] consists in possibility of derivation of expressions for the electron thermionic and field emissions by different ways. This allows to support the conclusion about universality of the formulas describing these emission mechanisms on the one hand. At the other hand, it also allows to study the possible deviations from the universal approach associated with strong electric fields, specifics of internal structure of the metals, boundary between two bodies and emission by photoelectron effect, etc.

2.2.6 *Individual Electron Emission*

Ecker and Muller [83, 84] were first to propose yet another type of emission, so-called, individual electron emission. This mechanism of electron emission was suggested to explain the nature of the cathode electron current in a vacuum arc without extremely large electric field, which necessary to use F-emission. The authors took into account that at the cathode plasma boundary, an ion approaching to the negatively charged surface induces a local field, which may influence the electron emission due to local fluctuation of the field E against its average value in the electric layer at the cathode. Using statistical methods of probability theory, it was obtained an expression for the probability distribution $W(E)$ of the ion density in space near the cathode surface. The current density of individual electron emission was proposed to calculate using Fowler–Nordheim dependence (2.2.14) of emitted electrons on electric field as

$$j = \int_0^\infty W(E) j(E) \mathrm{d}E \tag{2.2.19}$$

Integration of (2.2.19) can be performed by taking into account the lower limit for ion neutralization distance, the average electric field at the surface E_{ev}, the field distribution from the individual ions E_{in}, and the material work function φ. The numerical calculations were conducted for work function of 3.5–4.5 eV and for E_{ev}, used as parameter in range of $(0.4 - 4) \times 10^7$ V/cm. The result showed that the current density due to individual emission significantly exceeds that obtained using only average electric field. The role of individual emission was considered calculating the transmission coefficient for the potential barrier in approximation of (2.2.5) with adding the forces due to ion presence and its mirror image [85]. According to the model in works [83, 84], the potential barrier consists of two parts. The first part is usual and depends on electric field and electron image forces, while the second part determined by the potential distribution induced due to ion approaches the surface. An expression for probability distribution $W(E)$ of the ion density similar to one obtained in [83] was used. The calculation showed the relatively large electron emission caused by fields from separated ions, but the emission current was significantly lower than that obtained by Ecker and Muller [83], which is not real as indicated in [84].

The nonlinear behavior of the electric field at the cathode caused by approaching to the surface of the individual ions was considered in order to study the electron emission from the cathode of an electrical arc [86, 87]. The model predicts different result dependent on relation between cathode potential in the electrical sheath u_c and material work function φ. Thus, when $u_c < \varphi$ (low voltage discharge), the role of the individual ions can be considered, while for $u_c > \varphi$ (vacuum arc), the current density calculated taking into account the individual ions can be even lower than that obtained due to average electric field action. An influence of discrete ion fields on the electron emission from heated cathode to a temperature T was considered for electron emitted over the reduced barrier (by Schottky effect), named I-T-F-emission, and for tunneling electrons, named I-F-T-emission [88]. An analysis of the early published and recent (last decade) works indicates ambiguous effect of individual ions on field emission [89]. However, the general result of the above-mentioned works is that the electric field generated due to individual ion approaching to the emitting surface can influence only when the average field E_{ev} exceeds some critical value, which is about $(5–7) \times 10^7$ V/cm. But at such large values of E_{ev}, the cathode electron emission problem can be explained in frame of usual F-emission mechanism.

2.2.7 Fowler–Northeim-Type Equations and Their Correction for Measured Plot Analysis

As it was mentioned above, the standard Fowler–Northeim equation was derived for the case of a free-electron model and for a planar metal surfaces. Usually, the only simple type of test was conducted for consistency of field emission data with theory, i.e., by checking whether an experimental Fowler–Northeim plot in the logarithmic scale corresponds to a straight line. However, the described emission phenomena can be different from the emission in a real physical situation. According to Forbes [90], it should be taking into account significant effects resulting from: voltage drop in the measuring circuit, leakage currents; patch fields; field-emitted vacuum space charge, current-induced changes in emitter temperature; field penetration and band-bending; strong field fall-off, quantum confinement associated with small-apex-radius emitters, and field-related changes in emitter geometry or emission area or local work function. The metal can be with rough surface, including sharp emitters at which the field E_{loc} can be enhanced by geometric factor β_0 (which can be changed during the emission process) where $\beta_0 = E_{\text{loc}}/E$ and E is the field in the cathode–anode gap.

It was proposed [90] that the correlation between theoretical results and experimental data can be presented in ways, involving different choices of independent variable (e.g., barrier field, F_C; macroscopic field, F_M; device voltage, V_d; or measured voltage, V_m) and of dependent variable (e.g., characteristic local current density, J_C; macroscopic current density, J_M; device current i_d; or measured current, i_m). The need to distinguish between device parameters (i_d, V_d) and measured parameters (i_m, V_m) arises because measurement circuits may, in principle, have resistance in series and in parallel with the field emitter. In practice, parallel resistance can usually be made effectively infinite by improved system design, but, in some cases, it may be impossible to eliminate the series resistance. A quantitative test of Fowler–Northeim theory and theory related to Fowler–Northeim plots was developed in [91–93]. Reformulation of Fowler–Northeim theory presented in the cited papers makes it easier to generalize standard theory to treat the more realistic tunneling barriers and others leading to derive some universal Fowler–Northeim type equation by assumptions discussed in [90].

Fursey and Glazanov [94] modify the Fowler–Nordheim theory of field emission, for nanoscale emitters, considering the case when the protrusion apex curvature radius can become comparable or even less than the width of the potential barrier on the metal–vacuum boundary. In such case, the electric field near the surface is not homogeneous, and the barrier shape deviates from the traditional plane model. A numerical solution of Schrödinger equation for electron tunneling conducted, and the emission current density and energy distribution of emitted electrons for arbitrary barrier shape were calculated. Comparison of the obtained results with the prediction of conventional Fowler–Nordheim theory was discussed [68]. It was indicated that the traditional theoretical models such as “solid state,” “free electrons,” “band structure” need to be additionally thoroughly analyzed.

Recently also Kyritsakis and Xanthakis [95] derive analytically a generalized Fowler–Nordheim-type equation taking into account the curvature of a nanoscale emitter, so such solution is generally applicable to any emitter shape provided that the emitter is a good conductor and no field-dependent changes in emitter geometry occur. The ellipsoidal geometry was chosen for testing, where the beam width of electron beams emitted from the tip as a function of the tip-anode distance was calculated [96]. The developed approach allows to obtain the local field at the tip and the radius of curvature of the emitting surface.

It is well known that a protrusion growth takes place under high electric field applied to the metal surface. Zadin et al. [97] showed that nanoscale voids, presumably forming in the surface layers of metals during the technological processing, can be responsible for the onset of the growth of a surface protrusion. Finite element simulations were used to study the evolution of annealed copper, single crystal copper, and stainless steel surfaces that contain a void under high electric fields. The mechanical stress applied to the surface of each sample due to the electric field is calculated by using the Maxwell stress tensor for the electrostatic field equal $\varepsilon_0 E_n^2/2$ where, ε_0 the vacuum permittivity, and E_n is the electric field at the surface. According to the results, the stainless steel and single crystal copper surfaces demonstrate the formation of well-defined protrusions, when the external electric field reaches a certain critical value about (3–4) $\times$ 10^7 V/cm. The protrusion growth depends on the aspect ratio, defined as a ratio of depth to void radius ($h = r$), in the range from 0.2 to 2. Among the three materials, annealed copper surface starts yielding at the lowest electric fields due to the lowest Young's modulus and yield stress. However, it produces the smallest protrusions due to a significant strain-hardening characteristic for this material.

Timko et al. [98] determined the enhanced geometric factor β_0 measuring the electron field emission current as a function of external field E. Factor β_0 is then determined by a least-squares fit to the current deduced from Fowler–Nordheim equation corresponding to E. The local field E_{loc} was defined as $E_{\mathrm{loc}} = \beta_0 E$. The results indicated that for Cu the local field weakly depends on energy and the average of the local field was 96 MV/cm at breakdown, while the field enhancement factor increased from 50 to 80 with increasing energy in range 5–90 mJ. For Mo, the local field increased with increasing energy, while the field enhancement factor remained constant at 34. According to Descoeudres et al. [99] with copper electrodes, the local breakdown field is around 110 MV/cm, independent of the gap distance and β_0 was determined in range of 50–150 depending on surface modifications induced by the sparks result.

2.2.8 *Explosive Electron Emission*

This emission type initiated by rise of an electron current due to F-emission during high-voltage application to a needle-shaped emitter (cathode) at which the field E_{et} can be enhanced by geometric factor β_0. The emission was associated with Joule overheating of the emitter and its destruction. Dyke and Trolan [100] first reported the emitter overheating, explosion and transition phenomena to an arc discharge. Experimentally, it was shown that the time of vacuum arc formation was less than 5×10^{-8} s. During the formation of the arc, the current density j increased by approximately, two orders of magnitude. The observed current increase required either that j was considerably greater than current density, for which the space charge was necessarily neutralized, or that the effective emitting area was correspondingly increased, or both. The positive ions required by the former may have been supplied by the heated emitter whose surface material was evaporated as ions or as neutral atoms, which were ionized near the emitter surface in the high-density electron beam.

The follow-up study indicated that when the field current density from a tungsten emitter is continuously increased, the normal emission is terminated by an explosive vacuum arc [101]. The vacuum arc was initiated at a critical value of the field current density of the order of 10^8 A/cm^2. It was indicated that the current densities below the critical value supported by an electron emission, which apparently involved both by high electric fields and high temperatures. Dolan, Dyke and Trolan [102] provided an analysis of the transient heat conduction problem with resistively heating of used in experiments cone tungsten emitter whose half-angle was in the range 2.75°–15.5°, with a hemispherical tip of radius between 0.15 and 1.5 μm. Heat radiation is supposed to be negligible. Boundary conditions are based on the assumptions of zero temperature at a relatively great distance from the apex of the emitter, and no heat flow through the apex. The maximum steady-state temperature was obtained by the relation

$$T_{\max}(\deg\ \mathrm{C}) = 9.5 \times 10^{-4} j^2 r^2$$

where j is current density in A/cm^2, and r is emitter radius in cm. If $T_{\max}$ is to be held at less than 1000 °C, an arbitrary value used for illustration, it is clear that the product j^2r^2 must not exceed 10^6. Thus, in the range of radii from 10^{-4} to 10^{-5} cm, j for direct current operation may reach corresponding values from 10^7 to 10^8 A/cm^2 under the conditions here described in [102].

Flynn [103] studied interesting phenomena. He showed that after applying voltage, an intense field emission leads in a very short time to the initiation of the triggering arc, which causes a jet of vapor extended at high velocity from the cathode into the space between the anode and cathode. Flynn guessed that the vapor produced by the triggering arc is already ionized and that the jet actually consists of mixture of electrons and positive ions. Therefore, during the expansion from the cathode, advancing jet is effectively electrically conductive, and the applied voltage appears between the advancing ionized vapor front and anode. The limitation on the current will be set by electron space charge in the vacuum region. It proposed that this current could be drawn from the advancing plasma front. Flynn [103] indicates that the conducting plasma begins to traveling toward the anode with velocity v and the effective gap at any time will be $d - vt$, where d is the cathode–anode distance. Thus, the equation for the current density was obtained in frame of electron space charge limitation approach as [103]

$$j = k_{\rm sp} A U^{3/2} (d - vt)^{-2} \quad (2.2.20)$$

where $k_{\rm sp} = 2.34\ {\rm AV}^{-2/3}$ and A is the parameter described the plasma edge are expansion approximated function on vt. Using a value for v of 10^6 cm/s, measured previously by Easton et al. [104], the calculation conducted by Flynn showed that the predicted current grows agree well with measured current and voltage waveforms.

Considering the early works mentioned above, it can be noted that the phenomena studied with relatively low resolution in time for electrical pulses of about 1 μs. The improvement in experimental diagnostic systems with the highest resolution of about 10^{-10}–10^{-9}s happens in 60th and 70th decades of the last Century. This development led to new nanosecond pulse techniques [105–107]. The nanosecond experiments were conducted with a needle-shaped emitters and with a configuration of half-sphere disk. Approximately, 4–6 ns after applying of the voltage, a local luminescence at the cathode was observed that indicated the explosion points and that a plasma was formed. At the next few ns, the luminous region was extended showing plasma expansion in a vacuum. At this period, significant current rise was detected. The plasma velocity was determined as (2–3) × 10^6 cm/s (as by Easton et al. [104]) by measuring the displacement of the front of the plasma luminescence consistently during the breakdown development with nanosecond time resolution. It was shown that the electron emission from the cathode during an explosive transition of the metal into a plasma determines the rise of the electron current in the gap. This phenomenon was named as explosive electron emission (EEE), which was in detail investigated and widely presented since then [108–114]. Summarized results of EEE investigations were described in [115].

The results showed that the explosion phenomenon characterized by some delay time t_d between voltage application and the moment of tip explosion. Experimentally, for tungsten tip, a relationship was established as

$$\int_0^t j^2(t')\mathrm{d}t' = j^2 t_d = 4 \times 10^9 \frac{\mathrm{A}^2\mathrm{s}}{\mathrm{cm}^4} \quad (2.2.21)$$

It was shown that the determined value of $j^2 t_d$ was constant in a large range of the current densities and the time delays for tungsten tip. The relationship (2.2.21) can be understood considering the tip heating by equation

$$\gamma c \frac{\mathrm{d}T}{\mathrm{d}t} = \nabla(\lambda \nabla T) + j^2 \rho \quad (2.2.22)$$

where γ, c, λ, ρ are the specific density, heat capacity, heat conductivity, and specific resistivity of a metal, respectively. Taking into account linear dependence $\rho = \rho_0 T$ and neglecting the conductivity term due to very small time t_d is $\frac{\mathrm{d}T}{T} = \frac{j^2 \rho}{\gamma c}\mathrm{d}t$, and then after integration, the following relation can be obtained

Table 2.1 Parameter j^2t for different materials

Metal	Cu	Au	Al	Ag	Ni	Fe
$j^2td \times 10^{-9}$	4.1	1.8	1.8	2.8	1.9	1.4

$$j^2t_d = \frac{\gamma c}{\rho_0} Ln\left(\frac{T_{cr}}{T_o}\right) \tag{2.2.23}$$

where T_0 and T_{cr} are the initial and critical for the explosion temperatures, respectively. Calculation according to (2.2.23) generally indicates agreement with the measurement for the tungsten. It is also indicated that the calculated results depend on the tip geometry (cylinder, cone, radius of the cone apex curvature, Nottingham effect) [112, 114, 115]. Calculation using (2.2.23) for tungsten yields $j^2t_d = 2\times10^9$ A^2s/cm^4, when T_{cr} equals the liquid temperature, and it is 2.7×10^9 A^2s/cm^4, when T_{cr} equals the boiling temperature [108, 115].

As the measured time $t_d \sim 10^{-9}$ s, then $j \sim 10^9$ A/cm^2 and such current density was supported by tunneling emission from a tip with strong electric field ~10^8 V/cm. Since an average field was applied, the requested field can be reached by enhancement factor worked when the electrode gap is significantly larger than the tip radius of sub-microscale size.

It should be noted that earlier, the results of a calculation of resistance of explosion gold wire as a function of similar variable $\int_0^t I^2(t')dt'$ was called an action and it was reported by Tucker [116]. The thermal instability of resistivity heated microprotrusion was studied in [117] where the applicability of the instability mechanism for different cathode materials to the time delay calculation was discussed. Parameter j^2t_d as presented in Table 2.1 according to [115].

After explosion event, the electron flux is produced by emission from the expanding plasma-vacuum boundary at distance from the surface vt with gap distance d. Assuming that the plasma–surface at the shape of expanding boundary is $\pi r^2 = \pi v^2t^2$, the expression for current density was obtained in form similarly to Flynn's work of [103]

$$j(t) = A_{\exp 1}U^{3/2}(t)\frac{\pi(vt)^2}{(d - vt)^2} \tag{2.2.24}$$

The specifics of dependence (2.2.24) consist in that it was established experimentally. In a nanosecond time range and for half-sphere surface on the plane cathode, the constant was measured to be $A_{\mathrm{expl}} = 4.44 \times 10^{-6}$ A/V$^{3/2}$ [112]. The electrons emitted from the extended explosion plasma characterized the sharply discharge current rise due to effect of explosion emission. The explosion depends on many factors including destruction of the local emitter, and therefore, the current–voltage waveform cannot be reproduced as it was produced by tunneling mechanism at field emission. The phenomenon is enough complicated due to different physical processes appeared during the explosion and transition from the solid state to the highly velocity plasma flow.

Computer simulation of metal protrusion explosion was performed in quasitwo-dimensional [118] and in 2D [119] approximations. The numerical investigation of the protrusion thermal state was conducted in [120]. The later investigation was related to develop different heat model and to determine the critical temperature for protrusion explosion [115].

An attempt to describe the complex protrusion explosive processes was performed in work of [121]. A generalized model was developed considering transient and spatial inhomogeneity of the electromagnetic, thermal, gasdynamic, and the material properties arising during the energy dissipation due to field emission current in the protrusion. The model accounted the phenomena associated with the quick variation of induced magnetic field due to high rate of current rise (see below) by the explosion. The electrical current can be converged in the vicinity of the protrusion apex, and therefore, the non-uniform heating and metal destruction created.

Short time character of intensive energy deposition in the small volume causes formation of local regions with high pressure and abrupt dispersion of the substance, requiring the thermophysical data for the material, peculiarities of transport coefficients and equation of state for the metal in different phases. The main equations of electrodynamics of the protrusion explosion can be derived from equations of plasma motion, Maxwell's equations, and generalized Ohm's law [121]:

$$\frac{\partial \vec{H}}{\partial t} = \mathrm{rot}\left[\vec{V}, \vec{H}\right] - \mathrm{rot}\left(\nu_m, \mathrm{rot}\vec{H}\right) \tag{2.2.25}$$

$$\frac{\partial \vec{j}}{\partial t} = \nu_m \sigma \Delta \left\{ \frac{j}{\sigma} - \frac{1}{eN}(\nabla P) + \frac{R_T}{eN_e} - \frac{1}{c}\left[\vec{V}, \vec{H}\right] + \frac{1}{eNc}\left[\vec{j}, \vec{H}\right] - 4\pi \, \mathrm{grad}\, \rho_e \right\} \tag{2.2.26}$$

where $\vec{V}$ is the velocity vector, σ is the electrical conductivity, $\nu_m = c^2/4\pi\sigma$ is the magnetic viscosity, R_T is the thermoelectric force, ρ_e is the space charge, P_e and N_e are the electron pressure and density, respectively. In contrast to work in [118], there the effects of electron pressure and proper magnetic field were taken into account when Ohm's law is used. The "displacement current" can be neglected when $4\pi\sigma\tau \gg 1$, $\tau = 0.1$–1 ns is the characteristic time scale. It is possible to neglect the terms $\frac{1}{c}\left[\vec{V}, \vec{H}\right]$ and $\frac{1}{eNc}\left[\vec{j}, \vec{H}\right]$ in (2.2.26) when

$$\frac{v}{c} \ll \left\{ \begin{array}{ll} \frac{c}{4\pi\sigma r_0}, & r \geq v\tau \\ (4\pi\sigma\tau)^{-0.5} & v\tau > r_0 \end{array} \right\} \tag{2.2.27}$$

where r_0 is the emitting region radius, v is the hydrodynamic velocity. The boundary conditions included the equation of electron T-F-emission in general form (2.2.16). The total current from the emitter is

$$j = e\int j(T, E)\,\mathrm{d}S \tag{2.2.28}$$

The solid protrusion heating regime was studied by heat conduction equation in the following nonlinear form

$$\rho c(T)\frac{\partial T}{\partial t} = \nabla[\lambda(\rho, T)\nabla T] + \frac{j^2}{\sigma} + K_T(T)\left(\vec{j}, \nabla T\right) \tag{2.2.29}$$

In the right side of (2.2.29), the second term is Joule, and the third term is Thomson effects, where K_T, $\lambda(T)$, $c(T)$, and $\rho(T)$ are the Thomson coefficient, heat conduction coefficient, heat capacity, and density of the substance.

Boundary conditions:

$$\lambda(\rho, T)\frac{\partial T}{\partial n} = \frac{j}{e}W_n + \sigma_{\mathrm{SB}}T^4$$

W_n is the mean energy carried out by a single electron. The explosion phase and material properties were simulated in non-viscous 2D gas flow approximation [121]:

$$\frac{\partial \rho}{\partial t} + \mathrm{div}(\rho\vec{w}) = 0$$

$$\frac{\partial \rho u}{\partial t} + \mathrm{div}(\rho u \times \vec{w}) + \frac{\partial P}{\partial z} = 0 \tag{2.2.30}$$

$$\frac{\partial \rho v}{\partial t} + \mathrm{div}(\rho v \times \vec{w}) + \frac{\partial (\mathrm{Pr})}{r\partial z} = 0$$

$$\frac{\partial \rho\left(Ei + \frac{w^2}{2}\right)}{\partial t} + \mathrm{div}\left[\rho\vec{w}\left(Ei + \frac{w^2}{2}\right) + P\vec{w}\right] + \mathrm{div}Q_T - \rho\frac{j^2}{\sigma} = 0$$

Here, $\vec{w}$ is the velocity vector of expanding phase with components v and u. P and Ei pressure and internal energy.

The above system of equations was solved in cylindrical coordinates with axial symmetry. The protrusion was modeled as revolution-ellipsoid-like body with given long semi-axis "a" and short "b", placed on the surface of the plane massive aluminum cathode. Interelectrode gap was 25 μm, anode voltage 40 kV, curvature radius 0.1 μm. The numerical solution electrodynamic and heat conduction equations was conducted using not-implicit finite differential method with iterations. Hydrodynamic system of equations was numerically solved by the PIC code with artificial viscosity [121].

The simulations showed that the current moderately increased before the explosion while strongly raised with rate of 10^{10}–10^{11} A/s at the explosion moment (Fig. 2.4). The heating time until time of explosion depends on protrusion size, and this time reduced by Nottingham's effect. Maximum of the current density as calculated inside

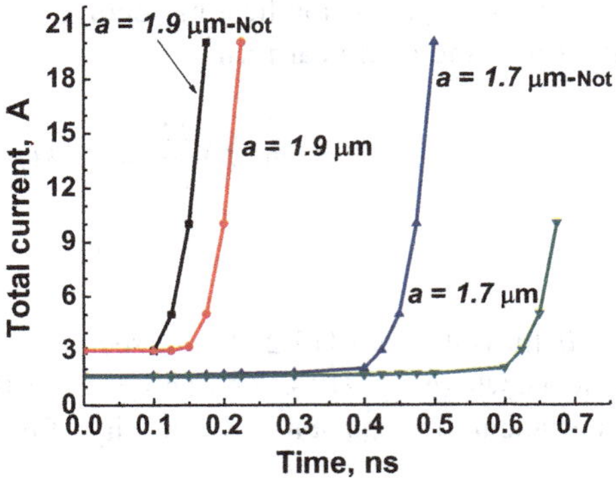

Fig. 2.4 Time dependence of the total current from the protrusion for short semi axis of 0.5 μm. Different sizes **a** of the long semi axis were shown near the graphs calculated with (index-"Not") and without Nottingham effect

near the apex of the protrusion body. At the same location was determined the pressure (200 kbar) and velocity (1–2 × 10^6 cm/s) maximums for protrusions of 1.6–0.4 μm.

References

1. Hertz, H. (1882). Ueber die Verdunstung der Flussigkeiten, insbesomdere des Quecksilbers, im luftleeren Raume. *Annalen der Physik und Chemie*, B27, N10, Online, *253*(10), 177–193.
2. Langmuir, I. (1913). The vapor pressure of metallic tungsten. *Physical Review*, *2*(5), 329–342.
3. Langmuir, I. (1916). The evaporation, condensation and reflection of molecules and the mechanism of absorption. *Physical Review, 8*(2), 149–176.
4. Langmuir, I. (1914). The vapor pressure of the metals platinum and molybdenum. *Physical Review, 4*(4), 377–386.
5. Knudsen, M. (1909). Eine Revision der Gleichgewichtsbedingung der Gase, Thermische Molekular stromung. *Annalen der Physik, 28,* 205–229.
6. Knudsen, M. (1911). Molekularstromung der wassestoffs durch bohren und das hitzdraht-manometer. *Annalen der Physik, 35,* 389–396.
7. Knudsen, M. (1915). Die Maximale Verdampfungsgeschwindigkeit des Quecksilbers. *Annalen der Physik und Chemie, 47,* 697–708.
8. Fonda, G. R. (1928). Evaporation of tungsten under various pressures of argon. *Physical Review, 31,* 260–266.
9. Fonda, G. R. (1923). Evaporation characteristics of tungsten. *Physical Review, 21,* 343–347.
10. Granovsky, V. L. (1951). On steady-state vaporization of fluid at different temperatures of vaporizer and condenser. *Soviet Technical Physics Letters, 21*(9), 1008–1013 (ref in Russian).
11. Afanas'ev, Yu V, & Krokhin, O. N. (1967). Vaporization of matter exposed to laser emission. *Soviet Physics, Journal of Experimental and Theoretical Physics, 25*(4), 639–645.
12. Anisimov, S. I., Imas, Yu. A., Romanov, G. S., & Khodyko, Yu. V. (1971). Action of high-power radiation on metals. *National Technical Information Service.* Virginia: Springfield.
13. Bhatnagar, P. L., Gross, E. P., & Krook, M. (1954). A model for collision processes in gases. I. Small amplitude processes in charged and neutral one-component systems. *Physical Review, 94*(3), 511–525.
14. Eldridge, J. A. (1927). Experimental test of Maxwell's distribution law. *Physical Review, 30,* 931–936.
15. Knake, O., & Stranski, I. N. (1956). Mechanism of vaporization. *Progress in Metal Physics, 6,* 181–235.

16. Kelly, R., & Dreyfus, R. W. (1988). On the effect of Knudsen layer formation on studies of evaporation, sputtering and desorption. *Surface Science, 198,* 263–276.
17. Anisimov, S. I. (1968). Vaporization of metal absorbing laser radiation. *Soviet Physics, Journal of Experimental and Theoretical Physics, 54*(1), 339–342.
18. Kucherov, R. Ya., & Rikenglaz, L. E. (1959). Slipping and temperature discontinuity at the boundary of a gas mixture. *Soviet Physics, Journal of Experimental and Theoretical Physics, 36*(6), 1253–1255.
19. Mott-Smith, H. M. (1951). The solution of the Boltzmann equation for a shock wave. *Physical Review, 82*(6), 885–892.
20. Luikov, A. V., Perelman, T. L., & Anisimov, S. I. (1971). Evaporation of a solid into vacuum. *International Journal of Heat and Mass Transfer, 14*(2), 177–184.
21. Anisimov, S. I., & Rakhmatulina, A Kh. (1973). The dynamics of the expansion of a vapor when evaporated into a vacuum. *Soviet Physics, Journal of Experimental and Theoretical Physics, 37*(3), 443–444.
22. Fisher, J. (1976). Distribution of pure vapor between two parallel plates under the influence of strong evaporation and condensation. *Physics of Fluids, 19*(9), 1305–1311.
23. Kogan, M. N., & Makashev, N. K. (1971). *Role of the Knudsen layer in the theory of heterogeneous reactions and in flows with surface reactions.* Izv. Akad. Nauk USSR, Mekhanika Zudkosti and Gaza, 913–920 I974 Consultants Bureau, a division of Plenum Publishing Corporation, 227 N.Y. 10011; Original in Russian: 6, 2–11.
24. Kogan, M. N., & Makashev, N. K. (1972). *Boundary conditions for a flow at a surface with chemical reaction.* Izv. Akad. Nauk USSR, Mekhanika Zudkosti and Gaza, I974 Consultants Bureau, a division of Plenum Publishing Corporation, 227 N.Y. 10011; Original in Russian: 1, 129–138.
25. Makashev, N. K. (1972). *Strong over condensation in one- or two-component of rarefied gases at arbitrary of Knudsen number.* Izv. Akad. Nauk USSR, Mekhanika Zhidkosti and Gaza, I974 Consultants Bureau, a division of Plenum Publishing Corporation, 227 N.Y. 10011; Original in Russian: 5, 130–138.
26. Hamel, B. B. (1965). Kinetic model for binary gas mixtures. *Physics of Fluids, 8*(3), 418–425.
27. Khight, C. J. (1976). Evaporation from a cylindrical surface into vacuum. *Journal of Fluids Mechanics, 75*(part 3), 469–486.
28. Khight, C. J. (1979). Theoretical modeling of rapid surface vaporization with back pressures. *AIAA Journal, 17*(5), 519–523.
29. Aoki, K., & Cercignani, C. (1983). Evaporation and condensation on two parallel plates at finite Reynolds numbers. *Physics of Fluids, 26*(5), 1163–1164.
30. Baker, R. L. (1991). Kinetically controlled vaporization of a polyatomic gas. *AIAA, 29*(3), 471–473.
31. Ytrehus, T., & Ostmo, S. (1996). *International Journal of Multiphase Flow, 22*(1), 133–135. Ytrehus, T. (1977) *Rarefied gas dynamics.* In L. Potter (Ed.) (Vol. 51, pp. 1197–1212). NY: AIAA.
32. Kogan, M. N. (1969). *Rarefied gas dynamics.* NY: Plenum.
33. Arthur, M. D., & Cercignani, C. (1980). Non-existence of a steady rarefied supersonic flow in a half-space. *Zeitschrift für angewandte Mathematik und Physik* (also translated in Journal of Applied Mathem. and Physics), *31*(5), 634.
34. Muratova, T. M., & Labuntsov, D. A. (1969). Kinetic analysis of vaporization and condensation. *High Temperature, 7*(5), 959–967 (ref in Russian).
35. Labuntsov, D. A., & Kryukov, A. P. (1979). Analysis of intensive evaporation and condensation. *International Journal of Heat and Mass Transfer, 22,* 989–1002.
36. Rebhan, E. (1985). Generalizations of the theorem of minimum entropy production to linear systems involving inertia. *Physical Review A, 32*(1), 581–589.
37. Rebhan, E. (1990). Maximum entropy production far from equilibrium: The example of strong shock waves. *Physical Review A, 42*(2), 781–788.
38. Ford, I. J., & Lee, T. L. (2001). Entropy production and destruction in models of material evaporation. *Journal of Physics. D. Applied Physics, 34,* 413–417.

39. Bronin, S. Ya., & Polischuk, V. P. (1984). Knudsen layer at vaporization and condensation. *High Temperature, 22*(3), 550–556 (Ref. in Russian).
40. Sibold, D., & Urbassek, H. M. (1993). Monte Carlo study of Knudsen layer in evaporation from elemental and binary media. *Physics of Fluids A, 5*(1), 243–256.
41. Meland, R., & Ytrehus, T. (2003). Evaporation and condensation Knudsen layers for nonunity condensation coefficient. *Physics of Fluids, 15*(5), 1348–1350.
42. Meland, R., Frezzotti, A., Ytrehus, T., & Hafskjold, B. (2004). Nonequilibrium molecular-dynamics simulation of net evaporation. *Physics of Fluids, 16*(2), 223–243.
43. Anisimov, S. I., & Khokhlov, V. A. (1995). *Instabilities in laser-matter interaction.* Boca Raton, FL: CRC Press Inc.
44. Rhim, H., & Hwang, S.-T. (1976). Knudsen flow under an external potential field. *The Physics of Fluids, 19*(9), 1319–1323.
45. Beilis, I. I. (1982). On the theory of erosion processes in the cathode region of an arc discharge. *Soviet Physics Doklady, 27,* 150–152.
46. Beilis, I. I. (1985). Parameters of kinetic layer of arc discharge cathode region. *IEEE Transaction Plasma Science, PS-13*(5), 288–290.
47. Beilis, I. I. (1986). Cathode arc plasma flow in a Knudsen layer. *High Temperature, 24*(3), 319–325.
48. Keidar, M., Boyd, I. D., & Beilis, I. I. (2001). On the model of Teflon ablation in an ablation-controlled discharge. *Journal of Physics. D. Applied Physics, 34,* 1675–1677.
49. Keidar, M., Boyd, I. D., & Beilis, I. I. (2004). Ionization and ablation phenomena in an ablative plasma accelerator. *Journal of Applied Physics, 96*(10), 5420–5428.
50. Elieser, S. (2002). *The interaction of high-power lasers with plasmas.* Bristol & Philadelphia: IoP Publishing.
51. Bogaerts, A., Chen, Z., Gijbels, R., & Vertes, A. (2003). Laser ablation for analytical sampling: What can we learn from modeling? *Spectrochimica Acta. B, 58,* 1867–1893.
52. Chen, Z., & Bogaerts, A. (2005). Laser ablation of Cu and plume expansion into 1 atm ambient gas. *Journal of Applied Physics, 97,* 063305.
53. Itina, T. E., Hermann, J., Delaporte, Ph, & Sentis, M. (2002). Laser-generated plasma plume expansion: Combined continuousmicroscopic modeling. *Physical Review E, 66,* 066406.
54. Gusarov, V., & Aoki, K. (2005). Ionization degree for strong evaporation of metals. *Physics of Plasmas, 12,* 083503.
55. Beilis, I. I. (2007). Laser plasma generation and plasma interaction with ablative target. *Laser and Particle Beams, 25,* 53–63.
56. Beilis, I. I. (1995). Theoretical modelling of cathode spot phenomena. In R. L. Boxman, P. J. Martin, & D. M. Sanders (Eds.), *Handbook of vacuum arc science and technology* (pp. 208–256). Park Ridge, N.J: Noyes Publ.
57. Beilis, I. I. (2012). Modeling of the plasma produced by moderate energy laser beam interaction with metallic targets: Physics of the phenomena. *Laser and Particle Beams, 30,* 341–356.
58. Sommerfeld, A. (1928). Zur Elektronentheorie der Metalle auf Grund der Fermischen Statistik. *Zeitschrift für Physik A, 47* (Tell I, Issue 1–2), 1–32: (Tell II), 43–60.
59. Richardson, O. W. (1915). The influence of gases on the emission of electrons from hot metals. *Proceedings of The Royal Society of London, Series A, 91,* 524–535.
60. Richardson, O. W. (1924). Electron emission from metals as a function of temperature. *Physical Review, 23*(2), 153–155.
61. Dushman, S. (1923). Electron emission from metals as a function of temperature. *Physical Review, 21*(6), 623–636.
62. Nordheim, L. (1928). Zur theorie der thermischen emission und der reflexion von elektronen an Metallen. *Zeitschrift für Physik, 46*(12), 833–855.
63. Richardson, O. W. (1901). On the negative radiation from hot platinum. Proceedings of the Cambridge Philosophical Society. *Mathematical and Physical Sciences, 11,* 286–295.
64. Schottky, W. (1923). Uber kalte und warme Elektronenentladungen. *Zeitschrift für Physik, 14,* 63–106; Schottky, W. (1925). Uber das Verdampfen yon Elektronen. *Zeitschrift für Physik, 34*(12), 645–675.

65. Millikan, R. A., & Eyring, C. F. (1926). Laws governing the pulling of electrons out metals by intense electrical field. *Physical Review, 27*(1), 51–67.
66. Millikan, R. A. (1912). The effect of the character of the source upon the velocities of emission of electrons liberated by ultra-violet light. *Physical Review (Series I), 35,* 74–76.
67. Sommerfeld, A., & Bethe, H. (1933). Elektronentheorie der Metalle. In *Handbuch der Physik* (Vol. 24–2, S. 333–622). Heidelberg: Springer Verlag.
68. Fowler, R. H., & Nordheim, L. (1928). Electron emission in intense electric field. *Proceedings of the Royal Society, 119,* 173–181.
69. Beilis, I. I. (1971). Analytic approximation for the Nordheim function. *High Temperature, 9,* 157–158.
70. Bauer, A. (1954). Zur Theorie des Kathodenfalls in Lichtbogen. *Zeitschrift für Physik, 138,* 35–55.
71. Guth, E., & Mullin, C. J. (1942). Electron emission of metals in electric fields III. The transition from thermionic to cold emission. *Physical Review, 61*(5–6), 339–348.
72. Murphy, E. L., & Good, R. H. (1956). Thermionic emission, field emission, and the transition region. *Physical Review, 102*(6), 1464–1473.
73. Rakhovsky, V. I. (1970). *Physical bases of the communication of electrical current in a vacuum.* Moscow: Nauka.
74. Elinson, M. I., & Vasil'ev, G. F. (1956). *Field emission.* Moscow: Fizmatgiz (in Russian).
75. Itskovich, F. I. (1966). On the theory of field emission from metals. *Soviet Physics, Journal of Experimental and Theoretical Physics, 23,* 945–953.
76. Wenzel, G. (1926). Verallgemeinerung Quantenbedingungen Zwecke Wellenmechanik. *Zeitschrift für Physik, 38*(7–8), 518–529.
77. Kramers, H. A. (1926). Wellenmechanik und halbzahlige Quantisierung. *Zeitschrift für Physik, 39*(10–11), 828–840.
78. Kemble, E. C. (1937). *The fundamental principles of quantum mechanics* (Vol. 4). N.Y.: McGraw-Hill Book Company Inc.
79. Dolan, W. W., & Dyke, W. P. (1954). Temperature-and-field emission of electrons from metals. *Physical Review, 95*(2), 327–332.
80. Beilis, I. I. (1974). Emission process at the cathode of an electrical arc. *Soviet Physics—Technical Physics, 19*(2), 257–260.
81. Brodskii, A. M., & Gurevich, Yu. Ya (1969). A threshold theory of electron emission. *Bulletin of the Academy of Science USSR, Ser. Fiz., 33,* 363–371.
82. Brodskii, A. M., & Gurevich, Yu. Ya. (1973). *Theory of electron emission from metals.* Moscow: Nauka.
83. Ecker, G., & Muller, K. G. (1959). Electron emission from the arc cathode under the influence of the individual field. *Journal of Applied Physics, 30*(9), 1466–1467.
84. Ecker, G., & Muller, K. G. (1959). Der Einfluß der individuellen Feldkomponente auf die Elektronenemission der Metalle. *Zeitschrift für Naturforschung, 14a,* 511–520.
85. Porotnikov, A. A., & Rodnevich, B. B. (1975). Field emission with allowance for individual ion field. *Soviet Physics—Technical Physics, 20,* 1403–1404.
86. Porotnikov, A. A., & Rodnevich, B. B. (1976). Field emission of electrons into a plasma. *Journal of Applied Mechanics and Technical Physics, 17*(6), 764–767.
87. Porotnikov, A. A., & Rodnevich, B. B. (1978). I-F emission. *Soviet Physics—Technical Physics, 23,* 740–741.
88. Porotnikov, A. A., & Rodnevich, B. B. (1978). Thermoautoelectronic emission taking account of individual ionic fields. *Journal of Applied Mechanics and Technical Physics, 19,* 164–166.
89. Petrosov, V. A. (1978). Effect of individual ions on field emission. *Soviet Physics—Technical Physics, 23,* 1109–1113.
90. Forbes, R. G. (2013). Development of a simple quantitative test for lack of field emission orthodoxy. *Proceedings of The Royal Society of London: A Mathematical, Physical and Engineering Science, 469,* 20130271.
91. Forbes, R. G., Deane, J. H. B. (2007). Reformulation of the standard theory of Fowler–Nordheim tunnelling and cold field electron emission. *Proceedings of The Royal Society of London: A Mathematical, Physical and Engineering Science, 463,* 2907–2927.

92. Forbes, R. G. (2008). On the need for a tunneling pre-factor in Fowler-Northeim tunneling theory. *Journal of Applied Physics, 103,* 114911.
93. Forbes, R. G. (2008). Physics of generalized Fowler-Nordheim-type equations. *Journal of Vacuum Science and Technology B, 26,* 788–793.
94. Fursey, G. N., & Glazanov, D. V. (1998). Deviations from the Fowler–Nordheim theory and peculiarities of field electron emission from small-scale objects. *Journal* of *Vacuum Science and Technology B, 16,* 910–915.
95. Kyritsakis, A., & Xanthakis, J. P. (2015). Derivation of a generalized Fowler-Nordheim equation for nanoscopic field-emitters. *Proceedings of The Royal Society, 471,* 20140811.
96. Kyritsakis, A., & Xanthakis, J. P. (2013). Beam spot diameter of the near-field scanning electron microscopy. *Ultramicroscopy, 125,* 24–28.
97. Zadin, V., Pohjonen, A., Aabloo, A., Nordlund, K., & Djurabekova, F. (2014). Electrostatic-elastoplastic simulations of copper surface under high electric fields. *Physical Review Accelerators and Beams, 17,* 103501.
98. Timko, H., Aicheler, M., Alknes, P., Calatroni, S., Oltedal, A., Toerklep, A., et al. (2011). Energy dependence of processing and breakdown properties of Cu and Mo. *Physical Review Accelerators and Beams, 14,* 101003.
99. Descoeudres, A., Levinsen, Y., Calatroni, S., Taborelli, M., & Wuensch, W. (2009). Investigation of the dc vacuum breakdown mechanism. *Physical Review Accelerators and Beams, 12,* 092001.
100. Dyke, W. P., & Trolan, J. K. (1953). Field emission: Large current densities, space charge, and the vacuum arc. *Physical Review, 89*(4), 799–808.
101. Dyke, W. P., Trolan, J. K., Martin, E. E., & Barbour, J. P. (1953). The field emission initiated vacuum arc I. Experiments arc initiation. *Physical Review, 91*(5), 1043–1054.
102. Dolan, W. W. & Dyke, W. P. (1953). The field emission initiated vacuum arc II. Resistivity heated emitter. *Physical Review, 91*(5), 1054–1057.
103. Flynn, P. T. G. (1956). The discharge mechanism in the high-vacuum cold-cathode pulsed X-ray tube. *Proceedings of The Royal Society B, 69*(7), 748–762.
104. Easton, E. S., Lucas, F. B., & Creedy, F. (1934). High velocity streams in the vacuum arc. *Electrical Engineering, 53*(11), 1454–1460.
105. Kassirov, G. M., & Mesyats, G. A. (1964). Breakdown mechanism of short vacuum gaps. *Soviet Physics—Technical Physics, 9,* 1141–1145.
106. Mesyats, G. A., Bugaev, S. P., Proskurovskii, D. I., Eshkenazi, V. I., Yurike, Ya. Ya. (1969). Study of the initiation and development of a pulsed breakdown of short vacuum gaps in the nanosecond time region. *Radio engineering* and *electronic physics, 14,* 1919–1925.
107. Mesyats, G. A. (1974). *Generation of high power nanosecond pulses.* Moscow: Soviet Radio. (In Russian).
108. Mesyats, G. A., & Proskurovskii, D. I. (1971). Explosive electron emission from metallic needles. *Soviet Physics, Journal of Experimental and Theoretical Physics Letter, 13*(1), 4–6.
109. Mesyats, G. A. (1971). The role of fast processes in vacuum breakdown. *Phenomena in Ionized Gases* (pp. 333–363), Sept. 13–18. England: Oxford.
110. Fursey, G.N., Antonov, A. A., & Zhukov, V. M. (1971). Explosive emission accompanying transition field emission into vacuum breakdown. *Vestnik LGY* (10), 75–78.
111. Fursey, G. N., & Vorontsov-Vel'yaminov, P. N. (1967). Qualitative model of initiation of a vacuum arc. *Soviet Physics—Technical Physics, 12,* 1370–1382.
112. Bugaev, S. P., Litvinov, E. A., Mesyats, G. A., & Proskurovskii, D. I. (1975). Explosive emission of electrons. *Soviet Physics Uspekhi, 18*(1), 51–61.
113. G.N. Fursey, Field Emission and Vacuum Breakdown, IEEE Transaction on Elect. Insul., E-20 N4, 659–679, (1985).
114. Mesyats, G. A., & Proskurovskii, D. I. (1989). *Pulsed electrical discharge in vacuum.* Springer-Verlag.
115. Mesyats, G. A. (2000). *Cathode phenomena in a vacuum discharge.* Moscow: Nauka.
116. Tucker, T. (1961). Behavior of exploding gold wires. *Journal of Applied Physics, 32*(10), 1894–1900.

117. Nevrovskii, V. A., & Rakhovsky, V. I. (1980). Time for thermal instability to develop in microscopic protrusion on the cathode during vacuum breakdown. *Soviet Physics.-Technical Physics, 25*(10), 1239–1244.
118. Loskutov, V. V., Luchinsky, A. V., & Mesyats, G. A. (1983). Magnetohydrodynamic processes in the initial stage of explosive emission. *Soviet Physics Doklady, 28,* 654–656.
119. Bushman, A. V., Leshkevich, S. L., Mesyats, G. A., Skvortsov, V. A., & Fortov, V. E. (1990). Mathematical modeling of an electric explosion of a cathode microppoint. *Soviet Physics Doklady, 35,* 561–563.
120. Glazanov, D. V., Baskin, L. M., & Fursey, G. N. (1989). Kinetics of the pulsed heating of field emission cathode points with real geometry by high-density emission current. *Soviet Physics—Technical Physics, 34*(2), 534–539.
121. Beilis, I. I., Garibashvily, I. D., Mesyats, G. A., Skvortsov, V. A., & Fortov, V. E. (1990). Electrodynamic and gasdynamic processes of the explosive electron emission from the metal spikes. In R. W. Stinnett (Ed.), *24 International Symposium on Discharge and Electrical Insulation in Vacuum,* Santa Fe, Sandia National Lab., USA, Sept. 548–551.

Chapter 3
Heat Conduction in a Solid Body. Local Heat Sources

In this chapter, the electrode heat regime under energy flux dissipation in the solid from a vacuum arc was analyzed. Specifics of electrode heat conduction and electrode heating by concentrate heat sources are studied. The local power density from the minute area arc spot is modeled. The mathematical formulation and methodology of solutions of heat conduction equation for different target configurations (plate, film, and bulk) are presented; as well, analytical expressions are derived.

3.1 Brief State-of-the-Art Description

Arc spot is a highly concentrated and intense heat source. The temperature distribution in the electrode arises as result from the spot energy action, which produced due to the processes of mass and charge generation necessary for operation of the arc. Therefore, the description of electrode and plasma phenomena primary related to the study of the thermal regime of the electrode and to the creation of the correct mathematical approach of its calculation.

In early works [1–6], thermal processes were studied using one-dimensional equations of heat conduction and sometimes with assumed surface temperature at the electrode surface, which correspond to temperature of the phase transition (melting or boiling point) of the electrode material. The transient electrode temperature was also calculated using heat flux density as boundary condition taking into account the energy losses by radiation and evaporation of the material [7–14]. These works made possible to get a general understanding of the temperature distribution in the electrodes and the scale of the thermal effect propagation by the arc heat flux. In spite of such studies of limited by the heat conduction equation in one-dimensional formulation, nevertheless the obtained results allow to estimate the contribution of separate thermal effects and to understand the temperature distribution over time in case of exposed the energy of a pulsed discharge to the electrode.

I. Beilis, *Plasma and Spot Phenomena in Electrical Arcs*, Springer Series on Atomic, Optical, and Plasma Physics 113, https://doi.org/10.1007/978-3-030-44747-2_3

The extensive use of the arc in technology for welding, EDM, arc plasma [15–19], electron beam [20], and laser [21, 22] treatment of the solid targets also led to appearance of works related to two- and three-dimensional solutions of the heat conduction equation. The main approach was developed by Rykalin [15, 16] who used experimental data indicating that current density (power density) in spot of the welding arc can be approximated by Gauss law. The temperature distribution was studied solving the equation of heat conduction using method of sources and source–sink method [15, 16, 23]. As a result, the temperature distribution in solids caused by moving arc spot and temperature behavior in vicinity of the spot in which the heat source was modeled as point, linear, and normally distributed source of the power was obtained. In the following, we will discuss several limited cases relevant to analysis of the cathode spot in vacuum arc.

3.2 Thermal Regime of a Semi-finite Body. Methods of Solution in Linear Approximation

Below we consider some examples of solutions of the heat conduction equation for different types of heat sources. In order to further analysis of electrode thermal regime in vacuum arcs, the solutions developed by Rykalin [16] are derived and using his approach our own studies of the thermal regime for specific finite body cases are described.

3.2.1 Point Source. Continuous Heating

In general, the power in a heat source was distributed either at a surface or in a volume. In case when this volume is significantly smaller than the area of the heat expansion, the temperature distribution can be determined using **point source** approximation. Let us consider the power density of such heat source as q(W), which continuously heats an infinite initially cold body during time t. At each elementary time t', the element of heat is $\mathrm{d}Q = q(t')\mathrm{d}t'$. Element of heat $\mathrm{d}Q$ diffuses into the body and causes at moment t an elemental temperature rise:

$$\mathrm{d}T\left(R, t - t'\right) = \frac{\mathrm{d}Q}{c\gamma[4\pi a(t - t')]^{3/2}}\mathrm{e}^{-\frac{R^2}{4a(t-t')}},$$
$$R^2 = x^2 + y^2; \tag{3.1}$$

where c is the heat capacity, γ is the mass density, a is the thermal diffusivity, x and y are coordinates in the system with center of the source, and R is the radius vector. According to the principle of superposition, the following expression can be used:

$$T(R,t) = \int_0^t (R, t - t')\mathrm{d}T \tag{3.2}$$

Using the above element of heat dQ and (3.1) for the elemental temperature rise, (3.2) can be represented in form

$$T(R,t) = \int\limits_0^t \frac{q(t')\mathrm{d}t'}{c\gamma[4\pi a(t-t')]^{3/2}} \mathrm{e}^{-\frac{R^2}{4a(t-t')}} \tag{3.3}$$

After integration of (3.3), the temperature distribution caused from action of the heat point source with constant power q in the infinite body is

$$T(R,t) = \frac{q}{4\pi\lambda R}\left[1 + \Phi\left(\frac{R}{\sqrt{4at}}\right)\right] \tag{3.4}$$

where λ is the heat conductivity and Φ is the Gaussian probability function. At $R = 0$, the $T \to \infty$ indicating the approximation of heat flux by point source cannot describe the heat regime in the region immediately close to the source. However, with $t \to \infty$, the temperature tends to a steady state at which it is inversely proportional to the distance from the source R:

$$T(R, t \to \infty) = \frac{q}{4\pi\lambda R} \tag{3.5}$$

3.2.2 *Normal Circular Heat Source on a Body Surface*

In the case of such source, the heat flux density $q(r)$ is given by

$$q(r) = q_m \mathrm{e}^{-kr^2} \tag{3.6}$$

where r is the distance of a point from the source axis (Fig. 3.1), q_m is the maximal heat flux density (W/cm^2) on the axis of the source, and k (cm^{-2}) is the coefficient characterizing the concentration of the normal distribution. To determine k, let us obtain the total source power q in form

$$q = \int\limits_0^\infty q_m \mathrm{e}^{-kr^2} 2\pi r \mathrm{d}r \tag{3.7}$$

Integrating (3.7), the total power q is

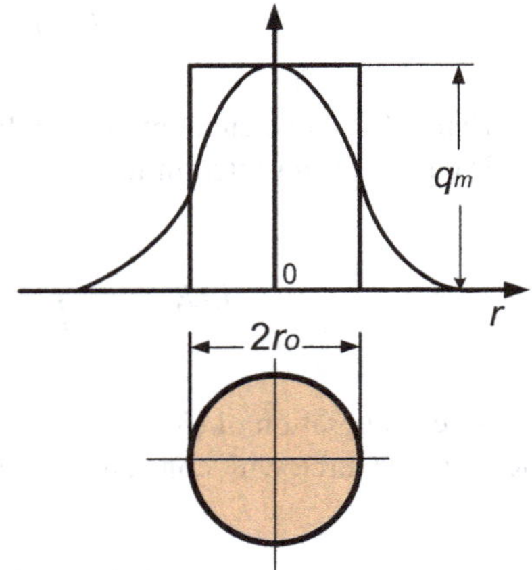

Fig. 3.1 Heat sources with equivalent total power and different distributions. One with normal circular distribution of heat flux and concentration coefficient, k, and the other with uniform heat flux within a circle of radius r_0

$$q = \frac{\pi}{k} q_m \tag{3.8}$$

In order to evaluate k, let us introduce an equivalent heat source with maximal power density q_m but distributed uniformly over a spot of radius r_0, and therefore, the total power flux is $q = q_m \pi r_0^2$. Comparing this expression with expression (3.8), the concentration coefficient is $k = r_0^{-2}$.

3.2.3 *Instantaneous Normal Circular Heat Source on Semi-infinity Body*

Instantaneous normal circular heat source with power q and concentration k acts upon the surface of semi-infinity body at moment $t = 0$. The source center was placed at the center 0 of rectangular system of coordinate x, y, z (Fig. 3.2). The heat distribution in the body at time t will be

$$q(r)\mathrm{d}t = q_m \,\mathrm{d}t \,\mathrm{e}^{-kr^2} \tag{3.9}$$

Using source method and element of surface $\mathrm{d}x'\mathrm{d}y'$, the process of heat expansion in the body can be expressed as

$$\mathrm{d}T(x, y, z, t) = \frac{2q(r')\mathrm{d}x'\mathrm{d}y'\mathrm{d}t}{c\gamma[4\pi at]^{3/2}} \mathrm{e}^{-\frac{R'^2}{4at}},$$
$$R'^2 = (x - x')^2 + (y - y')^2 + z^2 \tag{3.10}$$

$R'^2 =$ is the distance to some point in the body from a point (x', y', 0) and $r'^2 = (x - x')^2 + (y - y')^2$.

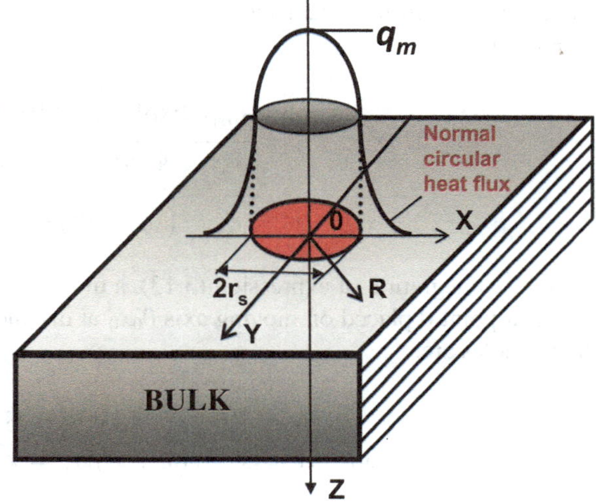

Fig. 3.2 Heating a semi-infinity body by a flat normal circular heat source (spot radius $r_s = r_0$)

Taking into account (3.9) and expressions for R'^2 and r'^2, integration of (3.10) can be provided at the all body surface in form

$$T(r, z, t) = \frac{2q_m \mathrm{d}t}{c\gamma(4\pi at)^{3/2}} \int\limits_{-\infty}^{\infty}\int\limits_{-\infty}^{\infty} \mathrm{d}x'\mathrm{d}y'\mathrm{e}^{\left[-\frac{\left(x-x'^2\right)+\left(y-y'^2\right)^2+z^2}{4at}-\frac{x'^2+y'^2}{4at_0}\right]} \tag{3.11}$$

where $t_0 = (4ak)^{-1}$ is the constant time that depends not only on the heat distribution of the source, but also on the metal properties.

Taking into account that $q = 4\pi at_0q_m$, after integrating of (3.11), the temperature distribution in the semi-infinity body from a normal circular heat source is

$$T(r, z, t) = \frac{2q\mathrm{d}t}{c\gamma}\frac{\mathrm{Exp}\left(-\frac{z^2}{4at}\right)}{(4\pi at)^{1/2}}\frac{\mathrm{Exp}\left(-\frac{r^2}{4a(t+t_0)}\right)}{4\pi a(t+t_0)} \tag{3.12}$$

3.2.4 Moving Normal Circular Heat Source on a Semi-infinity Body

Let us consider a continuous normal circular heat source with power q and concentration k moving at surface of semi-infinite body with velocity v. The time t during which the continuously moving source is divided into element $\mathrm{d}t'$. The heating of the semi-infinite body from such source will be considered as superposition of the heat $\mathrm{d}q\mathrm{d}t'$ acted by instantaneous normal circular heat source (3.12) applied at time

t' at distance vt' from initial location of the source center (x_0, y_0, z_0) and during time $t'' = t - t'$:

$$dT\left(r', z, t''\right) = \frac{2q dt'}{c\gamma} \frac{\mathrm{Exp}\left(-\frac{z^2}{4at''}\right)}{(4\pi a t'')^{1/2}} \frac{\mathrm{Exp}\left(-\frac{r'^2}{4a(t''+t_0)}\right)}{4\pi a(t''+t_0)} \tag{3.13}$$

$$r^2 = r'^2 = \left(x_0 - vt'^2\right) + y_0^2;$$

Before integration of expression (3.13), a moving coordinate system x, y, z with center at point 0 placed on moving axis 0_0x_0 at distance vt_0 is used, and therefore, the variables are:

$$x_0 = v(t + t_0) + x; \; y_0 = y; \, z_0 = z;$$
$$x_0 - vt' = x + v\left(t_0 + t''\right); \, r^2 = x^2 + y^2;$$

Taking into account the above definitions after integration of expression (3.13), the temperature distribution caused by continuously moving normal circular heat source at surface of the semi-infinite body is

$$T(x, y, z, t) = \frac{2q}{c\gamma(4\pi a)^{3/2}} \mathrm{Exp}\left(-\frac{vx}{2a}\right) \int\limits_0^t \frac{dt''}{\sqrt{t''}(t_0 + t'')}$$
$$\mathrm{Exp}\left[\frac{z^2}{4at''} - \frac{r^2}{4a(t_0 + t'')} - \frac{v^2}{4a}\left(t_0 + t''\right)\right] \tag{3.14}$$

Assuming $v = 0$ and $x = y = z = 0$, the temperature of the center of a fixed source will be:

$$T(0, 0, 0, t) = \frac{2q}{c\gamma(4\pi a)^{3/2}} \int\limits_0^t \frac{dt''}{\sqrt{t''}(t_0 + t'')} \tag{3.15}$$

The integral (3.15) can be expressed as [16]:

$$T(0, 0, 0, t) = \frac{q}{2\lambda(4\pi a t_0)^{1/2}} \frac{2}{\pi} \mathrm{arctg}\sqrt{\frac{t}{t_0}} \tag{3.16}$$

The temperature distribution in a bulk can be obtained solving the equations for normal circular heat source.

3.3 Heating of a Thin Plate

3.3.1 Instantaneous Normal Circular Heat Source on a Plate

A normal circular heat source with power q and concentration coefficient k is applied to the surface of a plate of thickness δ. The center C of the source coincides with the origin O of the coordinate system (Fig. 3.3)

As the plate is thin, the heat introduced through an element of area dF is almost instantaneously diffused through the thickness δ and heats uniformly the element $\delta \mathrm{d}F$. Using the source power distribution (3.6), the instantaneous temperature increase will be obtained as

$$\mathrm{d}T(r,t) = \frac{q_m \mathrm{d}t}{c\gamma\delta}\mathrm{e}^{-kr^2} \tag{3.17}$$

Taking into account that $k = (4at_0)^{-1}$ and source heat $Q = 4\pi a t_0 q_m \mathrm{d}t$ according the model [16], the temperature distribution of (3.17) can be obtained using an imaginary concentrated source with heat Q expanded in the plate during time t_0 in form

$$T(r,t_0) = \frac{Q/\delta}{c\gamma(4\pi a t_0)}\mathrm{e}^{-\frac{r^2}{4at_0}} \tag{3.18}$$

Thus, the heat ($Q = q\mathrm{d}t$) diffusion in a plate of an instantaneous normal circular source which is equivalent of heat diffusion of concentrated linear source during time to $t > t_0$ will be expressed as

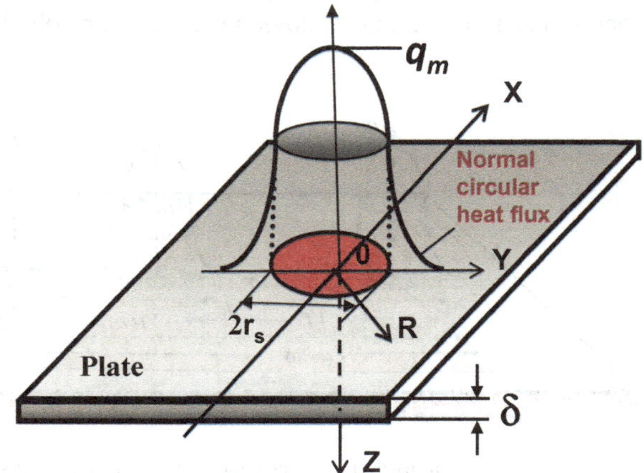

Fig. 3.3 Heating a thin plate by a flat normal circular heat source

$$T(r,t) = \frac{q\mathrm{d}t}{c\gamma\delta 4\pi a(t_0 + t)} \exp\left(-\frac{r^2}{4a(t_0 + t)}\right) \quad (3.19)$$

3.3.2 Moving Normal Circular Heat Source on a Plate

The heat diffusion from a continuously acting source will be considered using superposition principle, as a summation of an infinity number of superposed elementary heat diffusion process of corresponding instantaneous sources acting during an element of time *dt*. A continuous normal circular source of an effective power q and concentration coefficient k acts on the plate surface of thickness δ. According to Rykalin's [16] heating model (Fig. 3.4) the center C of the source, which is coincident with the initial time moment $t = 0$ and with the origin O_0 of the coordinate system $x_0y_0z_0$, moves along the x_0-axis on the plate surface with velocity v. The values of q, k, and v are assumed constant. At time t, the distance of the C from the origin is vt.

The time interval t will be considered as sum of element $\mathrm{d}t'$ and the corresponding distance as sum of elements $v\mathrm{d}t'$. An instantaneous normal circular source $\mathrm{d}Q = q\mathrm{d}t'$ with its center O' is introduced at time t' at distance vt' from initial location point O_0, and the heat expands in the plate during $t'' = t - t'$ and rises an element's temperature at some point A according to (3.19)

$$\mathrm{d}T\left(r, t''\right) = \frac{q\mathrm{d}t'}{c\gamma\delta 4\pi a(t_0 + t'')} \exp\left(-\frac{r'^2}{4a(t_0 + t'')}\right) \quad (3.20)$$

Here, $r'^2 = (\mathrm{AO}')^2$, $r'^2 = (x_0 - vt')^2 + y_0'^2$.

According to superposition principle, the temperature source produced by the continuous normal circular moving source can be obtained by integration (3.20):

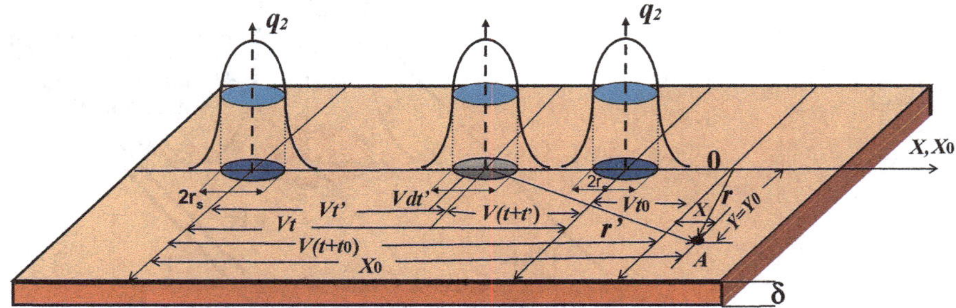

Fig. 3.4 Schema of a normal circular heat source action on the surface of a thin plate

$$T(x_0, y_0, t) = \int_0^t \frac{q\mathrm{d}t'}{c\gamma\delta 4\pi a(t - t' + t_0)} \exp\left[\frac{(x_0 - vt')^2 + y_0^2}{4a(t - t' + t_0)}\right] \tag{3.21}$$

Now can be taking into account a moving coordinate system xOy with origin at point O which is located on O_0x_0-axis at distance vt_0 ahead of the center C and that $x_0 = v(t + t_0) + x$; $y_0 = y$; $z_0 = z$, and it can be finally obtained:

$$T(x, y, t) = \frac{q}{4\pi\lambda\delta} \exp\left(-\frac{vx}{2a}\right)$$

$$\int_{t_0}^{t+t_0} \frac{\mathrm{d}t'}{(t - t' + t_0)} \exp\left[-\frac{r^2}{4a(t - t' + t_0)} - \frac{v^2}{4a}(t - t' + t_0)\right] \tag{3.22}$$

Here, $r^2 = x^2 + y^2$ is the radius vector of point A related to the origin of the moving coordinate system, O. The temperature on the central axis of the plate Oz can be obtained from (3.22) when $r = 0$ and $x = 0$:

$$T(0, t) = \frac{q}{4\pi\lambda\delta} \int_{t_0}^{t+t_0} \frac{\mathrm{d}t'}{(t - t' + t_0)} \exp\left[-\frac{v^2}{4a}(t - t' + t_0)\right] \tag{3.23}$$

3.4 A Normal Distributed Heat Source Moving on Lateral Side of a Thin Semi-infinite Plate

Different plasma systems consist of current carrying designs that used thin plate where moving arcs were occurred. Below the expressions for temperature distribution will be derived when an instantaneous or continuously moving normally distributed heat source acts on the lateral side of such thin plates (see also [24, 25]).

3.4.1 Instantaneous Normally Distributed Heat Source on Side of a Thin Semi-infinite Plate

Let us consider an instantaneous normal heat source with maximal power q_m, concentration k, and normally distributed power on side of a thin semi-infinite plate in form:

$$q(y_0) = q_m \mathrm{e}^{-ky_0^2} \tag{3.24}$$

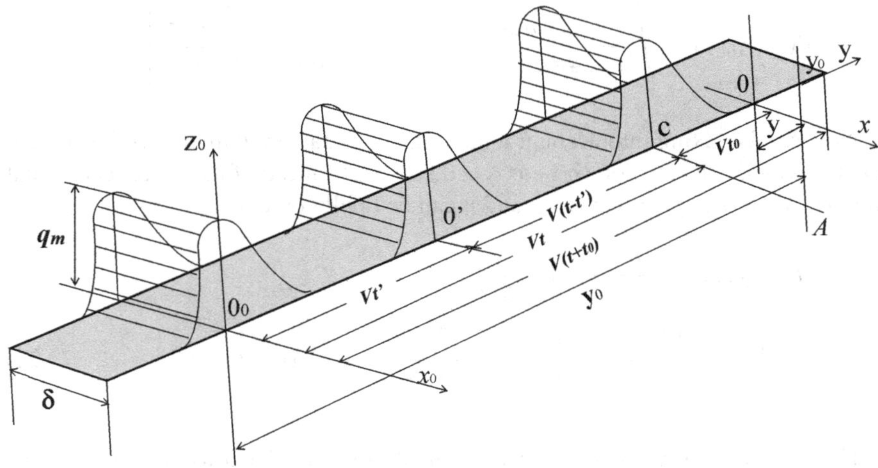

Fig. 3.5 Schema of later side of the plate with a moving normally distributed in y-direction over the plate thickness δ heat source

The maximum power is coincident with the origin O_0 of the coordinate system x, y_0, z_0 (Fig. 3.5). The plate was assumed thin that the heat from the source instantaneously was distributed uniformly through the plate thickness δ (i.e., in x-direction the width of source $\boldsymbol{x_0} = \boldsymbol{\delta}$), and the temperature is constant in the x-direction. In this case, the heat diffuses in y_0- and z_0-directions. In accordance with superposition principle, the heat diffusion process in semi-infinite plate from normally distributed source will be considered as sum of numerous elementary instantaneous sources with heat per unit length $\mathrm{d}Q_l$:

$$\mathrm{d}T(y_0, z_0, t) = \frac{\mathrm{d}Q_l}{c\gamma\delta 4\pi at}\exp\left(-\frac{y_0^2 + z_0^2}{4at}\right) \tag{3.25}$$

The source heat evolved in a surface element $\delta\mathrm{d}y$ during elementary time $\mathrm{d}t$ is:

$$\mathrm{d}Q_l = q_m\,\mathrm{d}y\mathrm{d}t\,\exp\left(-ky_0^{\prime 2}\right) \tag{3.26}$$

Here, $q = q_m\delta(\pi/k)^{0.5}$ is the heat per unit time and $Q_l = q\mathrm{d}t/\delta$. Substitute (3.26) in (3.25) and taking into account that plate is semi-infinity:

$$\mathrm{d}T(y_0, z_0, t) = \frac{2q_m\mathrm{d}y_0'\mathrm{d}t}{c\gamma 4\pi at}\exp\left[-ky_0^{\prime 2} - \frac{\left(y_0 - y_0^{\prime 2}\right) + z_0^2}{4at}\right] \tag{3.27}$$

Taking $k = (4at_0)^{0.5}$ and after integrating on y_0' from $-\infty$ to $+\infty$, the expression, described the temperature distribution in a thin plate caused by fixed instantaneous normal heat source, is obtained in form:

$$T(y_0, z_0, t) = \frac{Q_l}{2\pi\lambda} \frac{\exp\left(-\frac{y_0^2}{4a(t+t_0)}\right)}{\sqrt{t+t_0}} \frac{\exp\left(-\frac{z_0^2}{4at}\right)}{\sqrt{t}} \quad (3.28)$$

3.4.2 *Moving Continuous Normally Distributed Heat Source on Thin Plate of Thickness δ*

A continuous normal distributed in y-direction heat source of an effective power q and concentration coefficient k moves on lateral side of the thin plate surface of thickness δ. The width of source x_0 is equal to the plate thickness δ. The center C of the source (Fig. 3.5), which coincides with the initial time moment $t = 0$ and with the origin O_0 of the coordinate system x_0, y_0, z_0, moves along the y_0-axis on the plate surface with velocity v. The values of q, k, and v are assumed to be constant. At time t, the distance of the C from the origin is vt.

The time interval t will be considered as sum of element $\mathrm{d}t'$ and the corresponding distance as sum of elements $v\mathrm{d}t'$. An instantaneous normal circular source $Q_l = q\mathrm{d}t/\delta$ with its center O' is introduced at time t' at distance vt' from initial location point O_0, and the heat expands in the plate during $t'' = t - t'$ and rises an element's temperature at some point $\mathrm{A}(x_0, y_0)$ according to (3.28):

$$\mathrm{d}T\left(y_0, z_0, t''\right) = \frac{q\mathrm{d}t'}{2\pi\delta\lambda} \frac{\exp\left(-\frac{(y_0-vt')^2}{4a(t''+t_0)}\right)}{\sqrt{t''+t_0}} \frac{\exp\left(-\frac{z_0^2}{4at''}\right)}{\sqrt{t''}} \quad (3.29)$$

Let us introduce a moving coordinate system with origin at point O which is located on O_0y_0-axis at distance vt_0 ahead of the center C and that $y_o = v(t + t_0) + y$ or $y_0 - vt' = y + v(t'' + t_0)$; $z_0 = z$, $dt'' = -\mathrm{d}t'$, and the time-dependent temperature distribution can be finally obtained by integrating (3.29) in the following form:

$$T(y, z, t) = \frac{q}{2\pi\delta\lambda} \exp\left(-\frac{vy}{2a}\right) \int_0^t \frac{\mathrm{d}t''}{\sqrt{t''(t''+t_0)}}$$

$$\exp\left[-\frac{y^2}{4a(t''+t_0)} - \frac{z^2}{4at''} - \frac{v^2\left(t''+t_0\right)}{4a}\right] \quad (3.30)$$

3.4.3 Fixed Normal-Strip Heat Source with Thickness X_0 on Semi-infinite Body

Consider a semi-infinity body with surface $z = 0$. Instantaneous normal-strip heat source with concentrate k_c and with strip thickness x_0 acts upon the surface. At the surface element $dF = dxdy$, the elementary heat source $dq = q(y)dxdydt$ in a point $A = (x - x')^2 - (y - y')^2 - z^2$ increases the body temperature by

$$dT(x, y, z) = \frac{2q_m dx'dy'dt'}{c\gamma(4\pi at')^{3/2}} \exp\left[-\frac{(x - x')^2 + (y - y')^2 + z^2}{4at'} - k_c y'^2\right] \quad (3.31)$$

3.4.4 Fixed Normal-Strip Heat Source with Thickness X_0 on Semi-infinite Body Limited by Plane $X = -\delta/2$

Let us consider a body limited by plane $z = 0$ and ***by plane $x = -\delta/2$***. To this end, we introduce a new coordinate system $x_1 - \delta/2 = x$, $y = y_1$, $z = z_1$ in which origin O is placed on the adiabatic body boundary $z = 0$ (Fig. 3.6). Using the image principle [16], the body temperature will be increased with instantaneous acting of the normal-strip heat source with thickness x_0 by

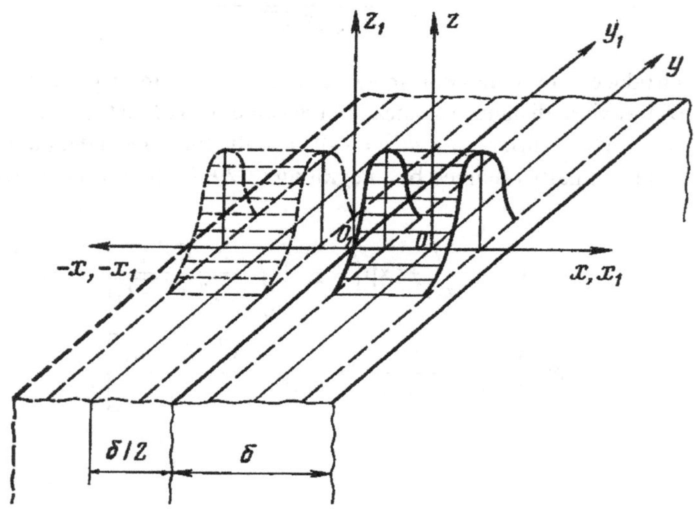

Fig. 3.6 Illustration of the image method to temperature field determination in a plate of finite thickness [25 Permission 4777140922619]

$$dT(x_1, y_1, z_1) = \frac{2q_m dx_1' dy_1' dt'}{c\gamma(4\pi at')^{3/2}} \left\{ \exp\left[-\frac{\left(x - x_1'^2\right) + \left(y - y_1'\right)^2 + z_1^2}{4at'} - k_c y_1'^2 \right] + \exp\left[-\frac{\left(x + x_1'^2\right) + \left(y + y_1'^2\right) + z_1^2}{4at'} - k_c y_1'^2 \right] \right\} \tag{3.33}$$

Integrating (3.33) in ranges $(\delta/2) \pm (x_0/2)$, $(-\delta/2) \pm (x_0/2)$, and $-\infty < y_1 < \infty$ and using the previous coordinate system with origin in the source center, the temperature distribution caused by continuous normal-strip heat source with strip width x_0 at moment t is

$$T(x, y, z, t) = \frac{q}{4\pi\lambda x_0} \int_0^t \frac{dt'}{\sqrt{t'(t' + t_0)}} \exp\left(-\frac{y^2}{4a(t' + t_0)} - \frac{z^2}{4at'} \right) \left[\begin{array}{c} \Phi\left(\frac{x + (x_0/2)}{\sqrt{4at'}}\right) - \Phi\left(\frac{x - (x_0/2)}{\sqrt{4at'}}\right) \\ +\Phi\left(\frac{x + \delta + (x_0/2)}{\sqrt{4at'}}\right) - \Phi\left(\frac{x + \delta - (x_0/2)}{\sqrt{4at'}}\right) \end{array} \right] \tag{3.34}$$

Expression (3.34) describes the temperature distribution in the finite body, which contacted with thermal isolated bodies caused by normal-strip heat source which center placed at distance $(\delta/2)$ from contact with the isolated bodies.

3.4.5 Fixed Normal-Strip Heat Source with Thickness X_0 on Lateral Side of Finite Plate ($X_0 < \delta$)

Let us consider a body limited by plane $z = 0$ and ***by two adiabatic planes $x = \pm\delta/2$***. Fixed normal-strip heat source of thickness x_0 acts the lateral plate side with thicknessδ for condition of $\boldsymbol{x_0 < \delta}$. For this case will be also used the new coordinate system $x_1 - \delta/2 = x, y = y_1, z = z_1$ in which origin O placed on the adiabatic body boundary $z = 0$ (Fig. 3.6). Considering the expressions from paragraphs 3.4.3 and 3.4.4 and the image principle the sum of elementary sources taking into account influence of the reflected sources due to the body limited by two surfaces, the following temperature field will be

$$dT(x_1, y_1, z_1) = \frac{q dt'}{c\gamma x_0} \frac{Exp\left(-\frac{y^2}{4at'}\right)}{\sqrt{4\pi a(t' + t_0)}} \frac{Exp\left(-\frac{z^2}{4at'}\right)}{(4\pi at')^{3/2}} X$$

$$\int \left[\sum_{-\infty}^{\infty} \begin{array}{c} \exp\left(\dfrac{\left(x_1 - x_1' - 2i\delta\right)^2}{4at'}\right) \\ + \sum_{-\infty}^{\infty} \exp\left(-\dfrac{\left(x_1 + x_1' - 2i\delta\right)^2}{4at'}\right) \end{array} \right] \mathrm{d}x_1' \quad (3.35)$$

Here, $(x_1 - 2i\delta)$ and $-(x_1 + 2i\delta)$ are the distances to the image sources, and i is the positive and negative numbers including zero. After integration (3.35) and using the previous coordinate system, the expression for temperature field can be obtained as

$$T(x, y, z, t) = \frac{q}{4\pi\lambda x_0} \int_0^t \frac{\mathrm{d}t'}{\sqrt{t'(t' + t_0)}} \exp\left(-\frac{y^2}{4a(t' + t_0)} - \frac{z^2}{4at'}\right) M(t') \quad (3.36)$$

$$M(t') = \sum_{-\infty}^{\infty} \left[\begin{array}{l} \Phi\left(\dfrac{x - 2i\delta + (x_0/2)}{\sqrt{4at'}}\right) - \Phi\left(\dfrac{x - 2i\delta - (x_0/2)}{\sqrt{4at'}}\right) \\ + \Phi\left(\dfrac{x + \delta(2i - 1) + (x_0/2)}{\sqrt{4at'}}\right) - \Phi\left(\dfrac{x + \delta(2i - 1) - (x_0/2)}{\sqrt{4at'}}\right) \end{array} \right]$$

3.4.6 Moving Normal-Strip Heat Source on a Later Plate Side of Limited Thickness ($X_0 < \delta$)

A continuous normal-strip (thickness x_0) source of an effective power q and concentration coefficient k moves on lateral side of the finite plate surface when condition $x_0 < \delta$ is fulfilled.

The center C of the source, which is coincident with the initial time moment $t = 0$ and with the origin O of the coordinate system x, y, z, moves along the y-axis on the plate surface with velocity v (Fig. 3.7). The values of q, k, and v are assumed constant. At time t, the distance of the C from the origin is vt.

The time interval t will be considered as sum of element $\mathrm{d}t'$ and the corresponding distance as sum of elements $v\mathrm{d}t'$. An instantaneous normal-strip source $Q = q\mathrm{d}t'/x_0$ with its center O' is introduced at time t' at distance vt' from initial location point O, and the heat expands in the plate during $t'' = t - t'$ and rises an element's temperature at some point:

$$\mathrm{d}T(x, y, z) = \frac{qM(t'')}{4\pi\lambda x_0} \frac{\mathrm{d}t'}{\sqrt{t''(t'' + t_0)}} \exp\left[-\frac{\left(y - vt'\right)^2}{4a(t'' + t_0)} - \frac{z^2}{4at''}\right] \quad (3.37)$$

Let us introduce a moving coordinate system with origin at point O_m which is located on y-axis at distance vt_0 ahead of the heat source center and that $y = v(t +$

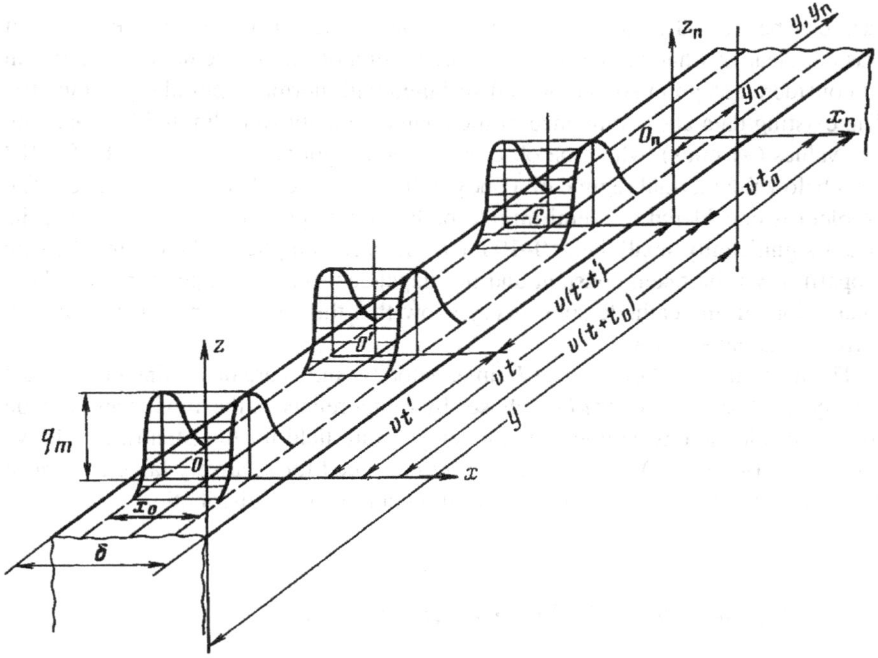

Fig. 3.7 Schema of later side of the plate with a moving normally distributed heat source [25 Permission 4777140922619]

$t_0) + y_m$ or $y - vt' = y_m + v(t'' + t_0)$; $z = z_m$, $dt'' = -dt'$. According to the principle of superposition, using (3.37) in the moving coordinate system, the temperature distribution in the finite plate will be increased with continuous normal-strip heat source action by the following expression:

$$T(x, y, z, t) = \frac{q}{4\pi\lambda x_0}\exp\left(-\frac{vy_m}{2a}\right)$$

$$\int_0^t \frac{dt''}{\sqrt{t''(t''+t_0)}}\exp\left[-\frac{y_m^2}{4a(t''+t_0)} - \frac{z_m^2}{4at''} - \frac{v^2(t''+t_0)}{4a}\right]M(t'') \quad (3.38)$$

3.5 Temperature Field Calculations. Normal Circular Heat Source on a Semi-infinite Body

One of most important issues relevant to arc heat transfer is to study the temperature fields in the case of three-dimensional heat regime of the electrode taking into account the transient, nonlinear phenomena caused by heat flux from the cathode spot. Such case can be treated as action of moving concentrated heat source. Previously, it was not clear to what degree of heat concentration of the source its heat action can be considered as point source or as distributed with normally circularly symmetry. The existing data about influence of the source velocity were limited by relatively low values (~1 cm/s), while the observed cathode spot velocity can reach 10^3–10^4 cm/s below the thermal regime of a body will be considered at different stages. The problem is considered for semi-infinite body with respect to arc spot in which its size is significantly small, i.e., 10–100 μm. Also, for simplicity, the thermophysical properties will be taken constant and not dependent on the temperature. The later calculations in the cathode spot theory show that this effect is not critical at spot current of about ~100 A.

The heat flux $q = Iu_{ef}$, current I, time, velocity, and t_0 are determined by current density $t_0 = (4ak)^{-1} = (4a\pi j/I)^{-1}$ chosen by values related to the moving arc cathode spot. The calculations demonstrate the temperature field for tungsten (u_{ef} = 15 V) and copper (u_{ef} = 10 V). The results were presented for relatively large current of heat source to obtain details of the field of the temperature distribution.

3.5.1 *Temperature Field in a Tungsten Body*

The calculations for tungsten were conducted according to (3.14) [26]. 3D configuration of isotherm curve of 3000 K temperature (close to tungsten melting temperature) is illustrated in Fig. 3.8. The result shows that the maximal values of the length (*b*) along the *x-axis* are larger than the width (*a*, *y*-axis) and depth (*h*, *z*-axis) of the isotherm configuration even at relatively low source velocity, v = 10 cm/s.

The temperature field as dependence on the source time action in *x-y* plane of the body surface for different temperatures of the isotherms is shown in Fig. 3.9. It can be seen that the sizes of the isotherms decreased with the temperature for constant

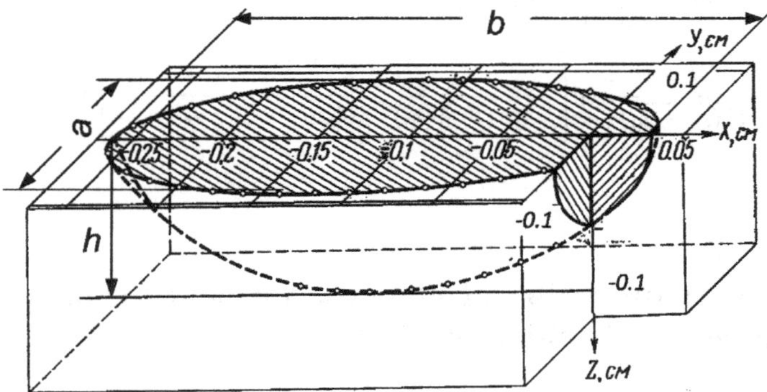

Fig. 3.8 Distribution of isotherm curve of 3000 K temperature at 5 ms after beginning of the heat source action. I = 1000 A; j = 10^6 A/cm^2; v = 50 cm/s [26]

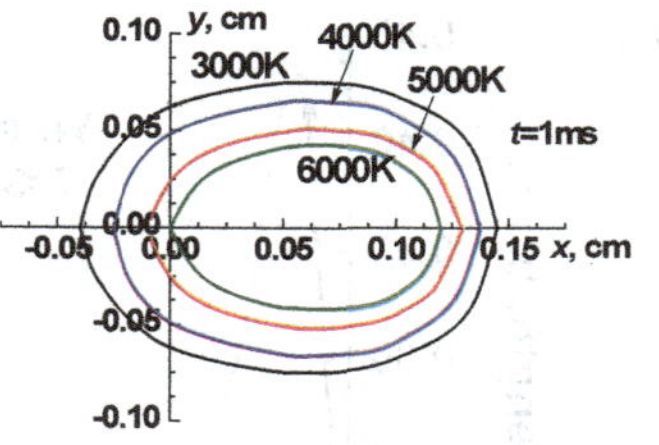

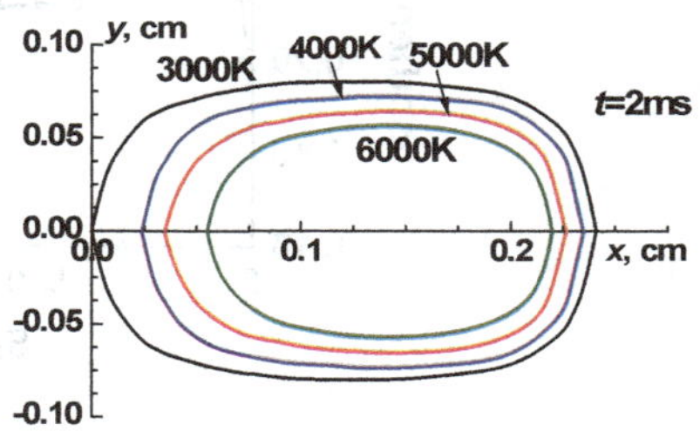

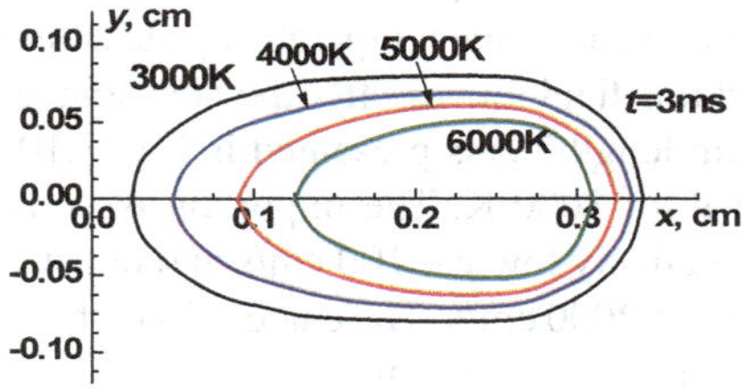

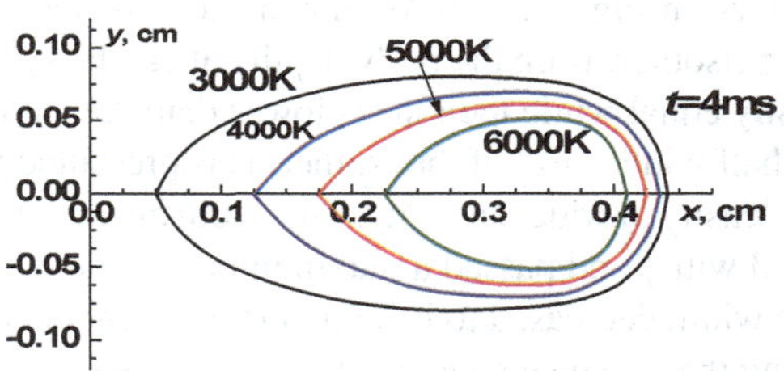

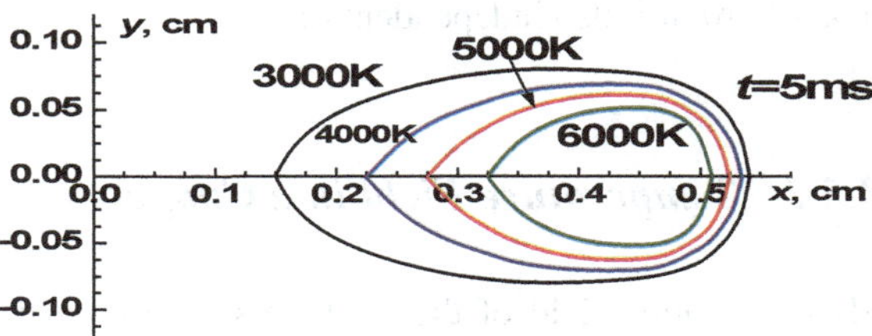

Fig. 3.9 Temperature field at surface of a semi-infinity body caused by moving spot with velocity 100 cm/s after time $t = 1$, 2, 3, and 5 ms from beginning source action for isotherm curves with 3000, 4000, 5000, and 6000 K temperatures, respectively, $I = 1000$ A; $j = 10^5$ A/cm^2

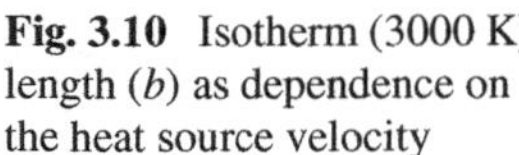

Fig. 3.10 Isotherm (3000 K) length (b) as dependence on the heat source velocity

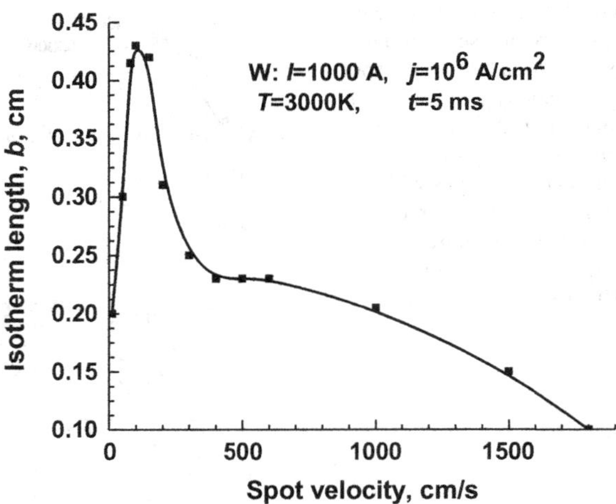

time. The isotherm length significantly increases with time in range of 1–5 ms, but its width changes weakly; the temperature field acquires elliptical (tear-shaped) form after 3 ms, and this field moves after the heat source.

The isotherm length (b) is presented in Fig. 3.10 as dependence on velocity v in wide range of v for 3000 K. The important issue is that this dependence passed a maximum at relatively low $v \sim 100$ cm/s and decreased to value lower than 0.01 cm at v larger of about 2000 cm/s. The calculations show that the sizes a and h decreased to about 0.01 cm at $v = 1500$ cm.

3D temperature field of the isotherms with temperature in range 3000–7000 K is presented in Fig. 3.11 for W in plane x-y (isotherms on the body surface) and in plane x-z (isotherms on the body depth). It can be seen that isotherms in both planes are mostly equal while their sizes lower than the isotherm length.

The half width (a/2) of the isotherms is presented in Fig. 3.12 as dependences on current density in wide range of j. The width mostly of all curves in these dependences increased with j and passed a maximum at in range of $j = 2\times10^4$–10^5 A/cm^2. After that, the width decreased to level that is independent on j. This is an important result indicating that the heat source acts as a point source. Isotherm depth (h) in range of $T = 3000$–6000 K (Fig. 3.13) increased with j reaching a level of (h) that after about 5×10^5 A/cm^2 also independent on j.

3.5.2 *Temperature Field in a Copper*

3D temperature field of the isotherms with temperature in range 500–2500 K is presented in Fig. 3.14 for Cu in plane x-y (isotherms on the body surface, a/2) and in plane x-z (isotherms on the body depth, h). It can be seen that isotherm's sizes

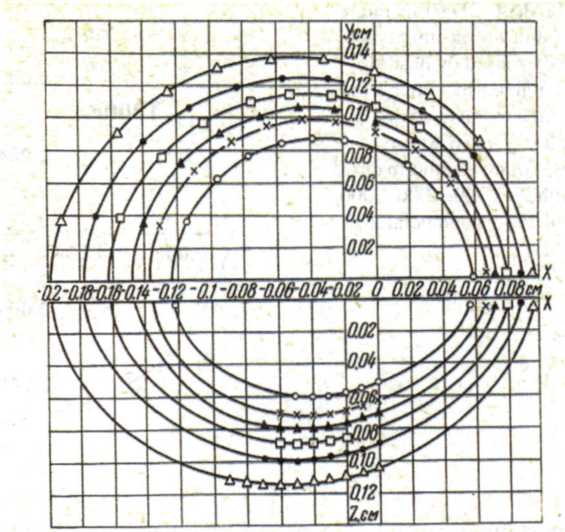

Fig. 3.11 3D temperature field in a semi-infinity W body caused by heat flux from moving spot with velocity 10 cm/s after 5 ms from beginning source action. $I = 1000$ A; $j = 10^5$ A/cm^2; isotherm curve with $T = 3000$, 4000, 5000, 6000, 7000, and 9000 K temperatures placed which sizes decreasing with T, respectively [From 26]

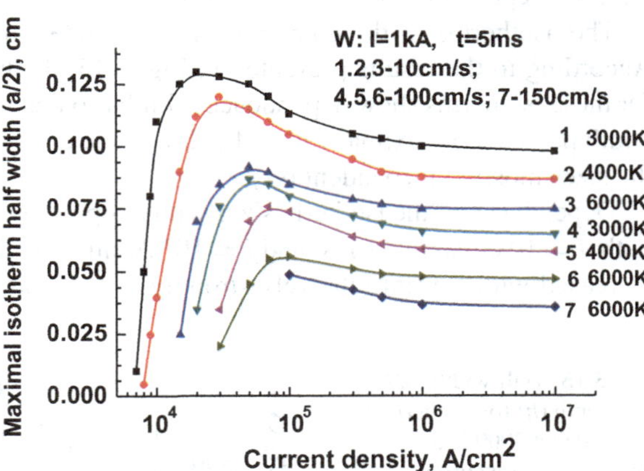

Fig. 3.12 Half width (a/2) of isotherm in range of $T =$ 3000–6000 K as dependences on current density, $v = 10$–150 cm/s

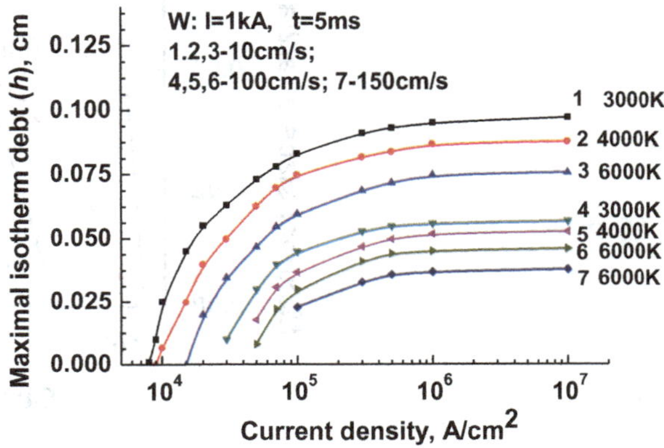

Fig. 3.13 Isotherm depth (h) in range of $T =$ 3000–6000 K as dependences on current density, $v = 10$–150 cm/s

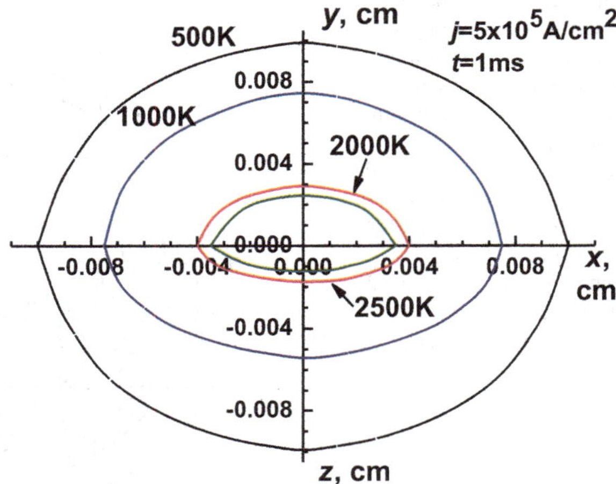

Fig. 3.14 3D temperature field in a semi-infinity Cu body caused by heat flux from fixed spot after 1 ms from beginning source action. $I = 30$ A; $j = 5 \times 10^5$ A/cm^2; isotherm curve with $T = 500$, 1000, 2000, and 2500 K temperatures

are symmetric in each planes, while they are not equal between the both planes, ($h < a/2$) except isotherm with $T = 500$ K.

The isotherms with temperatures $T = 1000$ and 3000 K were calculated. According to the results presented in Fig. 3.15, the half width ($a/2$) and depth (h) for these isotherms show dependences similar to that obtained for tungsten. The half width passed the maximal value of $a/2$ in range of $j = 2 \times 10^4$–10^5 A/cm^2 and then the sizes mostly independent on j.

Dependence of the isotherm sizes with I is presented in Fig. 3.16 for moving spot with $v = 10$ cm/s, $t = 1$ ms, and $j = 10^5$ A/cm^2. The half width ($a/2$) and depth (h) increased with current relatively strongly at $I = 300$ A, and then, this dependence

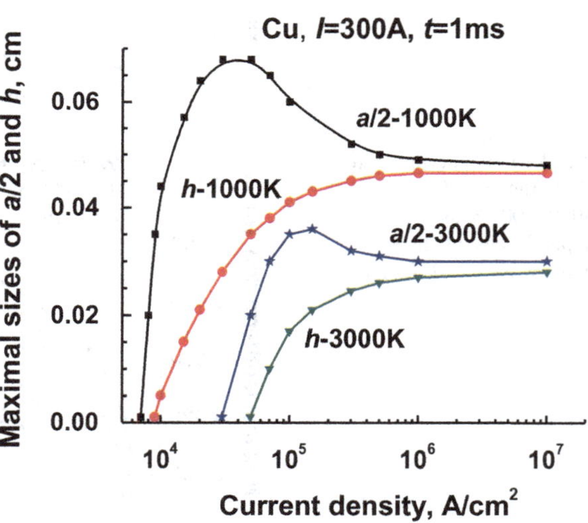

Fig. 3.15 Half width ($a/2$) and depth (h) for isotherm $T = 1000$ and 3000 K as dependences on current density, $v = 10$ cm/s

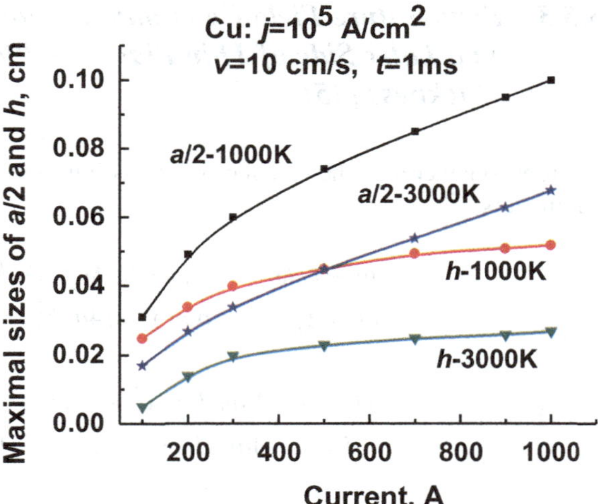

Fig. 3.16 Half width ($a/2$) and depth (h) for isotherm T = 1000 and 3000 K as dependences on current

indicates weaker increase of the isotherm sizes with I. Dependence of the isotherm sizes with heating time is presented in Fig. 3.17 for $I = 200$ A, and conditions are indicated in Fig. 3.16. It can be seen that the strongest increase of half width is in range of 10^{-5}–10^{-1}s and of depth is in range of 10^{-4}–10^{-2}s.

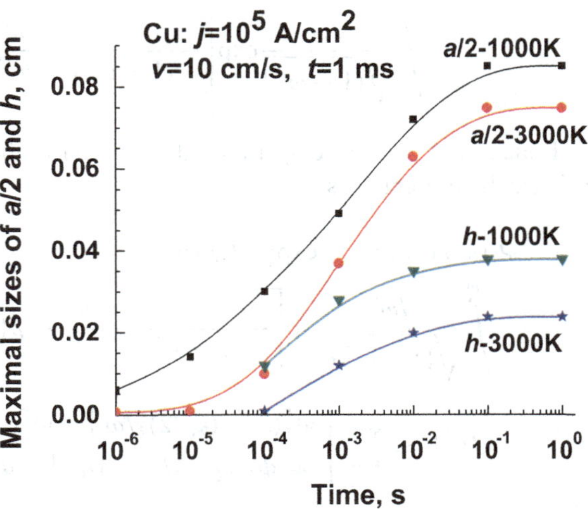

Fig. 3.17 Half width ($a/2$) and depth (h) for isotherm T = 1000 and 3000 K as dependences on spot time, I = 200 A

3.5.3 Temperature Field Calculations. Normal Heat Source on a Later Side of Thin Plate and Plate with Limited Thickness [25]

In order to generalize the calculation, let us introduce the following dimensionless parameters:

$$\omega = t/t_0, n = y_m^2/4at_0, t_0 = (4ak)^{-1},$$
$$m = z_m^2/4at_0, p = v^2t_0/4a, k_0^2 = x_0^2/4at_0,$$

$$D^2 = \delta^2/4at_0, l = x^2/4at_0, l_1 = l - 2iD,$$
$$l = x^2/4at_0, l_1 = l - 2(i-1)D,$$

$$\Psi(n, m, p, \omega) = T(x, y, v, t)2\pi\lambda\delta/q,$$
$$\Psi_1(n, m, p, \omega) = T(x, y, v, t)2\pi\lambda x_0/q$$

In this case, (3.30) for temperature distribution in the thin plate ($x_0 = \delta$) can be presented as:

$$\Psi(n, m, p, \omega) = \exp\left(-2\sqrt{np}\right) \int_0^{\omega} \frac{d\omega'}{\sqrt{\left(1+\omega'^2\right)}} \exp\left[-\frac{n}{\left(1+\omega'^2\right)} - \frac{m}{\omega'^2} - p\left(1+\omega'^2\right)\right] \tag{3.39}$$

Equation (3.38) for temperature distribution in the plate of finite thickness ($x_0 < \delta$) can be presented as:

$$\Psi_1(n, m, p, \omega) = \exp\left(-2\sqrt{np}\right) \int_0^{\omega} \frac{d\omega'}{\sqrt{\left(1+\omega'^2\right)}} \exp\left[-\frac{n}{\left(1+\omega'^2\right)} - \frac{m}{\omega'^2} - p\left(1+\omega'^2\right)\right] M(\omega') \tag{3.40}$$

$$M(\omega') = \sum_{-\infty}^{\infty} \left\{ \begin{array}{l} \Phi[(l_1 + (k_0/2))/\omega'] - \Phi[(l_2 - (k_0/2))/\omega'] \\ + \Phi[(l_2 + (k_0/2))/\omega'] + \Phi[(l_2 - (k_0/2))/\omega'] \end{array} \right\}$$

The transition from thin plate to plate of finite thickness was characterized by parameter $N = \delta/x_0$ which determines the relation between the degrees of concentration in y- and x-directions. Figure 3.18 shows dependence of Ψ on coordinate n for thin plate. It can be that relatively high level of heating ($\Psi \approx 1$) can be reached even on distance enough larger than the size y_0 ($n \gg 1$) at $\omega^2 \geq 10$ (this is time of

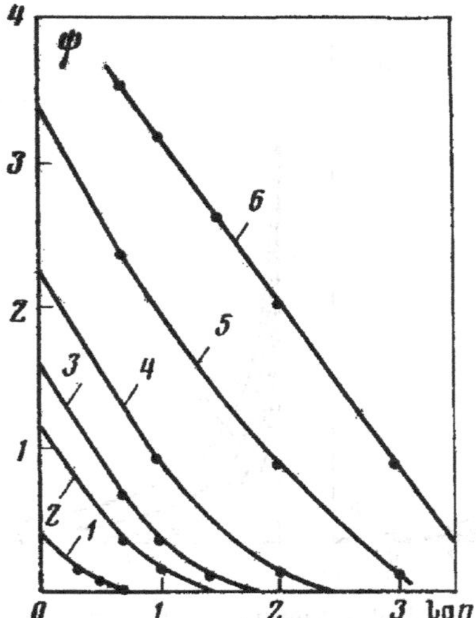

Fig. 3.18 Dependence of dimensionless temperature on coordinate n for thin plate, $p = 0$, $m = 0$; 1 $- \omega^2 = 1$, 2–10, 3–25, 4–10^2, 5–10^3, and 6–10^4. Figure taken from [25 Permission 4777140922619]

about 1 ms for $y_0 = 10^{-2}$ cm). For $\omega^2 = 10^3$–10^4, function Ψ increased by factor 3–4 when $n = 1$–10 and by factor up to 2 when $n = 10$–100.

The calculations showed that Ψ weakly depend on coordinate m for thin plate up to $m = 10^{-1}$, and at larger m, the dimensionless temperature sharply decreased depending on ω and n. The time-dependent Ψ was similar to the Ψ dependence on n, while for long time source action $\omega^2 = 10^3$, the most heating of the thin plate remained near y_0.

The heat source velocity (p) influence is illustrated in Fig. 3.19. In the range of p from 0 to1 ($v = 0$–5×10^2 cm/s), the plate surface temperature ($m = 0$) can be significantly decreased depending on coordinate n (about by order of magnitude). The obtained result shows that the influence of the source motion is significantly mostly on the distance that is comparable with the size of y_0.

The calculations for plate finite size when the heat source x_0 is lower than the plate thickness δ were conducted in order to determine the temperature of the source center for $p = 0$. The results are presented in Fig. 3.20 as dependence on k_0. This temperature is higher for lower source concentration in x-direction in comparison with its concentration in y-direction and decreases with parameter $N = \delta/x_0$, and the rise of the decreasing depends on k_0. However, for certain k_0, depending on $\omega =$ 1 and 10, the temperature is weakly changed (see Fig. 3.20). With furthermore k_0 increasing, the surface temperature is equal to the temperature calculated according to the thin plate regime ($N = 1$).

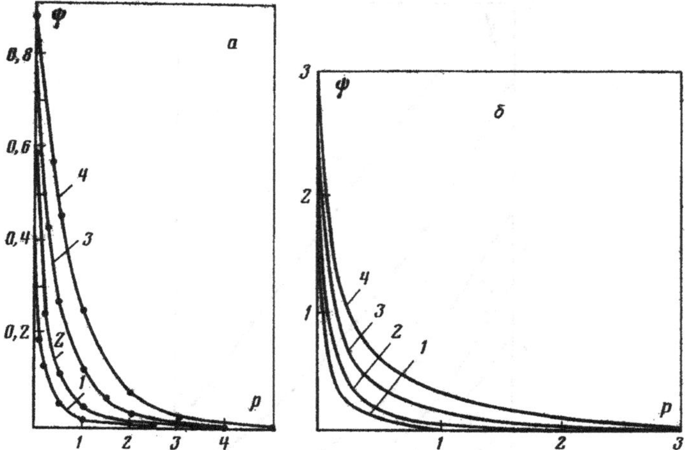

Fig. 3.19 Dependence of dimensionless temperature on velocity p for thin plate, $m = 0$; a – $\omega = 1$; b – $\omega = 10$; 1 – $n = 1$, 2–0.5, 3–0.1, and 4–0. [25 Permission 4777140922619]

Fig. 3.20 Dependence of dimensionless temperature in the center of fixed source on coefficient k_0 for plate finite size, I – $\omega^2 = 1$; II – 10, III – 10^2; 1 – $N = 10$, 2–5, 3–3. The horizontal dotted lines indicate the calculation using thin plate approximation. [25 Permission 4777140922619]

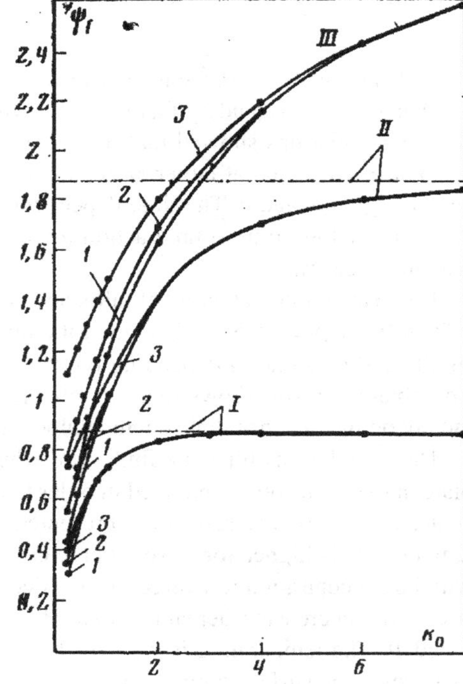

3.5.4 Concluding Remarks to the Linear Heat Conduction Regime

In general, the obtained results show that the temperature field configuration significantly depends on source parameters, allows indicate the limit values of j and v, the isotherm size of melting temperature which can be compared with size of the crater formed after the spot. In case of $v \leq 10$ cm/s, the configuration of this isothermal surface is spherical and can be similar to the crater form. In case of $v \geq 100$ cm/s, the elliptical configuration of the isotherms indicate that character of the spot motion is not continuous when the observed craters have spherical the form.

Considering the calculated temperature fields, it can be seen that the isotherm sizes relatively lower for metals like copper than for refractory materials like tungsten (see Figs. 3.11 and 3.16, $I = 1000$ A). The 3D temperature fields are mostly symmetric in both x-y and x-z planes for W, while it is not the case for Cu. For $v \leq 10$ cm/s and $t = 5$ ms, the isotherm of melting temperature is spherical form, while for $T \geq 2000$ K, the isotherm depth is lower than their width. The isotherm sizes of melting temperature exceed sizes of the boiling temperature by factor near two and significantly decreased with current and time. This difference decreased with the source power and its velocity increase. The size of $a/2$ is lower than 20 μm for Cu boiling temperature when $I = 30$ A and $t = 1$ ms.

The most important result is the dependence of the temperature field on spot current density. This dependence indicates that the isotherm width is maximal at $j = 10^5$ A/cm^2 and remain constant when the current density reaches about $j = 10^6$ A/cm^2 and more for both materials. This means that action of the heat source passed from the form of normal circular distribution of the heat to an action of a point heat source. In this case, the size of isotherm (e.g., melting temperature) is ambiguous function of j, i.e., it is impossible to determine the spot size from the comparison with experimental size of melting crater using autograph method. As example part of current can be collected at the surface around the spot size determined by it brightness.

The calculations of the temperature field for moving heat source showed that the isotherm changed their form from symmetric to tear-shaped (elliptical) depending on time and velocity of the heat source action. Although the maximal isotherm length is at $v \sim 100$ cm/s, the length size (as well the width and debt) strongly decreased at $v \sim 2000$ cm/s, as it was observed [16]. This result indicates that at some critical large value of the velocity or at some short time, the heat source action is weak, metal heating is not sufficiently high, and the arc spot cannot be supported self-consistently.

Regarding to the plate heating, as expected, the temperature distribution depends on the relation between the heat source and plate sizes. At relatively prolonged action of the heat source, the heating of the thin plate occurs primarily in the longitudinal direction (as opposed to the heating a semi-infinite body) and on sufficiently large distance compared with plate thickness and hence with the source size. The calculation for relatively large k_0 allowing use of the simple approach of $N = 1$ is important in case of a conjugated problem that is necessary to describe the self-consistent phenomena in the plasma–electrode systems. In general, it can be noted that although the relative level of heating of the plate finite thickness is lower than that heating level for a thin plate (Figs. 3.18 and 3.20), the absolute temperature value of the plate finite thickness is larger by factor N. The influence of k_0 increased with time rise of the heat source action.

3.6 Nonlinear Heat Conduction

The considered nonlinear process the solid heat conduction mainly include the loss heat due to the surface evaporation and radiation.

3.6.1 Heat Conduction Problems Related to the Cathode Thermal Regime in Vacuum Arcs

The heat conduction problems related to the thermal regime of electrodes, used in a developed technology, requested new simple engineering approaches. It also should be taking into account that the energetic parameters of the arc depend on the arc conditions due to interconnected phenomena between the electrode temperature and power dissipated in the arc spot. Thus, the spot problem is associated with a strong nonlinear heat conduction regime of the electrode. These circumstances stimulated use a heat balance assuming that the temperature uniformly distributed in the spot at the electrode surface and equal to the temperature in the spot center. The heat balance approach allows to study a simple form of the arc–electrode interaction using different theoretical models [27–30]. The method of heat balance, in essence, is similar to the channel approach for air arc with that difference that the arc current is concentrated on the electrode area named cathode spot and not in cross section of arc column, and the arc power dissipated in the electrode body and not in the surrounding gas. Although the indicated progress in 60th of the 20th century, approaches developed at that time did not consider three dimensionality, non-stationarity, and nonlinearity of the heat conduction processes simultaneously. Also, the information about temperature distribution inside the spot was absent, and therefore, the assumption about temperature uniformity in the spot was not verified and justified.

3.6.2 Normal Circular Heat Source Action on a Semi-infinity Body with Nonlinear Boundary Condition

A model which takes in account a general mathematical formulation of the heat conduction problem with nonlinear boundary condition was developed, and the calculation was presented for tungsten cathode [31]. This model used an assumption that the maximal temperature caused by normal circular heat source action cannot exceed the temperature T_c calculated from the energy balance, and the temperature below T_c can be obtained by linear heat conduction approach from (3.14). The temperature T_c was obtained using the heat conduction loss at center of the heat source. The question is how the temperature T can be different from T_c inside the source area far from the center and what is the deformation of temperature distribution comparing to that given by (3.14) in the transition region of low temperatures where the evaporation flux is also relatively weakly.

To understand the questions, the heat conduction problem should be formulated in general form (in differential form with nonlinear boundary condition), and result of the solution should be compared with the solution obtained from above described approximation [31]. Let us describe the details and some new study for copper cathode. This problem was considered for semi-infinity body because the spot size (10–100 μm) is much smaller than the electrode size ($\geq$1 cm). The normal circular source with power q having distribution is approximated by Gauss law with concentration coefficient $k = r_0^{-2}$, where $r_0 = r_s$ is the spot radius. The energy losses for metal melting q_{melt} were neglected comparing to the vaporization energy q_s for the case of copper group spot current $I = 100$ A, spot current density ~10^6 A/cm^2, erosion rate 10^{-2} g/s, depth ~10^{-3} cm, $t = 5$ ms, and λ_{melt}, λ_s, γ are the specifics weight, melting heat, and evaporation heat, respectively, due to condition

$$\frac{q_{\text{melt}}}{q_{ev}} = \frac{Ih\gamma\lambda_{\text{melt}}}{jq_{ev}G\lambda_s} < 1$$

Coinciding origin of the coordinate system ($r^2 = x^2 + y^2$, $z =$ zero, Fig. 3.2) with the source center and direction of the heat source motion with *x-axis,* the mathematical formulation of the problem includes the transient heat conduction equation with nonlinear boundary condition that is:

$$\frac{\partial T}{\partial t} = a\nabla^2 T + v\frac{\partial T}{\partial x} \tag{3.41}$$

Boundary condition:

$$\lambda_T \left.\frac{\partial T}{\partial z}\right|_{z=0} = -q_m \exp\left[-k\left(x^2 + y^2\right)\right] + \lambda_s W(T) + \sigma_{st} T^4 + \Phi(T) \tag{3.42}$$

Initial condition:

$$T|_{t=0} = T_0(x, y, z)$$

where λ_T, a are the heat conductivity and heat diffusivity, respectively, σ_{st} is the Stefan–Boltzmann coefficient of black-body radiation. $\Phi(T)$ is the energy loss by the macroparticle (MP) ejection which can be neglected according to the estimations taking into account the low rate of MP mass loss (relatively low MP velocity) comparing to the energy due to evaporation rate $W(T)$. In (3.42), a relation expressed as a function of evaporation rate on surface temperature the rate $W\ (T)$ was taken in form of Langmuir–Dushman:

$$LgW(T) = C - \frac{B}{T} - 0.5LgT \qquad (3.43)$$

C and B are empirical constants [32]. Formula (3.42) for evaporation mass rate describes the material evaporation into vacuum [33] and not applicable to determine the mass loss in vacuum arc from dense plasma of the cathode spot due to returned mass flux [34] (see Chap. 2). The measured cathode erosion rate was used to describe the rate of mass loss in vacuum arc [34]. However, the electrode erosion measurements are the result of some integral process over the eroded cathode areas and over the arc time and, consequently, cannot be used in form of a dependence on surface temperature as boundary condition. Therefore, below formula (3.43) was used to estimate the maximal rate in order to understand the effect of the material vaporization dependence on the temperature. In the same time, the obtained results can be useful for systems with the concentrate heat sources like low-power lasers or electron beams, where the returned mass flux to the heating area can be neglected. Some result obtained under electron beam action and consisting with the experiment, as example, can be seen in [35]. Equations (3.41)–(3.43) were solved for tungsten by heat source action with $j = 10^4$ A/cm^2, $t = 5$ ms, and $v = 10$ cm/s [31]. The result showed that the numerically obtained temperature distribution inside the source weakly changed and was like a plateau and near the value calculated from energy balance for the center, which is in form:

$$Iu_{\text{ef}} = \frac{(T_c - 300)c\gamma(4\pi a)^{3/2}}{\int\limits_0^t \frac{\text{Exp}\left[\frac{v^2}{4a}(t_0+t')\right]}{\sqrt{t'}(t_0+t')}\text{d}t'} + \lambda_s W(T) + \sigma_{\text{sb}}T^4 \qquad (3.44)$$

The above model was extended for calculations of the temperature field for a copper cathode [36]. The time-dependent 3D heat conduction (3.41)–(3.43) and (3.44) for fixed heat source were solved for a Gaussian distributed heat influx at the cathode surface with the characteristic Gaussian radius parameter equal to the effective spot radius r_s, Fig. 3.2. The nonlinear heat losses due to evaporation and radiation from the surface were taken into account in form of Langmuir–Dushman [32] and by Stefan–Boltzmann law.

3.6.3 Numerical Solution of 3D Heat Conduction Equation with Nonlinear Boundary Condition

3D heat conduction (3.41)–(3.43) together with the energy balance (3.44) for fixed heat source were solved numerically using finite difference approximation of the derivatives of (3.41)–(3.42) in explicit form and with variable mesh step.

The used numerical approach allowed to obtain the solution in the source region where the power rise was enhanced in direction to the spot center and is larger when the spot radius r_s is smaller. The process convergence and accuracy were controlled by variation of the mesh step and by comparison of the numerical solution in linear approximation with result calculated using the analytical expression (3.16) as dependence on time for source center. The results of the calculations are presented in Fig. 3.21 ($t = 150$ μs, $r_s = 150$ μm), Fig. 3.22 ($t = 37$ μs, $r_s = 150$ μm), and Fig. 3.23 ($t = 2.3$ μs, $r_s = 67$ μm). Non-stationary radial (r/r_s) temperature distribution on a

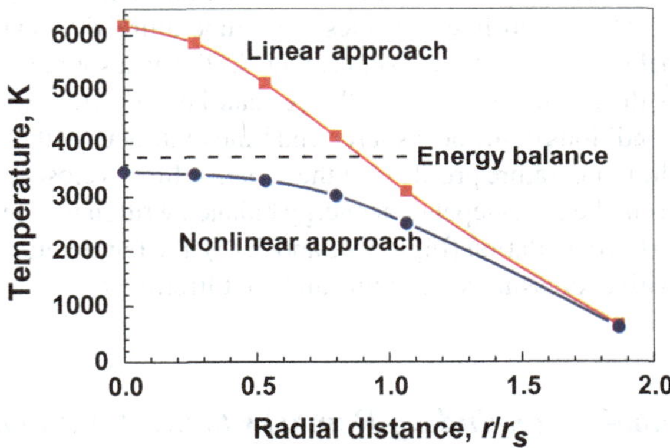

Fig. 3.21 Radially surface temperature distribution on a Cu cathode for $t = 150$ μs, $r_s = 150$ μm

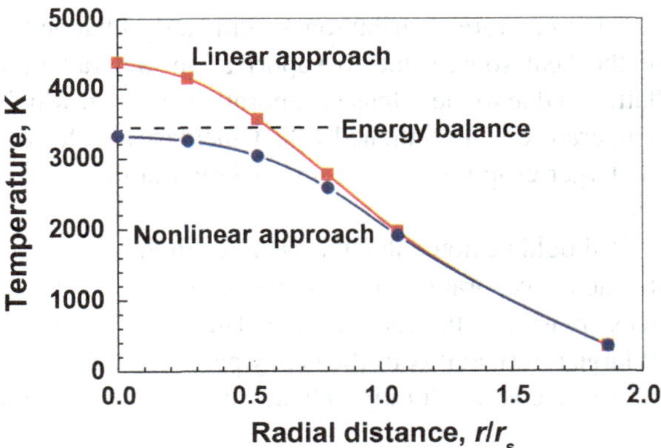

Fig. 3.22 Radially surface temperature distribution on a Cu cathode for $t = 37$ μs, $r_s = 150$ μm

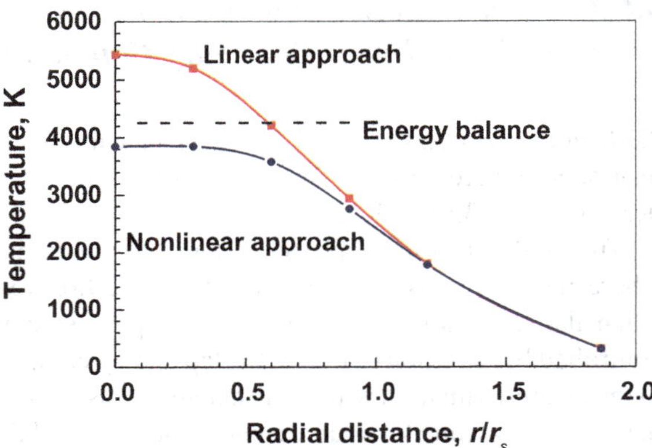

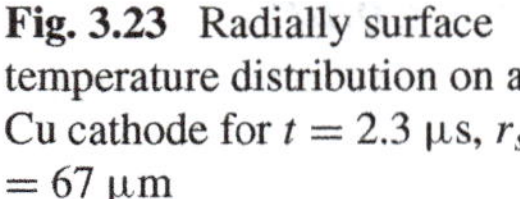
Fig. 3.23 Radially surface temperature distribution on a Cu cathode for $t = 2.3\ \mu\text{s}$, $r_s = 67\ \mu\text{m}$

Cu cathode under a heat source with spot power 1.7 kW is obtained in a linear and nonlinear approximation. t is the time elapsed from the start of heating.

The dotted line indicates the temperature T_c obtained from the cathode energy balance written for the spot center. As the heat source is fixed, the surface temperature is distributed symmetrically. As can be seen from the present results for different conditions of the heat source and time of its action, that there is a nearly flat portion of the temperature profile near the center, which is close to the temperature T_c calculated from the time-dependent energy balance written for the spot center (3.44). According to the calculation (Figs. 3.21 and 3.23), the temperature distribution is flattened more with decreasing spot radius and spot lifetime t.

3.6.4 Concluding Remarks to the Nonlinear Heat Conduction Regime

The temperature distribution significantly changed at the surface near the center of the heat source due to vaporization and radiation. The temperature profile is flattened due to the intense evaporative cooling, which varies exponentially with the temperature. The flattened effect increases with the power of the source because the larger evaporation heat loss is larger at the location of larger temperature at the surface.

It should be noted that the assumed in the above study exponential dependence of the metal evaporation rate with the temperature is acceptable also for the condition of dense plasma of the cathode spot. The observed lower electrode erosion rate is result of large returned flux of the heavy particles, which of course return their enthalpy to the surface and naturally enhance uniformity of the temperature distribution at the surface.

Thus, the simplified approach to description of the thermal processes as consequence of a uniform surface temperature in vicinity of the source center appears profitable, in view of the complexity of study of the self-consistent problems with nonlinear energy dissipation of the strongly heated solid. The complexity of the problem was discussed previously [36] indicating some difficulties related to appearance of radial plasma gradients by 2D study of the spot development.

References

1. Kulyapin, V. M. (1971). Quantitative theory of cathode processes in an arc. *Soviet Physics—Technical Physics, 16*(2), 287–291.
2. Amson, J. C. (1972). Electrode voltage in the consumable-electrode arc system. *Journal of Physics. D. Applied Physics, 5,* 89–96.
3. Williams, D. W., & Williams, W. T. (1972). Field-emitted current necessary for cathode initiated vacuum breakdown. *Journal of Physics. D. Applied Physics, 5,* 280–290.
4. Belkin, G. S., & Kiselev, V. Ya. (1978). Effect of the medium on the electrical erosion of electrodes at high currents. *Soviet Physics—Technical Physics, 23*(1), 24–27.
5. Struchkov, A. I. (1978). Erosion rate and properties of cathode spot of the first kind. *Soviet Physics—Technical Physics, 23*(2), 183–185.
6. Yuhimchuk, S. A. (1981). Destruction of the electrodes in plasmatrons by cathode spots. *Soviet Physics—Technical Physics, 51*(7), 1409–1414.
7. Osadin, B. A. (1965). Erosion of the anode in a heavy current discharge in vacuum. *High Temperature, 3*(6), 849–854.
8. Goloveiko, A. G. (1968). Elementary and thermophysical properties at the cathode in the case of a powerful pulse discharge. *Journal of Engineering Physics, 14*(3), 252–257.
9. Goloveiko, A. G., & Rzhevskaya, S. P. (1969). Limiting processes at the cathode in the case of a powerful pulse discharge. *Journal of Engineering Physics, 16*(6), 1073–1081.
10. Kulyapin, V. M. (1971). Some problems of heat conduction with phase transitions. *Journal of Engineering Physics, 20*(3), 497–504.
11. Jolly, D. C. (1982). Anode surface temperature and spot formation model for the vacuum arc. *Journal of Applied Physics, 53,* 6121–6126.
12. Prock, J. (1986). Time-dependent description of cathode crater formation in vacuum arcs. *IEEE Transaction on Plasma Science, PS-14*(4), 482–491.
13. Klein, T., Paulini, J., & Simon, G. (1994). Time-resolved description of cathode spot development in vacuum arcs. *Journal of Physics. D. Applied Physics, 27,* 1914–1921.
14. Rossignal, J., & Abbaoui, M. (2000). Numerical modelling of thermal ablation phenomena due to a cathode spot. *Journal of Physics. D. Applied Physics, 33,* 2079–2086.
15. Rykalin, N. N. (1947). *Warmetechnische Grundlagen des schweissen, Ausgabe derAcad.* Moskow: Wissenschaft UdSSR.
16. Rykalin, N. N. (1957). *Berechnung der warmevorgange beim schweissen.* Berlin: VEB, Verlag Tecnik.
17. DiBitonto, D. D., Eubank, P. T., Patel, M. R., & Barrufet, M. (1989). Theoretical models of the electrical discharge machining process. I. A simple cathode erosion model. *Journal of Applied Physics, 66*(9), 4095–4103.
18. Yeo, S. H., Kurnia, W., & Tan, P. C. (2007). Electro-thermal modelling of anode and cathode in micro-EDM. *Journal of Physics. D. Applied Physics, 40,* 2513–2521.
19. Kharin, S. N. (1986). Dynamic of arc phenomena at closure of electrical contacts in vacuum circuit breakers. *IEEE Transaction on Plasma Science, 33*(5), 1576–1581.
20. Rykalyn, N. N., Zuev, N. V., & Uglov, A. A. (1978). *Fundamentals of electron-beam treatment of materials.* Moscow: Machine Building Publishers.
21. Rykalin, N. N., Uglov, A. A., & Kokora, A. (1975). *Laser machining and Welding.* Pergamon Press–Mir Publisher.

22. Bogaerts, A., Chena, Z., Gijbelsa, R., & Vertes, A. (2003). Laser ablation for analytical sampling: What can we learn from modeling? *Spectrochimica Acta Part B, 58,* 1867–1893.
23. Carslaw, H. S., & Jaeger, J. C. (1959). *Conduction of heat in solids.* Oxford: At the Clarendon Press.
24. Beilis, I. I. (1979). Normal distributed heat source moving on lateral side of a thin semi-infinite plate. *Physics & Chemistry of a Material Treatment, 13*(4), 32–36.
25. Beilis, I. I. (1986). Thermal regime of metallic carcass of a combined electrode in MHD generators by contraction discharge. *High Temperature, 24*(6), 1173–1981.
26. Beilis, I. I., Levchenko, G. V., Potokin, V. S., Rakhovsky, V. I., & Rykalin, N. N. (1967). About heating of a body acted by a moving high-power concentrated heat source. *Physics & Chemistry of a Material Treatment, 1*(3), 19–24.
27. Lee, T. H., & Greenvood, A. (1961). Theory for the cathode mechanism in metal vapor arcs. *Journal of Applied Physics, 32,* 916–923.
28. Ecker, G. (1961). Electrode components of the arc discharge. *Ergebnisse der exakten Naturwissenschaften, 33,* 1–104.
29. Holm, R. (2000). *Electrical contacts.* Berlin, Reprint: Theory and applications. Springer-Verlag.
30. Hall, A. W. (1962). Cathode spot. *Physical Review, 126*(5), 1603–1610.
31. Beilis, I. I., & Rakhovsky, V. I. (1970). About nonlinear heat losses during the heating electrode by moving arc discharge. *Journal of Engineering Physics, 19*(4), 678–681.
32. Dushman, S. (1962). *Scientific foundation of vacuum technique.* NY-London: Wiley.
33. Langmuir, I. (1913). The vapor pressure of metallic tungsten. *Physical Review, 2*(5), 329–342.
34. Beilis, I. I., & Lyubimov, G. A. (1975). Parameters of cathode region of vacuum arc. *High Temperature, 13,* 1057–1064.
35. Beilis, I. I., Rakhovsky, V. I., Tkachov, L. G., & Smelyansky, B. Ya. (1969). Investigation of energy losses in a high-powered electron beam. *Soviet Physics—Technical Physics, 39*(9), 1650–1659.
36. Beilis, I. I. (2003). The vacuum arc cathode spot and plasma jet: Physical model and mathematical description. *Contributions to Plasma Physics, 43*(3–4), 224–236.

Chapter 4
The Transport Equations and Diffusion Phenomena in Multicomponent Plasma

This chapter considers transport phenomena of the plasma particle, and their fluxes produced due to contact of highly ionized dense plasma with the electrodes and study the hydrodynamic flow of the mass and energy of different plasma components.

4.1 The Problem

In general, a part of vaporized atoms from the electrode can be in an ionized state. At ion move to the cathode, they neutralize their charge at the cathode surface. Therefore, near the surface the plasma region is depleted by ions and an ion flux from the plasma volume to surface arises. Moreover, in a multicomponent plasma the particles flow arise due to their different gradients determined by different types of elementary collisions. The cathodic problem is how correctly determine the ion flux to the cathode taking into account the influence of other particles flow in the multi-component plasma environment. Early the transport and diffusion phenomena in multi-component plasma were studied in works of [1–5]. We use these results to describe the multicomponent phenomena for the condition of near-electrode plasma in a vacuum arc.

4.2 Transport Phenomena in a Plasma. General Equations

A layer near a wall exists in which a density gradient is produced due to ion and atom diffusion in the partially ionized gas. In general, the plasma components α are the different types of atoms, ions, and electrons, for which the fluxes of particles I_α, their enthalpy h_α, and diffusion coefficients $D_{\alpha\beta}$ should be obtained. For each of components, the transport equations in general form were derived using kinetic equations by the Grad method with 13-momentum approach [2, 3, 5]. The collision

I. Beilis, *Plasma and Spot Phenomena in Electrical Arcs*, Springer Series on Atomic, Optical, and Plasma Physics 113, https://doi.org/10.1007/978-3-030-44747-2_4

integrals were taken in Landau's form for Coulomb particles and in Boltzmann form for other interactions. General expressions for α particle mass fluxes of the particles J^{α} neglecting the terms for thermodiffusion, viscosity, and without magnetic field were presented as sum of mass fluxes due to the gradients of particle densities J_c^{α} and mass flux caused by electric field J_E^{α} in the following form:

$$\begin{aligned} J^{\alpha} &= J_c^{\alpha} + J_E^{\alpha}; \\ J_c^{\alpha} &= \sum_{\beta} \rho_{\beta} \frac{m_{\alpha} T_{\beta}}{m_{\beta} T_{\alpha}} G_{\alpha\beta} \nabla(\ln \rho_{\beta}); \\ J_E^{\alpha} &= \rho_{\alpha} \mu_{\text{mob}}^{\alpha} E_0 \end{aligned} \tag{4.1}$$

The expression for heat fluxes of the particles is obtained in the form:

$$h_{\alpha} = -\sum_{\beta} \lambda_{\alpha\beta} \nabla T_{\alpha\beta} + \sum_{\beta} \mu_{\alpha\beta} J^{\beta}; \tag{4.2}$$

where $J^{\alpha} = \rho_{\alpha}(v_0^{\alpha} - v_0)$; $v_0 = \sum_{\beta} c_{\alpha} v_0^{\alpha}$; $\rho_{\alpha} = m_{\alpha} n_{\alpha}$; $\rho = \sum_{\alpha} \rho_{\alpha}$; $p_{\alpha} = n_{\alpha} k T_{\alpha}$; $p = \sum_{\alpha} p_{\alpha}$; $c_{\alpha} = \frac{\rho_{\alpha}}{\rho}$.

The expression for energy equations is in the form:

$$\begin{aligned} \frac{3}{2} c_{\alpha} \frac{dT_{\alpha}}{dt} &= -c_{\alpha} T_{\alpha} \operatorname{div}(v) - \frac{m_{\alpha}}{k\rho} \operatorname{div} h^{\alpha} - \frac{T_{\alpha}}{\rho} \operatorname{div} J^{\alpha} + \frac{m_{\alpha}}{k\rho^2} J^{\alpha} \nabla P \\ &- \frac{5}{2\rho} J^{\alpha} \nabla T + \frac{J^{\alpha}}{k\rho} \left(e_{\alpha} - \sum_{\beta} \frac{m_{\alpha}}{m_{\beta}} e_{\beta} c_{\beta} \right) E \\ &+ \sum_{\beta} \frac{3 c_{\alpha} m_{\alpha} m_{\beta}}{(m_{\alpha} + m_{\beta})^2 \tau_{\alpha\beta}} (T_{\alpha} - T_{\beta}) \end{aligned} \tag{4.3}$$

Diffusion coefficients and mobility

$$\begin{aligned} D_{\alpha\beta} &= \frac{kT_{\alpha}}{m_{\alpha}} \frac{|a|_{\beta\alpha} - |a|_{\alpha\alpha}}{|a|}; \quad G_{\alpha\beta} = D_{\alpha\beta} - \sum_{\gamma} c_{\gamma} D_{\alpha\gamma} \\ \mu_{\text{mob}}^{\alpha} &= \sum_{\beta} \frac{n_{\beta}}{n_{\alpha} k T_{\alpha}} \left(e_{\beta} - \sum_{\gamma} \frac{m_{\beta}}{m_{\gamma}} c_{\gamma} e_{\gamma} \right) D_{\alpha\beta} \end{aligned} \tag{4.4}$$

$|a|$-nth order determinant with elements $(\alpha\beta)$ expressed as:

$$a_{\alpha\beta} - a_{\alpha\alpha} + \sum_{\gamma} b_{\alpha\gamma} (\mu_{\gamma\beta} - \mu_{\gamma\alpha}); \; \mu_{\alpha\beta} = \sum_{\gamma} \frac{\lambda_{\alpha\gamma} d_{\gamma\beta} \tau_{\gamma}^{*}}{\lambda_{\gamma}}; \; \lambda_{\alpha\beta} = \frac{\lambda_{\beta} |c|_{\beta\alpha}}{|c|}$$

$|c|$ is the determinant with elements $c_{\alpha\beta}$ and $|c|_{\beta\alpha}$ is the algebraic appending of elements $(\alpha\beta)$ to this determinant.

The coefficients $a_{\alpha\beta}$ and $b_{\alpha\beta}$, and characteristic collision times $\tau_{\alpha\beta}$ are obtained in the form:

$$a_{\alpha\alpha} = -\sum_{\beta\neq\alpha} \frac{m_\beta}{m_\alpha + m_\beta} \frac{1}{\tau_{\alpha\beta}}; \quad a_{\alpha\beta} = \frac{m_\beta}{m_\alpha + m_\beta} \frac{1}{\tau_{\beta\alpha}} \quad \alpha \neq \beta$$

$$b_{\alpha\alpha} = 0.6 \sum_{\beta\neq\alpha} \frac{m_\beta \gamma_{\alpha\beta}}{m_\alpha + m_\beta} \frac{1}{\tau_{\alpha\beta}}; \quad a_{\alpha\beta} = -\frac{0.6 m_\beta \gamma_{\alpha\beta}}{m_\alpha + m_\beta} \frac{1}{\tau_{\beta\alpha}} \quad \alpha \neq \beta$$

$$\frac{1}{\tau_{\alpha\beta}} = \frac{16}{3} n_\beta \left(\frac{1}{2\pi \gamma_{\alpha\beta}}\right)^{0.5} \sigma_{\alpha\beta}; \quad \frac{1}{\tau_{\beta\alpha}} = \frac{n_\alpha}{n_\beta} \frac{1}{\tau_{\alpha\beta}}; \quad \gamma_{\alpha\beta} = \frac{\gamma_\alpha \gamma_\beta}{\gamma_\alpha + \gamma_\beta}$$

Cross sections for Coulomb particle and for coefficients $c_{\alpha\beta}$, $d_{\alpha\beta}$:

$$\sigma_{\alpha\beta} = \frac{\pi}{2} \left[\frac{e_\alpha e_\beta \gamma_{\alpha\beta} (m_\alpha + m_\beta)}{m_\alpha m_\beta}\right]^2 \ln \Lambda_{\alpha\beta}; \quad \gamma_\alpha = \frac{m_\alpha}{kT_\alpha};$$

$\ln \Lambda_{\alpha\beta}$-Coulomb logarithm.

$$c_{\alpha\beta} = 1; \, c_{\alpha\beta} = -\frac{0.9 m_\alpha \gamma_\beta^2 (3 + 2\delta_{\alpha\beta}) \tau_\alpha^*}{(m_\alpha + m_\beta)(\gamma_\alpha + \gamma_\beta)^2 \tau_\beta} \alpha \neq \beta$$

$$d_{\alpha\beta} = -\frac{1.5 m_\alpha \gamma_{\alpha\beta} (1 + 2\delta_{\alpha\beta}) \tau_\alpha^*}{(m_\alpha + m_\beta) \gamma_\alpha^2 \tau_{\beta\alpha}} \quad \alpha \neq \beta \; \delta_{\alpha\beta} = \frac{1 - T_\beta / T_\alpha}{1 + m_\beta / m_\alpha}$$

For other types of collisions, the cross sections were expressed through Chapmen integrals $\Omega_{\alpha\beta}$ [2]:

$$\sigma_{\alpha\beta} = (2\pi \gamma_{\alpha\beta})^{0.5} \Omega_{\alpha\beta}$$

Here α, β, γ are each of the indexes listed the atoms "*a*", "*i*", and electrons "*e*", v_0 mass velocity.

4.2.1 Three-Component Cathode Plasma. General Equations

The above mathematical system describes the plasma for arbitrary number of components. In the case of vacuum arc, the discharge operates in vapor of the electrodes and density gradients were produced near the surface mainly in three-component plasma due to the diffusion of the ions, electrons, and atoms in the partially ionized metallic plasma. Let us derive the particle, energy fluxes, and diffusion coefficients, assuming only singly ionized atoms in the dense cathode plasma consisted of three components α that are ions, electrons, and atoms. Note that multiple ionized particles are formed in vacuum arc as well, and these issues will be addressed in Chaps. 5, 16 and 17.

The required system of equations was derived using above (4.1)–(4.4) and conditions that heavy particle's temperatures $T_a = T_i = T$, mass $m_a = m_i = m$, electron temperature $T_e > T$, and $\varepsilon/\theta \ll 1$, where $\varepsilon = m_e/m$, $\theta = T_e/T$.

The analysis brings to the following expressions for the particle fluxes:

$$I_i = -D_{ia}\frac{dn_i}{dx} - \theta(D_{ia} - D_{ie})\frac{dn_e}{dx} + \alpha D_{ia}\frac{dn_a}{dx} + D_{ie}\frac{en_i}{kT}E_p \tag{4.5}$$

$$I_e = -\frac{1}{\theta}D_{ea}\frac{dn_i}{dx} - (D_{ei} - D_{ea})\frac{dn_e}{dx} + \frac{1}{\theta}D_{ei}\frac{dn_a}{dx} + D_{ei}\frac{en_e}{kT_e}E_p \tag{4.6}$$

$$I_a = \theta(D_{ae} - D_{ai})\frac{dn_e}{dx} - D_{ai}\frac{dn_a}{dx} \tag{4.7}$$

Coefficients of diffusion of the particles after calculations of $|a|$-three-order determinant with elements $(\alpha\beta)$ for three-component plasma are expressed as:

$$\begin{gathered}
D_{ae} = \frac{\delta kT}{m}\left(\frac{a_1}{2\tau_{ia}} + \frac{a_2}{\tau_{ei}}\right); \quad D_{ea} = \frac{\delta kT_e}{m_e}\left(\frac{a_1}{2\alpha\tau_{ai}} + \frac{\varepsilon a_4}{\tau_{ie}}\right) \\
D_{ie} = \frac{\delta kT}{m}\left(\frac{a_1}{2\tau_{ai}} + \frac{a_3}{\tau_{ea}}\right); \quad D_{ei} = \frac{\delta kT_e}{m_e}\left(\frac{a_1}{2\alpha\tau_{ai}} + \frac{\varepsilon a_5}{\tau_{ae}}\right) \\
D_{ai} = D_{ia} = \frac{kT}{m}\frac{2\alpha\tau_{ai}}{a_1}; \quad \delta = \frac{2\alpha\tau_{ai}\tau_0}{a_1 a_0}; \quad \alpha = \frac{n_e}{n_e + n_a} \\
\frac{1}{\tau_0} = \frac{1}{\tau_{ei}} + \frac{1}{\tau_{ea}}; \quad a_1 - a_5 \approx 1 - 0.5;
\end{gathered} \tag{4.8}$$

Characteristic times of particle collisions for three-component plasma [3]:

$$\begin{gathered}
\tau_{ea}^{-1} = R_{01}\left(\frac{kT_e}{2\pi m_e}\right)^{0.5}\sigma_{ea}; \quad \tau_{ei}^{-1} = R_{02}\left(\frac{kT_e}{2\pi m_e}\right)^{0.5}\sigma_{ei}; \\
\tau_{ia}^{-1} = R_{01}\left(\frac{kT}{\pi m}\right)^{0.5}\sigma_{ia}; \quad \tau_{ee}^{-1} = R_{02}\left(\frac{kT_e}{\pi m_e}\right)^{0.5}\sigma_{ee}; \\
\tau_{aa}^{-1} = R_{01}\left(\frac{kT}{\pi m}\right)^{0.5}\sigma_{aa}; \quad \tau_{ii}^{-1} = R_{03}\left(\frac{kT}{\pi m}\right)^{0.5}\sigma_{ii}; \\
\tau_{ai}^{-1} = \frac{n_i}{n_a}\tau_{ia}^{-1}; \quad \tau_{ae}^{-1} = \frac{n_e}{n_a}\tau_{ea}^{-1}; \quad \tau_{ie}^{-1} = \frac{n_e}{n_i}\tau_{ei}^{-1}; \\
R_{01} = \frac{16}{3}n_a\left(\frac{k}{\pi}\right)^{0.5}; \quad R_{02} = \frac{16}{3}n_i\left(\frac{k}{\pi}\right)^{0.5}; \quad R_{03} = \frac{16}{3}n_e\left(\frac{k}{\pi}\right)^{0.5};
\end{gathered} \tag{4.9}$$

Cross sections for the charge particle's interactions [3]:

$$\sigma_{ee} = \frac{\pi}{2}\frac{e^4}{kT_e}\ln \Lambda_{ee}; \quad \sigma_{ei} = \frac{\pi}{2}\frac{z^2 e^4}{(kT_e)^2}\ln \Lambda_{ei}; \quad \sigma_{ii} = \frac{\pi}{2}\frac{z^4 e^4}{(kT_e)^2}\ln \Lambda_{ii}; \tag{4.10}$$

The expressions like (4.10) were analyzed in [11] in Chap. 1, and the cross sections of electron–ion interactions were presented by Granovsky formulas (1.23), which in simple form can be presented as

$$\begin{aligned}\sigma_{ei} &= 6.5 \times 10^{-14} T_e^{-2}(\text{eV}) \ln\left(1.4 \times 10^7 T_e(\text{eV}) n^{-1/3}\right)\\ \sigma_{ee} &= 2.6 \times 10^{-13} T_e^{-2}(\text{eV}) \ln\left(6.94 \times 10^6 T_e(\text{eV}) n^{-1/3}\right)\end{aligned} \tag{4.11}$$

The cross sections for charge-exchange collisions were calculated by Firsov expression (1.26), and atom excitations by electron impact were calculated by Dravin formula (1.35) from Chap. 1.

4.2.2 *Three-Component Cathode Plasma. Simplified Version of Transport Equations*

Nevertheless the above system of (4.5)–(4.9) is still complex, and to simplify the mathematical approach, let us calculated the main characteristics of plasma particle collisions. The results of calculations were presented in Table 4.1.

Then let us consider the following conditions

$$\begin{aligned}&\frac{\sigma_k}{\sigma_{ai}}\left(\frac{T_e m}{T m_e}\right)^{0.5} \gg 1\\ &\frac{n_i}{n_a} \le 10^{-2}\\ &\frac{n_i \sigma_k}{n_a \sigma_{ai}} \ge 1\\ &\frac{n_i}{N_{e0}}\left(\frac{2eu_c}{m_e}\right)^{0.5} \gg 1\end{aligned}$$

where σ_k is denoted as Coulomb cross sections, u_c cathode potential drop in the cathode sheath (see below chapter), and N_{e0} is the maximal electron emission flux at the cathode surface. Taking into account these conditions and the data of Table 4.1, the system (4.5)–(4.9) will be simplified as [6]:

$$I_i = (\theta D_{ie} - (1+\theta) D_{ia})\frac{\mathrm{d}n_i}{\mathrm{d}x} + \frac{D_{ie}}{kT}\left(n_i e E_p - f_i\right) + \alpha D_{ia} \Psi \tag{4.12}$$

Table 4.1 Cross-sections of charge particle interactions including ion-atom and recombination coefficient as dependence on electron energy

Electron energy (eV)	$N_i = 10^{18}$ (cm^{-3})		$N_i = 10^{19}$ (cm^{-3})		σ^* (cm^2)	σ_i^* (cm^2)	σ_{ai}^* (cm^2)	Recombination β (cm^6/s)
	σ_{ee} (cm^2)	σ_{ei} (cm^2)	σ_{ee} (cm^2)	σ_{ei} (cm^2)				
15	5.3×10^{-15}	1.5×10^{-15}	4.5×10^{-15}	1.3×10^{-15}	9×10^{-16}	8.7×10^{-16}	8.2×10^{-15}	–
11	9.3×10^{-15}	2.7×10^{-15}	7.6×10^{-15}	2.2×10^{-15}				
4	5.4×10^{-14}	1.6×10^{-14}	4.2×10^{-14}	1.3×10^{-14}	4×10^{-17}	1.3×10^{-17}	–	4×10^{-29}
2	1.7×10^{-13}	5.4×10^{-14}	1.2×10^{-13}	4.1×10^{-14}	–	–	–	10^{-27}
1	5×10^{-13}	1.7×10^{-13}	3×10^{-13}	1.2×10^{-13}	–	–	1.2×10^{-14}	2×10^{-26}

$$I_e = \frac{1}{\theta}(D_{ei} - (1+\theta)D_{ea})\frac{\mathrm{d}n_i}{\mathrm{d}x} - \frac{D_{ei}}{kT_e}\left(n_i e E_p - f_e\right) + \frac{\alpha}{\theta}(D_{ea} - D_{ei})\Psi \quad (4.13)$$

$$I_a = (D_{ai} + \theta D_{ae})\frac{\mathrm{d}n_i}{\mathrm{d}x} - \alpha D_{ai}\Psi \quad (4.14)$$

$$I_\alpha = \Gamma_\alpha - n_a v_0 \quad (4.15)$$

$$\Psi = \frac{1}{kT}\left[\frac{\mathrm{d}p}{\mathrm{d}x} - n_i\frac{\mathrm{d}(kT_e)}{\mathrm{d}x} - (n_a + n_i)\frac{\mathrm{d}(kT_e)}{\mathrm{d}x}\right]$$

Here p and v_0 are the pressure and velocity of flow, E_p is the electrical field in the plasma, and f_α is the volume force on α-component by the emitted electron beam from the cathode $N_e(x)$.

It was used the equation of plasma quasineutrality in the form

$$n_i = n_e + N_e(x)\left(\frac{2eu_c}{m_e}\right)^{-0.5}$$

For above-mentioned conditions, the diffusion coefficients are:

$$D_{ai} = D_{ea}; \quad D_{ea} = D_{ei}; \quad D_{ie}/D_{ei} \sim \varepsilon; \quad D_{ia} \gg D_{ie} \quad (4.16)$$

Eliminating the electrical field from (4.12) by (4.13) and consider (4.16) the expression for the ion flux Γ_i toward the cathode can be obtained in the form

$$\Gamma_i = -(1+\theta)D_{ia}\frac{\mathrm{d}n_i}{\mathrm{d}x} + n_i v + \alpha D_{ia}\Psi$$

$$D_{ia} = 0.4\frac{v_{iT}}{(n_a + n_i)\sigma_{ai}}; \quad v_{iT} = \left(\frac{8kT}{\pi m}\right)^{0.5} \quad (4.17)$$

Equation of electron energy:

$$\begin{aligned}\frac{5}{2}I_e\frac{\mathrm{d}kT_e}{\mathrm{d}x} &= -n_e kT_e\frac{\mathrm{d}v}{\mathrm{d}x} - \frac{\mathrm{d}h_e}{\mathrm{d}x} - kT_e\frac{\mathrm{d}I_e}{\mathrm{d}x} + eI_eE_p \\ &+ \frac{3n_e m_e}{\tau_{ei} m}k(T_e - T) + \frac{\mathrm{d}Q^*}{\mathrm{d}x} + eu_c\frac{\mathrm{d}N_e(x)}{\mathrm{d}x}\end{aligned} \quad (4.18)$$

Electron heat flux: $h_e = -\lambda_e\frac{\mathrm{d}T_e}{\mathrm{d}x} + \sum_\beta \mu_{e\beta} I^\beta$

Energy flux:

$$q_e = h_e + 2.5kT_e I_e$$

Electron specific heat conductivity:

$$\lambda_e = 5k^2 n_e \tau_{ee} \frac{T_e}{m_e}$$

where Q^* is the losses with inelastic collisions (ionization, recombination, excitation), radiation, and the last term in (4.18) is the energy influx from the emitted electron beam.

Equation of heavy particle energy:

$$\begin{aligned} V(n_i + n_e)\frac{5}{2}\frac{\mathrm{d}\left(kT + \frac{mv_0^2}{2}\right)}{dx} &= -\frac{\mathrm{d}(h_i + h_a)}{\mathrm{d}x} - kT\frac{d(I_i + I_e)}{\mathrm{d}x} \\ &+ \frac{m}{\rho}(I_i + I_e)\frac{\mathrm{d}p}{\mathrm{d}x} - \frac{5k}{2}(I_i + I_e)\frac{\mathrm{d}T}{\mathrm{d}x} \\ &+ eI_i E_p + \frac{3m_e k(T_e - T)}{\tau_{ei} m}\left(\frac{n_a}{\tau_{ae}} + \frac{n_i}{\tau_{ie}}\right) \end{aligned} \tag{4.19}$$

$$(h_i + h_a) = -(\lambda_i + \lambda_a)\frac{\mathrm{d}T}{\mathrm{d}x} + \sum_\beta \mu_{i\beta} I_\beta + \sum_\beta \mu_{a\beta} I_\beta$$

Taking into account that the heavy particle thermal velocity v_T is significantly larger than the direct velocity V of plasma flow near the cathode surface and that the term with electrical field can be negligible due to low ion current, (4.19) can be simplified to

$$\frac{5k}{2}\frac{\mathrm{d}T}{\mathrm{d}x}v(n_i + n_a) = -\frac{\mathrm{d}}{\mathrm{d}x}\left(-\lambda_a\frac{\mathrm{d}T}{\mathrm{d}x} + \mu_{ai} I_i\right) - \frac{3\varepsilon(T_e + T)}{\tau} \tag{4.20}$$

When (4.20) was obtained the following relations which were used:

$$\rho v\frac{\mathrm{d}v}{\mathrm{d}x} = -\frac{\mathrm{d}p}{\mathrm{d}x}, \quad \tau = \frac{n}{n}\tau_{ae} + \tau_{ie}, \quad \sum_\alpha I_\alpha = 0$$

Calculations of the contributions of the cross-terms bring to expressions

for electrons:

$$\mu_{ee} \approx 0.5\frac{kT_e}{m_e};$$

for ions:

$$\mu_{ai} \approx 0.17\frac{kT_i}{m_e}$$

The mathematical approach and the above system of equations describe the phenomena of the particle and energy transfer in ionized metallic vapor taking into account the difference between electron and heavy particle temperatures. The boundary conditions requested for numerical analysis should be formulated considering the spot model and cathode processes.

4.2.3 Transport Equations for Five-Component Cathode Plasma

When the arc with cathode spot operated in a surrounding gas, the additional particles except the vapor of cathode material should be also considered. One example is related to use an easy-to-ionized impurity in order to increase the electrical conductivity of the plasma. Such case was originated in systems for direct converting the gas enthalpy into electrical energy. In the similar systems, the plasma flow consists of ionized potassium or cesium impurities [7, 8]. As result, at least, additional two components (atom and ion) should be taken into account together with metallic neutrals and ions.

In order to understand the impurity role, let us consider plasma consisting from copper atom and ions, potassium atom and ions as well as from electrons with densities n_{am}, n_{im}, n_{ap}, $n_{ip,}$ and $n_{e,}$ index *m*-metal and *p*-potassium respectively. The phenomena in such plasma were analyzed and considering the mechanism of current continuity in cathode spots appeared at the electrodes of MHD generators [9]. Below the transport equations for this, five-component plasma will be derived.

First of all, let us consider the elementary characteristics for the particle collisions for conditions of electron temperature $T_e \sim 1$ eV different from heavy particle temperature and for $n_{am}/n_{ap} \sim 10^2$ [9]. Different collision groups can be denoted:

(i) Coulomb collisions with cross sections: $\sigma_{eim} \sim \sigma_{eip} \sim \sigma_{ee} \sim \sigma_{iim} \sim \sigma_{iip} \sim \sigma_{ipim} \sim 10^{-14}$ cm^2;
(ii) polarized particle interactions: $\sigma_{eap} \sim \sigma_{imap} \sim \sigma_{amap} \sim \sigma_{apap} \sim 10^{-14}$ cm^2;
(iii) exchange particle interactions characterized by charge exchange cross sections: $\sigma_{imam} \sim \sigma_{ipap} \sim \sigma_{amip} \sim \sigma_{ipam} \sim 10^{-14}$ cm^2;
(iv) elastic collisions for plasma particle and for emitted electrons with cross sections $\sigma_{eam} \sim \sigma_{ipam} \sim \sigma_{amam} \sim 10^{-15}$ cm^2;
(v) ionizations by emitted electrons with energy ~10–15 eV with cross sections ~10^{-14} cm^2; for potassium atom and ~10^{-16} cm^2 for copper atom;

Considering the above conditions, the analysis of diffusion equations for five-component plasma brings to the following expressions for the particle fluxes:

$$I_e = \frac{1}{\theta}\left\{ \begin{array}{l} [D_{eim} - (1+\theta)D^e]\frac{dn_e}{dx} + (D_{eam} - D^e)\frac{dn_a}{dx} + \left(D_{eap} + D_{eam}\right)\frac{dn_{ap}}{dx} \\ -\frac{e\theta E_p}{kT_e}\left(n_{ip}D_{eip} + n_{im}D_{eim}\right) \end{array} \right. \tag{4.21}$$

$$I_{ip} = \left[\theta D_{ipe} - (1+\theta)D^{ip}\right]\frac{dn_e}{dx} + D_{ipim}\frac{dn_{im}}{dx} + \left(D_{ipam} - D^{ip}\right)\frac{dn_a}{dx}$$

$$+\left(D_{ipap}-D_{ipam}\right)\frac{\mathrm{d}n_{ap}}{\mathrm{d}x}-\frac{eE_p}{kT}\left(n_eD_{ipe}-n_{im}D_{ipim}\right) \tag{4.22}$$

$$I_{im}=\left[\theta D_{ipe}-(1+\theta)D^{im}\right]\frac{\mathrm{d}n_e}{\mathrm{d}x}+D_{imip}\frac{\mathrm{d}n_{ip}}{\mathrm{d}x}+\left(D_{imam}-D^{im}\right)\frac{\mathrm{d}n_a}{\mathrm{d}x}$$
$$+\left(D_{imap}-D_{imam}\right)\frac{\mathrm{d}n_{ap}}{\mathrm{d}x}-\frac{eE_p}{kT}\left(n_eD_{ime}-n_{im}D_{imip}\right) \tag{4.23}$$

$$I_{ap}=\left[\theta D_{ape}-(1+\theta)D^{ap}\right]\frac{\mathrm{d}n_e}{\mathrm{d}x}+D_{apim}\frac{\mathrm{d}n_{ip}}{\mathrm{d}x}+D_{apip}\frac{\mathrm{d}n_{im}}{\mathrm{d}x}$$
$$+\left(D_{apam}-D^{ap}\right)\frac{\mathrm{d}n_a}{\mathrm{d}x}-D_{apam}\frac{\mathrm{d}n_{ap}}{\mathrm{d}x} \tag{4.24}$$

$$I_{am}=\left[\theta D_{ame}-(1+\theta)D^{am}\right]\frac{\mathrm{d}n_e}{\mathrm{d}x}+D_{amim}\frac{\mathrm{d}n_{im}}{\mathrm{d}x}+D_{amip}\frac{\mathrm{d}n_{ip}}{\mathrm{d}x}$$
$$-D^{am}\frac{\mathrm{d}n_a}{\mathrm{d}x}+D_{amap}\frac{\mathrm{d}n_{ap}}{\mathrm{d}x} \tag{4.25}$$

The flux with convective term is

$$\Gamma_\alpha=I_{am}+n_\alpha v$$

Here was taken in account $n_a=n_{am}+n_{ap}$, $n_e=n_{im}+n_{ip}$ and

$$D^\alpha=\sum_\beta c_\beta D_\alpha\beta\ \rho=\sum_\alpha\rho_\alpha;\ c_\alpha=\rho_\alpha/\rho;\ \rho_\alpha=n_\alpha m_\alpha;$$

$\alpha,\beta\rightarrow e, ip, im, ap, am;\ \alpha\neq\beta$.

The general form of coefficients of particle diffusion is

$$D_{\alpha\beta}=\frac{kT_\alpha}{m_\alpha}\frac{|a|_{\beta\alpha}-|a|_{\alpha\alpha}}{|a|}; \tag{4.26}$$

Coefficients of diffusion (4.26) of the particles after calculations of $|a|$-five-order determinant with elements $(\alpha\beta)$ for five-component plasma are expressed as:

$$D_{e\beta}=\frac{0.3\left(1-0.6\theta^{-1}\right)v_{eT}}{n_{ip}\sigma_{eip}}K_\beta^{(1)};$$
$$D_{ip\beta}=\frac{0.42v_{iT}\left(1-0.6\theta^{-1}\right)}{n_e\sigma_{ipim}+n_{am}\sigma_{imam}}K_\beta^{(2)}$$
$$D_{im\beta}=\frac{0.42v_{iT}}{n_e\sigma_{ipim}+n_{am}\sigma_{imam}}K_\beta^{(3)};$$

$$D_{ap\beta} = \frac{0.42v_{iT}}{n_{ip}\sigma_{apip} + n_{am}\sigma_{apam}} K_{\beta}^{(4)};$$

$$D_{am\beta} = \frac{0.42v_{iT}}{(n_{am} + n_{im})\sigma_{imam}} K_{\beta}^{(5)} \tag{4.27}$$

where v_{eT} and v_{iT} are the thermal velocities of electrons and heavy particles, respectively. Coefficients $K_{\beta}^{(n)}$ are functions of arguments consisting of n_{α}/n_{β} and $n_{\alpha}\sigma_{\alpha\beta}/n_{\beta}\sigma_{\beta\alpha}$. These corresponding expressions are very complicated and therefore are not presented here. Their estimation brings to the following relations:

$$K_{\beta}^{(1)} \sim 0.1;\ K_{\beta}^{(2)} \sim 1;\ K_{\beta}^{(3)} \sim 0.1;\ K_{\beta}^{(4)} \sim 0.1;\ K_{\beta\neq im}^{(5)} \sim 1;\ K_{im}^{(5)} \sim 0.1; \tag{4.28}$$

Analysis of expressions (4.27) and (4.28) allowed deducing the relations between diffusion coefficients of different particles in the form:

$$\begin{aligned} &D_{eip} \sim D_{eim} \sim D_{eap} \sim D_{eam} >> D_{ame} \sim D_{amip} \sim D_{amap} \\ &\sim D_{ipe} \sim D_{ipim} \sim D_{ipap} \sim D_{ipam} > D_{ape} \sim D_{apip} \sim D_{apim} \\ &\sim D_{apam} \sim D_{ime} \sim D_{imip} \sim D_{imap} \sim D_{imam} \sim D_{amim}; \end{aligned} \tag{4.29}$$

Equation of electron energy:

$$\frac{dq_e}{dx} = eI_eE_p + \frac{3n_em_e}{m}k(T_e - T)\sum_{\beta}\tau_{e\beta}^{-1} + \frac{dQ^*}{dx} + eu_c\frac{dN_e(x)}{dx} \tag{4.30}$$

$$q_e = \mu_{ee}I_e + 2.5kT_eI_e;\ \mu_{ee} = 0.5R;$$

$$R = \frac{1 + R_1}{1 + 1.3R_1/3};\quad R_1 = \frac{n_{ap}\sigma_{eap} + n_{am}\sigma_{eam}}{n_e\sigma_{ee}}$$

Energy losses due to fluxes of ionized atoms with potential ionization u_{ip} and u_{im} for potassium and metallic atoms, respectively:

$$Q^* = u_{ip}\left[\frac{dI_p}{dx} + \frac{d(vn_{ip})}{dx}\right] + u_{im}\left[\frac{dI_m}{dx} + \frac{d(vn_{im})}{dx}\right]$$

Estimations showed that $R \sim 2$–2.3 and $\mu \sim 1$–1.15. The energy (4.30) was derived for simplicity without heat conduction term.

The above system of equations for five-component plasma together with added equations described the cathode processes allows to calculate the cathode spot parameters in a complicating case of arc discharge originating at the cathode potassium film, which deposited from mixed plasma flow [10].

References

1. Aliyev'sky, M. J., & Zhdanov, V. M. (1963). Transport equations for non-isothermally multi-component plasma. *Journal of Applied Mechanics and Technical Physics, 5,* 11–17.
2. Polyansky, V. A. (1964). Diffusion and conductivity at low temperature partially ionized gas mixture. *Journal of Applied Mechanics and Technical Physics, 5,* 11–17.
3. Aliyev'sky, M. J., Zhdanov, V. M., & Polyansky, V. A. (1964). The viscous stress tensor and the heat flux in the two-temperature partially ionized gas. *Journal of Applied Mechanics and Technical Physics, 3,* 32–42.
4. Gogosov, V. V., Polyansky, V. A., Semenova, I. P., & Yakubenko, A. E. (1969). Equations of electrohydrodynamics and transport coefficients in a strong electric field. *Bulletin of the Akademii of Science of USSR, Fluid Dynamics* (2), 31–45.
5. Gogosov, V. V., & Shchelchkova, I. N. (1974). Derivation of the boundary conditions for concentrations, velocities and temperatures of the components in a partly ionized plasma with potential falls at the walls. *Bulletin of the Akademii of Science of USSR, Fluid Dynamics* (5), 76–88.
6. Beilis, I. I., Lyubimov, G. A., & Rakhovsky, V. I. (1972). Diffusion model of the near-cathode region of a high-current arc discharge. *Soviet Physics-Doklady, 17,* 225–228.
7. Coombe, R. A. (Ed.). (1964). *Magnetohydrodinamic generation of electrical power*. London: Chapman and Hall.
8. Frost, L. S. (1961). Conductivity of seeded atmospheric pressure plasmas. *Journal of Applied Physics, 32*(10), 2029–2035.
9. Beilis, I. I. (1982). On the role of metal electrode in the near-electrode processes of the arc discharge in MHD-generator's conditions. *Bulletin Sybirian Branch Academy of Science Series Technical Science* (3), 78–85.
10. Beilis, I. I. (1977). The near-cathode region of contracted discharge at the metal electrode of MHD generator. *High Temperature, 15*(6), 1269–1275 (Ref in Russian).

Chapter 5
Basics of Cathode-Plasma Transition. Application to the Vacuum Arc

Since the cathode is negatively charged with respect to the adjacent dense plasma, the electrons were repelled from the cathode, while a high-density ion flux toward the cathode is produced. As a result, a relatively large positive space charge in the vicinity of the cathode arises leading to cathode potential drop u_c formation and to occurrence of a strong electric field E at the cathode surface. The positive space charge formed in a layer is named as "electrical sheath." The sheath plays an important role in determining the heat and particle fluxes and supporting the cathode plasma generation. Therefore, the cathode is an active element of the electrical circuit providing the current continuity in the vacuum arc. Consequently, a proper determination of the electric field E at cathode surface is an important issue of the theory. This problem related to phenomena of plasma–wall transition which was reviewed in resent book [1] and where reported the main references to the works described this subject. Below the problem of cathode sheath and its electrical parameters are analyzed with respect to conditions of the cathode spot in a vacuum arc. However, firstly let us consider some early Langmuir studies of the current formation at the cathode plasma contact in order to understand the sheath problem in vacuum arcs for different cathode materials.

5.1 Cathode Sheath

The earliest works studied the characteristics of electrical discharges, including the electric field, potential distribution in the space charge regions, and voltage–current characteristics. The cathode region in a discharge with a hot CaO cathode was studied by Child in 1911 [2] taking into account that the space charge was produced only solely by positive ions. Langmuir studied the space charge due to electron current in order to understand the difference between measured thermionic current in a high vacuum and Richardson's saturated electron current [3]. As a result, the dependence

I. Beilis, *Plasma and Spot Phenomena in Electrical Arcs*, Springer Series on Atomic, Optical, and Plasma Physics 113, https://doi.org/10.1007/978-3-030-44747-2_5

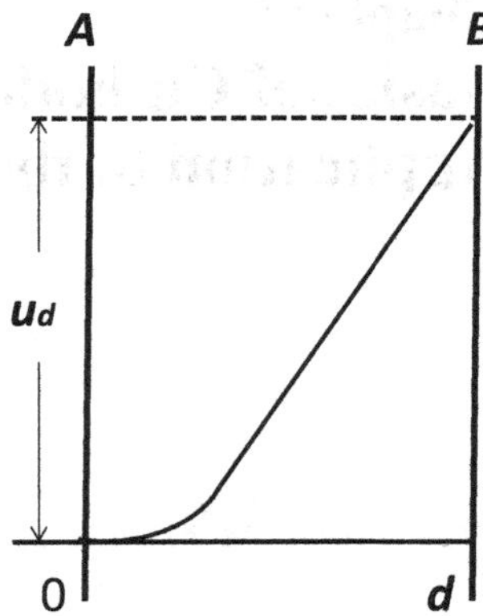

Fig. 5.1 Schematic potential distribution and voltage u_d between two parallel planes (A and B) with gap distance d

between current density, gap voltage, and gap distance was obtained solving the Poisson equation for two infinite parallel planes (Fig. 5.1):

$$\frac{\partial^2\varphi}{\mathrm{d}x^2} = 4\pi\rho \tag{5.1}$$

where x is the distance, φ is the potential and ρ is the space charge. Taking the relation between ion velocity and potential as $mv^2/2 = e\varphi$ (e and m are the electron charge and ion mass, respectively) and current density as $j = \rho v$, (5.1) will be:

$$\frac{\mathrm{d}^2\varphi}{\mathrm{d}x^2} = 2\pi\sqrt{\frac{2m}{e\varphi}}j \tag{5.2}$$

After multiplying (5.2) by $2d\varphi/dx$ and integrating one arrives to

$$\left(\frac{d\varphi}{dx}\right)^2 - \left(\frac{d\varphi}{dx}\right)_0^2 = 8\pi j\sqrt{\frac{2m\varphi}{e}} \tag{5.3}$$

Assuming the electric field at the external boundary $(\mathrm{d}\varphi/\mathrm{d}x) = 0$, voltage u_d and gap distance d, second integrating the expression known as Child–Langmuir formula:

$$j_i = \frac{\sqrt{e}}{9\pi}\left(\frac{2}{m}\right)^{0.5}\frac{u_d^{3/2}}{d^2} \tag{5.4}$$

For electron current density such expression is:

$$j_e = 2.34 \times 10^{-6}\frac{u_d^{3/2}}{d^2},$$

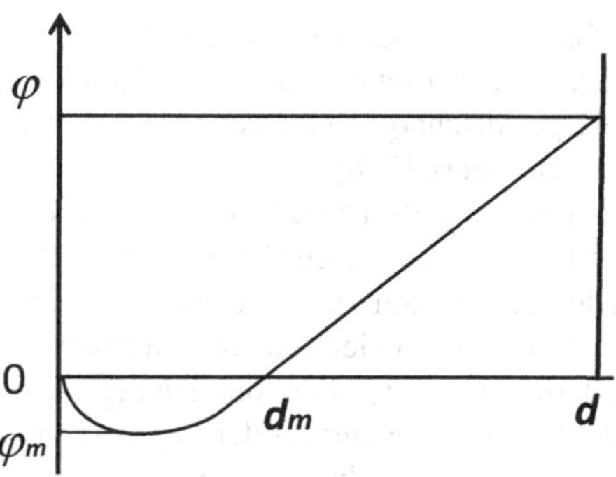

Fig. 5.2 Schematic potential distribution φ between two parallel planes with gap distance d and initial charge particle velocity

and for ion current density

$$j_i = 5.46 \times 10^{-8} \frac{u_d^{3/2}}{d^2}$$

Langmuir [4] investigated the effect of the initial electron velocity on the potential distribution and thermionic current assuming that the normal components of the velocities of the emitted electrons have the Maxwell distribution with cathode temperature T_c. In this case, the integration of Poisson equation between proper limits leads to

$$j_e = \frac{\sqrt{e}}{9\pi} \left(\frac{2}{m} \right)^{0.5} \frac{(\varphi - \varphi_m)^{3/2}}{(d - d_m)^2} \tag{5.5}$$

where φ_m is the minimal potential formed at distance d_m from the cathode surface due to own space charge of the emitted electrons. Figure 5.2 shows the potential distribution between the electrodes with the minimal potential close to the cathode. This potential distribution named also as distribution with "virtual cathode," which indicates that part of the emitted electrons with energies lower than $e\varphi_m \sim kT_c$ (k is the Boltzmann constant) are repelled back to the cathode. This effect explains the emission current limited by the space charge. Equation (5.4) can be obtained from (5.5) when $\varphi = u_d \gg \varphi_m$. Thermionic current limited by the space charge was also investigated for coaxial cylinders [5] and concentric spheres [6].

The above results related to the case when the space charge region length was comparable to the interelectrode gap distance. However, when a relatively small electrode (also for large electron density) is immersed in an ionized gas, then the size of space charge region surrounded the electrode is significantly smaller than the size of interelectrode gap filled with the ionized gas. Such space charge region surrounding closely to the electrode, Langmuir called as "sheath." He developed a method to investigate distribution of the potential, temperature, and density in the ionized gas of electrical discharges. The method consists in measure of a voltage–current characteristic of a small "collector" electrode (planar, cylindrical, or spherical) immersed in a gas discharge, interpretation of different parts of this characteristic and using the sheath theory [7, 8].

Langmuir [9] as well as Tonks and Langmuir [10] developed a theory of electron and positive ion space charge in the cathode sheath of an arc. To that end, the sheath edge and sheath structure were considered. Poisson equation solved considering that the plasma electrons have a Maxwellian velocity distribution corresponding to a temperature T_e. For ions having no initial velocity from plasma into sheath of thickness d, the current density j_i was taken from (5.4). The solution was obtained using Boltzmann distribution for plasma electron density and ion density from current density j_i as:

$$\mathrm{d}\eta/\mathrm{d}\lambda = 4/3\left\{\eta^{0.5} - \pi^{0.5}\alpha_m\left[K_E - \exp(-\eta)\right]\right\}^{0.5} \tag{5.6}$$

Here $\lambda = \frac{x}{d}$; $\eta = -\frac{e\varphi}{T_e}$; $\alpha_m = \frac{j_{eTO}}{j_i}\left(\frac{m_e}{m}\right)^{0.5}$, e-is the electron charge, x is the distance, φ is the potential, j_{eTO} is the plasma electron random current, K_E is the constant of integration which is equal to 1 with condition $\mathrm{d}\eta/\mathrm{d}\lambda = 0$ when $\eta = 0$ at the sheath boundary. The solution of (5.6) shows monotonic change of the potential in the sheath, when the parameter of currents ratio α_m increased in range from 0 to 0.88407 while for larger α_m this potential fluctuated, i.e., the sheath became unstable (see Fig. 5.2 in [9]). Studying the transition from plasma to sheath Langmuir first concluded [9] that the ion motion in the plasma cannot be arbitrary and should be determined by the electron temperature (this phenomenon was later studied by Bohm, see below) and by geometry of the ion source. The source of ion generation affects the sheath by the average velocity of the ions, which enter the sheath. The initial particle temperatures are also considered [10]. It was indicated that introducing the finite ion temperature is the only of the order of T/T_e, a very small quantity in most cases.

5.2 Space Charge Zone at the Sheath Boundary and the Sheath Stability

The role of particles temperature on the sheath stability expressing the ion current as $j_i = en_0v_i$ (n_0 is the plasma density at the sheath edge, v_i is the initial ion velocity determined by plasma potential at the sheath edge) was studied by Bohm [11]. The potential φ_0 in this work was assumed at the sheath edge. The Boltzmann distribution for plasma electron density and the simple form of ion density distribution as $n_i = n_0(\varphi_0/\varphi)^{0.5}$ in the sheath were taken in account. The Poisson equation was in form:

$$\frac{\partial^2\varphi}{\mathrm{d}x^2} = 4\pi e n_0\left[\sqrt{\frac{\varphi_0}{\varphi}} - \exp\left(-\frac{e(\varphi - \varphi_0)}{kT_e}\right)\right] \tag{5.7}$$

The solution obtained for condition $\partial\varphi/\partial x = 0$ when $\varphi = \varphi_0$:

$$\left(\frac{\partial\varphi}{\mathrm{d}x}\right)^2 = 8\pi e n_0\left[2\varphi_0\left(\sqrt{\frac{\varphi_0}{\varphi}} - 1\right) + \frac{kT_e}{e}\left[\exp\left(-\frac{e(\varphi - \varphi_0)}{kT_e}\right) - 1\right]\right] \quad (5.8)$$

Using expansion as a power series in ($\varphi = \varphi_0$) when φ is close to φ_0 (in the space charge zone at the sheath boundary), the following expression can be derived from (5.8):

$$\frac{\partial\varphi}{\mathrm{d}x} = \left[4\pi e n_0\left(\frac{e}{kT_e} - \frac{1}{2\varphi_0}\right)\right]^{0.5}(\varphi - \varphi_0) \quad (5.9)$$

As follows from (5.9), the monotonic potential distribution in the sheath (stable sheath) can be obtained when $\partial\varphi/\mathrm{d}x > 0$, i.e., when $\varphi_0 \geq kT_e/2e$ and which was known as Bohm criterion for ion initial ion velocity at the sheath entrance:

$$v_i = \left(\frac{kT_e}{m}\right)^{0.5} \quad (5.8)$$

This criterion is a consequence of different rates of decrease electron (according Boltzmann law) and ion (acceleration at constant current density) densities at zero boundary conditions for potential and electric field.

Ecker [12] continued the study of the influence of the plasma electron on the space charge zone at the sheath boundary. Considering the vacuum arc, he used the ion current density (constant) j_i of ion flux from the cathode plasma into the sheath. The ion density distribution was determined using the equation of current continuity conservation with ion velocity depended on the potential distribution in the sheath. Boltzmann distribution was assumed for returned to the cathode plasma electrons density (n_{e0} at the sheath edge). The problem was considered for the case of isothermal plasma with temperature T. The mathematical formulation of the Poisson equation is in form:

$$\frac{\partial^2\varphi}{\mathrm{d}x^2} = \frac{4\pi e}{\varepsilon_0}\left[\frac{j_i}{e\sqrt{\frac{2(kT/2 - e\varphi)}{m}}} - n_{e0}\exp\left(-\frac{e\varphi}{kT}\right)\right] \quad (5.10)$$

Twice integrating of (5.10) from $\varphi = 0$ to $\varphi = \varphi_d$ at $x = 0$ to $x = d$ and taking into account $n_{e0} = n_{i0} = \frac{j_i}{e}\sqrt{\frac{m}{kT}}$, the sheath thickness is expressed as

$$d = \int_0^{\varphi_d} \frac{d\varphi}{\sqrt{\frac{8\pi j_i}{e\varepsilon_0}\sqrt{kTm}\left(\sqrt{1 - \frac{2e\varphi}{kT}} - 1\right) + \left(e^{\frac{2e\varphi}{kT}} - 1\right) + E_0^2}} \quad (5.11)$$

And the total current density j defined by [12]

$$j = j_i - \frac{en_{e0}}{2}\sqrt{\frac{kT}{m}}\exp\left(\frac{e\varphi_d}{kT}\right) \tag{5.12}$$

As condition $j > 0$ and $j_i > 0$ are fulfilled and taken the above formula for plasma density n_{e0}, a minimal voltage at the sheath can be obtained for as

$$e\varphi_{\min} \geq kT\left[-Ln\left(\frac{1}{2}\sqrt{\frac{m}{m_e}}\right)\right] \tag{5.13}$$

The future study of the space charge region was conducted by analysis of the solution of (5.11) writing in dimensionless form. The requested input data for $d, j, T,$ and E_0 were taken as assuming that d is equal to the mean free path of the ion, j from the experiments, and E_0 as varied parameter. It was indicated the sheath voltage φ_d was obtained significantly lower than that calculated by Child–Langmuir formula, (5.4). This result was explained by influence of the negative space charge produced due to returned cathode plasma electrons. It should be noted that the calculated minimal voltage from (5.13) is about 1.6 V.

Later, Ecker [13] also determined the potential distribution through sheath and studied the influence of the boundary condition at sheath edge in case of $T_e > T$ on the sheath stability. The problem was formulated using (5.10) with dimensionless parameters as

$$\frac{\partial^2\phi}{dx^2} = \frac{1}{e\sqrt{1+\alpha\phi}} - \exp(-\phi) \tag{5.14}$$

Dimensionless distance x, potential ϕ, Debye length r_D, and are: $x = \frac{z}{r_D}$; $\phi = -\frac{e\varphi}{kT_e}$; $\alpha = \frac{2kT_e}{mv_0^2}$, $r_D = (\frac{kT_e}{4\pi n_0 e^2})^{1/2}$ and the boundary conditions

$$\phi(x=0) = \phi_0 = 0\,;\quad \frac{d\phi}{dx_e}(x=0) = \phi_0'$$

where v_0 is the initial ion velocity. The solution of (5.14) was obtained varying α and ϕ_0'.

The calculations of potential distribution in the sheath for a copper cathode is presented in Fig. 5.3. Note, that Langmuir obtained a similar potential distribution previously (Fig. 5.2 in [9]). According to the calculations, a critical field $\phi_0' = \phi_{cr}'$ is exists which indicate a solution that asymptotically approaches a constant $\phi_{\rm cr}$ for this field ϕ_{cr}' when $x \to \alpha$. When the entrance field $\phi_0' < \phi_{cr}'$, the solution shows a potential fluctuations in the sheath indicating that the Boltzmann distribution for the returned plasma electrons was no longer available. For fields $\phi_0' > \phi_{\rm cr}'$ and close to $\phi_{\rm cr}'$, the potential distribution continues to be constant at some distance x_c and then approaches the commonly known $\varphi^{3/2}$ law. The distance x_c decreases with ϕ_0' and $x_c = 0$ for $\phi_0' > 0.15$. The future analysis showed that the critical values of $\phi_{\rm cr}$ and ϕ_{cr}' increases with parameter α (both to about 0.6 for $\alpha = 9$), while $\phi_{\rm cr} = \phi_{cr}' = 0$ when

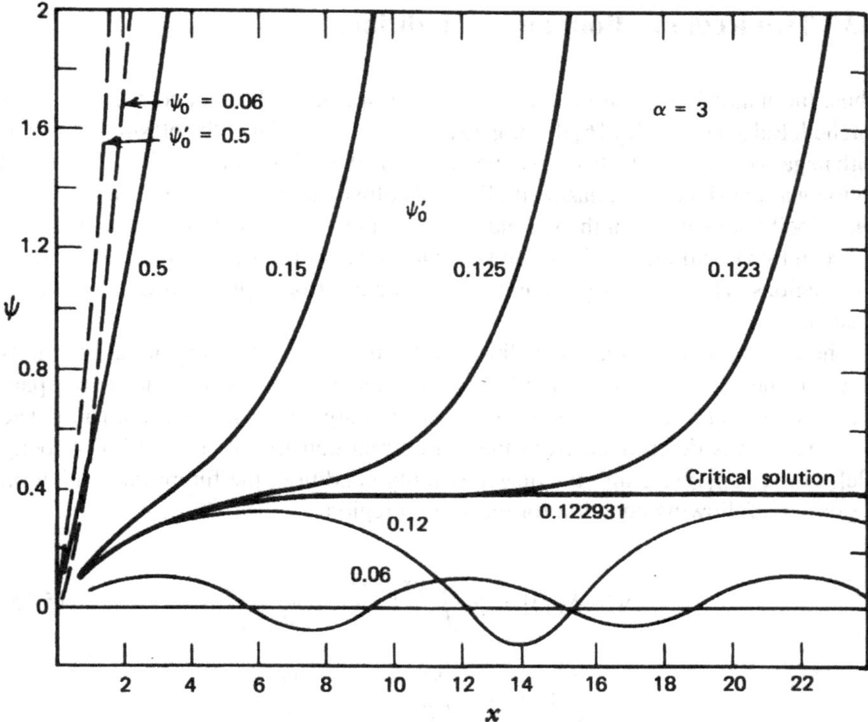

Fig. 5.3 Distribution of the potential ϕ in the sheath (solid lines) as function of the distance x from the plasma toward the cathode for different field ϕ_0' at the edge of the plasma (Cu, $\alpha = 3$). The dotted lines shows the calculations by Child–Langmuir formula (5.4) as found without returned plasma electrons [14]. (The figure is from [13] Ecker, G., Unified analysis of the metal vapour arc. Z. Naturf. 28a, 417–428, (1973). Used under a creative Commons 4.0 Attribution License https://creativecommons.org/licenses/by/4.0/legalcode.)

$\alpha = 2kT_e/mv_0^2 = 2$, which is the above-mentioned Bohm condition (5.8) for sheath stability. Thus, the above results show that the plasma near the cathode consists of two regions:

(i) the first, collisionless region, that closely adjacent to the surface in which potential distributed by Langmuir law $\varphi^{3/2}$;
(ii) the second region where the ion velocity and electrical field are formed in order to stable transition from the undisturbed collisional quasineutral plasma to the first region entrance.

Therefore, a correct study of the sheath can be provided using adequate boundary conditions.

5.3 Two Regions. Boundary Conditions

Thus, the major issue is how a solution of quasineutral plasma and sheath can be patched. Baksht et al. [15, 16] developed a two-region model of the plasma contacted with negative charged electrode. While the first is the collisionless region, the second region was considered as quasineutrality and collisional plasma. The ion current and ion velocity are studied in the second region to obtain the condition at the interface between two mentioned regions. Boltzmann distribution for the electrons is used in both regions. The x-axis beginning (zero) at the electrode surface is directed to the plasma.

The electric field increases in direction to the collisionless region and the ions move in the direction of the field. The flux depends on the ion mean free path l_i as $\exp[(x^{'}\text{-}x)/l_i]$, where $x^{'}$ is the point with charge-exchange ion collision. The ion density was determined from the ion current density divided by ion velocity $\{2e[\varphi(x)-\varphi(x^{'})]\}^{0.5}$. Using the quasineutrality condition, the future integration on $x^{'}$ gives the following equation for the second region:

$$\exp[(-\phi(\varsigma)] = \gamma_0 \int\limits_{\varsigma}^{\infty} \frac{\exp[-(\varsigma' - \varsigma)]d\varsigma}{\sqrt{\phi(\varsigma') - \phi(\varsigma)}} \tag{5.15}$$

$$\varsigma = \frac{x}{l_i}\,;\; \phi = -\frac{e\varphi}{kT_e}\,;\; \gamma_0 = \frac{v_i}{\sqrt{\frac{2kT_e}{m}}}$$

Parameter γ_0 was defined as the ratio of the directed ion velocity v_i to the ion velocity determined by T_e at point where assumed the potential $\phi = 0$. Equation (5.15) was solved numerically. The result shows that the electric field and $d^2\phi/d/\varsigma^2$ sharply increased when approaching to the collisionless region and at some point becomes infinite. This result indicates that the collisions are requested to support quasineutrality condition in the plasma. The potential distribution in the collisionless region is determined by the space charge and, therefore, this distribution should be described by Poisson equation. Taking into account the Boltzmann electron distribution and the above ion density, the potential distribution is

$$\frac{r_D^2(n_0, T_e)}{l_i^2}\frac{d^2\phi}{d\varsigma^2} = \gamma_0 \int\limits_{\varsigma}^{\infty} \frac{\exp[-(\varsigma\prime - \varsigma_0)]d\varsigma}{\sqrt{\phi(\varsigma) - \phi(\varsigma\prime)}} - \exp(-\phi) \tag{5.16}$$

Since parameter $(r_D/l_i)^2 \ll 1$, the contribution of the term in left side of (5.16) becomes large near the electrode where $d^2\phi/d/\varsigma^2 \gg 1$. At larger distance from the electrode, the potential is determined only by right side of (5.16). Therefore, (5.16) describes the potential distribution in both regions. The boundary conditions were chosen for ϕ and $d\phi/d/\varsigma$ at some distance where the space charge is relatively small and which can be determined by (5.15).

Thus, the above-described approach allows to solve the sheath problem and obtain monotonic potential distribution when the ion collisions with neutral atoms and the quasineutrality condition were taken in account in the plasma adjacent to the space charge region. In this case, the monotonic potential can be obtained for ion velocity at the entrance of the space charge region, which may not coincide with Bohm condition. This same conclusion regarding to the Bohm condition was reported by Valentini [17] considering collisions using hydrodynamic approach in order solving the sheath problem. Valentini and Herrmann [18] indicated that the Bohm condition was obtained from the solution for potential distribution and cannot be given as boundary condition. Scheuer and Emmert [19] investigated the plasma sheath transition for collisionless plasma with arbitrary ion temperature self-consistently. The Maxwellian velocity distribution of the ions in the absence of a potential gradient (far from the wall) was taken in account.

Franklin [20] reported a review that considered the problem of joining plasma and sheath solution over a wide range of physical conditions. Different approximations were considered. A method based on matched asymptotic approximations of different regions near the wall was analyzed [20–25]. It was shown that the asymptotic solution in the transition layer between plasma and sheath is suitable to match smoothly the plasma and sheath solutions. The relationship between the electric field and velocity at the plasma–sheath interface in the broad range of plasma parameters considered in [1, 26]. The plasma–wall transition in an oblique magnetic field was modeled in [27–30]. On another hand, kinetic approaches were developed to understand how the collisions in the adjacent plasma influence the sheath parameters. The results are considered in the next section.

5.4 Kinetic Approach

The feature of potential distribution in the plasma adjacent to the collisionless Langmuir region was solved considering different kinetic approaches. Harrison and Thomson [31] obtained an analytic solution with the kinetic model valid in the quasineutrality region. Simple expressions were obtained for the mean velocity and the mean square velocity of the ions, and for the wall potential with respect to the plasma. The conditions for the formation of a stable plasma–sheath boundary briefly examined and a general criterion is obtained without considering any specific mechanism of ion transport. No assumptions made regarding the energy distribution of the ions, and the result is therefore a refinement on Bohm's original criterion, which derived by assuming a monoenergetic ion flux. The equation for plasma–sheath transition was solved numerically with no separation into sheath and plasma [32]. It was deduced that the model of plasma region with sharply transition to a sheath is justified.

Using charge-exchange collisions in the plasma was considered by Petrov [33] and by Scherbinin [34,35]. The mathematical problem was formulated by kinetic equation with collision term *St* as

$$v\frac{\mathrm{d}f}{\mathrm{d}x} + \frac{eE}{m}\frac{\mathrm{d}f}{\mathrm{d}v} = St \tag{5.17}$$

Different typical collision *St* terms were used. As by Petrov [33]:

(i) $$St = \sigma_{ia} n_a [v_e f + \delta(|v|) j/e] \tag{5.18}$$

Here, v_e is the electron velocity, f is the ion function distribution, N is the neutral atom density, σ_{ia} is the charge exchange cross section, E is the electric field, j is the ion current density, and $\delta(|v|)$ is delta function. The first term in right part of (5.17) takes the ion disappearing as result of charge exchange and the second term describes the new ion generation with zero velocity. Since the total number of ions not changed then the integral on *St* equal zero. The method of characteristics was used to find solution of (5.17) [34, 35].

(ii) $$St = \frac{nf_0 - f}{\tau} \quad f_0 = \sqrt{\frac{m}{2kT\pi}}\exp\left(\frac{mv^2}{2kT}\right) \tag{5.19}$$

Here, τ is the characteristic time determined by the atom density and charge-exchange cross section. The following momentum $n = \int f\mathrm{d}v$ and $j = \int v\mathrm{d}v$ were determined and used for consistently solution of Poisson equation. The zero potential was given and small electric field (relatively to field at the electrode surface) was determined at the boundary with quasineutral plasma. The calculations [34, 35] allow obtaining the ion flux and ions and the ion energy flux. It was shown that these fluxes weakly different from that calculated using Maxwellian function when $eE_pl_i/4kT \leq 0.1$. According to work [33], the sheath thickness decreased when the rarefied charge exchange collisions were taken in account.

Riemann [36] developed a kinetic theory of boundary layer for weakly ionized gas discharge. The electron Boltzmann distribution ($T_e > T$) and negative charged absorbing wall were considered. The kinetic equation similar to (5.17–5.18) was studied to obtain in the sheath and in the plasma at sheath edge (named as presheath) solutions. The charge exchange collisions in the presheath were considered for cases with constant cross section and with constant collision frequency as well as collision-free case. Riemann also show that the monotonic potential distribution in the sheath-presheath system can be obtained taken in account the collision dominated case.

Later, Riemann reviewed the problem [37]. A rigorous kinetic analysis of the plasma in the vicinity of the presheath allows one to generalize Bohm's criterion according not only for arbitrary ion and electron distributions, but also for general boundary conditions at the wall. He showed that the obtained sheath condition marginally fulfilled and related to field singularity at sheath edge. Therefore, the smooth matching of the presheath and sheath solutions requires an additional transition layer. Based on this conclusion, Riemann [38] analyzed the influence of collisions on the plasma sheath transition using two-scale approach. A smooth matching of the presheath and sheath solutions was performed on an intermediate scale accounting as well for collisions and for space charge. This characteristic scale was given by a relation between the Debye length and mean free path length.

Scheuer and Emmert [39] developed a kinetic model for the collisional plasma presheath considering the ion temperature and Boltzmann distribution for the electrons. The ion collisions were taken in account by BGK approach [40]. An equation for potential distribution was obtained considering the ion–ion and ion–atom collisions. The calculated results showed the potential variation, wall potential, ion distribution function, and ion particle and energy fluxes into the sheath.

Procassini et al. [41] studied a bounded plasma by a particle-in-cell technique as a tool for examine a collisionless plasma using a distributed or volumetric particle source. The particle simulations of the plasma–sheath region with partial-width-distributed source indicate that there is no electric field present in the source-free region. The self-consistent description of the plasma-sheath region, which does not require the use of particle refluxing or the assumption of Boltzmann electrons, is possible if one includes the effects of Coulomb collisions.

Sternovsky [42] developed a kinetic model of the sheath and presheath that includes the effects of charge exchange and ionization collisions. In this treatment, accelerated ions disappear as result of charge-exchange collisions forming also cold ions. The electron density is modeled by the Boltzmann relation. Solutions are obtained by numerical integration of Poisson equation from a point near the plasma midplane to the wall. The collisions are found to reduce the current density at the wall to a value significantly below the usual ion saturation current. Solutions also found for the energy distribution of ions hitting the wall. The calculated results (plasma density, plasma electric field, etc.) were compared with that presented in other works [20, 35] indicating the agreement for similar conditions. The ion flux is reduced by approximately a factor of 2 when there are ten mean free paths from the plasma midplane to the wall.

Analysis of the above works show that, when the wall potential $\varphi_0 \gg kT_e$ the Langmuir collisionless region weakly depends on boundary condition and the Bohm condition or plasma collisions mainly determine the plasma presheath parameters depending on the model of the boundary conditions. However, the thickness of the volume space charge region can be shortened depending on ratio of the Debye length r_D to the ion mean free path $s_c = r_D/l_{ia}$. This important issue is detailed by Beilis et al. [43] considering the charge exchange collision. To this end, the kinetic equation was studied in form (5.17–5.18) for velocity function distribution f_i for the ions with a boundary condition (Fig. 5.4) as

$$f_i(v, 0) = \frac{j_i}{e}\frac{m}{kT}\exp\left(-\frac{mv^2}{2kT}\right) \quad v \geq 0 \tag{5.20}$$

The neutral atom velocity was assumed significantly lower than the ion velocity. The corresponding ion function distribution was in form:

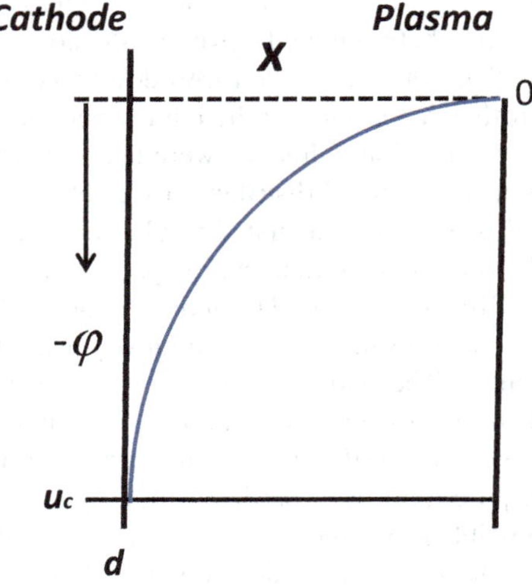

Fig. 5.4 Schematic potential distribution in the near cathode layer

$$f_i(v,x) = \begin{cases} \frac{j_i}{e}\frac{m}{kT}\exp\left[-\frac{m\left(v^2+\frac{2e\varphi(x)}{m}\right)}{2kT}\right]\exp(-\sigma_{ia}n_a x) \quad for \quad v \geq \sqrt{-\frac{2e\varphi(x)}{m}} \\ \frac{j_i}{e}\sigma_{ia}n_a\frac{m}{eE(x')}\exp\left[-\sigma_{ia}n_a(x-x')\right] \\ \varphi(x) = \varphi(x') - \frac{mv^2}{2e} \quad for \quad 0 < v < \sqrt{-\frac{2e\varphi(x)}{m}} \\ \frac{dx'}{d\varphi(x')} = -\frac{1}{E(x')} \end{cases} \tag{5.21}$$

Here, x is the axis directed to the cathode with zero at the plasma boundary, x' is the coordinate dependent on the electric field $E(x')$. The electron's function distribution f_e was taken as

$$f_e(v,x) = \frac{j_e}{e}\frac{m}{kT_e}\exp\left[-\frac{m\left(v^2-\frac{2e\varphi(x)}{m}\right)}{2kT_e}\right]\exp(-\sigma_{ia}n_a x)$$
$$for \quad v \geq \sqrt{\frac{2e[u_c+\varphi(x)]}{m}} \tag{5.22}$$

Here u_c is the cathode potential drop, σ_{ia} is the cross section of the charge exchange collision, j_e is the electron current density flowing from the plasma, and n_a is the neutral atom density. Taking (5.21) and (5.22), the Poisson equation will be:

$$\frac{\partial^2\varphi}{dx^2} = \frac{4\pi e}{\varepsilon_0}\left[\int f_e(x,v)dv - \int f_i(x,v)dv\right] \tag{5.23}$$

Using the following dimensionless variables

$$\phi = -\frac{e\varphi(x)}{kT};\quad \phi_c = -\frac{eu_c}{kT};\quad \theta = \frac{T_e}{T}\quad y = \frac{x}{r_D};\quad y' = \frac{x'}{r_D}\quad s_c = \frac{r_D}{l_i} = \sigma_{ia}n_a r_D$$

Poisson (5.23) can be written as

$$\frac{\partial^2\phi}{dy^2} = \exp(\phi - s_c y)\left[1 - \Phi(\sqrt{\phi})\right] + \frac{s_c}{\sqrt{\pi}}\int_0^y \frac{\exp\left[-s_c(y-y')\right]dy'}{\sqrt{\phi(y)-\phi(y')}}$$
$$- \frac{\exp(-\frac{\phi}{\theta})}{1+\Phi(\sqrt{\frac{\phi_c}{\theta}})}\left[1 + \Phi(\sqrt{\frac{\phi_c - \phi}{\theta}})\right]$$
$$\Phi(z_0) = \frac{2}{\sqrt{\pi}}\int_0^{z_0}\exp(z)dz \tag{5.24}$$

Equation (5.24) solved by Runge–Kutta method. It was used iteration procedure with method of successive approximations. As for zero approximation, the (5.24) at $s_c = 0$ (collisionless case) was considered in form:

$$y = \int_0^{\phi(y)} \frac{d\phi}{\{E^2 + 2[D(\phi, K) + C(\theta, \phi_c\phi)]\}^{0.5}}$$

were :

$$D(\phi) + \frac{1}{1-\Phi(\sqrt{K})}\left[e^{\phi} - 1 - \Phi(\sqrt{K+\phi})e^{\phi} + \Phi(\sqrt{K}) + \frac{2}{\pi}(\sqrt{K+\phi} - \sqrt{K})e^{-\phi}\right]$$
$$C(\theta, \phi_c\phi) = \frac{\theta}{1+\Phi(\sqrt{\frac{\phi_c}{\theta}})}\left\{1 - e^{-\frac{\phi}{\theta}} + \Phi(\sqrt{\frac{\phi_c}{\theta}}) - \Phi(\sqrt{\frac{\phi_c-\phi}{\theta}})e^{-\frac{\phi}{\theta}} + \frac{2e^{-\frac{\phi}{\theta}}}{\pi}\left[\sqrt{\frac{\phi_c-\phi}{\theta}} - \sqrt{\frac{\phi_c}{\theta}}\right]\right\} \tag{5.25}$$

Here, K is the parameter (order of unit) introduced for correcting the boundary ion function distribution at $v \geq \sqrt{\frac{2kT}{m}K}$ (like Bohm condition) needed to obtain the solution in the collisionless case.

The results of calculations are presented as two dependencies [43]: (i) dimensionless electric field at the cathode $eE_c r_D/kT$ and near-cathode layer (sheath and presheath) thickness d/r_D (d is the x when $\varphi = u_c$) as function on $s_c = r_D/l_{ia}$ (Fig. 5.5) and (ii) dimensionless potential drop eu_c/kT as function on d/r_D (Fig. 5.6).

The calculations show that during the transition from collisionless regime to collisional, the layer thickness decreased but the electric field is increased (Fig. 5.5). These results coincide with that obtained in [16] and it can be explained by presence of the charge-exchange collisions, which decrease the ion velocity and therefore increase the ion density.

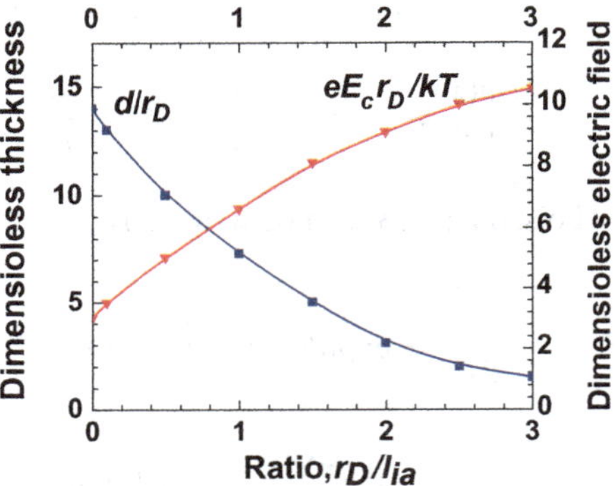

Fig. 5.5 Dependence of the near-cathode layer d and cathode electric field on the parameter indicated the collisional regime

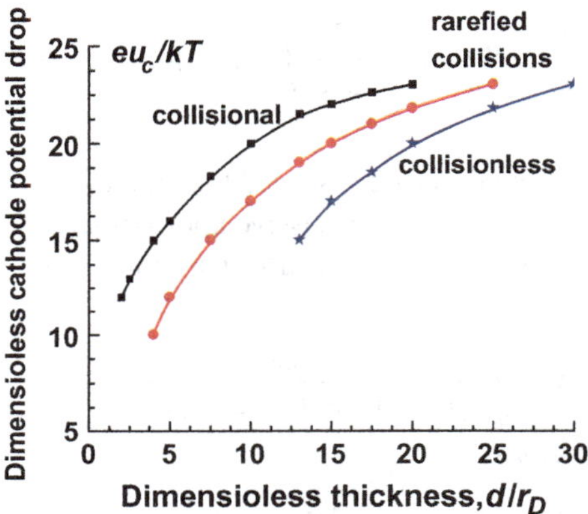

Fig. 5.6 Dependence of the cathode potential drop on thickness of the near cathode layer with the collisional regime as parameter

The dependence of the dimensionless cathode potential drop on dimensionless thickness of the near-cathode layer with the collisional regime as parameter is shown in Fig. 5.6. It can be seen that the dependence for collisionless case was shifted to larger values of d/r_D. For all cases the cathode potential drop have tendency to some saturation at larger d relatively r_D. The calculations also indicate that influence of the difference between electron and ion temperatures in the plasma was larger for collisionless regime in comparison with the collisional regime. When the parameter s_c increased from 1 (mostly collisionless regime) to 3 (collisional regime) the electric field increases only by factor two. This fact indicates that zero approach is a good approximation to describe the cathode electric field.

5.5 Electrical Field

Different approaches used to determine the electric field at the cathode surface are considered below.

5.5.1 Collisionless Approach

McKeown [44] studied the electrical field E at the cathode surface in the sheath. The positive space charge in the sheath is due to ion current density j_i originated from the plasma and the negative charge is due to electron emission current density j_{em} from the cathode. Plasma electrons and initial velocities were not considered. Poisson equation in this case has the following form:

$$\frac{\partial^2\varphi}{dx^2} = \frac{4\pi e}{\varepsilon_0}\left[\frac{j_i}{\sqrt{\frac{2e\varphi}{m}}} - \frac{j_e}{\sqrt{\frac{2e(u_c-\varphi)}{m_e}}}\right] \tag{5.26}$$

After integration the electrical field E at the cathode surface can be obtained in form

$$E^2 = 16\varepsilon_0^{-1}\left(\frac{u_c m_e}{2e}\right)^{1/2}\left[j_i\left(\frac{m}{m_e}\right)^{1/2} - j_{em}\right] + E_0^2 \tag{5.27}$$

where ε_0 is the dielectric permittivity of vacuum, m_e and m are the electron and ion mass, respectively, and E_0 is the electric field at the sheath boundary on, plasma side. Equation (5.27) is named in the literature as McKeown's equation and obtained assuming that the mean free path l_i is larger than the sheath thickness d in which the cathode potential drop is u_c. The last assumption is analyzed for cathode sheath in a vacuum arc neglecting the small contribution from the emitted electrons [45]. It was shown that the condition $l_i/d > 1$ fulfilled with current density $j \geq 5\times10^4$ A/cm^2.

5.5.2 Electric Field. Plasma Electrons. Particle Temperatures

According to analysis of the sheath (Sects. 5.2–5.4), the velocity of the ions ejected from non-disturbed quasineutral plasma increased in the presheath before ion entering in the sheath. Let us consider a case, which takes in account ion Maxwellian velocity distribution and the initial ion velocity is equal to mean velocity determined by this distribution. The density of the returned plasma electrons to the cathode is distributed according to the Boltzmann law ($T_e > T$). The potential drop is u_c. The space charge in Poisson equation is determined by ion density n_i ($Z = 1$), electron density n_{em} from the electron emission current density j_{em} and by Boltzmann electron density n_e. The mentioned densities of the charged particles are determined by the continuity and motion equations. In this case, Poisson equation is [46, 47]:

$$\frac{\partial^2\varphi}{dx^2} = 4\pi\left[\frac{j_i}{\sqrt{\frac{kT}{2\pi m}}\left[1+\frac{4\pi e(u_c-\varphi)}{kT}\right]^{0.5}} - \frac{j_{em}}{\sqrt{\frac{kT_c}{2\pi m}}\left[1+\frac{4\pi e\varphi)}{kT_c}\right]^{0.5}} - \frac{j_i}{\sqrt{\frac{kT}{2\pi m}}}\exp(-\frac{e\varphi}{kT})\right] \tag{5.28}$$

Integrating (5.28) from $\varphi = 0$ to $\varphi = u_c$ and $E = E_0$ at the external boundary, the electric field at the cathode surface E is

$$E^2 - E_0^2 = A_E\sqrt{u_c}\left\{\begin{array}{l} j_i(\frac{m}{m_e})^{0.5}\left[(1+\frac{kT}{4\pi eu_c})^{0.5} - \sqrt{\frac{kT}{4\pi eu_c}} - \frac{\sqrt{\pi}kT_e}{\sqrt{eu_c kT}}(1-\exp(-\frac{eu_c}{kT_e}))\right] \\ -j_{em}\left[(1+\frac{kT_c}{4\pi eu_c})^{0.5} - \sqrt{\frac{kT_c}{4\pi eu_c}}\right] \end{array}\right\} \tag{5.28}$$

Here $A_E = 16\pi\sqrt{\frac{m_e}{2e}} = 7.57 \times 10^5$ V/cm, T_c-cathode temperature. According to (5.28) the sheath will be stable when $T > \frac{\pi(kT_e)^2}{u_c}$ and with this condition, the monotonic potential distribution can be obtained near the sheath entrance. When the $\ll u_c \gg (kT, kT_e)$, the cathode surface field is given by McKeown (5.27). Taking the initial ion velocity according to the Bohm condition, i.e., $(kT_e/m)^{0.5}$, the cathode electric field can be obtained in form:

$$E^2 - E_0^2 = A_E\sqrt{u_c}\left\{\begin{array}{l} j_i(\frac{m}{m_e})^{0.5}\exp(0.5)\left[(1+\frac{kT_e}{2eu_c})^{0.5} - \sqrt{\frac{kT_e}{2eu_c}} - \frac{kT_e}{\sqrt{eu_c}}(1-\exp(-\frac{eu_c}{kT_e})\right] \\ -j_{em}\left[(1+\frac{kT_c}{4\pi eu_c})^{0.5} - \sqrt{\frac{kT_c}{4\pi eu_c}}\right] \end{array}\right\} \tag{5.29}$$

5.5.3 *Refractory Cathode. Virtual Cathode*

The rate of atom evaporation from cathodes of refractory materials is significantly lower than the rate of electron emission. Therefore, the ionized atom density can be lower than the electron density near the cathode surface. In addition, the ions accelerate toward the surface in the sheath and the emitted electrons are of low temperature equal to the cathode temperature. As result, a not compensated negative electron space charge is formed near the cathode surface. In this case, a minimum potential $\varphi = \varphi_m$ appears at distance $x = x_m$ producing so called *virtualcathode* (determined by Langmuir [4], Sect. 5.1) and the electron current will be limited by own space charge (Fig. 5.7) [48–50].

Let us consider the regime of virtual cathode studying Poisson equation following the approach developed in [48]:

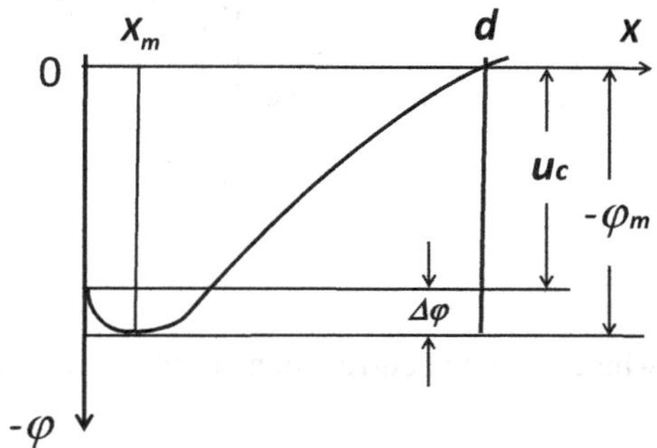

Fig. 5.7 Schematic presentation of the potential distribution in the cathode sheath for refractory material. At the distance $x = x_m$, a virtual cathode is produced due to negative space charge with minimal potential

$$\varepsilon_0 \frac{\partial E}{dx} = e[n_i(\varphi) - n_{er}(\varphi) - n_e(\varphi)] \tag{5.30}$$

Here, $n_i(\varphi)$ is the potential dependent density of ions flowed from the plasma toward the cathode, $n_{er}(\varphi)$ is the Boltzmann distributed density of returned electron from the plasma, and $n_e(\varphi)$ is the electron density determined by the electron emission current j_{em} reduced due to retarding potential $\Delta\varphi$.

5.5.3.1 Single Charged Ions

Let us define the particle densities taking into account the simple case of single-charged ion current extracted from the plasma boundary toward the cathode.

$$\begin{gathered}
n_{er}(\varphi) = n_{er0}\exp(-\frac{\varphi}{\varphi_e}); \quad n_e(\varphi) = n_m \exp(\frac{\varphi - \varphi_m}{\varphi_c})\Phi^*(\frac{\varphi - \varphi_m}{\varphi_c}); \\
n_i(\varphi) = \frac{j_i}{e\sqrt{\frac{2e(\varphi_i - \varphi)}{m}}} = \frac{n_{i0}}{\sqrt{1 - \frac{\varphi}{\varphi_i}}}; \quad n_{i0} = \frac{j_i}{e\sqrt{\frac{2e\varphi_i}{m}}}; \quad n_m = \sqrt{\pi}\frac{j_{eb}}{e\sqrt{\frac{2e\varphi_c}{m_e}}} \\
j_{eb} = j_{em}\exp(-\frac{\Delta\varphi}{\varphi_c}); \quad n_{e0}(\varphi) = n_e(0) = n_m \exp(\frac{\varphi_m}{\varphi_c})\Phi^*(\frac{\varphi_m}{\varphi_c});
\end{gathered} \tag{5.31}$$

where $\Phi^*(\xi) = (1 - \Phi(\xi))$, $\Phi(\xi)$ is the probability integral, $e\varphi_i = kT$. $e\varphi_e = kT_e$ and $e\varphi_c = kT_c$, T_c is the cathode temperature, n_{e0} is the electron density of emitted from the cathode at $x = d$. Substitution (5.31) in (5.30) and taking into account $E(x = x_m) = 0$ and $E(x = \mathrm{d}) = E_0$ the following expression is obtained:

$$e\int_{\varphi_m}^{0}\left[\frac{n_{i0}}{\sqrt{1 - \frac{\varphi}{\varphi_i}}} - n_{er0}\exp\left(-\frac{\varphi}{\varphi_e}\right) - n_m\exp\left(\frac{\varphi - \varphi_m}{\varphi_c}\right)\Phi^*\left(\sqrt{\frac{\varphi - \varphi_m}{\varphi_c}}\right)\right] = \frac{\varepsilon_0 E_0}{2} \tag{5.32}$$

Integrating (5.31) from $\varphi = \varphi_m$ to $\varphi = 0$ the following expression is obtained:

$$\varepsilon_0 E_o^2 = 4en_{i0}\sqrt{\frac{\varphi_m}{\varphi_i}}e\varphi_i\left[\begin{array}{l}\left(\sqrt{1+\frac{\varphi_i}{\varphi_m}}-\sqrt{\frac{\varphi_i}{\varphi_m}}\right)-\sqrt{\pi}s_r\sqrt{\frac{\varphi_e}{\varphi_m}}(1-\mathrm{e}^{-\frac{\varphi_m}{\varphi_e}})\\ -\sqrt{\pi}s_e\left(1+\exp(\frac{\varphi_m}{\varphi_c})\right)\Phi^*\left(\sqrt{\frac{\varphi_m}{\varphi_c}}\right)-\sqrt{\frac{\pi\varphi_m}{4\varphi_c}}\end{array}\right] \tag{5.33}$$

where j_{eT} is the current density of the returned electrons from the plasma and

$$s_e = \frac{j_e}{j_i}\varepsilon; \quad s_r = \varepsilon j_{eT}\exp(-u_c/\varphi_e); \quad j_{eT} = n_{er0}\sqrt{\frac{e\varphi_e}{2\pi m_e}}; \quad \varepsilon = \sqrt{\frac{m_e}{m}}; \tag{5.34}$$

5.5.3.2 Multiple Charged Ions. Quasineutrality

According to the study of the cathode spot parameters [49, 50], relatively large electron temperature occurred in the case of refractory cathode materials, like tungsten. In this case, a number of highly charged (Z) ions are produced. Taking into account this fact, other relation between parameters s_e and s_r is obtained taking into account the quasineutrality at the sheath–plasma boundary. Although that the electric field at this boundary $E \neq 0$ nevertheless the condition $(n_i\text{-}n_e)/n_e < <1$ is fulfilled. Below these phenomena are studies by considering the sheath structure in the spot on the refractory cathode.

Poisson equation reads as:

$$\varepsilon_0\frac{\partial E}{\mathrm{d}x} = e\left[\sum_Z Zn_{iZ}(\varphi) - n_{er}(\varphi) - n_e(\varphi)\right] \tag{5.35}$$

The densities $n_{er}(\varphi)$ and $n_e(\varphi)$ were determined by formulas (5.31). The density of multi-charged ions:

$$n_{iz}v = n_{iz0}v_0; \quad \frac{mv^2}{2} = \frac{mv_0^2}{2} + eZ\varphi; \quad v_0^2 = \frac{T_e}{m}; \quad v^2 = \frac{T_e}{m}(1+\frac{2eZ\varphi}{T_e}) \tag{5.36}$$

Using (5.35) and (5.36) and taking into account $E(x = x_m) = 0$ and $E(x = \mathrm{d}) = E_0$ the following expression can be obtained:

$$e\int_{\varphi_m}^{0}\left[\sum_Z \frac{Zn_{iZ0}}{\sqrt{1-\frac{2Z\varphi}{\varphi_e}}} - n_{er0}\exp(-\frac{\varphi}{\varphi_e}) - n_m\exp(\frac{\varphi-\varphi_m}{\varphi_c})\Phi^*(\sqrt{\frac{\varphi-\varphi_m}{\varphi_c}})\right] = \frac{\varepsilon_0 E_0}{2} \tag{5.37}$$

Integrating (5.37) from $\varphi = \varphi_m$ to $\varphi = 0$ and denoting $f_{iz} = \frac{Zn_{iZ0}}{\sum\limits_Z Zn_{iZ0}}$, the following expression is obtained:

$$\varepsilon_0 E_o^2 = e\sum_Z Zn_{iZ0}e\varphi_e\left[\sum_Z \frac{f_{iz}}{Z}(\sqrt{1+2\frac{Z\varphi_m}{\varphi_e}}-1)\right] - n_{er0}e\varphi_e(1-\mathrm{e}^{-\frac{\varphi_m}{\varphi_e}}) -$$
$$-n_m e\varphi_c\left[\exp(\frac{\varphi_m}{\varphi_e})\Phi^*(\sqrt{\frac{\varphi_m}{\varphi_c}}) - 1 + \sqrt{\frac{4\varphi_m}{\pi\varphi_c}}\right] \tag{5.38}$$

Assuming that all charged ions move with velocity v_0 the ion current density is $j_i = e_N^{-0.5}\sum\limits_Z eZn_{iZ0}v_0$.

Passing $\sqrt{\frac{4\varphi_m}{\pi\varphi_c}}$, the last brackets in (5.38) and taken n_m, n_{ir0} and v_0 from (5.31) and (5.36) equation for the boundary electric field is:

$$E_o^2 = \frac{4j_i}{\varepsilon_0}\sqrt{\frac{e\varphi_m m}{2}}\left\{\begin{array}{l} e_N^{0.5}\sum\left[\frac{f_Z}{\sqrt{Z}}(\sqrt{1+\frac{\varphi_e}{2Z\varphi_m}} - \sqrt{\frac{\varphi_e}{2Z\varphi_m}})\right] - \beta_{er}\sqrt{\frac{\pi\varphi_e}{\varphi_m}}(1-e^{-\frac{\varphi_m}{\varphi_e}}) \\ -\beta_e\left[1+\frac{1}{2}\sqrt{\frac{\pi\varphi_c}{\varphi_m}}e^{\frac{\varphi_m}{\varphi_c}}\Phi^*(\sqrt{\frac{\varphi_m}{\varphi_c}}) - \frac{1}{2}\sqrt{\frac{\pi\varphi_c}{\varphi_m}}\right] \end{array}\right\} \tag{5.39}$$

Now, the quasineutrality at the sheath–plasma boundary is considered in form.

$$e\sum_Z Zn_{iZ0} = n_e(0) + n_{er0}$$
$$\frac{e_N^{0.5}j_i}{\sqrt{\frac{e\varphi_e}{m}}} = \frac{j_{eb}}{\sqrt{\frac{2e\varphi_c}{\pi m_e}}}\exp(\frac{\varphi_m}{\varphi_c})\Phi^*(\sqrt{\frac{\varphi_m}{\varphi_c}}) + \frac{j_{er}}{\sqrt{\frac{e\varphi_e}{2\pi m_e}}} \tag{5.40}$$

After division of (5.40) by $j_i\sqrt{\frac{m}{e\varphi_e}}$:

$$e_N^{0.5} = \beta_e\sqrt{\frac{\pi\varphi_e}{2\varphi_c}}\exp(\frac{\varphi_m}{\varphi_c})\Phi^*(\sqrt{\frac{\varphi_m}{\varphi_c}}) + \sqrt{2\pi}\beta_{er} \tag{5.41}$$

or after multiply both sides of (5.41) by $\sqrt{\frac{2\varphi_m}{\varphi_e}}$:

$$e_N^{0.5}\sqrt{\frac{2\varphi_m}{\varphi_e}} = \beta_e\sqrt{\frac{\pi\varphi_m}{\varphi_c}}\exp\left(\frac{\varphi_m}{\varphi_c}\right)\Phi^*(\sqrt{\frac{\varphi_m}{\varphi_c}}) + \beta_{er}2\sqrt{\frac{\pi\varphi_m}{\varphi_e}} \tag{5.42}$$

Let us denote:

$$b = \sqrt{\frac{\pi\varphi_e}{\varphi_m}}(1 - e^{-\frac{\varphi_m}{\varphi_e}}); \qquad C_{\Sigma 1} = \sum_Z \frac{f_Z}{\sqrt{Z}}\sqrt{1 + \frac{\varphi_e}{2Z\varphi_m}};$$

$$C_{\Sigma 2} = \sum_Z \frac{f_Z}{\sqrt{Z}}\sqrt{\frac{\varphi_e}{2Z\varphi_m}}; \qquad F = \sqrt{\frac{\pi\varphi_m}{\varphi_c}}\exp(\frac{\varphi_m}{\varphi_c})\Phi^*(\sqrt{\frac{\varphi_m}{\varphi_c}})$$

and then (5.39) and (5.42) are

$$E_o^2 = \frac{4j_i}{\varepsilon_0}\sqrt{\frac{e\varphi_m m}{2}}\left\{e_N^{0.5}\sum[C_{\Sigma 1} - C_{\Sigma 1}] - \beta_{er}b - \beta_e\left[1 + F - \frac{1}{2}\sqrt{\frac{\pi\varphi_c}{\varphi_m}}\right]\right\} \tag{5.43}$$

$$e_N^{0.5}\sqrt{\frac{2\varphi_m}{\varphi_e}} = \beta_e F + \beta_{er}2\sqrt{\frac{\pi\varphi_m}{\varphi_e}}; \tag{5.44}$$

5.6 Electrical Double Layer in Plasmas

Until now, we considered a positive volume charge formed in a single sheath between plasma and insulating or conducting wall. However, in certain plasma configurations or for non-uniform parameters of the plasma (sharply changed geometry of the plasma flow, or plasma properties) a space charge layer can be produced between two separate plasma regions. Such layer is named as "*double layer*" or "*double sheath.*" Double layers consist of two adjacent charge plasma layers of opposite charge. This double sheath produced at the junction between two plasmas with different parameters (densities, temperatures) similar to potential barrier between to solid materials with different conductivity. The mechanism of double layer operation is an important issue as a possible similar layer support the current continuity at the cathode plasma region in some specific vacuum arc discussed below. Let us discuss, therefore, the published works in order to study the knowledge of this subject.

A double layer in a high-voltage discharge with a low concentration of charged particles is observed experimentally [51]. The potential profile of expanding collisionless plasma is obtained experimentally [52]. Plasma potential structures are measured by emissive probes. The results showed that the quasineutrality condition breaks and the ions accelerated by the charge separation into the plasma.

A review of double-layer experiments was presented by Hershkowitz [53, 54]. The formation and stability of double layers are analyzed for different potential steps $e\varphi/T_e$- ~ 1, < 10, or > 10, and ≫10. Double layers differ from sheaths in that they are removal from the plasma boundaries. Generally, the thickness of the plasma double layer is on the order of the mean free path. Free ions can enter from the right

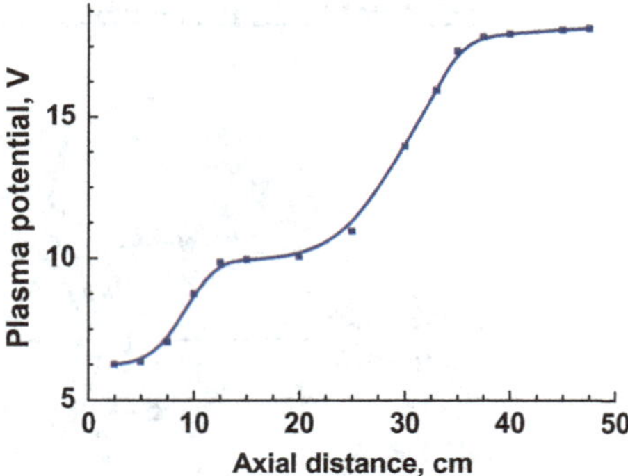

Fig. 5.8 Example of potential distribution with two step double layers

plasma side and be accelerated by the potential and electrons can be accelerated from the left plasma side by the potential of the double layer. It was indicated that multiple double layers occurred. Two- and three-dimensional structures including multiple double layers were described according to Hershkowitz [53], which can be shown by Fig. 5.8.

The laser-induced fluorescence was used for measuring the ion accelerations in helicon plasma devices. Spatially resolved non-invasive measurements were conducted of supersonic ion flows created in helicon-wave-heated plasmas without the use of accelerator or auxiliary-heating techniques. These studies include a range of plasma parameter in which double layers can be appeared [55]. The measurements of ion velocities were made along the axis of a helicon-generated Ar plasma column with radius changed by spatially separated mechanical and magnetic apertures. Ion acceleration to supersonic speeds was observed for both aperture types, simultaneously generating two steady-state double layers [56]. The electron populations in a magnetic nozzle were measured by electric probe and spectroscopic methods were used to determine the electron energy. It was indicated that the measured super thermal electrons are suggested to be as a source for the large potential drop of a double layer, which accelerates the ions [57].

Charles and Boswell [58] experimentally studied an electric double layer with $e\varphi/kT_e \sim 3$ and a thickness of less than 50 Debye lengths in an expanding, high-density helicon sustained rf discharge. It was observed that the plasma itself self-consistently generates the potentials, and there is no current flowing through an external circuit. The plasma electrons are heated by the *rf* fields in the source, provide the power to maintain the double layer, and hence ions were accelerated. Charles [59] reviewed the developments of double layers from the late 1980s to the spring of 2007. The double layer devices and properties are presented with an emphasis on current-free double layers. Applications of double layers are discussed for the field of plasma processing and for electric propulsion.

Andrews and Allen [60, 61] presented the early theoretical description of the double layer. They considered a double sheath between two plasmas when sharp changes of the discharge channel cross section occur in form represented in Fig. 5.9. The model was presented for gas discharge tube in which the double sheath develop

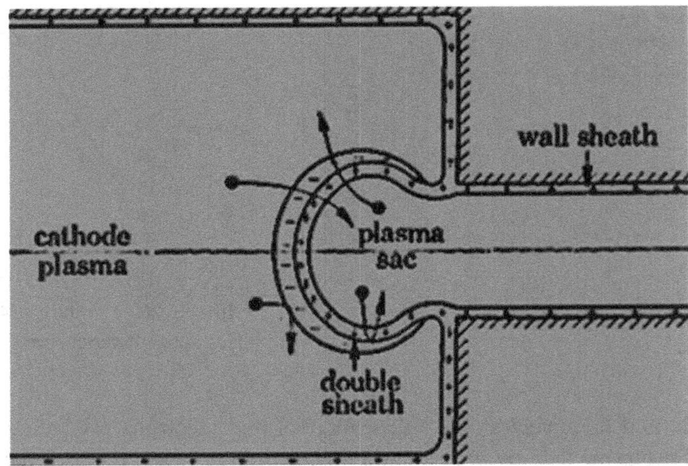

Fig. 5.9 Double sheath at a constriction in gas discharge tube

over the cathode side of a constriction in the tube and influences the flow of electrons and ions through the constriction. The electric field inside the double sheath accelerates electrons from the cathode plasma into the plasma sac that is a bright blob of the second plasma protruding out of the cathode region of the discharge tube. Ions from the second plasma are accelerated in opposite direction into cathode plasma.

The model takes into account the initial velocities and reflected particles. Four groups of charged particles in the double sheath were taken in account. There are the mentioned accelerated ions and electrons as well also returned thermal ions emitted from the cathode plasma and returned thermal electrons emitted from the second plasma. The thermal particles have Botzmann distribution in the sheath. The charge density of thermal ions with temperature T_{i} in the cathode plasma 1 is given by

$$en_{i1} = a\exp(-\frac{\varphi}{kT_i}) \tag{5.45}$$

The charge density of ions emitted from the plasma 2 is obtained from conservation of mass and energy assuming absence of collisions and ionization in the sheath with potential drop φ_s and is given by

$$en_{i2} = b\left[1 + \frac{2e(\varphi_s - \varphi)}{mv_i^2}\right]^{-0.5} \tag{5.46}$$

The charge density of electrons emitted from the plasma 1 is given by

$$en_{e2} = c\left[1 + \frac{2e\varphi}{mv_e^2}\right]^{-0.5} \tag{5.47}$$

The charge density of thermal electrons with temperature T_1 in the cathode plasma 2 is given by

$$en_{e2} = d\exp(-\frac{(\varphi_s - \varphi)}{kT_2}) \tag{5.48}$$

Using (5.45–5.48) and following denotes:

$$\Psi_e = \frac{mv_e^2}{2e\varphi_s} \qquad \Psi_i = \frac{mv_i^2}{2e\varphi_s} \qquad \tau_e = \frac{kT_e}{e\varphi_s} \qquad \tau_i = \frac{kT_i}{e\varphi_s} \qquad \eta = \frac{\varphi}{\varphi_s}$$

The total charge density for study Poisson equation is given by

$$en_{ch} = a\exp(-\frac{\eta}{\tau_i}) + b\left[1 + (\frac{1-\eta}{\Psi_i})\right]^{-0.5} + c\left[1 + \frac{\eta}{\Psi_e}\right]^{-0.5} + d\exp(-\frac{(1-\eta)}{\tau_e}) \tag{5.49}$$

where *a, b, c, and d* are the constants.

The boundary conditions assumed ionization in the plasma and a case where the sheath thickness was vanishingly small compared with the dimensions of plasmas. The plasma quasineutrality at both sides of the sheath is given

$$en_{ch} = \rho = 0; \quad \frac{d\rho}{d\eta} = 0 \tag{5.50}$$

Substituting (5.49) into (5.50) the coupling relations were obtained for double sheath and the above problem was solved numerically. As a result, the dependencies of Ψ_i on τ_e and Ψ_e on τ_I were calculated. It was obtained that $mv_i^2 > kT_e$ and $mv_e^2 > kT_i$. It implies that ion gains initial kinetic energy in the plasma sac before the sheath boundary with velocity, which is in excess of the Bohm velocity. Another important results is the ratio of the current densities of the electrons and ions accelerated across the double sheath

$$\frac{j_e}{j_i} = \alpha_d\sqrt{\frac{m}{m_e}}; \quad \alpha_d \sim \frac{\Psi_e}{\Psi_i} \tag{5.51}$$

According to the authors [61], α_d can be significantly lower than 1 shown in Fig. 5.10. The authors [61] compared this result with that obtained by Langmuir for a simple case (other Langmuir's approaches were not considered) of a double sheath studied assuming zero initial particle velocity, no reflected particles and zero electric field at both sides of the double sheath [9]. It was concluded that in the developed model, the ions from the plasma sac and electrons from the cathode plasma arriving at the two plasma–sheath boundaries must satisfy certain condition in order to support monotonic potential distribution (like Bohm condition).

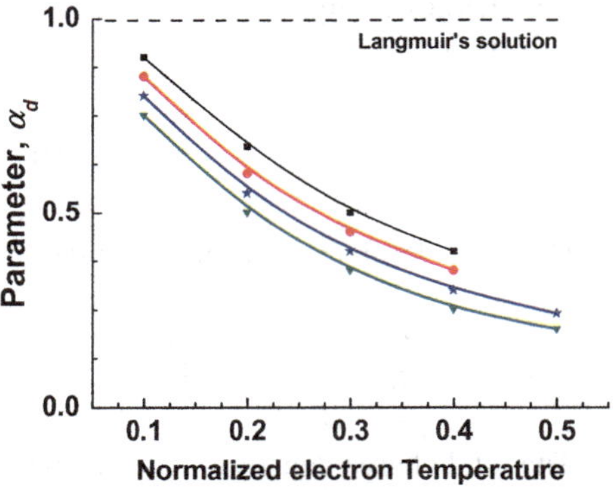

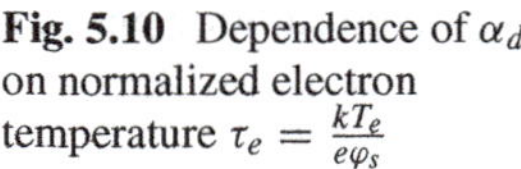

Fig. 5.10 Dependence of α_d on normalized electron temperature $\tau_e = \frac{kT_e}{e\varphi_s}$

Torven [62] solved Poisson equation considering the double layer as small shock amplitude. The ions were assumed to have a vanishingly small temperature, and the motion is considered on the ion acoustic time scale. The electron distribution function was approximated by a steady-state distribution function. As calculated for different time, the potential profiles give a spatial scaling of the shock wave. It was noted that these profiles equal roughly the experimentally observed steepening of initially monotonic profiles.

Several explanations for double layers are mentioned by Chen and co-authors [52]. In steady state, they can be explained in frame of BGK solutions, which depend entirely on the distribution functions of both ions and electrons at boundaries far from potential profile of the double layer. Instabilities and turbulence can play an important role in the formation of the double layers. Another possibility is due to ionization of plasma on the high-potential side, which can be enhanced by the energy gain that electrons receive. An attempt to understand the physical mechanism of current-free double layers observed in plasmas expanding along magnetic fields considered by Chen [63]. It was shown that the diverging magnetic field lines cause the presheath acceleration of ions, producing a potential jump resembling that of a double layer. The process stops when it runs out of energy.

The models considered above indicate that, in spite of extensive study of the double layers, the general mechanisms of different processes for double-layer formation are still unclear. However, experimental and theoretical studies of widely characterizing plasma double layers allow using these events to understand the mechanism of continuous electrical current in the cathode spot of a vacuum arc with mercury cathode.

The main problem of the arc with mercury is the small temperature at which the electron emission is negligible, while the vapor density is considerably large. As result, no electron beam was from the cathode for vapor ionization and therefore no reproduction of the plasma as with a usual case of copper cathode [64]. This problem is well known from beginning of twentieth century and numerous unsuccessful attempts to explain the mechanism were previously developed using classical approaches to explain the electron emission from the mercury. Therefore, the idea of a double layer arises which separates the cathode plasma in two regions, and the first adjacent to the cathode surface serves as plasma cathode from which the necessity electron beam emitted.

Beilis [65–67] developed the double sheath model for mercury arc which allows to understand the observed mercury cathode spots dynamics and unusual specifics of the arc with measured two step of cathode potential drop. The corresponding model is described in Chap. 16, in which the cathode spot theory is presented for different cathode materials.

References

1. Keidar, M., & Beilis, I. I. (2016). *Plasma Engineering*. Elsevier, London-NY: Acad Press.
2. Child, C. D. (1911). Discharge from hot CaO. *Physical Review, 32*(5), 492–511.
3. Langmuir, I. (1913). The effect of space charge and residual gases of thermionic currents in high vacuum. *Physical Review, 2*(6), 450–486.
4. Langmuir, I. (1923). The effect of space charge and initial velocities on the potential distribution and thermionic current between parallel plane electrodes. *Physical Review, 21*(4), 419–435.
5. Langmuir, I., & Blodgett, K. (1923). Current limited by space charge between coaxial cylinders. *Physical Review, 22*(4), 347–356.
6. Langmuir, I., & Blodgett, K. (1924). Current limited by space charge between concentric spheres. *Physical Review, 24*(1), 49–59.
7. Mott-Smith, H. M., & Langmuir, I. (1926). The theory of collectors in gases discharges, Physical Review, *28*(N4), 727–763.
8. Compton, K. D., Turner, L. A., & McCurdy, W. H. (1924). Theory and experiments relating to the striated glow discharge in mercury vapor. *Physical Review, 24*(6), 597–615.
9. Langmuir, I. (1929). The interaction of electron and positive ion space charge in the cathode sheath. *Physical Review, 33*(6), 954–989.
10. Tonks, L., & Langmuir, I. (1929). A general theory of the plasma of an arc, Physical Review, *34*, 876–922.
11. Bohm, D. (1949). In A. Guthry & R. K. Wakerling (Eds.), *The characteristics of electrical discharges in magnetic field*. McGraw-Hill: New York.
12. Ecker, G. (1953). Die Raumladungszone an der Grenze des Bogenplasmas. *Z fur Physics, 135,* 105–118.
13. Ecker, G. (1973). Unified analysis of the metal vapour arc. *Zeitschrift Naturf*, *28a*, 417–428.
14. Ecker, G. (1980). Theoretical aspects of the vacuum arc. In J. M. Lafferty (Ed.), *Vacuum arcs. Theory and application* (pp. 228–320). John Wiley & Sons.
15. Baksht, F.G., & Moyzhes, B. Y. (1968). On the theory of near electrode layer in a low temperature plasma, *Soviet Physics Technology Physics*, *38*(N4), 724–736, Russian.
16. Baksht, F. G., Moyzhes, B. Ya., Nemchinsky, V. A. (1969). On the calculation of near electrode layer in a low temperature plasma, *Soviet Physics Technology Physics*, *39*(N4), 558–556, Russian.
17. Emmert, G. A., Wieland, R. M., Mense, A. T., & Davidson, J. N. (1980). Electric sheath and presheath in a collisionless, finite ion temperature plasma. *Physics of Fluids, 23*(4), 803–812.
18. Valentini, H. B., & Herrmann, F. (1996). Boundary value problems for multi-component plasmas and a generalized Bohm criterion. *Journal of Physics. D. Applied Physics, 29,* 1175–1180.
19. Scheuer, J. T., & Emmert, G. A. (1988). Sheath and presheath in a collisionless plasma with Maxwellian source. *Physics of Fluids, 31*(12), 3645–3648.
20. Franklin, R. N. (2003). The plasma-sheath boundary region. *Journal of Physics. D. Applied Physics, 36,* 309–320.

21. Chekmarev, I. B., Sklyarova, E. M., & Kolesnikova, E. N. (1983). The Knudsen ion layer problem in the theory of the collisionless sheath interaction of completely ionized gas with a wall. *Beitr Plasmaphysics, 23*(4), 411–421.
22. Chekmarev, I. B., & Chekmarev, O. M. (1987). Asymptotic theory of a non-thermal collisional layer. *Technical Physics, 57*(3), 440–445.
23. Sklyarova, E. M., & Chekmarev, I. B. (1994). Asymptotic model of the interaction of completely ionized gas with a wall. *Technical Physics, 39*(7), 649–652.
24. Benilov, M. S. (2003). Method of matched asymptotic expansions versus intuitive approaches: Calculation of space-charge sheaths. *IEEE Transactions on Plasma Science, 31*(4), 678–690.
25. Sternberg, N., & Godyak, V. (2003). On asymptotic matching and the sheath edge. *IEEE Transactions on Plasma Science, 31*(4), 665–677.
26. Keidar, M., & Beilis, I. I. (2005). Transition From plasma to space-charge sheath near the electrode in electrical discharges. *IEEE Transactions on Plasma Science, 33*(5), 1481–1486.
27. Beilis, I. I., Keidar, M & Goldsmith, S. (1997). Plasma-wall transition: The influence of the electron to ion current ratio on the magnetic presheath structure, *Physics of Plasmas, 4*(N10), 3461–3468.
28. Beilis, I. I., & Keidar, M. (1998). Sheath and presheath structure in the plasma wall-transition layer in an oblique magnetic field. *Physics of Plasmas, 5*(5), 1545–1553.
29. Ahedo, E. (1997). Structure of the plasma-wall transition in an oblique magnetic field. *Physics of Plasmas, 4*(12), 4419–4430.
30. Ahedo, E. (1999). Plasma-wall transition in an oblique magnetic field: Model of the space-charge sheath for large potentials and small Debye lengths. *Physics of Plasmas, 6*(11), 4200–4207.
31. Harrison, E. R., & Thompson, W. B. (1959). The low pressure plane symmetric discharge. *Proceedings of Physical Society London, 74*(2), 145–152.
32. Self, S. A. (1963). Exact solution of the collisionless plasma-sheath equation. *Physics of Fluids, 6*(12), 1762–1768.
33. Petrov, V. G. (1973). Near cathode region by taken in account the charge-exchange collisions. *Soviet Physics-Technical Physics, 43*(5), 1083–1086.
34. Scherbinin, P. P. (1972). Function distribution of the ions in an electrical field near the electrode in case of resonance charge exchange collisions. *High Temperature, 10*(2), 255–264.
35. Scherbinin, P. P., & Chervyakov, I. B. (1977). Function distribution of the ions and their drift velocity in a constant electrical field near the boundary. *Soviet Journal of Plasma Physics, 3*(4), 841–847.
36. Riemann, K. U. (1981). Kinetic theory of the plasma sheath transition in weakly ionized plasma. *Physics of Fluids, 24*(N12), 2163–2172.
37. Riemann, K. U. (1991). The Bohm criterion and sheath formation. *Journal of Physics. D. Applied Physics, 24*, 493–518.
38. Riemann, K. U. (1997). The influence of collisions on the plasma sheath transition. *Physics of Plasmas, 4*, 4158–4166.
39. Scheuer, J. T., & Emmert, G. A. (1988). A collisional model of the plasma presheath. *Physics of Fluids, 31*(6), 1748–1756.
40. Bhatnagar, P. L., Cross, E. P., & Krook, M. (1954). A model for collision processes in gases. I. Small amplitude processes in charged and neutral one-component systems, *Physical Review, 94*(N3), 511–525.
41. Procassini, R. J., Birdsall, C. K., & Morse, E. C. (1990). A fully kinetic, self-consistent particle simulation model of the collisionless plasma-sheath region. *Physics of Fluids, B2*(N12), 3191–3205.
42. Sternovsky, Z., Downum, K., & Robertson, S. (2004). Numerical solutions to a kinetic model for the plasma-sheath problem with charge exchange collisions of ions. *Physical Review E, 70*, 026408.
43. Beilis, I. I., Yu., Kukharenko, A., Rakhovsky, V. I., & Chumakova, L. F. (1988). Kinetic calculation on vacuum arc electrode layer parameters. *Measurement of Technology, 31*, 517–520.

44. McKeown, S. (1929). The cathode drop in an electrical arc. *Physical Review, 34,* 611–614.
45. Beilis, I. I., Lyubimov, G. A., & Rakhovsky, V. I. (1969). Electric field at the electrode surface near the cathode spot of an arc discharge. *Soviet Physics Doklady, 14,* 897–900.
46. Beilis, I. I. (1977). Theoretical analysis of cathode phenomena in vacuum arc discharge In 3rd *International Symposium, Switching Arc Phenomena 1977* (pp. 194–200). Part I, Lodz: Poland, Thechnical University.
47. Beilis, I. I. (1980). The effect of electric field near the cathode surface upon parameters on the near-electrode region of arc discharge. *IX ISDEIV* (65–68). Netherlands: Eindhoven.
48. Granovsky, V. L. (1971) (Chap. 8). In L.A. Sena & V.E. Golant (Eds.), *Electrical current in a gas. Steady-State current*, Nauka, Russia.
49. Beilis, I. I. (1988). Model of steady-state arc spot on a refractory cathode. *Soviet Technology Physics Letters, 14*(2), 494–495.
50. Beilis, I. I. (1988). On a mechanism of vacuum arc spot function on refractory cathodes. *High Temperature, 14,* 1224–1226.
51. Lutsenko, E. I., Sereda, N. D., & Kontsevoi, L. M. (1976). Study of the charge volume sheath creation in plasmas. *Soviet. Journal of Plasma Physics*, *2*(N1), 39.
52. Chan, C., Hershkowitz, N., Ferreira, A., Intrator, T., Nelson, B., & Lonngren, K. (1984). Experimental observations of self-similar plasma expansion. *Physics of Fluids, 27*(1), 266–268.
53. Hershkowitz, N. (1994). How does the potential get from A to B in a plasma? *IEEE Transactions on Plasma Science, 22*(1), 11–21.
54. Hershkowitz, N. (1995). Review of recent laboratory double layer experiments. *Space Science Reviews, 41,* 351–391.
55. Cohen, S. A., Siefert, N. S., Stange, S., Boivin, R. F., Scime, E. E., & Levinton, F. M. (2003). Ion acceleration in plasmas emerging from a helicon-heated magnetic-mirror device. *Physics of Plasmas, 10*(6), 2593–2598.
56. Sun, X., Cohen, S. A., Scime, E. E., & Miah, M. (2005). On-axis parallel ion speeds near mechanical and magnetic apertures in a helicon plasma device. *Physics of Plasmas, 12,* 103509.
57. Cohen, S. A., Sun, X., Ferraro, N. M., Scime, E. E., Miah, M., Stange, S., et al. (2006). On collisionless ion and electron populations in the magnetic nozzle experiment (MNX). *IEEE Transactions on Plasma Science, 34*(3), 792–803.
58. Charles, C., & Boswell, R. (2003). Current-free double-layer formation in a high-density helicon discharge. *Applied Physics Letters, 82*(9), 1356–1358.
59. Charles, C. (2007). A review of recent laboratory double layer experiments. *Plasma Sources Science and Technology, 16,* R1–R25.
60. Andrews, J. G., & Allen, J. E. (1969). Theory of double sheath between two plasmas. In *Proceedings of 9th International Conference Ioniz. Phenomena in Gases, 158.* Bucharest: Romania.
61. Andrews, G. J., & Allen, J. E. (1971). Theory of a double sheath between two plasmas. In *Proceedings of Royal Society, Vol. A320* (pp. 459–472).
62. Torven, S. (1981). Modified Korteweg-de Vries equation for propagating double layers in plasmas. *Physics Review Letters, 47*(15), 1053–1056.
63. Chen, F. F. (2006). Physical mechanism of current-free double layers. *Physics of Plasmas, 13,* 034502.
64. Beilis, I. I. (2019). Vacuum arc cathode spot theory: history and evolution of the mechanisms. *IEEE Transactions on Plasma Science, 47*(8), 3412–3433.
65. Beilis, I. I. (1990). The nature of an arc discharges with a mercury cathode in vacuum. *Soviet Technology Physics Letter, 16*(5), 390–391.
66. Beilis, I. I. (1991). Model of a vacuum arc spot on a mercury arc cathode. *High Temperature, 29*(1), 30–34.
67. Beilis, I. I. (1996). Current continuity and instability of the mercurv vacuum arc cathode spot. *IEEE Transactions on Plasma Science, 24*(4), 1259–1271.

Part II
Vacuum Arc. Electrode Phenomena. Experiment

Chapter 6
Vacuum Arc Ignition. Electrical Breakdown

Before an electrical discharge occurs, two electrodes are insulated by the vacuum from each other. In a vacuum, an arc can ignited when a conducting material will be appeared in the electrode gap, to which a voltage is applied. Various methods are possible to excite the conducting media in the gap [1, 2]. Three main methods can be indicated such as: (i) triggering of the arc using additional trigger electrode; (ii) initiation of the arc by contact breaking of the main electrodes; (iii) by electrical breakdown using of high-voltage supply. The first method can use the additional electrode as contacting mechanical trigger or as high-voltage trigger; the second is the contact method, while the third method supports the plasma excitation by influence of the strong electric field at electrode surface.

6.1 Contact Triggering of the Arc

6.1.1 Triggering of the Arc Using Additional Trigger Electrode

When an additional electrode is used, a local contact with one of main electrodes is occurred. As example, the arc can be initiated by momentarily touching the cathode with a mechanical trigger a tungsten electrode, attached electrically to the anode through a current limiting resistor [3]. This local contacting current bring to the cathode surface modification and arc ignition between the main electrodes. In addition, an explosion of a Cu wire placed between the cathode and the anode can trigger the arc [4]. The wire explosion occurs when the voltage applied to the main electrodes.

I. Beilis, *Plasma and Spot Phenomena in Electrical Arcs*, Springer Series on Atomic, Optical, and Plasma Physics 113, https://doi.org/10.1007/978-3-030-44747-2_6

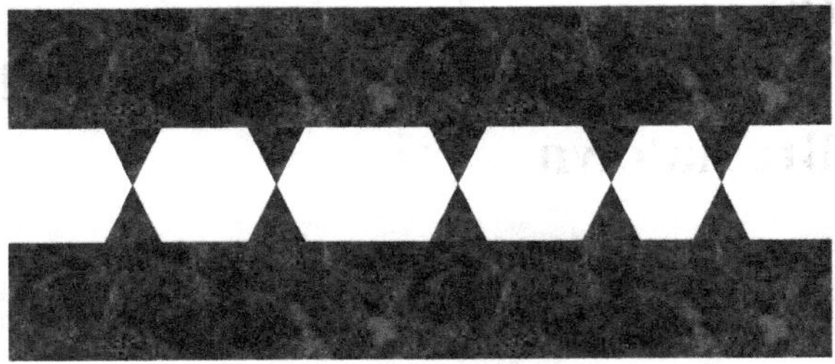

Fig. 6.1 Schematic presentation of the contact of not ideally smooth surfaces

6.1.2 Initiation of the Arc by Contact Breaking of the Main Electrodes

The contact area of two bodies depends on state of their surfaces. The surface state is determined by the surface roughness. Since the electrode surfaces cannot be ideally smooth, the contacting area is significantly lower than the electrode area. Schematically, the contact of roughness surfaces indicated in Fig. 6.1. When a local contacting current appeared, the current action brings to the cathode surface modification and resulting in an arc ignition between the main electrodes.

6.1.3 Contact Phenomena

Different phenomena (by bridging) appeared by contact closure or by contact breaking. It is heating, melting, bridging, evaporation, compression, restitution, and arcing at bouncing. The details of bridging phenomena can found in [1, 5, 6]. The parameter that usually characterized the contacting surfaces is "constriction resistance." According to Holm [1], a spherical model of the constriction resistance can be developed.

A circular contact surface with some semi-infinity contact materials was assumed (Fig. 6.2). The radius of the contact surface is b which is significantly smaller than the radius of the contact materials, B. Considering the constriction resistance of one of the materials (resistivity ρ), the resistance $\mathrm{d}R$ between the hemispheres with radius r and $r + \mathrm{d}r$ is

$$\mathrm{d}R = \frac{\rho \mathrm{d}r}{2\pi r^2} \tag{6.1}$$

The constriction resistance can be obtained by integrating (6.1) taking into account two hemispheres and that $b \gg B$:

$$R = \frac{\rho}{2\pi} \int\limits_{b}^{\infty} \frac{2\mathrm{d}r}{r^2} = \frac{\rho}{\pi b} \tag{6.2}$$

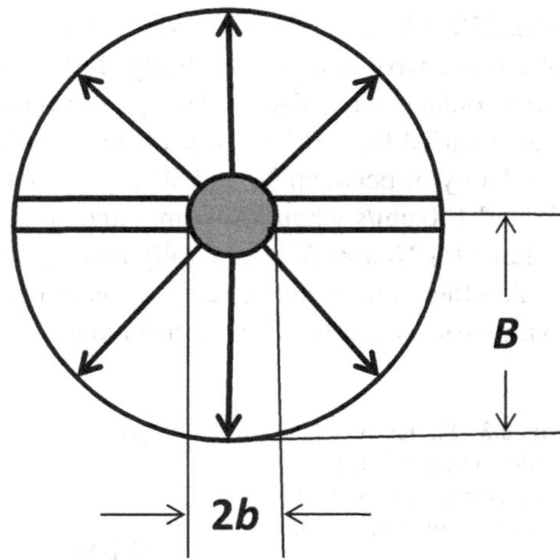

Fig. 6.2 Schematic illustration of the parallel current flow and of the spherical model of the constriction resistance

Using power I^2R dissipated in the constriction resistance with electrical current I, the following energy balance determines the temperature ΔT increasing:

$$I^2 Rt = m_R c \Delta T \tag{6.3}$$

Considering ideal spherical symmetry of the constriction volume ($4\pi b^3/3$) with material heat capacity c, mass density γ and (6.2) for R the heating time t of the constriction material is:

$$t = \frac{4c\gamma\pi^2 b^4}{3I^2\rho}\Delta T \tag{6.4}$$

Taking for Cu $\gamma = 8.9$ g/cm^3, $\rho = 1.6$ μOhm cm, $c = 0.38$ W/g/grad, $b = 10$ μm, $I = 10$ A the time is $2.8 \times 10^{-5}\Delta T/I^2$. When $I = 10$ A and the melting temperature 10^3C is reached through t of about 0.3 ms. Since the contact material is deformable during the heating, in reality the contact surface is more complicated by contact closure or breaking. The liquid metal and a metallic bridge were produced at the contact breaking. The bridge diameter, length, surface geometry, and the lifetime depend on power heating, rate of the contact breaking, and liquid material properties.

According to the review of [5], the bridge diameter obtained for Pt by high-speed photo registration in range 20–80 μm when the current increased from 20 to 80 A. The length is in range 10–100 μm and 0.2 cm for breaking velocity 20 cm/s and 0.1 cm/s, respectively. The liquid metal was transferred from one to other electrode by contact breaking.

Slade and Nahemow [7] investigated an initial separation of high current of rapidly opening contacts. Four pure metals, namely, Cu, Ag, W, and Ni, were investigated in air at atmospheric pressure and close to maximal arc current of 1 kA. Effect of the magnitude of the current on the formation time of the bridge investigated at 4,

310, 550, 790, and 900 A for Ni electrodes. Three voltage-time characteristics were observed, two of which show bridge formation and the third showed a rapid transition from contact to arc. The used image-converter camera with the exposure time of each frame varied from 0.1 to 10 μs (Fig. 6.3). In these experiments, the electrodes had a velocity of between 30 and 50 cm/s when the molten bridge formed and between 75 and 100 cm/s when it ruptured and an arc formed. It was shown that the melting voltage for Ni and W is generally much greater than the quasistatic regime.

Another important effect is associated with electrodynamic repulsion in symmetric contacts. The electrodynamic mechanism produces mechanical force

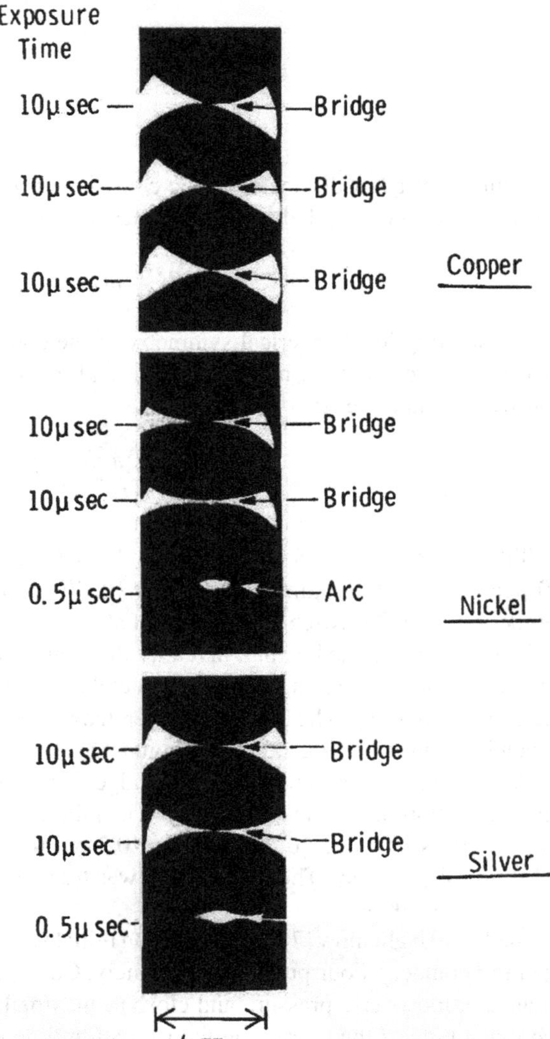

Fig. 6.3 Molten-metal bridge between Cu, Ni, and Ag electrodes. Figure taken from [7] with permission

generated by the interaction of the electric current with its own magnetic field which lines are concentric circles around the electrode cylinder axis [1]

$$H \sim \frac{I}{2\pi r} \tag{6.5}$$

The force change $d f_r$ of an element chosen at an angle with respect to cylinder axis is proportional to magnetic permeability μ_0:

$$\mathrm{d} f_r \sim \frac{\mu_0 I^2}{2\pi r} \mathrm{d}r \tag{6.6}$$

After integration (6.6) at constriction region, the force was obtained by Holm [1] is:

$$f_c \sim 10^{-7} I^2 \ln \frac{B}{b}\ (N) = 10^{-8} I^2 \ln \frac{B}{b}\ (kG) \tag{6.7}$$

The electrodynamic force reaches significant value for relatively large current at the contact point increasing up to about 5 kg when the current increased up to 5 kA [5]. This force decreased with number of contact points at the electrode surface because the circuit current distributed between the contact points. The electrodynamic force also depends on the mechanical contact load due to change of the number and geometry of the contact points at the surface.

According to Holm [1] in most cases for relatively long bridges, the hottest part of the bridge displaced toward the cathode leading to reversing the current. Therefore, the metal transfer was found to be directed from cathode to the anode. The bridging metal transfer is due to asymmetrical heating caused by three main effects including Thomson, Peltier, and Kohler effects discussed qualitative and quantitative by Holm [1].

Thomson effect means that heat is given off by current carriers coming from warmer to cooler regions (positive effect) of the current path in a conductor. The heat is proportional to the current and to the temperature difference as $\sigma_T I \Delta T$, where σ_T is Thomson coefficient that is determined experimentally.

Kohler effect associated with current by tunnel electron penetration through thin film at the contact surface. The electrons do not alter their energy level by tunneling. Since they land in an anode with a lower negative potential than the cathode the electrons have a surplus kinetic energy there. As result, the kinetic energy is given off as heat in the anode. In case of bridging, Kohler process was associated with assumption that an oxygen film deposits and chemisorbs on the contact material are remain in the closed contacts. As such during a subsequent opening the current density becomes high in the last contact area, the Kohler effect develops heat in the anode.

Peltier effect produces a heat at junction of different metals by thermoelectric process due to electrical current through two junctions and depends on the direction of current. The produced heat power is proportional to the current by Peltier coefficient, which is an effective voltage. In case of the liquid bridge, an opening contact has one boundary (1) against solid metal where the current enters the bridge and another boundary (2) where the current leaves the bridge. If Peltier effect positive, then (2) is heated and (1) is cooled. This means that the cathodic side becomes hotter than the anode side. A contribution of mentioned above different effects in the bridge thermal regime was discussed in [1, 5].

The experimental and theoretical study of the dynamic of arc phenomena including touch compression and bridging was presented at closure contacts in [8] and at opening contacts in [9–11]. Different initial stages of contact opening were described by the mathematical model considering the contacts before the melting, as well as evolution, and rupture of the liquid metal bridge, the arc ignition, and spreading in the bridge metallic vapors until the transformation of the metallic arc phase. The mathematical formulation was extended concerning the non-stationary phenomena and non-ideal electrical contact, when the contact opening is so rapid that the above Holm relationship will no longer be valid. Also a modification of approach is associated with a filament or roughness between the electrodes, connecting the contact with the filament resistance $R_f = 2\rho l/\pi r^2$, where $2l$ is the length of filament, that has to be added to the constriction resistance R. Further heating of this filament leads to its melting, creation of molten bridge, and arc ignition after bridge boiling. The lifetime, energy of each stage, contact voltage, and temperature were calculated. It is shown, that arc duration depends on opening velocity, contact scale roughness, energy, and length of liquid bridge.

Electrical contact at closure can appear due to breakdown when a strong electric field was produced at the electrode surface (gap width of 1–100 μm [6]) or due to touch and thermal destruction of a microspike and thermal atom ionization. One or other contact mechanism depends on the speed of closure electrodes. In experiment [8], the breakdown mechanism (gap voltage < 400 V) was observed for slow contact speed <0.1 m/s while at contact speed 0.5 m/s and greater the electrical contact was occurred by the mechanical touch of molten microspikes on the surface.

6.2 Electrical Breakdown

Electrical breakdown is an irreversible process, since a high-voltage gap will sharply drop to a much lower voltage and the insulating vacuum sharply changed to relatively high conductive medium in the interelectrode gap while current increasing. Despite that, the developed electrical arc is a relatively high-current discharge, which operates at low voltage, arc initiation requires much higher initial voltage, and an appropriate power supply will be required [12]. The required voltage is determined by phenomena of the trigger gap breakdown. Three main ways of triggered vacuum arc by electrical breakdown can be indicated as: i) electrical breakdown of the gap between the main electrodes; ii) electrical breakdown of the gap using additional triggered electrode; (iii) electrical breakdown at an insulator surface used in a triggered electrode. In essence, the first two ways have similar mechanisms of arc triggering while the last should be analyzed separately.

6.2.1 Electrical Breakdown Conditions

The presence of a gas in the gap is an important condition allowing the electrical breakdown development. When a voltage is applied such gas can be appeared due to the atom desorption. If the cathode is under an electric field and it is warm, the atom evaporation and electron emission take place also. When the gap pressure approaches a value so that the mean free path of electrons becomes less than the electrode gap distance d the conditions were created for avalanche ionization development. At the initial stage, when the gas is relatively cold the charged particle increase can be occur according to classical mechanism by the atom ionization due to electron acceleration in the applied voltage. The condition of breakdown development was obtained in following form [13].

$$N_{\text{ed}} = \frac{N_{\text{ed}} \exp(\alpha d)}{1 - \gamma\left[\exp(\alpha d) - 1\right]} \tag{6.8}$$

where N_{e0} is the initial electron flux at the cathode surface, α is the coefficient characterized the charge particle generation in the volume, γ is the coefficient characterized the charge particle generation at the cathode surface, N_{ed} is the amplified electron flux in direction to the anode. The electrical field of breakdown and transition to a spark in presence of a residual gas with pressure p depends on gap product pd according to Paschen law and the details can be found in [13].

In vacuum the breakdown development depends critically on specifics of the above mentioned coefficients, which characterize also the generation of a conducting vapor and maintaining a hot area located at the electrodes. The mechanism of electrical breakdown in a vacuum is determined by mechanism of the charged particles generation in the electrode gaps.

6.2.2 General Mechanisms of Electrical Breakdown in a Vacuum

The most studies of electrical breakdown for a fixed gap showed that the discharge could be initiated by the charged particle generation either at cathode or at the anode or by particles in the interelectrode gap. A review of the breakdown phenomena and historic review of the study development can be found in [2, 5, 14–17]. Below the main points of developed theories will be indicated.

The surface condition of the electrodes has important influence on vacuum breakdown. The surface microroughness is determined by mechanical treatment of the surface, it is heating or cooling. The surface microrelief can be significantly changed by prolonged application of a voltage as well as during the breakdown process. As result, a number of microirregularities including microprotrusions, whiskers, metal and dielectric particles, grooves, cracks and craters, characterizes the electrode surface microstructure.

The presence of the surface irregularities caused enhance of the local electric field E_l compared to the averaged field E determined by an applied voltage U to gap distance as U/d. The field enhancement is characterized by coefficient $\beta = E_l/E$. This coefficient depends on the protrusion geometry, electrode gap and is proportional to the ratio h_p/r_p, where h_p and r_p are the protrusion height and radius, respectively [16, 18, 19] and see Chap. 2.

The presence a strong electric field at the cathode surface stimulates use the field electron emission (F-emission) to explain the breakdown phenomena in a vacuum [20] and that the breakdown is determined primarily by conditions at the cathode. It was suggested [21] that breakdown involves a rupturing of the cathode surface under the action of local heating and mechanical strain associated with the electric field. Also, it was indicated that the highest electric field that could be applied to a tungsten cathode without breakdown occurring was about 4.7×10^6 V/cm. However, Chiles [22] observed for a number of metals that the luminosity always appeared at the anode before it appeared at the cathode. His evidence is given to show that positive ions from the anode with velocity in range from 5×10^5 to 8.9×10^5 cm/s have sufficient time to cross the gap during the average observed time intervals. This experimental result continues the existing problem related to the location of breakdown initiation. The problem was clarified using methods with high sensitivity measurements of the radiation brightness from the gap indicating that the observed brightness at the cathode side rather weaker than at the anode side [17].

A model developed using the secondary ion and electron emission suggested weak effect due to low coefficients of these processes. Another approach used weakly coupled and charged macroparticles at the cathode surface. These particles can be removed from the surface due to force of the strong electric field ~$E^2/8\pi$ or under heat action of the emitted electrons from the cathode. The model based on their acceleration in the gap with further transfer the surface energy $W_{sf} = \sigma_s U$ (where $\sigma_s = U/d$) to the anode, heating it and causes the surface vaporization. This approach was applicable to relatively long time of breakdown and cannot explain the phenomena occurred at short time (<10 ns).

The mechanism of breakdown initiation by field electron emission from the single crystal tungsten cathode was further demonstrated by experiments of Dyke with co-authors [23–25]. The following results can be summarized using suggested by the authors' experiments and the theoretical study (see also Chap. 2). The experiment showed effects of bright ring appearance and spontaneous growth of the emission current in time. The breakdown was initiated at a critical value of the field current density of the order of 10^8 A/cm^2 due to significant field enhancement by geometry factor. Above this limit value, an *explosive* vacuum arc [23–25] occurred between electrodes. At current density below the critical value, an electron emission was observed which apparently involved both high temperature and high electric field. The emitter temperature increased due to resistive heating of the microemitters. Electron micrographs of the surface microprofiles were used to obtain the emitter geometry, which allows calculating both the electric field and current density taking into account the temperature increasing. The volumetric electron space charge was taken in account and this effect found to be effective at a current density of the order of 10^7 A/cm^2. Evaporated emitter material appeared to be the source of the positive ions, which are required to neutralize space charge during the large increase in current accompanying the transition from field emission to vacuum arc. It was indicated that the bombardment of the cathode by ions formed at the anode or in the residual gas was judged negligible during an arc initiation.

In contrary, Chatterton's [26, 27] calculations of the critical fields for anode or cathode primary melting (and surface processes which become important at temperatures lower than the melting point) showed that both cathodic and anodic mechanisms of breakdown are possible. The work presents the results of calculations for different metals on the onset of breakdown due to field emission from an arbitrary protrusion at the cathode in an attempt to resolve the dilemma presented by the above work. The calculations take into account the effects of high-field intensification at the protrusions, Nottingham heating, electron backscattering and space charge. The anode temperature was calculated taking into account the extent electron beam at the anode, the radius of the bombarded area that depends on the electron velocity due to field emission, the shape of the field lines near the emitting protrusion at high-current densities, the space charge effects. The model gives the critical cathode field, which will cause melting at the cathode or anode. The anode or cathode melting can occur when fields exceed $(3 - 5) \times 10^7$ V/cm, which was in agreement with the experimental evidence. The differences between predicted fields for primary cathode and anode melting are small.

The further studies of the cathodic-anodic breakdown initiation dilemma were provided theoretically by Charbonnier et al. [28] and experimentally tested by Bennette et al. [29]. The experiments conducted for gap distances varied from a few tenths to a few thousandths of a centimeter at gap voltages up to 30 kV applied either continuously or in single pulses of 1–100 μsec duration. The breakdown mechanisms studied considering thermal processes initiated at both the anode and the cathode by the prebreakdown field-emitted electron current. The electric field was enhanced at the tip of microscopic cathode protrusions. Also, it was analyzed the mechanical processes resulting from yield of one of the electrode surfaces under the action of electrostatic stress produced by the electric field in the gap. In the case of thermal breakdown initiation, a numerical parameter γ_B was obtained. This parameter is a ratio where the numerator depends only on conditions at the protrusion tip and the denominator only on conditions at the anode. γ_B can indicate either an anode or a cathode-initiated arc depending on their thermal conditions.

The experiments [29] showed that for all electrodes (W, Mo, Cu), except aluminum, breakdown is caused by excessive field emission from the sharpest protrusion and consequent thermal instabilities. For aluminum, the electrostatic stresses induce irreversible changes at lower gap voltages than a voltage would be required for thermal instability, and it is cause the primary electrical breakdown. Joule heating in prebreakdown stage usually more localized at the cathode than at the anode. In case of pulsed gap voltages, short pulses tend to favor cathode initiation while long pulses favor anode initiation. The theoretical predictions [28] were in agreement with the results of the measurements.

Williams and Williams [30, 31] reported results of measurements of prebreakdown current as a function of electric field applied for molybdenum electrodes in a vacuum (10^{-9} Torr) for a fixed gap separation (0.05 cm). Two sets of electrodes were used: unpolished anode, polished cathode; and polished anode, unpolished cathode. The results indicate that the state of the anode surface governs the breakdown voltage and influence the microgeometry of the cathode surface. At fixed cathode surface, the surface finish of the anode plays an important part in determining the type of microgeometry obtained on the cathode after repeated sparking. Amplification factor with sparking indicates that the anode surface finish has a greater effect on breakdown voltage than the cathode surface finish. When the anode is either polished or unpolished, the protrusions for the first few breakdowns are conical at the polished cathode and the discharge was cathode initiated for the first few breakdowns. With a number of breakdowns, the cathode geometry eventually becomes cylindrical, indicating anode breakdown. Also, in a conditioned gap, both anode and cathode are involved in the discharge initiation. Anode initiation was obtained for a certain set of electrodes.

According to Lafferty [12] as the voltage across the gap is increased, a small current produced by F-emission begins to flow. When these electrons bombarded the anode, additional charged particles and radiation were produced. These in turn produce enhanced electron emission on the cathode. When the applied voltage exceeds a critical value, these effects become cumulative and breakdown occurs with the release of vapor from the electrodes. The breakdown then develops into a metal-vapor arc with the formation of a discharge. It indicated that the presence of both electrons and ions is an essential requirement for producing complete breakdown of the vacuum gap. This mechanism, however, difficult to applied for relatively large gaps when U/d can be low for enough electron emission.

To understand the breakdown evaluation, a luminescence of the interelectrode gap at the anode, cathode, and in the center of the gap was determined by means of a photomultiplier [32]. An apparatus with high resolution of ~1 ns was used, which allows establish a correspondence between the rise in current and the output of the radiation in the spark stage of breakdown. The luminescence location and the onset of the rise in current were compared during a DC spark vacuum breakdown.

The copper or molybdenum electrodes were used. The cathode made in the form of a hemisphere 10 mm in radius and the anode was a plane disk 20 mm in diameter with rounded edges and used a sharp needle as cathode and a plane as anode. It was reported that the luminescence at the cathode started at exactly the same moment as the rise in current. The luminescence at the anode appears on average 10 ns after the onset of the rise in current. This fact also agrees with the results of electron-optical investigations into the pulsed breakdown, which showed that evaporation from the anode takes place under the influence of bombardment by electron fluxes emitted from the cathode jets, this being a secondary process relative to the act of breakdown initiation. The influence of the anode material generation on the breakdown voltage increase was shown in different other experiments [5, 16].

6.2.3 Mechanisms of Breakdown Based on Explosive Cathode Protrusions

Basic principles of the Dyke's model [23–25] of cathodic field emission thermal instability and explosive destruction of cathodic microprotrusion creating the medium necessary for the subsequent development of the discharge were further developed by Fursey et al. [33–37] and Mesyats et al. [38–42]. They took in account that the plasma expands from the explosive protrusion. An intense critical field at the pointed cathode leads to critical field emission and to the cathode microinhomogeneity explosive destruction.

According to Fursey and Coworkers [34, 36], upon explosive destruction of the point (like to wire electrical exploding), a dense plasma sphere is formed which expands into the vacuum. This dense plasma also formed a positive space charge (in polarized plasmoid with minus in direction to the anode) providing a strong field at surface vicinity of the cathode points. The stronger field increased the field emission from the protrusion and this may lead to their destruction initiating new plasmoid. The process may be repeated several times, expanding in width around the surface and forming a plasma cloud near the cathode. As result, a melting layer on the cathode surface was produced and an extraction of new points occurs (under the action of ponderomotive forces) with their subsequent destruction on attaining a critical current density. Extraction and destruction cause of the electron emission and regeneration of the interelectrode medium necessary to the discharge development. It showed that the explosive emission current density distribution on the anode surface (obtained by field emission microscope) reflected the structure of the electron beam. The beam has a shape of symmetric spot in the form of rings. A similar form of the anode erosion was observed.

According to Mesyats group [16, 38–41] the mechanism of breakdown initiation was based on the electron explosive emission (see Chap. 2). The current increase in the beginning was observed by local luminescence, which is a plasma cloud named also cathodic flare. During the development current grows sharply and plasma flame expands at a high speed (~2×10^6 cm/s) in direction to the anode. The experiments and the models of Mesyats group concluded that in the sequence of the processes involved in a pulsed breakdown, the explosion of microprotrusions accompanied by the generation of cathode plasma and by the appearance of intense electron emission from the cathode that should be treated as a breakdown initiation. The local field was 10^8 V/cm and the local current density was over 10^8 A/cm^2.

The electrons emitted by the cathode plasma were accelerated in the vacuum gap and their energy passed to the anode surface (about 10^9 W/cm^2 during ~14 ns) [17].

As result, the energy was accumulated in the anode producing a flux of the anode material plasma, which can be in form of anodic flare. This flare was several order larger than expanding cathodic flare and therefore the anode luminescence larger. So, at the final stage of a discharge, the anode may be the principal supplier of the conducting material requested for the gap breakdown and discharge development [16, 17]. In the first stage of current increase (short time), an anode mass gain was observed due to a deposit from the cathode plasma fluxes [39]. The difference of breakdown at steady voltage from pulse breakdown is in the processes leading to cathode surface variation and its emission properties. The anode material transfer to the cathode in prebreakdown stage suggested to be considered as one of such processes.

In essence, the role of the anode plasma in the cathode–anode breakdown mechanism was published in the literature (see above). However, this mechanism and the field emission mechanism in the mentioned works were studied with more detail interpretation considering the different stages of breakdown by use the high-voltage technique of nanosecond pulse, high-speed oscilloscopic registration and electron-optical imaging [35]. The used technique and corresponding experimental results allowed develop a physical model for better understanding the evolution of the breakdown and transition to the vacuum arc.

Other investigations of vacuum electrical breakdown showed different mechanisms of gap triggering by critical vapor density that generated by the detachment of an anode macroparticle from the hottest region on the anode surface and its subsequent evaporation during its transit to the cathode [43] and by additional plasma injection of a hydrogen density [12]. The electrode polarity influence was considered where indicate support theories of electrical breakdown in vacuum which postulate that the initiating event occurs at the cathode [44].

6.2.4 Mechanism of Anode Thermal Instability

Thus, the above considered publications indicate two main mechanisms of electrical breakdown. One mechanism is limited by the material supply due the protrusion explosion. This mechanism is confirmed by the observation of the light emission from the cathode material vapor prior to the anode vapor in narrow gaps formed by electrodes of dissimilar metals with short impulse voltage. The second mechanism was supported by the directly anode evaporation due to anode local heating with the energy of accelerated cathode electrons in the gap. There is important parameter the time delay of the breakdown onset.

A study of vapor time evolution into the gap was carried out in the work of [45]. Measurements of the breakdown delay time following the application of the impulse in the range 1–200 μs for an electrode separation of 0.1 cm have been correlated with the optical resonance radiation, characteristic of each electrode material, emitted from the gap during the early stages of breakdown. The data have been obtained from three spatial regions: near the cathode, middle of the gap, and near the anode. The time delays of the breakdowns at was obtained for various overvoltages values that characterized by a coefficient K_{ov} which is a ratio of the impulse voltage amplitude to the DC breakdown voltage.

The results show that, when high overvoltages were applied ($K_{ov} > 1.3$), the breakdown delay times <20 μs are characterized by the initial emission of cathode vapor radiation, whereas for times 20–200 μs and $K_{ov} < 1.2$, anode vapor radiation is emitted prior to cathode vapor radiation. It concluded that a transition in the impulse breakdown mechanism occurs from one involving initiation in anode vapor to one in which initiation takes place in cathode vapor, as the magnitude of the applied voltage is increased above the DC breakdown value.

So, the cathode material generation was correlated with short-time delays of the breakdown, is in agreement with the model of explosive electron emission, whereas the anode material generation and longtime delays was proposed to understand on the basis of the thermal anode phenomena [46]. It is important to study the phenomena occurring at relatively large delay times, especially when also the impurity materials in the gap can be produced due to gas desorption from the anode surface.

A model of the anode thermal instability, which caused by local heating of the anode by beam of the field emission electrons and by the following exchange of ions and electrons between cathode and anode has been suggested [47–49]. It is shown that uncontrolled growth of the anode temperature T_a is possible in the case of the electron beam bombardment when simultaneous action of the above two processes is taken into account. It was also taken into account that ions are formed due to ionization of vapor of the anode material and desorption products [50]. The local anode temperature grows because of an additional heating of the anode due to an increase in the electron current, caused by the vapor ionization, and due to secondary processes on the cathode. Taking into account electro-ion exchange processes the heat flux density $q(T_a)$ was approximated by following expression [47]:

$$q(T_a) = \frac{q_{a0}}{\left[1 - \frac{\alpha\gamma}{T_a}\exp\left(-\frac{L_a}{T_a}\right)\right]}$$

where α, γ are the coefficients of ionization by electron impact and secondary ion-electron emission of the cathode material, respectively, q_{a0} is the heat flux density without anode vapor ionization, L_a is the specific heat of the anode vaporization. The heat model considers the anode as semi-infinite body and the heat flux distributed uniformly on the circular anode area with a given radius R_a. The appearance of thermal instability is controlled, on the one hand, by the ionization of the anode vapor and desorption material and the intensity of secondary processes on the cathode, and on the other hand, by heating and cooling of the anode. The temperature of the instability onset is close to the melting point of the anode material, which is in good agreement with experimental data [29].

A relation between $\alpha\gamma$ and $q_{a0}R_a/\lambda L_a$ determined the onset of anode thermal instability. The time delay was determined as duration between the moment of flux q_{a0} applied and the onset of the sharply temperature increase. This time for Cu anode and q_{a0} in typical range $10^4 - 10^5$ W/cm^2 [29] is indicated as $10^{-5} - 10^{-3}$ s and it is depends on an overvoltage, i.e., on the depth of beam penetration.

An experimental study of gas evaporation from the electrode surface and the decisive effect of these processes on prebreakdown characteristics of the gap and the vacuum breakdown initiation was demonstrated in [48, 50]. The total pressure before the beginning of the experiments was 8×10^{-10} Torr. The gap consisted of a molybdenum single-crystalline cathode and interchangeable cylindrical anodes: single-crystalline copper and molybdenum and polycrystalline nickel. The content of impurities in the materials employed was less than 10^{-5}%. The cathode made in the form of a rounded cone with a half-angle of 30°. The cathode diameter was 4.65 mm. The gap between the electrodes varied from 0.05 to 1.5 mm. A DC voltage up to 100 kV in 10 kV steps was applied across the vacuum gap.

According to mass-spectrometric data, an increase of the pressure in the chamber was caused by the growth of the concentration of all the components of residual gases, the growth of H_2, CO/N_2, CO_2, and water group dominating. Molecular hydrogen was dominated in the spectrum of residual gases. Experimental dependences $U_B = f(d)$, obtained for three anode materials are shown in Fig. 6.4. Thus, there is a clear tendency to the saturation of the curve $U_B = f(d)$ associated with decrease in the mean field intensity E_0, on the cathode surface with the increasing d. The dependence of $E_0(d)$ is obtained by simulating the electrical field [48]. As shown in Fig. 6.4 the dependences $U_B = f(d)$ for gaps, whose anodes are made of various materials, coincide almost completely. This coincidence was explained by the fact that impurities are desorbed from the surface of anodes of various materials with approximately the same intensity, rather than to the fact that the breakdown initiation is determined by the electrode material properties (ultimate tensile strength of Cu and Mo).

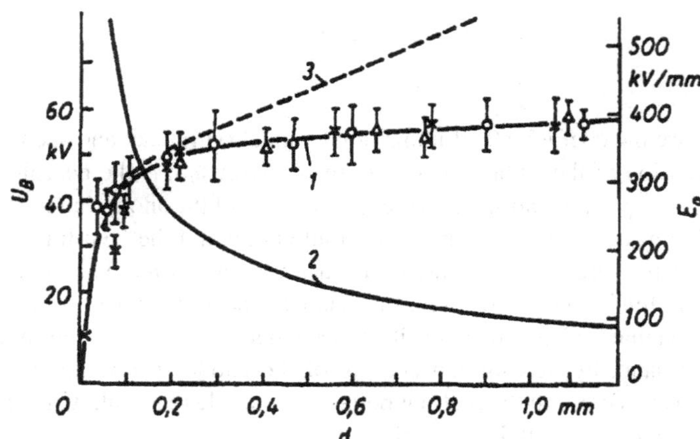

Fig. 6.4 To the total voltage effect. 1 experimental dependence $U_B = f(d)$, 2 electric field at the cathode, E_0 versus interelectrode gap d, 3 breakdown voltages reduced to plane parallel electrodes, x—Mo anode, Δ—Ni anode, o—Cu anode. Figures taken from [48] with permission 501550145

The authors [48] indicate that behavior of the dependence U_B on distance d is due to the fact that the anode local heating and thermodesorption are determined by the total energy of bombarding electrons, that is, by the total voltage and applied across the gap. It should be noted, if the thermal effects were important then the anode material properties should influence the breakdown phenomena because the heat- and thermoconductivity significantly different for such metals as Cu and Mo. However, this point was not discussed.

6.2.5 Electrical Breakdown at an Insulator Surface

The breakdown voltage of a vacuum gap is reduced by the presence of solid dielectric inclusion in the gap and it decreases with increase of the permittivity of the material. The electrical breakdown of a vacuum gap with a solid insulating material between the electrodes (flashover discharges) has been extensively investigated by various workers [2, 5, 17]. The initiation of the electrical breakdown at an insulator surface always takes place in the cathode in contact with the insulator [42, 51, 52]. The electric field strength was amplified because of the microroughness of the metal and dielectric surface in the contact area. The field amplification is proportional to dielectric permittivity ε. According to Bugaev [53] the amplified field is

$$E = -\frac{\varepsilon E_0}{\left[\frac{\varepsilon \delta_0}{\delta} + 1\right]}$$

where E_0, δ are the applied field and the dielectric thickness, δ_0 is the gap between dielectric and cathode. For conditions of the work [53], the field E at smooth surface is about 10^6 V/cm. For real cathode surface with roughness presence, the electric field reaches significant value, which causes explosion electrons emission. The results of spectral studies in the initial stage of discharge (up to 10 ns) showed protrusion explosion due to Joule heating and plasma producing with simultaneously heating of dielectric under electron bombardment, desorption of adsorbed gases, and evaporation of the ceramic material. The plasma propagates along the dielectric surface in the thin gas layer under the action of the tangential component of the electric field with subsequent ionization of the gas and vapor-like propagation of the streamer [13]. Therefore, for insulators with the large dielectric constant, the breakdown electric field was relatively low (~10^3 V/cm) in comparison with the vacuum gap breakdown.

Boxman [54] studied the mechanism of arc ignition depending on polarity of the main electrode. The main electrodes were 7 cm diameter copper disks. The gap could be varied using movable upper (independent) main electrode, while the lower main electrode was attached mechanically and electrically to the stainless steel chamber. The trigger gap consisted of two hollow copper cylinders insulated from each other by a ceramic disk. The upper cylinder connected electrically and mechanically to the lower main electrode. The lower cylinder used as the trigger anode (the trigger electrode) and was connected electrically to the trigger circuit. The trigger operated by applying a high potential across the ceramic disk causing a metal triggering arc. For positive polarity of the independent electrode, the trigger arc establishes its own cathode spots, which then supply current to the main discharge, dividing to form new spots as needed. For negative polarity of the independent electrode, the breakdown occurred due to high electric field in the thin sheath adjacent to the surface of this electrode due to plasma produced by the trigger arc and because all the potential difference dropped across the positive sheath.

Govinda Raju et al. investigated the triggering and breakdown properties including the time delay to the firing of a triggered vacuum gap containing dielectric materials of different resistance. The trigger gap as dependencies on the main gap voltage, main gap length, trigger pulse duration, trigger current, and trigger voltage determined for electrode materials as silicon carbide, steatite ceramic, boron nitride [55], and barium titanate dielectric [56]. Three electrodes used from copper for triggering the gap. The main electrodes were of 25 mm in diameter. Increasing either the trigger voltage or trigger current decreases the time delay. Variations of the main gap length and main gap voltage affect the time delay only slightly. Two groups of time delays, long (>100 μs) and short (<10 μs), were observed simultaneously before electrode "conditioning" while only short-time delays was present after conditioning. The presence of dielectric significantly decreases the triggering voltage. For example, for a fixed trigger voltage of 1 kV and trigger current of 28 A, the time delay decreases from 1.4 μs to 0.6 and then to 0.4 μs with increasing anode voltage, respectively, from 10 to 15 and 23 kV. The resistance of the dielectric is found to decrease with increasing trigger current and this is attributed to the deposition of metal vapor on the surface.

Kamakshaiah and Rau [57] studied a triggered vacuum gap for diameter of the main electrodes of copper in 45 mm with a dielectric gap (trigger gap) of 0.5 mm between the trigger electrode and the cathode. They noted that use a zirconated titanate as dielectric material has an extended life of 2000 firings without much noticeable deterioration of the electrical properties for main discharge currents up to 3 kA and is much superior to these made with silicon carbide used in the previous work. It was shown that probability of firing (determined as the ratio of the number of successful firings of the main gap to the total number of pulses applied to the trigger gap) increases with increase of trigger pulse energy, trigger pulse duration, main gap voltage and aging, but it decreases with increase of the main gap separation.

The threshold trigger voltage required for reliable firing appears to decrease with increase of the permittivity of the materials. It is 1.2 kV for zirconated titanate with $\varepsilon = 1750$, whereas this value is lower 0.725 kV for silicon carbide with lower $\varepsilon = 285$. Nevertheless, other investigators [57] also observed (in magnetic field or with polymer dielectrics) a reduction in breakdown voltage with increase of the permittivity of some insulators. In view of these results, the authors of work [57] conclude that it is difficult to say positively that permittivity has a dominant control over the trigger voltage.

Anders et al. [58] showed that the arc initiation can be provided by applying the relatively low voltage of the arc power supply to the cathode–anode-separating

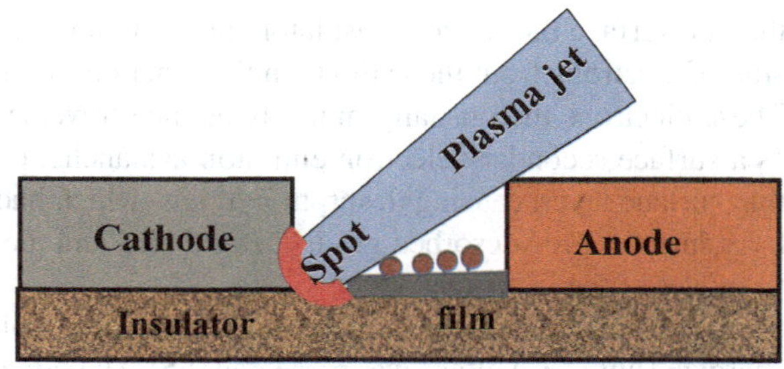

Fig. 6.5 Schematic presentation of the film deposited at the insulator on the vacuum arc triggering. Figure taken from [59]

insulator was coated with a conducting layer. This "triggerless" principle has been tested successfully with a large number of cathode materials.

Zolotuchin and Keidar [59], Teel et al. [60] demonstrated a pulsed microcathode triggering vacuum arc consisting of two rectangular electrodes on an alumina ceramic plate with a variable interelectrode gap. The insulator covered by a conductive film of carbon paint shown in Fig. 6.5. The effect of dielectric material, microroughness, gap and magnetic field on number of ignition was investigated. Overall over million pulses can be achieved as result of optimizing the interelectrode gap, the electrical power and the value of the magnetic field parallel to the film surface.

The coated boron nitride insulator was used for ignition of a vacuum arc with black body electrode assembly in [61, 62]. Material eroded by the arc from cathode spots (metallic plasma and MPs) was reflected from the hot anode, and the metallic plasma was generated in the vessel (Fig. 6.6). The closed vessel operated as a black body for the MPs while the plasma will be emitted through the anode apertures.

However, when the arc extinguished the plasma deposits the boron nitride (BN) shield. For first, an explosion of a Cu wire placed between the cathode and the anode triggered the arc and then the next arc ignition was triggered by explosion of the cathode material condensed on the BN insulator from the previous arcing.

Miller [63, 64] presented a review of flashover of insulators in vacuum. The voltage breakdown along the surface of insulators in vacuum and the theories relating to surface flashover and related experimental results were discussed. It concluded

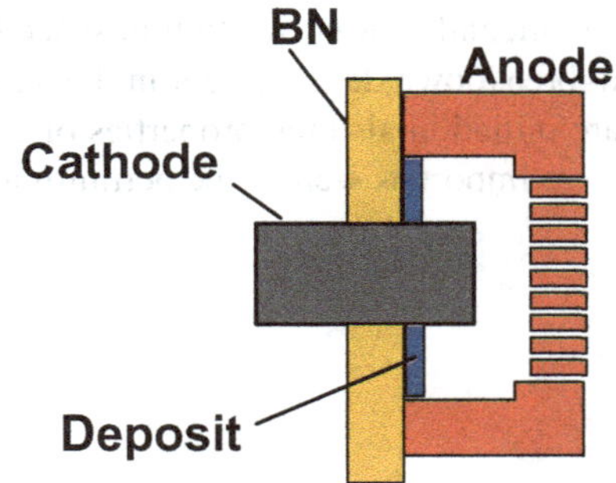

Fig. 6.6 Schematic presentation of cathode material deposited on the BN insulator (from previously burning of the arc), which serve for triggering the next arc in a vacuum

that the surface flashover of insulators in vacuum is generally initiated by the emission of electrons from the cathode in the junction of electrode-insulator in vacuum. These electrons then usually multiply as they traverse the insulator surface, either as a surface secondary electron emission avalanche, or as an electron cascade in a thin surface layer, causing desorption of gas, which had been adsorbed on the insulator surface. This desorbed gas is then ionized and the ionized gas leads to surface flashover of the insulator.

A dynamic model that computes the flashover voltages of polluted insulators energized with DC voltage was presented [65]. The salient feature of this model is that it takes into account the configuration of the insulator profile at every instant, which plays an important role in the flashover process of the DC polluted insulators. The relatively recent investigations of surface flashover discharges along an insulators can be considered in [66–68] including insulator high-voltage flashover with electrons produced by laser interaction [69].

6.3 Conclusions

Cathode mechanism explained pulse voltage breakdown initiation with short-time delay (~10^{-9} s) when anode cannot be heated. In some conditions (relatively long time), the amount of the material due to protrusion explosion is not enough for discharge development. Therefore, the authors of cathode mechanism initiation considered the further stage when the anode was heated by the electron beam from the explosion plasma supporting the necessary density of the medium for the discharge. The anode mechanism of breakdown initiation explained the breakdown with long-time delay and for statically applied voltage in gaps of d ~ 1 mm. The anode mechanism occurred by current density on the cathode protrusion lower than for cathode explosion emission mechanism. However, for all cases, the current density and the electric field at the surface should be $>10^8 - 10^9$ A/cm^2 and ~10^8 V/cm in order to support the cathode thermal instability leading to explosion of the protrusion and explosive electron emission. These conditions were required also in case of flashover breakdown along the insulator surface. Nevertheless, the necessary voltage can be significantly lower due to thin space layer at the contact cathode-dielectric and due to breakdown development in the desorbed material. When a dielectric was used to arc initiation also the properties of electrode metal deposit on the insulator surface were import as well as the permittivity for increasing the firing probability.

References

1. Holm, R. (2000). *Electrical contacts. Theory and Application.* Reprint of the four rewritten edition, 1967. Springer.
2. Lafferty, J. M. (Ed.) (1980). *Vacuum arcs: Theory and application.* Wiley: New York.
3. Beilis, I. I., Grach, D., Shashurin, A., & Boxman, R. L. (2008). Filling trenches on a SiO_2 Substrate with Cu using a hot refractory anode vacuum arc. *Microelectronic Engineering, 85,* 1713–1716.
4. Beilis, I. I., Koulik, Y., & Boxman, R. L. (2014). Cu film deposition using a vacuum arc with a black-body electrode assembly. *Surface & Coatings Technology, 258,* 908–912.
5. Rakhovskii, V. I. (1970). *Physical bases of the commutation of electric current in a vacuum. Translation NTIS AD-773 868 (1973) of Physical fundamentals of switching electric current in vacuum.* Moscow: Nauka Press.
6. Namitokov, K. K. (1978). *Electroerosion Phenomena.* Energy: Moscow (in Russian).
7. Slade, P. G., & Nahemow, M. D. (1971). Initial separation of electrical contacts carrying high currents. *Journal of Applied Physics, 42*(9), 3290–3297.
8. Kharin, S. N., Nouri, H., & Amft, D. (2005). Dynamics of arc phenomena at closure of electrical contacts in vacuum circuit breakers. *IEEE Transactions on Plasma Science, 33*(5), 1576–1581.
9. Kharin, S. N. (1996). Post bridge phenomena in electrical contacts at the initial stage. *IEEE Transactions on Components Packaging, and Manufacturing Technology, 19*(N3), 313–319.
10. Kharin, S. N. (2000). Influence of the pre-arcing bridging on the duration of vacuum arc. In *Proc.* XIXth *International Symposium on Discharges and Electrical Insulation in Vacuum* (Vol. 1, pp. 278–285). China: Xian.
11. Kharin, S. N. (1999). Modeling of transition arc phenomena in opening electrical contacts. In *Proceedings of International Conference on Electrical Contacts, Electromechanical Components and their Applications* (pp. 133–139). Japan: Nagoya.
12. Lafferty, J. M. (1966). Triggered vacuum gaps. *Proceedings of the IEEE, 54*(1), 23–32.
13. Raether, H. (1964). *Electron avalanches and breakdown in gases.* London: Butterworths.
14. Latham, R. V. (1978). Initiation of electrical breakdown in vacuum. *Physics in Technology, 9*(1), 20–25.
15. Latham, R. V. (Ed.) (1995). *High voltage vacuum insulation. Basic concepts and technological practice.* London, N.Y.: Academic Press.
16. Mesyats, G. A., & Proskurovskii, D. I. (1989). *Pulsed electrical discharge in vacuum.* Berlin: Springer.
17. Mesyats, G. A. (2000). *Cathode phenomena in a vacuum discharge: The breakdown, the spark and the arc.* Moscow: Nauka.
18. Nevrovskii, V. A., & Yaroslavskii, V. N. (1982). Dependence of the β coefficient from microscopic cathode projections on the width of the electrode gap. *Soviet Physics—Technical Physics, 27*(2), 180–183.
19. Miller, H. C. (1984). Influence of gap length on the field increase factor β of an electrode projection. *Soviet Physics—Technical Physics, 55*(1), 158–161.
20. Millikan, R. A., & Eyring, C. F. (1926). Laws governing the puling of electrons out metals by intense electric field. *Physical Review, 27*(1), 51–67.
21. Ahearn, A. J. (1936). The effect of temperature, degree of thoriation and breakdown on field currents from tungsten and thoriated tungsten. *Physical Review, 50*(1), 238–253.
22. Chiles, J. A. (1964). A photographic study of the vacuum spark discharge. *Journal of Applied Physics, 8*(9), 622–626.
23. Dolan, W. W., Dyke, W. P., & Trolan, J. K. (1953). Field emission: Large current densities, space charge, and the vacuum arc. *Physical Review, 89*(4), 799–808.
24. Dyke, W. P., Trolan, J. K., Martin, E. E., & Barbour, J. P. (1953). The field emission initiated vacuum arc. I. Experiments on arc initiation. *Physical Review, 91*(N5), 1043–1054.
25. Dolan, W. W., Dyke, W. P., Trolan, J. K. (1953). The field emission initiated vacuum arc. II. The resistively heated emitter. *Physical Review, 91*(N5), 1054–1057.

26. Chatterton, P. A. (1966). A theoretical study of field emission initiated vacuum breakdown. *Proceedings of the Physical Society of London, 88*(1), 231–245.
27. Chatterton, P. A. (1966). Further calculations on field emission initiated vacuum breakdown. *Proceedings of the Physical Society of London, 89*(1), 178–180.
28. Charbonnier, F. M., Bennette, C. J., & Swanson, L. W. (1967). Electrical breakdown between metal electrodes in high vacuum. *Journal of Theoretical and Applied Physics, 38*(2), 627–633.
29. Bennette, C. J., Swanson, L. W., & Charbonnier, F. M. (1967). Electrical breakdown between metal electrodes in high vacuum. II. Experimental. *Journal of Applied Physics, 38*(2), 634–640.
30. Williams, D. W., & Williams, W. T. (1973). Initiation of electrical breakdown in vacuum. *Journal of Physics. D. Applied Physics, 6*(6), 734–743.
31. Williams, D. W., & Williams, W. T. (1972). Effect of electrode surface finish on electrical breakdown in vacuum. *Journal of Physics D: Applied Physics,* 5(N10), 1845–1854.
32. Yurike, Y. Y., & Puchkarev, F., & Proskurovskii, D. I. (1973). Observation of the luminescence between electrodes during the raise in the spark current in vacuum breakdown at constant voltage. *Soviet Physics Journal, 16*(3), 293–297 [Translated from *Izvestiya VUZ, Fizika* N3, 12–16 (1973)].
33. Fursey, G. N., & Vorontsov-Vel'yaminov, P. N. (1968). Qualitative model of initiation of a vacuum arc. I. Breakdown mechanism. *Soviet Physics. Technical Physics, 12*(N10), 1370–1376.
34. Fursey, G. N., Vorontsov-Vel'yaminov, P. N. (1968). Qualitative model of initiation of a vacuum arc. II. Field emission mechanism of a vacuum arc onset. *Soviet Physics. Technical Physics, 12*(N10), 1377–1382.
35. Fursey, G. N., Kartsev, G. K., Mesyats, G. A., Proskurovskii, D. I., & Rotshtein, V. P. (1969). Field emission initiated vacuum arc in an extremely strong field at high current densities. In *IX International Conference on Phenomena in Ionized Gases*. Bucharest, Romania, 88.
36. Fursey, G. N., Zhukov, V. M., Gulin, B. F., & Ventova, I. O. (1972). Experimental studies of the transmission process from field emission to vacuum breakdown. In *Proceedings* of *V International Symposium on Discharges and Electrical Insulation in Vacuum* (pp. 43–46). Poznan, Poland, September 1972.
37. Fursey, G. N. (1985). Field emission in vacuum breakdown. *The IEEE Transactions on Dielectrics and Electrical Insulation, E-20*(N4), 659–679.
38. Bugaev, S. P., Iskol'dskii, A. M., Mesyats, G. A., & Proskurovskii, D. I. (1967). Electron-optical observation of initiation and development of a pulsed breakdown in a narrow vacuum gaps. *Soviet Physics—Technical Physics, 12*(12), 1625–1627.
39. Mesyats, G. A., Bugaev, S. P., Proskurovskii, Eshkenazi, D. I., & Yurike, Y. Y. (1969). Study of the initiation and development of a pulsed breakdown of short vacuum gaps in the nanosecond time region. *Radio Engineering* and *Electronic Physics, 14,* 1919–1925 (*Radiotechnika and Electronika, 14*(12), 2222–2230).
40. Mesyats, G. A. (1971). The role of fast processes in vacuum breakdown. In *X International Conference on Ionized Gases*. Oxford, England, September 13–18, pp. 333–357.
41. Mesyats, G. A., & Proskurovskii, D. I. (1971). Explosive emission electrons from metallic needles. *Soviet Physics—JETP, 4*(4–6) (*Pis'ma Zh. Exp. Teor. Fiz, 13*(N1), 7–10).
42. Bugaev, S. P., Litvinov, E. A., Mesyats, G. A., Proskurovskii, D. I. (1975). Explosive emission of electrons. *Soviet Physics Uspekhi, 18*(N1), 51–61. [*Usp. Fiz. Nauk, 115*(N1), 101–120 (1975)].
43. Davies, D. K., & Biondi, M. A. (1968). The effect of electrode temperature on vacuum electrical breakdown between plane-parallel copper electrodes. *Journal of Applied Physics, 39*(7), 2979–2990.
44. Miller, H. C., & Farrall, G. A. (1965). Polarity effect in vacuum breakdown electrode conditioning. *Journal of Applied Physics, 36*(4), 1338–1344.
45. Yen, Y. T., Turna, D. T., & Davies, D. K. (1984). Emission of electrode vapor resonance radiation at the onset of impulsive breakdown in vacuum. *Journal of Applied Physics, 55*(9), 3301–3307.
46. Nevrovskii, V. A. (1978). Thermal anode instability in the prebreakdown stage of vacuum breakdown. *Soviet Physics—Technical Physics, 23,* 1317–1321.

47. Nevrovskii, V. A. (1982). Dynamics of anode thermal instability development s electrical breakdown in a vacuum. *Bulletin of the Account of Science of SIB Brunch, Service Technology Science, N2*(1), 130–132; Nevrovskii, V. A. (1982). Development of thermal instability of the anode in the prebreakdown stage of vacuum breakdown. *Soviet Physics. Technical Physics, 27*(N2), 176–179.
48. Nevrovskii, V. A., Rakhovsky, V. I., & Zhurbenko, V. G. (1983). Dc breakdown of ultra-high vacuum gaps. *Beitr Plasmaphys, 28*(4), 433–458.
49. Nevrovskii, V. A., & Rakhovskii, V. I. (1986). Electrode material release into a vacuum gap and mechanisms of electrical breakdown. *Journal of Physics D: Applied Physics, 60*(1), 125–129.
50. Nevrovskii, V. A., Rakhovsky, V. I., Zhurbenko, V. G., & Zernaev, V. A. (1981). Gas desorption from electrodes and electrical breakdown in vacuum. *Journal of Physics D: Applied Physics, 14,* 215–220.
51. Bugaev, S. P., & Mesyats, G. A. (1966). Nanosecond time development of a pulsed discharge at a dielectric vacuum interface. *Soviet Physics. Technical Physics, 10*(7), 930–932.
52. Bugaev, S. P., & Mesyats, G. A. (1968). Investigation of the pulsed breakdown mechanism at the surface of a dielectric in vacuum. I. Nonuniform field. *Soviet Physics. Technical Physics, 12*(N10), 1363–1369.
53. Bugaev, S. P., Iskol'dskii, A. M., & Mesyats, G. A. (1968). Investigation of the pulsed breakdown mechanism at the surface of a dielectric in vacuum. I. Uniform field. *Soviet Physics. Technical Physics, 12*(N10), 1358–1362.
54. Boxman, R. L. (1977). Triggering mechanisms in triggered vacuum gaps. *IEEE Transactions on Electron Devices, ED-24*(N2), 122–128.
55. Govinda Raju, G. R., Hackam, R., & Benson, F. A. (1976). Breakdown mechanisms and electrical properties of triggered vacuum gaps. *Journal of Applied Physics, 37*(N4), 1310–1317.
56. Govinda Raju, G. R., Hackam, R., & Benson, F. A. (1977). Time delay to firing of a triggered vacuum gap with barium titanate in trigger gap. *Proceedings of IEE, 124*(N9), 828–831.
57. Kamakshaiah S., & Rau, R. S. N. (1977). Low voltage firing characteristics of a simple triggering vacuum gap. *IEEE Transactions on Plasma Science, PS-5*(N3), 164–170.
58. Anders, A., Brown, I. G., MacGill, R. A., & Dickinson, M. R. (1998). 'Triggerless' triggering of vacuum Arcs. *Journal of Physics D: Applied Physics, 31*(N5), 584–587.
59. Zolotuchin, D., & Keidar, M. (2018). Optimization of discharge triggering in micro-cathode vacuum arc thruster for CubeSats. *Plasma Source Science and Technology, 27,* 074001.
60. Teel, G., Fang, X., Shashurin, A., & Keidar, M. (2017). Discharge ignition in the micro-cathode arc thruster. *Journal of Applied Physics, 121,* 023303.
61. Beilis, I. I., Shashurin, A., & Boxman, R. L. (2008). Anode plasma plume development in a vacuum arc with a "black-body" anode–cathode assembly. *IEEETransaction on Plasma Science, 36*(N4), 1030–1031.
62. Beilis, I. I., Koulik, Y., & Boxman, R. L. (2014). Cu film deposition using a vacuum arc with a black-body electrode assembly. *Surface and Coatings Technology, 258,* 908–912.
63. Miller, H. C. (1989). Surface flashover of insulators. *IEEE Transactions on Electrical Insulation, 24*(5), 765–786.
64. Miller, H. C. (1993). Flashover of insulators in vacuum. Review of the phenomena and techniques to improve holdoff voltage. *IEEE Transaction on Electrical Insulation, 28*(N4), 512–527.
65. Sundararajan, R., & Gorur, R. S. (1993). Dynamic arc modeling of pollution flashover of insulators under dc voltage. *IEEE Transaction on Dielectric & Electrical Insulation, 28*(N2), 209–218.
66. Benwell, A., Kovaleski, S. D., & Gahl, J. (2007). Dielectric flashover with triple point shielding in a coaxial geometry. *Review of Scientific Instruments, 78*(N11), 115106.
67. Zhan, J.-Y., Mu, H.-B., Zhang, G.-J., Huang, X.-Z., Shao, X.-J., & Deng, J.-B. (2012). Cathode-like luminescence from vacuum-dielectric interface induced by self-stabilizing secondary electron emission. *Applied Physics Letters, 101*(N4), 041604.

68. Nakano, Y., Kojima, H., Hayakawa, N., Tsuchiya, K., Okubo, H. (2014). Pre-discharge and flashover characteristics of impulse surface discharge in vacuum. *IEEE Transactions on Dielectric & Electrical Insulation, 21*(N1), 403–410.
69. Krasik, Y. E., & Leopold, J. G. (2015). Initiation of vacuum insulator surface high-voltage flashover with electrons produced by laser illumination. *Physical Plasmas, 22*(N8), 083109.

Chapter 7
Arc and Cathode Spot Dynamics and Current Density

In this chapter, the arc definition and the experimental data characterizing the cathode spot including arc voltage, cathode voltage, threshold (minimal) arc current, spot definition, and problem of spot dynamics and the current density are described.

7.1 Characteristics of the Electrical Arc

7.1.1 Arc Definition

Usually, the electrical discharge is defined by volt–current characteristics changed at arc stage by transition from glow discharge. The electrode gap voltage was dropped from about 400–600 V and mA's for current range (glow) to about 20–30 V and currents >1 A in 1 Torr argon discharge. In addition, local luminous hot areas were observed. The last fact indicates that production of the charged particles changed from the sputtering in glow discharge (large voltage) at whole cathode surface to the phenomena arising at localized cathode areas. As it was described below, these phenomena include the electron emission and plasma generation due to intense local cathode heating by the energy flux produced by the plasma interaction with the cathode, and therefore, the arc can be supported by low gap voltage.

The methods of direct electrical arc initiation in a deep vacuum were described in Chap. 6: (i) high-voltage electrical breakdown, (ii) arcing contacts, and iii) by momentary contact of an additional small triggering electrode. The contact phenomena were studied by Holm [1] (see Chap. 6). There the main point is significantly lower contact area with respect to the real contact area due to surface irregularities. Therefore, the constriction resistance was located in the immediate neighborhood of small spikes. Under electrical current, the electrode's locally constricted areas were heated forming metallic bridges that melt and vaporize electrode material [2]. The bridging phenomena were the transient stage for arc initiation in breaking contact producing the metallic plasma. The main purpose of the listed methods is

I. Beilis, *Plasma and Spot Phenomena in Electrical Arcs*, Springer Series on Atomic, Optical, and Plasma Physics 113, https://doi.org/10.1007/978-3-030-44747-2_7

producing the primary plasma from the cathode material in order to initiate arc and then develop it. Therefore, the vacuum arc was operated in an electrode vapor, and it discharges high current (relative to glow). *Thus, the vacuum arc was defined as a relatively high-current discharge with low gap voltage, local heating of the cathode, and which is operated in vapor from the electrode material.* The oscillated voltage and the arc stability depend on the current.

7.1.2 Arc Instability

Usually, the arc is an unstable discharge, and the arc voltage fluctuates. Kesaev [3] widely investigated the arc stability, i.e., arc duration that is limited due to its spontaneous extinguish for mercury in detail [4] and for low-melting cathode materials (Bi, Pb, Zn, In, Sn, Ga, Al, and Cu) [3, 5, 6]. The relatively large number of the experiments showed stochastic character of the arc duration, which is described by following function

$$N_t = N_\vartheta \exp(-t/\theta) \tag{7.1}$$

where N_t is the number of readings with arc time larger than t and θ is the mean arc duration. It was obtained that the arc stability (the mean arc duration θ in (7.1)) increased with arc current, magnetic field, and surrounding gas pressure. The arc stability for a solid cathode at moderate currents (10–50 A) was significantly higher than the stability of a liquid cathode arc (except Al cathode). The results of arc stability were explained in frame of proposed so-called "internal instability of metallic arcs" [7, 8]. According to the internal instability, the stationary arc operation was considered as a number of old disappeared and new regenerating of a cathode spots, when $\mathrm{d}I/\mathrm{d}t < 10^7$ A/s.

Figure 7.1 demonstrates amplitude of the oscillations which significantly decreased when current rise from 2 to 10 A for cathodes from Ag, Cu, and Au [3]. Some larger oscillations for current 25 A were measured also by Nazarov et al. [9]. The result [9] was obtained for plane Cu cathode (25 × 40 mm) and 1 mm W anode separated by 1 mm.

The fluctuations of the arc current, ion current, light intensity, and arc voltage of a single cathode spot in a vacuum arc were investigated by [10, 11]. Cleaned Cu cathode and anode with diameter of 10 mm [10] and 30 mm in diameter and 3 mm thickness [11] were used. A circular current probe of 10 mm in diameter was placed at distance of 36 mm from cathode. DC arc initiated after electrode separation, when an electrode distance of 5 mm was reached. Measurements of light intensity were carried by observation of the cathode spot in a direction, normal to the cathode. Time integrated spectroscopy used to determine the spectral composition of the light emitted by the cathode spot.

The spontaneous fluctuations of interelectrode voltage at arc current 50 A and rate of current rise in the order of $\mathrm{d}I/\mathrm{d}t = 5 \times 10^6$ A/s were observed showing relatively

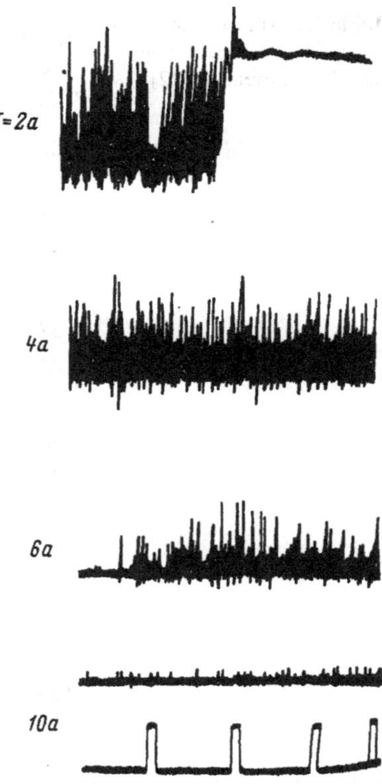

Fig. 7.1 Waveform of arc oscillations for Au electrodes at different arc current. The scale of marks (down) indicates amplitude of 50 V and frequency of 10 kHz [3]

stable discharge with variations of only some volts. However, lower currents (<40 A) and voltage peaks of some hundreds of volts characterize the occurrence of instabilities. It has shown that voltage fluctuations occur simultaneously with an excess of mass production at the cathode spot causing a synchronous increase in light intensity and an increase in probe ion current.

So, the amplitude and frequency of the oscillation for low currents can depend on discharge conditions that follow considering the results of above different experiments and presented in Fig. 7.1. The arc stability and amplitude of oscillations depend also on the rate of current rise dI/dt. The oscillation amplitude increases with the rate of current rise and decreases with arc current. At dI/dt about of 10^8 A/s (characterized a spark discharge), a spontaneous formation of new spots was observed.

Paulus et al. [12] investigated the transition of a vacuum arc from one steady state to another by superposition of current step for cylindrical cathodes from Cu, Al, Pb, and Zn of 25 mm diameter and 5 mm electrode gap. It was observed that at low rate of current rise (dI/dt = 5×10^6 A/s), the magnitude of oscillations is relatively weak and the arcing voltage slightly increased. At high rate of current rise (dI/dt > 10^8 A/s), a significant increase in the interelectrode voltage can be produced. This voltage depends on initial arc current I_{in}, current step ΔI, and transition time Δt. The

Table 7.1 Transient arc voltage U_{di} as function on initial arc current I_{in} [12]

I_{in} (A)	ΔI (A)	U_{di} (V)	$\mathrm{d}I/\mathrm{d}t \times 10^8$ A/s
28	150	235	7.5
28	100	180	5.0
28	50	40	2.5
32	150	70	7.5
32	100	63	5.0
37	150	70	7.5
37	100	40	5.0
37	50	29	2.5
42	150	42	7.5
42	100	33	5.0
45	150	40	7.5
45	100	34	5.0
50	150	37	7.5
50	100	30	5.0
50	50	25	2.5
80	200	40	10
80	150	32	7.5
120	50	25	2.5

transient arc voltage U_{di} dependence for Cu cathode and $\Delta t = 0.2$ μs was presented in Table 7.1. As follows from the results, the value of U_{di} decreases with I_{in}, while U_{di} increases with larger values of $\mathrm{d}I/\mathrm{d}t$. Also the value of U_{di} increases with melting temperature of cathode material, Al (350 V, 15 A), Pb (500 V, 5 A), and Zn (2100 V, 5 A) for $\mathrm{d}I/\mathrm{d}t = 7.5 \times 10^8$ A/s.

7.1.3 Arc Voltage

The electric field in the electrode gap of an arc was not distributed uniform and is substantially large at the cathode surface, while it is relatively low in the expanding cathode plasma. The potential distribution is schematically shown in Fig. 7.2. The

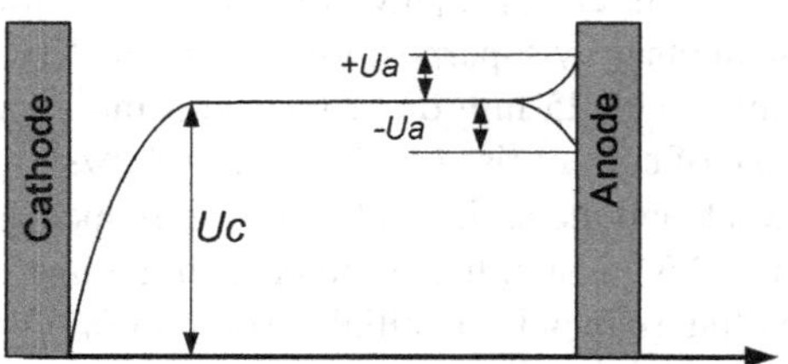

Fig. 7.2 Schematic potential distribution between cathode and anode in an arc

large part of the arc voltage is dropped in the adjacent to the cathode sheath named ***cathode potential drop*** u_c [3, 13, 14]. A small part of potential drop is in the conductive gap plasma and small positive or negative ***anode potential drop*** u_a (depends on arc current and plasma density and gap distance) in the anode region.

Early Duddell 1904 [15] showed that when the arc is maintained at constant current I, the potential difference u_{arc} between electrodes can be represented by an equation $u_{arc} = e + RI$, where e is the ponderomotive force and R is the resistance. Duddell studied experimentally the arc resistance and electromotive force and found that the arc parameters had a statistical character even at constant conditions.

During last few decades, in general, the total arc voltage u_{arc} was measured considering the minimal level at the oscillogram of voltage oscillations [3, 4]. This voltage consists of several parts, one of which corresponds to the space charge sheath u_c and determines the main cathode processes (heat, ionization, etc.). The other parts corresponds to the potential drop in the near-cathode dense plasma u_{pl} (short region that is not shown in Fig. 7.2) and in the anode region u_a. So, it is:

$$u_{arc} = u_c + u_{pl} \pm u_a \tag{7.2}$$

The voltage u_{arc} significantly depends on the arc current I at its threshold level, in moderate range (10–100 V) and at high-current level of few kA and higher. Figure 7.3 illustrates the voltage u_{arc} dependence in wide range of arc current I shown by Mitchel [16].

According to Mitchel [16], the voltage u_{arc} is practically weakly changed up to arc current of about 1 kA. The voltage u_{arc} in this current range is largely independent of separation and electrode geometry, and is determined by the cathode material. For copper, the mean value is 20 V, and oscillations of 1–2 V peak to peak at frequencies from a few kilohertz to a few megahertz have been observed on this mean value. From 1 to 6.5 kA, significant increases are apparent, and the arcing voltage rises linearly with current from 20 to about 40 V at 6.5 kA. For currents in excess of 7 kA, a grossly unstable voltage occurs; the mean value of this voltage is almost proportional to the current and may give arcing voltages in excess of 120 V.

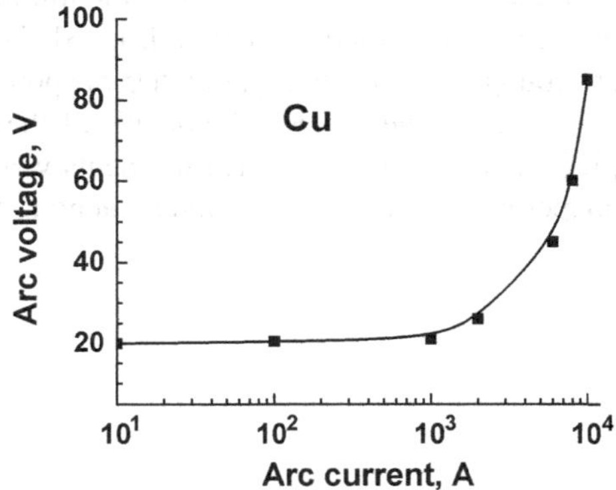

Fig. 7.3 Arc voltage as function on arc current for 25 mm diameter Cu electrodes at 5 mm separation

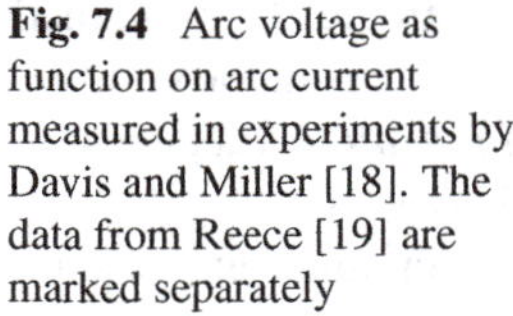

Fig. 7.4 Arc voltage as function on arc current measured in experiments by Davis and Miller [18]. The data from Reece [19] are marked separately

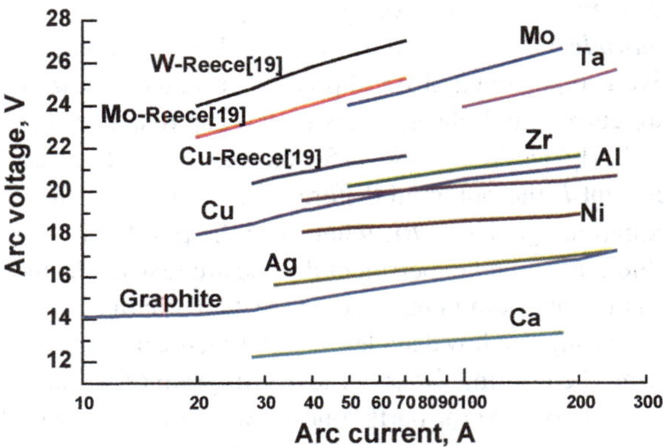

The arc voltage u_{arc} was widely investigated, see [17] as an example. Figure 7.4 demonstrates the relatively weak dependence on arc current in range lower than 250 A measured by Davis and Miller [18] and by Reece [19]. It can be seen that u_{arc} linearly increased with logarithm of current I. The measurements for current range $I \leq 1$ kA are presented in Table 7.2. Some dispersion of the value of u_{arc} obtained in different experiments was observed. It could be understood taking into account that different conditions of arc burning can be occurred including arc time, cathode surface state, current, etc.

The arc voltage dependence on current for high-current arcs is presented in Fig. 7.5 [22]. It can be seen that u_{arc} mostly linearly increased with current up to 6–7 kA for different cathode materials.

7.1.4 *Cathode Potential Drop*

The voltage–current characteristic was used to study the cathode potential drop. The experiment showed that at low arc currents ($I < 10$ A), the arc voltage weakly oscillates and peaked above the some minimal value. The minimal arc voltage was defined as the cathode potential drop [3]. This value was obtained by measuring the arc voltage for very short gaps when the potential drop in the plasma and in the anode region can be neglected. The details of the experimental methods can be found in [4, 5], and the results of measurements were presented in Table 7.3. In the case of copper bulk cathode, the cathode potential drop [3] $u_c = 15–16$ V.

Table 7.2 Arc voltage measured in different works

Cathode element	Arc voltage, u_{arc} (V)								
	Anders [20] $I = 200–300$ A	Brown [21] $I = 200$ A	Agarwal [22] $I < 1$ kA	Daalder [23] $I = 40–200$ A	Plutto [24] $I = 100–300$ A	Paul [25]	Kantsel' [26] $I = 700$ A	Kimblin [27] $I = 100$	Garcia [28]
Li	23.5	–	–	–	–	9.0	–		
Na	–	–	–	–	–	7.0	–		
K	–	–	–	–	–	–	–		
Cs	–				–	–	–		
Cu	22.7	20.5	19.5	18–20.0	19.2–20.0	15.5	20.0	20	16.3
Ag	22.8	19.0	–	17.5	16.5–17.8	12.5	–	20	
Au	19.7	18.0	–	–	–	12.9	–		
Bi	14.4	–	9.0	–	–	7.5	–		
Mg	18.6	16.0	12	–	13.4–15.0	11.0	–		
Ca	20.5	–	–	–	10.3	8.0	–		
Sr	18.5	–	–	–	8.8	7.0	–		
Y	19.9	–	–	–	–	–	–		
Zn	17.1	14.0	10.0	12	12–12.5	9.1	–	13	
Ge	20.0	–	–	–		–			
Cd	14.7	–	9.0	–	10.5–11.1	8.1		11	
Hg		–	–	–	–	7.5	–		
Er	19.2	–	–	–	–	11.5	–		
Al	22.6	20.5	18.0	20.0	20–20.8	15.5			
Si	21.0	–	–	–	–	–	–		
In	16.0	14.0	–	–	–	9.5	–		

(continued)

Table 7.2 (continued)

Cathode element	Arc voltage, u_{arc} (V)								
	Anders [20] I = 200–300 A	Brown [21] I = 200 A	Agarwal [22] I < 1 kA	Daalder [23] I = 40–200 A	Plutto [24] I = 100–300 A	Paul [25]	Kantsel' [26] I = 700 A	Kimblin [27] I = 100	Garcia [28]
V	22.7	–	–	–	–	–	–		
Ti	22.1	20.5	–	–	–	15.5	–	20	15.0
Zr	22.7	22.0	–	–	–	18.5	–		16.3
Gd	20.4	16.5	–	11.0	–	11–11.5	–		
Tb	19.6	–	–	–	–	–	–		
Dy	19.8	–	–	–	–	11.5	–		
Ho	20.0	18.0	=	–	–	–	–		
Hf	23.3	–	–	–	–	15–20.0	–		
C	31	–		–	–	12.0	–	20	33.3
Sn	17.4	14.0	12.5	13.5	–	10.5	–		
Ba	16.5	–	–	–	–	6.3–6.5	–		
La	18.7	–	–	–	–	10.5	–		
Ce	17.6	–	–	–	–	–	–		
Pr	20.5	–	–	–	–	–	–		
Nd	19.2	–	–	–		11.5			
Sm	18.8	–	–	–	–	–	–		
Pb	17.3	12.5	–	10.5	10.3	8.0	–		
V	22.7		–		–	–	–		
Nb	27.9	25.5	–		–	19–24.0	27.0		
Ta	28.6	24.5	–		–	20–23.0	–		

(continued)

Table 7.2 (continued)

Cathode element	Arc voltage, u_{arc} (V)								
	Anders [20] I = 200–300 A	Brown [21] I = 200 A	Agarwal [22] I < 1 kA	Daalder [23] I = 40–200 A	Plutto [24] I = 100–300 A	Paul [25]	Kantsel' [26] I = 700 A	Kimblin [27] I = 100	Garcia [28]
Bi	14.4	–	–		–	8.4–8.7	–		
Cr	22.7	20.0	–		–	15.5	–	18	
Mo	29.5	24.5	–	25–26.5	–	23.0	28.0	28	25.1
Ru	23.8		–		–	–	–		
Pd	23.5	18.0	–		–	16.0	–		
Rh	23.8	24.0	–		–	–	–		
W	28.7	28.0	−1	28.0	–	23.2	–	25	40.1
Ir	25.5	–	–		–	–	–		
Fe	21.7	20.5	–		–	15.5	19.0	20	
Co	21.8	20.5	–		–	18.9	–		
Ni	21.7	20.0	–	18.0	18-19.6	16.0	21.0		17.6
Pt	23.7	20.5	–		–	13.5	–		
Ga			–		–	10.0	–		
Yb			–		–	6.5–7.1	–		
Te			–		–	11.0	–		
Sb			–		–	8.8	–		
Mn			–		–	12.0	–		
Th		16.0							
U		19.0							
L8-59			–		14–14.5		–		

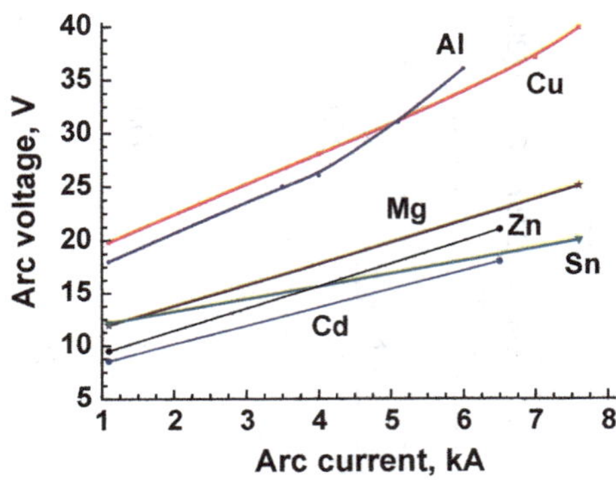

Fig. 7.5 Arc voltage as function on arc current in range of I from 10 A to about 7 kA

7.1.5 *Threshold Arc Current*

The minimal arc current at which the arc exist was defined as threshold current I_0. As follows from Table 7.3, I_0= 1.6 A for often used Cu cathode. The measurements of the threshold arc current [3] showed that I_0 is linearly dependent on the combination of boiling temperature T_b and heat conductivity λ_T of the cathode materials in form:

$$I_0 = 0.52 \times 10^{-3} T_b \sqrt{\lambda_T} \tag{7.3}$$

As it was explained in [3], the dependence (7.3) directly indicated the preferable role of the cathode vaporization process to characterize the threshold arc current.

7.2 Cathode Spots Dynamics. Spot Velocity

When an arc of a relatively moderate current was ignited, very small local luminous regions were observed. The observed luminous cathode area traditionally called "the cathode spots." The spot dynamics is determined by their current, velocity, lifetime, and size. The methods used to determine the spot parameters and the results of the observations are analyzed and discussed below.

7.2.1 *Spot Definition*

The cathode spots move along the cathode producing characteristic erosion trace on the surface. The trace indicates that the cathode was sufficiently hot and eroded under the luminous area, which in essence is a plasma radiation due to local electrical energy dissipation. *Thus, the cathode spot can be defined as an arc-constricted region included the local hot cathode body and a dense plasma clot supported the current*

Table 7.3 Cathode potential drop measured on body cathodes u_c, on film cathodes u_{cf}, and the threshold arc current I_0 (u_i is the ionization potential)

Cathode element	u_i (eV)	u_c (V) [3]		I_0 (A) [3]	u_c (V) [14]	u_{cf} (V) [29]	Substrate [29]
		Air	Vacuum				
Li	5.39		15.0		11.1–11.7		
Na	5.14				8.7–9.0		
K	4.34				6.7–7.4		
Cs	3.89		6.2		6.2		
Cu	7.72	16.0	16.0	1.6	14.7–15.4	11.2	Cu
Ag	7.57	15.3	13.0	1.2	12.1–13.6	9.1	W
Au	9.22	15.5	15.0	1.4	13.1–14.8	9.7	Au, W
Be	9.32	14.4	17.0	0.9	18.6–19.2	14.0	W
Mg	7.64		12.5		11.6–13.0	8.8	Mg
Ca	6.11				10.8–11.4	10.3	W
Sr	5.69				8.4–9.2	8.8	W
Zn	9.39	10.0	10.0	0.3	9.8–11.1	9.6	W
Cd	8.99	9.8	11.0	0.19	8.6–10.2	8.3	W
Hg	10.4		*8–9.5*; 19	0.04–0.07	8.0–9.5		
Er	6.20				13.1–14.0	8.6	W
Al	5.98	14.4	15.5	1.0	17.2–18.6	4.2	
Ga	6.00	12.8	15.0	0.45	15.0		
In	5.78	11.8	13.0	0.45	9.5–11.9	4.1	W
Tl	6.11		10.5–11.5(L)		10.5–11.5		
Ti	6.82	14.3		2.0	16.8–17.6	4.0	Ti
Zr	6.84				17.7–18.5		
Hf	6.8				16.9–17.4	3.5	Hf
C	11.26				15.2–18.9		
Sn	7.34	12.0	12.5	0.6	10.6–13.0	4.3	W
Pb	7.42				8.8–10.2		
V	6.74				17.3–18		
Nb	6.88				19.9–21.6		
Ta	7.88	13.5		1.5	16.8–21.4	4.1	Ta
Bi	7.29	12.3	9.0–12.5	0.27	8.4–8.7	4.4	W
Cr	6.76	14.8		2.5	16.7–17.4	4.6	Cr
Mo	7.10	16.0		1.5	16.6–17.2	4.4	Mo
W	7.98	16.1		1.6	16.2–22.6	4.5	W
Te	9.01				11.0–12.4	5.0	W
Fe	7.87	15.1	17.0	1.5	17.1–18.0	4.7	W

(continued)

Table 7.3 (continued)

Cathode element	u_i (eV)	u_c (V) [3]		I_0 (A) [3]	u_c (V) [14]	u_{cf} (V) [29]	Substrate [29]
		Air	Vacuum				
Co	7.86	15.2	16.0	3.2	16.8–17.7	4.2	Co
Ni	7.63	15.0	18.0	2.36	16.3–17.3	4.9	Ni
Pt						5.3	Pt

continuity at the cathode region. Below, the cathode spot parameters for different metals was considered and was preferable for Cu as that of the most studied and typical case.

7.2.2 *Study of the Cathode Spots. General Experimental Methods*

Two general experimental methods were used to study the cathode spot characteristics and spot behavior: (i) study the luminous area and (ii) autograph method—study the traces on the cathode surface. The first approach is widely used for study of spots appeared in the arc with bulk cathodes while the second was used for spots appeared in the arc with bulk [30] and film cathodes [3]. The luminous regions of chaotic moving spots were observed mainly by high-speed optical photography, see [30, 31], where some spot behavior was reported. Below, the main characteristics and the spot dynamics based on the literature data are summarized.

7.2.3 *Method of High-Speed Images*

Kesaev [3, 4] studied the spot behavior on solid and liquid mercury cathodes using mirror scanning of the spot images. This approach allowed observing the chaotic spot motion and spot splitting by generating new sub-spots named cells which carried certain current I_{cell}. The new cells were produced from an old cell when its current increased to $2I_{\mathrm{cell}}$. The arc current equal to minimal I_{cell} was named as threshold current of the arc. This result obtained for mercury cathodes nevertheless was used for understanding the observed similar spot dynamics by other investigators for different cathode materials, which *was not considered by Kesaev.*

Different types of cathode spots were detected by the experimental study of the cathode surface luminosity in vacuum arcs [30, 31]. Spot types were classified according to the spot velocity v_{sp}, the spot lifetime t_s, and the spot current I_s. The spot behavior depends on cathode surface state (irregularities, oxide film impurities, etc.), arc current, arc duration, gap plasma density, and gap length. The experimental investigation indicates that the cathode spot can disappear and again appear with a characteristic spot lifetime t_s on bulk or film cathodes [3, 4, 31]. Let us consider the main literature data regarding to the spot dynamics and then summarize the experimental results of the published works.

7.2.3.1 Early Observations of Spots on Different Cathodes

A short duration arcs between various metals (Cu, Al, Sn, Cd) in air have been studied by Somerville et al. [32] by means of a Kerr cell camera and by microscopically observing the tracks left on the electrodes. The arc was usually struck by moving the electrodes together until sparking occurred. The transient arcs with durations ranging from one microsecond to one millisecond were considered with currents ranging up to 200 A. It was indicated that the cathode spot becomes multiple soon after the initiation of the arc at relatively large rate of current rise of 10^7–10^8 A/s, and the increase in track area is due to an outward motion of the spots. The spot appearance depends somewhat on the electrode material and on the distance of the anode from the cathode.

Froome [33–35] studied the spot behavior with arc current in range 40–450 A (duration up to 100 μs) on cathodes from Hg in air and sodium–potassium alloy in argon under 10^{-2} Torr pressure considering the photographs obtained by means of the Kerr cell shutter with exposition 100 ns. The spots expanding with velocity up to 10^4 cm/s radially away from the initiated point was observed when the rate of current rise was reached 10^8 A/s. For lower rate and a given current, the spots location takes the form of a line, or broken line of total length proportional to the current.

Reece [19] studied near-electrode plasma brightness in high-current vacuum arcs of switching devices during separation of the electrodes. The investigations showed two types of discharges with Cu electrodes. The first (named type A) occurs at currents of below 10 kA that consists of group of entirely separate cathode spots each carrying about 100 A. The cathode spots split up, and the arcing spreads across the whole cathode surface with velocity of ~1 m/s. The observed relatively small electrode damage was caused by type A since the spot power input was over small period of milliseconds. At higher currents (12 kA), the arc (named type B) was appeared similar to high-pressure arc indicating cathode and anode melting. The brightness of the arc is much greater than in the type A and does not split into multi-spot regime but spreads on the surface.

Initially, the spots are classified on two types as fast-speed and low-speed spots given by Grakov and Hermoch [36]. The pulse arc was ignited in air of atmosphere pressure with current 400 A in maximum and duration 180 μs. The spot images were obtained with exposition of 8 μs.

It was observed that the spots are produced initially at the minimal distance from the anode, and then, they move at the cathode surface with different velocities (fast or slow moving) depending on the cathode materials and arc duration. During the spots motion, their splitting changes the number of spots. Therefore, the motion trace was branching which can be seen in Fig. 7.6 indicating the erosion tracks leaving on the Al cathode surface after extinguishing the arc.

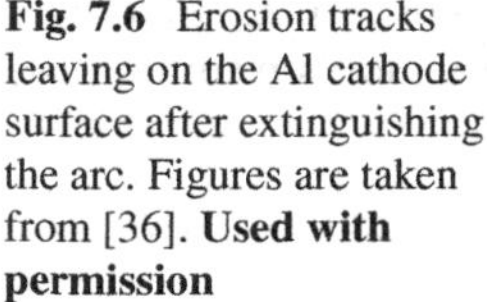

Fig. 7.6 Erosion tracks leaving on the Al cathode surface after extinguishing the arc. Figures are taken from [36]. **Used with permission**

The fast-moving (speed) spots (*named as type* 1) are produced on cathodes of relatively low volatile from metals (Ni, Cu, Al) while the both fast and slowly (*named type* 2) moving spots exist at metals of relatively high volatile (Cd, Sn, Zn). No data about velocity values were reported. The behavior of spots of type 1 was similar for all cathode materials, and these spots were appeared from the arc beginning and preferable at scratch with abrasion on the surface. The spot of type 2 represents a group spot, which consisted of a number of sub-spots located closely to each other. They were always produced at the location where previously the spots of type 1 were located, and their lifetime can reach 100 μs. Although the experiments [36] were conducted in air, the results of spot dynamics indicate a behavior that was observed in vacuum mentioned above by Kesaev [3, 4] and will consider below.

The more extended study with fast-speed image camera of the spot behavior was conducted (for Sn, Zn, Ag, Cu, W) in experiments [37–39]. The cathode surface was polished with diameter 6–10 mm for $I \leq 100$ A and 20–24 mm for $I > 100$ A. The pointed anode was used placed at distance of 2.5 mm from the cathode center. The arc duration (with rate d$I/t = 5 \times 10^6$ A/s) was continued up to 25 ms. Different spot types were observed at different time stage of arc development. In the beginning (<100 μs) of 100 A arc only, fast-speed spots (also *named type* 1) that radially (a *ring shape*) expanded from the initiated point at the cathode surface were detected. These spots were characterized by velocity of (1–5) × 10³ cm/s, current per spot of I_s= 5–20 A, and lifetime t_s= 5–20 μs. After about 100 μs, the low-speed spots (also *named type* 2) with velocity <100 cm/s, I_s= 10–30 A, and t_s= 50 μs were occurred. These spots exist separately when low pressure of an external gas (0.1–1 Torr) presented in the arc gap.

In high vacuum, several spots (type 2) moved together on the cathode surface (velocity ~10 cm/s and t_s= 1–10 ms) with distance between them equal or smaller than the overall spot size. This collection of spots is called a "*group spot.*" The velocity of spots (fragments) was significantly larger inside the group than velocity of the group spot. For arc currents ≤1000 A, the group spot current was in the range of 100–1300 A with current per spot inside of the group 10–120 A depending on cathode materials like Cu, Ag. Sometimes the group spots can be appeared for $I \leq$ 100 A, especially for volatile cathode materials like Sn.

Similar spot behavior at the cathode surface was observed by W, Mo, Cu, Ni, Al, and Sn contacts separation or its closing with current in range of 0.2–14.5 kA in an experimental system modeling an interrupter [40]. The contact out-gassing was conducted by their heating to 800–1900 °C and then cleaning by many time of arcing with current 500 A. The experiment showed two arc types with fast-moving arc channels (*named type* 1) and practically stable channels (*named type* 2) that corresponded high-speed and weakly moving group spots (with current 200–1500 A as example for W), respectively. The fast-moving spot (type 1) was detected for relatively large contact gap while the low-moving spot (type 2) appeared when the contacts approached closely one to other in relatively short gaps. The last effect indicates that low-speed spots were produced when a relatively large vapor pressure arise in a short arc gap. The lifetime of spots type 2 increased with current by power law with exponent equal to 1.2 for W (40 μs at 400 A), to 1 for Mo (60 μs at 400 A), and to 0.7 for Al (20 μs at 400 A), i.e., by most linear dependence.

Djakov and Holmes [41] investigated the distribution of spot numbers depending on arc current (5–150 A, arc duration of few ms) for different metals (Bi, Zn, Pb, Al, Cu) using an image converter camera. The cylindrical cathode with diameter 25 mm and from 25 to 50 mm long and annual anode with an inner diameter of 40 mm (outer diameter 60 mm) separated by 20 mm were used. The exposure time was 0.21 μs, and the time interval between exposures was in range 0.1–1 ms. It was shown that the spot moves randomly with spot displacement of 0.5-1.5 mm in 1 ms. This displacement indicates that the spot velocity was in range 50–100 cm/s. At small arc current, the average spot number is more often than not equal to one while the average spot number increased linearly with arc current. The spot current changed from about 5 A for Bi to about 75–100 A for Cu cathodes.

In their follow-up work [42] the cathode spot named as fragments in a copper vacuum arcs has been studied using a high-magnification optical system. The cathode was 2 mm thick copper strip sandwiched between two 1 mm molybdenum sheets. The experiment showed that the spot usually remained on copper insert. The anode consists of two flat parallel copper strips, and the gap distance was varied up to 30 mm. The arc current was used in range 25–100 A and being maintained constant over periods of about 10 ms. The authors showed that the observed spots were a group of fragments and the number of fragments increased from 1 to 4 with arc current. The arc consists of two spot types that are defined by their current density. The spot with current density (0.5–1) $\times$ 10^6 A/cm^2 *named as type* 1, *while for spot type* 2 it was 5 $\times$ 10^6 A/cm^2 (there the type was defined according to different spot current density). The number and size of fragments fluctuated, and a transition from one spot type to second spot type and vice versa occurs on an average of every 100 μs according to their type definition [42].

Cooper arcs were photographed during separation of a 1.3 cm radius anode from a 1.7 cm radius cathode to a spacing of 2.5 cm in time 150 ms using high-speed camera at 5000 frames/s [43]. The arc currents were 30, 100, and 350 A, and nitrogen pressure in the range 10^{-3}–200 torr was used. The measurements showed that at pressures below 0.5 torr, the arcs were similar to vacuum arcs with a multiplicity of mobile cathode spots with spot current of about 100 A. At higher pressures in range 0.5–200 torr, the current of each individual spot decreased to 15 A. The specific spot behavior was investigated in high-current vacuum arcs (few kAs). A ring shape expansion of the spots was indicated as common main result of the high-current experiments. Persky et al. [44] observed a ring expansion of the group spots in vacuum arc with current in range of 1–10 kA and duration up to 4 ms. Molybdenum cathodes with diameters 25–50 mm were cleaned by a number of pulse arcs before the measurements with time resolution of 50 ns. The spot velocities were measured in range of 10^3–10^4 cm/s. The ring shape expansion of the spots on the cleaned cathodes for different metals by the pulse arcs with current 7–10 kA was observed by Sherman et al. [45] and Agraval et al. [46]. The spots were formed in a group spot (named here "*cluster*") with current per spot of 100 A. The spot velocity was varied in range of (1–2.5) $\times$ 10^3 cm/s, and this velocity was not changed significantly on the gap length in range 5-20 mm.

Siemroth et al. [47, 48] observed the spot structure in a vacuum arc with sinusoidal current pulse (peak 3–5 kA, duration 0.5–1 ms) using image-converting high-speed framing camera. It was indicated that a single cathode spot size was about 100 μm and consists of a number of simultaneously existing sub-spots, each with a diameter of about 10 μm and a mean distance of 30–50 μm between them. The lifetime of the sub-spot on cathodes from Ti, Cu, and C was in the range from 100 ns to several μs. For Cu, this time was about 3 μs.

7.2.3.2 Spot Types on Fresh and Cleaned Cathode Surfaces

Achtert et al. [49] observed extremely fast-moving cathode spots (named type 1) in experiments with DC arcs ignited between different metallic electrodes in the arc beginning on fresh cathodes. After cathode surface cleaning by the discharges of themselves also fast-moving spots were detected but slower than that on fresh cathodes (named spots of type 2). The arcs with contaminated cathode surfaces (fresh) and arcs with clean surfaces were studied by Juttner [50] and Bushik et al. [51–53]. Two kinds of experiments were conducted (i) with current 20–60 A DC and (ii) with current 8–20 kA pulse duration about 1 ms. The cathode consisted of two adjacent cylinders from Cu and Mo and another from Mo stripe which was covered by 10 μm Cu or Al film.

A fast-moving spots of type 1 with branch-like structure were observed in DC arc. The velocities of these spots changed in range of (0.5–1) $\times$ 10^5 cm/s and lifetime of about 4 μs. The spot type 1 burns mainly in desorbed contamination and on oxide films [54]. They ignited again when air was exposed in the system after previously discharges cleaned the surface. No cathode erosion was observed, and the characteristic trace depth was about 0.1 μm. The velocities of spot type 1 appeared at Cu and Mo are different when the arc burn on two adjacent cathodes from Cu and Mo [55].

At large arc current with one cathode material, the spot appeared in form of expanding ring. During some discharge time, the density of the branches decreases until the spot type 1 changed to the type 2 with velocities of about (0.4–1) $\times$ 10^4 cm/s and lifetime of larger than 10 μs. It was also indicated that at 200 μs, the low-velocity spots type 2 were detected inside of the expanding ring. The spot type 2 appeared immediately after arc ignition with velocity 3×10^3 cm/s in case of clean cathode surface by preliminary arcing. The spot type 2 moves randomly as a bright single spot which caused a macroscopically visible erosion. The motion of this spot type was characterized by spot association, splitting, conjugation, and extinguish. The current per spot 2 can be larger by factor 2–4 than for spot 1. The spot 2 preferably appeared at the boundary line when the arc burn on two adjacent Cu and Mo cathode materials. Thus, the spot velocities on fresh and cleaned cathode surfaces remained relatively large, and the spot named by type 2 exceeds the velocity of spots named as type 2 and defined in [37–39].

The group spot (like defined in [37–41]) behavior was observed by Beilis, Djakov, and Juttner [56] in a 40–1500 A vacuum arc for Cu (65 mm diameter) and CuCr (50 mm diameter) separated electrodes. At initial stage with a non-arced cathode, it was found about 8 A per spot. After some arcing, this value significantly increased. At stage of arc time larger than 3 ms after arc ignition, the group spot was observed and the number was found to increase linearly with current. The average current per resolvable spot amounted to 100 ± 30 A for CuCr and 150 ± 70 A for Cu. The group spot consists of fragments that conduct currents <10 A. A random spot displacement R was observed, having mean square values of $R^2/\Delta t \approx 1 \times 10^{-3}$ m^2/s for Cu and $R^2/\Delta t \approx 4 \times 10^{-4}$ m^2/s for CuCr. For characteristic displacement time 25 μs, the spot velocity for Cu can be calculated as 5×10^2 cm/s. This velocity is larger than obtained in experiments [37] and [40] due to their very short gap distance (<2.5 mm) and electrode configuration (point anode [37]) in comparison with the ones used here (10 mm). The observation also showed that the Cu spots brightness fluctuated with intervals of 17 ± 3 μs.

The above random spot motion [56] was in accordance with such results obtained by Daalder [57]. He studied the spot displacement using a framing camera recorded 224 frames of cathode spot movement at 20–10 μs exposure time. Interframe times were 90 or 180 μs. The cathode was movable, and an arc was struck by separating the cathode 1 mm from the mesh anode. The cathode surface was slightly curved so that the arc was always ignited at the cathode center. The anode was a grid of stainless steel, which had an optical transparency of 46%. The cathode surface contaminants were removed by arcing prior to each experiment. A laser beam, projected on the film, was produced a datum from which spot coordinates were measured with a spatial resolution of 50 μm.

The random walk was studied assuming that during each time interval Δt, there is an equal probability of the spot displacement at a step $\pm s$ along the $\boldsymbol{x}$- and the $\boldsymbol{y}$-axis. After n sets of displacements, the spot position is (x, y) *and* will take time $t = n\Delta t$. For a one-dimensional random walk, the probability after n steps at time t at the spot position was given by the normal density function:

$$p(x) = \frac{1}{(2\pi Dt)^{0.5}} \mathrm{Exp}(-\frac{x^2}{2Dt})$$

or in two-dimensional random walk, the probability is

$$p(x, y) = \frac{1}{2\pi Dt}\mathrm{Exp}\left(-\frac{x^2 + y^2}{2Dt}\right)$$

where $D = (s^2/\Delta t)$. Taking into account that $R^2 = (x^2 + y^2)$, $x = R\cos\varphi$, and $y = R\sin\varphi$, after integration over the domain of the angle φ, the probability will be

$$p(R) = \frac{R}{Dt}\mathrm{Exp}\left(-\frac{R^2}{2Dt}\right)$$

It follows that the mean value is

$$R_{ev} = \left(\frac{\pi}{2}Dt\right)^{1/2}$$

The most probable value of R is

$$R_m = (Dt)^{1/2}$$

The parameter D has the same dimensions (m^2/s) as the classical diffusion coefficient.

Taking into account the above probability of random spot walk, the parameter R_{ev} was determined measuring the average distances between cathode craters, which were 2–3 times of the crater diameter by Daalder's [57]. Table 7.4 gives the values of D and the calculated average distance R_{ev} for different cathode metals. For copper D was found to be independent of current, and for cadmium, D increased with current. The difference between 7.9 and 4.6 A is relatively small. It was noted that according to data in Table 7.4 for cathode metals with a wide range of different thermophysical properties, the D values do not differ much (by about a factor two).

Table 7.4 Parameter D and R_{ev} for different cathode metals [57]

Metal	Current (A)	D (m^2/s)	$R_{ev}/t^{0.5}$ ($m/s^{0.5}$)
Copper	45	6.5×10^{-4}	3.2×10^{-2}
	18	6.5×10^{-4}	3.2×10^{-2}
	9	6.5×10^{-4}	3.2×10^{-2}
Aluminum	21.6	1.4×10^{-3}	4.7×10^{-2}
Molybdenum	59.5	7.4×10^{-4}	3.4×10^{-2}
Cadmium	7.9	5.9×10^{-4}	3×10^{-2}
	4.5	5.1×10^{-4}	2.8×10^{-2}
	1.8	8.3×10^{-4}	3.6×10^{-2}

Table 7.5 Average spot velocities on different cathode metals as a function of time [57]

Metal	v_{ev} (m/s) (t = 0.1 ms)	v_{ev} (m/s) (t = 1 ms)	v_{ev} (m/s) (t = 10 ms)
Copper	3.2	1	0.32
Aluminum	4.7	1.5	0.47
Molybdenum	3.4	1.1	0.34
Cadmium (4.5)	2.8	0.89	0.28

Using the average distance travelled by the spot and the crater formation times derived from the random walk data (see Table 7.4), the spot average velocity $v = (R_{ev}/t)$ was calculated in range from 0.3 to about 5 m/s for Cu, Cd, Al, and Mo cathodes, for arc time in range of 10–0.1 ms (Table 7.5).

The cathode spot dynamics on 30 mm diameter Cu, CuCr, and stainless steel contacts were studied in 1–5 kA, with duration 0.5–5 ms vacuum arc using by high-speed photography camera by Chaly et al. [58]. The results show that after arc initiation by contact opening, the fast-moving spots expand on clean cathodes in a ring form with radius increased with arc time and are dependent on arc current. In contrary, at the composite cathodes (CuCr), a slowly moving group spot were formed in short gaps. The group spot diameter depends on arc current and the contact gap.

7.2.3.3 High Temporal and Spatial Resolution of Spots on Arc-Cleaned Cathodes

Anders et al. studied the cathode spots on arc-cleaned copper and molybdenum electrodes in vacuum by fast image converter framing and streak camera photography with high temporal and spatial resolution [59], the discharge ignition in breakdown stage and the future arc phase with high time resolution using laser absorption photography [60], and the brightness distribution of vacuum arc cathode spots [61].

The electrode gap in range 20–100 μm, needle Cu cathode of 280 μm diameter, and rod Cu anode were used with arc duration of 800 ns and current 50–150 A. The current rise time was 5 ns [61]. The authors indicate that the plasma of a vacuum-arc cathode spot emits continuum radiation from its small central region, while the line radiation at larger distances from the spot center. This experimental result was explained by assuming a possible spot structure with sub-spots (fragments).

It was found that the ignition phase is about 50 ns, and in arc phase, the spot consists of fragments with current 20-40 A. The authors indicate that the fragment size up to 10 μm and that the size of the brightness structure can be obtained between 20 and 40 μm for spots 232 ns and 360 ns after arc ignition, respectively [60]. Figure 7.7 is selection of streaks at different sweeps times, which demonstrate fluctuations depending on time scale [62]. Figure 7.8 demonstrates photographic picture of luminescence brightness through the spot [63]. This distribution seems like to Gauss distribution and, respectively, current density distribution, which was assumed by calculation of the electrode thermal regime [64].

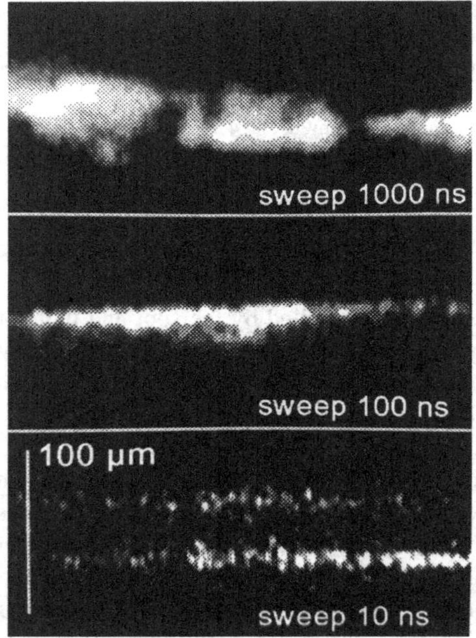

Fig. 7.7 Nanoseconds brightness fluctuations and displacements in streaks by using the UV-optics with different time intervals [62]

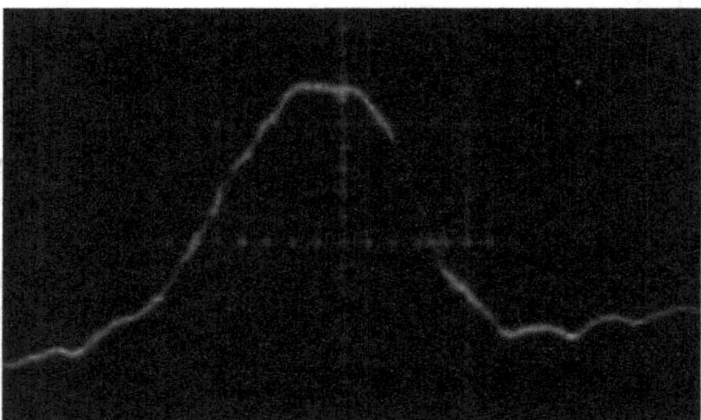

Fig. 7.8 Luminescence brightness distribution through the spot [63]

Cathode spots of vacuum arcs (duration 2–20 ms) on cleaned electrodes have been studied by Djakov [65, 66] using image convertor photographs providing 0.5–10 μs exposures separated by 0.5 ms in frame mode or 200 μs exposure in the streak mode. The electrode surfaces were cleaned by repeated vacuum arcing up to state when the transition from fast-moving spots to only slowly moving cathode spots with velocity about or less 100 cm/s was observed. The experimental setup and conditions were used as it was described in [42]. The observation showed that spherical spot plasma was generated on fresh (oxidized) surfaces. The slowly moving spots on cleaned surfaces, at which two groups of spots with areas different by factor about 2 were observed, generated the conical plasma shape. The last result agrees with the spot modes reported in experiment of [41]. By using space and time resolution (>100 ns), the continuous motion or jump-like spot fragments displacement and immobile single fragment spot existing for about 10–50 μs or longer (do not show significant changes for several μs) and radius 10–30 μm have been detected. The cleaning procedure changes the initially smooth cathode surface into randomly rough surface having pits with dimensions 10–100 μm. The thermal analysis provided in the work showed that the local cathode surface relief with the irregularities strongly affects the cathode spot size and velocity of motion. It was indicated that a furthermore study also with nanosecond resolution like work of [59] is needed.

The dynamics of spot brightness using streak camera for arc current waveform 1, 15 ns, and about 3 μs with amplitude 90 A was studied by Juttner [67] with the cathode cleaned by the prior arcing. The cathode was Ti wire of 300 μm diameter facing anode of 10 mm diameter at distance <100 μm. At first 400 ns, several spots with current 20–40 A, mean time for spot disappearance of 14 ns, and size in range 5–10 μm were observed. Several types of long time fluctuations were observed at 34 μs after arc beginning with periods 154 ns, 1.4 μs, and about 11 μs. These long time fluctuations were interpreted as result of random spot displacement and spot back to its origin location.

The further detailed experiments using high-speed framing camera "IMACON 468" were conducted by Juttner [68] for cathodes from Cu and Ti wires 470 and 300 μm, respectively, with arc current 70 A and gap distance of 100–300 μm. Other Juttner's experiments [69] used cathode with the rounded end of a Cu wire of 0.5 mm diameter and experiments with cathode from Cu wire 0.4-0.8 mm at arc current 30-70 A [70]. The results were reviewed [71]. Arc spots were ignited at the apex of the cathode by electrical breakdown. The pictures of the spots were chosen being delayed with respect to arc ignition by 3–100 μs. Prior to the measurements, the cathode surface was cleaned by numerous arcs, so that it was completely covered by protrusions and craters.

The high-speed observations of spot dynamics for different time after arc ignition and with different exposure time were presented in Figs. 7.9, 7.10, 7.11a, b. The results show that the spots were structured (spot splitting) consisting of a number of fragments, and spot motion is random. However, the total spot dimension can be about or even exceeds 50 μm for all the above experiments indicating weak dependence on time after arc ignition when the observations were initiated. The displacement of the spots were obtained as 2×10^{-3} m^2/s for time >100 ns (in accordance with such results obtained by Daalder [57]). The brightness changes were detected in all ranges of used exposure times. Luminous circles with diameter 30–60 μm and non-circles of sizes 50–100 μm with fragment sizes up to 20 μm were observed. In 70 A, arcs can be detected up to seven fragments. The fragments seem periodically to attract and repel one another with different range of velocities <10 m/s or >100 m/s depending on the surface relief. When the fragments attain their smallest mutual distance, the

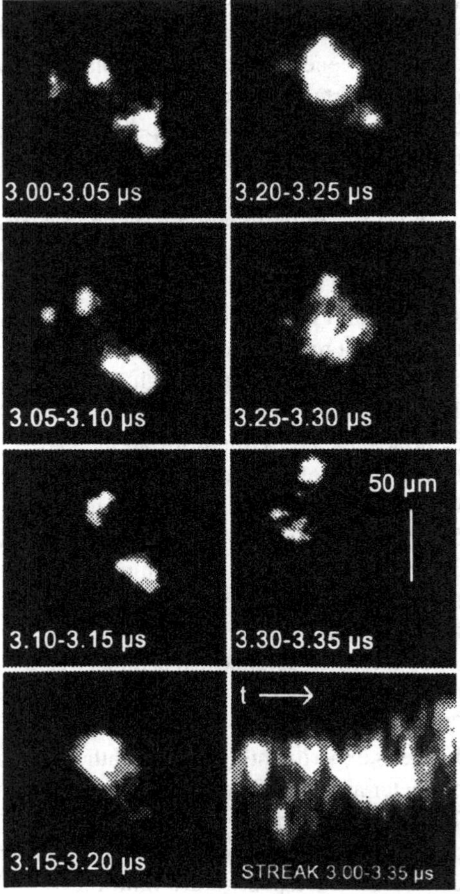

Fig. 7.9 Frames and streak taken three 3 μs after ignition with frame exposure time 50 ns. No pause between the frames. Streak sweep 350 ns and indicate upward motion of the spot with velocity reaching up to 10^4 cm/s [62]

luminous seems as whole spot. This effect was used by the author to explain various periodical fluctuation times (1–3 and 10–20 μs) obtained in [67] and reported in the literature.

At currents of 30–70 A and sufficiently long time after ignition in range of 3–300 μs, the spots consist of fragments with diameters of 10–20 μm [70]. These fragments appear and then disappear with formation time of 50 ns and residence time of 100 ns. Consecutive fragment formation appears as displacement with momentary velocities up to 1000 m/s. The fragment dynamics leads to random displacement of the spot center with a ratio of mean square displacement $<R^2>$ to the observation time t of $<R^2>/t = (2.3 \pm 0.6) \times 10^3$ m^2/s. This holds down to $t = 0.1$ μs. Juttner indicated that the components of the spot could be distributed over the whole crater rim, but the spot center changes its position from frame to frame (Fig. 7.11a), resulting in a net motion around the crater with velocity even larger than 1000 m/s.

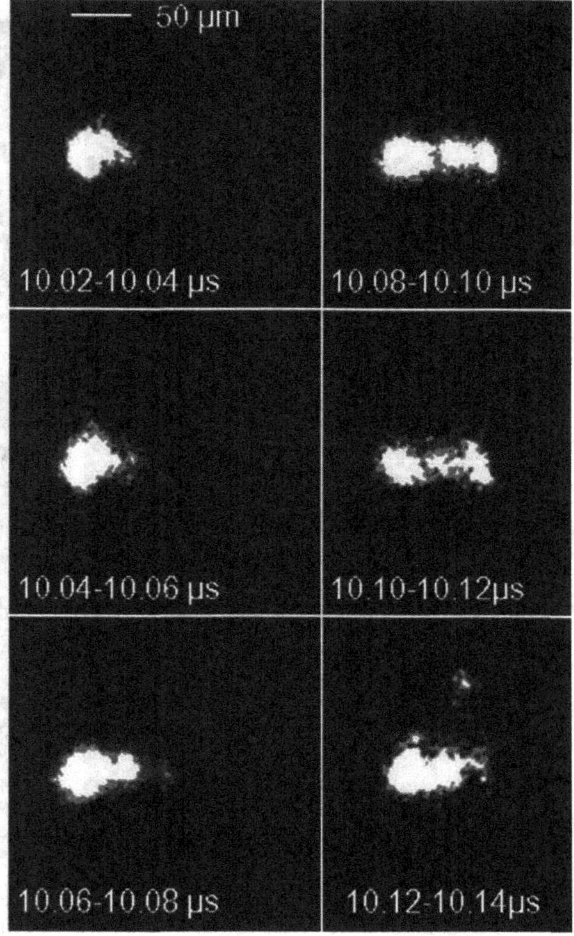

Fig. 7.10 Six frames taken with the glass optics10 μs after ignition. Exposure time 20 ns. No pause between frames [62]

The diffusion character of the motion of cathode spot fragment with 10–50 ns exposure timescale was observed at 200–300 μs after arc ignition with current 30–40 A on Cu strip cathode of size 1×5 cm^2 [73]. The distance between fragments (or spot size) was 50–100 μm with fragment size ~20 μm and sub-fragments <20 μm. The probability of spot with one fragment was large ~70%, and while finding two fragments, the probability was about 20%, and for 3 fragments, it was low, 4–5%. The experiment [74] showed that in gases, the fragments were more separated than in a vacuum which was similarly to that observation in [37].

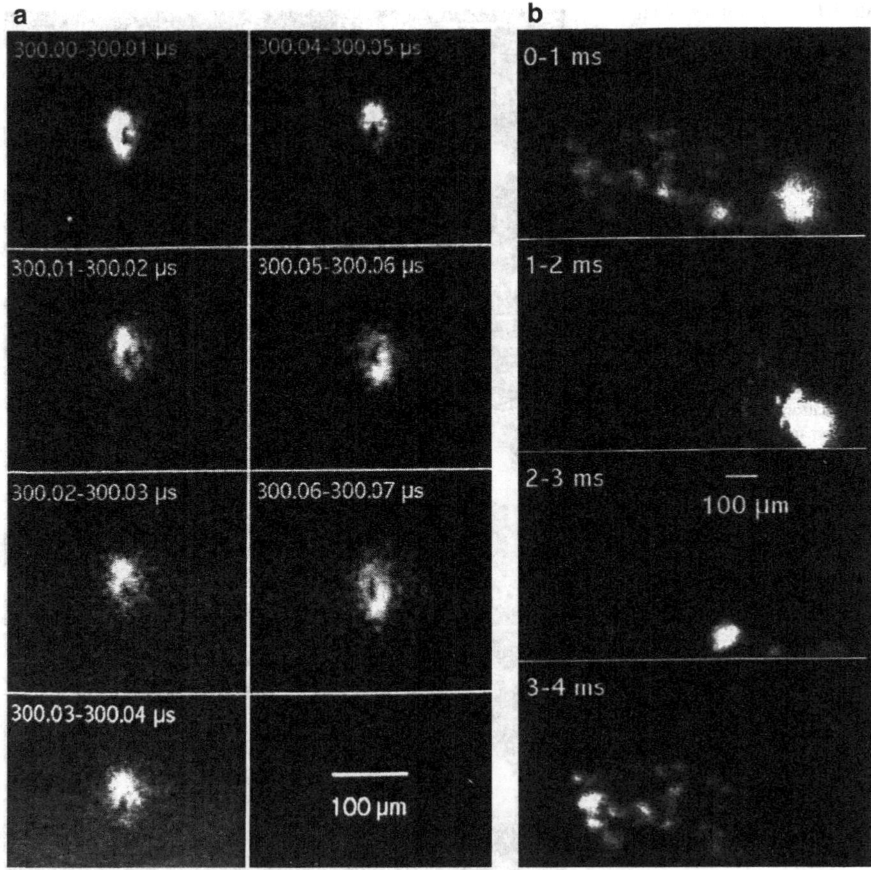

Fig. 7.11 High-speed photographs of an arc current 30 A started 300 μs after arc ignition, **a** with 10 ns exposure time per frame and **b** with 1 ms exposure time per frame. No time break between the frames [72]

7.2.4 Autograph Observation. Crater Sizes

The approach employs with the erosion tracks produced by spot heat action on the cathode surface. In frame of this approach, the track or crater sizes on bulk cathodes and tracks on thin film cathodes leaving after arc extinction were studied. Cobine and Gallagher [75], one of first researches, measured the width of tracks left by a moving spots on oxidized Cu, Al, and W cathodes in air at atmosphere pressure with currents 2.6, 10, and 140 A. The motion in a straight line was supported by applied external magnetic field. The single tracks were detected at relatively low arc current ($\leq$5 A), while for larger currents the overlapping or multi-branching tracks were observed. The appearance of the track suggests that the oxide film was removed from the base metal under the spot action by direct evaporation, and therefore, the track indicated pure cathode material. The time passed a track was about 5 μs.

Table 7.6 Width of tracks as function on cathode materials and arc current [75]

Metal	Current (A)	Width of tracks (μm)	Current density (kA/cm^2)
W	2.6	67	73.75
	5.0	83	92.41
	10	164	47.34
Pb	5	156	26.16
Al	2.6	105	30.03
Cu	2.6	54	113.53
Cu–Hg	2.6	127	20.5
	5.0	172	21.52
	10	275	16.84

It was indicated for Al multiple spots formed at currents as low as 2 A, while for W there was apparently but a single spot at currents up to at least 10 A. Assuming circular geometry, the results of measurements are illustrated in Table 7.6 for currents where mostly single spot was appeared. The lower width was observed for Cu and W cathodes at 2.6 A arc, while for 140 A arcs the tracks like branching tree were observed.

Daalder [76] observed a crater size distribution on copper cathode surfaces in breaker. The average opening speed of the breaker was 1.5 ms. The anode was acted as the movable contact. The cleaning process of the electrodes was by an etching with acids degreasing in vacuum at 700 C. The maximal length of the gap between contacts in a breaker reached was 6 mm. After polishing, an emery paper was used to obtain a surface with unidirectional ridges in order to investigate a large discharge area that increases with increasing of the cathode surface roughness. After arcing, the cathode surface was studied by microscopic photographs with magnification 50–250.

The observations indicate that most of craters were almost circular in cross section, and the single crates were detected at low arc current about 4–100 A. The crater diameter distributions were determined for different current in range 4–230 A. The results [76] illustrated dependence of the crater number on crater sizes with current as parameter. Figure 7.12 indicates that the most probable crater diameter increased with the arc current. The total number of craters was in range of 170–440 depending on current. The crater sizes (at maxim number) increased from 4 to 12 μm, and the maximal number (pick in the distribution) of craters was decreased from about 40 to 7% when arc current increased from 4 to 105 A. The multi-sport arc regime was observed in current range of 120–230 A when spot splitting in the discharge occurred. A small fraction of larger craters up to 30 μm also observed. The results showed asymmetrical distribution with respect to peak of crater number and have a character, which presumably is due to the overlapping of the two traces observed.

Using scanning microscope, Guile and Hitchcock [77–80] widely investigated the damage regions on 2.5–340 nm thickness copper oxide films in atmospheric-pressure air arc with duration from few ns to 40 μs. The cathode surface was flat end of a

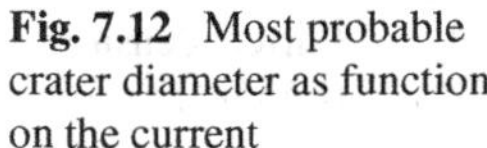

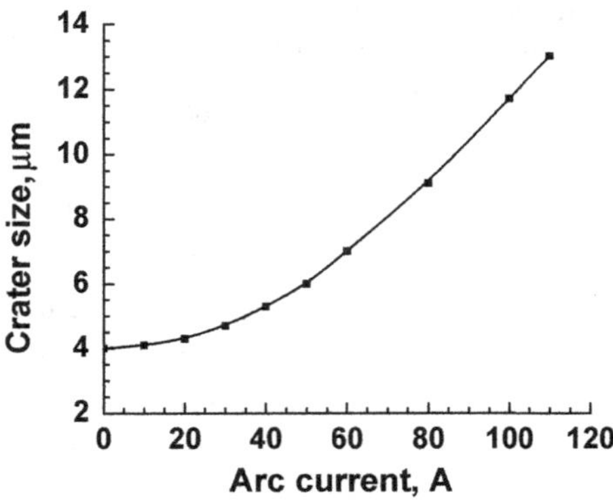

Fig. 7.12 Most probable crater diameter as function on the current

cylindrical rod of 3–6 mm diameter. The arc was initiated by breakdown of 30 μm electrode gap at voltage 600 V. After 4–14 A arcing, it was found that a patch of the oxide layer was stripped from the underlying metal, and within the stripped area, multiply craters were formed. The crater numbers was proportional to arc current and only approximately proportional to time of the arc duration. At 25 ns arc, the most common crater diameter was 0.1 μm, and for larger time, it was 0.16 μm for current 4.5 A. The crater numbers increased from 10^3 to 10^4 when arc duration increased from 10 to 100 ns, while the total stripped (damage) area (also in form of a crater) increased from about 6 μm at 4 ns to about 60 μm at 300 ns of arc duration for 4.5 A arc.

The research of Cu cathode craters was extended by Guile and Hitchcock [81] for reduced air and nitrogen pressures up to 1 torr, arc currents up to 1.1 kA, and Cu cathode with two oxide (Cu_2O) films thicknesses 2.5 and 100 nm. The experiments were conducted with 1.5–2.5 mm gap distance for 100 nm film thickness and mainly 2 mm gap (separately with 7 mm gap) for 2.5 nm film thicknesses. The measurements showed that the cathode crater diameters and surface densities have been similar to those obtained previously for low arc current in atmosphere pressure. This result [81] partially demonstrated in Fig. 7.13, which shows crater diameter distribution for arcs of different currents in different pressures of air and nitrogen with 100 nm oxide film thickness. This same similarity has been observed for crater rate production on 2.5 nm film in arcs ignited at reduced nitrogen.

The dependencies of cathode track width on arc velocity and on current have been studied for 2.5 nm oxide thickness and were shown a difference from that for thicker films. For moving arcs, the cell arc and cell lifetime were obtained using data for arc velocity, crater density, arc current, and cathode track width obtained by optical microscopy. According to [80, 81], the arc cell was 6–10 mA and cell lifetime was 10–16 ns for 100 nm Cu oxide film. Mau [82] also observed spot tracks on Cu oxide film produced on rotating disk cathode in an arc with current of 15–60 A and spot

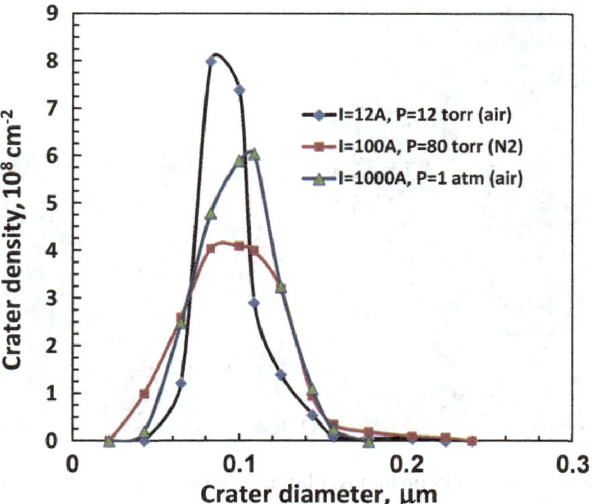

Fig. 7.13 Dependence of the crater density at the cathode surface on crater diameter at Cu oxide films with thickness of 100 nm

lifetime of ~10 µs. The tracks has branching form with width of about 2 µm and current of about 0.1 A.

Guile and Hitchcock [81] indicated that the previous optical investigations, for example in [37, 43], cannot describe the arc root structure since the crater sizes was about 0.1 µm. They also stated that the cells are so close together (1 µm) that they could not to be separately resolved by high-speed photography. Therefore, the optical method could only see groups of cells, and these groups can be detected separately. According to and Hitchcock [81], such groups named cathode spots were observed in [37], and the cathode area engaged by first type of spots with lifetime of 5–20 µs spreads with velocity 1–50 m/s at first time after arc ignition. The measured area in [81] containing all cells expanded during arcing and the radial velocity was reduced from about 800 m/s (100 nm) and 300 m/s (2.5 nm) at 15 ns to range 10–40 m/s at ~1 µs similarly to what indicated in [37].

Kesaev [3] studied the spots (cells) behavior in an arc burning on cathodes from thin Cu films deposited on a glass and Bi films deposited on a glass and Mo substrates. The spot moved and left a clear erosion track in which all the metal thin film was removed which is shown in Fig. 7.14a–c. The branch tracks in Fig. 7.14a for Al film and in Fig. 7.14b for Cu film indicated the spot splitting process when the arc current exceeds the value of spot current. For thin films (<0.6 µm for Cu), the spots leave continuous tracks, and width of the tracks can be increased before the spot division. However, for tick films >1 µm, the spots leave discontinuous tracks (Fig. 7.14c). It was obtained that the number of spots increased proportionally to the arc current.

The thickness of copper thin films was varied in the range of d*f* ~0.01–0.1 µm and arc duration up to 100 µs. From observation of the Cu tracks, the following parameters were determined: spot velocity v_s ~10^3–10^4 cm/s, erosion rate G_f ~46–64 mg/, spot current I_f ~0.1–0.9 A, and track width δ ~2–15 µm for d*f* <0.14 µm (Table 7.7) [3]). The similar data (from [3]) for Bismuth films deposited on glass were

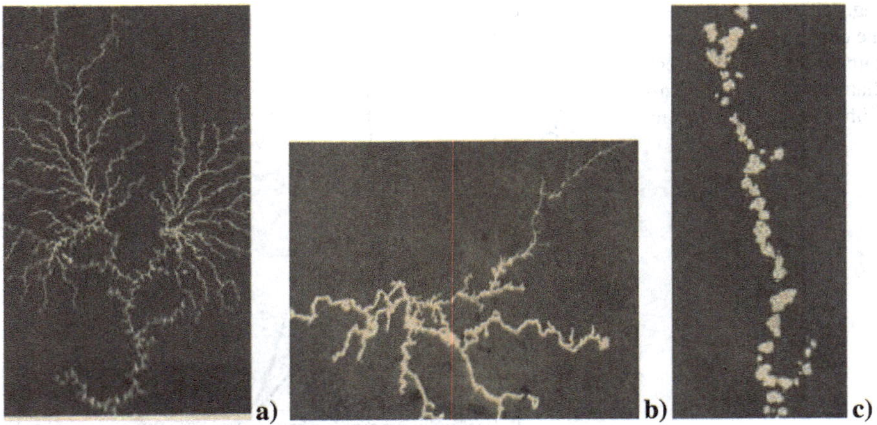

Fig. 7.14 Tracks on thin Al film (**a**), tracks on thin Cu film (0.6 μm) on glass (**b**) and track on thick Cu film (2.5 μm) on glass (**c**) [3]

Table 7.7 Copper films on glass substrate [3]

Film thickness *df* (μm)	Cells velocity *vs* (10^3 cm/s)	Cells current *If* (A)	Track width *δ* (μm)	Erosion rate *Gf* (mg/s)	Erosion rate *Gf* (mg/C)
0.017	13.0	0.1	2.5	0.046	0.46
0.025	10.1	0.15	4.0	0.085	0.56
0.034	8.0	0.3	6.0	0.137	0.46
0.06	5.7	0.4	9.0	0.26	0.64
0.12	3.4	0.9	16	0.56	0.62
0.12	3.8	0.9	15	0.58	0.64
0.14	3.4	1.2	20	0.82	0.68
0.60	–	3.0	32	–	–
0.66	2.4	3.2	70	9.33	2.9

presented in Table 7.8 and on molybdenum substrate were presented in Table 7.9.

The data in Tables 7.7–7.9 show that Bi films have the spot current *If* ~0.05–0.25 A and track width *δ* ~5–17 μm for d*f* <0.6 μm. The measured average erosion rate for Cu (0.58 mg/C) is lower by factor 6 lower than for Bi film on glass substrate (3.34 mg/C), and the erosion rate sharply decreased for Bi on metallic substrate (1 mg/C). It was due to relatively low boiling temperature for Bi in comparison with that for Cu and by larger energy dissipation in metallic in comparison with that in glass substrate.

Zykova et al. [37] indicated that the track depth on the cathode surface after the high-speed spots varied from 2 μm on refractory to 20 μm on low*temperature melting metals. The total damage surface area corresponded to the expanding image

Table 7.8 Bismuth films on glass substrate [3]

Film thickness *df* (μm)	Cells velocity, *vs* (10^3 cm/s)	Cells current *If* (A)	Track width δ (μm)	Erosion rate *Gf* (mg/s)	Erosion rate *Gf* (mg/C)
0.055	3.7	0.05	5	0.1	2.00
0.068	3.2	0.06	6.0	0.135	2.14
0.11	2.7	0.08	8.0	0.23	2.90
0.15	2.65	0.10	10.0	0.39	3.90
0.19	2.15	0.10	10.0	0.40	4.00
0.23	2.0	0.14	12	0.54	3.86
0.34	1.57	0.2	15	0.78	3.90
0.45	1.21	0.25	17	0.90	3.60
0.52	1.3	0.3	17.0	1.16	3.87

Table 7.9 Bismuth films on Mo substrate [3]

Film thickness *df* (μm)	Cells velocity, *vs* (10^3 cm/s)	Cells current *If* (A)	Track width δ (μm)	Erosion rate *Gf* (mg/s)	Erosion rate *Gf* (mg/C)
0.027	4.0	0.07	5.0	0.053	0.76
0.034	4.0	0.07	5.0	0.066	0.94
0.041	3.0	0.07	5.0	0.06	0.86
0.048	2.5	0.07	5.0	0.059	0.84
0.068	3.0	0.09	6.0	0.12	1.33
0.12	1.5	0.12	7.5	0.13	1.08
0.11	1.5	0.12	7.5	0.123	1.00
0.20	1.0	0.15	10	0.20	1.30
0.59	0.5	0.25	10	0.46	1.85

area is 10^{-2}–10^{-1} cm^2 after the high-speed spots. After slow-speed spot type 2 (defined in [37]), the crater size was 5×10^{-3} cm; after slow speed group spot, the crater area was in range of 10^{-3}–10^{-2} cm, and their depth was in range of 10^{-3}–10^{-2} cm that depended on surrounding gas pressure.

Juttner [50, 83] observed that craters size was increasing from 5 to 20 μm on copper and molybdenum cathode surfaces when current increased from 10 to 130 A of nanosecond arcs. It was indicated that pulse and DC arcs produce similar crater forms. Juttner [69] indicated a large craters size of diameters up to 100 μm for Ti and up to 50 μm for Cu produced by few spot fragments.

The crater size can be changed when a new spot ignited on previously hot cathode surface. Reece [19] indicated that individually melted area on Cu cathode was greater than 10^{-5} cm^2 (considering micrographs of cathode 30 A arcing surface) and which can grow with cathode background temperature. According to [84] with increasing

Al cathode temperature in range 300–770 K, the sizes of craters increased by a factors in range of 2–4.

The arc craters and traces on virgin tungsten cathode wires of 0.1–0.2 mm in diameter were studied [85]. The cathode spot was initiated by a trigger pulse of 30 ns duration with current about 3 A. Rectangular arc pulse with duration of 1.2 ns in the current range of 2–50 A was applied. The cathode–anode distance was small 0.1–0.2 mm. To identify the track width, the cathode spot motion was directed by the magnetic field varied in the range 0.1–0.7 T. The cathode was cleaned prior to the experiment by heating at 2000 K.

Varying the magnetic field in used range does not really affect the mean crater diameter or the track width. It was indicated that cathode spot division and the simultaneous existence of several cathode spots could explain the increase in the track width as compared with the mean crater size. The cathode spot has been found to start dividing at a current of a few amperes. For currents of 20–50 A, two parallel tracks are observed. Track splitting was also observed on a hot cathode (1800 K) at a current of 45 A. The crater diameter weakly depends (5–8 μm with lifetime of about 30–50 ns) on arc current (2–50 A) and increases to about 40 μm and lifetime to about 200 ns at 1800 K heated cathode [85]. Figure 7.15 shows the traces for different arc currents [86].

In the absence of a magnetic field, the spot moves randomly. For the most probable crater radius, the ratio r^2/t increases slowly $(1.2–4) \times 10^{-4}$ m^2/s with current in range 5–45 A. For the maximal radius, the ratio r^2/t increases $(2.5–25) \times 10^{-4}$ m^2/s with current for a cold cathode, and the ratio r^2/t increases $(3.6–30) \times 10^{-4}$ m^2/s with current for a hot cathode. The cathode spot division [85] could explain this ratio difference.

7.2.5 Summary of the Spot Types Studies

Let us summarize the above cathode spot's high-speed image dynamics and the data of craters and tracks at the cathode surface left during the spots action.

7.2.5.1 Spot Image Dynamics

The most of experiments were conducted with cathodes treated before the cathode spot study by the arc in order to clean them from the surface oxide films or impurities. Also some experiments were conducted with cathodes not treated by the arcing. In essence, two main kinds of experiments can be distinguished:

(i) Observations in arcs with relatively large electrode diameters (~1 cm), durations in range of 1–10^3 μs, with current from 100 to 1000 A and up to few kA.

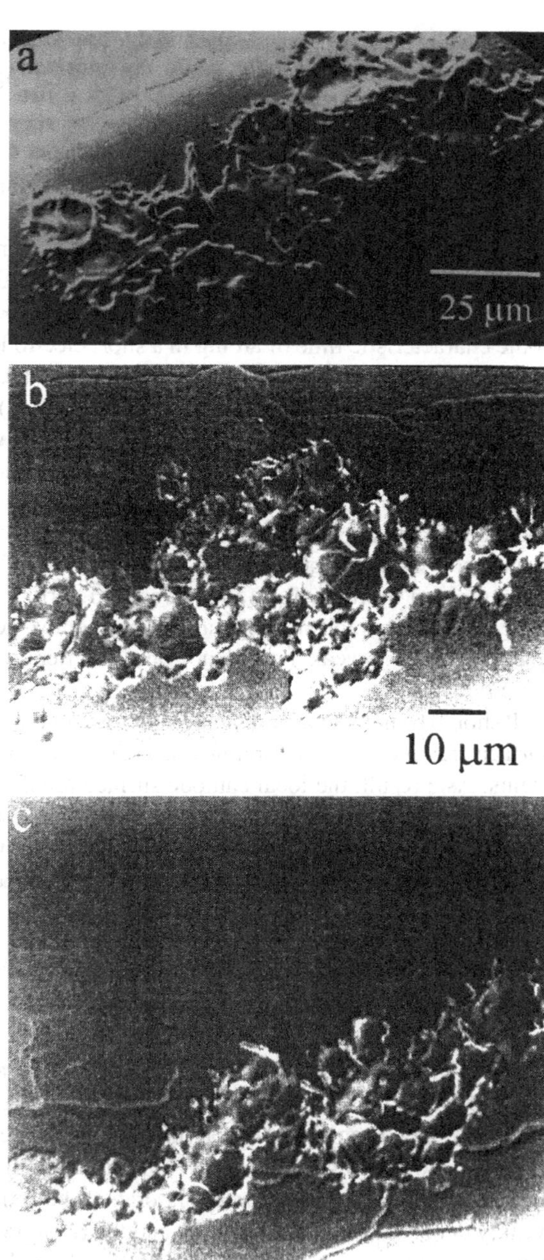

Fig. 7.15 Craters and track on tungsten cathode from vacuum arc of 1250 ns duration and $H = 0.75$T, **a** 45 A, **b** 23 A, and **c** 10 A [86]

(ii) Observations in arcs with relatively low electrode diameters (about 300–500 μm) and arc current ≤70–150 A with high space and nanosecond time resolution.

The spot parameters were tested at few ns and up to few hundred μs after arc initiation. Considering the spot dynamics in vacuum arcs, the results of first kind of experiments indicated that fast speed spot motion was observed always in the arc beginning (>10^3 cm/s). The spot velocity at treated cathode surfaces can be lower by order of magnitude than at the not treated surfaces remaining at enough high level (up to 10^3 cm/s). The low-speed spots (<100 cm/s) were observed after some characteristic time of arcing in a short electrode gaps when the cathode can be heated and the plasma pressure was increased in the gap and also in a low pressure of a surrounding gas. The low-velocity (≤10 cm/s) group spots were appeared in good vacuum (≤10^{-4} Torr) and arc current ≤1 fkA while the velocity of group spot significantly increased to order of 10^3 cm/s for high-current arcs (~10 kA).

The results of second kind of experiments indicated that mostly all fast-moving spots and spots appeared after ~1-100 μs consist of moving fragments similar to group spots which consists of a sub-spots number. A fragment size of about 10–20 μm with minimal lifetime 10 ns and lower, and current per fragment of about 10-20 A were found. The total spot size can be about or even exceeds 50 μm (Juttner's experiment) indicating weak dependence on time after arc ignition when the observations were initiated.

It should be noted that another point is the cathode cleaning by preliminary cathode surface treatment by the arcing related to both kind of above-mentioned experiments. As a result, the local cathode surface relief substantially changed producing different sizes of protrusion including sub-micron spikes. The produced irregularities strongly affect the cathode spot size, lifetime of the fragments, and velocity of motion possible to observe as a distribution of cathode surface luminosity with high space and nanosecond resolutions.

7.2.5.2 Summary of the Autographs Study

Observations of the autograph showed very wide range of craters and tracks sizes (from 0.1 to 30 μm). The crater characteristic (the number, size distribution, and lifetime) depends on cathode type (films or bulk), arc currents, arc duration, and time of observation during the arcing. While for oxide film cathode having ns lifetime of the cells (spots), the most probable crater size is relatively small (~0.1 μm) and weakly depends on arc current [81], and the most probable crater diameter of 12 μm was measured in millisecond (100 A) arcs and for cleaned cathodes [76]. Separately, craters can be obtained for relatively small arc current and arc time that characterized only low-life and fast-speed type of spots according to image observation. Kesaev's data in experiments with film deposited on substrates (using the magnetic field to obtain separately tracks) can give a possibility to describe the spot mechanism on film cathodes as a self-sustained object. The respective approach will be described below in Chaps. 16 and 17.

7.2.5.3 Comparison of Imaging and Autograph Results

The above description of the spot behavior showed different spot sizes when it was studied by high-speed imaging and by autograph method. Thus, the high-speed photographs of emission from the arc plasmas yield r_s= 10-100 μm in vacuum and more in the presence of ambient low-pressure gases [39, 87, 88]. The absorption photographs of short duration vacuum-arc spots yielded $r_s \leq$ 10 μm [58]. The observation of the craters size at the cathode surface after the arc indicates small radii 0.1-1 μm for film of an oxide cathodes and is increasing from 5 to 20 μm (Cu and Mo cathode) for DC arcs and for nanosecond arcs [37, 71, 76, 83, 85]. Hence, the image observation sizes during the spot emission cannot agree with the damage sizes observed on the cathode surface after arcing. This difference in the results indicates different specifics of the mentioned experimental methods, which cannot adequately reflect the active main electro-conductive plasma zone which supported the current continuity at plasma–solid cathode interface and interpreted as the "spot area."

According to Hantzsche et al. [89], the craters do not constitute a direct image of the active spot. The active spot zone is characterized by strong electron emission and evaporation. This zone may be surrounded by a less active area of plasma–surface interaction, ionization, and radiation, especially in an ambient gas. Harris (see in [30]) indicated that the image size could be enlarged up to 100 μm due to motion of an excited ion at the length of his relaxation in active zone. This same size can be estimated considering the length of luminous plasma expansion (by particle excitations and de-excitations) with velocity 10^6 cm/s at time 10 ns (r_s= vt = 100 μm) after the spot initiation. For spot lifetime 1 μs at certain location, the spot luminous size can reach 100 μm at plasma expansion velocity of 10^4 cm/s. In general, the ejected plasma velocity depends on spot current and heavy particle density in the expanding region. For heavy particle density ~10^{20} cm^{-3}, the plasma velocity of active zone can be low ~10^3 cm/s with following acceleration by future expansion where the density and therefore the plasma brightness significantly decreased [90]. Also, the plasma expansion was influenced by the background gas pressure [91] and by background plasma in a short gap arcs. Thus, the spot imaging depends on expansion of dense active zone that in turn determined on the spot plasma parameters.

The research and assumptions in works [61] conclude that the brightness distribution of a single cathode spot at rest is a superposition of continuum radiation from the spot core plasma and line radiation from ions and mainly atoms. The diameter of the continuum radiation is related to the current carrying area at the cathode since the very dense plasma in the spot center transfers the largest part of the current. The line radiation not reflected the cathodic current area and is mainly depended on the density distribution of radiating particles. The emission photographs reflect the process of plasma expansion rather than the active spot zone where electrons are emitted, atoms are evaporated, and the plasma is formed [88]. Emission photographs can enlarge the spot size. According to research in [92], the absorption photographs refer directly to the dense plasma in the spot.

On the other hand using the autograph method at large current, the cathode surface observation showed that the tracks and crater were superimposed. It can be noted about some difficulties to find a correspondence between the damage zone and current carried area (cathode spot) as well about a time of their appearing and lifetime. Another weak point is related to fact that the cathode surface studied after the arcing (without magnetic field) indicates an integral picture of the crater number produced during all arc time, and the time-dependent number distribution remained unknown. Therefore, the mentioned problems limit the autograph approach to use it for understanding the structure of cathode current continuity and mechanism of transient spot operation on a bulk cathode. So, the existing approaches of the spot observation do not reflect unambiguous data on its size in spite of the importance of this size to determine the active spot area.

7.2.5.4 An Argument Against the Numerical Classification of Spot Types

Nevertheless, the high-speed imaging determines the transient spot behavior, the influence of the low external gas pressure, magnetic field action of the cathode geometry, cathode treatments, and the rate of current rise. As mentioned above, different types of local brightness areas, named as cathode spot, were observed depending on the listed conditions. Mainly they differ by the velocity, numbers, or current local spot, and lifetime. So, the dynamics of different spots available is to be classified and determined by their types, which at present is characterized by the classification published in the literature described in Sect. 7.2.3.

However, some misunderstanding was occurred in the literature using the Arabic numerals (1, 2 …) in order to recognize the different spot types. While some of the literature labels the spot types by the numbers 1, 2, 3, this characterization confusingly groups together spots with very different characteristics, and the numerical names do not convey any meaning. A few such examples can be listed as:

(1) Fast velocity spots (10^4–10^3 cm/s) were named as type "1" in [35, 36, 38], while the spots with similar velocity (~10^3 cm/s) appeared at cathode and cleaned by arcing were named as type "2" in [49, 50];
(2) The spots occurred on fresh (with impurities or oxide films) cathodes named as type "1" in works [49, 50];
(3) Spot type "2" was identified in works of [35, 36, 38] with relatively low velocity (10–100 cm/s), such spots appeared separately after some arcing time under low pressure of external gas. At same time, the fast-moving spot appeared on clean cathodes according to works [49, 50] also named as type "2" while in [36] the spot of type 2 represents a group spot.

(4) The fast spot occurred at the film cathodes [3] also named by type "1," while this definition cannot be related to the spot at a bulk cathode.
(5) The spot types with different large velocities in [30] were denoted by Roman numerals (I, II). The references are mentioned as examples, while numerous other works used the type number mixing the real spot type characteristics.

Thus, different characteristics, moreover, different spot mechanisms related to the types are defined by same spot number. Such above-mentioned definitions confused understanding the object's physical state and properties. It should be noted that even on cleaned cathode after surface treatment with arcs duration more than 10–100 ns, the produced irregularities influence the type of high-speed spots appearing. Different other definitions such as "emission site (ES)" and "emission center (EC)" (appeared under external high-voltage source (Chap. 6), which is absent in low-voltage arc discharges) at the cathode are also named historically as cathode spot. Thus, it is important to distinguish the spots by different form of their appearing in order to study in further the mechanism taking into account the common properties obtained experimentally. Therefore, below it is presented an attempt to classify the spot types by observations of their typical properties and conditions of appearing.

7.2.6 *Classification of the Spot Types by Their Characteristics*

It is established that the prevalence of different types of spots depended on arc current, cathode material, and cathode surface characteristics (e.g., oxide film and roughness). Also, it is commonly accepted that for relatively low current (100–300 A) in the initial stage of an arc, superfast, fast-moving (up to 10^3–10^4 cm/s, depending on surface cleaning), and short life (<1 μs) cathode spots appear on all metals. With higher current, higher vapor or gas pressure, and longer arc pulse duration, the spot velocity is less, and the spot lifetime is longer.

Under some arc conditions, a few spots can be located close to one another, producing a low-velocity "group spot." A high-velocity group spot was observed in presence of a magnetic field. The special type of spot appeared on metallic films deposited on a substrate which occurred at very low current ~0.1 A due to operation at low-ionized metal atoms (in comparison with oxide films) and very low energy due to film heat conduction.

Therefore, it is preferable to spot types distinguished by their adequate parameters. In general, the spot parameters are velocity v_{sp}, spot current I_s, lifetime t_s, and their disposition which mainly depends on the state of cathode surface. For clarity, the spot classification can be defined directly by parameters which characterized their appearance. For example, all fast-moving spots can be distinguished as *superfast spot*-**SFS** (at surface with impurity, oxides and roughness), *moderate fast-moving spot*-**MFS** (cleaned and flat surface), and film cathodes-**FiS** (metal films on glass or metals). There is also the separated *slowly independent (or individual) spots*-**SIS** and *group spots*-**GS**, which represent association of sub-spots, fragments, or cells. The main spot types classified by such manner with their main characteristics are presented in Table 7.10. The data are mostly related for Cu cathode. Note that, some types of spot depend on gap distance or on the self- or applied magnetic field. For example, the usually slow group spot can be appeared as fast-moving group (v_{sp}~10^3 cm/s) at roof-shaped cathode in transverse magnetic field [93].

Table 7.10 Classification of the spot types by their characteristics and conditions of appearance

Abbreviated name	Extended name*	Spot motion	Conditions of appearance	Spot parameters		
				v_{sp} (cm/s)	I_s (A)	t_s (μs)
SFS	Superfast spot, (including fragments)	Fast	On surface with impurity, oxides, and no uniformity	10^3–10^4	≤10	0.01–10
MFS	Moderate fast spot (including fragments)		Cathode surface cleaned by preliminary arc (roughness) or by cathode heating	10^3	≤10	<10
FiS	Film cathode spot		Metal films deposited on glass or metals substrates	≥10^3	~0.1	1–100
GS	Fast group spot		Usually under magnetic field	up to 10^3	Few of ten	1–10
	Slowly group spot	Slow	In a vacuum for relatively high arc current	≤10	100–300	10^2–10^3
SIS	Slowly independent spots		Electrode gap filled by an electrode vapor or low-pressure gas	10–100	10	~100

*All above-listed spot types change their motion from random displacement to the direct motion in presence of a magnetic field

7.3 Cathode Spot Current Density

Let us take in account the above data of spot dynamics to consider the spot current density problem in vacuum arcs.

7.3.1 *Spot Current Density Determination*

The spot current density cannot be measured directly in the spot determining the electrical current distribution (e.g., by electrical or another probe type) due to extremely small spot size. The generally accepted approach consists of its study by determining the spot radius r_s and spot current I_s. Assuming that, the current was distributed uniformly in the spot, and the current density j was calculated as

$$j = \frac{I_s}{\pi r_s^2}$$

Usually, the parameters r_s and I_s were detected from the dynamics of spot luminous or studying the cathode surface damage.

7.3.2 *Image Sizes with Optical Observation*

It has been commonly accepted that in the initial stage of a discharge, fast-moving cathode spots appeared on all metals. Also, this type of spots was mainly investigated optically with high time and space resolution for a numbers of years [70]. Most of such results (in ns and μs time intervals) reported relatively small spot sizes ($\leq$10μm) indicating a large spot current density of 10^7–10^8 A/cm^2 [59–61, 67–71, 91]. According to Juttner [64], the probability of spot changes in some time interval τ due to disappearance of spots which increases from 0.2 to about 0.9 when τ increase from 10 to 100 ns, respectively, at spot current of 20–40 A. This means that the rate of current rise during the spot life decreased in range (2–0.2) $\times$ 10^9 A/s and (4–0.4) $\times$ 10^9 A/s for currents 20 and 40 A, respectively.

The large spot current density was also obtained by less commonly used methods. The characteristic spot size was obtained considering region of the line splitting by the Zeeman effect and assuming that the radiation screening was determined by the surface roughness with dimension comparable with the size of a crater [94].The obtained current density was 10^8 A/cm^2. The Cu cathode was used with gap distance about 0.5 mm. Also, the spot current density was proposed to determine measuring a rate of current rise and spot velocity at crossing of the arc spot over a slit, which divide the cathode into two electrically insulated sections. The spot radius was obtained from $2r_s = v_s\Delta t$, and the time Δt was determined by the time of current change behind the slit [95, 96]. The obtained current density was 10^8 A/cm^2. It can be seen in the above-mentioned studies a tendency to explain the large values of current density by explosion electron emission.

On the other hand, the low-current density in fast-moving cathode spots on polished copper, obtained by measuring the luminous regions, have been reported for rates of current rise of 5×10^5, 5×10^5, and 6×10^5 in vacuum arc of 400 μs duration [97]. The results indicate 4×10^4, 4.4×10^4, and 5×10^4 A/cm^2, spot velocity 4×10^3, 3×10^3, and 2×10^3cm/s for arc current 100, 50, and 24 A, respectively. The maximum current density ~10^5 A/cm^2 was observed usually immediately after division of a spot, and the minimum value immediately before division.

Another similar research was conducted by Smith et al. [98] for arc current 700 A with $dI/dt = 3.6 \times 10^6$ A/s using 12 framing photograph at arc beginning from 12.5 up to 40 μs with 2.5 μs of each frame. The data from this experiment indicate area of 2×10^{-4} cm^2 for single spot, current per spot of 17.5 A with current density from 1.3×10^5 to 7.8×10^4 A/cm^2. The spot velocity was reported as 5×10^{-3} cm in 30 μs, i.e., about 1.7×10^2 cm/s. The relatively lower current density of (0.5–5) × 10^6 A/cm^2 was also reported by Djakov [41] and of 10^6–10^7 A/cm^2 by Simroth [47, 48, 99]

7.3.3 *Crater Sizes Observation*

Cobine and Gallagher [75] reported about cathode spot area measurements from the width of tracks left on metals by the cathode spot of an arc burned in air. The obtained current densities were 1.24×10^5 A/cm^2 for Cu, 3×10^4 A/cm^2 for Al, and 7.4×10^4 A/cm^2 for W.

The large current density 2×10^7–10^8 A/cm^2 were obtained considering the craters dimensions indicated in [76, 77]. From these measurements, Fig. 7.16 shows that the current density increases with current up to 50 A, then passes through maximum, and

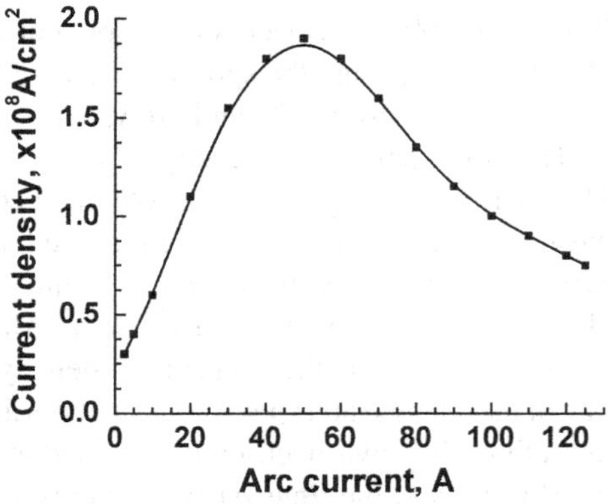

Fig. 7.16 Spot current density as function on arc current determined by Daalder using the observed crater sizes

decreases with larger arc current. The future decreasing of current density with arc current can be expected due to increase of the crater size and spot current decreasing.

Puchkarev and Murzakayev [85] used the autograph method to investigate the arc spot over a virgin tungsten cathode under a magnetic field of 0.1–0.7 T in the current range 2–50 A. They demonstrate the dependence of spot current density on discharge condition. The cathode spot has been found to start dividing at a current of a few amperes. The current density measured at the moment of cathode spot death is 1.2 $\times$ 10^7 A/cm^2 for a cold cathode and 3 $\times$ 10^6 A/cm^2 for a hot one. The cathode spot lifetime is 25–50 ns for a cold cathode, and it increases to 150–200 ns for a heated one. Neither the current density nor the cathode spot lifetime depends on current. For extremely short arc (10 μm) at closure gold electrical contacts, the current density obtained using the crater sizes (5–10 μm) was reported about (2–6) $\times$ 10^6 A/cm^2 [100]. The current density in arcs with oxide films and metallic films on the glass and metallic substrate using the craters and traces sizes was obtained also in large range of 10^6–10^8 A/cm^2 [3, 80–82].

7.3.4 Influence of the Conditions. Uncertainty

Considering the published data, the spot current density obtained by size measurements of the optical images, and the crater size of the arc varied in wide range of 10^5–10^8 A/cm^2 on bulk cathodes and about 10^5–10^6 A/cm^2 on thin film cathodes. The problem of determination of the spot current density arises with significant uncertainties that stimulate large number of investigations and publications cited above.

The results of optical experiments depend on time and space resolutions of the amplification factor of the light and used spectral region. The different experimental data were both due to the results characterized different spot types and by the dispersion of the measurement characterized similar spot types, in particular, fast-moving spot. The dispersion of the measurement also can be due to used significant values of rate of current rise >10^8 A/s that is related to the case of a spark rather than of an arc. Sometimes, the current density was reported, but no values for spot current were presented, and therefore, value of the current density cannot be directly approved by the data on the luminous or crater characteristic sizes.

Using the autograph method, the question is what crater area is related exactly to the current carrying area. To determine the current I_s even in a single discharge at the cathode, one has to know if there are one or more current active areas at the same time [76]. As an example, here it is relevant to mention the experiment [39] which reported the result of surface observation after the arcing. A detailed study of the trail by means of scanning and conventional microscopy proved that it consisted of single craters. In general, these craters accumulated in areas less in size than the area of a single spot (optical size). According to analysis of the experiment [101], the area occupied by the craters amounts only to about 1% of the area of the cathode spot before extinction or splitting (Fig. 7.17).

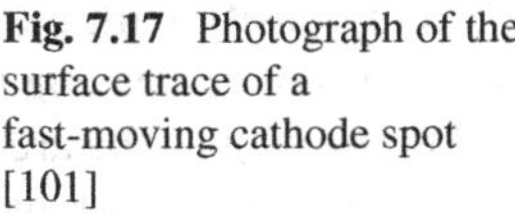

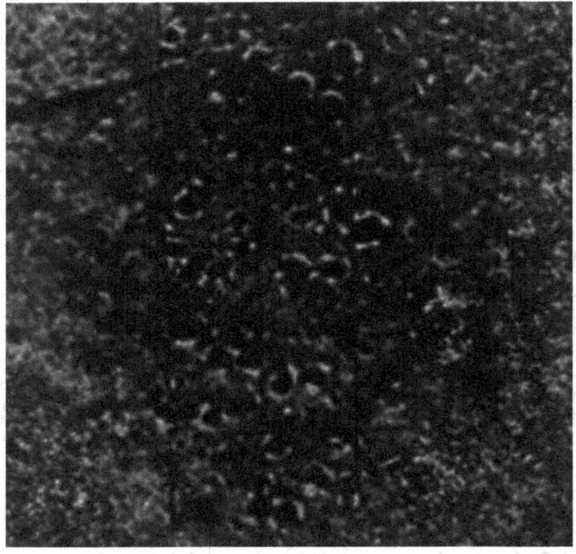

Fig. 7.17 Photograph of the surface trace of a fast-moving cathode spot [101]

Another misunderstanding is the discrepancy between sizes of macroparticles and corresponding craters. Thus, Daalder [76] observed macroparticles whose diameter exceed by one order of magnitude of the diameter of craters. This point also requests to study the relation between macroparticles production and the process of crater formation [102].

7.3.5 Interpretation of Luminous Image Observations

The most important question is the interpretation of the measured luminous objects. According to works [59–61], the current density was determined by spot area where the plasma emits continuum radiation, i.e., by spot central region, at which the plasma is optically thick in the visible region. The main argument to base the measured current density of 10^8 A/cm^2 is to satisfy this value to the model of explosive electron emission as a mechanism of a cathode spot, according to which the plasma is significantly dense to support the observed if optically thick. It further stated that, in contrary, the evaporation–ionization model is associated with a relatively low-current density and a low plasma density.

The last statement is a simple misunderstanding conclusion because the evaporation-ionization approach reported also this result (large plasma density possible to be as optically thick) without any explosion. It follows from study of the structure of high-current arc cathode spots in vacuum as dependence on spot

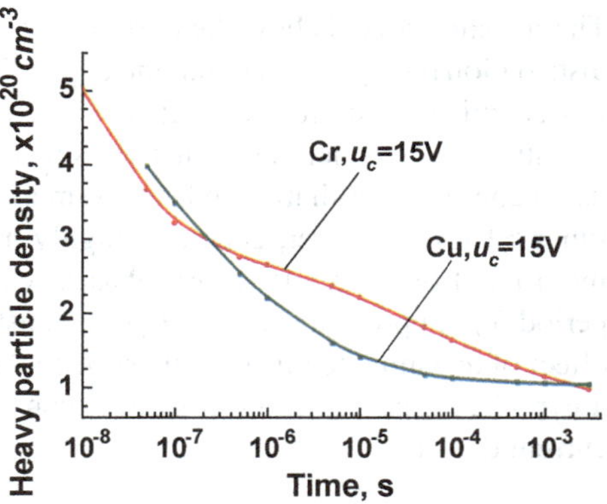

Fig. 7.18 Plasma density as a function of spot lifetime with cathode potential drop as a parameter for Cr and Cu

time [56]. The results [56] of time-dependent density calculation for Cu and Cr cathodes are demonstrated in Fig. 7.18. Besides, the calculated current density is about (5–7) × 10^6 A/cm^2 when the time approaches 10–100 ns. The electron temperature for nanosecond cathode spot operation can exceed 5 eV in case of cathode potential fluctuation (see [103]). So, there is no necessary to use the explosion emission to reach the plasma parameters indicated in [60, 61, 67] and to obtain the continuum radiation from the explosive plasma. The recently developed new similar models applied to spots burning on microprotrusion and on the bulk cathodes will be presented in the theoretical part of this description [104, 105]. The advantages of these theoretical approaches will be discussed, and the measuring continuum radiation area [59–61] that interpreted as active spot radius, which amounts to 5–10 μm, can be understood according to the mentioned models.

In addition, the interpretation luminous areas as sharply active spot radius should be conducted with enough degree of discretion and accuracy. For example, the section results of [60] reported "These structures, having a size of about 5–10 μm, can be called microspots or cathode spot fragments," and then, "we can deduce a maximum current per microspot of 20–40 A." Taking the reported data, the current density is calculated as 2.5 × 10^7 A/cm^2 using current 20 A and the size 10 μm. Also, it was stated that "can appear and disappear within one or several nanoseconds; their typical residence time is a few nanoseconds." If it is true, the rate of current rise is ~10^{10} A/s. At this condition, the discharge was similar to spark, and the current density is possible to be significantly large as described for nanosecond cathode spot in [103]. Note that, the observed luminous fragments located very close one to the other are influencing the current distribution and an effective area.

As it was indicated in [89], the plasma particles are excited at the edge of the dense spot core, but they radiate at some distance due to the finite lifetime of the excited levels and the plasma expansion. It should be noted that the high-speed plasma expansion was occurred in the direction normal to the cathode surface. The radial expansion is less than the normal, and the velocity of dense plasma is significantly smaller than in the developed jet [90]. The expanding exited particles are also of ions and electrons, which contribute the spot current increasing and the active area.

The question arise is how the current was distributed and what is integral over this distribution taking into account the expanding plasma area (transition region) before it was optically thin for possibility of line radiation?

Juttner [66] concludes that high-speed pictures insight into the cathode spots using continuum radiation indicate a small size of the spot plasma also for discharge times >1 μs. However, generally highly non-stationary spot nature is periodically interrupted by more stationary phases with increased brightness by their average period T_n, approximated by $(T_n)^n$= const. Although that it was explained as an effect of accumulated heat at the surface during the random walking of the spots as a hypothesis, nevertheless this effect can be related to the sizes determined by the current density.

7.3.6 *Effects of Small Cathode and Low-Current Density. Heating Estimations*

It should be noted that the discussed experiments were conducted with small Ti cathode diameter ($2R = 300$ μm) for 90 A arc and very low gap distance <100 μm in [57] and with copper cathode in form a thin wire (diameter 280 μm) separated 20–50 μm from the copper anode in [60]. As the heat conductivity for Ti is relatively low ($\lambda_T = 0.155$ W/cm), this cathode can be sufficiently heated. Assuming that heat flux (due to very short gap) is formed by the used arc current and an u_{ef} cathode-effective voltage is 10 V (can be larger due to fluctuations in the transient process) the average cathode surface temperature T_s can be estimated from

$$T_s = \frac{I u_{\text{ef}} \sqrt{at}}{\lambda_T S},$$

where a is the thermal diffusivity, t is the time after arc ignition, and $S = \pi R^2$ is the cathode surface area.

The estimation shows that T_s for 90 A arc and Ti cathode can reach ~1600 K for arc time 400 ns and ~3900 K for arc time 3 μs. For large Cu heat conductivity cathode [68–70] with $u_{\text{ef}} = 10$ V, $I = 100$ A, and $R = (1.5\text{–}4) \times 10^{-2}$ cm, the temperature reaches T_s~10^3 C when t increased up to ≥100 μs. These temperatures can be larger taking into account non-uniform current distribution in the arc. Thus, the background temperature of the surface relatively small cathodes and gaps can reach a value that influences the spot ignition and its development and therefore the current density. Indeed, according to Juttner [67, 71], the brightness increases when a fragment is formed at a location that is still hot due to the action of a predecessor. Also, there the rate of current rise can sufficiently be larger than 10^8 A/s.

The above-mentioned low-current density (4×10^4 A/cm^2) obtained by the measurements done by taking the pictures of a spot in the light of atomic and ion emission lines [96] is problematic due to their limited heat regime. At the velocity 4×10^3 cm/s and spot diameter 0.015 cm, the time of spot residence is ~3.75 μs. For these conditions and u_{ef}= 10 V in 3D heat conduction approximation, the cathode temperature in the spot reached of about 600 K assuming that the ion heat flux was determined even by the measured value of spot current density. So, the spot operation cannot be supported self-consistently. According to Daalder [76] the main reason for the differences in the current density reported in [97] and the values found in other studies can be ascribed to the difference of arc regions observed which directly follows from the different used measuring techniques. Another interpretation of detected the large luminous area with radius of about 100 μm [97] can be understandable taking into account the accelerating plasma expansion with velocity 2×10^6 cm/s during 10 ns [101].

7.3.7 Concluding Remarks

Thus, the above consideration of the experimental study of the spot current density indicates unclear and questionable points, and the results meet some vagueness and complications to construe of the measurements. In order to obtain an adequate interpretation of the spot experiment in different conditions, it is preferable to understand the main spot phenomena by interconnection with the measurements.

An attempt to explain the spot nature by electron explosion emission indicates only a single action with duration of a few nanoseconds, and then, the plasma disappears by high-velocity expansion, i.e., only cyclic regime. As it was shown, in some vacuum discharges, the fast cathode processes were developed without explosive emission [106]. The spot lifetime can be significantly larger than the nanoseconds explosion and plasma expansion times. For example, to reach a few amperes per spot in case of current rise of about 10^7 A/s (characterized the arc type discharge) the spot lifetime, which sometime appeared during the arc pulse (not on DC current), should be at least in range of 0.1 μs [97]. Obviously, negligible current reached at nanoseconds for such rate of current rise. In addition, the spots leave continuous tracks on thin film cathodes indicating complete evaporation of the metal from the glass substrate without any signs of an explosion process.

Therefore, in general the cathode spot operation should consider the interconnected processes in a self-sustained object by a closed model used minimal number of input parameters. The analysis shows that it available also at DC arc when spot life is in range of 10–100 ns. This means that in case of electrical breakdown of the gap (e.g., by explosions of a protrusion), the further development of the spot should describe the phenomena occurring in the cathode body and in the initial near-cathode plasma in order to continue to generate the plasma to support the current continuity. Chapter 15 presents a review of the developed different theoretical approaches with an analysis of the existing cathode spot mechanisms. The explosion model and models based on the traditional approaches [107, 108] were considered. Also, the author's approaches are systematized below by spot modeling and theoretical description of the surface and plasma phenomena by taking into account the modern plasma theory. In this framework, the spot initiation and development occurs by arising the amount of preliminary plasma as will be described later (Chaps. 16 and 17).

References

1. Holm, R. (2000). *Electric contacts. Theory and applications*. Berlin: Springer.
2. Kharin, S. N., Nouri, H., & Amft, D. (2005). Dynamic of arc phenomena at closure of electrical contacts in vacuum circuit breakers. *IEEE Transactions on Plasma Sciences, 33*(N5), 1576–1581.
3. Kesaev, I. G. (1968). *Cathode processes in electric arcs*. Moscow: Nauka Publishers.
4. Kesaev, I. G. (1964). *Cathode processes in the mercury arc*. NY: Consultants Bureau.
5. Kesaev, I. G. (1963). Stability of metallic arcs in vacuum. I *Soviet Physics—Technical Physics, 8*(N5), 447–456.
6. Kesaev, I. G. (1963). Stability of metallic arcs in vacuum. II *Soviet Physics—Technical Physics,* 8(N5), 457–462.
7. Kesaev, I. G. (1960). Internal instability of an arc with a mercury cathode. Transient processes in the cathode region of the arc. II *Soviet Physics—Technical Physics, 5*(N6), 635.
8. Kesaev, I. G. (1960). Phenomena of internal instability of an arc with a mercury cathode. Destruction and re-formation of a cathode spot in the condition of a stationary arc. III *Soviet Physics—Technical Physics, 5*(N7), 758.
9. Nazarov, S. N., Rakhovsky, V. I., & Zhurbenko, V. G. (1990). Voltage drop over a vacuum arc and the cathode-spot brightness. *IEEE Transactions on Plasma Science, 18*(3), 682–684.
10. Smeets, R. P. P., & Schulpen, F. J. H. (1988). Fluctuations of charged particle and light emission in vacuum arcs. *Journal of Physics. D. Applied Physics, 21*(2), 301–310.
11. Smeeth, R. P. P. (1986). Stability of low current vacuum arcs. *Journal of Physics. D. Applied Physics, 19*(4), 575–587.
12. Paulus, I., Holmes, R., & Edels, H. (1972). Vacuum arc response to current transient. *Journal of Physics. D. Applied Physics, 5*(1), 119–133.
13. Kesaev, I. G. (1964). Law governing the cathode drop and threshold currents in an arc discharge on pure metals. *Soviet Physics—Technical Physics, 9*(8), 1146–1154.
14. Grakov, V. (1967). Cathode fall of an arc discharge in a pure metal. *Soviet Physics—Technical Physics, 12,* 286–292.
15. Duddell, W. (1904). On the resistance and electromotive forces of the electric arc. *Philosophical Transactions of the Royal Society of London. Series A, Containing Papers of a Mathematical or Physical Character, 203*, 305–342.
16. Mitchel, G. R. (1970). High current vacuum arcs. Part 1—An experimental study. *Proceedings in Instruments and Electrical Engineeringd, 117*(N12), 2315–2326.
17. Kutzner, J., & Miller, H. C. (1992). Integrated ion flux emitted from the cathode spot region of diffuse vacuum arc. *Journal of Physics. D. Applied Physics, 25,* 686–693.
18. Davis, W. D., & Miller, H. C. (1969). Analysis of the electrode products emitted by dc arcs in a vacuum ambient. *Journal of Applied Physics, 40,* 2212–2221.
19. Reece, M. P. (1963). The vacuum switch. *Proceedings IEE, 110,* 793–811.

20. Anders, A., & Yushkov, G. (2002). Ion flux from vacuum arc cathode spot in absence and presence of a magnetic field. *Journal of Applied Physics, 91*(8), 4824–4832.
21. Brown, I. G., Feinberg, B., & Galvin, J. E. (1988). Multiply stripped ion generation in the metal vapor vacuum arc. *Journal of Applied Physics, 63*(10), 4889–4898.
22. Agarwal, M. S., & Holmes, R. (1984). Arcing voltage of the metal vapor vacuum arc. *Journal of Physics. D. Applied Physics, 17,* 757–767.
23. Daalder, J. E. (1977). Energy dissipation in the cathode of a vacuum arc. *Journal of Physics. D. Applied Physics, 10,* 2225–2234.
24. Plyutto, A., Ryzhkov, V., & Kapin, A. (1965). High speed plasma streams in vacuum arcs. *Soviet Physics JETP, 20,* 328–337.
25. Paul, M. O. (1967). Vacuum arc voltage. *Nature, 215,* 1474–1475.
26. Kantsel', V. V., Kurakina, T. S., Potokin, V. S., Rakhovskii, V. I., & Tkachev, L. G. (1968). Thermophysical parameters of a material and electrode erosion in a high current vacuum discharge. *Soviet Physics—Technical Physics, 13,* 814–817.
27. Kimblin, C. W. (1973). Erosion and ionization in cathode spot regions of vacuum arcs. *Journal of Applied Physics, 44*(7), 3074–3081.
28. García, L. A., Pulzara, A. O., Devia, A., & Restrepo, E. (2005). Characterization of a plasma produced by pulsed arc using an electrostatic double probe. *Journal of Vacuum Science and Technology, A23*(3), 551–553.
29. Grakov, V. (1967). Cathode fall in vacuum arcs with deposited cathodes. *Soviet Physics—Technical Physics, 12,* 1248–1250.
30. *Vacuum arcs. Theory and applications.* J. M. Lafferty (Ed.), Wiley, NY (1980).
31. *Handbook of vacuum arc science and technology*, R. L. Boxman, P. J. Martin, & D. M. Sanders (Eds.) Noyes Publ. Park Ridge, N.J. (1995).
32. Somerville, J. M., Blevin, W. R., & Fletcher, N. H. (1952). Electrode phenomena transient arcs. *Proceedings of Physical Society. B, 65*(12), 963–970.
33. Froome, K. D. (1948). The rate of growth of current and the behaviour of the cathode spot in transient arc discharges. *Proceedings of Physical Society. B, 60*(5), 435–464.
34. Froome, K. D. (1949). The behaviour of the cathode on an undisturbed mercury surface. *Proceedings of Physical Society. B, 62*(12), 805–812.
35. Froome, K. D. (1950). The behaviour of the cathode spot on an undisturbed liquid surface of low work function. *Proceedings of Physical Society. B, 63*(6), 377–385.
36. Grakov, V., & Hermoch, V. (1963). Cathode spots behaviour in a high current electrical discharge. *Czechoslovak Journal of Physics, B13,* 509–517.
37. Zykova, N. M., Kantsel, V. V., Rakhovsky, V. I., Seliverstova, I. F., & Ustimets, A. P. (1971). The dynamics of the development of cathode and anode regions of electric arcs I. *Soviet Physics—Technical Physics, 15*(11), 1844–1849.
38. Rakhovsky, V. I. (1970). *Physical bases of the commutation of electric current in a vacuum.* Translation NTIS AD-773 868 (1973) of Physical fundamentals of switching electric current in vacuum, Moscow, Nauka Press.
39. Rakhovsky, V. I. (1976). Experimental study of the dynamics of cathode spots development. *IEEE Transactions on Plasma Science, 4,* 87–102.
40. Selicatova, S. M., & Lukatskaya, I. A. (1972). Initial stage of a disconnection vacuum arc. *Soviet Physics—Technical Physics, 17,* 1202–1208.
41. Djakov, B. E., & Holmes, R. (1971). Cathode spot division in vacuum arcs with solid metal cathodes. *Journal of Physics. D. Applied Physics, 4,* 504–509.
42. Djakov, B. E., & Holmes, R. (1974). Cathode spot structure and dynamics in low current vacuum arcs. *Journal of Physics. D. Applied Physics, 7,* 569–580.
43. Kimblin, C. W. (1974). Cathode spot erosion and ionization phenomena in the transition from to atmosphere pressure arcs. *Journal of Applied Physics, 45*(12), 5235–5244.
44. Persky, N. E., Sysun, V. I., & Kromoy, Y. D. (1989). The dynamics of cathode spots in a vacuum discharge. *High Temperature, 27*(6), 832–836.
45. Sherman, J. C., Webster, R., Jenkins, J. E., & Holmes, R. (1975). Cathode spot motion in high current vacuum arcs on copper electrodes. *Journal of Physics. D. Applied Physics, 8,* 696–702.

46. Agarwal, M. S., & Holmes, R. (1984). Cathode spot motion in high current vacuum arcs under self-generated azimuthal and applied axial magnetic fields. *Journal of Physics. D. Applied Physics, 17,* 743–756.
47. Siemroth, P., Schuke, T., & Witke, T. (1995). Microscopic high-speed investigations of vacuum arc cathode spots. *IEEE Transactions on Plasma Science, 23,* 919–925.
48. Siemroth, P., Schuke, T., & Witke, T. (1997). Investigation of cathode spots and plasma formation of vacuum arcs by high speed microscopy and spectroscopy. *IEEE Transactions on Plasma Science, 25*(4), 571–579.
49. Achtert, J., Altrichter, B., Juttner, B., Peach, P., Pusch, H., Reiner, H. D., et al. (1977). Influence of surface contaminations on cathode processes of vacuum discharges. *Beitrage Plasma Phys, 17*(6), 419–431.
50. Juttner, B. (1981). Cathode phenomena with arcs and breakdown in vacuum. *Beitrage Plasma Phys, 21*(3), 217–232.
51. Bushik, A. I., Juttner, B., & Pusch, H. (1979). On the nature and the motion of arc cathode spots in UHV. *Beitrage Plasma Phys, 19*(3), 177–188.
52. Bushik, A. I., Shilov, V. A., Juttner, B., & Pusch, H. (1986). Behaviour of the spots and local heating of the cathode surface. *High Temperature, 24*(3), 326–332.
53. Mitskevich, M. K., Bushik, A. I., Bakuto, I. A., Shilov, V. A., & Devoino, I. G. (1988). *Electroerosion treatment of metals.* Minsk: Nauka & Technika. in Russian.
54. Bushik, A. I., Bakuto, I. A., Mitskevich, M. K., & Shilov, V. A. (1990). About evolution of phenomena on cathode surface in the process of repeated pulse vacuum arc action. In *Proceedings XIV International Symposium on Discharges and Electrical Insulation in Vacuum* (pp. 208–212).
55. Bushik, A. I., Juttner, B., & Pusch, H. (1980). Dynamics of the cathode processes on bimetal electrode under ultrahigh vacuum. *High Temperature, 18*(4), 555–560.
56. Beilis, I. I., Djakov, B. E., Juttner, B., & Pursch, H. (1997). Structure and dynamics of high-current arc cathode spots in vacuum. *Journal of Physics. D. Applied Physics, 30,* 119–130.
57. Daalder, J. E. (1983). Random walk of cathode arc spots in vacuum. *Journal of Physics. D. Applied Physics, 16,* 17–27.
58. Chaly, A. M., Logatchev, A. A., & Shkol'nik, S. M. (2005). Cathode spot dynamics on pure metals and composite materials in high current vacuum arcs. *IEEE Transactions on Plasma Science, 25*(4), 564–570.
59. Anders, A., Anders, S., Jüttner, B., Pursch, H., Bötticher, W., & Lück, H. (1992). Vacuum arc cathode spot parameters from high-resolution luminosity measurements. *Journal of Applied Physics, 71*(10), 4763–4770.
60. Anders, A., Anders, S., Juttner, B., Botticher, W., Luck, H., & Schroder, G. (1992). Pulsed dye laser diagnostics of vacuum arc cathode spots. *IEEE Transactions on Plasma Science, 20*(4), 466–472.
61. Anders, A., Anders, S., & Juttner, B. (1992). Brightness distribution and current density of vacuum arc cathode spots. *Journal of Physics. D. Applied Physics, 25,* 1591–1599.
62. Juttner, B. (1998). Displacement times of arc cathode spots in vacuum. In *Proceedings of XVIIIth International Symposium on Discharges and Electrical Insulation in Vacuum* (pp. 194–197). Endhoven, The Netherlands.
63. Rakhovsky, V. I. (1972). Recent advances in the study of vacuum electric arc. In *Proceedings of XVIIIth International Symposium on Discharges and Electrical Insulation in Vacuum* (p. 179). Poznan, Poland.
64. Beilis, I. I., Levchenko, G. V., Potokin, V. S., Rakhovsky, V. I., & Rykalin, N. N. (1967). About heating of a body acted by a moving high-power concentrated heat source. *Physics. & Chemistry of a Material Treatment, 1*(N3), 19–24.
65. Djakov, B. E. (1993). Cathode spot phenomena in low current vacuum arcs on arc cleaned electrode surfaces. I. Spot size. *Contributed Plasma Physics, 33*(4), 307–316.
66. Djakov, B. E. (1993). Cathode spot phenomena in low current vacuum arcs on arc cleaned electrode surfaces. II. Spot dynamics. *Contributed Plasma Physics, 33*(4), 201–207.

67. Juttner, B. (1995). The dynamics of arc cathode spots in vacuum. *Journal of Physics. D. Applied Physics, 28*(3), 516–522.
68. Juttner, B. (1997). The dynamics of arc cathode spots in vacuum: New measurements. *Journal of Physics. D. Applied Physics, 30*(2), 221–229.
69. Juttner, B. (1998). The dynamics of arc cathode spots in vacuum. Part III: Measurements with improved resolution and UV radiation. *Journal of Physics. D. Applied Physics, 31,* 1728–1736.
70. Juttner, B. (1999). Nanosecond displacement times of arc cathode spots in vacuum. *IEEE Transactions on Plasma Science, 27*(4), 836–844.
71. Jüttner, B. (2001). Cathode spots of electric arcs. *Journal of Physics. D. Applied Physics, 34,* R103–R123.
72. Juttner, B. (2004). The influence of the surface structure on the behaviour of arc cathode spots in vacuum. In *Proceedings XXIth International Symposium on Discharges and Electrical Insulation in Vacuum* (pp. 147–151). Yalta, Crimea.
73. Djakov, B. E., & Juttner, B. (2002). Random and directed components of arc spot motion in vacuum. *Journal of Physics. D. Applied Physics, 35,* 2570–2577.
74. Juttner, B. (1997). Properties of arc cathode spots. *Journal of Physics France, 7,* C4-31-C4-45.
75. Cobine, J. D., & Gallagher, C. J. (1948). Current density of the arc cathode spot. *Physical Review, 74*(10), 1524–1530.
76. Daalder, J. (1974). Diameter and current density of single and multiple cathode discharges in vacuum. *IEEE Transactions on Power Systems, PAS-93*(N5), 1747–1758.
77. Guile, A. E., Hitchcock, A. H., & Stephens, G. W. (1977). Emitting site lifetimes, currents and current densities on arc cathodes with 100 nm thick copper-oxide films. *Proceedings IEE, 124*(3), 273–276.
78. Guile, A. E., Hitchcock, A. H., & Barlow, J. M. (1977). Transition in size and number of emitting sites with increase in arc speed over a Cu cathode. *Proceedings IEE, 124*(4), 406–410.
79. Hitchcock, A. H., & Guile, A. E. (1977). Effect of copper-oxide thickness on the number and size of arc cathode emitting sites. *Proceedings IEE, 124*(5), 488–492.
80. Hitchcock, A. H., & Guile, A. E. (1977). A scanning electron microscope study of the role of copper oxide layers on arc cathode erosion rates. *Journal of Materials Science, 12,* 1095–1104.
81. Guile, A. E., & Hitchcock, A. H. (1978). Arc cathode craters on copper at high currents and with reduced gas pressures. *Proceedings IEE, 125*(3), 251–256.
82. Mau, H. J. (1964). Uber die Katodenfleck-Structur des bewegten electrischen Lichbogens, IX Int. Colloquium, TH Ilmenau, 1964, Berlin, 71, Berlin Akad. Verl.
83. Juttner, B. (1979). Erosion craters and arc cathode spot in vacuum. *Beitrage Plasma Physics, 19*(1), 25–48.
84. Niirnberg, A. W., Fang, D. Y., Bauder, U. H., Behrisch, R., & Brossa, F. (1981). Temperature dependence of the erosion of Al and TiC by vacuum arcs in a magnetic field. *Journal of Nuclear Materials, 103,* 305–308.
85. Puchkarev, V. F., & Murzakayev, A. M. (1990). Current density and the cathode spot lifetime in a vacuum arc at threshold currents. *Journal of Physics. D. Applied Physics, 23*(1), 26–35.
86. Bochkarer, M. B., & Murzakaev, A. M. (2004). Investigations of vacuum arc cathode spots with high temporal and spatial resolution. In *Proceedings XXIth International Symposium on Discharges and Electrical Insulation in Vacuum* (pp. 244–251). Yalta, Crimea.
87. Anders, S., & Juttner, B. (1987). Arc cathode processes in the transition region between vacuum arcs and gaseous arcs. *Contributions to Plasma Physics, 27,* 223–236.
88. Anders, S., & Juttner, B. (1991). Influence of residual gases on cathode spot behavior. *IEEE Transactions on Plasma Science, 19,* 705–712.
89. Hantzsche, A., Juttner, B., & Ziegenhagen, G. (1995). Why vacuum arc cathode spots can appear larger larger than they are. *IEEE Transactions on Plasma Science, 23*(1), 55–64.
90. Beilis, I. I. (1995). Theoretical modeling of cathode spot phenomena. In R. L. Boxman, P. J. Martin, & D. M. Sanders (Eds.), *Handbook of vacuum arc science and technology* (pp. 208–256) Noyes Publ. Park Ridge, N.J.

91. Drouet, M. G., & Meunier, J. L. (1985). Influence of the background gas pressure on the expansion of the arc plasma. *IEEE Transactions on Plasma Science, PS-13*(N5), 285–287.
92. Anders, S., & Anders, A. (1991). Emission spectroscopy of low-current vacuum arcs. *Journal of Physics. D. Applied Physics, 24*(11), 1986–1992.
93. Beilis, I. I., Sagi, B., Zhitomirsky, V., & Boxman, R. L. (2015). Cathode Spot motion in a vacuum arc with a long roof-shaped cathode under magnetic field. *Journal of Applied Physics, 117*(N23), 233303.
94. Vogel, N., & Juttner, B. (1991). Measurements of the current density in arc cathode spots from the Zeeman splitting of emission lines. *Journal of Physics. D. Applied Physics, 24,* 922–927.
95. Jüttner, B., Pursch, H., & Anders, S. (1984). On the current density at the cathode of vacuum arcs. *Journal of Physics. D. Applied Physics, 17*(8), L111–L113.
96. Juttner, B. (1985). Current density in arc spots. *IEEE Transactions on Plasma Science, PS-13*(N5), 230–234.
97. Golub, V. I., Kantsel, V. V., & Rakhovsky, V. I. (1972). Current density and behaviour of cathode spots on copper during an arc discharge. In *Proceedings of 2nd International Conference on Gas Discharges (London)* (pp. 224–226).
98. Smith, G. P., Dollinger, R., Malone, D. P., & Gilmour, A. S. (1980). Relative cathode spot and cell areas and currents in a copper cathode. *Journal of Applied Physics, 51*(7), 3657–3662.
99. Siemroth, P., Schultrich, B., & Schtilke, T. (1995). Fundamental processes in vacuum arc deposition. *Surface and Coatings Technology, 92*(96), 74–75.
100. Boyle, W. S., & Germer, L. H. (1955). Arcing at electrical contacts on closure. Part IV. The anode mechanism of extremely short arcs. *Journal of Applied Physics, 26*(N5), 571–574.
101. Beilis, I. I., Kanzel, V. V., & Rakhovsky, V. I. (1975). On explosive model of fast moving cathode spot. In *Proceedings of Union Conference on Electrical Contacts and Material, 1973; Chapter in book: Electrical contacts* (pp. 14–16).Moscow, Nauka.
102. Rakhovsky, V. I. (1984). Current density per cathode spot in vacuum arcs. *IEEE Transactions on Plasma Science, PS-12*, 199–203.
103. Beilis, I. I. (2001). A mechanism for nanosecond cathode spot operation in vacuum arcs. *IEEE Transactions on Plasma Science, 29*(N5), Part 2, 844–847.
104. Beilis, I. I. (2011). Continuous transient cathode spot operation on a microprotrusion: Transient cathode potential drop. *IEEE Transactions on Plasma Science, 39*(N6), Part 1, 1277–1283.
105. Beilis, I. I. (2013). Cathode spot development on a bulk cathode in a vacuum arc. *IEEE Transactions on Plasma Science, 41*(N8), Part II, 1979–1986.
106. Mazurek, B., & Cross, J. D. (1988). Fast cathode processes in vacuum discharge development. *Journal of Applied Physics, 63*(10), 4899–4904.
107. Ecker, G. (1961). Electrode components of the arc discharge. *Ergeb. Exakten Naturwiss, 33,* 1–104.
108. Lee, T. H., & Greenwood, A. (1961). Theory for the cathode mechanism in metal vapor arcs. *Journal of Applied Physics, 32*(5), 916–923.

Chapter 8
Electrode Erosion. Total Mass Losses

Vacuum arc is a discharge of relatively high current (>1 A for bulks) where the conductive material in the interelectrode gap arises in course of mass loss from the electrodes during the arcing. Different phenomena of the electrode degradation occurred due to the local and high intensive electrical energy dissipation in the arc gap and at the electrode surfaces. The experiments [1–5] showed that total mass loss of the electrodes (cathode or anode) occurs in the form of plasma (neutral atoms, ions, and electrons), liquid droplets, and solid particles.

The observed cathode surface after arcing was represented by damage pattern including small craters, traces, and relatively large cavities depending on arc current. The craters about 300 μm were observed even for graphite cathodes [6]. The erosion mass and erosion rate significantly depend on the arc current, gap distance, and arc duration. On the other hand, during the electrode erosion, a high pressure near the electrodes was produced and therefore causing macroparticle ejection.

8.1 Erosion Phenomena

8.1.1 General Overview

The phenomena of electrical erosion are an extremely important issue in a number of practical applications of the arc. On one hand, it plays harmful role leading to intense damage in different devices like high-voltage switchgears, relay contacts, contacts in electrical apparatus, and other discharge. At the same time, the positive role of the erosion phenomenon is extensively used for cutting, drilling, electrical discharge machining (electrospark treatment) of metals, thin film deposition, coating, and ion beam generation and ion implantation.

Duffield [7] reported about first experiments studied the electrode erosions in the electrical arcs. According to his brief review, some interesting results were obtained

I. Beilis, *Plasma and Spot Phenomena in Electrical Arcs*, Springer Series on Atomic, Optical, and Plasma Physics 113, https://doi.org/10.1007/978-3-030-44747-2_8

already in nineteenth century. Silliman in 1825 found that the lost weight from positive electrode is more rapid than from the negative carbon electrode. The same result reported by Matteucci in the "Comptes Rendus, vol. 30, p. 201 (1850)", concluding from his experiments. (1) The loss is chiefly dependent upon elevation of temperature by other factors being constant. (2) For carbon and iron, the positive electrode mass loses more than the negative, the ratio of the losses varying according to the arc length from 2:1 to 5:1 for carbon, the ratio being smaller for iron. For zinc, copper, tin, lead, brass, and gold, the negative loses more than the positive. (3) For electrodes that are more refractory, the quantity of mass lost was less. W. S. Weedon (Trans. Electrochem. Soc, vol. 5, p. 171 1904) has measured the losses from the electrodes of certain metallic arcs. For copper electrodes, he found that, if the current increased by 2.5 times, the cathode loss was five times as great. When an iron arc burned in hydrogen using water-cooled electrodes, the anode lost and the cathode gained.

In the mentioned review [7] Duffield also described his early experiments, which have been carried out to determine the amount of material lost by carbon electrodes of a continuous current of an arc under different arc current and length. The experiments dealt with arcs whose lengths varied from 0.5 to 15 mm or more and the current from 1 to 10 A. The carbons employed were initially all of the same diameter (10 mm.), but, after having been burnt to shape, their diameters in the neighborhood of the arc gap were much smaller, the electrodes having so adjusted themselves that there was a characteristic contour for each value of the current strength and arc length.

Two methods were used to assess the erosion rate. In one, the carbons were held in special clips, which could be removed and weighed. The second method uses some arrangement, which obviated the necessity for touching the electrode at all when the weighing made. Table 8.1 presents the results.

The measurements show that the mass loss per coulomb for a given current increases with increasing arc length until a nearly constant value is reached at about 8 mm. This is true for both the anode and cathode, but there is a difference in the initial rates of grows. The anode loss per coulomb is at first gradual and then increases more rapidly as the electrodes are further separated, whereas the cathode slope is very steep at the outset and shows no point of change. The obtained dependences explained by the increasing oxidation of the hot electrodes in the air with increase of the arc length. The greater anode loss than the loss from the cathode under similar conditions of arc length and current suggested to explain by the observed larger of a crater on the anode.

While the cathode absolute mass loss increases with the current, the loss of material per coulomb for long arcs decreases with increasing current. For long arcs, the rate of loss at first decreases rapidly with increasing current but subsequently more slowly until some minimum value is reached for large currents. For short arcs, the cathode loss of material per coulomb is practically constant. The author has not presented any reasonable explanation of the dependence for long arcs. However, it is possible to understand this dependence taking into account the specifics of returned heavy particles at atmosphere of air arc with current increasing. This specific also follows from the data at short arcs indicated constant erosion per coulomb.

Table 8.1 Rate of carbon electrode mass loss as dependences on arc length and arc current [7]

Anode erosion rate (μg/C)						Cathode erosion rate (μg/C)						
	Current (A)											
Length (mm)	2	4	8	10	100	1	2	4	6	8	10	100
0.1			89.0		77					31.2		32
0.2	165	132		92			33.4	33.8	34.2		31.6	
0.5					10		102	71	48	45	41	
1	253	176	128	110		202		92	65	55	54	
2	266	181	131	120		270	192	107	77	61	59	
3	276	194				351		136				
4	296	200	138			374	217	145	103		74	
5		213				377		155				
6	346		152			363	239	157	114	101	85	
7		214						158				
8	345	225	155			351	246	157	120	97	85	
10		222	166				251	161	121		89	
12		228	159					160	123			
17							258	163		101		
30					132							102

The electrode erosion widely studied by Ragnar Holm and colleagues in period of 30–50 years of the twentieth century for understanding the processes in electrical contacts in light of development of automatic and telemechanic devices. These early investigations described in last edition of his book [8]. Important results obtained including effect of electrode polarity, mass transfer from one electrode to another, the role of thermophysical, mechanical, and chemical properties in erosion processes.

The investigation of the erosion of contacts showed a possibility to use this phenomenon for machining of a metallic targets in order to hardening, deposition of a protective layer, change the surface properties or geometry [9]. The machining, which is used in practice, includes processes of doping, deposition or piece treatment using repeated current pulsed electrical arcs to melt and evaporate material from an anode and to transport it to the workpiece, which is held in close proximity. The workpiece serves as the cathode, and the electrical discharge action at its surface removes surface contaminants so that, for example, a coating forms. The applications of the contact systems known also as "electroerosion", "electrosparking", or "electrical discharge machining (EDM)" systems were operated usually in gas atmosphere or in liquids to minimize the erosion rate. When a part of electrode material should be removed due to electrical erosion, the dielectric liquids were used for workpiece treatment. The gas media was used for electrospark doping or coatings when a material transferred from one electrode to other. The contact erosion phenomena occur at very short electrode gaps and can be significant both on the anode and on the cathode. The short distance between the electrodes together with the high pressure of the generated metallic plasma excludes the used media from the electrode gap and allows to minimize its influence on the arc. Below the electrode erosion will be considered for the arcing in different media and at both electrodes in order to understand the phenomena with respect to the erosion rate in vacuum arcs.

8.1.2 Electroerosion Phenomena in Air

An electric arc arises at closure or breaking contacts. According to early work of Germer and Haworth [10], the vaporization of anode metal is one of the major factors in the erosion of telephone relay contacts. The observed crater of about 5 μm at positive platinum electrode was produced by discharge of energy 11 g cm^2 s^{-2}, while on the negative electrode surface was showed the presence of metal spattered. The crater formed on the anode is result of the melting and boiling of the metal caused by electron bombardment before the electrodes touch. The mass transfer in closure and breaking contact systems with very small arc current around 1 A and in extremely short gap (0.1 μm) was studied by Germer and Haworth [11] and with arc current (1–7) A by Boyle and Germer [12]. The measurements were conducted by microscopic observation of the electrode surfaces and by the optical method of the mound metal, which develops on negative electrode. The electrode surfaces were observed after the discharges. It was found that craters were developed in the positive and mound on negative electrodes. The metal was transferred from anode to the cathode. It was reported that the measured transferred metal volume was proportional the discharge energy and the values for palladium and platinum contacts were 4.5×10^{-14} and 4.0×10^{-14} cm^3/erg, respectively, at 10^5 closures. Germer and Boyle [13] studied the transfer of cathode or anode material from one to other electrode.

The following erosion mechanism was proposed [10]. When an arc occurs at the metal surface, the energy from charged condenser is dissipated almost entirely upon the positive electrode and melts the metal forming a crater from which then the metal can be vaporized. Part of the melted metal landed on the negative electrode and, with repeated operation, results in a mound of metal transferred from anode to the cathode. It was also noted that bridge erosion can be also as a mechanism of the transfer of metal from one electrode to the other which occurs when an electric current is broken in a low voltage circuit [14]. It is resistive metal heating at some local contacts when the electrodes were pulled apart, and a molten bridge was formed between the electrodes before the contact is finally broken.

Mandel'shtam and Rayskiy [15] presented an experimental research of the erosion mechanism of metals in a spark discharge. Accordingly, the authors determine the fundamental difference of a spark discharge, high voltage or low voltage, from an arc discharge. The key difference lies in the fact that in a spark the discharge channel does not expand due to the short duration of the current pulses. Due to this fact, the current density in a spark discharge channel reaches significantly higher values, than that in an arc discharge. This property of a spark discharge forms the specific features of the high temperature discharge channel, leading to excitation of "spark lines," and the specific features of the discharge on the electrode. These features result in the extremely large energy dissipated at the electrode surface, and it does not propagate over large distance in the metal body due to the short discharge duration. As result, the energy was passed into a thin subsurface layer of the metal, and an intensive vaporization arises. It is known from studies of the spark discharge that the formation of electrode metal vapors occurs in the form of plasma flares escaping perpendicular to the electrode surfaces at a high velocity.

Experiments [15] with copper and other electrodes showed that the magnitude of erosion with a large electrode gap (4 mm) was significantly less than with a small gap (0.1 mm). With a small gap, the anode undergoes considerably more severe damage than the cathode. With a large gap, the cathode was damaged due to formation of the flare, while a cathode material was deposited locally at the anode center.

It was concluded that the mechanism of electric spark erosion of metals as a secondary process conditioned by the destructive effect of the plasma flares ejected from the opposite electrode. Movement of the flares at a supercritical velocity is an essential condition for realization of this mechanism. It follows from the fact that the flares expand to the sides immediately after electrode damage that attests to the presence of great pressures in the flare. The damage of the electrode itself ejected the plasma was significantly less in intense damage to the opposite electrode. The explanation for this phenomenon apparently lies in the fact that the buildup of the plasma generation on the electrode occurs more slowly than stopping of moving flares by the opposite electrode.

Pravoverov and Struchkov [16] investigated the erosion rate of a number of electrode materials. The arc was initiated in air of atmosphere pressure by high voltage breakdown with 3 ms pulse and pick current of 9.5 A, gap distance of 0.3 mm. Each experiment uses 10^4 pulses with frequency of 1 Hz of arc between electrodes of 6 mm diameter and 3 mm thickness. The erosion mass was determined by weight method. The total amount of the charge per pulse was obtained as 1.94×10^{-2} C taking into account effective pulse duration of 2.5 ms. The measured erosion rate G in μg/s and erosion rate G_r in μg/C (named also as erosion coefficient) data of different metals are presented in Table 8.2 using the mentioned total charge value. The erosion rate is relatively low except the mass loss for low melting materials such as Zn, In, Sn, and Pb.

Zhouy and Heberlein [17] developed a specially designed thermal plasma reactor system for the investigation of arc cathode erosion. Argon gas and mixtures of argon and hydrogen gases have been used in this investigation. The cathode materials used include pure tungsten and 2% thoriated tungsten. Experiments with cathodes of various shapes and sizes have been performed (cone shape and truncated cone shape with 1 mm diameter of the truncated surface, both with 60° included angle). Cathode diameters have been 6.4 and 3.2 mm. The chamber pressures have been 340 and 760 Torr, and the current levels have been in the range 80–350 A. Observation of cathode spot behavior has been carried out simultaneously by employing a microscope and a high-speed vision system. Cathodes have been examined by SEM and EDX after arcing. For pure tungsten cathodes, the initial cathode geometry has

Table 8.2 Erosion rate in μg/C of different metals according to data of [16]. The gain mass on Ag, Pd, Sn, and Pb anodes is indicated by "+"

Material	Erosion rate (μg/s)		Erosion rate (μg/C)	
	Cathode	Anode	Cathode	Anode
Ag	9.2	+4	1.186	+0.515
Cu	18	17	2.32	2.19
W	30	<1	3.866	<0.129
Nb	36	−31	4.639	4.00
Pd	56	+36	7.216	4.639
Ni	61	2.4	7.861	0.31
Zr	4	62	0.515	8.00
Al	32	68	4.124	8.763
Fe	108	76	13.92	9.794
Cr	84	110	10.825	14.175
Y		220		28.350
Zn	440	120	56.70	15.464
In	520	800	67.0	103.0
Sn	1800	+240	231.96	+30.93
Pb	6000	+8000	773.2	+1031

almost no effect on the cathode spot's behavior due to the molten state of the metal under the cathode spot.

Various mechanisms of arc cathode erosion were presented, and the factors affecting the cathodes erosion were discussed. The major erosion mechanism is the ejection of liquid droplets from the cathode spot. The erosion rate is dependent not only on the cathode tip's condition, namely its temperature and the depth of the molten state, but also on the fluid dynamics of the plasma gas flowing along the cathode. Turbulent pressure fluctuations lead to oscillations of the molten metal and the formation of smaller droplets. However, the initial cathode geometry has a certain influence on the cathode's erosion for 2% thoriated tungsten cathodes. This influence and relatively low erosion rate are demonstrated in Table 8.3. A non-uniform erosion pattern will occur if the cathode is overcooled, probably due to ion bombardment in the low-temperature regions of the arc attachment spot.

Peters et al. [18] reported on erosion studies of a cathode as used in a plasma torch operating at of 200 A. A hafnium insert in a water-cooled copper sleeve serves as the cathode. The experiments allowed the measurement of material loss from the cathode during different phases of an operating cycle. The first phase is the plasma start phase. This phase begins with arc ignition (by high voltage), continues with the arc transfer to the anode and with the gas flow change. The second phase is the steady running phase begins with the change in gas flows and continues until the arc is shutdown. Arc shutdown is the final phase of operation. It used two different current stop mechanisms on the power supply. During the controlled ramp-down process,

Table 8.3 Cathode's weight loss for thoriated tungsten cathodes under various operating conditions at 200 A arc current [17]

Cathode diameter (mm)	Electrode shape	Gas	Weight loss (μg/s)	
			5 min of arcing	60 min of arcing
3.2	Cone shape	Ar	5.33	0.778
3.2	Cone shape	Ar:H_2 (1:1)	8.33	
3.2	Truncated cone	Ar		1.53
6.4	Cone shape	Ar		0.556
6.4	Truncated cone	Ar		0.0278

the arc current operates from its steady value to zero over a period of ~150 ms. For the rapid current shutdown, this drop occurs in ~1 ms. Erosion has been found [18] to be predominantly due to ejection of molten material droplets.

The erosion rate significantly increases when an arc with relatively large current. Essiptchouk et al. [19] showed this increase in experiment with arc current in range of 100–500 A. The measurements were conducted in a system, which was equipped with water-cooled commercial copper-ring electrodes placed in an axial magnetic field. The cathode was isolated from the adjacent parts of the setup by thermal- and electro-insulating spacers. The working gas was axial flow air in the interelectrode gap with velocity of 7.6 m/s. Most of the experiments were made with a cathode ring thickness of 10 mm (outer ring diameter of $2R = 60$ mm). The water-cooling flow rate was maintained constant. In order to increase the cathode surface temperature, some experiments were carried out with a different cathode ring, with $2R = 120$ mm. After each 10 min arc experiment, the cathode ring insert was extracted and weighed to obtain the average mass erosion rate. The results of erosion measurement showed two main characters in the dependence of erosion rate on the arc current. These regimes are determined by the magnetic field, namely (i) regime with relatively low increasing erosion rate and (ii) regime with relatively large increasing erosion rate. The dependencies [19] can be illustrated in the form of Fig. 8.1.

Erosion measurements on a copper cathode were reported in works of [20, 21]. The 100 A arc, driven by a magnetic field, runs continuously for up to 30 min between two concentric cylindrical electrodes having an interelectrode gap of 4 mm. Argon–nitrogen gas mixtures in various proportions are blown through the electrode gap. The erosion rate in argon is drastically reduced by the addition of only 1% nitrogen and is further reduced as the nitrogen content increases in the gas mixture. The decrease in erosion rate is found to be correlated to an increase in arc velocity. The erosion rates dropped from 9.0 to 1 μg/C as the arc velocity was increased from 15 to 135 m/s.

The electroerosion phenomena of gas discharges occurring in technological devices for different applications such as machining of target, electrical interrupters, electrosparking treatment, between contacts in electrical apparatuses were investigated by Namitokov [22]. The contact surface properties, erosion mechanisms in closure and breaking contacts, of safety fuses, and erosion processes that appeared at different discharge stages were reviewed and analyzed. Theoretical investigations

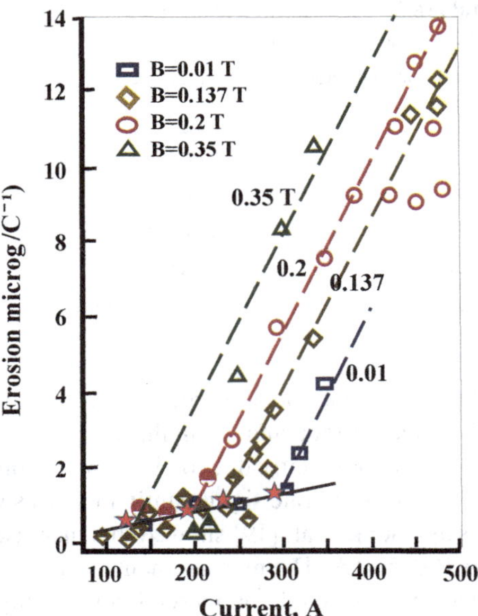

Fig. 8.1 Erosion rate of a copper cathode as function on current I, for a cathode with inner diameter $2R =$ 40 mm, outer diameter $2R =$ 60 mm and for four different magnetic field values. Lines are linear approximations of the experimental points for each value of B. The half-painted points relate to the relatively low increasing erosion regime. Stars mark the transition points from low increasing to a large increasing erosion regime

were presented, which concentrated on models developed to understand the dynamics of the electroerosion processes in spark and pulse arc discharges.

Electrode erosion has been considered in a number of works in relation to electro-spark treatment or also named as electro-discharge machining [23–27]. Parkansky et al. [28] reviewed the applications of electrosparking technique (named there as pulsed air arc deposition (PAAD)), and then, experiments related to the influence of stress on the maximum coating thickness were presented. It was indicated that is a process of coatings which use a chain of high current (10–100 A) short-duration (1–100 μs) pulsed electrical arcs. The melted and evaporated material from a source anode is transported to the workpiece served as cathode, which is held in vicinity to the anode. This coating method has the advantageous characteristics as superior adhesion, high coating density, ability to apply alloy coatings, simple facilitation of selective coatings, minimal workpiece heating and thus no heat deformation nor loss of prior heat treatments, and no need some specific chamber to place the workpiece.

Problems were studied, which related to the hard coatings, heat-resistant coatings, corrosion resistance, reduction in contact resistance, residual tensile stress, effect of mechanical stress applied for the electrodes during the deposition [28]. It was concluded that the maximum coating thickness, which can be applied, about 100 μm, is limited by residual tensile stress, which ultimately causes surface damage and material loss.

The erosion of contact electrodes can be reduced and controlled, by injecting parallel to the contact surface an additional transverse electrical current with a density less than 1 A/mm^2 [29]. In addition to the reduction of electrode erosion, the additional current affects the localization of the arc root on the cathode and affects the morphology of the electrodes surface left after extinguishing the arc. The effect reported here may be applied in real systems to prolong the life of electrical contacts but also to increase the erosion in some other cases.

The main issue at the electroerosion treatment of the metals is the mass transfer from one electrode to another. The electrode mass loss depends on the thermophysical characteristics of the electrode materials. Palatnik [30] proposed a criterion in which the electrode temperature reaches a destruction temperature in an estimated characteristic time. The model assumed that the anodic and cathodic heat fluxes are equal. Taking into account that the cathode and anode spot mechanisms are different, a general criterion was developed, and the heat losses into an infinitely long electrode body (required to reach a certain phase temperature T of the electrode surface) were determined as [31]:

$$q = \sqrt{at}(T - T_0)c\rho \tag{8.1}$$

where the temperature T can be as the melting T_m or boiling T_b, T_0 is the background temperature, a is the thermal diffusivity of the electrode, c is the heat capacity, ρ is the mass density, and t is the time. The erosion process was determined by a ratio of the heat losses into the electrodes of the same geometry. In case anode q_a and cathode q_c heat losses, the ratio is

$$Cr = \frac{q_a}{q_c} = \frac{(T_a - T_0)}{(T_c - T_0)}\sqrt{\frac{\lambda_a c_c \rho_c}{\lambda_c c_a \rho_a}} \tag{8.2}$$

According to (8.2), the mentioned ratio described by parameter Cr depends only on the thermophysical constants of the electrode materials, and Cr can be estimated without consideration of complicated plasma phenomena. In order to understand the dependence of anode erosion, the heat losses expressed by (8.1) were considered for pure Al, Ti, Cu, Fe, W, and WC-based hard alloys contacts with respect to that loss for W material [31]. The calculations using the anode material thermal constants indicated good correlation with the measured series of anode erosion rate.

8.1.3 Electroerosion Phenomena in Liquid Dielectric Media

Kimoto et al. [32] measured the anode and cathode erosion rate in kerosene as dependence on discharge long time 100–700 μs, arc current 10–200 A electrode distance < 0.1 mm by weight method. Anode—hollow Cu tube 20 mm OD and 5 mm ID. The carbon steel cathode was 60 mm diameter and 5 mm thick. The effect of arc time on the erosion rate was determined for time range of 10–100 μs. Also, the experiments were conducted in tap water with gap 0.01 and 0.03 mm for pairs: carbon steel anode–Cu cathode and tungsten carbide anode–soft steel cathode. The cathode crater areas were determined as function on time for arc currents 18, 35, and 50 A.

The obtained results showed that crater areas increase in range (2–9) × 10^{-4} cm^2 when the time increased from 3 to 500 μs, respectively. The current density was obtained using arc current and the crater sizes in range 5 × 10^4–10^6 A/cm^2. The measured copper anode erosion rate G_a was much smaller than that for carbon steel cathode G_c. G_a increases from 1.33 to 500 μg/s, and G_c increases from 170 to 1300 μg/s when current rises from 20 to 200 A for 500 μs arc. Similar result was obtained for 700 μs arc. Thus, the anode erosion rate increases with current by more than 100 times. However, in range of time 10–100 μs, the rate G_a decreases more 100 times with time but still remains smaller than G_c. When the discharge takes place in pure water or in dilute electrolyte solution, the increase in time results in decrease in erosion rate of the copper cathode comparing to the steel anode. The influence the electrode polarity on the erosion rate was explained by the difference of the current density, which depends on the electrode materials of corresponding pairs.

Zolotykh [33] studied the electrode erosion in kerosene by a single discharge pulse with maximal current 1000 A and maximal voltage 200 V, space 30 μm, and energy 2.5 J. Anode is Cu plate—0.1 mm thick, cathode Cu wire 1 mm diameter. The optical observation was conducted by high-speed camera. The photograph showed calm evaporation and gas bubble is being formed after first 80 μs. The following dynamics was observed of the erosion mass evacuation. After 180 μs, the metal was evaporated in shape of flares, and about 85–90% of the metal remained in the crater. After 240 μs ejected vapor cool off, the shape of flares was stopped, and a growing cone was started. After 320 μs, a stream of metal particles or droplets was ejected from the crater, and after 450–500 μs, the crater was cleaned out.

A hydrodynamic model was developed to compare with the observed dynamics of metal evacuation from the crater. The obtained agreement allows to conclude that hydrodynamic nature of forces plays important role for metal ejection that determined the mechanism of electrode erosion.

Zingerman and Kaplan investigated the anode [34] and cathode [35] erosion as dependence on short gap distance in range of 3–500 μm in an oil with initial voltage of 1 kV and relatively high arc current 2 kA. The electrode materials were steel, copper, aluminum, and brass. A 30 × 45 mm plate 2 mm thick was used as anode, and the cathode was a rod 4 mm in diameter in [34], while polarity vice versa was used in [35].

As the arc power Iu_{arc} was enough large, the crater sizes were observed in mm ranges. In general, the measurements showed that for increasing distances d (μm) between electrodes, the diameter L of the craters increased for electrodes of the different materials and polarity. In semilogarithmic coordinates, these dependences become a straight line except steel for cathode (Fig. 8.2a) and brass for anode (Fig. 8.2b). The brass anode diameter remained constant up to $d = 100$ μm, and then, its value increased with d. The degree of the lines inclination k_d can be obtained from dependence $L(d) = L_0 + k_d \times \text{Log}\,(d/d_{in})$ taking into account the initial L_0 (mm), initial d_{in} (mm), and maximal d_{max} distances (in μm), respectively, from Fig. 8.2a and Fig. 8.2b.

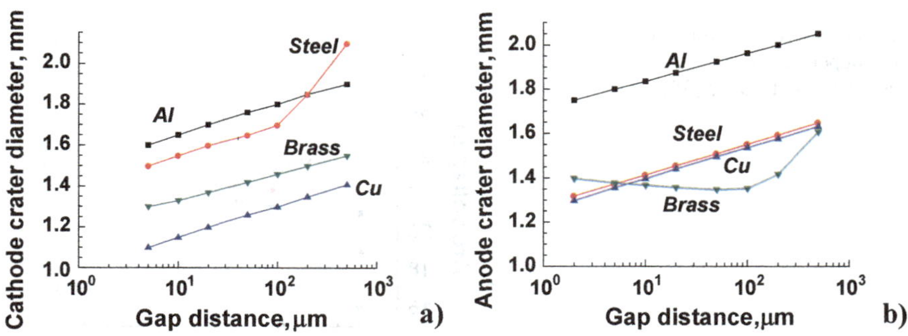

Fig. 8.2 Crater diameter on the cathode (**a**) [34] and on the anode (**b**) [35] as function on interelectrode distance for different materials

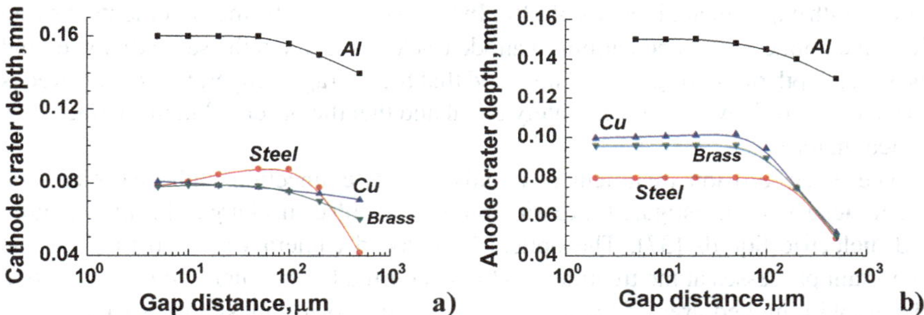

Fig. 8.3 Crater depth on the cathode (**a**) [34] and on the anode (**b**) [35] as function on interelectrode distance for different materials

The crater depths h as function on electrode distances presented for cathode in Fig. 8.3a and for anode Fig. 8.3b for different electrode materials. The figures show that these dependences of h in semilogarithmic coordinates mostly constant up to $L = 75$ μm for anodes and 100 μm for cathodes, and then, h decreased with distance more significant for anodes than for cathodes. Simultaneously, the voltage for arc burning in oil was measured. Figure 8.4 indicates that the voltages are relatively low, mostly constant for all used cathode materials up to $d = 100$ μm. The voltage increases with d for larger gap distances and exceeds that values measured for arcs in vacuum that can be explained by the oil media and relatively large arc current.

A comparison of the electrode damages showed that the calculated volume of the craters for cathode and anode was close one to other for all considered materials. It was indicated that the difference between the diameters of the craters in the anode and in the cathode does not exceed 15%. Also, the difference between arc voltages for positive and negative polarity of the flat electrodes was small, within 3–5%. Taking into account these results, the authors [34, 35] concluded that the erosion phenomena are similar for cathode and anode for high current short discharge, and erosion mechanism was determined by the electrode heat regime. This conclusion

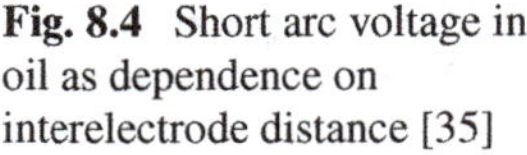

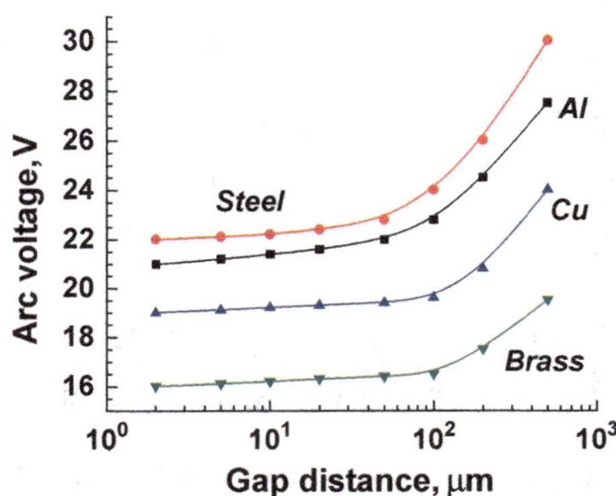

Fig. 8.4 Short arc voltage in oil as dependence on interelectrode distance [35]

agrees with experimental results of [36]. In that work [36], the measurements showed that tin cathode tracks and cadmium anode tracks have much the same diameter and the same depth of melting. It was indicated that the average temperatures of the tracks on each electrode were approximately equal and that the observed melting tracks are of heat nature.

The electroerosion treatment of metals and its connection with near-electrode phenomena was investigated experimentally for pulse discharges in air, vacuum, and dielectric liquids [37]. The optical high-speed camera was used to study the important processes at the treatment, which occurred by the interaction of the active plasma with melted area of the electrode surface (erosion mass removal) and with surrounding liquid (limited the plasma expansion). The main pulse discharge was studied as result of its initiation by an additional triggered discharge. It was observed that the main discharge developed in previously created gas cavity by the triggering discharge. The electrode erosion was in form of melted metal and as metallic vapor. The erosion products were cooled penetrating deep in the moving surrounding liquids due to the hydrodynamic phenomena. The characteristic time of discharge processes (microseconds) was significantly shorter than the processes determined by hydrodynamic (millisecond) forces. This fact determines the parameters of the discharges during their repetitions with some frequency.

Parkansky et al. [38] studied the erosion of the anode and cathode dependence on material combination of the electrode pairs, i.e., on the thermophysical properties of the electrode materials for submerged arc. Low voltage, low energy pulsed arcs with a pulse repetition rate of 100 Hz, energy of 48 mJ and duration of 20 μs were tested to determine the electrode erosion rate during treatment of 10 mg/l methylene blue dissolved in 40 ml of deionized water, with and without the addition of 0.5% H_2O_2. A wide kind of material properties was used for formation of the anode/cathode pairs of Fe/Fe, Ti/Ti, Cu/Cu, Cu/Fe, Fe/Cu, Ti/Fe, Fe/Ti, Cu/Ti, and Ti/Cu.

The anode G_a and cathode G_c erosion rate sequences were measured. In general, the cathode erosion obtained in the treated solution without adding of H_2O_2 increased from 0.0053 to 0.0212 g/C for a certain sequence. The anode erosion obtained in the same conditions increased from 0.0021 to 0.0237 g/C. Smaller cathode erosion was measured in the solutions with copper cathodes than with other cathodes. The largest anode erosion was observed for a Cu anode.

The decrease of the G_a is opposite to the increase of G_c with heat flux change. The correlations of measured anode/cathode erosion ratio G_a/G_c with ratio of heat losses q_a/q_c (using criterion of (8.2)) for various electrode materials were found. The experimental result [38] shows in form presented in Fig. 8.5. The correlations as dependences of ratio G_a/G_c on heat fluxes q_a/q_c (related to melting temperature) for various anode (q_a) and cathode (q_c) materials were illustrated in Fig. 8.5a for submerged arc in water with H_2O_2 and in Fig. 8.5b without H_2O_2 [38]. G_a/G_c was obtained for the three groups of electrode pairs with one of the three (Ti, Fe, and Cu) anodes in each group noted by 1, 2, and 3. For each group, the pairs were used with different cathodes. Four points of G_a/G_c in each group changed from largest to smallest using, respectively, the cathodes Cu, Fe, Al, and Ti.

The obtained results are important for application of the submerged arc, in particular, in the electro-discharge machining of pieces in a liquid. The piece was usually an anode. In this case, the anode erosion should be larger for its effective treatment, but the cathode (instrument) erosion should be lower in order to increase the cathode lifetime. The electrode pair can be chosen considering change of ratio q_a/q_c for a

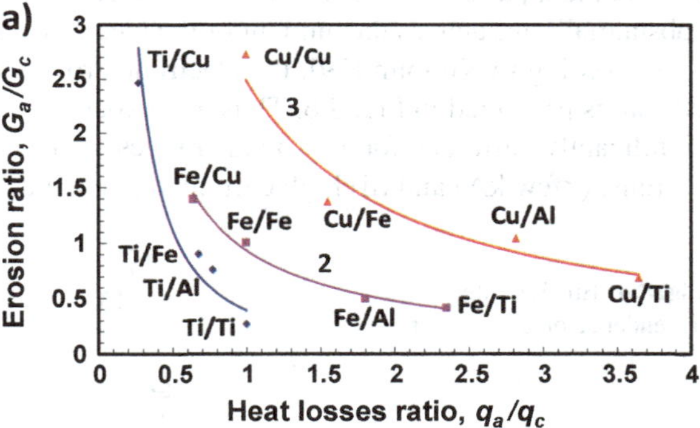

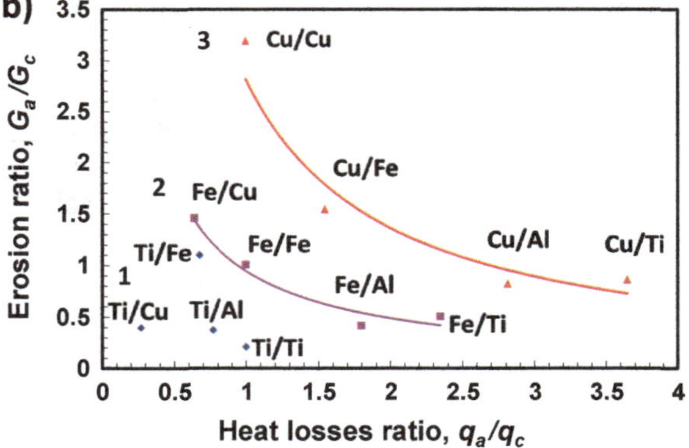

Fig. 8.5 **a** Dependences of G_a/G_c versus q_a/q_c obtained by using four materials of cathodes (Cu, Fe, Al, and Ti) with each of the Ti, Fe, Cu anodes for submerged arc in water with addition of the 0.5% H_2O_2. **b** Dependences of G_a/G_c versus q_a/q_c obtained by using four materials of cathodes (Cu, Fe, Al, and Ti) with each of Ti, Fe, Cu anodes for submerged arc in water without H_2O_2 [38]

number of electrode material pairs. The lower heat flux ratio corresponds to the larger anode erosion, but the larger heat flux ratio corresponds to large cathode erosion.

8.2 Erosion Phenomena in Vacuum Arcs

The above- described electrode mass loss was related to specific condition of relatively short vacuum arc interelectrode distance and at times submerged in liquids when the plasma generation was influenced by the gap and surrounding media, for example, reducing the spot mobility at the electrode surface. In case of vacuum arc, the gap distance is not limited by the arc application, for particular, for plasma deposition, coatings, ion source, and vacuum interrupters, etc. In these cases, plasma expands relatively free especially in vacuum arcs with ring anodes. Therefore, the electrode erosion is mainly related to the cathode and depends on processes including heating and plasma continuity mechanisms in the spot for moderate arc currents.

The investigations have shown that many factors such as geometry of the cathode, vacuum level, interelectrode distance and gas, surface chemistry and electrode microstructure, surface microrelief affect the nature and rate of cathode erosion [1–5]. The main parameter that determines degree of the cathode heating and therefore substantially influences the amount of the electrode erosion is the arc current.

According to Kimblin [39], the electrode erosion dependence influenced by arc current is presented in Fig. 8.6. Thus, the experiment show that the cathode erosion significantly different for two main ranges of the electric current: (i) moderate currents (<few kA) and (ii) high current vacuum arcs of few kA and larger.

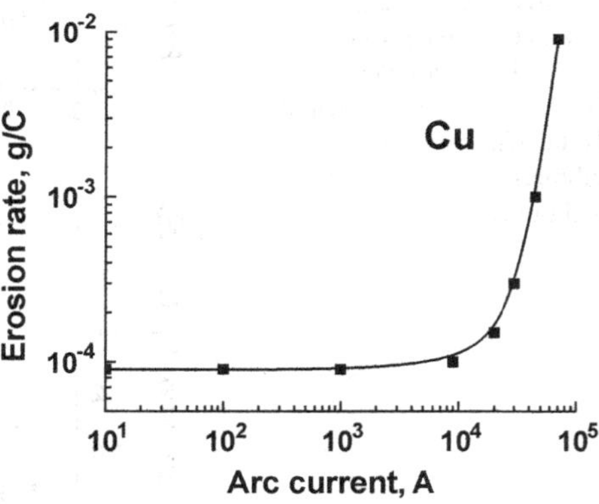

Fig. 8.6 Erosion rate dependence on arc current

8.2.1 Moderate Current of the Vacuum Arcs

According to Kesaev's review [40], the first study of the cathode erosion (from 1910 upwards) was related to rate of vaporization and liquid metal spraying of the mercury cathode. When the liquid metal spraying was detected separate, the erosion rate of Hg was close to that determined later in period 1928–1938 for solid cathodes like Cu in range of 0.015–0.007 mg/C [40]. The measured total erosion rate (with macroparticles, MP's) for Hg cathode presented Kesaev [40] weakly depends on arc current (1–18 A) and was 4 mg/C for moving and 0.6 mg/C for fixed (at Ni rod) cathode spot. It was indicated that the difference in the data determined by difference of the MP's fractions. We think that it can be caused also by difference of the cathode temperatures measured for both cases. The MP's fraction significantly depends on the cathode material properties and on conditioning of the metal, amount of the gas or impurities in its volume. Robertson [41] has observed about 87 μg/C for Cu cathode in 15 A arc of 15 s duration, 10 mm electrode gap and in 1 Torr pressure of nitrogen.

Reece [42] has reported an erosion rate on copper of about 80 μg/C and cadmium of about 400 μg/C at a current of 5 A (see Tables 8.7 and 8.8). The erosion rate was measured by Plyutto et al. [43] for different cathode metals by weighing and analysis of the condensed materials for arc currents in range of 100–300 A. The results with and without MPs are presented below in Tables 8.7 and 8.8. It was observed that no MP's were observed for brass and Mg cathodes at currents up to 100 A and for arc time below 5 s, while the large MP's fraction was detected in case of Al, Cu, and Ag cathodes.

Klyarfeld et al. [44] studied the cathode erosion components in a vacuum arc by separating of the mass loss by the droplets and by evaporation using centrifugal force generated in rotation of a cylindrical cathode around its axis. Depending on the direction of this force, the droplets were whether ejected or were prevented from the expansion, but the vaporized mass remained unchanged. The erosion of fixed cathode (32 mm diameter) for a number of metals was measured also by weighing method. The cylindrical anode material was used from steel. The initial pressure of the residual gas was 10^{-4} Torr. Two groups of metals were tested that characterized by (i) low melting temperature Zn, Cd, Sn, Pb, and Bi studied for arc current in range 10–60 A and (ii) higher melting temperature Mg, Be, Ti, Fe, and Cu studied for arc current in range 20–130 A.

The measurements demonstrated the weak dependence of the erosion rate on current used and relatively high erosion rate for metals of first group that was about 1000 μg/C for Bi, Sn, Pb and about 2–6 μg/C for Cd, and Zn. The metals of second group were characterized by lower level of mass loss, and it was in range of 40–100 μg/C for Cu, Fe and 30–50 μg/C for Ti, Mg and close 10 μg/C for Be. The comparison erosion data obtained under fixed or rotating cathode was presented in Table 8.4. The cathode erosion products consist mainly of ejection of droplets of liquid metal for first group of the cathodes (see data for Sn in Table 8.4). For this group, an increase of the cathode mass was obtained due to the radial mass acceleration directed normally to the cathode surface and outward.

Table 8.4 Erosion rate measured under different condition of spot location and cathode state depending on arc current [44]

Cathode spot location	Cathode state	Erosion rate (μg/C)					
		Tin		Copper		Magnesium	
		40 (A)	80 (A)	40 (A)	80 (A)	40 (A)	80 (A)
On external surface of cylinder	Fixed	1200	1500	77	120	24	28
	Rotating	3700	47	78	120	23	26
On internal surface of cylinder	Fixed	600	650	35	38	10	12
	Rotating	390	410	31	38	12	13

With reverse of this direction, the cathode mass loss decreased by factor one and a half. The metals of the second group do not change the erosion mass with a radial mass acceleration regardless of whether this acceleration is directed outward or toward the metal indicating that the cathode mass loss consists mainly of material evaporated by the cathode spot. A small amount of droplet was detected, and this mass constituted low fraction of the total cathode mass loss.

Kutzner and Zalucki [45] measured the electrode erosion rate in a chamber with pressure 10^{-5}–10^{-6} Torr for current range of 30–450 A and gap distance in range of 1–7 mm. The arc was initiated by electrode separating with velocity of 1–2 m/s. Plane-parallel electrodes were investigated with cathode diameter of 20 mm and anode of 30 mm for metals Cu, Ag, Ti, and Ta. The results showed that the dependence of cathode erosion rate (in μg/C) on arc current I was not monotonic (Cu, Ag, Ti). For example of Cu, the erosion increased from 40 μg/C at $I = 30$ A, passed a maximum of 90–130 μg/C at $I = 100$–130 A, and then, decreased with arc current up to 70–90 μg/C at 400 A depending on gap distance. For Ag and Ti metals, this dependence was nonlinear and more complicated, and erosion rate was in range of 25–100 μg/C for Ti and in range of 60–220 μg/C for Ag. In case of Ta, the erosion rate increased monotonic from 10 to 60 μg/C.

The erosion rate decreased with the increase of gap distance, and the smallest change was for high melting metal. A ratio of mass changed at the anode to that mass at the cathode $\Delta m_a/\Delta m_c$ was measured as function on the gap distance. The following principal conclusions were made: (i) $\Delta m_a/\Delta m_c < 1$ indicating that some material from the cathode was condensed on the shield, (ii) $\Delta m_a/\Delta m_c$ decreases with the increase of gap of electrodes for all the applied metals and arc currents, (iii) the tested metals, except Ta, were characterized by decrease $\Delta m_a/\Delta m_c$ in the function of arc current, (iv) with the increase of gap of electrodes, values $\Delta m_a/\Delta m_c$ were higher for high melting metals (Ta, Ti), in comparison with the results for Ag and Cu.

The measured cathode erosion dependencies were explained by the cathode spot dynamics, their location at the surface, increase of the spot numbers, and increase of the arc plasma pressure by a force induced due to the self-magnetic field. It was also shown that the measured anode mass produced due to condensation of the metallic plasma ejected from the cathode decrease with increase of the gap distance. The ratio of the anode mass gain to the cathode mass loss decreased from 0.6 to 0.2 with gap increase from 1 to 7 mm, for Cu and Ag.

According to the [2], the erosion rate of Cu, Ni, and W cathodes significantly decreased in the presence of transverse magnetic field up to 0.08 T. This reduction can be by the factor of six in comparison with the erosion rate measured without magnetic field for arc currents up to 4 kA. The MPs fraction also decreased. This effect was explained by increase of the spot velocity and lower time life at the spot location due to thermal character of the erosion mechanism.

During the arcing, the heated gases stimulate spraying of the liquid metal. In order to measure the cathode erosion rate mainly in vapor phase in work of [46, 47], the electrodes were degassed in vacuum system by strong and prolonged heating and by continuing of the pumping.

The erosion rate was determined by cathode weighing before and after a number of arcing (10–20 ms of each arc time [2]). The experiments were conducted in range of currents from 200 to 700 A detecting a nonlinear dependence of G on the arc current. The erosion products were condensed on probe plates located around the cylindrical electrodes. The observation of the deposited probes showed structureless films indicating that the cathode erosion was in general as metal vapor. Figure 8.7 shows the erosion rate (in mg/s) in semilogarithmic scale measured for Cu, Ni, and W cathodes in mentioned range of current.

The erosion rate for Fe, Nb, and Mo cathodes was measured only for arc current of 700 A, but the corresponding lines were extrapolated like to the curves for Cu, Ni, and W. The measured dependencies were approximated by following equation [47]

$$G = A_{er} 10^{B_{er} I} \tag{8.3}$$

where B_{er} is the constant indicated a slope of the curves which is equal 1.95×10^{-3} for all cathode materials and A_{er} is the constant indicated the metal properties which is equal 8.6, 4.7, and 2.6 for Cu, Ni, and W cathodes. Constant A_{er} was determined from the extrapolated dependencies as 2.444, 138, and 0.777 for Fe, Nb, and Mo, respectively.

Study of cathode erosion by Zykova et al. 1971 [48] in the presence of a surrounding gas pressure demonstrated that the erosion rate depends significantly

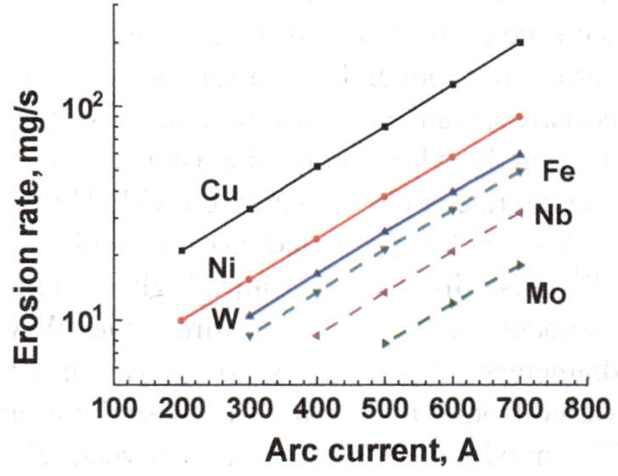

Fig. 8.7 Dependence of the cathode erosion rate (mg/s) on arc current for different materials [47]

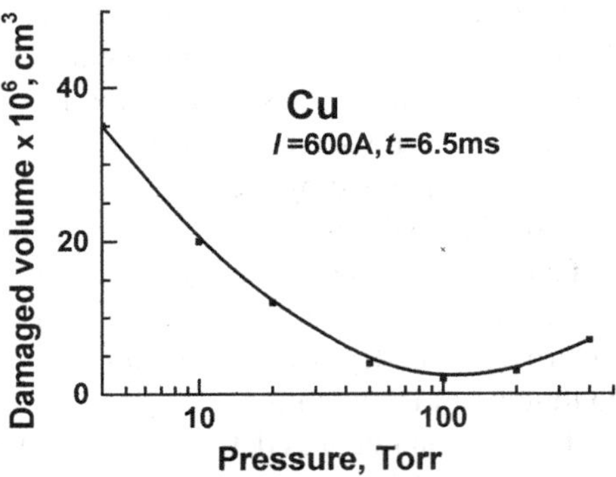

Fig. 8.8 Damaged cathode volume as function of surrounding He pressure. The data below 10 Torr are presented quality, and they should be continued up to pressure 10^{-4} Torr (not shown) at which the volume is about 4×10^{-5} cm^3 [48]

on the type of cathode spot. The spot type and therefore the erosion rate changed with gas pressure observed by change of the cathode damaged state. The dependence of the crater volume on He gas pressure is presented in Fig. 8.8.

As follows from this experiment, the volume passed a minimum at 100 Torr. The result of Fig. 8.8 was explained by the fact that in range of pressure from 400 to 100 Torr, only slowly independent spots (SIS) with current of 10 A were detected, which of short time and relatively lower damaged the cathode. In comparison with low pressure of 10^{-4} Torr, a slowly group spot (SGS) with current of 200 A was appeared, and the large cathode damage occurred, (see Table 7.10 of Chap. 7).

The erosion rate for Cu cathode was reported by Rakhovsky [49] for arcs operated at relatively small current (I < 100 A). In this case, only super fast spot (SFS) or moderate fast spots (MFS) with low current per spot of 5–10 A with large velocity were observed [50] (Chap. 7). The cathode was polished (but not cleaned), and after pulse arc, the erosion mass was determined from the crater volume, which was obtained by studying the crater area from a photography by raster and conventional electron microscopes as well as by determining the crater depth by profilometry [49, 51]. The typical crater diameter was about 1 μm with this same size of the depth crater. The cathode surface is between craters mostly not damaged. The average value of the electric transfer rate for copper and spots of the fast-moving type was found to be 0.45 μg/C. According to the experiment, the number of spots grows linearly with the current. When the arc current increased, keeping dI/dt constant, then the current per spot and the current density stay constant. As result, the erosion per spot is also constant, and therefore, the erosion rate per Coulomb is a constant for constant dI/dt in the whole range of existence of fast-moving spots. It was also indicated that erosion rate could be increased with dI/dt.

Juttner [52] presented erosion results for electrodes that were cleaned by about 10^4 pulse discharges in ultra-high vacuum (UHV) before the measurements. The cathodes were in form of wires from W with diameters 20 and 40 μm, Cu with diameters 50 μm, and strips of Al foil of 10×400 μm. The gap distance was varied from 2 to 20 mm. The pulse duration was 10, 40, 80, 120, 500, and 1000 ns. The maximum current varied between 20 and 100 A. It was indicated that within the accuracy of the measurements (about 20%), the erosion rate did not depend on current, pulse duration number of discharges, pulse voltage gap distance and depend only on pure material constant. The results are presented below in Tables 8.7 and 8.8.

The cathodic erosion of Cu, Cr, and Cd has been measured by Rondeel [53] for DC arc. The material loss or gain of the electrodes measured by weighing the electrodes. The contacts separated with a velocity of approximately 1 m/s in a chamber with a pressure of about 10^{-6} Torr. Two types of copper were used, one containing about 100–200 ppm gas-forming impurities (OFHC), the other about 10 ppm. Cr and Cd contained about 10 ppm and 50 ppm gas-forming impurities, respectively. Cylindrical electrodes with a diameter of 15 mm were used. The contact gap was about 1 mm during the erosion measurements.

Some measurements were carried out at other contact separations, giving the same results for the cathodic erosion rate as long as the arc voltage remained at the specific value of that metal. The influence of the oxide layer and gross heating of the contacts on the erosion rate were thoroughly checked. With a total charge through the arc of about 300 C for each measurement, and for maximum arc duration of 20 ms, the mentioned effects were of no influence on the erosion rate.

The cathodic erosion rates (μg/C) of the metals Cu, Cr, and Cd as a function of the arc current in the current range 300–2000 A were presented in Fig. 8.9. As can be seen, the erosion rate is only slightly dependent on the arc current. A somewhat higher increase observed for Cd than for copper and chromium. The observed erosion rates were 76, 22, and 400 μg/C for Cu, Cr, and Cd, respectively, for 500 A arc. Measurements on different kinds of copper indicate that the crystal grain size is of influence on the erosion rate. The marked difference between the erosion rates of Cu and Cr compared with that of Cd was explained by the fact that the boiling point of Cd is much lower than that of the Cu and Cr. It seems justified to assume that a high boiling point is one of the properties connected with a low erosion rate.

Kimblin [54] demonstrated measurements indicating that the cathode erosion rate dependent on different variables, such as arc current, arcing time, and cathode size. The erosion rate was determined by weighing the cathode prior to installation

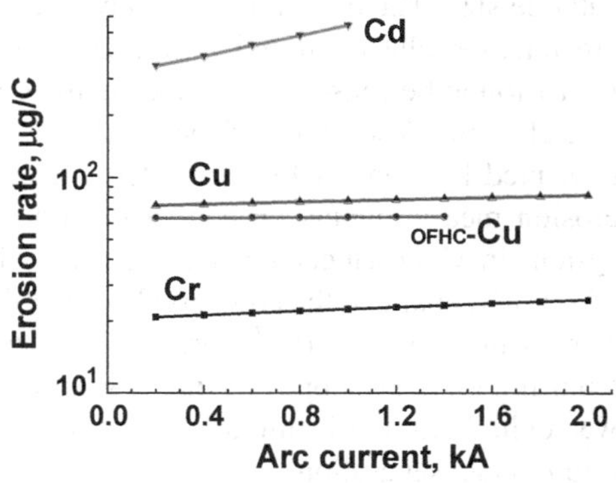

Fig. 8.9 Erosion rate as a function of the arc current for arc time of 20 ms [53]

in vacuum chamber and after the arc. The cathode was repeatedly arced for with duration 3–5 s for total time between 10 and 250 s and arc current of 100 A. The total charge passed through the arcing was detected. The arc was unstable in case of W cathode, and therefore, the erosion was measured at current of 250 A for total time arcing of 10 s. The arc initiated by separation of the current carrying electrodes. Since the average arc duration was relative large, the mass loss due to the bridge phenomena was negligibly small (<1%). The measured rate of erosion for a number of cathode metals is presented below in Tables 8.6 and 8.7 were results could be compared with other published data.

At future Kimblin's study [38], the cathode erosion was measured as a function of ambient gas pressure in range 10^{-3}–100 Torr for 100 A arc. A Cu cathode radius of 1.7 cm was drawn to spacing of 2.5 cm from 1.3 cm radius Cu anode during 2 s of maximum separation. The cathode was repeatedly arced for an arced time of about 100 s. The measurements show that the cathode erosion rate of about 190 μg/C was remained constant up to 10^{-2} Torr in nitrogen, while this erosion value saved up to about 10^{-1} Torr in helium.

It is important to make a note about an analogy study presented by Meunier and Drouet [55]. Their experiments were performed in a spherical vacuum chamber, 25 cm in diameter, which was pre-evacuated at a residual pressure of 10^{-6} torr then filled with He, Ar, at pressures from 10^{-3} to 760 torr. The arc, 2–10 mm long, was ignited using a laser pulse focused on the anode surface. For the erosion rate measurements, a commercial copper cathode 1 cm in diameter with a conical tip was used, while the anode consisted of a flat copper disk, 2 cm in diameter. The arc current $I = 300$ A remained constant during the discharge, which lasted less than 10 ms for the cathode erosion rate study. Each arcing period consisted of 100–500 short arcs with 20 s intervals between the pulses, which gave a cumulative arcing time of up to 5 s. The measurements showed that the erosion rate is not affected by the gas pressure in the chamber from 10^{-6} Torr to the 0.1–10-torr, indicating a constant value at 70 μg/C.

Daalder [56] presented an interesting and detail investigation. Similarly, to work in [54], he showed that the erosion rate is dependent on arc current, arcing time, and cathode size. The experimental study offers an explanation of the large variation in erosion rate values as found by earlier experiments. An analysis showed the variation in erosion to be caused by changes in the output of neutral species.

Below the description of the erosion data and its analysis will be done as it was presented by Daalder [56] in order to exactly understand the interpretations. The erosion measurements on copper cathodes were conducted with an experimental system, in which a mechanical device imparting the movable contact an average speed of 1.5 m/s separated the contacts. The chamber was evacuated prior to an experiment at pressures of 5×10^{-7} Torr or less. The electrode material was manufactured from high purity copper having a gas content of less than 10 ppm. Its geometry was cylindrical (3–4 mm height), having a slightly convex end surface to achieve mid-contact separation.

These contacts were brazed on stainless steel studs so that they could be mounted in the breaker by a clamping system. This method ensured a negligible change of weight by fastening or releasing the electrodes. Throughout the range of measurements, a cathode mass loss of about 10 mg was achieved by choosing an appropriate number of arcing sequences in each experiment at a fixed current and a specific discharge time.

The mass loss due to arcing was measured by weighing. Two cathode sizes having a diameter D of 25 mm and 10 mm were used. In both cases, the anode diameter was 25 mm and the gap distance 1.5 mm. The current values were chosen from 33 to 200 A. For a fixed current of 200 A was found the erosion rate to vary from less than 50 μg/C to about 190 μg/C, i.e., a variation of about a factor four, by choosing increasing values of the discharge time from few 2–3 ms to 2 s (Fig. 8.10). The dependencies are similar for other currents, but the erosion rate being less for lower currents. Also, the average slopes of the curves become less for diminishing currents.

Figure 8.11 shows the data of a 25 mm cathode as a function of the number of arcing coulombs (It) together with the results obtained with a 10 mm cathode. In

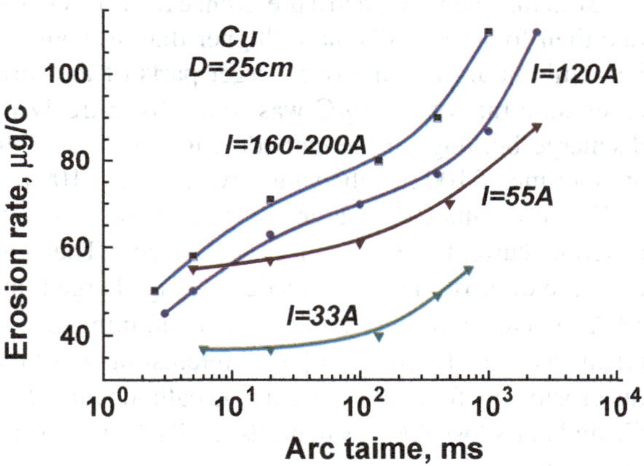

Fig. 8.10 Erosion rate of copper as a function of arcing time for 25 mm diameter electrodes [56]

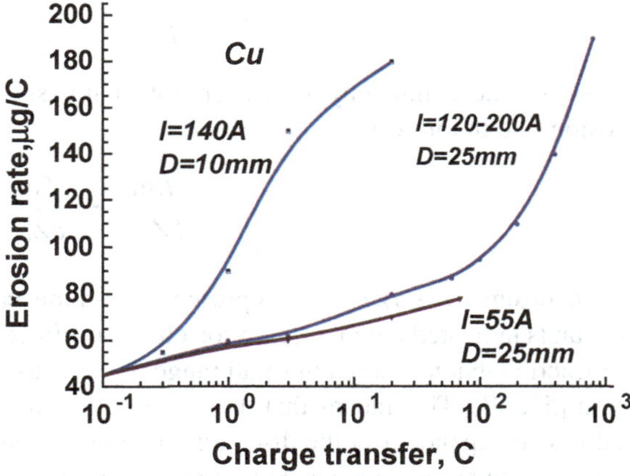

Fig. 8.11 Erosion rate of copper as a function of charge transfer by the arc for two cathode sizes [56]

the case of a 25 mm cathode, two curves indicate the erosion rate dependency of approximately 50–120 and 160–220 A. For high values of charge transfer, there is a divergence of these curves, while for lower charge transfer, the curves merge into one within the accuracy of the measurements. The fact that the current range investigated actually is a transition region from a single to a multiple cathode spot discharge probably explains why the erosion rate is not solely dependent on charge transfer. The dependence of the erosion rate on the size of the cathode is clearly demonstrated when comparing 10 and 25 mm cathodes. The small cathode shows a much stronger increase in erosion rate for values of charge transfer, while the 25 mm cathode only has a moderate gain, again indicating the importance of the electrode geometry under arcing in erosion rate measurements.

The velocity of the discharge and its manner of movement over the cathode surface are of importance for the resulting erosion. It was revealed experimentally that sometimes the discharge movement was concentrated in a relatively small part of the total available cathode area. It was observed that the arc, when initiated on the cathode edge (25 mm cathode), had a preference to burn here for prolonged times. The erosion rate then found was distinctly higher than it would be under "normal" conditions, i.e., with an arc moving over larger parts of the surface. For a 200 A, 20 ms arc, an erosion rate of 72 μg/C was found for a freely moving arc, while in case of a discharge burning on the cathode edge, a value of 90 μg/C was observed. At an arcing time of 100 ms, the values were 77 and 100 μg/C, respectively.

For low values of charge transfer, there is a convergence of data obtained at different current values and cathode sizes. There is a tendency of a continuing decrease of erosion rate with a decreasing charge transfer. The few data collected at 33 A indicate a more or less constant and minimum value of about 40 μg/C. These results were analyzed in [56] in an interesting way taking into account that generally the erosion coefficient represents the cathode mass loss ΔM_i in form of ions erosion G_{ri} and mass loss ΔM_n as neutrals (MP's + atoms) erosion G_{rm}. So, the total erosion rate G_r is

$$G_r = \frac{\Delta M_\mathrm{i}}{I\Delta \mathrm{t}} + \frac{\Delta M_\mathrm{n}}{I\Delta \mathrm{t}} \tag{8.4}$$

As Δm_i determined by ion current ion charge state Z and atom mass m_i then the erosion rate in ion form

$$G_{ri} = \frac{I_i \mathrm{m_i}}{I Z_{ef} e} = \frac{f \mathrm{m_i}}{e Z_{ef}} \tag{8.5}$$

According to (8.5), G_{ri} that represents a minimal cathode erosion rate depends on constants indicated ion mass, electron charge e, effective ion charge state Z_{ef}, and the ion fraction that measured in small range of $f = 0.8$–1.2 for different cathode metals by in [56, 57]. This means that the erosion rate represented by the ions is constant and thus independent of the discharge time or cathode size for arc currents up to at least a few hundred amperes. For large currents, the variation of the total erosion rate should entirely be due to the neutrals leaving the cathode either as neutral vapor or as droplets. The ion erosion rate therefore is a lower limit of the total erosion rate, which will be reached in case no neutral matter is leaving the cathode. For copper, the value of the ion erosion rate is found by inserting the numerical values obtained in (8.5) from which can be found G_{ri}= 38 μg/C. Figure 8.11 shows that this limit is about reached for small charge transfers, i.e., for a low energy input into the cathode. Table 8.5 shows the calculated results of ion erosion rate and the minimal erosion rate measured in the literature. The comparison of the calculated G_{ri} with directly measured minimal erosion rate obtained by weight measurements shows reasonable agreement.

Table 8.5 Cathode erosion rate data by ion flux using $f = 1.1$ and experimental data of minimal erosion rates for different metals [58]

Metal	Calculated ion erosion rate (μg/C)	Z_{ef} [42, 56, 57]	Measured minimal erosion rates (μg/C)	References
Pb	236	1		
Bi	238	1		
Cd	128	1	130	[59]
Sn	135	1		
Zn	74.5	1	76	[59]
Be	10	1	14.8	[59]
C	13.2	1.04	16–17	[39]
Ag	90.4	1.36	108	[59]
Al	22	1.4		
	19.5	1.58	25	[59]
Mg	19	1.45	25	[44, 58]
Ca	31	1.47		
Ni	43.7	1.53		
	48.9	1.37	49	[59]
Fe	40	1.6		
	42.5	1.5	50	[42]
Cu	39.2	1.85	37–59	[56]
	42.7	1.7	35–40	[44, 58]
Ti	30.4	1.8	45	[44, 58]
			52	[39]
Cr	29.7	2	22	[58]
Zr	47.9	2.2		
Mo	55	1.99		
	37.8	2.9	47	[39]
Ta	72	2.87	62	[39]
W	90	2.3	50–100	[58]

For moderate arc currents and arcing times, the erosion loss of the refractory metals is predominantly in ionized form, whereas for similar conditions, the low melting metals largely produce droplets. It was noted that ions and neutrals differ in origin of generation, the cathode spots being the centers of fully ionized metal vapor, while neutrals are formed in surrounding areas. An increase of the energy input means an increase of the erosion rate of neutral mass for wide range of investigated currents. This increase is also determined by cathode geometry and the velocity and way of movement of the arc spots.

Therefore, conditions of energy input per unit cathode area and heat loss by conduction into the electrode metal will have an important effect on the local surface temperature, which determines the ensuing evaporation of atoms and/or freeing of particles from melted areas. The wide range of electrode size influence on the erosion rate was measured. Table 8.6 shows the erosion rates measured for degassed small cathode sizes, and these values are compared with data obtained for similar condition using larger cathodes.

Thus, the erosion rate consists of two major particle fluxes: that due to ions and that due to neutrals. The ion erosion rate is independent of arcing time, cathode size, or arc movement, whereas the erosion rate of neutrals depends on these (external) parameters. We conclude that the origin and the mechanism of emission of these species are not the same. For sufficiently short arcing times, mass loss in the form of ions is dominant. For large arcing times, neutral mass is also eroded. These specific variables should be accounted mainly for the different values of erosion rate found in the literature.

Table 8.6 Cathode erosion rate of different metals and varying cathode sizes [58]

Metal	Diameter cathode (mm)	Arc current (A)	Arc time (ms)	Total erosion rate (μg/C)	Erosion rate by ion flux (8.5) (μg/C)
Cu	0.5	130	7	3600	39
	1	195	65	51.30	39
	1	200	28	4780	39
	10	140	20	150	39
	25	200	20	70	39
Al	1	200	23	2060	20
	1	200	27	2390	20
	2	215	19	1320	20
	3	390	6.4	400	20
	12	300	20	40	20
Pb	2	170	16	15900	236
	2	230	20	17300	236
	30	221	31	500	236

Daalder [60] has made an analysis of cathode surfaces eroded by vacuum arcs carrying currents up to a few hundred amps. A variety of structural changes was observed on Cd, Cu, Mo, and W cathodes. The different erosion structures have been interpreted as representing a number of stages with different degrees of erosion. The type of erosion pattern generated is dependent upon local conditions of energy input. The droplet production is mainly due to the heat flow, which is equivalent with the ion flow incident on the cathode surface. This process is a surface heating effect in which the properties of the cathode metal, spot lifetime, the velocity of the spots, and the character of their moving over the surfaces are essential parameters determining the cathode surface melting and mass loss.

Tuma [61] experimentally investigated the erosion products ejected from Cu cathode spot in an 80 A arc of 2 s duration. The optical and spectroscopic measurements have been used to determine the total ion, macroparticle (MP) fluxes, and flux of neutral atoms. It was shown that the erosion material consists mainly of ions and MPs. The emission flux of the neutrals from the cathode was less than 1% of total erosion flux. The total flux distribution was observed peaked in the direction of the cathode plane, while the ion flux distribution was forward peaked. The MP's sizes were detected as 9 μm and less, while most of the particle number per cm^2 was in range of 1–2 μm. The measured data indicate a collisional character of the cathode plasma flow. An analysis showed that the neutral atom flux was generated by the MP's vaporization. The main mass distribution was presented as function on volume angle.

The erosion yields by vacuum arcs of 16.6 A in a magnetic field of 0.05 T have been measured by Nurnberg et al. [62] as a function of temperature for Al, Al + 3%Mg, and TiC which was deposited by chemical vapor deposition in a 5 μm thick layer on Mo and stainless steel. The tokamak condition was simulated using plasma as anode. After inserting a new cathode material, the chamber was generally baked at 550 K for 6 h, and a vacuum of 10^{-8} mbar was reached after cooling down. The different cathode materials were heated to temperatures up to 770 K. The erosion after being subjected to about 30 to 200 arcs was determined by weight loss. The observed values for Al + 3%Mg are the same as for Al, and there is no difference within the experimental errors. Similarly, it was concluded between the erosion for TiC deposited on Mo and on stainless steel.

For Al and TiC (also for SS-316), the erosion rate was increased from 50 to 200 μg/C and from 55 to 110 μg/C, respectively, with cathode temperature. For Al, the erosion increases steeply above 0.8 of melting temperature, arc velocity decreases with temperature, and the arc craters become larger. The data for Cu are in Table 8.7. Additional experiments have been conducted in a vacuum arc apparatus for stainless steel SS-316 and Ti cathode, and the anode was pure graphite [63]. The vacuum arc was ignited by contacting and separating the electrodes. The interelectrode distance was from 2 to 4 mm. Arc current was 11.6 A. For SS-316, and for Ti cathodes, the erosion was 60 and 42.5 μg/C, respectively.

Aksenov et al. [64, 65] studied the erosion rate for a number of cathode materials in stationary vacuum arcs for cathode diameter 64 mm and length about 15 mm (except 50 mm for Ti). The arc current for all metals was 110 A, but it was 160 A

Table 8.7 Erosion rate measured for Cu cathode

Erosion rate (μg/C)										
Reece [41] 5 A	Plyutto [43] I = 100/300 A	Kantsel' [47] I = 0.2–0.7 kA	Kimblin [54] I = 100 A	Rondeel [54] I = 0.2–2 kA	Daalder [56] I = 200 A	Rakhovsky [51] I = 100 A	Juttner [52] I = 100 A	Nurnberg [62] T = 300–800 K	Brown [66] I = 100 A	Rao and Munz [71] I = 125 A
80	130/65	105–290	115	73–82	50–190	0.45	50	55–30	35	30–90

for Mo and Nb. The results were presented in Table 8.8. For electrolytic copper, the erosion rate was measured as 65 μg/C. When a surrounding gas was presented, the erosion rate not decreased up to 10^{-2}–10^{-1} Pa for Ti and Zr, 1 Pa for Cu and Cr and up to 10 Pa for Mo.

Brown [66] has measured the erosion rates of cathode materials in vacuum arcs for a number of different cathode elements. The cathodes are in the form of 6.35-mm diameter rods of a length of about 5 cm. The flat face of one end of the rod is used as the cathode surface. Those experiments employed a source that operated in a pulsed mode, with a pulse length of typically about 250 μs and repetition rate in the range 1–100 pps. The arc current was 100 A. The cathode mass was measured using an accurate beam balance before and after operating the source for a known number of discharges, typically several thousand. The results are presented in Tables 8.7 and 8.8.

Table 8.8 Erosion rate measured in different works

Cathode element	Erosion rate (μg/C)							
	Reece [41] 5/30 A	Plyutto [42] I = 100 and 300 A	Kantsel' [46] I = 200–700 A	Kimblin [53] I = 100 A	Daalder [55] I = 200 A	Aksenov [63, 64] 110–160 A	Juttner [51] I = 100 A	Brown [65] I = 100 A
		Total/as ions						
Ag	35	140/72						
Au				150	78			
Mg	40/70	36/25			17			31
Y								45
Zn	230	32/16		215	68			
Cd	400	62/31		655				
Hg								
Al		120/60			19	110	100	28
Ti				52		39		30
Zr						51		53
Gd								55
Tb					217			
C				16		27		
Sn					123			295
Pb	1100				215			510
Nb			21–45			38		
Ta					65			56
Bi					218			
Cr				40	22–27	36		20

(continued)

Table 8.8 (continued)

Cathode element	Erosion rate (μg/C)							
	Reece [41] 5/30 A	Plyutto [42] I = 100 and 300 A	Kantsel' [46] I = 200–700 A	Kimblin [53] I = 100 A	Daalder [55] I = 200 A	Aksenov [63, 64] 110–160 A	Juttner [51] I = 100 A	Brown [65] I = 100 A
Mo			15–26	47	50	52		36
W			35–85	62	64		136	55
Ir								
Fe			28–71	73				48
Co								44
Ni		100/50	50–130		44			47
LS-59		84/78						

Vacuum arc of 50–400 A and duration controlled from 1 to 20 ms have been studied [67] on Cu–Cr electrodes used in commercial interrupters in a vacuum of 10^{-7} Torr. Thousands arcs between flat-surface electrodes of 26 mm diameter and 4 mm gap distance were used for weighing the mass loss. The results showed that the erosion rate weakly depends on the arc current and the duration. The average erosion rate for Cu-Cr was measured as 36 μg/C.

Shulman et al. [68] studied erosion rates for contacts from vacuum interrupters used for low voltage contactors for contact materials, Ag–WC, Cu–Cr and Cu–Cr–Bi. The specifics of the erosion rates were considered for two gap distances 2.2 and 8.5 mm. The contacts were 23 mm in diameter. The contacts were then cleaned and weighed to within 10^{-5} g.

In case of gaps of 2.2 mm, the erosion rates were determined from the erosion of both contacts with 60 Hz half-cycle currents (from 630 to 685 A peak). The contacts were opened at a preset time in a half cycle. The starting polarity of the applied AC current (fixed number of half cycles) was alternated to permit the AC contactor to open alternately in a positive or negative last half cycle. This promoted balanced erosion and deposition on both contacts. So, the polarity of the current automatically changed for each operation to ensure uniform erosion history for both contacts during many thousands of operations.

In case of gaps of 8.5 mm, the absolute cathodic erosion rates were determined with applied DC current pulses on the order of 100 A. The arcing time in each operation was 1.4 s. A repetition generator automatically cycled through the arcing sequence every 40 s until the desired total charge transfer was achieved. Table 8.9 shows the results of experiments for different gap distances indicated the type of erosion process. According to the measurements for arc operating at a gap of 2.2 mm, a significant fraction of the ions and neutral vapor leaving one contact deposits on the other contact, so the effective erosion rate is markedly smaller than the cathode erosion rate for arc operated at large gap of 8.5 mm.

It was noted [68] that the magnitude, duration, and repetition rate of the DC current pulses used in 8.5 mm erosion measurement were kept small enough so as

Table 8.9 Data of effective erosion rates of closely spaced contacts and absolute values for cathode at large gap measured in [68]

Gap distance erosion type	Ratio gap/diameter	Material	Current (A)	Erosion rate (μg/C)
Absolute DC cathode Erosion at 8.5-mm	0.37	Ag–WC (40/60 wt%)	93.5 and 103.5	23.6
		Cu–Cr (60/40 wt%)	86.7	38
		Cu–Cr–Bi	137.5	35
Effective AC erosion at 2.2-mm	0.087	Ag–WC (40/60 wt%)	684	2.9
		Cu–Cr (60/40 wt%)	633	3.1

not to overheat the cathode. This ensured that the erosion rate would not be increased by heating effects. Also, small enough the heating of the anode, i.e., the very small amount of erosion of the passive anode could be ignored in comparison with the cathode erosion.

Shulman et al. in their later work [69] have been extended measurements of the effective electrode erosion at relatively low gap distance for larger contacts and gaps and the, respectively, results of the measurements were presented in Table 8.10. The important parameter of the cathode–anode assembly is the ratio of the gap distance to the electrode diameter indicated the angle between the center of the cathode plane and the edge of the anode. When this angle is >25°, most of droplets and particle flux is lost from the gap. This means that the erosion rates increase with the mentioned ratio increasing which is followed from the presented measured data.

Early, in 1942, Cobine and Gallagher [70] observed that surface irregularities, like scratches, affect cathode spot movement in vacuum arcs. Vacuum arc erosion measurements were performed by Rao and Munz [71] on copper cathodes having different surface roughness and surface patterns in 10^{-5} Torr vacuum (1.3324 mPa),

Table 8.10 Effective AC erosion rates for different gaps and diameters [69]

Material	%	Gap (mm)	Current peak (A)	Contact diameter	Ratio gap/diameter	Erosion rate (μg/C)
Cu–Cr	75/25	4	848	27.5	0.149	3.1
	75/25	6	848	27.3	0.22	7.5
	(75/25)–(25/75)	10	1000	30	0.33	7.9
Ag–WC	50/50	4	848	26.9	0.149	8.2
	50/50	6	848	26.9	0.223	8.4
	50/50	8	848	27.1	0.295	8.9
	50/50	4	848	27.0	0.148	11.8

in an external magnetic field of 0.04 T. Oxygen-free high electrical conductivity pure copper (99.99%) was used in this study. Different surface patterns and surface roughnesses were created on copper strips using two methods, namely grit blasting and grinding with emery paper. Two different sizes of alumina grit, namely, 190 and 708 μm, were blasted with nitrogen, to create an isotropic surface roughness and pattern. The erosion rates of these cathodes were obtained by measuring the weight loss of the electrode after 135 arc pulses, each of which was 500 μs duration at an arc current of 125 A.

Results obtained indicate that both surface roughness and surface patterns affect the erosion rate, namely the erosion rates measured indicate that erosion rates decrease with decreasing roughness levels. The erosion rate increased from 30 to 50 μg/C when grit size increased from 25 to 200 μm for grit blasting and from 30 to 90 μg/C when grit size increased from 200 to 700 μm for grinding with emery paper. The slope of these two dependencies was equal. Isotropic surfaces give lower erosion rates than patterned surfaces at the same roughness.

8.2.2 Electrode Erosion in High Current Arcs

At low arc currents, the electrode erosion was caused mainly by the spot action, and the rate in μg/C is weekly changed, while for moderate region of current, this parameter slightly deviated from constant values due to increase of neutral erosion component from the spot areas. However, in case of high current arcs, the electrode was melted on large area of the surface between the spot due to significant heat flux from the spot and especially from the relative dense plasma around the spots. Therefore, it was expected a large contribution of the liquid metal in the total electrode mass loss. Different forces cause the liquid metal ejection including convective gas flow, electromagnetic force, plasma pressure from the spot, gravity, and others. Not all liquid material can be ejected due to viscosity, Laplace force, and friction, and therefore, a ratio of ejecting material to the total melted material was characterized by a coefficient of ejection. At high current arcs, both cathode and anode are significantly losing their mass.

The dependence of electrode erosion on current in discharge of a capacitor bank was measured by Belkin and Kiselev [72, 73] for current range of $I = 70$–800 kA and arc pulse in range of $t = 35$–200 μs. Sectionalized metallic screens deposited by the eroded metals surrounded the discharge gap. Each section was weighed before and after the discharge. A microscope observation indicates the screen traces of evaporated and droplets phases of the materials. The amount of metal ejected in vapor phase was estimated removing the droplets from the screen. At this way, it was found that about 85–90% was eroded in liquid phase for Cu. The total electrode mass M_{er} loss was presented as dependence on total charge determined as [72]

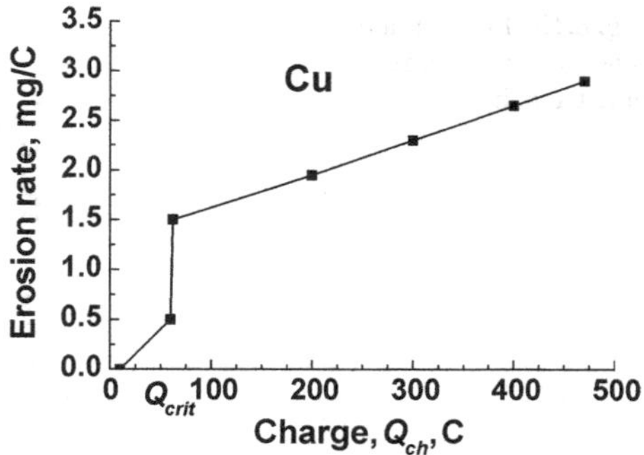

Fig. 8.12 Erosion rate as function on charge for Cu anode in a vacuum [74]

$$Q_{\mathrm{ch}} = \int_0^t I \mathrm{d}t \tag{8.4}$$

It was shown a sudden increase of M_{er} from 0.1 to 220 mg with Q_{ch} increase from 2 to 100 C. This increase of M_{er} passed from one linearly dependence through jump at some critical value of Q_{crit} to another linearly dependence. The value of Q_{crit} increases as 8, 33, and 35 C for electrodes from Cu, Mo, and Ti, respectively [73].

The similar dependence of electrode erosion was obtained in [74] for currents 2–25 kA in case of discharge in vacuum circuit breakers. The electrodes were separated with velocity of 30 cm/s up to distance of 10 mm Fig. 8.12. The jump in erosion dependence takes place at currents 8–11 kA and was caused by sharply rise of liquid metal ejection. At currents larger than 11 kA, the mass loss increases, and the anode erosion rate exceeds that for cathode by factor about 1.5–2. At current 16–20 kA, the erosion rate is ~2.1 mg/C, and for $I = 23$–25 kA, it is 3.2 mg/C. It was indicated that for high currents, the erosion rate in vacuum close to that measured in air. The coefficient of liquid metal ejection increases with surrounding gas pressure due to crater depth increasing.

In vacuum circuit breakers, a ceramic-metals materials, which do not have refractory components, were preferable to use due to relatively low erosion of such contacts with remaining other useful properties inherent to component a composition. The value of erosion rate for pure copper and ceramic-metal of Fe–Cu with 30% of Cu in vacuum were investigated by Belkin and Kiselev [75]. The pressure was kept at level of $(1\text{–}3) \times 10^{-5}$ Torr. The cylindrical contacts with diameter of 28 mm and gap of 10 mm were used. The erosion from pure Cu contact was performed at currents with amplitude up to 25 kA while up to 18 kA for ceramic-metal. The difference of the erosion rates for mentioned contact pairs [75] is presented in Figs. 8.13 and 8.14. The comparison shows that the erosion rate for pure Cu exceeds that measured for Fe–Cu pair by factor of one order of magnitude.

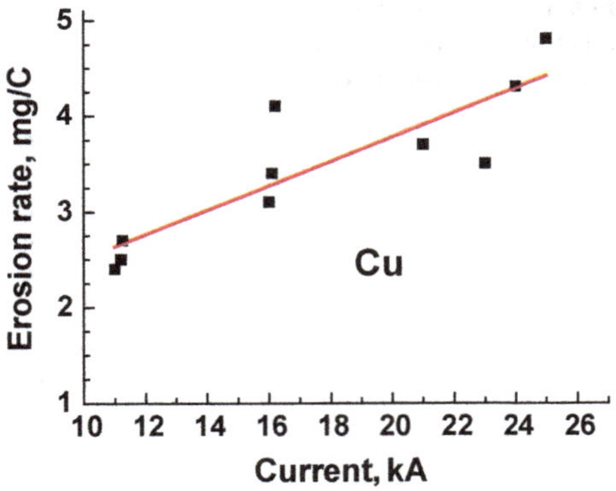

Fig. 8.13 Erosion rate as function on arc current for pure Cu pair

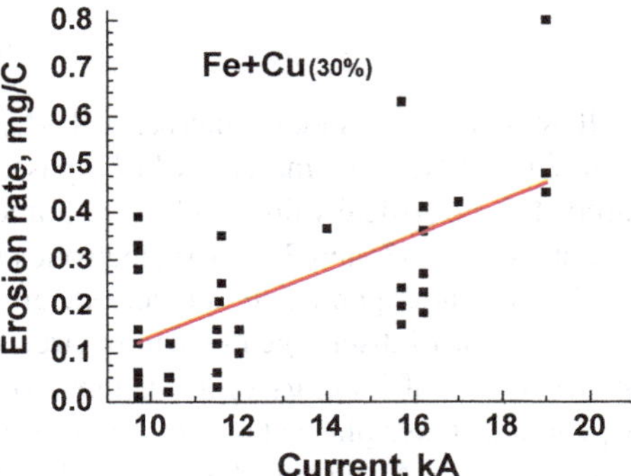

Fig. 8.14 Erosion rate as function on arc current for pure Fe–Cu (30%) pair

A discontinuity character of erosion has been observed in the current range from 9.6 up to 11.5 kA. This dependence causes by growth of an intensity of the liquid copper ejection. It was found a lot of Cu of different form and sizes including hardened metal in the form of drops, threads. In case of Fe–Cu, the hardened metal drops of different form were detected at current of 9.6 kA. They are well attached by a magnet.

Mitchel [76] has been studied experimentally the vacuum arc between butt electrodes of 6–75 mm diameter, at separations up to 40 mm, throughout a range of sinusoidal current loops up to 100 kA peak. Studying the images of vacuum arc at 6 kA, arising between 25 mm-diameter o.h.f.c. copper electrodes with 18 mm apart, showed a uniform cathode-plasma expanding up to the anode. Similar images were produced significantly later by Batrakov et al. [77] studying high current Cu arcs in millisecond range. Single frame from a high-speed camera record of a typical vacuum arc is presented in Fig. 8.15. The erosion rate has been studied experimentally for the vacuum arc between Cu electrodes of 45 mm diameter.

The erosion rate was measured over a range of sinusoidal currents up to a maximum of 40 kA peak [76]. The total material loss was obtained at 10 and 5 mm electrode separation, and the results are summarized in Fig. 8.16. When the peak current was less than about 10 kA, only the cathode was eroded; the erosion rates

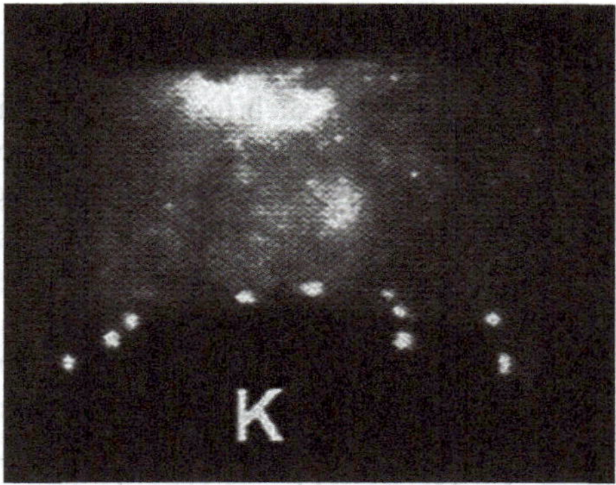

Fig. 8.15 Images of high current arc gap with copper electrodes [77]

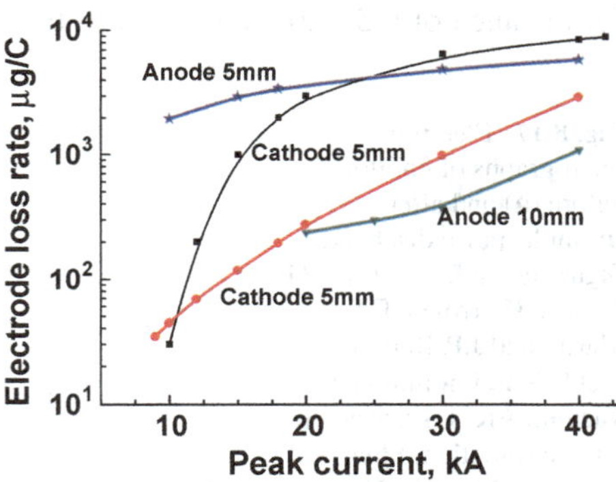

Fig. 8.16 Erosion rates of 45 mm-diameter copper electrodes at 5 and 10 mm separation [76]

measured being of the order indicated in the literature for arcs of a few hundred amperes. When the arc current was greater than about 10 kA peak, both the anode as well as the cathode lose their mass.

The anode and cathode loss rates at 10 mm separation are greater than the low-current cathode spot erosion rate by some 30 times at a current of 39 kA peak. At 5 mm separation and currents between 10 and 40 kA peak, the erosion rates of both electrodes are between three and ten times that experienced at the same currents at 10 mm separation.

8.2.3 Erosion Phenomena in Vacuum of Metallic Tip as High Field Emitter

The arc initiation, operation, and spot behavior significantly determined by the cathode surface relief. The protrusions, spikes, tips, and others enhancing the cathode electric field increase the electron emission. According to Dyke et al. [78, 79], an explosive vacuum arc was initiated between electrodes at critical values of the electrical field of 10^8 V/cm and current density of 10^8 A/cm^2.

They showed that such field and current density were produced by applying of high voltage to field emission emitter in form of a tip of submicron size. It was also demonstrated that the tip was destroyed after applying a pulse voltage. The tip was significantly eroded after multiple pulse discharges changing the tip geometry as it is shown in Fig. 8.17.

Considering the phenomena of explosion emission, the tip erosion rate was investigated in detail in [80, 81]. These results were presented also later in [82, 83]. Let us consider the condition of the experiments and the results presented in these works. A single voltage pulses of duration $t_p = 5, 20, 40$ and 80 ns were used in the experiments for Mo, Cu, Al, and Ni cathode tips. The curvature radius of tip point was in range of 0.1–0.5 μm. The pulse voltage amplitude was varied from 10 to 40 kV for distance of 0.5, 1.0, and 2 mm to the plane anode. The emitter mass loss was

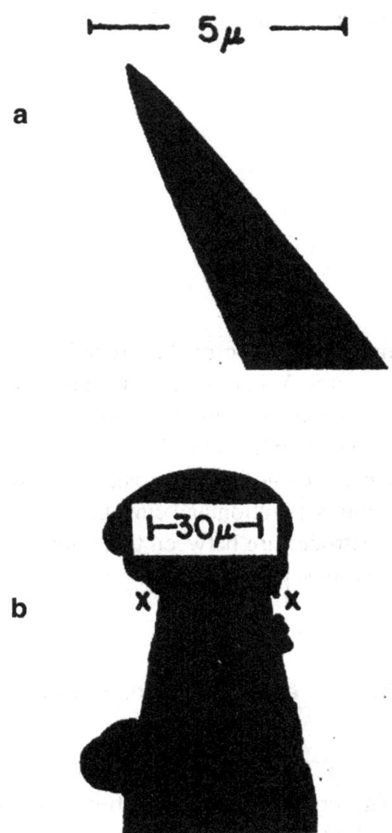

Fig. 8.17 Electron micrographs of emitter before (**a**) and after (**b**) multi-pulse discharge. Figure taken from [78]. W.P. Dyke, J.K. Trolan, E.E. Martin and J.P. Barbour. The Field Emission Initiated Vacuum Arc. I. Experiments on Arc Initiation, Phys. Rev. 91(5) 1043 (1953). **Used with Permission**

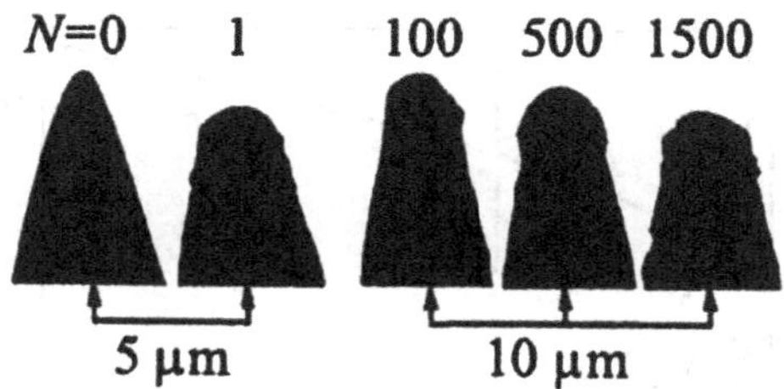

Fig. 8.18 Micrographs of geometry change of the molybdenum tip before nanosecond pulse discharge ($N = 0$), after one $N = 1$ and multiple $N = 100$, 500, and 1500 pulses (from [82] G.A. Mesyats and D.I. Proskurovskii, Pulsed Electrical Discharge in Vacuum. Berlin: Springer, (1989) **Springer publisher**)

determined calculating the volume change from the micrographs of the tip before and after the discharge pulse.

The observation showed that up to $t_p = 40$ ns, the material removed only from peak of the tip, and the lateral surface was not eroded, while this surface was damaged at $t_p = 80$ ns. The plasma was appeared through 1 ns after the pulse beginning at condition that the electric field at the tip point exceeds 10^8 V/cm. The rise of the current was at rate of 10^9 A/s. Figure 8.18 demonstrates profiles of points of molybdenum tip after the action of one or a number pulse discharges. The larger amount of mass loss was detected after first pulse, while the losses at next pulses can be lower by an order of magnitude.

The eroded mass significantly depends on the cone angle θ of the tip. When this angle is increased from 2° to 40°, the mass loss decreased by about one or two orders of magnitude. The erosion rate decreases with gap distance. For example, for Mo tip and voltage 30 kV, the rate of mass loss decreases from 10^{-2} g/s at 1 mm gap to 10^{-3} g/s at 2 mm gap.

The removable mass from the tip increases with applied voltage and discharge pulse duration. Figure 8.19 presents the mass loss per pulse duration as function on number of pulses at multiple actions for different materials of the tip and the dependence of the peak tip radius on the pulse numbers for Mo. Two regions were indicated in the present dependences. In region *I*, the mass loss strongly decreased, while region *II* (R_e= 8–2 μm) indicates weakly mass change with N and with angle θ. It can be seen that mass M_1 is the smallest in amount for Cu tip, and differences between masses M_1 for other materials are relatively small.

According to region *II*, the removed material measured for various pulse duration at $N =$ constant give the change of the rate of mass with tip. Using such detailed dependence, the authors of [82] concluded that the material was removed not only at the first moment from the primary tip but also at subsequent times from the new tips produced during the pulse t_p which was varied up to 300 ns. Another estimation using the volume of eroded t_p part indicates that the linear size of the removed mass was smaller than the radius R_e. Taking into account this estimation, it was assumed that the mass loss at region *II* corresponds to that occurred at a quasiplane massive

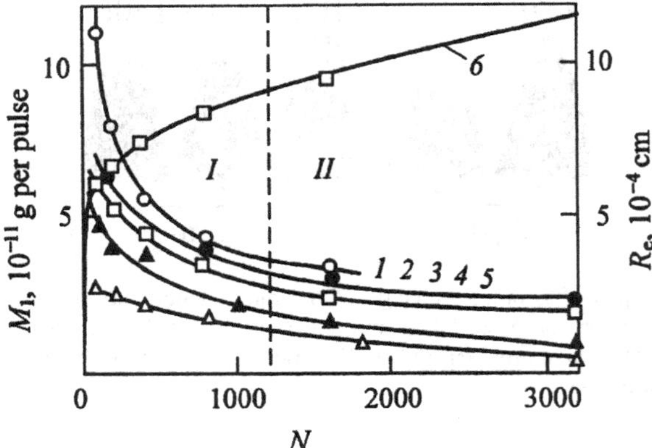

Fig. 8.19 Dependence of cathode mass loss M_1 for Ni(1), Mo(2), Al(2, 3), and Cu(5) on number of discharge pulses (N), voltage 30 kV, gap distance of 1 mm with $t_p = 5$ ns(1, 2, 4) and 10 ns(3, 5); $\theta = 10°$ (1, 2, 4) and $\theta = 24°$ (3, 5); The curve (6) indicates change of tip radius R_e on number N for Mo with $t_p = 5$ ns, $\theta = 12°$ (Figure is from [82] G.A. Mesyats and D.I. Proskurovskii, Pulsed Electrical Discharge in Vacuum. Berlin: Springer, (1989) Springer publisher)

cathode. It should be noted that the above conclusions could be valid if the high voltage character of the pulse is remained by producing an electric field of order of 100 MV/cm at a peak of the tip.

Using the measured removed mass, the estimated erosion rate was indicated between 10–100 μg/C. It was indicated that the new protrusions could not be stretched from the liquid metal under the action of the strong electric field because their sizes exceed the region of large electric field and their direction was not perpendicular to the cathode surface. An attempt was to use this result to explain the observed droplet ejection in direction parallel to the cathode plane in the arcs should take in account that the droplets were detected at large distance from a dense plasma region at the cathode.

While studying the behavior of an arc operating on W point cathodes at currents 1–5 A, it was measured a high erosion rate of (2–20) × 10^{-4} and (1–8) × 10^{-3} g/C for cold (300 K) and heated cathodes (1800 K), respectively. It was noted that the spread in values caused by fluctuations in arc duration [84, 85]. The weighing method was used to determine the erosion rate for a clean W cathode at currents of 2–3 A and arc duration of 1.3 μs. The mass loss was measured for a duration of (1–5) × 10^4 pulses. The erosion rate ~10^{-4}g/C was found. The weighing method showed a typical value of the erosion rate for a low-current arc. Thus, the local erosion rate (from a tip determined by change of tip geometry) shows higher value than that obtained by weighing. It can be understood taking into account that the weighing method shows a difference between the evaporated mass and the mass returned to the cathode surface at area larger than the crater size.

8.3 Summary and Discussion of the Erosion Measurements

Considering the results of electrode erosion, it can be noted that this phenomenon remains a problem that affects numerous applications. Different reasons exist for electrode erosion, and different erosion mechanisms take place under different conditions and used technique for determination of the mass loss. This is the cause that the published works report contradictory results about the cathodic erosion rate. For example, Reece [42] obtained for Cd the same value 400 μg//C as Rondeel [53] measured at arc current of 500 A. The large dispersion of the data was because the arc was operated in different electrode geometry, different media including gases and air at different pressures as well in different liquids as oil or kerosene. In case of short gap arcs, the important issue is mutual influence of the erosion products between cathode and anode resulting in mass loss or gain mass of a metallic of electrodes of different polarity.

Mainly widely used is the weight method to measure of the erosion, which contains of some uncertainness due to mass transport and redeposition in regions away from the arc. In case of vacuum interrupters, Shulman et al. [69] indicated an important parameter which is the gap distance ratio to the electrode diameter that means the relation between the metallic mass extracted in surrounding (method) and that condensed at opposite electrode.

Rondeel's [53] analysis for copper contacts at the same value of the DC current and for an identical geometry showed that measurements conducted by Cobine and Vanderslice [86] reported an erosion much higher than that found by Kantsel et al. [47] and by Kutzner and Zalucki [45]. The dependences [86] of the erosion rate on the number of Coulombs are contrasted with all other data also. Daalder [56] indicated that their high values obtained were most probably due to long durations of current flow through the electrodes prior to separation and low contact opening speeds. Also, the work of [87] showed that erosion rate rises when the cathode material consists of a large amount of absorbed gases.

A constructive discussion of the published data and detailed comparison of their results were given by Daalder [56]. Firstly, a number of discrepancies in data on copper erosion rates have been explained by the different arcing times and cathode sizes. The combination of small cathodes, high currents, and short arcing times is the cause for producing a relatively high mass loss from the cathode obtained in [47]. Rondeel [53] also explains the data of [47], which were obtained using complicated measuring technique. The material loss was determined indirectly by measuring the amount of metal condensed on a pair of plates located outside the electrode gap. When the current increased, more cathode spots move to the periphery of the cathode influencing the condensed material distribution. This will introduce an error in the measurements due to a considerable amount of metal was lost by sputtering [45]. This makes the estimation of the erosion rate even more complicated.

Secondly, Daalder [56] proposed to explain the published data considering their dependence on the carrying charge comparing the results either by lowering the current at constant arcing time or by varying both current and arcing time during the experiments. In other words, the dependence of the erosion rate on current as found by the investigators is not a single-valued function and is only valid for a specific combination of arcing times and current levels. The erosion rate can also be found at much higher current levels if one chooses appropriate (short) arcing times.

Most of data indicated that erosion rate not depends on the arc current within the current range 20–100 A. Kutzner and Zalucki [45] and Klyarfeld et al. [44] both observed a strong decrease of the erosion rate (by factor of three) for currents between 30 and 100 A. It can be understood taking into account the minimal mass loss due to outflow of the ion fraction in the plasma jet of a vacuum arc proposed in [56] and calculated for Cu as about 40 μg/C (see 8.5). The weakly change of the cathode erosion rate at low current range was caused by presence super fast spot type which characterized by weak contribution of very small fraction of droplets [49]. However, the Cu erosion rate of 0.45 μg/C with craters <1 μm reported in experiments with discharges of long duration in [48, 50] provided to be significantly lower than that indicated by the minimal value (40 μg/C) which request to vacuum arc supporting by the ion current. This large difference in the cathode erosion investigations (also comparing to erosion rates in arc with oxide and film cathodes) cannot be explained by a simple experimental error.

Juttner et al. [87] attempt to explain the low erosion rate obtained in works of [2, 49] indicating the presence of large fraction of contamination in the fresh cathode material resulting in appearing of the super fast spots producing low erosions. They state that contamination of the surface can have a great effect on the character of erosion. After some arcing, the cathode was cleaned, and the erosion rate increases by the spots of lower velocity. If the value 0.45 μg/C will be accepted, then, according to 8.5, the ion fraction should be reduced by two orders of magnitude comparing to that measured by Kimblin ($f \approx 0.1$) to showing adequate minimal erosion rate due to small ion current. However, the quantity $f \approx 0.001$ was not detected experimentally. Such difference cannot be explained by difference of used rate of current rise, dI/dt (see above [82]).

References

1. Kesaev, I. G. (1968). *Cathode processes in electric arcs*. Moscow: Nauka Publishers. (in Russian).
2. Rakhovsky, V. I. *Physical bases of the commutation of electric current in a vacuum*. Translation NTIS AD-773 868 (1973) of *Physical fundamentals of switching electric current in vacuum*. Nauka Press: Moscow (1970).
3. Lafferty, J. M. (Ed.). (1980). Vacuum arcs. Theory and applications. Wiley publisher: NY.
4. Boxman, R. L., Martin, P. J., & Sanders, D. M. (Eds.). (1995). *Handbook of vacuum arc science and technology*. Noyes Publ.: Park Ridge, N.J.
5. Anders, A. (2008). *Cathodic arcs: From fractal spots to energetic condensation*, Springer.
6. Restrepo, E., Garcia, L. A., Castro, J. J., & Devia, A. (2005). Craters formation in a graphite cathode produced by pulsed arc at low pressure. *Applied Surface Science, 252,* 1276–1282.
7. Geoffrey Duffield, W. (1915). The consumption of carbon in the electric arc. I. Variation with current and arc-length. II. Influence upon the luminous radiation from the arc. In *Proceedings*

of the Royal Society of London. Series A, Containing Papers of a Mathematical and Physical Character (Vol. 92, No. 636, pp. 122–143).
8. Holm, R. (2000). Electric contacts. Theory and applications, Springer: Berlin.
9. Lazarenko, B. R., & Lazarenko, N. I. Electrospark machining of metals, Gosenergoizdat, Moscow, Issues in (1946) and in (1950) (in Russian).
10. Germer, L. H., & Haworth, F. E. (1948). A low voltage discharge between very close electrodes. *Physical Review, 73*(9), 1121–1122.
11. Germer, L. H., & Haworth, F. E. (1949). Erosion of electrical contacts on make. *Journal of Applied Physics, 20*(11), 1085–1109.
12. Boyle, W. S., & Germer, L. H. (1955). Arcing at electrical contacts on closure. Part IV. The anode mechanism of extremely short arcs. *Journal of Applied Physics, 26*(5), 571–574.
13. Germer, L. H., & Boyle, W. S. (1955). Two distinct types of short arcs. *Journal of Applied Physics, 27*(1), 32.
14. Lander, J. J., & Germer, H. (1948). The bridge erosion of electrical contacts. Part I. *Journal of Applied Physics, 19*(10), 910–928.
15. Mandel'shtam, S. L., & Rayskiy, S. M. (1949). The mechanism of electrical erosion of metals, Human translation, FTD-ID(RS)T-0560–87 23, Sept. 1987; С.Л. Мандельштам, С.М. Райский О механизме электрической эрозии металлов, Известия АН СССР, сер физич. Т.ХШ, Т5, 549-565 (in Russian).
16. Pravoverov, N. L., & Struchkov, A. I. (1976). Erosion of clean metals in an electrical arc. *Electrotechnics, 100*(1), 46–47 (in Russian).
17. Zhou, X., & Heberlein, J. (1998). An experimental investigation of factors affecting arc-cathode erosion. *Journal of Physics. D. Applied Physics, 31,* 2577–2590.
18. Peters, J., Yin, F., Borges, C. F. M., Heberlein, J., & Hackett, Ch. (2005). Erosion mechanisms of hafnium cathodes at high current. *Journal of Physics. D. Applied Physics, 38,* 1781–1794.
19. Essiptchouk, A. M., Marotta, A., & Sharakhovsky, L. I. (2003). The influence of the arc current on the cold electrode erosion. *Physics of Plasmas, 10*(9), 3770–3773.
20. Szente, R. N., Munz, R. J., & Drouet, M. G. (1987). Effect of the arc velocity on the cathode erosion rate in argon-nitrogen mixtures. *Journal of Physics. D. Applied Physics, 20*(6), 754–756.
21. Szente, R. N., Munz, R. J., & Drouet, M. G. (1988). Arc velocity and cathode erosion rate in a magnetically driven arc burning in nitrogen. *Journal of Physics. D. Applied Physics, 21*(6), 909–913.
22. Namitokov, K. K. (1978). Electroerosion phenomena, energy, Moscow.
23. Jilani, S. T., & Pandey, P. C. (1982). Analysis and modelling of EDM parameter. *Precision Engineering, 4,* 215–221.
24. Yeo, S. H., Kurnia, W., & Tan, P. C. (2007). Electro-thermal modelling of anode and cathode in micro-EDM. *Journal of Physics. D. Applied Physics, 40,* 2513–2521.
25. Schulze, H.-P., Herms, R., Juhr, H., Schaetzing, W., & Wollenberg, G. (2004). Comparison of measured and simulated crater morphology for EDM. *Journal of Materials Processing Technology, 149,* 316–322.
26. Sharakhovsky, L. I., Marotta, A., & Essiptchouk, A. M. (2006). Model of workpiece erosion for electrical discharge machining process. *Applied Surface Science, 253,* 797–804.
27. Izquierdo, B., Sanchez, J. A., Plaza, S., Pombo, I., & Ortega, N. (2009). A numerical model of the EDM process considering the effect of multiple discharges. *International Journal of Machine Tools and Manufacture, 49,* 220.
28. Parkansky, N., Boxman, R. L., & Goldsmith, S. (1993). Development and application of pulsed-air-arc deposition. *Surface & Coatings Technology, 61,* 268–273.
29. Parkansky, N., Boxman, R. L., Goldsmith, S., & Rosenberg, Yu. (1997). Arc erosion reduction on electrical contacts using transverse current injection. *IEEE Transactions on Plasma Science, 25*(4), 543–547.
30. Palatnik, L. S. (1953). Phase transformation at electrospark treatment of the metals and criterion of the observed data. *Doctrate Academy of Science U.S.S.R. 89*(3), 455 (in Russian).
31. Parkansky, N., Beilis, I. I., Boxman, R. L., Goldsmith, S., & Rosenberg, Yu. (1998). *Surface & Coatings Technology, 108–109,* 253–256.

32. Kimoto, Y., Tamiya, K., & Hirata, K. (1969). Electrode erosion due to transient arc discharge in liquids. *Electrical Engineering in Japan, 89*(1), 42.
33. Zolotykh, B. N. (1960). The mechanism of electrical erosion of metals in liquid dielectric media. *Soviet Physics—Technical Physics, 5*(12), 1370–1373.
34. Zingerman, A. S., & Kaplan, D. A. (1958). Electric erosion of an anode as a function of interelectrode distance. *Soviet Physics—Technical Physics, 3*(2), 361–367.
35. Zingerman, A. S., & Kaplan, D. A. (1960). The dependence of cathode erosion on the length of the discharge. *Soviet Physics—Technical Physics, 5*(7), 792–795.
36. Somerville, S. M., Blevin, W. R., & Fletcher, N. H. (1952). *Electrode Phenomena Transient Arcs-Proceedings of Physical Society, B65*(12), 963–970.
37. Mitskevich, M. K., Bushik, A. I., Bakuto, I. A., Shilov, V. A., & Devoino, I. G. (1988). *Electroerosion treatment of metals*. Minsk: Nauka & Technika. (in Russian).
38. Parkansky, N., Yakubov, V., Beilis, I. I., Boxman, R. L., & Berkh, O. (2015). Electrode erosion during submerged arc treatment of methylene blue water solution. *Journal of Physics. D. Applied Physics, 48,* 225202.
39. Kimblin, C. W. (1974). Cathode spot erosion and ionization phenomena in the transition from vacuum to atmospheric pressure arc. *Journal of Applied Physics, 45*(12), 5235–5244.
40. Kesaev, I. G. (1964). *Cathode processes in the mercury arc*. NY: Consultants Bureau.
41. Robertson, R. M. (1938). The force on the cathode of a copper arc. *Physical Review, 53*(7), 578–582.
42. Reece, M. P. (1963). The vacuum switch. *Proceedings of IEE, 110,* 793–811.
43. Plyutto, A. A., Ryzhkov, V. N., & Kapin, A. T. (1965). High speed plasma streams in vacuum arcs. *Soviet Physisc—JETP, (U.S.S.R.), 20,* 328–337.
44. Klyarfeld, B. N., Neretina, N. A., & Druzhinina, N. A. (1969). Metal sputtering by the cathode spot of a vacuum arc. *Soviet Physics—Technical Physics, 14,* 796–799 (Zhurn Techn Phys, 39 1061–1065).
45. Kutzner, J., & Zalucki, Z. (1970). Electrode erosion in the vacuum arc. In *Proceedings in International Conference on Gas Discharge* (pp. 87–94). IEE: London.
46. Rakhovskii, V. I. (1965). Study of the erosion of high melting electrodes in a heavy current arc discharge in vacuo. *Soviet Physics—Technical Physics, 9*(11), 1593–1597.
47. Kantsel', V. V., Kurakina, T. S., Potokin, V. S., Rakhovskii, V. I., & Tkachev, L. G. (1968). Thermophysical parameters of a material and electrode erosion in a high current vacuum discharge. *Soviet Physics—Technical Physics, 13,* 814–817 (Zh. Tekh. Fiz. 38, 1074–1078).
48. Zykova, N. M., Kantsel, V. V., Rakhovsky, V. I., Seliverstova, I. F., & Ustimets, A. P. (1971). The dynamics of the development of cathode and anode regions of electric arcs I. *Soviet Physics—Technical Physics, 15*(11), 1844–1849.
49. Rakhovsky, V. I. (1976). Experimental study of the dynamics of cathode spots development. *IEEE Transactions on Plasma Science, 4,* 87–102.
50. Rakhovskii, V. I. (1972). Recent advances in the study of vacuum arc. In *Proceedings V International Symposium on Electrical Discharges and Insulation in Vacuum* (pp. 215–223). Poznan, Poland.
51. Rakhovsky, V. I. (1975). Erosion of electrodes in a contracted discharge. *Izv. Sibirian Branch of Academy of Science Series and Technology N3* (1), 11–27 (in Russian).
52. Juttner, B. (1979). Erosion craters and arc cathode spot in vacuum. *Beitrage Plasma Physics, 19*(1), 25–48.
53. Rondeel, W. G. J. (1973). Cathodic erosion in the vacuum arc. *Journal of Physics. D. Applied Physics, 6*(14), 1705–1711.
54. Kimblin, C. W. (1973). Erosion and ionization in the cathode spot regions of vacuum arcs. *Journal of Applied Physics, 44*(7), 3074–3081.
55. Meunier, J.-L., and Drouet, M. G. (1987). Experimental study of the effect of gas pressure on arc cathode erosion and redeposition in He, Ar and SF6 from vacuum to atmospheric pressure. *IEEE Transactions on Plasma Science PS-15,* 515–519.
56. Daalder, J. E. (1975). Erosion and the origin of charged and neutral species in vacuum arcs. *Journal of Physics. D. Applied Physics, 8*(14), 1647–1659.

57. Davis, W. D., & Miller, H. C. (1969). Analysis of the electrode products emitted by dc arcs in a vacuum ambient. *Journal of Applied Physics, 40,* 2212–2221.
58. Daalder, J. E. (1981). Cathode spots and vacuum arc. *Physica, 104C,* 91–96 (North-Holland Publishing Company).
59. Daalder, J. E. (1978). Thesis, Eindhoven University of Technology.
60. Daalder, J. E. (1979). Erosion structures on cathodes arced in vacuum. *Journal of Physics. D. Applied Physics, 12*(10), 1769–1779.
61. Tuma, D. T., Chen, G. L., & Davies, D. K. (1978). Erosion products from the of cathode spot of a copper vacuum arc. *Journal of Applied Physics, 49*(7), 3821–3827.
62. Nurnberg, A. W. (1981). Temperature dependence of the erosion of Al and TiC by vacuum arcs in a magnetic field. *Journal of Nuclear Materials, 103 & 104,* 305–308.
63. Nurnberg, A. W., Bauder, U. H., Mooser, C., & Behrisch, R. (1981). Cathode erosion in vacuum arcs and unipolar arcs. *Contributions to Plasma Physics, 21*(2), 127–134.
64. Aksenov, I. I., Konovalov, I. I., Padalka, V. G., & Khoroshikh, V. M. (1984). *High Temperature, 22*(4), 517–521.
65. Aksenov, I. I., Konovalov, I. I., Pershin, V. F., Khoroshikh, V. M., & Shpilinskii, L. F. (1988). *High Temperature, 26*(3), 315–318.
66. Brown, I. G., & Shiraishi, H. (1990). Cathode erosion rates in vacuum—arc discharges. *IEEE Transactions on Plasma Science, 18*(1), 170–171.
67. Malik, N. K, Edgley, P. D., & Malkin, P. (1984). Experiments on low current vacuum arcs. In *Proceedings XIth International Symposium on Discharges & Electrical Insulation Vacuum* (pp. 147–150) Germany, Berlin.
68. Schulman, M. B., Slade, P. G., & Bindas, J. A. (1995). Effective erosion rates for selected contact materials in low-voltage contactors. *IEEE Transactions on Computer packaging Technology Part A, 18*(2), 329–333.
69. Schulman, M. B., Slade, P. G., Loud, L. D., & Li, W. (1999). Influence of contact geometry and current on effective erosion of Cu-Cr, Ag-WC, and Ag-Cr vacuum contact materials. *IEEE Transactions on Computer packaging Technology, 22*(3), 405–413.
70. Cobine, J. D., & Gallagher, C. J. (1948). Current density of the arc cathode spot. *Physical Review, 74*(10), 1524–1530.
71. Rao, L., & Munz, R. (2007). Effect of surface roughness on erosion rates of pure copper coupons in pulsed vacuum arc system. *Journal of Physics. D. Applied Physics, 40,* 7753–7760.
72. Belkin, G. S., & Kiselev, V. J. (1966). Erosion of the electrodes in high current pulse discharges. *Soviet Physics—Technical Physics, 11*(2), 280–283.
73. Belkin, G. S., & Kiselev, V. J. (1967). Influence of electrode material on erosion at high currents. *Soviet Physics—Technical Physics, 12*(5), 792–793.
74. Belkin, G. S., & Kiselev, V. J. (1978). Effect of the medium on the electrical erosion of electrodes at high current. *Soviet Physics—Technical Physics, 23*(1), 24–27.
75. Belkin, G. S., & Kiselev, V. J. (1974). *Metal ejection electrodes under action of a vacuum arc* (pp. 259–264). Swansea: Proceedings VI International Symposium on Discharges & Electrical Insulation Vacuum.
76. Mitchel, G. R. (1970). High current vacuum arcs. Part 1—An experimental study. In *Proceedings of Institute of Electrical Engineering, 117*(12), 2315.
77. Batrakov, A., Schneider, A. V., Rowe, S. W., Sandolache, G., Markov, A., & Zjulkova, L. (2010). Observation of an anode spot shell at the high current vacuum arc. In *Proceedings of* XXIVth *International Symposium Discharges Electrical Insulation in Vacuum,* Braunschweig, Germany (pp. 351–354).
78. Dyke, W. P., Trolan, J. K., Martin, E. E., & Barbour, J. P. (1953). The field emission initiated vacuum arc. I. Experiments on arc initiation. *Physical Review, 91*(5), 1043.
79. Dolan, W. W., Dyke, W. P., & Trolan, J. K. (1953). The field emission initiated vacuum arc. II. The resistively heated emitter. *Physical Review, 91*(5), 1054.
80. Bazhenov, G. P., Litvinov, E. A., Mesyats, G. A., Proskurovskii, D. I., Shubin, A. F., & Yankelevich, E. B. (1973). Metal supply in the cathode burst in explosive electron emission from metal tips. I. First explosion in a tip with field emitter geometry. *Soviet Physics—Technical Physics, 18*(6), 795–798.

81. Bazhenov, G. P., Litvinov, E. A., Mesyats, G. A., Proskurovskii, D. I., Shubin, A. F., & Yankelevich, E. B. (1973). Metal supply in the cathode burst in explosive electron emission from metal tips. I. First explosion in a tip with field emitter geometry. *Soviet Physics—Technical Physics, 18*(6), 799–802.
82. Mesyats, G. A., & Proskurovskii, D. I. (1989). *Pulsed electrical discharge in vacuum.* Berlin: Springer.
83. Mesyats, G. A. (2000). *Cathode phenomena in a vacuum discharge: The breakdown, the spark and the arc.* Moscow: Nauka.
84. Puchkarev, V. F., Proskurovskii, D. I., & Murzakaev, A. M. (1988). Unsteady processesin the cathode spot of a vacuum rac at currents near the threshold. Spot on a needle cathode. *Soviet Physics—Technical Physics, 33*(1), 51–54.
85. Puchkarev, V. F., & Chesnokov, S. M. (1992). Erosion rate and voltage distribution in contracted (with cathode spot) and diffuse (spotless) low-current vacuum arcs. *Journal of Physics. D. Applied Physics, 25,* 1760–1766.
86. Cobine, J. D., & Vanderslice, T. A. (1963). Electrode erosion and gas evolution of vacuum arc. *IEEE Transactions Communication and Electronics, 66*, 240–245.
87. Achtert, J., Altrichter, B., Juttner, B., Peach, P., Pusch, H., Reiner, H. D., et al. (1977). Influence of surface contaminations on cathode processes of vacuum discharges. *Beitrage Plasma Physics, 17*(6), 419–431.

Chapter 9
Electrode Erosion. Macroparticle Generation

During the vacuum-arc burning, a high-pressure plasma is produced in the spot near the electrode surface causing an ejection of electrode material in form of the macroparticle (MP) as a liquid droplets or solid particles. The erosion mass in form of MPs is substantially determined by the arc current, thermophysical properties of the electrode material, and arc duration.

9.1 Macroparticle Generation. Conventional Arc

According to Kesaev [1, 2] first study of the cathode mass loss was performed considering the mercury cathode. It observed that relatively large cathode eroding mass was due to presence of significant fraction of droplets. This result was right for spot moving randomly and for fixed spot. The most part of the mercury was removed from the cathode in the form of droplets with a diameter greater than 20 μm. The number of droplets and their velocity increases when the diameter was reduced. The spots ejected about 10^5 droplets per second at arc current of 6 A. The observed velocities were in range of 3–8 m/s.

Plyutto et al. [3] have studied presence of a fraction of MPs in the total cathode mass loss. This work reported that the geometrical distribution of the plasma and MP fluxes significantly different. While the plasma flux was axially directed, the MP flux was oriented in the plane weakly inclined to the cathode surface. The amount of the MP depends on the cathode material, arc current, and arc time. According to Plyutto et al. [3], the MPs not observed for cathodes from LS-59 and Mg up to current 100 A, while for Al, Ag, and Cu, the MP fraction reached about 60–80%. Klyarfel'd et al. [4] also informed about the MP's contamination by using a centrifugal force to separate the amount of generated melting metal (see Chap. 8, Table 8.4).

Udris [5, 6] presented one of primary experimental study of the molten-metal particles emitted by vacuum arc for wide range of electrode materials. The particles were detected by their impressions leaved on glass covered with a layer of black

I. Beilis, *Plasma and Spot Phenomena in Electrical Arcs*, Springer Series on Atomic, Optical, and Plasma Physics 113, https://doi.org/10.1007/978-3-030-44747-2_9

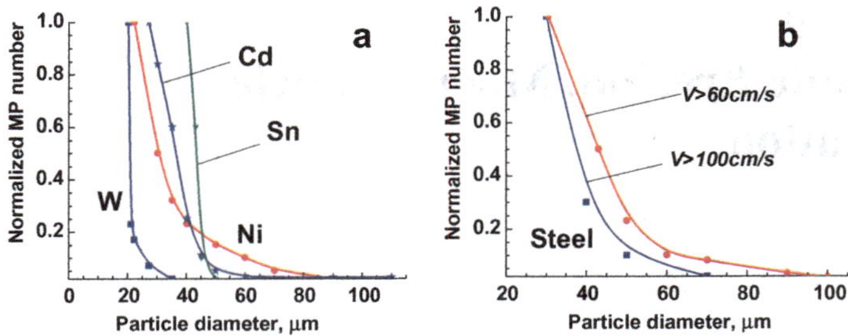

Fig. 9.1 Relative number of particle distribution as dependence of their diameter in microns (Sn, Cd, Ni, W, left figure, (**a**)) and velocity (steel cathode, right figure, (**b**))

graphite. It was determined size distribution of the particles larger than 1 μm and their velocity for a number of low- and high-melting cathode materials. Figure 9.1 shows that the number of particles decreased with their diameter for Sn, Cd, Ni, W, and steel [6]. For Mo the normalized MP numbers in range of 0-1 arise at about of 20 μm particle diameter. The significant drops are at the minimal diameter in the distribution for each metal. The smaller particle size was ejected with higher velocity that measured in range of 0.1–100 m/s.

The impressions left by the particles were circular or elliptical form and indicate a liquid state of the metals, while solid state for graphite particles. The number of emitted particles was 100 for W and 10^5 for Hg per Coulomb. Udris [6] noted that at Hg cathode, the molten particles were ejected due to force produced by an electric field $E^2S/8\pi$. Taking the characteristic surface particle S, their diameters, and velocities, the electric field required for particles departure (neglecting the viscosity) was determined as (1–2.4) × 10^7 V/cm. This is high electric field but the estimation is relatively arbitrary and not applicable for other metals. There should be mentioned another mechanism considered by McClure [7] for metallic cathodes of vacuum arcs. A pressure P_{mc} on the molten area of the cathode surface which produced due to plasma jet reaction ejected from the cathode spot, was calculated using measured both the current density j and the cathode force F per unit current. Assuming that this pressure causes the MPs emission from the liquid pool with a kinetic energy depending on MP mass density ρ, the particle velocity can be obtained in following form:

$$v_{\mathrm{mp}} = \left(\frac{2P_{\mathrm{mc}}}{\rho}\right)^{0.5} = \left(\frac{2Fj}{\rho}\right)^{0.5} \tag{9.1}$$

Using (9.1) McClure showed that the mechanism of pressure action allows to generate MPs at velocities in range of 20–200 m/s from copper cathode surface.

Similar mechanism was indicated by Gray and Pharney [8] for various types of low-current (5 A) arcs at atmospheric pressure. Material transfer is considered as result of molten-metal ejection out of the electrode craters by the recoil force, which follows the abrupt cessation of ion bombardment. Once ejected, the liquid droplets re-solidify and condense on the reactor walls, or are entrained by the gas flow to the exhaust line. The droplet formation during the discharge was provided by a force F_{ion}, due to ion bombardment, acting on the molten pool at the crater bottom. Following the abrupt cessation of the discharge filament, the unbalanced recoil force is then directed outward from the metal bulk into the interelectrode gap. The surface tension force F_{st} of the molten metal will act against this recoil force F_{ion} and for $F_{ion} > F_{st}$, the molten metal pool will be displaced causing the separation from the bulk in the form of a molten droplet. The momentum becoming due to the cathode sheath potential drop determined the force F_{ion}. However, the momentum of ions accelerated towards the cathode surface in the sheath was compensated by the momentum of the cathode electric field and cannot to be considered as force acted on the cathode molten pool in the spot region, which was indicated early by Beilis [9, 10].

It should be noted that Jenkins et al. [11] observed a free flight liquid copper droplets ejected by the cathode spots and attached to the quartz windows after around 1 ms from extinction of high-current vacuum arc of 2–11 kA arc. Later, these authors [12] measured the droplet size distribution produced by 4.3 kA arc with copper electrodes. It was observed that the number of droplets decreased monotonically by seven orders of magnitude with their impacted diameter in range of 0.15–90 μm.

Utsumi and English [13] study included the size distribution of the molten particles, the volume carried by these particles as function of the volume carried by the metal vapor, and the velocity distribution for Au, Pd, and Mg electrodes in the range of relatively small currents of 2–6 A. Vacuum arc was initiated by opening the electrodes. The average single-arc duration was varied from 100 μs to 10 ms depending on the material and a number of arcs were used to reach the total arcing time of about 20 s. The size and the distribution were measured by analyzing the deposit on the glass substrate which examined by both optical and a scanning electron microscope. The particle velocity distribution was measured by mechanical filter, which was made of multiple blades mounted around the perimeter of a rotating drum driving by vacuum-light dc motor through a magnetic coupling.

The results showed that the total particle number emitted from Au and Pd arcs was about 5×10^7 per Coulomb and the diameter was varied from lower than 0.2 to about 5 μm. The ratio of volume carried by the particles and vapor depends on the material and the corresponding percentages were varied for Mg, Au, and Pd approximately 80, 50, and 10%, respectively. The particle velocity was in rage of $(2–4) \times 10^2$ m/s for Au and Pd and 40–50 m/s for Mg. The majority of the volume carried by the particles was due to the particles whose diameter was greater than 1μm for both Au and Pd arcs. The number of ejected particles decreased monotonically as their size increased.

An investigation of the particles and plasma ions to the total cathode mass loss for Cu, Cd, and Mo at arc currents 50–200 A was conducted by Daalder [14, 15]. According to used methodology, the electrodes were surrounded by a condensing shield cylinder with diameter of 75 mm and length of 55 mm. Inside the cylinder was placed four target disks with diameter of 10 mm and a surface with mirror appearance. The discs collect the particles emitted from the cathode and allowing analyzes their shape, size, number, and angle distribution. The arc current levels were between 50 and 200 A. The cathode (Cu, Cd, and Mo) diameter was 25 mm. maximum electrode spacing was varied from 4 to 10 mm.

It was shown that generally the particles have a circular shape, but the large particles have nearly flat middle part which is surrounding by higher ring-shaped shoulder and for sizes of few microns, the structures tend to become hemispherical shape. Minimum resolved particle diameter was about 0.1 μm detected for all materials.

The observed number of particles was normalized by their diameter, charge transfer, and by distance from the target discs to the cathode plane (L_{pc}). It was showed that this parameter decreased from 50 to 200 (particle number/μm/C/mm) to about 0.1 when the particle sizes increased from minimal resolved diameter up to 40–70 μm depending on L_{pc} (0–26 mm). However, the volume of deposited particles increased with their outer diameter by few orders of magnitude in the mentioned range of particle sizes. Table 9.1 shows increase by factor three in the particle mass with charge transfer for copper. This increase for Cd is significantly larger than for Cu, which most likely is due to a lower melting temperature. An exception is the experiment Cd-1 at which for lower charge transfer, a larger particle mass loss was obtained. This result was due to arc movement. In the experiment with high-melting molybdenum cathode, the particle mass loss was measured only of about few μg/C indicating strong dependence of the particle amount on the fusion temperature of the cathode metal.

Angular distribution. The particle erosion rate per solid angle as viewed from the cathode center was obtained as function of the angle with the cathode plane. The maximum opening angle was in range of 39–63°. The results [15] were illustrated in Fig. 9.2 indicating that the droplet mass was expelled in a narrow region having angle of 20–30°. For copper, maximum of ejected MP is approximately 10° from the cathode plane, and for cadmium, it is 20°–30° and the MP size decreases with increasing angle from the cathode plane. Kutzner et al. [16] also observed a flux mass directed along the cathode surface.

Table 9.1 Contribution of particles and ions in the cathode erosion depending on charge transfer obtained in a number of experiments for Cu and Cd cathodes [15]

Material and experiment number	Charge transfer $Q = It$ (C)	Mass loss in particle form (μg/C)	Mass loss as ion flux (μg/C)
Cu-1	2.4	25	37
Cu-2	20.8	37	39
Cu-3	200	76	39
Cd-1	0.21	245	128
Cd-2	0.47	166	128
Cd-3	45	250	128
Cd-4	135	360	128

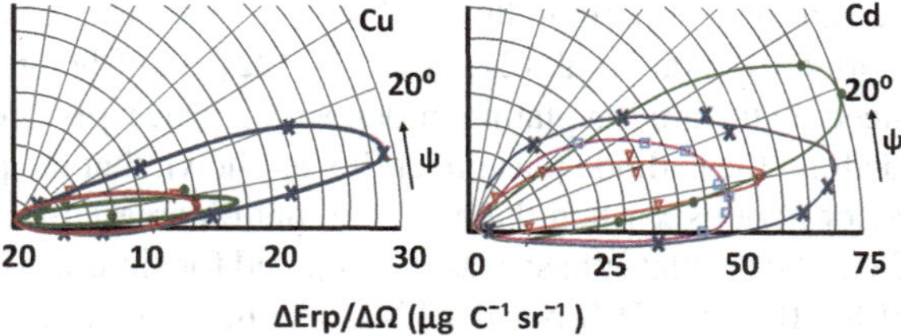

Fig. 9.2 Angular distribution of particle mass in μg/C and per solid angle in sr for different charge transfer indicated by numbers of experiments in Table 9.1 for copper and cadmium cathodes. Left: ●-Cu-1;∇-Cu-2; ×-Cu-3. Right: ●-Cd-1; ∇-Cd-2; □-Cd-3; ×- Cd-4

The relative contribution of particle with certain diameter to the total ejected particle flux was also obtained by Daalder [15] and presented in form shown in Fig. 9.3. The distributions of mass composition are about the same for both experiments Cu-1 and Cu-3. The dominant mass output occurring at small angles is due to particle sizes in range of 10–40 μm. The smaller particle sizes are occurred at larger angles with the cathode plane and for lower charge transfer (Cu-1). A similar result was obtained for cadmium for which the particle diameter of 10–40 μm was dominant in the mass output. It was also noted that a particle flux was returned to the cathode. This effect may be existed due to particle reflection from the anode which was discussed by Daalder [17] and detailed by Anders [18] (see also a data below).

Aksenov et al. [19, 20] conducted experiments for studying of droplets characteristics as well as their angle distribution ejected from Ti cathode covered by TiN. The measurements showed that in high vacuum, the main fraction of the droplets was with 5–15 μm diameter, which was moved at relatively small angle to the cathode plane. More uniformly was distributed droplets with diameters 2–5 μm. With nitrogen pressure up to 0.1 Pa, the total number of droplets decreased due to decrease of large diameter droplets moving under small angles. The small droplet sizes also decreased with future increase of the nitrogen pressure up to 1 Pa.

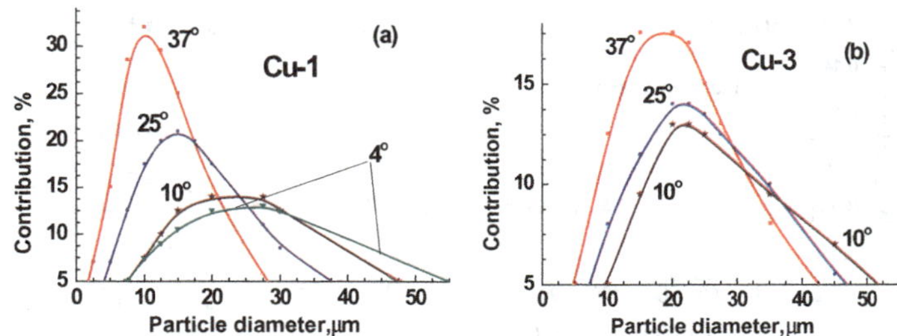

Fig. 9.3 Contribution of particles with different charge transfer sizes in total mass output for a specific angle with the cathode plane and for experiments indicated by the numbers in Table 9.1 for copper cathodes. Linear approximations of the measured data: **a** Cu-1 for 4°, 10°, 25°, and 37°. **b** Cu-3 for 10°, 25°, and 37°

A reduction of number of particle flux and the sizes of particles emitted from Ti cathode was reported by Akari et al. [21], Tai et al. [22], and Cheng et al. [23] using TiN films produced by the arc ion plating or deposition processes due to presence of a magnetic field along the cathode plasma jet. It was shown that the particle number decreased by factor about six when the magnetic field increased from 0 to 0.022 T for 100 A arc [21]. The droplets phase was investigated for sintered titanium cathode with porosity of 8–10% by PTES powder [24]. The investigations were conducted with an arc burning in a Hall accelerator with steady arc of (0.9–3.6) $\times$ 10^3 s duration and current in range of 80–120 A. The obtained condensate was studied by optical microscope. A condensate volume was detected with a few numbers of droplets of 5–3 μm, 3–18 droplet of 3–1 μm and 27–60 of droplets of 1–0.5 μm. No information about relatively larger particles (10–60 μm) was published and no discussion of the result was presented.

Kandah and Meunier [25] studied the particle number and size distribution on the covered substrate using graphite cathode of 12.7 mm in diameter and 25.4 mm length with circular graphite anode thickness 9.5 mm and of 50.8 mm in diameter. Arc current was varied between 44 and 110 A, the electrode gap was 5 mm, arc time varied between 10 μs and 100 ms and time between pulses was 15–20 s. Distance between the cathode and the silicon substrate was 75–130 mm. Some results [25] are presented in form shown in Fig. 9.4. The number of particle density increases linearly with the total electric charge passing through the cathode, which was varied by changing the number of arc pulses (Fig. 9.4, left). The particle density increases linearly also with the arc pulse duration for constant of the total electrical charge (Fig. 9.4, right).

For given pulse duration, the particle density increases with arc current. The research showed that the structure of the particles was found to be graphite, the average particle size was observed in range of 0.2–2.0 μs, and numbers of density of the particles decreased with distance to the substrate. Study of the particle morphology indicated presence of liquid phase. Considering the phase diagram for carbon, this result can be understood assuming that the graphite droplets may

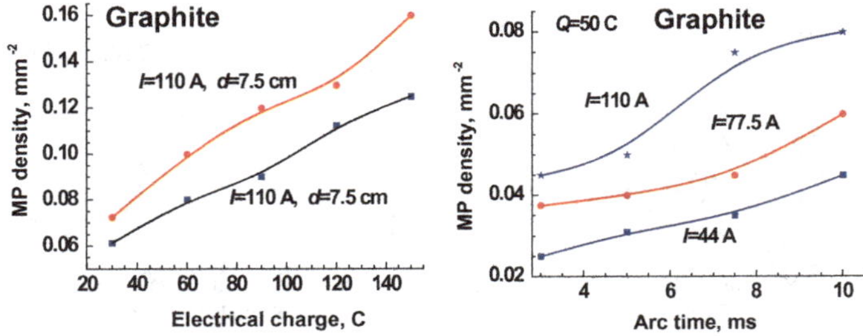

Fig. 9.4 Number density of particles as function of electrical charge through cathode (left): for current 44 and 110 A, pulse duration time 14 ms; and of the arc pulse duration time (right): for electrical charge 50 C, current 44, 77.5, and 110 A

be produced under relatively large plasma pressure and surface temperature in the region of the cathode spot.

The particles emitted from the cathode vacuum arc in solid state were characterized for refractory cathodes. Such case for tungsten and molybdenum cathodes was reported in [26, 27]. A hyperbolically decreasing of the particle number with their size was observed. The collected particles showed no signs of melting, but rather appeared to be cleaved along crystalline planes. The detected solid particles were caused by thermoelastic stress produced due to large temperature gradients. The effect was occurred when the thermoelastic stress exceeds the breaking point of the material which take place at rate of current rise of $dI/dt \sim 10^8$ A/s of the arc pulse. With decreasing of this dI/dt the thermoelastic stress effect decreased and mostly disappeared at of $dI/dt < 10^6$ A/s. However, at large $dI/dt > 10^9$ A/s, the destruction of the cathode with emission of the solid particles can be even on plastic materials like Cu.

A series works studying the particle mass transport, size distribution, velocity, and spatial distribution were presented by Boxman and Goldsmith and co-authors considering the applications of vacuum-arc depositions (see reviews [28–31]). Macroparticle dynamics in multi-cathode-spot (MCS) vacuum arcs were studied by utilizing laser-Doppler anemometry (LDA) methods for in situ measurement of the cathodic particle velocities and relative emission rates [32, 33]. Arc current pulses having peak values of 1–2 kA were investigated. The arc was sustained between two cylindrical copper electrodes, 14 mm in diameter and spaced 4 mm apart. The two electrodes were coaxial and spaced 5.5 mm apart. The stainless steel particle collectors of 4-mm-diameter collected the Mps, emitted by the arc. The collectors were polished to a mirror finish and cleaned before the arcing sequence with degreasing materials. Photographs of the collector surface were taken using an optical microscope and scanning electron microscope. Two current waveforms, with rise times to peak currents of 1 and 6 ms and pulse duration of about 5 and 30 ms, respectively, were used in the experiment. The electrodes were fabricated from Cd, Zn, AI, Cu and Mo. Broad velocity component distributions, ranging from about 10 m/s up to about 800 m/s, were measured.

It was found that the macroparticle velocity increased with the melting temperature of the cathode metal, distance from the cathode surface, time after arc initiation and spatial location in the discharge volume and the instantaneous value of the arc current, and decreased with macroparticle size and the rise time of the current waveform [34]. The particle emission rate was found to reach its peak a few milliseconds after the occurrence of peak current.

Time- and space-resolved measurements of the axial and transverse velocity components of molybdenum particles produced in a 1 kA peak current, 30 ms duration vacuum arc were conducted using laser-Doppler anemometry [35]. Both axial (v_z) and transverse (v_x) velocity components of at least one hundred macroparticles were measured at each "measuring point," defined by its (x, z) location in the inter-electrode region and its time after arc initiation. Three "measuring points" were fully analyzed. The average values of the axial velocity component at the points near the cathode, near the anode, and outside the interelectrode cylinder are bounded by 290–165, 310–190, and 170–100 m/s, respectively. The average values of the transverse component at the same points are bounded by 315–200, 310–200, and 355–230 m/s, respectively. The rms values of total velocity were obtained in the range 400–700 m/s from the single-component averages assuming an isotropic velocity distribution in the plane normal to the discharge axis. Comparing the average values of the radial and axial components indicates that typical emission angles are at approximately 30° ± 5° with respect to the cathode surface.

The values of the Mo macroparticle velocity was found to be much greater than the values found for Cu, AI, Zn, and Cd macroparticles (decreasing order) measured under identical experimental conditions. The obtained results suggest that the macroparticle emission rate is an increasing function of both the arc current and the average temperature of the cathode surface. The particle axial velocity component, increases with distance from the cathode surface and with the arc current, possibly demonstrating momentum transfer from the arc plasma to the ejected droplets.

Weighing the anode and the cathode showed that the cathode lost a total of 140 μg/C and the anode gained a total of 15 μg/C [36]. The macroparticle erosion rate was determined to be 105 μg/C, and together with ionic emission, accounted for most of the cathodic erosion. Thus, approximately 2/3 of the cathodic mass loss is emitted in the form of macroparticles. The number of macroparticles emitted decreased exponentially with macroparticle diameter, with 20–80 μm macroparticles carrying the bulk of the mass transport. Macroparticles are emitted preferentially at an angle of 20° with respect to the cathode surface. In comparison to previous investigations, higher macroparticle erosion rates, a larger proportion of large macroparticles, and a higher emission angle are observed, and the differences are attributed to the large current density occurred in the present experiment. The minimal size of the macroparticles was fixed as 1 μm, while the larger particles reached a size even exceeding 100 μm.

Measurements of size and velocity of the particles produced during and after high-current vacuum arc were conducted using ultrafast laser shadow technique. The current pulses of up to 12 kA peak and 5 ms duration between plane Cu [37] or 24 kA with CuCr [38] contacts of 25 mm diameter spaced by 9 mm were applied. The measurements showed strong activity at the Cu cathode, which begins with decreasing current before current zero and until 600 μs after current zero. The liquid droplets were generated with diameter <200 μm and velocity of a few m/s. The solidification of the melts take place at <1 ms. The anode melt was of mm depth and big droplets generation was late as 8 ms after the arc. Anodic particles were ejected nearly tangentially, whereas cathodic particles do not have any direction preferred.

For CuCr contacts, smaller depth of the melt was indicated which is due to higher melting point of Cr. The molten region has only a depth of 100–200 μm compared to a depth of order of mm for Cu under similar conditions. Therefore a smaller particle sizes were detected that shown in Fig. 9.5. Maximal particles were observed at about 5 μm diameter was found at lower current, whereas larger particles occur after the anode spot mode with a second peak around 7 μm for a peak current of 5 kA. The number of particle dependent on time after current zero was presented in Fig. 9.6.

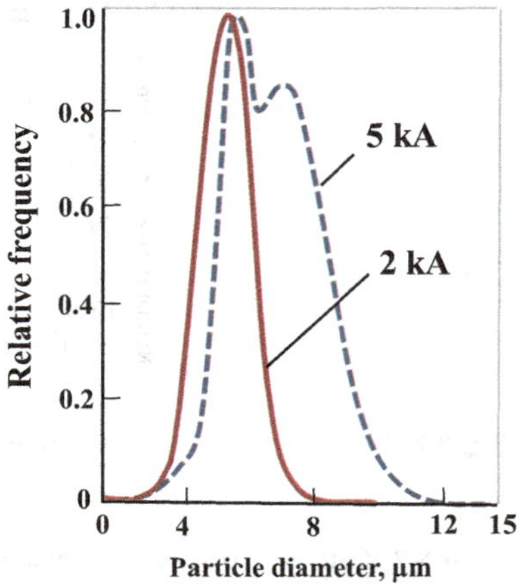

Fig. 9.5 Normalized particle distribution on particle diameter integrated over a time period of 5 ms after current zero [38]

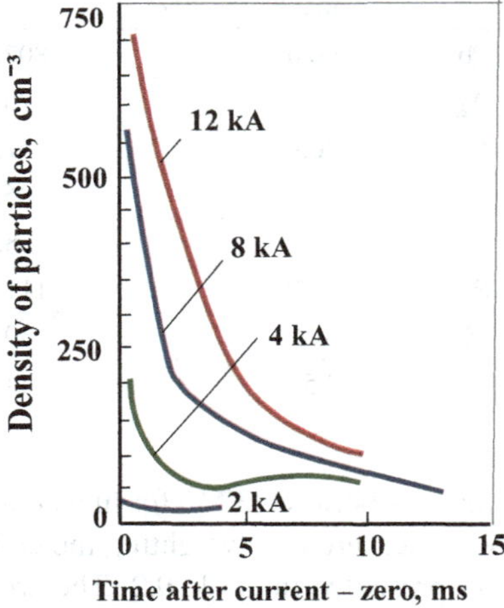

Fig. 9.6 Droplet density as function of time after current zero for various peak currents (sinusoidal) [38]

The large particle numbers were detected at relatively larger currents more than 4 kA when the anode spot appeared, whereas at lover currents (2 kA) less than 30 particles per cm^3 were found.

Anders [39] studied the particle contamination in thin films deposited by a vacuum arc generated in arc plasma source with axial magnetic field in direction near the discharge axis. Cathode diameter 6.4 mm, aluminum anode inner diameter 19 mm, distance between cathode and anode exit planes 15 mm. Rectangular current pulse of 250 μs duration with current amplitude varied between 50 and 250 A. Figure 9.7 shows the distribution of the particle diameter normalized to the charge transferred through the cathode in Coulombs and the sample area in mm^2.

The total mass deposited in the form of macroparticles was compared with the mass of the film deposited by ions at the sample. This ratio presented in Table 9.2

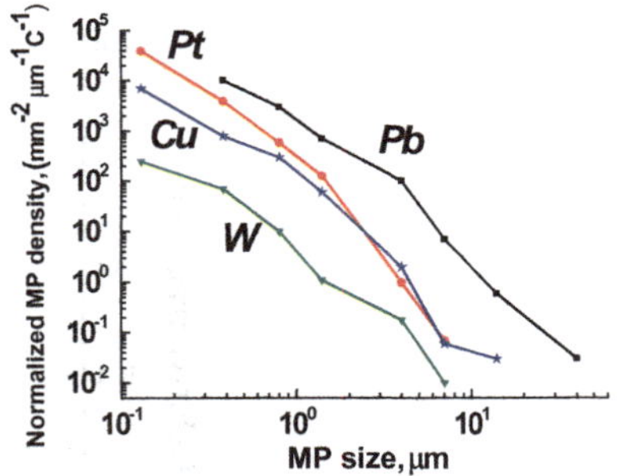

Fig. 9.7 Normalized particle distribution as function on their size (diameter)

Table 9.2 Cathode erosion rate without magnetic field (MF) and with MF of 170 mT [39]

Material	Mass fraction of droplet compared to that by ions (%)	Erosion in μg/C without MF	Erosion in μg/C with MF	Erosion increase due to MF (%)
Pb	90.0	807.0	975.0	21
Ag	9.9	76.3	83.7	10
Cu	20.9	51.1	56.0	10
Au	–	86.2	100.9	17
Ni	61.1	38.4	49.1	28
Pt	43.0	111.1	116.4	5
Mo	–	30.9	38.8	26
W	3.5	44.4	50.0	13

changes between 3.5% for tungsten to 90% for lead. The erosion rate of the cathode was measured by weighting the cathode before and after discharges and the results are presented in Table 9.2. The erosion rate increases slightly with axial magnetic field.

9.2 Macroparticle Charging

The macroparticle properties are determined not only by their density, sizes, velocities but also by their electrical charge. The particle moving in the cathode quasineutral plasma jet becomes floating potential U_p (negative with respect to the plasma) due to much higher random electron velocity (T_e is the temperature) than either the direct or random velocity of the ions v_i [40]. This potential is determined from condition when both fluxes electron and ions to the particle are equal:

$$U_p = \frac{kT_e}{2e} \ln(\frac{v_i^2 \pi m_e}{2kT_e}) \tag{9.2}$$

The electric field at the macroparticle surface may be estimated using the characteristic Debye length L_d as $E_p = U_p/L_d$. Using macroparticle charge definition as $Q_p = \varepsilon_0 E_p 2\pi R_p^2$ and electron density as $n_e = fI/A_c e v_i$ the value of Q_p in steady state case is [40]

$$Q_{pe} = \left(\frac{\varepsilon_0 k T_e f I}{2A_c e v_i}\right) 2\pi R_p^2 \ln\left(\frac{v_i^2 \pi m_e}{2kT_e}\right) \tag{9.3}$$

where ε_0 is the permittivity of free space, n_e is the electron density, R_p is the macroparticle radius, f is the ion current fraction, I is the current, and A_c is the cathode area. The fast-moving electron determined the negative particle potential. However, in quasineutral plasma, due to relatively slow ion motion, the time of particle charging depends on the ion flux. When the ion flux to the MP is equal to the electron flux, a steady state can be reached. When the time of particle charging is larger than characteristic time of the discharge, the particle charge process will be *nonstationary*. Hazelton and Yadlowsky [41] investigated experimentally the effect of MP radius on the non-stationary charging in a plasma with a relatively large fixed MP density This charging result was obtained under conditions in which the MP flight time was smaller than the full charging time.

Also, the charging is in *equilibrium* when during the charging process, the plasma is only slightly disturbed, and remains in equilibrium and the MP becomes fully charged to a value dependent on the its size. In contrast, we will consider non-stationary *non-equilibrium* MP charging, which depends not only on the MP size but also on the MP density and the plasma density distribution along the MP track.

Close to the cathode in vacuum arcs, the electron density is very high, and thus, the Debye length L_D is very small, and usually a condition $R_p \gg L_D$ was fulfilled. This is the case considered in [40]. When expanding plasma of the vacuum arc was utilized far from the cathode spots, typically in a direction normal to the cathode surface, the electron density is greatly reduced as a function of distance from the cathode. A smaller MPs are increasingly favored in the MP size distribution in this direction with condition $R_p \ll L_D$ should be analyzed.

The non-stationary particle charging was considered by Keidar et al. [42] taking into account the kinetics of the MP charging controlled by the ion and electron fluxes to the MP which in turn depends on the potential distribution in the sheath. Poisson's equation and equation of time dependent electric field at the MP surface were solved

$$\varepsilon_0 \nabla U(r) = e[n_e(r) - Z_i n_i(r)] \tag{9.4}$$

$$\varepsilon_0 \frac{\partial E(r)}{\partial t} = -[j_e(r) + j_i(r)] \tag{9.5}$$

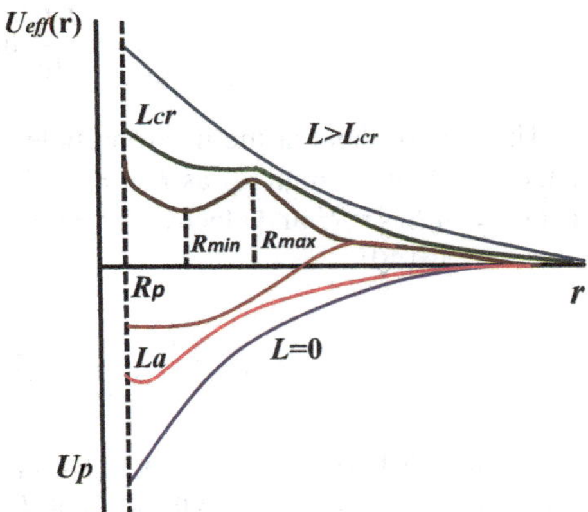

Fig. 9.8 Effective potential lines for ions with different angular momentum L in the neighborhood of a macroparticle

$$U(R_\infty)\frac{\partial U(R_\infty)}{\partial r} = 0 \tag{9.6}$$

where $n_e(r)$ and $n_i(r)$ are the electron and ion densities, respectively, $j_e(r)$ and $j_i(r)$ are the electron and ion current densities, respectively, Z_i, is the charge number of the ions, $R_p < r < R_\infty$, R_∞ is the outer boundary of the sheath around the particle.

The electron and ion parameters were obtained in [42]. The problem was solved at the radius R_p using approach of orbital motion limit (OML) [43] which is illustrated by Fig. 9.8. A simple estimation indicates that the characteristic time for the electric field to reach steady state decreases with radius and has its maximum time at the MP surface.

The radial flux of a charge carrier is governed by the effective potential energy $U_{eff}(L, r) = L^2/2mr^2 + qU(r)$, where q, m are the particle charge and mass, respectively, $L = mv_\theta r$ is the angular momentum relative to the center of force, v_θ is the tangential projection of the particle velocity vector, r is the distance from the center of force, and $U(r)$ is the local electrical potential. The resulting sign of U_{eff} depends on the relative magnitude of L and $U(r)$. In general, we have a family of $U_{\text{eff}}(r)$ lines, which correspond to different values of L (Fig. 9.8). When L_a is the value of L at which $U_{\text{eff}} < 0$ for all r and when L_{cr} is the value of L at which U_{eff} has no extremum and $U_{\text{eff}} > 0$ for all r.

In the case of $L > L_{cr}$ a particle approaching from far away will be repelled at radius r if $E < U_{\text{eff}}(r)$, and in the case $L < L_a$, the approaching particle will collide with the MP. For an arbitrary L in the interval $L_a < L < L_{cr}$, U_{eff} has local extrema at $r_{\min}$ or $r_{\max}$. The charge which accumulates on the MP is given by $Q = 4\pi R_p^2 \varepsilon_0 E_p$. Figure 9.9 shows the time evolution of the MP charge in the case of a titanium plasma [42]. The maximum charging time of 10 μs occurs for the minimum values of n_e and R_p. The charge, which accumulates on the particle, depends on the electron temperature T_e.

Let us consider the non-equilibrium MP charging when the particle density is relatively large for the case of $L_D > R_p$. According to the model used, the OML

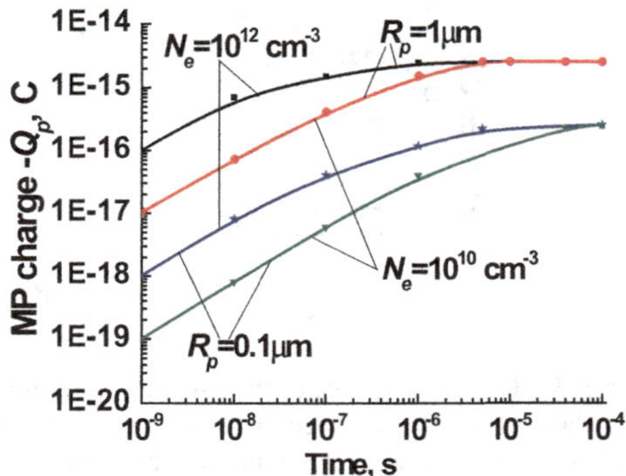

Fig. 9.9 Charge on the MP. $T_e = 6$ eV; $T_i/T_e = 0.1$; 1 – $n_e = 10^{10}$ cm^{-3}; 2 – N,= $n_e = 10^{11}$ cm^{-3}; *a* – $R_p = 0.1$ µm; *b* – $R_p = 1$ µm

approach [44], the MP charge evolution $Q_p(t)$ was determined by

$$\frac{\mathrm{d}Q_p(t)}{\mathrm{d}t} = -4\pi R_p^2[j_p e^{-\phi(t)\gamma} + j_{e0}(t)e^{-\phi(t)} - j_{i0}(t)(1 + Z_i\phi(t)\beta] \tag{9.7}$$

$$Q_p = 4\pi\varepsilon_0 R_p \varphi_p \tag{9.8}$$

where $\phi = e\varphi/k_B T_e$; $\gamma = k_B T_e/W_p$; $\beta = T_e/T_i$ and k_B is the Boltzmann's constant, W_p is the MP kinetic energy in the plasma bulk, φ_p is the MP potential relative to the plasma, j_{e0} and j_{i0} are the random electron and ion current densities, respectively. The ratio of n_p density to the ion density n_i was defined by parameter $\kappa = n_p/n_i$. This parameter characterizes the degree of non-equilibrium MP charging in the case of uniform plasma. For κ => 0.01 the charging process will be a strongly nonequilibrium, while $\kappa \to 0$, it will be equilibrium charging. It follows that for a short time (non-stationary MP charging), the MP charge increases with radius [44] (Fig. 9.10). But for a longer time ($t \to \infty$) and for MP radii larger than some critical radius an effect of MP charge saturation with radius was observed. The saturation occurs because the electron and ion fluxes are limited by the quasineutrality condition.

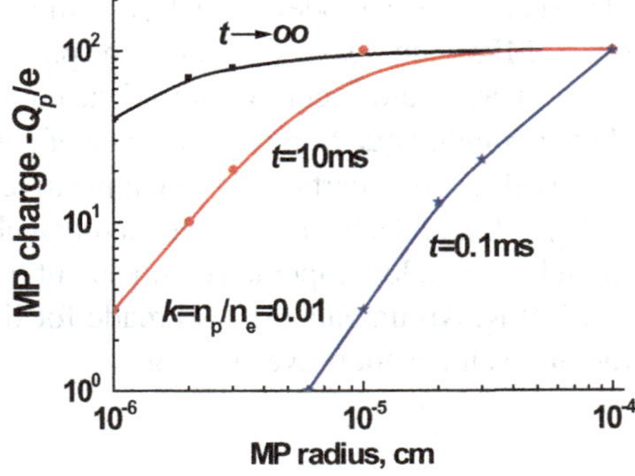

Fig. 9.10 Dependence of the MP charge on the MP radius with time as a parameter for plasma with $\kappa = 0.01$, $n_i = 10^{12}$ m^{-3}, $T_e = 2$ eV, and $\beta = 10$

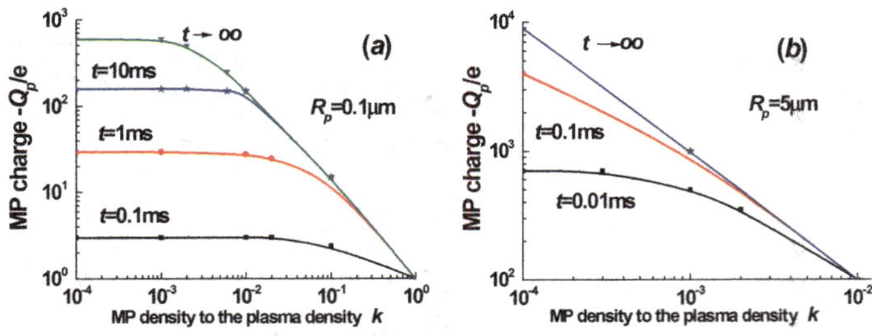

Fig. 9.11 **a** Dependence of the MP charge on the parameter $\kappa = n_p/n$ with time as parameter for a plasma with $R_p = 0.1$ μm, $n_i = 10^{12}$ m^{-3}, $T_e = 2$ eV, and $\beta = 10$.) The dependence of the MP charge on the parameter κ with time as parameter for a plasma with $R_p = 5$ μm, $n_i = 10^{12}$ m^{-3}, $T_e = 2$ eV, and $\beta = 10$

The variation of the MP charge with MP density is shown in Fig. 9.11a, b [44]. Generally, MP charge decreases with MP density (parameter κ) when parameter κ is larger than a critical value. For the case $R_p = 0.1$ μm, the critical parameter κ changes from 0.1 to 0.001 when the MP residence time increases from 0.1 to 10 ms. The MP charge reduction is significant for in the case $\kappa \geq 0.0001$ and in the case $R_p = 5$ μm. It may be also seen that MP charge is relatively independent of the MP residence time for an MP residence time greater than 0.1 ms (see Fig. 9.11b).

As for MPs traveling through a non-uniform plasma jet, the plasma density dependence with distance is approximated as $n_i = n_{i0}(x_0/x)^m$, where m is the non-uniformity parameter and x_0 is the distance during which charging occurred is larger than the characteristic linear scale of the plasma density changes [44]. It was obtained that for MP traveling in the plasma jet, the MP does not reach the equilibrium charge if $m \geq 2$ and x_0 ~ 1 cm. For the rarefied part of the vacuum arc plasma jet, it was obtained that the axial plasma density distribution may be characterized by $m \leq 2$ [45, 46], while for the radial density distribution m ~ 3 [47].

The velocity, mass, and charge of copper MPs emitted by a 100 A arc were experimentally measured and compared to a model based on thermionic electron emission [48]. The MP velocity was determined by using a time-of-flight velocity filter. The charge was calculated by measuring particle deflection in a transverse electric field. The model predicts, and the experimental results verify, that the charge on the MPs becomes positive once the plasma is extinguished, and the MP travels in a vacuum, as would occur in a pulsed vacuum arc, versus a dc arc. Experimental results show a roughly quadratic dependence of particle charge on the particle diameter ($q \sim D^2$), with a 1 μm particle having a positive charge of ~1000 electronic charges (1.6 × 10^{-16} C), and a 5 μm particle having a charge of ~25,000 electronic charges. The model is particle temperature dependent, and gives $q \sim D^2$ at 1750 K and $q \sim D^{1.7}$ at 2200 K. Arguments are also made for limitations on particle temperature due to radiative and evaporative cooling.

9.3 Macroparticle Interaction

The charging effect allows control the particle motion in systems of different applications, in particular, by their interaction with a wall, substrate, and surrounding plasma. Below these phenomena will be considered.

9.3.1 Interaction with Plasma

The interaction between the interelectrode plasma and macroparticles (droplets) produced by a multitude of cathode spots in a vacuum arc between Cu electrodes was analyzed by Boxman and Goldsmith [49], using experimental data of the MP size distribution and erosion rate and a flowing plasma model. The effect of the plasma on the MPs is considered by treating the MPs as floating probes and calculating the particle, momentum, and energy fluxes to them. It was shown that the momentum imparted on them by collisions with high-velocity ions (ion bombardment) accelerated the MPs. The rate of momentum transfer to the MP is given by the ionic particle flux multiplied by the momentum of each ion, and taking into account the acceleration of the ions by the MP sheath with negative floating potential ϕ_p [49]:

$$m_p \frac{\mathrm{d}v_{pz}(t)}{\mathrm{d}t} = \frac{I_p m_i}{eZ}\left[v_i e^{-\phi(t)\gamma} + \left(-\frac{2Ze\phi_p}{m_i}\right)^{0.5}\right] \tag{9.9}$$

where m_p and v_{pz} are the MP mass and axial velocity, respectively. The acceleration of the MP's from ion impact is in the axial direction. The effect of MP acceleration was determined by comparison of the final axial velocity v_{pz} with the initial radial velocity of the particle v_{pi} as function on the current density and the particle radius R_p, using $I_p = fJ\pi R_p^2$. The result indicated that the axial deflection is significant (i.e., v_{pz}/v_{pi}; is comparable to unity) for used range of the current density (1–100) $\times$ 10^2 A/cm^2, when the lower value of MP velocity is invoked. However, when the higher value of v_{pi} is used, the deflection is only significant for the smaller particles and only at the upper values of the current range.

The steady-state MP temperatures of 2000–2600 K were obtained considering the balance of the energy influx (primarily from ion bombardment). As result, the particle evaporation provides a distributed vapor source. The calculated neutral density is proportional to the current density and is in the range 0.02–0.5% of the electron density, depending upon electron temperature and MP velocity. In the 10^3 A/cm^2 Cu arc, the calculated neutral density was in the range (1–25) $\times$ 10^{11} cm^{-3}.

Neutral vapor was ionized (in a period much smaller than the time of flight) producing an ion population initially characterized by a slow, random velocity distribution (in contrast to the fast directed velocity of the cathode-spot produced ions). The MP ion production rate approaches significant levels in Cu arcs for $j > 10^3$ A/cm^2, and may be significant in Cd arcs for the entire current density range (10^2–10^4 A/cm^2).

Anders [50] analyzed the problem of mass gain or mass loss of macroparticle immersed in the vacuum arc plasma when the ions collided the MP. The mass and energy transfer from the plasma to the macroparticle were considered. The mass balance was taken in form:

$$\frac{\mathrm{d}m_{mp}(t)}{\mathrm{d}t} = \{f_i[n_i(t)T(t)v_i] - f_a[T_{mp}(t)^{0.5}]\}m_a A_{mp}(t) \tag{9.10}$$

where the first term in right side of (9.10) is the flux of ions arriving at the MP which depends on density n_i, ion temperature T, the velocity of the directed ion motion v_i, etc. The second term is the flux of atoms evaporated, m_a is the atom mass, and A_{mp} is the surface area of the MP. To determine MP growth, we assume that each ion incident on the MP will stick to it and contribute to its mass gain. The solved energy balance was included ion bombardment heating, plasma electron heating, electron emission cooling, radiation, and evaporation cooling.

The estimated constant temperatures which needed to complete MP evaporation, were obtained higher than the boiling temperatures of the materials (C-4800 K; Ti-4500 K; Cu-3800 K; and W-7200 K). It was noted, however, that numerous publications indicate that MPs are liquid when ejected from metal cathodes, and solid at large distances from the cathode. Based on the observations, it assumed that a typical initial MP temperature is higher than the material's melting temperature but lower than its boiling temperature. Therefore, it concluded that complete MP destruction by MP–plasma interaction is difficult or impossible in any case. This is due to a reduction of ion bombardment heating by the electrically equivalent thermionic electron emission, and an increase of the loss mechanisms with increasing MP temperature, when the MP approaches the temperature region of strong evaporation. Note that it is true for detected MPs but some of MPs could be disappeared before.

Zalucki [51] developed a model of calculation of Cu and Cd droplets vaporization and neutral atom generation assuming that initial temperature of the droplets weakly changed during their flight through the electrode space at the current zero or after current zero. The initial droplets temperature (ejected by the cathode) ratio to the boiling temperature was given in range of 0.6–1. The calculated atom density was presented as dependencies on the mentioned ratio, distance from the cathode and discharge time taking into account the droplets diameters and velocities distribution. The influence of the evaporating droplets (according to the obtained results) on the development of electrical breakdown in vacuum gap after arcing was discussed.

Peters et al. [52] indicate that the molten droplet ejection seems to be the major mechanism of hafnium cathode erosion. Most ejection events are associated with changes in the conditions of the plasma, e.g., during start-up, change of gas flow, and shutdown. The ejection was explained by imbalances of the forces acting on the molten surface. The balance include: (i) force P_{st} due to surface tension; (ii) force P_{ion} associated with the arc ion current flow in the near-cathode electrical sheath; and (iii) the electromagnetic force $P_{j\times B}$, which caused by the interaction of the radial component of the current density j_r with the self-induced magnetic field in the azimuthal direction. This force was important with small-arc diameters and high-current densities.

The analysis [52] was indicated, if the forces on the molten pool due to the plasma ($P_{j\times B}$ and P_{ion}) are assumed to be balanced by the surface tension forces (P_{st}), the radius of droplets formed at the edge of the molten pool can be shown to be inversely proportional to the forces resulting from the plasma. This implies that small droplets will be ejected during the arc ignition when the ion pressure is large and large droplets will be ejected at other times during operation. The magnitudes of these forces on the molten cathode pool due to the plasma were calculated using the measured current density values obtained from the optical experiments. It was obtained that droplet diameter was 3 μm at during start-up ($I = 20$ A), and 51 μm at shutdown ($I = 200$ A). The problem however is that the force due to ion action cannot to be described by the momentum acquired in the sheath because the electric field momentum compensates this momentum.

9.3.2 Interaction with a Wall and Substrate

The model of motion of the individual electrically charged MP in the straight and quarter-torus plasma guides was developed, taking into account MP charging by interaction with the plasma [53]. The influence of the electric field in the plasma, which depends on the magnetic field, on the charged MP motion was studied. The fraction of the MPs transmitted through the toroidal plasma guide was obtained as a function of the wall potential and MP velocity. The guide wall potential has a strong effect on the transmission of MPs having velocities in the 25–100 m/s range. In the duct at floating potential, the fraction of the MPs transmitted through the torus approaches 100% for 0.1 pm Ti MPs having an initial velocity parallel to the duct wall. The main mechanism of MP transmission through curved ducts is repeated electrostatic reflection of the charged MP from the wall.

The justification of this mechanism was tested experimentally [54]. In the experimental system, the vacuum arc metallic plasma source consists of a 90 mm diameter Ti cathode and of a 122 mm i.d. annular anode. The gap distance was 10 mm. A 250 A DC discharge was ignited by a mechanical trigger electrode in a vacuum of 3×10^{-5} Torr. The plasma jet emitted by the cathode passed through the annular anode into a magnetized quarter-torus duct (macroparticle filter) with a 240 mm major radius and 80 mm minor radius. The duct wall was smooth, i.e., no baffles or corrugations were placed in the duct. The distance from the cathode to the torus entrance was 60 mm. The anode was electrically grounded while the quarter-torus body and the chamber were floating. The toroidal magnetic field in the center of the duct of 6 mT per 1 A in the torus coils was operated up to a maximum value of 10 mT.

Coatings were deposited on 25 mm diameter steel samples placed in the entrance and exit planes of the filter at three radial positions: near the outer wall (samples 1 and 4), at the center of the duct (samples 2 and 5), and near the inner wall (samples 3 and 6). The samples were cleaned by immersion in alcohol prior to mounting, and exposed to the Ti plasma flux for 15–120 s. Four random photomicrographs per sample were acquired using a 100 lens, with a field of view of 63 × 47 m. The smallest detectable particle had an area of 0.03 m^2 or a diameter 0.2 m. The particles overlapping the frame boundary were counted as 50%. The photomicrographs were acquired with a video camera, digitized, and stored, after which automatic particle detection and measurement algorithms were employed to find the total number particles, the average particle size, and the area fraction covered by the particles. The results reported are the average over the four photomicrographs for each sample. The SEM micrographs were taken at 1000 magnification in the secondary electrons mode, with an acceleration voltage of 25 kV.

Two significant results were obtained in this experiment. First, the MP density on samples placed near the outer wall decreased with magnetic field, and second, the MP density on samples placed near the outer wall is larger than that on the center of the torus exit plane. The average MP flux in the central region of a quarter-torus plasma duct from the entrance to the exit was decreased by factor of 50, while near the outer wall, the flux decrease was only by a factor of 15. The average MP flux at the exit plane adjacent to the outer wall is a factor of 3 greater than the centerline MP flux. With increasing toroidal magnetic field (up to 10 mT), the average MP flux in the center of the filter exit was not changed substantially, while adjacent to the outer wall, the average flux decreased by a factor of 2.

Comparison of the experimental [54] and theoretical [53] results leads us to the main conclusion that MP transmission in the near outer wall region is not related to mechanical bouncing or entrainment in the plasma beam, but rather is the result of electrostatic reflection, while the MP transport in central region is due to mechanical bouncing.

An experiment was carried with the vacuum arc source of 64 mm diameter cathode to study the effect of MP electrostatic reflection [55]. A 150 A d.c. discharge was run for 90 s with Ti, Zr, and Cr cathodes and for 15 s on Cu cathode. Coatings were deposited on 25 mm diameter steel samples. Two samples were mounted inside the chamber in a way that one would be under floating potential and other under −1000 V with respect to the anode. The distance between the cathode and samples was about 25 cm. Four random photomicrographs per sample were acquired using optical microscope. The micrographs were acquired with a video camera, digitized, and stored, after which automatic particle detection and measurement algorithms were employed to find the total number particles, the average particle size, and the area fraction covered by the particles.. The SEM micrographs were taken at 500× magnification in the secondary electrons mode.

The obtained results showed that the floating potential depends on the cathode material and lies in range of (5–10) V. The results of the image analyses are summarized in Table 9.3, where presented number of MPs within the frame with area about $4.1 \times 10^4\ \mu m^2$, fraction area covered by MP's and MP density. It can be seen that the values of all above parameter are generally smaller in the case of sample deposited under −1000 V bias compared to the sample deposed under floating potential. The maximal effect was obtained for copper cathode.

Table 9.3 Results of the MP image analyses [55]

Sample bias	Material	Number of MPs	Area fraction (%)	Density (1/μm^2)
Floating	Copper	3418	32.19	0.085
Floating	Chromium	260	1.5	0.04
Floating	Titanium	3275	11.52	0.08
Floating	Zirconium	847	5.19	0.02
−1000 V	Copper	725	6.66	0.02
−1000 V	Chromium	113	0.6	0.02
−1000 V	Titanium	1334	11.52	0.03
−1000 V	Zirconium	235	2.47	0.01

The MP number significantly decreases with substrate biasing. This effect is more essential for those MPs having small size. The comparison of the MP number in the case of floating samples and the samples deposited under –1000 V easily shows significant MP reduction with –1000 V biasing. In the case of copper MPs, the number of small MP's (<0.5 μm) decreases by factor of 3, while number of larger MPs decreases less. Opposite can be observed in the case of titanium MPs. In this case, the number of small MPs (<0.5 μm) also decreases by factor of 3, while number of larger MPs decreases more substantially. The experimentally observed reduction of the MP number with sample biasing under –1000 V can be understood in term of model based on the MP charging and transport in the quasineutral plasma and near-substrate sheath. The substrate bias causes to appear electric field in the quasineutral plasma and sheath.

The temperature of MP traveling from the cathode to the substrate during about 100 μs was studied [56]. The MP temperature dynamics were considered assuming that the MP thermal diffusivity time is smaller than the MP flight time. This means that MP has a uniform temperature throughout its volume. The energy balance equation for the MP which takes into account ion heat flux and radiation, evaporation heat losses, and heat transfer to gas atoms has following form [56]:

$$\frac{4}{3}\pi R_p^3 c\gamma \frac{\mathrm{d}T}{dt} = \pi R_p^2 n_e v_i \left(\frac{m v_i^2}{2}\right) - 4\pi R_p^2 [\varepsilon \sigma_B T^4 + \Gamma (W_{ev} + 2k_B T) + S] \tag{9.11}$$

where c is the heat capacity, γ is the MP mass density, R_p, is the MP radius, σ_B is the Stefan–Boltzmann constant, v_i is the ion directed velocity, m, is the ion mass, ε is the emissivity, Γ is the rate of evaporation loss per unit area, W_{ev} is the heat of vaporization, k_B is the Boltzmann constant, and S is the heat flux transferred to the gas.

Equation (9.11) requires an initial value of the MP temperature T_0, which is not known a priori and used as introduced parameter in the range between the cathode melting and boiling points. The calculated [56] MP temperature as function of time

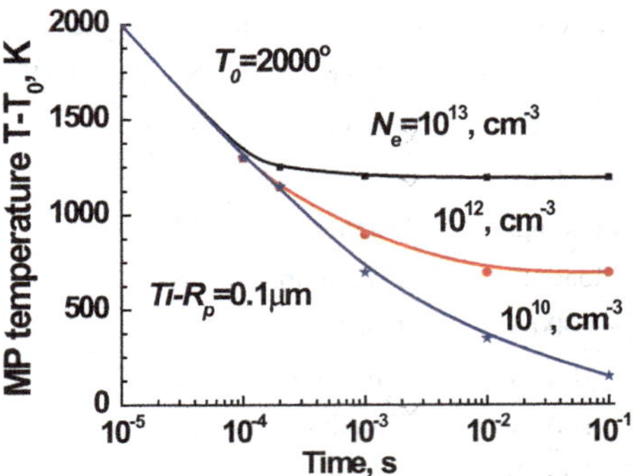

Fig. 9.12 Temporal evolution of titanium macroparticle temperature for different plasma densities. $T_0 = 2000$ K, $R_p = 0.1$ μm

for different plasma density is presented at Fig. 9.12. It follows from this result that the MP was cooled and the initial MP temperature was decreased at times smaller than 10^{-4}–10^{-3} s. In typical devices with characteristic length of 0.5 m and characteristic MP velocities 10–100 m/s (for MP radii of 0.1 μm), the characteristic time is 10^{-3}–10^{-2} s, and thus, the MP temperature at the substrate does not depend on the initial MP temperature.

It has been experimentally observed that the MP content in coatings decreases with increasing gas pressure [20, 21]. It was experimentally obtained [57] that the coating MP content reduction with gas pressure was more significant when the plasma is incident upon the glancing angle. The study in work [57] suggests a mechanism, based on increasing electrostatic repulsion of the MPs in a background gas, which explains the angular dependence of the MP reduction in the coating.

A model of the electrostatic sheath at the substrate, which is electrically isolated in a background, was developed [56]. In the steady state, the substrate assumes the floating potential, which depends on the electron temperature and ion velocity in the case of beam-like ion flow. In the vacuum arc, a fully ionized metal plasma jet interacts with a background gas through elastic (drag) and inelastic (charge exchange) collisions.

The voltage drop in the substrate sheath as a function of the background pressure was calculated taking into account charge exchange between ions and gas atoms. The MP motion in the sheath was studied considering MP charging and MP reflection from the substrate. It was obtained that the voltage drop in this sheath increases with increasing background pressure by factor of 2. A negatively charged MPs were reflected from a substrate, which also has a negative potential if the potential energy is larger than the MP kinetic energy:

$$Q_p \varphi_p \geq \frac{m_p v_p^2}{2} \tag{9.12}$$

where Q_p, m_p, v_p are the MP charge, mass, and velocity component normal to the substrate (in case when the substrate oriented by angle β with respect to the axial

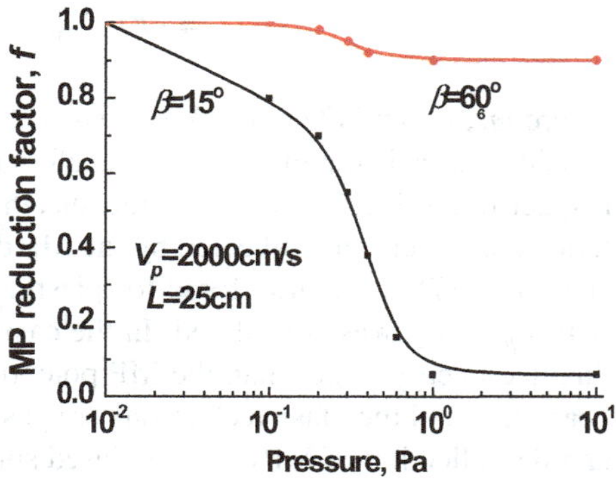

Fig. 9.13 Titanium MP reduction factor f as a function of a background gas pressure (in Pa) for different substrate angles. $L = 25$ cm, $v_p = 2000$ cm/s [56]

direction of the MP flux), and φp is the substrate floating potential relative to the plasma.

Taking into account the measured size distribution of MPs [22], it is possible to calculate the probability of MP reflection from the substrate. A function f was defined as the ratio of the number of MPs in the coating with gas pressure P_g to the number when $P_g = 0$. The background gas reduces MP contamination in the substrate by raising the substrate potential, and thus reflecting more MPs. The ratio f as a function of the gas pressure, for Ti MP velocity $v_p = 20$ m/s, distance from the cathode to the substrate $L = 0.25$ m, and with the substrate angle β as parameter is shown in Fig. 9.13. It may be seen that f substantially decreases at the substrate placed with a 15° angle while for a 60° angle the decrease is small. In the case of a 90° angle, no significant dependence of parameter f on the gas pressure was calculated. Thus, an increase of the background gas density caused a decrease in the ion flow velocity, due to collisions of the ions with the gas atoms. This results in an increased substrate floating potential, and hence a high probability of electrostatic repulsion of the MPs by the substrate, particularly for glancing incidence with the substrate. The probability of reflection decreases with MP size, velocity, angle of the MP velocity vector with respect to the substrate, and increases with the background pressure.

Above consideration was related to the MP charging in a quasineutral plasma. Generally, the MP charging in the sheath is different from that in the quasineutral plasma [58]. Keidar and Beilis [59] developed a model of MP charging in the sheath of a substrate, which is biased negatively with respect to the plasma. A negative substrate biasing leads to the appearance of the sheath, which, in turn, affects the MP charging and motion. The near-substrate sheath and its influence on the MP trajectory were studied solving the Poisson equation with the volume charge due to plasma ion and electron density and equation of MP motion. The equation for MP trajectory may be written as:

$$m_p \frac{\mathrm{d}^2 x}{\mathrm{d}t^2} = \gamma_{pl} \sigma (v_i - v_p) - Q \frac{\mathrm{d}\varphi}{\mathrm{d}x}$$

$$Q = 4\pi\varepsilon R_p\varphi_p; \quad \frac{dQ}{dt} = I_i - I_e \tag{9.13}$$

where m_p, v_p, and Q are the MP mass, velocity, and charge; γ_{pl} is the mass plasma density, and σ is the momentum transfer cross section, φ_p is the MP potential with respect to the local potential in the sheath, I_i is the total ion current, and I_e is the total electron current collected by the MP depended on φ_p. The first term on the left side of (8.17) is the drag force for plasma with a large directed velocity. The case with $R_p \ll L_D$ was considered. In the case of vacuum-arc plasma jet the ions have directed energy larger than the MP potential and therefore can be considered to be beam-like and the plasma electrons are gas-like. The titanium MPs were considered and the reflection of MPs from a biased substrate was taken in account.

An interesting result was obtained [59] (Fig. 9.14), which show that the MP can be charged not only negative but also positive depending on ion current density. x is the length coordinate having its origin at the plasma–sheath interface and directed normal to the wall. Initially, at the plasma–sheath interface, the MP has a negative charge, similar to steady state charging in the quasineutral plasma. In both cases of smaller- and higher-current density, the MPs acquire a positive charge. Therefore, the charged MP can be either reflected or attracted to the substrate and this effect depend strongly on the time and sheath voltage drop U [59] (Fig. 9.15). In the case of ion current density $j = 10$ A/m^2, the MP was always repelled from the wall if U > 100 V.

The obtained results [59] strongly deviate from those obtained without consideration of MP charging in the sheath, especially in the low MP velocity regime. Saturation of the MP repulsion effect with negative bias voltage, observed experimentally, can be obtained only through study of the MP charging and motion it in the sheath.

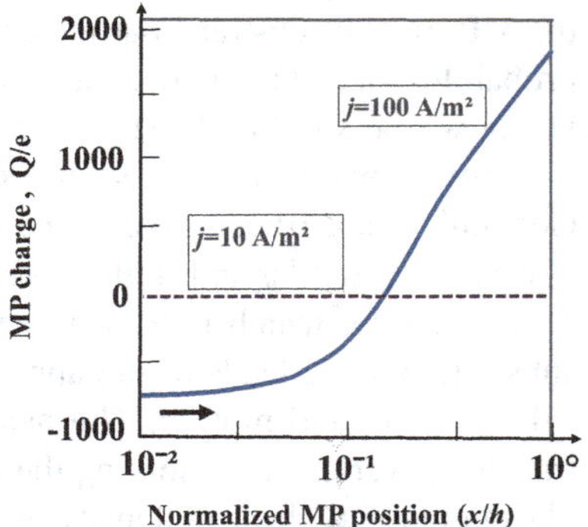

Fig. 9.14 Spatial distribution of MP charge in the sheath with current density as a parameter. $R_p = 0.1$ μm, $v_p = 30$ m/s, and $U = -1000$ V. The arrow indicates direction of MP charge evolution. h is the sheath thickness

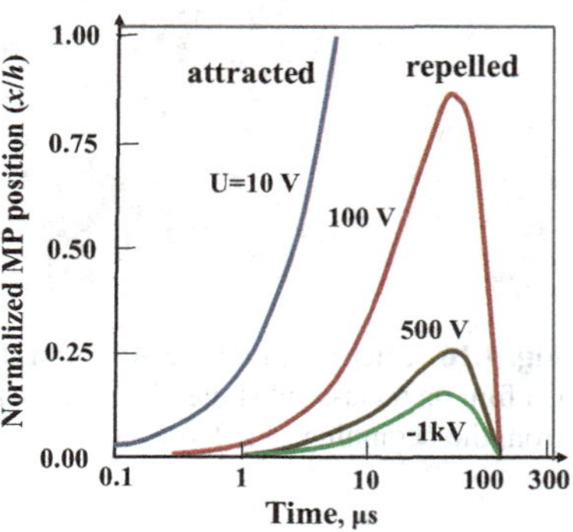

Fig. 9.15 MP trajectories in the sheath with voltage drop as a parameter. $R_p = 0.1$ μm, $v_p = 30$ m/s, and $j = 10$ A/m^2. h is the sheath thickness

9.4 Macroparticle Generation in an Arc with Hot Anodes

Generation of a clean metallic plasma from a vacuum arc is problematic due to producing numerous numbers of droplets. An attempt to solve the problem was made using the Hot Anode Vacuum Arc (HAVA). The plasma in HAVA is diffusely attached to the hot anode and the plasma was generated by evaporation of the hot anode material [60, 61]. When a refractory anode was used, the plasma can be generated from the cathode material. Two types of vacuum arc plasma sources with refractory anodes were developed in last decade. These arc plasma sources have potential for relatively clean metallic deposition with good characteristic that will be described separately in chapter of applications (Chap. 23). Below only the results of some detected particle distribution are detailed.

9.4.1 Macroparticles in a Hot Refractory Anode Vacuum Arc (HRAVA)

The metallic plasma from cathode was produced in a HRAVA that recently developed and investigated experimentally [62]. The HRAVA is constructed with a conventional water-cooled cathode, but the anode is made from a non-consumable refractory material. It was shown that the HRAVA operated initially as a conventional vacuum arc sustained by cathode spot emitted plasma together with MPs; however, at a later stage, the interelectrode region contained cathodic metal vapor re-evaporated from the hot refractory anode. The re-evaporated condensed cathode material (including MPs) was strongly ionized, creating an anodic plasma. The HRAVA served as a radially expanding plasma source with low-MP contamination. Two deposition regions, facing the plasma emitted from the cathode (C-region) and the anode (A-region), were observed [63] (Fig. 9.16).

The MP size distribution function for both regions illustrated in Fig. 9.17 for deposited glass substrate by an arc, which sustained between a 30 mm diameter

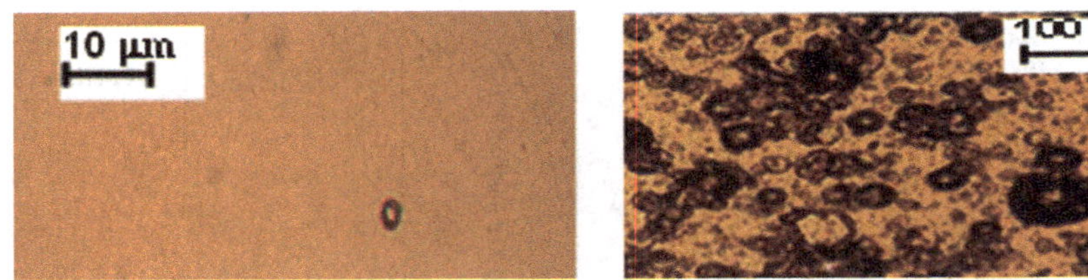

Fig. 9.16 Micrograph of the A-region (left) and C-region (right) showing MP contamination in a Cu film on a glass substrate placed at distance from the arc axis $L = 110$ mm and exposed for 60 s from the beginning of a 200 A arc

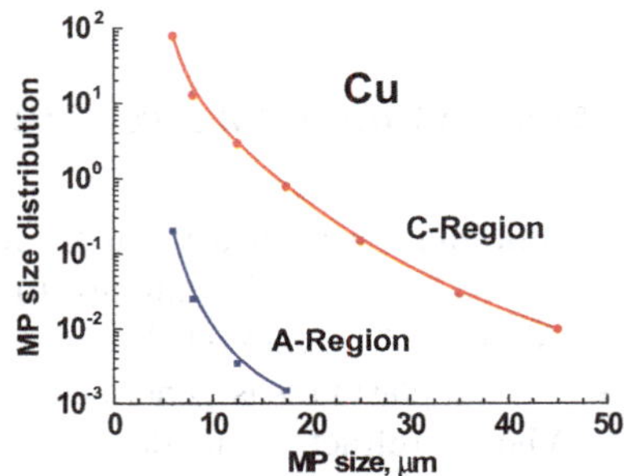

Fig. 9.17 Macroparticle size (diameter D) distribution ($\frac{\Delta N_{MP}}{\Delta D_{MP} \Delta C \Delta z}$ μm^{-1} C^{-1} mm^{-1}) in the C-region (solid curve) and A-region (dotted curve) of the glass substrate placed at distance from the arc axis $L = 110$ mm and deposited at C-region for 10 s, and at A-region for 30 s, beginning 60 s after 200 A arc ignition with gap of $h = 10$ mm

cylindrical, water-cooled, copper cathode and a molybdenum anode of 32 mm diameter and 30 mm height [63]. It can be seen in both Figs. 9.16 and 9.17 that the region facing the cathode contained numerous MPs while the region facing the anode was deposited with significantly reduced MP's.

The dependence of copper MP flux density on I with tungsten and molybdenum anodes is shown in Fig. 9.18 (for distance $L = 110$ mm) [64]. It was observed that MP flux density in the A-region decreased approximately linearly with the arc current

Fig. 9.18 Cu MP flux density versus I (L = 80 mm, h = 10 mm, 30 s deposition time starting 1 min after arc ignition, W and Mo anodes)

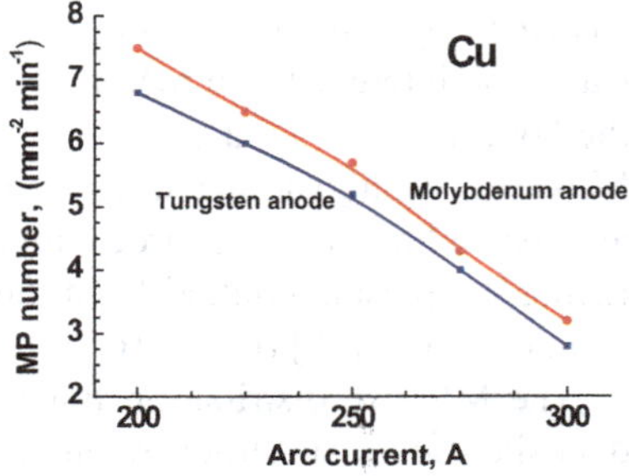

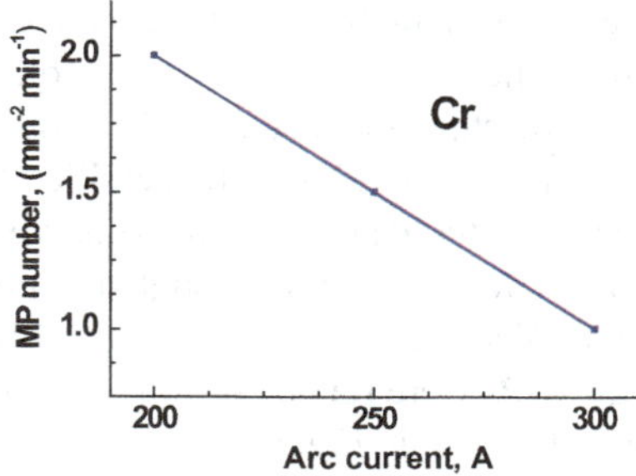

Fig. 9.19 MP flux density as function on I for Cr ($L = 110$ mm)

and reached ~3 $\text{mm}^{-2}\ \text{min}^{-1}$ for current $I = 300$ A. The dependence of the MP flux density on I with a Cr cathode and $L = 110$ mm was investigated in [65]. It was observed that the MP flux density in the A-region decreased approximately linearly with the arc current and reached ~1 $\text{mm}^{-2}\text{min}^{-1}$ for $I = 300$ A (Fig. 9.19).

9.4.2 Macroparticles in a Vacuum Arc with Black Body Assembly (VABBA)

In VABBA a water-cooled cathode and non-consumable hot anode are configured to form a closed vessel, which is schematically shown in Fig. 9.20a with a hollow shower anode [66, 67]. Material eroded by the arc from cathode spots (metallic plasma and MPs) is condensed on the initially cold anode and then is re-evaporated from the anode at stage when the anode is hot due to energy flux from the arc. The metallic plasma behavior in the closed vessel is similar to ordinary gas plasma. The closed vessel operated as a black body for the MPs while the clean plasma will be emitted through the anode apertures at the hot anode stage, forming a clean plasma plume.

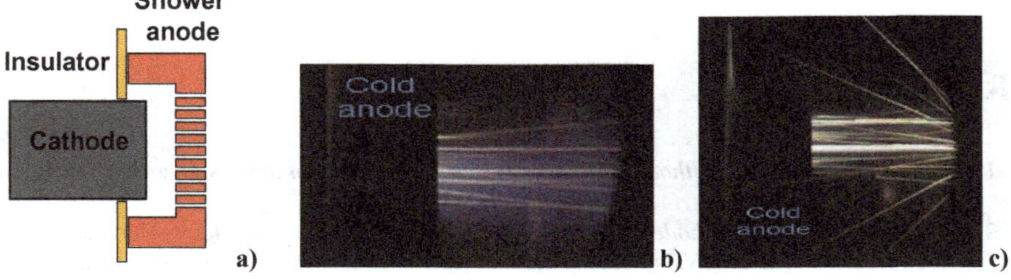

Fig. 9.20 Schematic cathode-anode presentation (**a**), MP's ejection from the shower anode at initial cold stage (**b**), MP's reflection from the shutter placed between the anode front surface and a substrate (**c**)

A shutter placed between the anode front surface and a substrate, which was used to prevent the substrate from the MP's condensation at cold stage of the anode. The shutter is open at the hot anode stage that occur in about 60 s after arc ignition and a clean plasma is produced which was used for thin metallic film deposition. However, at first few second of arc ignition (cold anode), numerous MPs were produced, which demonstrated in Fig. 9.20b the MP's ejection through the shower anode having small holes with 1 mm diameter. When the shutter was placed between the anode front surface and a substrate, the MPs were reflected from the obstacle and they returned in direction to the anode that shown in Fig. 9.20c.

9.5 Concluding Remarks

MP velocity components are detected in the range 10–700 m/s. The MP-axial velocity component increases with distance from the cathode surface and with the instantaneous value of arc current demonstrating momentum transfer from the arc plasma to the ejected droplets [32]. The droplet density depends on cathode material and arc current significantly decreasing with melting temperature and substantially increasing with current larger than 100 A indicating increase of the cathode erosion rate by large contribution of the particle mass. However, the number of ejected droplets was considerable decreased in an electrode assembly with refractory anode in a stationary vacuum arc.

A larger proportion of large macroparticles is observed, with the predominant mass transport effected by macroparticles having diameters in the 20–80 μm range and particle erosion of 105 μg/C for 1–2 kA arc [32, 36], in comparison to 10–40 μm and cathode erosion rate of 25–70 μg/C in Daalder's experiments with current 100–200 A [56]. In plasmas where the MP density is sufficiently high, the MP's capture a substantial fraction of the plasma charges, while the individual MP charge is less than in a comparable MP-free plasma. The MP charge decreases with parameter dependent on the MP residence time in the discharge. Macroparticles passing through highly nonuniform plasmas with sufficiently high velocity cannot attain their equilibrium charge value.

References

1. Kesaev, I. G. (1968). *Cathode processes in electric arcs*. Moscow: Nauka Publishers. (in Russian).
2. Kesaev, I. G. (1964). *Cathode processes in the mercury arc*. NY: Consultants Bureau.
3. Plyutto, A. A., Ryzhkov, V. N., & Kapin, A. T. (1965). High speed plasma streams in vacuum arcs. *Soviet Physics—JETP, 20*, 328–337.
4. Klyarfeld, B. N., Neretina, N. A., & Druzhinina, N. A. (1969). Metal Sputtering by the cathode spot of a vacuum arc. *Soviet Physics—Technical Physics, 14*, 796–799.

5. Udris, Y. (1963). Disintegration of materials by an arc cathode spot. *Radio Engineering and Electronic Physics, 8,* 1050–1056.
6. Udris, Y. (1970). On the emission of cathode material particles in low pressure arc discharge. In *Proceedings of International Conference on Gas Discharge, London* (pp. 108–112). IEE: London.
7. McClure, G. W. (1974). Plasma expansion as a cause of metal displacement in vacuum-arc cathode spots. *Journal of Applied Physics, 45*(5), 2078–2084.
8. Gray, E. W., & Pharney, J. R. (1974). Electrode erosion by particle ejection in low-current arcs. *Journal of Applied Physics, 45*(2), 667–671.
9. Beilis, I. I. (1982). On the theory of the erosion processes in the cathode region of arc discharge. *Soviet Physics—Doklady, 27*(2), 150–152.
10. Beilis, I. I. (1985). Parameters of the kinetic layer of arc-discharge cathode region. *IEEE Transactions on Plasma Science, PS-13*(5), 288–290.
11. Jenkins, J. E., Sherman, J. C., Webster, R., & Holmes, R. (1975). Measurement of the neutral vapor density decay following the extinction of a high-current vacuum arc between copper electrodes. *Journal of Physics. D. Applied Physics, 8,* L139–L143.
12. Jenkins, J. E., Sherman, J. C., Webster, R., & Holmes, R. (1977). The sizes of copper droplets emitted by cathode spots in a high current vacuum arc. University of Liverpool Arc Research Project, Report ULAP-T47.
13. Utsumi, T., & English, J. H. (1975). Study of the electrode products emitted by vacuum arcs in form of molten particles. *Journal of Applied Physics, 46*(1), 126–131.
14. Daalder, J. E. (1975). Erosion and the origin of charged and neutral species in vacuum arcs. *Journal of Physics. D. Applied Physics, 8*(14), 1647–1659.
15. Daalder, J. E. (1976). Components of cathode erosion in vacuum arcs. *Journal of Physics. D. Applied Physics, 9,* 2379–2395.
16. Kutzner, J., Seidel, S. T., Zalucki, Z. (1970). Distribution of neutral particle flux in vacuum arcs. In *Proceedings of IV International Symposium on Discharges & Electrical Insulation Vacuum* (pp. 276–285). Waterloo, Canada.
17. Daalder, J. E. (1977). Reflections of ions in metal vapor arcs? *Journal of Physics. D. Applied Physics, 10,* L87–L90.
18. Anders. (2008). *Cathodic arcs: From fractal spots to energetic condensation*, Springer.
19. Aksenov, I. I., Kudryavzeva, E. E., Kunchenko, V. V., Padalka, V. G., Popov, A. I., Khoroshikh, V. M. (1984). Investigation of gas pressure on droplet phase parameters of cathode erosion in a steady state vacuum arc. Preprint, Central Nauch-Issledov. Inst. Information and Tekhniko-Ekonom. Issled.- (CNIIatominform), Moscow (in Russian).
20. Aksenov, I. I., Konovalov, I. I., Kudryavzeva, E. E., Kunchenko, V. V., & Padalka, V. G. (1984). Droplet phase of cathode erosion in a steady state vacuum arc. *Soviet Physics—Technical Physics, 29,* 893–894.
21. Akari, K., Tamagaki, H., Kumakiri, T., Tsuji, K., Koh, E. S., & Tai, C. N. (1990). Reduction in macroparticles during the deposition of TiN films prepared by arc ion plating. *Surface & Coatings Technology, 43*(44), 312–323.
22. Tai, C. N., Koh, E. S., & Akari, K. (1990). Macroparticles on TiN films prepered by the arc ion plating process. *Surface & Coatings Technology, 43–44,* 324–335.
23. Cheng, Z., Wang, M., & Zou, J. (1997). Thermal analysis of macroparticles during vacuum arc deposition of TiN. *Surface Coatings and Technology, 92*(1–2), 50.
24. Antsiferov, V. N., Kosogor, S. P., Semenov, Yu L, & Gurbich, V. L. (1988). Investigation of droplet phase of erosion of a porous cathode in a vacuum arc. *Soviet Physics—Technical Physics, 33*(8), 922–924.
25. Kandah, M., & Meunier, J.-L. (1995). Study of microdroplet generation from vacuum arcs on graphite cathode. *Journal of Vacuum Science and Technology, A13*(5), 2444–2450.
26. Rakhovskii, V. I., & Yagudaev, A. M. (1969). Electrode emission mechanism in a pulsed vacuum arc. *Soviet Physics—Technical Physics, 14,* 227–230.
27. Rakhovsky, V. I. (1975). Erosion of electrodes in a contracted discharge. *Izv. Sibirian Branch of Academy of Science Series and Techology, N3*(1), 11–27 (in Russian).

28. Boxman, R. L., & Goldsmith, S. (1987). Cathode spot arc coatings: Physics, deposition and heating rates and some examples. *Surface & Coatings Technology, 33,* 153–167.
29. Boxman, R. L., & Goldsmith, S. (1989). Principles and applications of vacuum arc coatings. *IEEE Transactions on Plasma Science, 17*(5), 705–712.
30. Boxman, R. L., Goldsmith, S., Ben-Shalom, A., Kaplan, L., Arbilly, D., Gidalevich, E., et al. (1995). *IEEE Transactions on Plasma Science, 23*(6), 939–944.
31. Boxman, R. L., Zhitomirsky, V., Gidalevich, E., Beilis, I. I., Keidar, M., & Goldsmith, S. (1996). Recent progress in filtered vacuum arc depositions. *Surface & Coatings Technology, 86*(87), 243–253.
32. Shalev, S., Boxman, R. L., & Goldsmith, S. (1983). In situ determination of macroparticle velocities in a copper vacuum arc. *IEEE Transactions in Plasma Science, PS-11*(N3), 146–151.
33. Shalev, S., Boxman, R. L., Goldsmith, S., Einav, S., & Avidor, J. M. (1984). Laser Doppler anemometry: A tool for studying macroparticle dynamics in a vacuum arc. *Journal of Physics E: Scientific Instruments, 17,* 56–61.
34. Shalev, S., Boxman, R. L., & Goldsmith, S. (1986). Macroparticle dynamics during multi-cathode-spot vacuum arcs. *IEEE Transactions on Plasma Science, PS-14*(N1), 59–62.
35. Shalev, S., Boxman, R. L., & Goldsmith, S. (1985). Velocities and emission rates of cathode-produced molybdenum macroparticles in a vacuum arc. *Journal of Applied Physics, 58*(7), 2503–2507.
36. Disathik, G., Boxman, R. L., & Goldsmith, S. (1987). Characteristics of macroparticle emission from a high-current-density multi-cathode spot vacuum arc. *IEEE Transactions on Plasma Science, PS-15*(N5), 520–523.
37. Gellert, B., Shade, E., & Dullni, E. (1986). Measurement of particle and vapor density after high current vacuum arcs by laser techniques. In *Proceedings of XII International Symposium on Discharges & Electrical Insulation Vacuum* (pp. 209–213) Shoresh Tel Aviv, Israel.
38. Gellert, B., & Shade, E. (1990). Optical investigation of droplet emission in vacuum interrupters to improve contact materials. In *Proceedings of XII International Symposium on Discharges & Electrical Insulation Vacuum* (pp. 450–454). Santa Fe, New Mexico, USA.
39. Anders, S., Anders, A., Yu, K. M., Yao, X. Y., & Brown, I. G. (1993). On the macroparticle flux from vacuum arc cathode spots. *IEEE Transactions on Plasma Science, 21*(5), 440–446.
40. Boxman, R. L., & Goldsmith, S. (1992). Macroparticle contamination in cathodic arc coatings: generation, transport and control. *Surface & Coatings Technology, 52*(1), 39–50.
41. Hazelton, R. C., & Yadlowsky, E. J. (1994). Measurement of dust grain charging in a laboratory plasma. *IEEE Transactions on Plasma Science, 22*(2), 91–96.
42. Keidar, M., Beilis, I. I., Boxman, R. L., & Goldsmith, S. (1995). Non-stationary macroparticle charging in an arc plasma jet. *IEEE Transactions on Plasma Science, 23*(6), 902–908.
43. Bernstein, I. B., & Rabinovich, I. N. (1959). Theory of electrostatic probe in a low density plasma. *Physics of Fluids, 2,* 112–120.
44. Beilis, I. I., Keidar, M., Boxman, R. L., & Goldsmith, S. (1997). Non-equilibrium macroparticle charging in low-density plasmas. *IEEE Transactions on Plasma Science, 25*(2), 346–352.
45. Ivanov, V. A., Juttner, B., & Pursch, H. (1985). Time-resolved measurements of the parameters of arc cathode plasmas in vacuum. *IEEE Transactions on Plasma Sciience, PS-13,* 334–337.
46. Beilis, I. I., & Zektser, M. P. (1991). Calculation of the cathode jet parameters of the arc discharge. *High Temperature, 29,* 501–504.
47. Keidar, M., Beilis, I. I., Boxman, R. L., & Goldsmith, S. (1996). 2-D expansion of the low-density interelectrode vacuum arc plasma jet in an axial magnetic field. *Journal of Physics D, Applied Physics, 29,* 1973–1983.
48. Rysanek, F., & Burton, R. L. (2008). Charging of macroparticles in a pulsed vacuum arc discharge. *IEEE Transactions on Plasma Science, 36*(5), 2147–2162.
49. Boxman, R. L., & Goldsmith, S. (1981). The interaction between plasma and macroparticles in a multi-cathode-spot vacuum arc. *Journal of Applied Physics, 52*(1), 151–161.
50. Anders. (1997). Growth and decay of macroparticles: a feasible approach to clean vacuum arc plasmas? *Journal of Applied Physics, 82*(N8), 3679–3688.

51. Zalucki, Z. (1985). Estimation of post-arc neutral vapor density in the gap volume generated by evaporating macroparticles in a diffuse vacuum arc. *IEEE Transactions on Plasma Science, PS-13*(N5), 321–326.
52. Peters, J., Yin, F., Borges, C. F. M., Heberlein, J., & Hackett, C. (2005). Erosion mechanisms of hafnium cathodes at high current. *Journal of Physics D: Applied Physics, 38*, 1781–1794.
53. Keidar, M., Beilis, I. I., Boxman, R. L., & Goldsmith, S. (1996). Transport of macroparticles in magnetized plasma ducts. *IEEE Transactions on Plasma Science, 24*(1), 226–234.
54. Keidar, M., Beilis, I. I., Aharonov, R., Arbilly, D., Boxman, R. L., & Goldsmith, S. (1997). Macroparticle distribution in a quater-torus plasma duct of a filtered vacuum arc deposition system. *Journal of Physics. D. Applied Physics, 30*(21), 2972–2978.
55. Keidar, M., Aharonov, R., & Beilis, I. I. (1999). Influence of an electrical field on the macroparticle size distribution in a vacuum arc. *Journal of Vacuum Science and Technology, A17*(5), 3067–3073.
56. Keidar, M., Beilis, I., Boxman, R. L., & Goldsmith, S. (1996). Macroparticle interaction with a substrate in cathodic vacuum arc deposition. *Surface & Coatings Technology, 86*(87), 415–420.
57. Jilani, S. T., & Pandey, P. C. (1982). Analysis and modelling of EDM parameters. *Precision Engineering, 4,* 215–221.
58. Winske, D., & Jones, M. E. (1994). Particulate dynamics at the plasma-sheath boundary in DC glow discharge. *IEEE Transactions on Plasma Science, 22,* 454–464.
59. Keidar, M., & Beilis, I. I. (1999). Macroparticle Reflection from a Biased Substrate in a Vacuum Arc Deposition. *IEEE Transactions on Plasma Science, 27*(3), 810–812.
60. Ehrich, H., Hasse, B., Muller, K. G., & Schimidt, R. (1988). The anodic vacuum arc. Experimental study of arc plasma. *Journal of Vaccum Technology, A6*(N4), 2499–2503.
61. Ehrich, H., Hasse, B., Mausbach, M., & Muller, M. (1990). The anodic vacuum arc & its application to coating. *Journal of Vacuum Science and Technology, A8*(3), 2160–2164.
62. Beilis, I. I., & Boxman, R. L. (2009). Metallic film deposition using a vacuum arc plasma source with a refractory anode. *Surface & Coatings Technology, 204,* 865–871.
63. Beilis, I. I., Shashurin, A., Arbilly, D., Goldsmith, S., & Boxman, R. L. (2004). Copper film deposition by a hot refractory anode vacuum arc. *Surface & Coatings Technology, 177–178*(1–3), 233–237.
64. Beilis, I. I., Snaiderman, A., Shashurin, A., Boxman, R. L., & Goldsmith, S. (2007). Copper film deposition and anode temperature measurements in a vacuum arc with tungsten anode. *Surface & Coatings Technology, 202*(4–7), 925–930.
65. Beilis, I. I., Snaiderman, A., & Boxman, R. L. (2008). Chromium and titanium film deposition using a hot refractory anode vacuum arc plasma source. *Surface & Coatings Technology, 203*(5–7), 501–504.
66. Beilis, I. I., Koulik, Y., & Boxman, R. L. (2014). Cu film deposition using a vacuum arc with a black-body electrode assembly. *Surface & Coatings Technology, 258,* 908–912.
67. Beilis, I. I., Koulik, Y., Yankelevich, Y., Arbilly, D., & Boxman, R. L. (2015). Thin-film deposition with refractory materials using a vacuum arc. *IEEE Transactions on Plasma Sciences, 43*(N8), Part I, 2323–2328.

Chapter 10
Electrode Energy Losses. Effective Voltage

The near-cathode plasma is relatively dense and hot. The plasma generation supported by thermal phenomena in the cathode body and in region of the plasma cathode transition. The conductive plasma expanding up to the anode executes the electrical contact needed for vacuum arc burning. As result, a large heat fluxes outputted by the both cathode and anode bodies. The ratio of the heat power (in Watt) to the arc current is named as *equivalent* or *effective* voltage for cathode u_{cef} and for anode u_{aef}. The measurements of electrode heat losses and the effective voltages were conducted by different methods. One of them is a calorimetric approach measuring the maximal temperature of water and its mass flux in systems of water cooling electrode or measuring the temperature of electrode having a high thermal resistance lead. The energy input in the electrode was obtained using the mass and heat capacity of the water flux or electrode. Another approach involved calculating the heat conduction flux using an analytical expression for the temperature distribution in one-dimensional approximation and corresponding electrode temperature distribution measured by thermocouples. Let us consider the measuring effective electrode voltages by different approaches and conditions.

10.1 Measurements of the Effective Voltage in a Vacuum Arc

Fay and Hogen [1] reported the results of heat transfer to the W electrode by measuring the rate of change of electrical resistance of the electrode in a channel with low-pressure (0.1 atm) ionized gas flow, 250 A. This change was observed in range of 10^2–10^3 A/cm^2 and was found that the heat flux was proportional to the current. The heating rate divided by the total current was measured to be 4.9 V for anode and 3.2 V for cathode in case of short gap distance. Noted that the resistance of the used ionized gas flow should be accounted with the gap increasing in range from 0.1 to 30 inches.

I. Beilis, *Plasma and Spot Phenomena in Electrical Arcs*, Springer Series on Atomic, Optical, and Plasma Physics 113, https://doi.org/10.1007/978-3-030-44747-2_10

The energy input into the electrodes was studied by Reece, using a calorimetric method and temperature measuring by thermocouples in both electrodes of a vacuum arc [2]. It was indicated that at an arc current of 30 A the arc voltage for Cu could be divided into an effective cathode voltage drop of 8 V, and an effective anode voltage drop of 13 V. The effective voltage drops are defined as the ratio between the energy dissipated in the electrodes and the discharge current, assuming that the energy loss to the surroundings can be neglected.

Osadin [3] determined the effective anode voltage from the anode energy balance using the measured erosion rate of the anode. Liquid electrodes from Sn and PbBi eutectics were used in the experiment with frequency (10^3–10^4) pulse discharge at pressure of 10^{-5} Torr. It was obtained effective voltage of 2 V for the eutectics and 7.3 V for Sn. It was assumed that the heat was preferable dissipated in the electrode considering the energy balance. Osadin [4] also reported the results of calorimetric investigation in order to obtain the energy distribution between different component of the erosion products.

Konovalov [5] studied the heat flux into the electrodes mounted in the opposite walls of a channel with additive ionized gas (Ar or He) flow. A pressure of 10^{-4} Torr was supported in the vacuum channel, and the ionized plasma jet with velocity of 10^6 cm/s was generated by an additional pulse vacuum discharge. The experiments were conducted in a discharge of 100–300 μs pulse duration and current of 30–300 A, with cylindrical copper electrodes of 2.4, 1.25, 0.86 mm, and of 80 mm length.

The heat flux was determined using calorimetric method and solution of heat conduction equation in one-dimensional approximation neglecting the energy loss by the electrode erosion. Figure 10.1 shows dependence of the energy density flux by electrode heat conduction on charge transfer through the electrodes (2.4 mm diameter). It can be seen that the dependences are linearly and the ratio of the electrode energy density to the charge transfer density indicate the effective cathode voltage drop as 2.5 V, and an effective anode voltage drop as 12.5 V. It should be noted that the

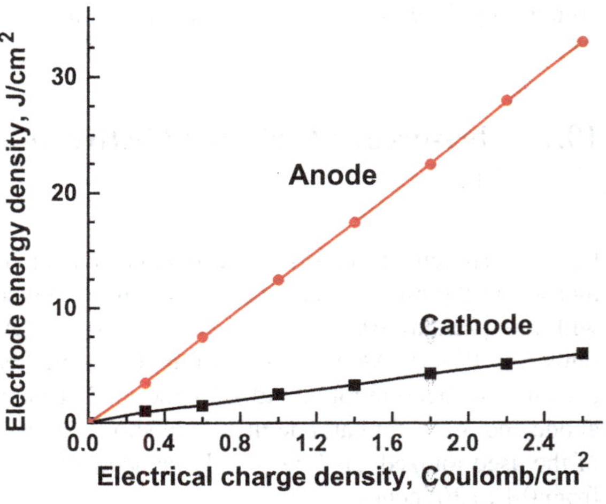

Fig. 10.1 Energy density flux into cathode and anode as dependence on charge transfer through the Cu electrodes [5]

real cathode heating process could be different from used one-dimensional approach when a number of spots supports the current. The author formulated his cathode heat model assuming that the surface temperature flattened due to rapid motion of the spots. The heat model of the anode (mostly fixed spot) was based on assuming that the time reaching the uniform temperature at the electrode surface is more shortly in comparison with the time of reaching of the heat wave the thermocouple location.

Konovalov [6] measured the heat flux to the electrodes for high-current arc using the experimental system mentioned above [5]. The erosion less approach of the anode energy balance in one-dimensional approximation was considered. Experiments were conducted with Cu and steel electrodes of 3–3.5 mm diameters in a pulse discharge of 25 μs arc and at current amplitude up to 10 kA.

Calorimetric method was used measuring the temperature of the external side of a sensor by a thermocouple. The results of measurements of conductive energy flux density W_{con} as dependence on charge density Q_{ch} are presented in Fig. 10.2 for Cu and in Fig. 10.3 for steel electrodes. This dependence for Cu anode is linear while for other cases the dependences weakly declined from linear function indicating that a small energy spending with electrode erosion should be taken in account. The effective voltages are represented by the ratio of W_{con}/Q_{ch}. It was obtained that the cathode-effective voltage in linear part of the dependences is 2 V for Cu and 6 V for steel while for anode is about 9 V for Cu and 11 V for steel. Note that the weakness of the results is due to using the one-dimensional electrode heat model that should be considered as an approximation approach, in spite of the mentioned above assumptions.

Belkin and Danilov [7] determined the electrode energy loss for vacuum arc calorimetrically using a fine current lead with high thermal resistance. The electrode temperature was measured with chromel–alumel thermocouple. Using the measured maximal change of the temperature, the effective voltage u_{ef} was obtained as

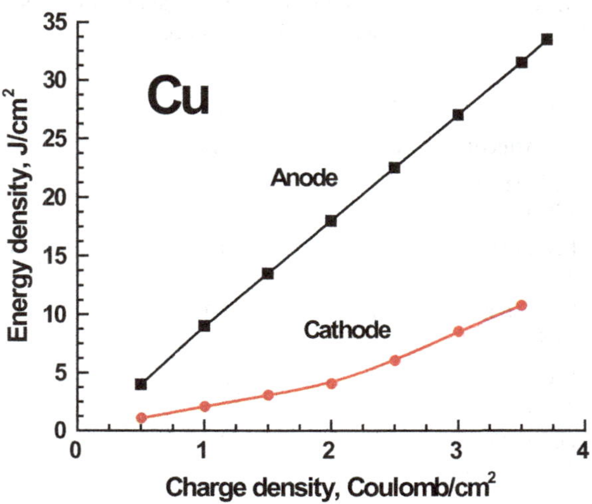

Fig. 10.2 Density of the energy flux into electrodes as function on total charge density for copper electrodes [6]

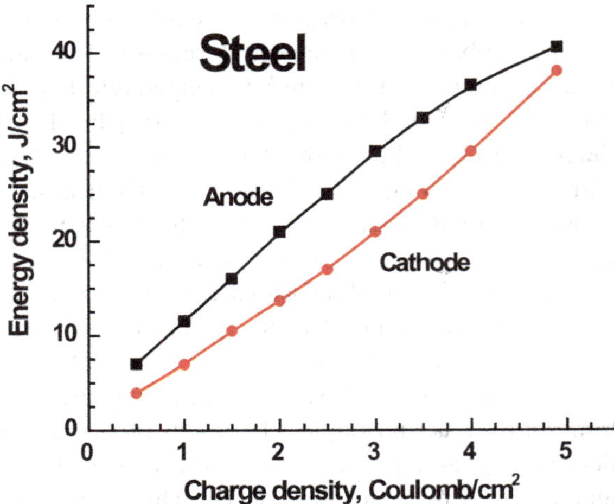

Fig. 10.3 Density of the energy flux into electrodes as function on total charge density for steel electrodes [6]

$$u_{ef} = mc\Delta T/Q_{ch} \tag{10.1}$$

The electrode mass m was obtained by it weighing, and the heat capacity c was assumed constant. The experiment conducted at a current of 2.5 kA. Table 10.1 shows the results obtained for a number of electrode materials.

It was found that the equivalent electrode potential drop (effective voltage) is roughly twice great for anode as for the cathode. The effective voltages of the clean materials are greater than those voltages for the metalloceramic electrodes. The

Table 10.1 Effective voltages for different electrode materials and the arc voltage between cathode and anode [7]

Material	Anode-effective voltage, u_{aef}, V	Cathode-effective voltage, u_{cef}, V	Sum ($u_{aef} + u_{cef}$), V	Arc voltage, V
W	11.2	6.5	17.7	28
Cu	12.4	6.1	18.5	19
Fe (Armco)	8.0	6.8	14.8	19
FeCu (87% + 13%)	8.7	4.5	13.2	14
FeCu (75% + 25%)	9.3	4.9	14.2	16
FeCu (69% + 31%)	8.8	5.1	13.9	16
WCu (57% + 43%)	8.8	5.0	13.8	15
FeCu (70% + 26% + 4%)	6.6	3.6	10.2	15

sum of cathode- and anode-effective voltage is lower than the arc voltage, and this difference indicates some energy loss at the surrounding.

Rondeel [8] measured of the energy dissipated in the anode and the cathode using temperature measuring by thermocouple inserted into the electrodes. This was measured in an arc current of 600 A, 1 mm contact separation, and the arc voltage was 20 V. It was obtained that about 25% is transferred to the cathode, and 75% of the total energy dissipated in the arc was divided between the losses to the surroundings and input to the anode. The energy for the erosion rate of the cathode material was only a small fraction (10%) of the total energy dissipated in the arc. From this fact (low evaporation loss), Rondeel [8] concluded that it will be extremely difficult to determine the cathodic erosion from the electrode energy balance explicitly.

Rondeel [9] demonstrated the influence of an axial magnetic field on the electrode effective voltage. Arc current was about 500 A, and arc duration was between 20 and 30 ms.

The energy input to both electrodes and a cylindrical shield outside the electrodes was determined by measuring the temperature rise with three copper–constantan thermocouples mounted in resistive components. The electrodes of 20 mm diameter were from copper with 5 mm gap length. Figure 10.4 presents the results for anode and cathode. It can be seen the effective voltage increase from about 4 V (cathode) and 10 V (anode) at zero magnetic field to about 8 V (cathode) and 15 V (anode) at field of 0.35 T. The effective voltage of about 2 V was detected due to energy loss to the shield. It can be assumed that the increase of the electrode effective voltages is due to the increase of the arc voltage (from 20 to 35 V) with magnetic field that measured here.

Detailed research was conducted by Daalder [10, 11]. He presented the results of an experimental study of the thermal state of a vacuum arc cathode by measuring an accumulated heat during the arcing. The cathode was constructed as a thin disk with diameter of 30 mm and was placed at a massive holder by three small supports in order

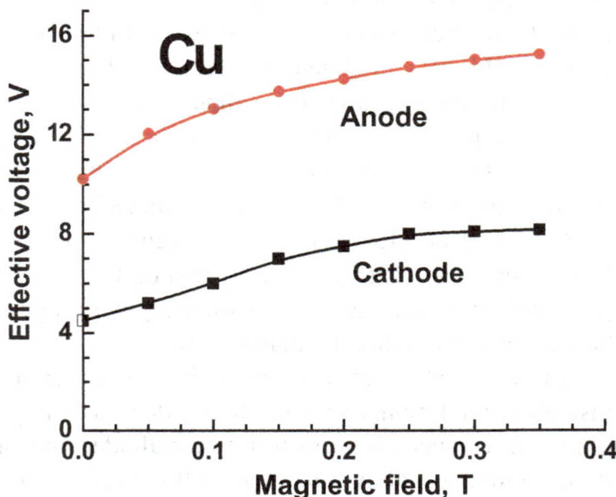

Fig. 10.4 Anode- and cathode-effective voltages as function on magnetic field strength [9]

to minimize heat conduction losses. Before each series of experiments, the cathode surface was additionally cleaned by drawing a number of arcs, thus removing any absorbed layers of gas and oxidized surface layers if present. During the experiment, the temperature of the rear surface of the disk was measured by thermocouples of iron–constantan and the temperature was recorded during a time of 200 s after arc extinguishment. The arc time t_{arc} was varied (at constant current) from a few ms to a maximum of half a second. The value of the maximum temperature T_{max} was used to determine the accumulated heat W_{con} under an assumption that the heat conduction and radiation losses were small:

$$W_{con} = m_d c_p T_{max} \tag{10.2}$$

where m_d is the disk mass and c_p is the specific heat of the disk materials. As it indicated by Daalder for each metal, the choice of the current range was such that conditions of both one cathode spot and several spots were investigated. For the low-melting metals, Pb, Cd, Zn, and Sn, the current was varied between 10 A (one spot) and 200 A (in case of Cd, Pb, and Zn about 15–20 spots were counted). For the higher-melting metals (Ag, Ni, Al, Cu, and Mo), the current range was extended and reached more than 600 A (for Cu and Mo 8–9 spots were observed). Most measurements were taken at an electrode spacing of 1 mm. For Cd and Al, the gap distance varied from approximately 0–5 mm up to 10 mm. Daalder noted that no dependency on this parameter was observed indicating that cathode heating is due to processes occurring in the direct vicinity of the vacuum metal interface and/or in the metal itself. The cathode-effective voltage was obtained by

$$u_{cef} = \frac{W_{con}}{I t_{arc}} \tag{10.3}$$

The results of measurements are presented in Table 10.2, which show that in case of the metals Cu and Mo the value of u_{cef} increases with arc current. This increase is due to the increase of arc voltage with current for considered cathode metals. For most metals considered here, the increase of arc voltage with current is in the order of a volt in a current range from about 10 A up to several hundreds of Amperes. This increase is particularly noticeable in the transition region of one to a few cathode spots. For Cu and Mo, the increase in arc voltage was a few volts, and it was found that this rise was related to an increase in cathode-effective voltage.

Zektzer et al. [12] showed an agreement of calculation results with Daalder's [10] experimental data using cathode spot theory of Beilis [13], which described the plasma flow in near-cathode region using hydrodynamic approach and the cathode heating by the ion flux bombardments.

An investigation has been made by Paranin et al. [14] of an arc on a thermally insulated gadolinium cathode. The anode was a molybdenum radiation-cooling disk placed at distances 1–6 cm from the cathode. The cathode temperature was determined with a visual pyrometer as well with photoelectric pyrometer. The main heat

Table 10.2 Cathode-effective voltages and arc voltage for a number of cathode metals with the arc current of 100 A, unless specified otherwise [10]

Material	Arc voltage, u_{arc}	Cathode-effective voltage, u_{cef}	u_{cef}/u_{arc}
Pb	10.5	2.6	0.25
Cd	11.0	2.7	0.25
Zn	12.0	3.0	0.25
Sn	13.5	3.9	0.29
Ag	17.5	5.25	0.30
Ni	18	5.35	0.30
AI	20	6.2	0.31
Cu	18 (40 A)	5.4	0.30
Cu	20 (100 A)	6.2	0.31
Mo	25 (50 A)	8.2	0.33
Mo	26.5 (200 A)	8.7	0.33
Mo	28 (600 A)	9.25	0.33
W	28	9.5	0.34

losses (90%) from the insulated cathode were due to the emission from the surface. Therefore, a specific method was developed to measure the heat flux into the cathode by directly calibrating using a heating by an electron beam. The cathode was eroded in the vapor phase with saturated pressure of 1–100 Pa. The cathode was heated by electron beam energy in temperature range of 1900–2300 K. The arc current was $I < 200$ A. As the heating power and correspondingly cathode temperature increased, the heat flux into the cathode Q_{ef} decreased and here was the continuous transition from self-heated regime with $Q_{ef} > 0$ to the externally heating with $Q_{ef} < 0$. It was obtained that the, respectively, cathode-effective voltage decreased from positive 5 V to negative value of −3 V when the cathode temperature increased up to 2200 K.

Dorodnov et al. [15] studied the heat flux to the electrodes of a pulse-type plasma accelerator. The working body of the discharge is the material of the vaporizing central face electrode. The material of the central electrode is copper. The heat flux was determined using the results from measurements of a maximum temperature of the electrode with thermocouple, which was embedded at distance 15 mm from electrode working surface. The measurements were conducted for central and outer electrodes of the accelerator in range of 3–15 kA with different polarities. The dependencies of the heat flux to the outer electrodes on the discharge current (averaged data shown by lines) are presented in Fig. 10.5.

The results showed that the heat flux to the central electrode for both polarities was proportional to the discharge current and determined by the electrode material (Cu, Mo, Nb). This flux to a central anode (15–18.5%) was somewhat greater than that flux to a central cathode (12.5–14% of the supplied power). The total losses in both electrodes were 25–45% of the total supplied electric power. It was noted that the relatively fraction of the electrode losses decreased with a rise in discharge current and this fact indicated an increase in the efficiency of the accelerator.

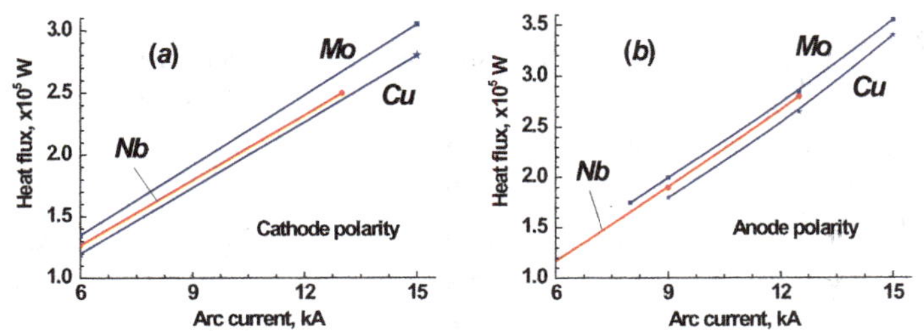

Fig. 10.5 Dependence of the heat flux to the outer electrodes on the discharge current, **a** as cathode polarity, **b** as anode polarity; material of the outer electrode are indicated as Cu, Mo and Nb [15]

10.2 Effective Electrode Voltage in an Arc in the Presence of a Gas Pressure

Rieder [16] reported one of the early works regarding to the investigation of heat energy loss by the free air arc into the electrodes. The data of energy losses in the electrodes were obtained using the volt-ampere characteristics for air low-current arc and results of calorimetric measurements. According to the calorimetric results for copper, it was shown linearly dependences of the power (in Watt) dissipated in the cathode and in the anode on arc current (3–35 A) for different arc length as parameter and saturated dependences on arc length (0–12 mm) for different arc current as parameter. Considering the electrode balances the effective anode fall of 2.3–3.3 V and cathode fall of 8.9–9.9 V were determined.

The power dissipated at the wire anode of a DC arc discharge (10 A) has been measured, for arcs in argon at atmospheric pressure, by noting the length of the anode which is melted [17]. The measurement of the power transferred by the electrons energy and the work function gives rise to a useful means of determining the anode fall. The non-stationary heat conduction equation was solved by given the melting temperature as boundary condition and by assuming that the energy input to the anode is balanced only by the wire heat conduction. The assumption of one-dimensional heat conduction in the anode may be justified because of the large ratios of length to diameter.

Experimentally it was fixed the time when the anode melting was reached. In the experiment, a graphite cathode and Cu wire anode with diameters of 4.0, 3.0 and 2.2 mm (length of 8 cm) were used at a constant temperature T_0, fixed in a supporting rod of brass, which was cooled in a water bath. As result, the effective anode fall indicated as 4.22 V, but not explained how it was obtained. In this same time, the heat flux into the anode wires were found to be 312 and 580 cal cm^{-2} s^{-1} for wire diameters of 3.0 and 2.2 mm, respectively. Using these data and arc current of 10 A, the anode-effective voltage can be calculated as 9.19 and 9.18 V, respectively, assuming uniform distribution of arc current at the anode.

Experiments were conducted by Capp [18] to elucidate energy transfer processes in electrode-dominated arcs with a cadmium or a zinc cathode with diameter of 8 mm

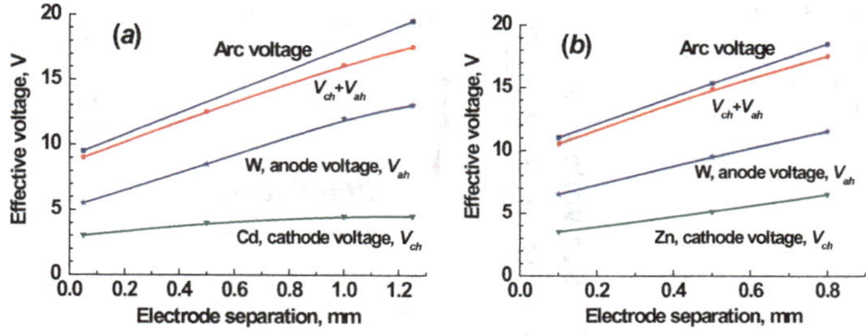

Fig. 10.6 Anode- and cathode-effective voltages as function on electrode gap for Cd cathode (a) and Zn cathode (b)

and a tungsten anode. The energy transfer to the anode and cathode was measured for known arc power and duration in nitrogen at atmospheric pressure. The thermal capacities of the electrodes were measured by heat exchange with hot water in a calorimeter, and this measurement was checked by calculation taking into account the electrode mass. The average power going into each electrode was calculated from its thermal capacity and temperature rise during the arc time.

The obtained power divided by the arc current determined an effective voltage, which denoted by Caap [18] as V_{ah} for the anode and V_{ch} for the cathode. These denotes were used according to Caap's dependences which are presented in form shown in Figs. 10.6, where the results of effective voltage V_{ah} and V_{ch} together with their sum and measured arc voltage were shown as function (averaged data shown by lines) on electrode gap distances for a cadmium (Fig. 10.6a) and a zinc cathode (Fig. 10.6b).

As can be seen over the range of currents (2, 3, and 4 A) and used gap distance, the results show that the arc voltage (arc power) dependence is close to the dependence of heat conducted into the electrodes ($V_{ah} + V_{ch}$). An increasing proportion of the arc power, however, goes into the anode as the gap increases. The power into the cathode increases slowly (especially for Cd) with the gap, compared with the corresponding increase into the anode. It was noted that the power losses by evaporation and radiation are not important.

The effective voltage for water-cooled or radiative-cooled anodes were investigated for arc current in range of 50–350 A in argon of pressure 20–760 Torr [19]. The discharge was ignited between W rod cathode of 3 mm diameter and water-cooled or radiation-cooled flat anode, made from W plate of 0.2–0.5 mm thickness. The heat flux into the anode was determined by calorimetrical method measuring the cooling water parameters. The anode temperature was reached values in range of 2000–3000 K depending on gas pressure and arc current. The results showed that the energy losses for radiation-cooled anode were lower (by ~2 V) than that for water-cooled case and it is increased from 7–9 to 10–12 V with gas pressure in the indicated above range for arc current of 200 A.

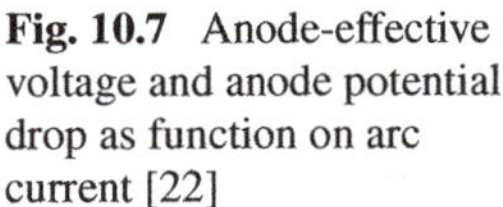

Fig. 10.7 Anode-effective voltage and anode potential drop as function on arc current [22]

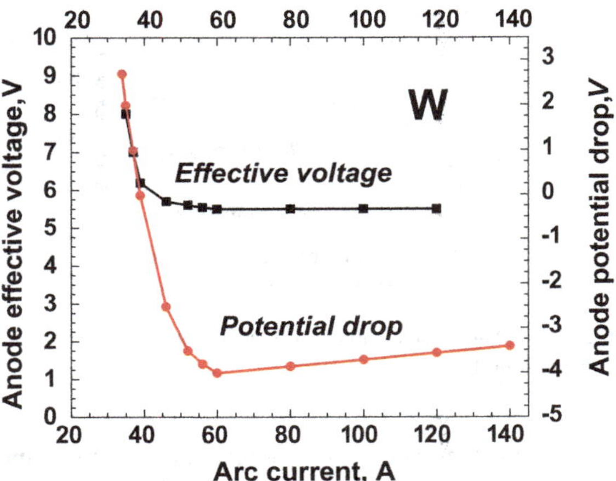

Merinov et al. [20] studied a discharge with radiation-cooled anode in inert gases. Using photosensor, the authors have determined the anode-effective voltage calorimetrically, which slightly increase from 5.5 to about 6 V with arc current increase from 80 to 130 A. In the experiments [21], which continue the study of work in [19] arc current was used up to 200 A and inert gas pressure varied in range of 2–120 Torr. The calorimetric measurements of radiation-cooled anode plate at arc current 100 A showed that the effective voltage increase from 5 to 6.4 V when the argon pressure increased from 5 to 120 Torr, respectively. The probe measurements at 15 Torr pressure of He, Ar, and Xe showed the presence of negative anode drop that was about −1.5, −2.5, and −5 V, respectively, to the mentioned gas type. Using the calorimetric and probe methods of work in [21], Merinov and Petrosov [22] studied the anode-effective voltage and anode voltage drop for tungsten in argon discharge at pressure 5 Torr. The results are shown in Fig. 10.7 as dependence on arc current.

Analyzing the experimental data, the authors of [22] indicated that it is not possible to obtain data regarding the sign and magnitude of the anode potential drop from the anode energy balance using the result of the calorimetric measurements. Energy transport close to and at the anode (contraction zone) in high intensity arcs at atmospheric pressure was studied by Pfender [23]. It was shown that the anode potential drop was negative in the range from one to three volts, which depend on the anode current density and plasma temperature.

Salihou et al. [24] studied experimentally the power dissipated by heat conduction into the cathode of an electric arc by measuring the steady-state temperature reached during arcing in a system with argon at 1 atm flowing through it at a low rate. The arc was ignited by electrode separation. The rear side of the cathode was maintained at constant temperature (the temperature of the water-cooled brass). The electrodes were connected through an adjustable resistor to a regular DC power supply which provided a constant current up to 5 A. The electrodes of different materials (copper, silver, and tungsten with a tungsten anode) had a diameter of 3 mm, length of 8 cm with an electrode spacing of 0.4–1 mm. The temperatures were measured by five

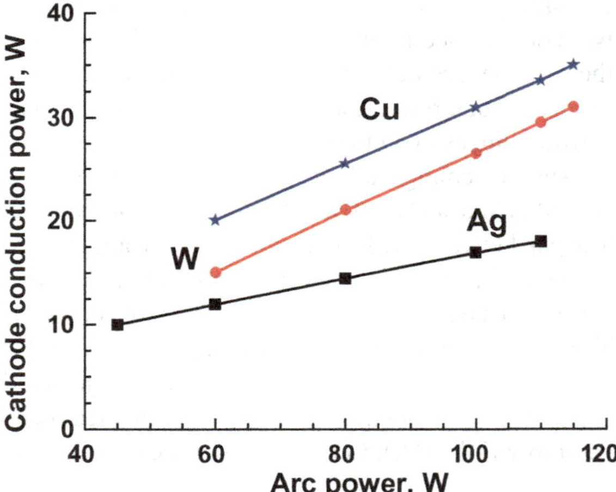

Fig. 10.8 Power lost by cathode conduction versus the electrical power input (Iu_{arc})

thermocouples (chromel–alumel, type K) inserted into holes of same depth (1.5 mm) along the cathode axis in order to obtain a function of temperature distribution along the electrode axis.

Using heat conduction equation and the measured temperature distribution along the electrode axis, the average power going into the cathode by conduction was determined. The results are presented in Fig. 10.8 as function on arc power [24]. The arc voltage u_{arc} was also measured ranges of 22–24 V for Cu, 18–20 for Ag, and 22–25 for W. It can be seen that cathode loss increases linearly with increasing the power input and it found to be material sensitive. Thus, for a given electrical power input of 100 W, the power lost by heat conduction is about 17%, 26%, and 29%, respectively, for Ag, W, and Cu.

10.3 Effective Electrode Voltage in a Vacuum Arc with Hot Refractory Anode

This type of DC arc with duration of few minutes sustained between a thermally isolated refractory cylindrical anode and a water-cooled cylindrical copper cathode. The interelectrode distance varied around value of 10 mm. The arc started as a conventional cathode spot vacuum arc. In the initial stage, the cathodic plasma jet deposits a film of the cathode material on the anode as well as the anode was heated by jet energy. The temperature of the thermally isolated anode begins to rise, reaching eventually a sufficiently high temperature to re-evaporate the previously deposited material, which subsequently ionized in the plasma of the interelectrode gap. The arc at this stage characterized by plasma evolution and named as hot refractory anode vacuum arc (HRAVA). The transition to the HRAVA mode completed when the density of the interelectrode plasma consists mostly of ionized re-evaporated atoms expanding radially as anode plasma. The transitions period of the HRAVA mode were determined by the propagation of a luminous plasma plume from the anode to the cathode (see below). The HRAVA mode is a discharge-instrument, which allows to clearly determine the structure of the heat fluxes into the cathode, anode and surrounding as evolution in time.

Another configuration consists of a closed volume, which formed by a water-cooled cylindrical relatively low-melting cathode and a cup-shaped refractory anode. The gap between the front cathode surface and the inside flat anode surface was about 10 mm. The cathode material was emitted into the gap between the front cathode surface to the inside flat anode surface of the closed volume. The arc heats the anode. The closed volume confined and evaporated droplets while the plasma was extracted through small anode apertures. The plasma ejected through an array of holes. Such arc with closed electrode configuration named as a vacuum arc with a black body assembly (VABBA). Let us consider the experiments for both configurations.

10.3.1 Effective Electrode Voltage in a Hot Refractory Anode Vacuum Arc (HRAVA)

Heat transfer to a thermally isolated graphite anode in a long duration (up to 200 s) vacuum arc was investigated by Rosenthal et al. [25]. The anode surface temperature was optically determined. The anode bulk temperatures near the front and rear surfaces were measured as a function of time using two high-temperature thermocouples for arc currents 175 and 340 A. A one-dimensional nonlinear heat flow model for the anode was developed and was solved to determine the time-dependent effective anode voltage as the ratio between the input power to the anode and the arc current. The measured anode temperature distribution and characteristic time of observed expansion of the anode plasma plume were used. The effective cathode voltage was determined for copper by measuring the temperature of the cathode cooling water. The net power input to the cathode was determined as

$$W_{\text{con}} = c_{\text{w}} \Delta T \frac{\Delta m_{\text{w}}}{\Delta t_{\text{w}}} \tag{10.4}$$

where c_{w} is the heat capacity of water, $\Delta m_{\text{w}}/\Delta t_{\text{w}}$ is the mass flow rate of cooling water. In the case of a 175 A arc with a mass flow rate of 36.5 g/s, the temperature near the beginning of the arc was approximately 8 °C above ambient water temperature, reaching a value of 8.4 °C near the end of the arc. It was indicated that this temperature rapidly increased after arc ignition and then slowly increased during the course of the arc.

The time-dependent input flux $q_{\text{in}}(t)$ to the anode was approximated by the following expression:

$$q_{in}(t) = q_{ss} + q_0 \exp\left(-\frac{t}{\tau}\right) \tag{10.5}$$

where $q_{ss}(t)$ is the input flux at steady state, $(q_o + q_{ss})$ is the initial input flux (at $t = 0$), and τ is the characteristic time for the development of the anodic plume. For q_{ss}, we take the value obtained at steady-state arc (see Table 10.3), while for q_o we seek a value yielding the best fit to the measured temperatures. Figure 10.9 characterizes time evolution of the anodic plume for Cu cathode and graphite anode.

Figure 10.10 characterizes time evolution of the time-dependent input flux calculated using (10.5) and measured arc parameters presented in Table 10.3. The behavior of the anode-effective voltage was explained by the HRAVA theory [26]. The initial relatively large anode-effective voltage is determined by the large kinetic and potential energy of the cathode spot jet, which freely interacts the anode surface.

With time, this energy jet was dissipated by interaction with the generated anode plasma, and therefore, the dissipated energy was lost in a surrounding space by radial heated anode plasma expansion. As result, the heat flux to the anode decreases and determined by energy flux determined by the electron temperature and anode work function in the steady-state HRAVA stage.

The steady-state effective heating potentials are approximately 7, 6, and 12 V for the cathode, anode, and surrounding structures, respectively, while in the initial stage energy loss in surrounding is about 3 V. The anode thermal analysis for the energy losses in the steady state showed that the energy flux input was balanced mainly by radiation from the front surface of the anode, and in lower value by radiation from the side surfaces through the space between the anode and the shields, as well as from the bottom surface of the anode.

Table 10.3 Pairs of values indicate the initial and steady-state values, respectively. I is the arc current. T_{sur}, K is the anode surface temperature at steady state. τ is the time that anodic plume filled the all electrode gap. t_{ss}, is the temperatures and arc voltage steady-state time. q_{in} is the input heat flux to the anode; u_{cef} is the effective cathode voltage. u_{aef} is the effective anode voltage. u_{sef} is the effective surrounding voltage. u_{arc}, is the arc voltage [25]

I, A	T_{sur}, K	τ, s	t_{ss}, s	q_{in}, W/cm^2	u_{cef}, V	u_{aef}, V	u_{sef}, V	u_{arc}, V
175	1900	30	100	218–140	6.6–7.2	10–6.4	4–10.4	20.5–24
340	2300	15	60	510–260	6.6–7.2	12–6.2	2.4–12.6	21–26

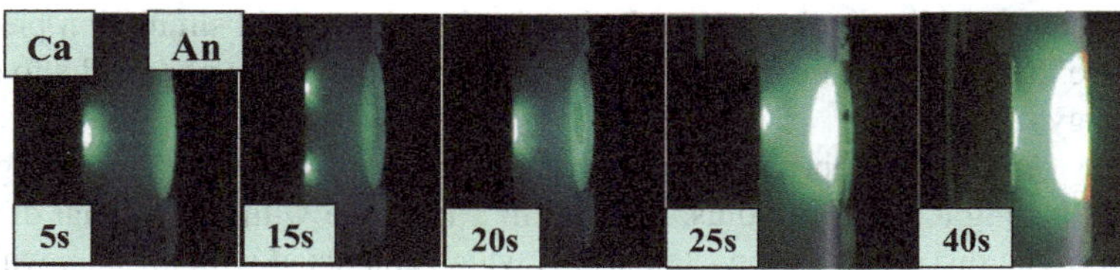

Fig. 10.9 Plasma plume development in time in HRAVA with a disk Cu cathode (Ca) and graphite anode (An). Electrode gap is 18 mm, arc current-175 A

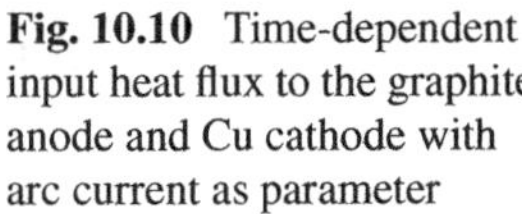
Fig. 10.10 Time-dependent input heat flux to the graphite anode and Cu cathode with arc current as parameter

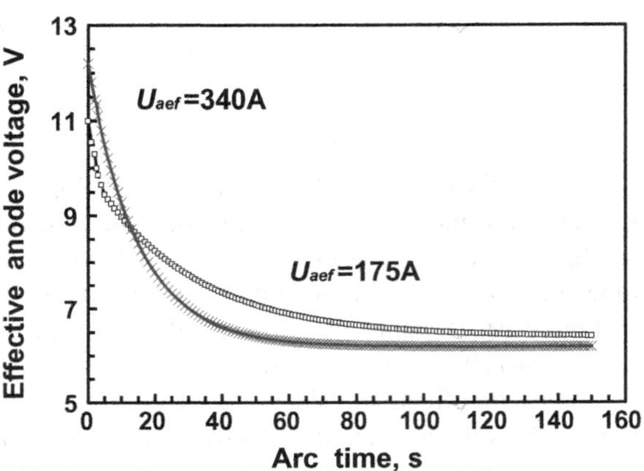

10.3.2 Energy Flux from the Plasma in a Hot Refractory Anode Vacuum Arc (HRAVA)

As the cathode jet dissipated in the gap plasma, an important issue is understood how the interelectrode anode plasma was heated and what is the energy flux from the anode plasma. Beilis et al. [27] experimentally studied this flux. Two types of thermal probes were used: (1) a *thermal resistor*, where the incoming plasma energy flux was determined by measuring the temperature distribution along the probe length; and (2) a *water-cooled probe*, where the energy flux was determined from the water temperature change and flow rate.

The thermal resistor probe was constructed from a tungsten rod with a diameter of 1 mm and a length of 120 mm. A BN cylinder with an outer diameter of 3 mm shielded the lateral surface of the 20-mm-long distal part of the rod, and the proximal part of the rod passed through a boron nitride (BN) disk, which shielded it from the plasma heat flux. The proximal end of the rod was maintained at room temperature. The temperature of the rod was measured at two points (at the distal end of the probe and 20 mm from the distal end) with high-temperature thermocouples. The thermocouple was made from W + Re5% to W + Re26% alloy wire, which were welded together at their tips. A thermal model for the probe was formulated which considered heat conduction in the W rod and BN insulator, and radiation from the distal end of the rod and BN insulator and from the lateral surface of the insulator. The temperature distribution was calculated using the Q-field simulation program, and the experimentally measured temperatures were fitted to these results to obtain the energy flux.

The water-cooled probe was designed and built, consisting of a 20-mm-long tungsten rod having a 0.5 mm radius, inserted in a BN tube with an external diameter of 3 mm. The distal part of tungsten probe, which contacted with plasma, while the BN tube insulated the lateral surface. The distal part of the probe had a diameter of 2.5 mm and a length L of 8 mm. The proximal end of the tungsten probe was cooled by water flowing with a rate G_w. The water temperature was measured by a J-type thermocouple. The increase in water temperature was measured during the arc. The net energy flux to the probe was calculated from the measured temperature increase ΔT and the water flow rate G_w from $Q_w = c_w \Delta T G_w$. The heat flux per unit area can be obtained as $q = Q_w/F$, where $F = 6.8 \times 10^{-5}$ m^2 is the area of plasma-probe contact.

The steady-state analysis of the thermal resistor probe showed that the energy flux at the distal end of the probe for $t = 120$ s, obtained by fitting the measured temperatures with the calculated model, is plotted in Fig. 10.11. The energy flux decreased with distance from the axis and increased approximately linearly with the arc current. For 175 and 340 A, the energy flux is 1 and 2 MW/m^2, respectively.

The measured values of the energy flux density q are presented in Table 10.4. It may be seen that this flux increased linearly with arc current and the dependence is similar to results obtained using the thermal resistor probe. The small difference of q for $I = 175$ A indicated here (90 s) in comparison to the energy flux presented in Fig. 10.11 is due to the time to reach steady state of the HRAVA is larger than 90 s.

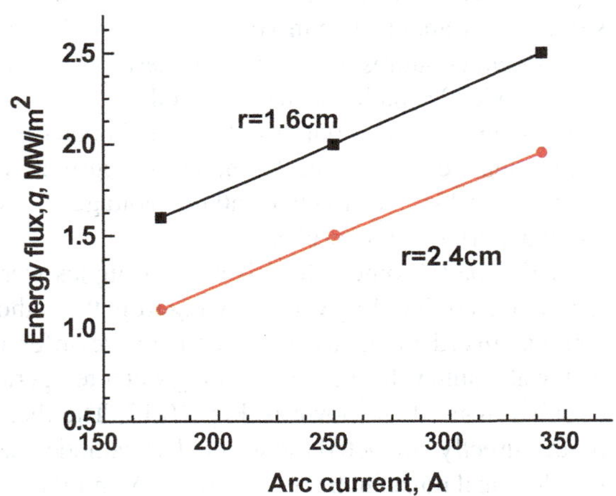

Fig. 10.11 Energy flux from the plasma to the front surface of the resistor probe as function of the arc current at the electrode gap edge $r = 1.6$ cm and at distance from the electrode axis $r = 2.4$ cm

Table 10.4 Energy density flux q measurements for different arc currents I at arc duration 90 s by the water-cooled probe ($r =$ 24 mm). G_W (g/s) is the water flow rate, and $\Delta T°$ is the water temperature increase. From [27]

Current, A	G_w (g/s)	ΔT, °C	$q \times 10^{-2}$, MW/cm^2
175	0.167	37	40.1
	1.67	5	51.37
	0.33	26	53.44
250	0.667	24	98.76
	2	9	111.3
340	1.786	20	220.4

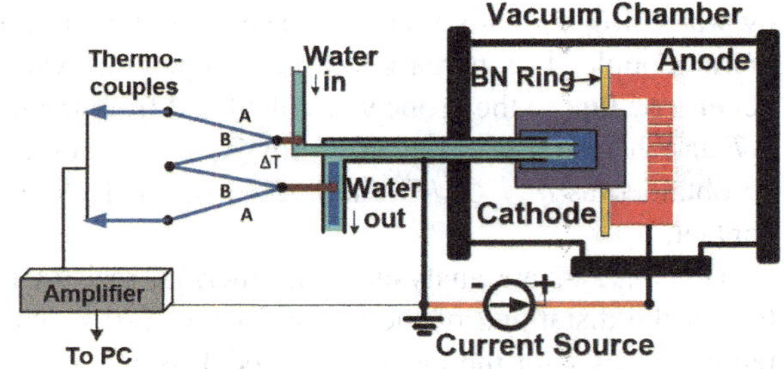

Fig. 10.12 Schematic diagram of the experimental setup with a shower anode and with differential thermocouple circuit

10.3.3 Effective Electrode Voltage in a Vacuum Arc Black Body Assembly (VABBA)

Beilis et al. [28] investigated the energy dissipated in the cathode of arc with this type of electrode configuration. A water-cooled cylindrical Cu cathode with 30 mm diameter and a cup-shaped W or Ta anode with 50 mm outer diameter formed a closed volume. The gap between the front cathode surface and the inside flat anode surface was about 10 mm (Fig. 10.12). Both materials are refractory with close boiling temperatures. Cu cathode material was emitted into the closed volume and the cathode plasma jet heated the anode. At the hot anode stage, the plasma ejected through an array of 250 holes of 1 mm diameter in the W anode or through a single 4 mm diameter hole in the Ta anode. Arc currents were $I = 175$–250 A, and the arc time was 150 s. The effective cathode voltage U_{cef} was determined calorimetrically, using a thermocouple probe.

A K-type (chromel–alumel) thermocouples were used to measure the temperature increase of cooling water flowing from the cathode, relative to water entering the cathode. In order to measure the temperature difference directly, without dependence on the absolute value of the incoming water temperature, a differential thermocouple circuit was used as shown in Fig. 10.12. The thermocouple junctions (output and input) directly contacted the water. The cathode was grounded in order to minimize the electrical noise in the probe circuit. A and B correspond to the positive and negative legs of the thermocouples. BB junction remained outside the water. An electrically floating amplifier was used to amplify the thermocouples signal. The measured data was collected on a personal computer using an analog–digital converter and LabVIEW software. The water mass flow rate was $F = 0.193$ or 0.290 kg/s. The effective cathode voltage was determined as:

$$u_{cef} == \frac{c_p F \Delta T}{I} \tag{10.6}$$

where ΔT is the temperature difference between the output and input cathode cooling water.

Typical time dependence of ΔT and u_{cef} when using a W showerhead anode with $I = 200$ A and $F = 0.290$ kg/s is shown in Fig. 10.13. When the arc was ignited, u_{cef} increased to 6–7 V, where it remained for ~40 s (both anodes) while the anodes were cold. For hot anodes (heated by the arc, which visually the anode brightness increased with time and then strongly radiated), u_{cef} increased up to ~11–12 V and reached a steady state. The time to reach steady state is shorter with arc current.

Figure 10.14 shows the dependence of the average steady-state u_{cef} on arc current for the W anode. When I increases from 175 to 250 A and $F = 0.290$ kg/s u_{cef} increases from about 11.1 to 11.6 V. For $F = 0.193$ kg/s, when I increases from 200 to 250 A the effective cathode voltage increased with from about 11.4 to 12.0 V. However, the measured u_{cef} for two different F are in range of error bar. This point and small absolute difference of u_{cef} indicate that u_{cef} practically independent on the rate of water flow cooling the W cathode.

Figure 10.15 shows the dependence of steady state (average) u_{cef} on arc current for the Ta anode with $F = 0.193$ kg/s. u_{cef} initially increases with I from about 10.6 to 11.3 V when I increases from 175 to 200 A and then slightly decreases to about

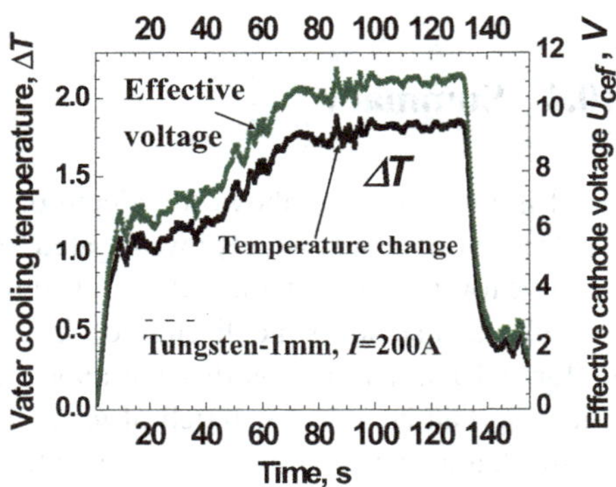

Fig. 10.13 Time dependence of water temperature increase ΔT and effective cathode voltage u_{cef} for a W showerhead anode, $I = 200$ A, $F = 0.290$ kg/s

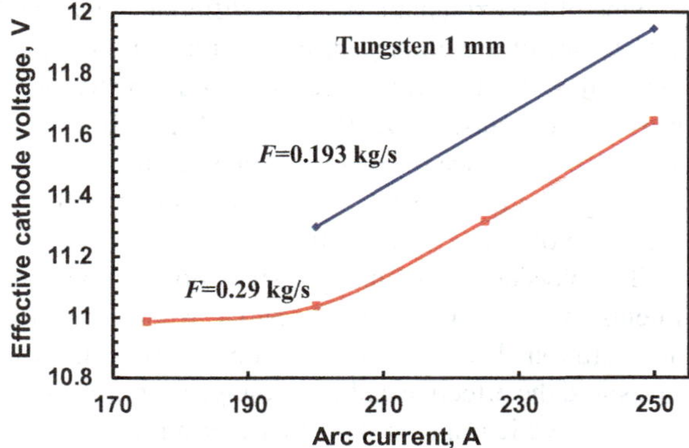

Fig. 10.14 Steady-state effective cathode voltage, u_{cef} dependence on arc current with a W showerhead anode and with water flow rate F as a parameter

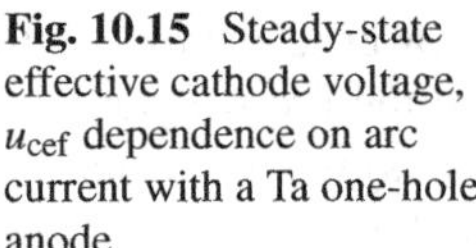

Fig. 10.15 Steady-state effective cathode voltage, u_{cef} dependence on arc current with a Ta one-hole anode

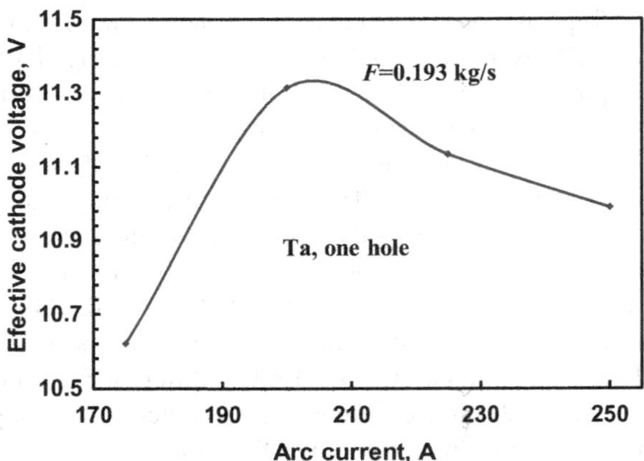

11 V for $I = 250$ A. The results in Figs. 10.13, 10.14 and 10.15 indicate very weak u_{cef} changes with I and with F demonstrated for W showerhead anode.

10.4 Summary

As it is follows from above the electrode energy loss and corresponding effective electrode voltage was widely investigated for arcs in different condition: in vacuum, in additive low pressure of ionized gas, in atmosphere pressure, at low and high currents, for pulse and steady-state discharges. The above-mentioned results are summarized in Table 10.5. The fact is that the anode voltage exceeds the cathode voltage and often u_{aef} can be twice as much of u_{cef}, ~12 and 6 V, respectively, for conventional arc discharge. The cathode-effective voltage for arcs in vacuum is high compared to those found in gases. Also, the heat losses in the presence of even low-pressure gas are lower than that measured in vacuum.

One of the explanations of this difference may be the influence of ambient gas in which part of arc energy dissipated reducing the power lost by conduction to both electrodes. Another influence can be due to change the spot types causing different erosion mechanism [29, 30, See also Chap. 10.7]. The low u_{cef} can be due to rough surface or presence of oxide or deposited films. Estimation [31] for Cu thin films in the range of thickness 0.17–1 μm shows that cathode-effective voltage is low and in the range of 1.5–2 V, respectively.

The physics of arc energy distribution can be understood considering the experiments in arcs with refractory anode where the anode temperature increased with arc duration. It is easy to show considering the VABBA experiment [28]. It was measured the effective cathode voltage as $u_{cef} = 6$–7 V during the initial stage (about 30–40 s) which agrees with that measured in the conventional cathodic arc [10]. Similar value was obtained using theory of the cathode spot which took into account the heat flux formed incident on the cathode spot surface from ion and electron energy

Table 10.5 Summarized data of measurements of electrode effective voltages

Material	u_{cef}, V	u_{aef}, V	Reference	Notes
W	3.2	4.9	[1]	Additive plasma, 0.1 atm, $I = 250$A
Cu	8	13	[2]	Vacuum, $I = 30$ A
Sn	–	7.3	[3, 4]	10^{-5} Torr,
Cu	2.5	12.5	[5]	10^{-4} Torr, additive Ar, He
Cu	2	9	[6]	10^{-4} Torr, additive Ar, He $I = 10$kA
Steel	6	11	[6]	10^{-4} Torr, additive Ar, He $I = 10$kA
Cu	6.1	12.4	[7]	Vacuum, $I = 2.5$ kA
Cu, Cr, Cd	4	10	[8, 9]	10^{-6} Torr, $I = 600$ A
Cu	6.2	–	[10]	10^{-7} Torr, up to $I = 600$A
Gd	From 5 to −3	–	[14]	$I < 200$ A, Cathode $T =$ 1900–2300 K
Cu, Mo, Nb	12.5–14%	15–18.5%	[15]	5×10^{-5} Torr, $I =$ 3–15 kA
Cu	–	9.2	[17]	Ar, atm
Cd	3–4	6–13	[18]	2–5 A, Nitrogen, atm
Zn	3.5–3.5	6–11	[18]	2–5 A, Nitrogen, atm
W	–	From 7–9 to 10–12	[10]	Ar, 20–760 Torr, $I =$ 50–350 A
W	–	9–6	[22]	He, Ar and Xe, 2–120 Torr
Cu	6.6–7.2	12	[25]	Vacuum, $I = 175$ and 340 A

fluxes toward it [32]. In the HRAVA (open gap with planar electrode surfaces), u_{cef} was approximately one third of the arc voltage, while the measured effective anode voltage ***in the initial stage*** (~12 V) was approximately two thirds of the arc voltage (20 V), and it was resulted from the energy conveyed by the cathodic plasma jets onto the anode [10, 26]. The nature of energy flux flows with the cathodic plasma jet will be analyzed below in Chap. 12, [33]. HRAVA ***steady state*** with a planar anode surface, the anode-effective voltage was about 6 V, while the other power part dissipated in the anode plasma radially expanding from the gap [10, 26].

When the VABBA anode is hot, the relatively large u_{cef} of ~11–12 V at steady state can be understood taking into account the part of cathode plasma jet energy returned to the cathode surface (due to closed assembly) and dissipated in its body. This returned energy flux is due to an anode plasma flux (where the cathode jet energy was dissipated and was transferred to the anode plasma) and radiation from the hot interior surface of anode toward the cathode surface. The hot exterior anode surface radiation as well as the relatively low power in the escaping plasma flux (compared to the interior power flux) is dissipated in the surrounding space. Thus, u_{cef} in the developed VABBA with a hot anode of 11–12 V consists of two parts: (i) from the cathode spots as in conventional arc (~6 V), and (ii) by returned cathode plasma jet energy (~5–6 V) which remains in the closed volume formed by the VABBA electrodes.

Finally, the above overview shows that the energy loss by cathode heat conduction determined by the cathode spot mechanism, while the energy loss by anode heat conduction depends by the cathode plasma jet energy incident the anode surface. The energy flux to the anode varied from energy flux of the cathode plasma jet in the initial stage to an energy flux streaming from the anode plasma to the anode surface in steady-state arc stage. The anode plasma energy was determined by the anode work function and electron temperature [5, 6, 18, 26]. The sum of the effective voltage drops of the anode and cathode has to be somewhat lower than the total arc voltage due to surrounding energy loss. This energy loss was determined by the gap arc plasma expansion, which depends on interelectrode distance in a vacuum or by the pressure of filled gas.

References

1. Fay, J. A., & Hogan, W. T. (1962). Heat transfer to cold electrodes in a flowing ionized gas. *Physics of Fluids, 5*(8), 885–890.
2. Reece, M. P. (1963). The vacuum switch. *IEE Proceedings, 110,* 793–811.
3. Osadin, B. A. (1965). Erosion of the anode in a heavy-current discharge in vacuum. *High Temperature, 3*(6), 849–854.
4. Osadin, B. A. (1965). Energy release in a high current vacuum discharge. *High Temperature, 3*(7), 952–956.
5. Konovalov, A. E. (1967). Investigation of the electrode heating of an arc burning in a flowing plasma. In *Izvestiia Sibirskogo otdeleniia Akademii nauk SSSR.* Seriia tekhnicheskikh nauk (Vol. 3, No. 1, pp. 135–138) (In Russian).
6. Konovalov, A. E. (1969). Heat fluxes to the electrodes in condition of high current discharge. *High Temperature, 7*(1), 91–96.
7. Belkin, G. S.,& Danilov, M. E. (1973). Measurements of the energy introduced into the electrodes during the burning of an arc in vacuum. *High Temperature, 11*(3), 533–536.
8. Rondeel, W. G. J. (1973). Cathodic erosion in the vacuum arc. *Journal of Physics. D. Applied Physics, 6*(14), 1705–1711.
9. Foosnaes, J., & Rondeel, W. G. J. (1976). The energy balance of a vacuum arc in an axial magnetic field. In *Proceedings of the VIIth International Symposium Discharges on Discharges and Electrical Insulation in Vacuum* (pp. 312–316). Russia, Novosibirsk.
10. Daalder, J. E. (1977). Energy dissipation in the cathode of a vacuum arc. *Journal of Physics. D. Applied Physics, 10*(16), 2225–2254.
11. Daalder, J. E. (1978). A cathode spot model and its energy balance for metal vapor arcs. *Journal of Physics. D. Applied Physics, 11*(12), 1667–1682.
12. Zektser, M. P., & Lyubimov, G. A. (1979). Electrode heating by the cathode spot of a vacuum arc. *Journal of Physics. D. Applied Physics, 12*(5), 761–763.
13. Beilis, I. I. (1974). Analysis of the cathode spots in a vacuum arc. *Soviet Physics-Technical Physics, 19*(2), 251–256.
14. Paranin, S. N., Polishchuk, V. P., Sychev, P. E., Shabashov, V. T., & Yartsev, I. M. (1986). Thermal conditions in hot evaporating cathode in a stationary vacuum arc with diffuse cathode emission. *High Temperature, 24*(3), 307–313.

15. Dorodnov, A. M., Ivashkin, A. B., Koslov, N. P., Reshetnikov, N. N., & Chursin, M. M. (1969). Heat fluxes to electrodes and erosion of electrodes in a quasisteady-state plasma accelerator of end-face type. *High Temperature, 7*(5), 881–887.
16. Rieder, W. (1956), Leistungsbilanz der Elektroden und Charakteristiken frei brennender Niederstrombogen. *Zeitschrift fur Physik, 146*, 629–643.
17. Sugawara, M. (1967). Anode melting caused by a D.C. arc discharge and its application to the determination of the anode fall. *British Journal of Applied Physics, 18*(12), 1777–1781.
18. Capp, B. (1972). The power balance in electrode-dominated arcs with a tungsten anode and a cadmium or zinc cathode in nitrogen. *Journal of Physics. D. Applied Physics, 5*(12), 2170–2178.
19. Dorodnov, A. M., Koslov, N. P., & Reshetnikov, N. N. (1973). Investigation of energetic characteristics of radiative cooling anode. In *Proceedings of the II All-Union Conference on Plasma Accelerators* (p. 388) Minsk (in Russian).
20. Merinov, N. S., Ostretsov, I. I., Petrosov, V. A., & Porotnikov, A. A. (1973). Measurements of the anode potential drop in the inert gas medium at high anode temperatures. In *Proceedings of the II All-Union Conference on Plasma Accelerators* (p. 386), Minsk (in Russian).
21. Merinov, N. S., Ostretsov, I. I., Petrosov, V. A., & Porotnikov, A. A. (1976). Anode processes with a negative potential drop at the anode. *Soviet Physics-Technical Physics, 21*(4), 467–472.
22. Merinov, N. S., & Petrosov, V. A. (1976). Calorimetric measurement of the anode potential drop. *Soviet Physics-Technical Physics, 21*(6), 724–726.
23. Pfender, E. (1980). Energy transport in thermal plasmas. *Pure and Applied Chemistry, 52*(7), 1773–1800.
24. Salihou, H., Abbaoui, M., Lefort, A., & Auby, R. (1995). Determination of the power lost by conduction into the cathode at low current arc. *Journal of Physics. D. Applied Physics, 28*(12), 1883–1887.
25. Rosenthall, H., Beilis, I. I., Goldsmith, S., & Boxman, R. L. (1995). Heat fluxes during the development of hot anode vacuum arc. *Journal of Physics. D. Applied Physics, 28*(2), 353–363.
26. Beilis, I. I., Boxman, R. L., & Goldsmith, S. (2002). Interelectrode plasma evolution in a hot refractory anode vacuum arc: Theory and comparison with experiment. *Physics of Plasmas, 9*(7), 3159–3170.
27. Beilis, I. I., Keidar, M., Boxman, R. L., & Goldsmith, S. (2000). Interelectrode plasma parameters and plasma deposition in a hot refractory anode vacuum arc. *Physics of Plasmas, 7*(7), 3068–3076.
28. Beilis, I. I., Koulik, Y., & Boxman, R. L. (2013). Effective cathode voltage in a vacuum arc with a black body electrode configuration. *IEEE Transactions on Plasma Science, 41*(8, Part II), 1992–1995.
29. Rakhovsky, V. I. (1976). Experimental study of the dynamics of cathode spots development. *IEEE Transactions on Plasma Science, 4,* 87–102.
30. Boxman, R. L., Martin, P. J., & Sanders, D. M. (1995). Handbook of vacuum arc science and technology. Park Ridge, N.J.: Noyes Publ.
31. Beilis, I. I., & Lyubimov, G. A. (1976). Theory of the arc spot on a film cathode. *Soviet Physics-Technical Physics, 21*(6), 698–703.
32. Beilis, I. I. (1977). Cathode spots on metallic electrode of a vacuum arc. *High Temperature, 15,* 818–824.
33. Davis, W. D., & Miller, H. C. (1969). Analysis of the electrode products emitted by dc arcs in a vacuum ambient. *Journal of Applied Physics, 40,* 2212–2221.

Chapter 11
Repulsive Effect in an Arc Gap and Force Phenomena as a Plasma Flow Reaction

Electrode spots support the current continuity in vacuum arcs. Due to extremely high-density energy dissipation in the spots, a large energy flux to the electrode is formed. As result, the electrode is intensely vaporized and conductive metallic plasma being generated. The electrodes are locally damaged and an erosion mass flux extracted in form of directed plasma jet was produced on the one hand. On the other hand, the reaction of the jet, in turn, invokes a force acting on the electrode. The nature of electrode force is important for understanding the vacuum-arc phenomena. In this chapter, the results of measurements illustrating the principle of the force phenomena and early experimental data about vapor pressure and it expansion near the electrode will be presented (mainly in chronological order and with description close to the published experiments to better understand the original works).

11.1 General Overview

The works regarding the cathode plasma jet phenomena were widely reviewed [1–7]. Mainly, those reviews discussed the results obtained in the second half of the twenty century. This chapter presents additional qualitative and quantitative results of measurements of expansion of the electrode plasma, which obtained in the end of nineteen century and in first half of the twenty century. The data published at this period about the force and velocity of the plasma jet produced by the cathode spot remain actually up today. Therefore, below we try giving details of the first observations, about early used methodology and the results as well present an analysis of the last corresponding researches to clarify the state and physics of measured data and specifics of this problem.

I. Beilis, *Plasma and Spot Phenomena in Electrical Arcs*, Springer Series on Atomic, Optical, and Plasma Physics 113, https://doi.org/10.1007/978-3-030-44747-2_11

11.2 Early Measurements of Hydrostatic Pressure and Plasma Expansion. Repulsive Effect upon the Electrodes

Dewar 1879 [8–11] studied applications of a carbon electrical arc and Dewar 1881 [12] (see also, "An electric arc is passed between carbon poles in an atmosphere… Ann. Chim., 1888") one of first reported about the hydrostatic pressure developed at electrodes named as poles in the papers. Dewar measured a hydrostatic pressure within the arc developed at hollow carbon electrodes (drilled centrally along their length) using especially developed water manometer to measure the pressure in condition of arc burning. The hollow electrodes are coupled through two tubes tubing with the two manometers, whose construction is described as follows:

Two glass cylinders 50 mm in diameter have each a uniform horizontal tube open at both ends, 2 mm in diameter, passing through the corks, and fitted inside apertures. When fluid was added to a fixed level, these vessels constituted the manometers. The tubes leading from the hollow poles have been made of metal or thick India rubber, and to prevent heating of the tubes and manometer by radiation from the arc, hollow tin screens through which a current of water flowed continuously have carefully guarded them. The lengths of tube between the manometers and poles have been varied, and in some cases, the tube made into a spiral form has been immersed in water to guard against unequal heating. The little glass stoppers were convenient for the alteration of the zero point by the addition or withdrawal of fluid from the vessels. In the experiments, water, ether, and alcohol have been used in the manometers, but the largest number of the experiments has been made with ether. This fluid is most convenient because of its mobility.

It was found that when the arc passes between two pointed carbon poles, two very different arc forms arise. In one case, the envelope of the intensely heated materials is well defined, almost spherical in appearance, surrounding the whole of the end of the positive pole, but touching the negative only at a single point. At other times, the arc is very unsteady, noisy with apparent blasts of flame-looking ejections. These blasts are invariably associated with a great increase of intensity in the hydrocarbon and cyanogen spectrum. In this unstable condition, manometric observations are impossible.

Many experiments were made to ascertain if a local heating of the carbon tube caused any permanent pressure by taking the arc at right angles to a carbon tube placed in a block of magnesia. This experiment showed that repulsion of the enclosed gas in the tubes through an electric charge had no effect on the manometer. During the maintenance of the steady arc, the manometer connected with the positive pole exhibits a fixed increase of pressure, while the manometer connected with the negative pole shows no increase of pressure, but rather on the average a diminution.

According to Dewar [12] "when the negative carbon tube is about 1 mm in diameter, and the point sharp and the tube short, so as to diminish any air friction, at the negative electrode the manometer seems also to give a positive pressure. Using intermittent (i.e., pulse) arc, the manometer measurements showed an increase of pressure at both electrodes. It appears from the above experiments that the interior of the gaseous envelope of the electric arc always shows a fixed permanent pressure, above that of the surrounding atmosphere. This looks as if the well-defined boundary of the heated gases acted as if it had a small surface tension. This pressure may be due to various causes including transit of material from pole (electrode) to pole, or a succession of disruptive discharges."

If assume that the luminous area near the electrode related to arise of a spot the positive pressure at the anode can be explained by interaction with the plasma ejected from the observed luminous points at the cathode and transported through the gas. The pressure, which sometimes absent at the cathode, can be due to small spot location, whose action was not detected.

Feddersen in 1861 [13] photographed sparks passing between different metallic electrodes after reflection of spark light from rotating mirrors. Shuster studied the electrical discharge in experiments to understand the nature of the electricity in gas discharges [14, 15]. Considering Federson's work [13] later, Schuster and Hemsalech in 1898 [16] stated that it was insufficient to only observe the bearing of the subject, but that the light should be sent though a spectroscope to distinguish between the luminosity of the air molecules and the metallic atoms. They investigated macroscopic expansion of the discharge electrode vapor into atmospheric pressure air. Their electrical discharges were obtained from a battery of six Leyden jars, having total capacity of 0.033 μF, and charged from an induction machine. Different inter-electrode gaps were used and the discharge images were projected onto the slit of the spectroscope at a distance at which the image size was equal to the spark size. When the electric spark pass between metallic electrodes the spectrum of metals appeared and stretched from one to other electrode. The method that determined the vapor velocity consists in fixing a photographic film round the rim of a rotating wheel.

The investigated sparks have enough powerful to give a good impression of its spectrum on the film. The metal lines were found to be inclined and curved when the wheel rotates and their inclination serves to measure the rate of diffusion of the metallic particles. The lines of the air, on the other hand, remain straight, through slightly widened. The authors found that low atomic weight electrode materials, viz., Al and Mg, had the highest velocities: the average velocity of the Al was over three times greater than that of Zn.

From a number of measurements, a certain figures were deduced for velocities at different and equidistant points on photographs, which were taken the average of all these figures as the mean velocity of the vapor. In Table 11.1 the data refer to the mean velocities between the pole and a point 2 mm obtained for zinc. The influence of the capacity and change in the gap length was investigated. It was also obtained the average velocity taken for different distances.

A comparison of the velocities was conducted for discharges with different capacities. It was found that when the discharge gap is small, there a diminution of velocities was measured than when the capacity increases. This was not what should have been expected at first sight, as with the large number of jars, it should be expected higher temperatures, and therefore greater velocity of metallic material expansion. When the gap is 1 cm, the experiments do not reveal any marked change due to capacity

Table 11.1 Average velocity of vapor for zinc electrode [16]

Gap length (cm)	Wavelength	Number of jars		
		2	4	6
		Velocity of vapor (m/s)		
0.51	4925	814	556	416
	4811	1014	668	529
1.03	4925	400	499	415
	4811	501	548	545
1.54	4925	723	1061	435
	4811	1210	1526	492

values. It was noted that when the gap was increased, the discharge become irregular and unsteady, and therefore, no certain conclusion can be drawn from the measurements and the values are doubtful. When six jars were used practically, identical values are obtained for all discharge gaps, but for small capacity, the centimeter gaps seem to give a lower velocity than in the two other cases.

Comparing Zn and Cd with each other, it was obtained almost identical values for both corresponding doublet and triplet lines. In spite of the high atomic weight of Bi, a remarkable average velocity of 1.4×10^5 cm/s was measured. The temperature of cadmium vapor was calculated as 2700 K using the measured velocity of 5.6×10^4 cm/s.

In general, the experiment [16] allows obtaining the following approximate data. The first spark passing through the air will give rise to sound speed, which during the complete time of the discharge, will only travel a few millimeters. Therefore, it was considered that its own pressure as into the vacuum expanded the mass of metallic vapors suddenly set free. Assuming that there is not much difference between the expansion in a vacuum and in the hot air, the measured velocity can be taken approximately as the sound speed in the metallic vapors. This gives a relation between the vapor temperature and vapor density. According to work [16], temperature of cadmium vapor was calculated as 2700 K using the measured velocity of 560 m/s. The observed luminous caused by the first discharge remains for a time of about 0.5 μs near the surface; the metallic vapor then begin to diffuse and reach the center of the spark (the gap being 1 cm) in a time which in case of cadmium was about 6 μs. The metallic vapor remains luminous in the center of the spark for a longer period than near the electrodes; the duration of the time during which some luminosity can be traced with a discharge supported by six Leyden jars is about 15 μs.

Duddell 1904 [17] experimentally studied the arc resistance indicating that the values of the current and the size and the electrode vapor configuration determined the arc potential drop in spite of the fact that the length, the nature of the electrodes, and the other conditions may be kept constant. This research opened the way for an interpretation mechanism upon the electromotive forces within the electric arc considering an effect of liberation (i.e., emission) of electrons from that electrode.

Duffield 1915 [18] measuring the carbon electrode erosion rate in a consumption carbon arc (see Chap. 8) also studied presence of a hot cathode spot and a cathode stream. However, before describing the corresponding experiment, Duffield has given a resume of two existing point of view to the cathodic arc process.

In 1890, Fleming [19] put forward the view that "the negative carbon is projecting off a torrent of negatively electrified carbon molecules, and these impinging against the positive carbon wear out a crater in it by a sandblast-like action. The electromotive force is thus able to keep up a projection of negatively charged carbon molecules from the end of the negative carbon, which molecules are loosened from the mass by heat, and then move away from the surface in virtue of the electric charge which they retain." Dealing with the potential difference necessary to maintain an arc, Fleming suggested that a certain fraction of the working electromotive force might be employed in detaching carbon molecules from the mass of the poles.

Sir J. J. Thomson (Thomson, Conduction of Electricity through Gases, p. 613.) put forward a different view (actually up today): "The cathode is bombarded by positive ions, which maintains its temperature at such a high value that negative corpuscles come out of the cathode. These, which carry by far the larger part of the arc discharge, bombard the anode and keep it at incandescence. They ionize also either directly by collision or indirectly by heating the anode, the gas or vapor of the metal of which the anode is made producing in this way the supply of ions, which keep the cathode hot. It will be seen that the essential feature of the discharge is the hot cathode".

The important part played by the high temperature of the cathode and arise of cathode stream were illustrated by Duffield in the following experiment. Two carbon rods *A* and *B* form the positive and negative poles (electrodes) of an arc. When the arc springs from the end of negative *B* to the rounded surface of *A*, then *B* can be moved rapidly up and down the length of *A* without extinguishing the arc, whereas if the poles are reversed, the arc at once goes out when the positive is moved. The general effect is such as though the hot spot upon the negative pole acts as a nozzle through which a stream of negative electricity is discharged. The arc develops whenever this stream falls upon another conductor connected to the opposite terminal of the source of current supply. The experiment was modified by mounting one pole so that it can be rotated about its long axis, and by forming an arc between its rounded surface and the end of a second carbon rod. When the latter is the cathode, the former can be rapidly rotated without affecting the arc, but not when it is the anode unless the motion is extremely slow, when it is possible to wrap the arc completely round the circumference of the cathode, the hot spot upon it remaining fixed and turning with the pole.

It should be noted that the above-mentioned description of cathode processes is remain close to the modern theories up today, which used the modern plasma knowledge, in spite of that past of more than 100 years.

Suspending one electrode of a carbon arc and a keeping the other electrode fixed, Duffield et al. 1920 [20] found an apparent repulsion between them. There was in fact a pressure upon each electrode, which tends to separate them. Duffield has observed the repulsion between electrodes in preliminary experiments in 1912 and the original data were presented in a paper entitled "A pressure upon the poles of a carbon arc" at British Association Meeting, Australia, 1914. Three series of measurements were conducted in the period up to 1920 with this same experimental setup with differences in the disposition of the carbons in different sets of experiments.

A stirrup was suspended by a torsion fiber, or by two fibers *F*, as in the illustration (Fig. 11.1), in this was placed a copper rod *E* to whose extremity was fixed at right angles, a short carbon rod *C*, which was balanced by counterpoise *W* at the other end. The arc was formed between this carbon rod and other (*D*) fixed ether as shown in Fig. 11.1 or in some another manner. In its zero position, the Cu rod swung freely between two stops *S* placed close to one end. The sensitivity of the suspension and the

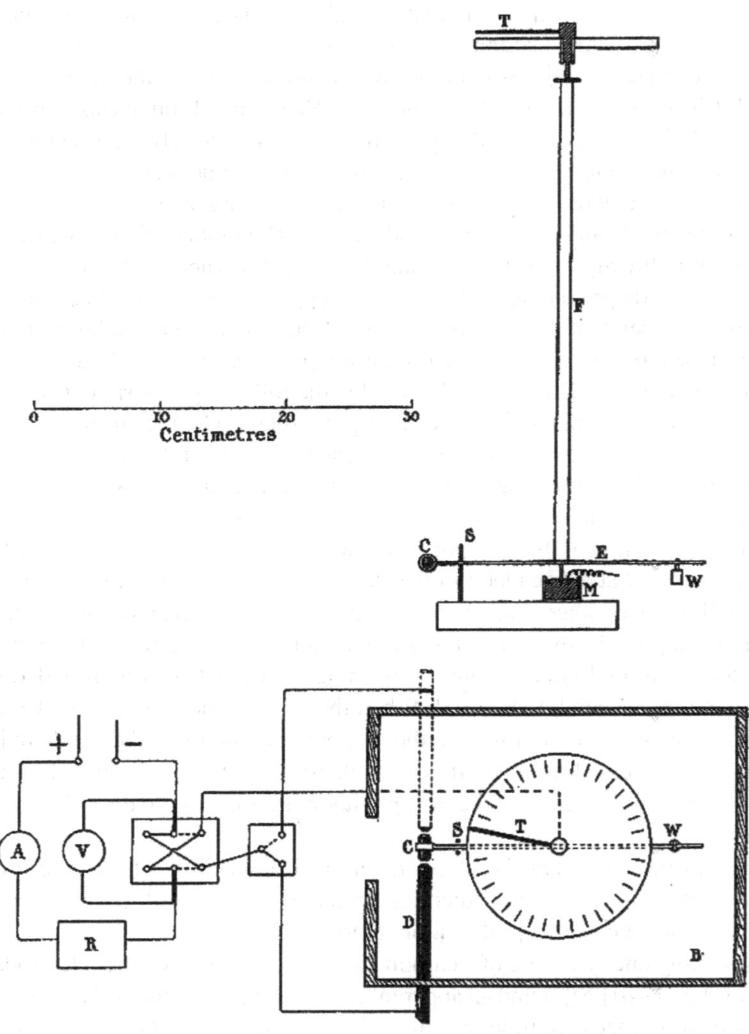

Fig. 11.1 Schematic presentation of the system for force measurement in an air electrical arc [20]

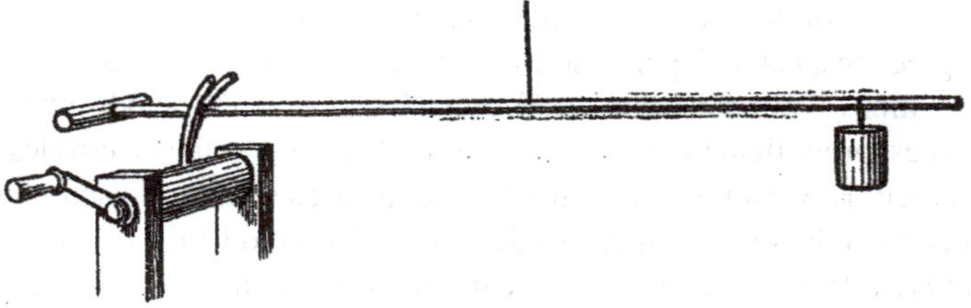

Fig. 11.2 V-grip presentation [20]

long period of swing necessitated some simple means for bringing the rod back to the zero position, and the V-grip device illustrated in Fig. 11.2 was ultimately adopted in place of the stops *S*. The adjustment was made by twisting the torsion head until on turning down the V-grid; the suspended wire remained stationary. This control also enabled the arc length to be maintained nearly constant during an experiment. The difference between the readings of the torsion head *T* when the current was on and off measures the couple acting upon the suspended Cu rod if the constant of the suspension are known.

The movable parts of the system were completely enclosed in a box *B* with a glass top to prevent disturbance from air currents in room and appropriate windows and holes in it to enable observations to be made. The torsion fiber was enclosed in a vertical tube. A lens focused an image of the arc upon a screen facilitate the measurement of the arc length. The observed couple was due to total pressure upon the electrodes determined forces within the arc. These forces include different effects such as air flow, electrostatic, electromagnetic, and earth's magnetic field.

The force on a carbon anode (a) and cathode (b) is presented in Fig. 11.3 as a function of the arc current, with arc length L_{arc} as a parameter. The graphs are not straight lines, and the total force is somewhat greater at the anode than at the cathode. The dependence on arc length is not monotonic. In both cases, the measured force significantly reduced when the current approaches to 3 A. The authors postulated that the force was due to electrons projected from the cathode, which moved with a speed large compared with that which they would acquire in the field itself.

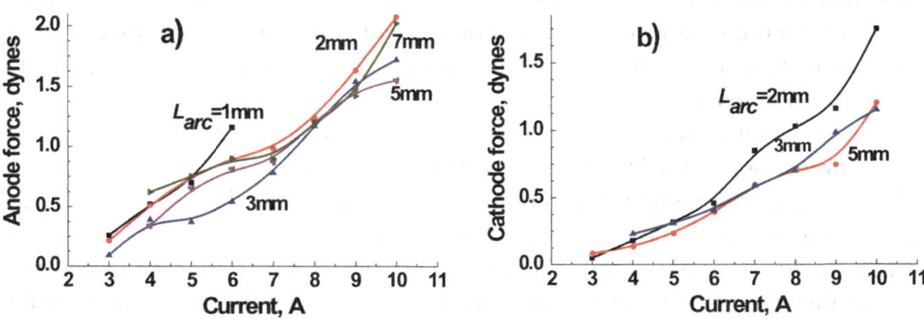

Fig. 11.3 Variation of total force at anode (**a**) and cathode (**b**) with DC current and arc length *L* [20]

Duffield [20] pointed that in the previous measurements of Dewar 1881 [12], no data were given respect to dependencies on current and arc length and the hydrostatic pressure (1 mm of water is about 100 dynes/cm^2) near the anode was about one hundred times larger than the total pressure (in dines) upon the electrodes measured in this research. However, this conclusion and the area were not discussed.

To understand the measured electrode effects, Duffield [20] assumed that a pressure could arise from the expulsion of carbon atoms in the process of evaporation. Any reaction due to evaporation he calculated using the measured mass per second (8.5 g/s for 10 A arc) which leave the electrode and atom velocity determined as thermal velocity by electrode boiling (400 K) temperature (2.97 $\times$ 10^5 cm/s). As result, it was found a disagreement between observed and calculated data.

Tyndall 1920 [21] and Beer and Tyndall [22] provided experiments to understand the role of ion fluxes from the electrodes in formation of the force at the electrodes. The experiments were made with a horizontal straight arc of soft carbons of 1.3 cm diameter, through each of which a hole of 3.5 mm diameter had been bored by removing the soft core. All experiments were carried out at atmospheric pressure. The pressure at either anode or cathode was by connecting it to one limb or a tilting bubble gauge of the usual type. In general, the other limb was open to air to some point in or near the body of the arc at which the pressure was that of the surrounding atmosphere. The particular instrument used responded to a difference of pressure of about 1/1000 mm of water. An image of the arc was focused on a screen so that the distance between the carbon electrodes could be maintained at required values and simultaneously reading the manometer pressure for various currents and lengths of arc.

The preliminary observations in [21, 22] showed that the Dewar effect [12] was readily obtained, namely an excess pressure at the anode for a silent arc and a comparatively negligible pressure, which was sometimes negative at the cathode. The anode pressure was greatly reduced when the arc was hissing. An observation showed that the anode pressure was decreased with time. It was found that this result depends on carbon contained absorbed gases. This was solved in a series of experiments with carbon electrodes from which the gas had been desorbed by running the arc at a high current for 5 min before the measurements were made. The current used for desorption being in all cases greater than that of which pressure measurements were subsequently taken. The electrodes used were with a diameter of about 8 mm, partly to ensure a more complete desorption of gas and still more to minimize as far as possible the tendency of the arc to leave its central position for effective gauge use in the measurements.

Another experimental difficulty was "hissing," which considerably restricted the scope of the observational work. It was particularly noticeable with postwar carbons. When the arc was hissing, the luminous patch on the anode was quivering and travelling about, with the result that the effective pressure was greatly reduced on that electrode owing to the departure from centrality, if from no other cause as well. On the other hand, hissing caused the cathode pressure, which was generally negative for a large hole, to become positive.

Anode pressures and length of arc for various currents with holes of different sizes were obtained. The observations show that the pressure at the anode is practically, if not wholly, independent of the length of the arc. This result strongly suggests that the pressure originates in the region quite close to the anode itself, and experiments with different sizes of the holes and current afford additional support for this view.

The linear dependence of the anode pressure on arc current is demonstrated in Fig. 11.4. The readings for this dependence were taken under conditions in which the greatest care was observed in ensuring centrality of the arc, the hole (1.5 mm) being always smaller than the luminous spot. For other hole sizes (except perhaps at the largest currents), the holes were always larger than the luminous patch and the centering consequently not perfect. Therefore, the obtained anode pressures were significantly lower (<5 dynes/cm^2) at arc currents up to 10 A and linearly depended then up to 15 A increasing from about 5 to 50–100 dynes/cm^2 depending on hole sizes (see Table 11.2).

Quantitative measurements at the cathode were difficult, owing to the fact that the luminous spot was much smaller and will rarely remain centered on a small hole in the electrode long enough to obtain readings which are definite except in sign. In the case of the large holes, the luminous spot was appeared right inside the hole at a point on its wall. In the case of the small holes, the spot was about half on the hole. The careful measurements were taken with a hole of 1.5 mm diameter, selecting only

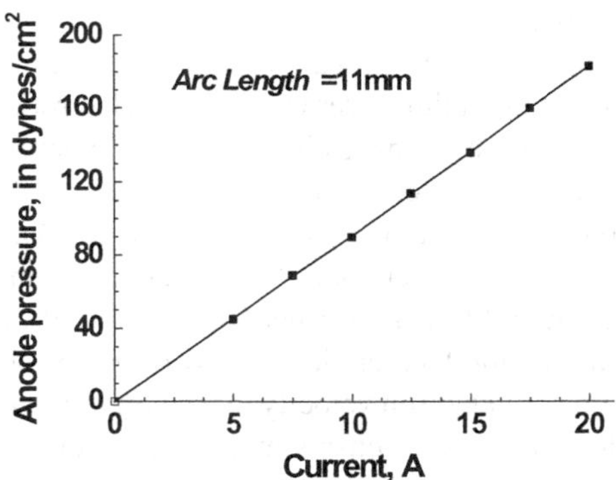

Fig. 11.4 Anode pressure as function on arc current [22]

Table 11.2 Anode pressures for different hole diameters [22]

Arc current (A)	7.5	10.5	14.5
Hole diameter (mm)	Anode pressures (dyne/cm^2)		
1.7	23	52	85
2.3	28	53	82
3.0	25	48	72
4.0	32	55	89
5.0	35	59	96

Table 11.3 Pressure at the cathode of 9 mm diameter [22]

Hole = 1.5 (mm)			Hole = 1.1 (mm)		
Length (mm)	Current (A)	Pressure (dyne/cm^2)	Length (mm)	Current (A)	Pressure (dyne/cm^2)
–	13.5	6.9	7	13.5	24.7
			7	17.0	13.2
1.3	17.0	18.6	5	18.1	5.7
			6	19.5	31.8

the rare occasions on which the arc was really central. In every such case, a relatively large positive pressure was recorded. This result was confirmed by a few readings taken with a smaller hole 1.1 mm in diameter. The order of magnitude of the effect is given by the results shown in Table 11.3.

According to Table 11.3, the pressure was given as the force per unit area. The force can be estimated assuming that the area was determined at least by the hole radius of 1 mm. In this case, the larger force from the table will be about 0.05 dyne/C, which is very small value. As the authors [22] noted that with the large holes used by Dewar [12], he did not observe the positive cathode pressure of a perfectly central arc but was dealing with negative pressures similar to those observed in the present work. They also supposed that the pressure was caused with the effects of space charge near the electrode.

The pressure observed at the anode was produced by the motion of a space charge layer of negative electrons in a thin layer close to the electrode while at the cathode was produced by the motion of a space charge layer of positive ions. The thickness of space charge layer was estimated as 10 μm for anode.

Tyndall 1921 [23] developed a model to analyze the mechanism of the effects of the electrostatic forces within the arc, and for re-examining the theoretical basis of the Duffield effect in the light of the further information of the Dewar hydrostatic pressure has provided. The model included a study of pulses obtained in the electric field by different ions and of electron velocity in the arc. It was noted that the in order to produce a Duffield effect, the electrons must retain their original momentum in travelling across the arc. However, they could exert any drag on the gas and therefore they will very rapidly be pulled up quite close to the cathode, thereby setting up a suction there, which will balance the recoil of the electrode and so reduce the Duffield effect to zero. In the opinion of the author, the theory of pulse from electrons moving across the arc either from the cathode itself or from its immediate neighborhood is not in accordance with some experimental data. It was therefore concluded that (i) Duffield's explanation of the mechanical pressure he observed is doubtful and (ii) a pressure of the order of magnitude observed can be studied with a theory of the striated discharge only when a method for measuring the mobility of the ions will be improved in the arc conditions.

A Sellerio 1916 [24] and 1922 [25] studied experimentally a repulsive effects upon the carbon electrodes in an atmospheric pressure electrical arc. The system

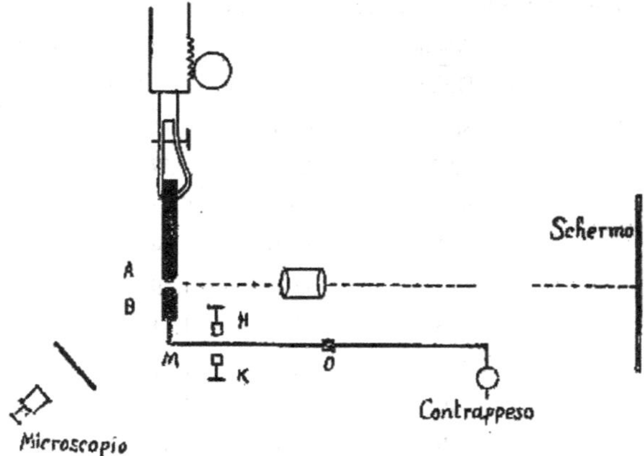

Fig. 11.5 Schematic presentation of the system for the force measurements [24]

of measurements was presented in Fig. 11.5. A sort of torsion balance, which was connected to the horizontal rod, a suspension fiber normal to the plan hanging down from a graduated (Schermo) torsion head, and a little iron style dipping into a mercury trough. Therefore, the current flowed through the mercury either in the direction OMBA, or in the opposite one. When the arc is started, the arm OM tends to recoil, and to hold it stationary it is necessary to give to the fiber a certain torsion α, corresponding to a force as

$$F = 0.036\,\alpha\ \text{(dyne)} \tag{11.1}$$

The force F there is to be distinguished: (i) the true repulsion rising within the arc gap; (ii) some disturbance occasioned from the heat, as air convection currents, etc.; (iii) the influence $M_F = 0.296I$ of the earth's magnetic field; and (iv) the electrodynamic action $E_{ed} = 0.03I^2$ between OMB and the fixed circuit. It was taken OM = 15 cm, MB = 4 cm, and $H = 037$. Taking in account the earth's magnetic field and electrodynamic actions, the corrected force P^+ for anode and for cathode P^- are

$$P^+ = F - 0.2961 - 0.03I^2 \tag{11.2}$$

$$P^- = F + 0.2961 - 0.03I^2 \tag{11.3}$$

In the experiments, almost all the readings were taken by keeping the arc length L_{arc} constant and varying the current. The observations showed that the carbon quality has a far greater influence than L_{arc} on the results. When the carbon rods are very close together ($L_{\text{arc}} \sim 0$), the repulsions become evidently greater. It would be useless to relate here the individual series of measurements… Therefore only the mean values of P upon anode and upon cathode for arc length $L_{\text{arc}} = 1$–4 mm were presented in Table 11.4 by given the current and after the corrections with (11.2) and (11.3).

Table 11.4 Electrode force for given arc currents [24, 25]

	Anode			Cathode		
Current (A)	Deflexion α	F (dyne)	P^+ (dyne)	Deflexion α	F (dyne)	P^- (dyne)
3	36	1.30	0.14			
4	62.2	2.23	0.57			
5	88	3.16	0.93			
6	114	4–10	1.24			
7	140	5.04	1.49			
8	166	5.90	1.61			
9	204	7.5	2.26			
10	240	8.65	2.69			
11	290	10.4	3.49	16	0.58	0.21
12	336	12.1	4.23	48	1.73	0.96
13	380	13.7	4.77	80	2.87	1.66
14	436	15.7	5.68	112	4.03	2.1
15	480	17.3	6.07	144	5.19	2.87
16	540	19.4	6.97	176	6.35	3.38
17	600	21.6	7.88	210	7.55	3.87
18				260	9.35	4.95
19				318	11.50	6.35
20				376	13.60	7.42

The diameter of the electrodes has no great influence in the present research, of course only while it remains large relatively to the crater size. For, putting a carbon rod 12 mm. in diameter against a similar one of 3 mm, when the latter is acting as anode, the arc hums and the repulsion becomes greater. Further remarks made also with thick carbon rods either by increasing the current strength over the hissing point or by shortening the length has generally shown that when the arc is not quite steady and silent, the forces acting upon the electrodes become greater.

In order to explain the observed mechanical pressure, many hypotheses have been suggested [25]. One of them was take in account that the arc loss of matter by each electrode and a transport from anode to cathode occur, thus—on the anode at least—a recoiling effect must be occasioned. Whatever the nature of the forces propelling the particles may be, the recoil can be due to evaporation electrode material. In the case, the relation between the measured force P of an evaporated mass flux per second N (particle mass m) from an electrode crater of are A_c with the mass velocity v can be estimated as follows:

$$P = Nmv \tag{11.4}$$

Table 11.5 Specific pressures at the anode and cathode for different arc currents [25]

Current I (A)	2	4	6	8	10
Force, P (dyne)	0.1	0.44	0.9	1.43	2.12
Anode crater area (cm^2)	0.017	0.088	0.16	0.24	0.31
Cathode crater area (cm^2)	0.005	0.027	0.05	0.073	0.096
Anode pressure (dyne/cm^2)	5.8	5.0	5.6	5.9	6.8
Cathode pressure (dyne/cm^2)	20	16.5	18	19.5	22

In the electric arc, there is a repulsive effect upon the electrodes, increasing with the current. The range of P is less than 10 dyne with currents up to 20 A. The carbon quality has a great influence on P. It seems that metallic impurities cause an increase in the pressure on the cathode and diminish that on the anode [24]. With uncured carbons, the repulsion on the cathode appears smaller than that on the anode. On the contrary, the specific pressure per unit crater surface is greater on the cathode [24]. The propulsive velocity of carbon particles has been estimated as 280–400 cm/s [24]). The specific pressure p was obtained as ratio of the force P to the crater area A_{crat}. Taking the values P from Duffield's work [20] and the values A_{crat} for a circular crater from Sellerio's work [24], the specific pressures can be estimated at the anode and cathode. The results are presented in Table 11.5.

The author noted that the data in Table 11.5 are only recorded in order to give a rough estimate of the specific pressure. To pursue accurately this inquiry, coherent values of the force and crater area for the same carbons and the same arc length are required. Sellerio 1922 [25] also indicated that the manometric observations of [22] of the mechanical pressure with drilled carbons have shown for current strength under 20 A a hydrostatic pressure up to 30 dyne/cm^2, an order of magnitude not far away from that of the mechanical pressure. However, in a research on the electric arc Hg+/C~ between Hg as anode and a thin carbon rod as cathode, a pressure of 6500 dyne/cm^2 upon the positive pole Hg was calculated. In fact, it is known that in a Hg arc, a cavity of 1 mm or more in depth has been often observed, corresponding to a pressure range over 1300 dyne/cm^2, with currents of a few amperes. With a carbon–carbon arc, the specific pressure is, as shown, a hundredfold smaller (See [24, 25]).

Sellerio [25] indicated that the Duffield's (see above) assumption about the thermal velocity used for explain the electrode force was hardly defensible. He showed that the atom expansion from the electrode could be analogical to an expansion process from a compressed gas in bulb. By opening a tap, the gas escapes impressing a reaction upon the bulb as in turbines. The velocity with which the gas departs from the bulb is a function of the pressure difference and therefore can be different from the thermal velocity. Sellerio [25] noted that it could not be concluded with full knowledge whether the observed effect associated with electrical processes of the arc, or whether the force was appeared as result of an ordinary way the evaporation of electrodes at high temperature.

11.3 Primary Measurements of the Force at Electrodes in Vacuum Arcs

The above-considered early works has been provided mostly in air and experimental results demonstrated the existence of a repulsion of between the electrodes. This (observed even at primary arc study) effect indicates that the electrical arc exerts a force on the electrodes and appearing of a plasma flow with relatively large velocity. The forces on the electrodes were observed for low currents (<20 A) in atmospheric arcs and therefore the results showed relatively low values (about 1 dyne). Also, the number of experiments with electric arcs in a vacuum indicated that a jet of high-speed vapor was ejected from the cathode [26] and the forces were obtained in vacuum significantly larger than that measured in atmospheric arcs in similar range of currents. Let us consider the further quantitative data. Tanberg [27] measured the velocity of an expanding Cu plasma jet by: (1) the force of reaction of the vapor on the cathode and the rate of vaporization of the cathode material and (2) determining the force exerted by the vapor on a vane suspended in front of the cathode spot and the rate of vapor condensation on the vane.

Thus, the experiments were conducted measuring the force of reaction of the vapor on the cathode and by determination of the momentum imparted by the vapor to a vane suspended in front of the cathode. The experimental system is shown in Figs. 11.6a, b. Also by measuring, the amount of vapor condensed on the vane per unit time by weighing method, the vapor velocity was obtained. The average jet velocity from the cathode was calculated from the two methods, using, in the first case, the reaction force and mass loss of the cathode, and in the second case, the deflection and mass increase of the vane.

The arc current was in the range of 11–32 A, and the vacuum chamber pressure was ~10^{-4} torr. Cu cathode was used as a short cylinder of 0.6 cm diameter around which a quartz tube fitted in order to limit the spot motion and on the end surface of which the cathode spot was located. The deflection was measured on a scale, located

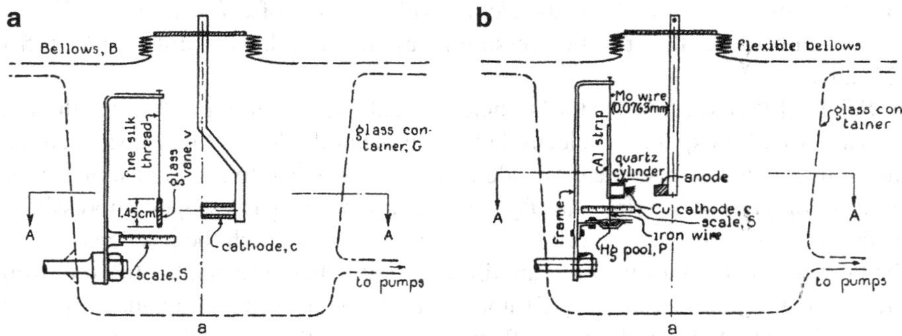

Fig. 11.6 Schematic presentation of force measurements by reaction of the vapor on the cathode (**a**) and determining the momentum imparted by the vapor to a vane suspended in front of the cathode (**b**). Figure taken from [27]. Permission number RNP/20/FEB/023149

Table 11.6 Typical data determined by reaction force from the cathode [27]

Current (A)	Reaction force (10^{-3} g)	Force (dyne/A)	Evaporation from cathode (10^{-4} g/s)	Jet velocity (10^6 cm/s)
11	199.3	17.8	1.7	1.60
16	263.6	16.2	2.5	1.46
19	363.1	18.75	3.0	1.68
32	462.0	14.16	4.9	1.31

underneath the cathode. The instrument was calibrated for cathode deflection against the corresponding force on the cathode surface. The arc was formed by contact between the cathode and anode. As soon as the arc ignited, the anode was moved to about 1.5 cm away from the cathode (Fig. 11.6a).

The series of typical data are presented in Table 11.6. Electrostatic and electromagnetic effects showing very small contributions indicated only in gram in [27] corrected the results of force measurements. These data of the force M given in gram divided by current and multiplied by is the acceleration due to gravity, 981 cm/s^2, given forces in dyne/A (third column). According to Table 11.6, the average reaction force from the freely swinging cathode is ~16.73 dyne/A and the average velocity is 1.5×10^6 cm/s.

According to the second method in the apparatus shown in Fig. 11.6b, the Cu cathode, *c*, surrounded by quartz tube, was fastened to a metal rod, which could be moved from outside the vacuum by means of the flexible bellows. The arc was formed by moving the cathode to touch the anode. As soon as the arc was ignited, the cathode was moved back in front of the glass vane, consisting of a square piece of Pyrex glass. Two fine silk threads from a supporting frame suspended this. The deflection of the vane was fixed directly on the scale, which determined the force as a function of the deflection. During a test, the arc would play, while the vapor would be projected from the cathode spot directly against the vane deflecting this according to the momentum imparted to it by the impinging particles.

The measurements using glass vane method were presented in Table 11.7. Using data from this table, it can obtain the average force as 4.73 dyne/A. Although the force measured at the vane was lower, the amount of mass collected was also proportionately less, hence about the same similar value of average velocity was obtained as 1.9×10^6 cm/s.

Table 11.7 Typical data determined by deflecting vane [27]

Current (A)	Force on vane (10^{-3} g)	Force (dyne/A)	Condensed vapor (10^{-5} g/s)	Jet velocity (10^6 cm/s)
14.2	64	4.42	4.26	2.08
16.0	64	3.93	5.37	1.66
18.0	94	5.12	6.50	2.01

Using the measured rate of cathode evaporation, Tanberg [27] estimated the cathode temperature assuming that the kinetic energy and the absolute temperature T of the atoms should be justified to use the equation applying to three degrees of freedom of the single atom as $mV^2/2 = 3kT/2$ (m is the atom mass in grams, V is the jet velocity). The calculations showed that cathode temperature for Cu atoms varied in range (4.4–7.3) $\times$ 10^5 K. As was noted, in this work, these temperatures were far in excess of even the most extreme temperatures ever measured in connection with any physical phenomenon of any duration. As was indicated by Tamberg [27], the second method was used due to very high temperature, calculated by measured jet velocity, compared with what could be expected from conservative estimates of the temperature existing at the cathode.

Another work of Tanberg and Berkey 1931 [28] showed that the temperature of a copper cathode is measured by an optical pyrometer and found to be about 3000 K in a 20 A arc. Spectroscopic examination of the cathode spot shows only a faint continuous spectrum indicating that the temperature of the cathode is not high and it is sufficient to give the rate of vaporization required under arc conditions. The high speed of the vapor stream issuing from the cathode region cannot be due to high temperature of the cathode itself.

Kobel 1930 [29] measured the force on a mercury cathode. In order better understand the experiment, let us describe the method as it presented by Kobel. Figure 11.7

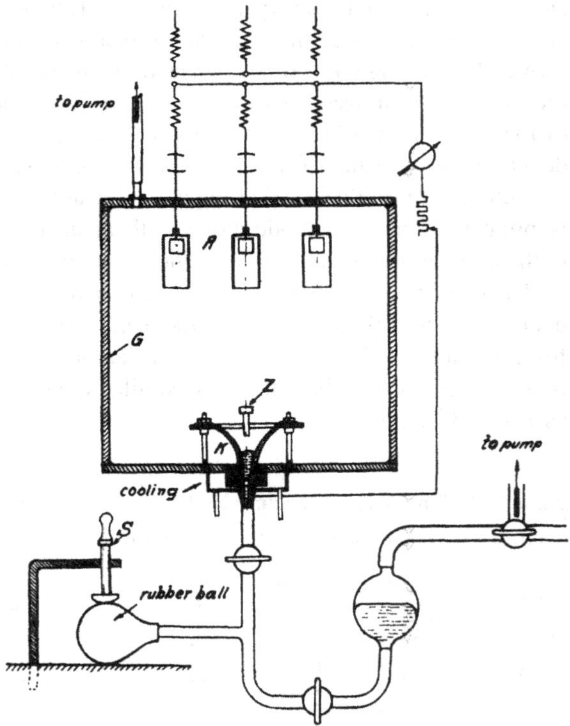

Fig. 11.7 Diagram of measurements. G is a glass chamber, K is the cathode, A is the three iron anodes, equally spaced in the form of a triangle. Figure taken from [29]. Permission number RNP/20/FEB/023159

shows diagrammatically a glass container of about 5.5 L capacity, in the floor of which is the cathode K, and in the cover, the three iron anodes A, equally spaced in the form of a triangle. The conical- and cylindrical-shaped tungsten insertion-piece is extended in an upward direction by a quartz cone so that the arc may be ignited in it by touching the mercury with the ignition anode Z.

After ignition, the mercury is lowered into the cylindrical part of the tungsten insertion-piece by rotating the screw *S*. When increasing the cathode current, the mercury pressure must be adjusted at the same time by means of the screw *S*, so that the surface of the mercury, which is entirely covered by the cathode spot, neither rises nor sinks. The pressure is indicated on a slanting capillary glass tube attached to the bottom of the vessel containing the mercury. The amount of mercury vapor was determined separately by decreasing the mercury level in the tungsten insertion-piece at a constant temperature of the container and with the mercury feed pipe closed.

When the cathode spot is moving freely, the surface of the mercury is much larger than the area of the spot, and in this case, the vaporization is greatly dependent upon the mean temperature of the surface of the mercury. When the cathode spot is fixed and the spot covers the entire surface of the mercury, the vaporization remains at a certain value, which is no longer greatly dependent upon the cooling of the cathode. This is also the case with a strongly cooled copper-mercury cathode. The copper disk used as a cathode that contained holes of about 2 mm diameter filled with mercury. A cooling liquid with a temperature of 10 °C was circulated round a channel in the disk during the test.

The arc only burned until all the holes covered by the cathode spot were used up. A subsequent examination of the copper disk showed that it had been scarcely burned by the arc, since an arc current of only 14 A was used. The gas pressure in the chamber, arc current, force on the cathode spot, and vapor velocity are given in Table 11.8. The data in column 3 give only the gas pressure in the chamber, but not the mercury vapor pressure. No corrections for electrostatic and electrodynamic forces were made because they do not affect the results.

As it can be seen the average force measured by Kobel [29] for Hg is about 40.7 dyne/A and larger by factor more than 2 that force obtained by Tanberg [27]. This same is related to the average velocity, 3.4×10^6 cm/s.

Berkey and Mason [30, 31], determined the vapor velocity by measuring the kinetic energy of the vapor stream. Also Mason 1933 [32] was given a resume of previous observations of the forces and jet velocities obtained from measurements

Table 11.8 Data of measurement [29]

Current (A)	Current density (A/cm^2)	Pressure (mm Hg $\times 10^{-3}$)	Force (dyne)	Force (dyne/A)	Jet velocity (10^6 cm/s)
30	1700	0.5	1470	49.0	4.2
37	2090	1.5	687	18.56	1.6
35	1980	0.5	1470	42.0	3.6
32	1810	0.5	1700	53.13	4.33.4

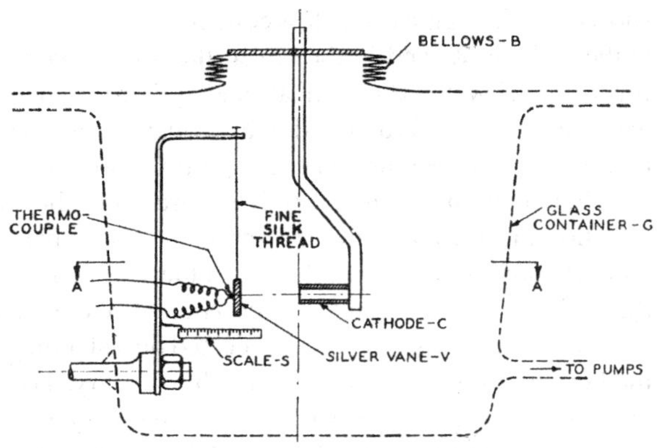

Fig. 11.8 Schematic presentation of the measurements on cathode vapor jet. Figure taken from [32]. Used under a Creative Commons 4.0 Attribution license https://creativecommons.org/licenses/by/4.0/legalcode

of momentum in the way in which the balancing force was registered. Figure 11.8 shows a sketch of the used setup with Cu electrodes mounted in a glass desiccator. The arc was ignited by a contact with the anode. A quartz cylinder surrounding the copper rod cathode restricted the cathode spot to the end of the 0.55 cm diameter rod. A silver vane 2.5 cm square and about 0.25 mm thick was suspended 3 cm directly in front of the cathode by a silk thread.

Below the vane, a scale was mounted which allowed the deflection of the vane to be observed. A number of thermocouples were to measure the temperature of the vane in degrees centigrade. The gas pressure of about 6×10^{-4} mm, an arc was struck and run for a few seconds. During a test, the pressure would rise to 20 or 30×10^{-3} mm. The mass of the copper deposit on the vane was determined by weighing method. Arc current and arc voltage were 18.5 A and 30 V, respectively. Distances from cathode to anode and to vane were 3 and 2 cm, respectively. The force acting on the vane was found from the deflection of the vane, the length of support, and the mass of the vane. The energy absorbed in the vane was obtained from the rate of temperature rise of vane when the arc was running. The results of several tests are given in Table 11.9. The values of velocity calculated from the energy input show close agreement from test to test. The mean velocity calculated from the force shows greater variation than the velocity, but still is of the same order of magnitude. In cases where the deflection could not be determined exactly, upper and lower limits are given for the deflecting force.

As it is followed from the measurements, the force is relatively low. As the authors due to the smallness of vane deflections, the distance from scale to eye and the uncertain retarding force of the coiled thermocouple leads noted it, the force on the vane could not be determined with very great accuracy. The average velocity of the vapor, computed from data obtained by either method, is over 10^6 cm/s, confirming Tanberg's momentum measurements. Also, it was indicated that the vapor is ionized to a considerable degree.

Creedy et al. 1932 [33] investigated the welding arc in order to understand what force phenomena in the arc produced an ejection of the welding droplets. They

Table 11.9 Parameters of vapor stream from cathode [32]

	(1)	(2)	(3)	(4)	(5)	(6)
Number of test	Power to vane (ergs/s × 10^7)	Force (dynes)	Force (dynes/A)	Mass Cu deposited on vane (g/s × 10^{-5})	Velocity from columns (1) and (4) (× 10^6 cm/s)	Velocity from columns (1) and (4) (× 10^6 cm/s)
1	7.26	–		2.38	2.46	–
2	4.64	–		2.38	1.98	–
3	3.59	53–70.5	2.86–3.81	2.38	1.78	2.23–2.96
4	–	70.5	3.81	2.38		2.96
5	–	70–88.2	3.8–4.77	2.38		2.96–3.7
6	–	53.6	2.9	2.22		2.41
7	4.36	35.8	1.94	2.22	1.98	1.62

developed an experiment to study the near-electrode region as project it intense brilliance image on a screen magnified by means of a lens because this arc region cannot be studied directly. In this case, a series of remarkable phenomena at once becomes apparent. In first, a low-current arc of four or five amperes considered between electrodes of about the same size as welding wire.

This experiment showed that the two ends of the electrodes each become fused into a globule, the positive electrode being materially hotter than the negative. In some cases, the globule may be exclusively on the positive electrode while the negative electrode contains only a small incandescent area corresponding to the cathode spot. On each electrode, there is a brilliant incandescent spot and frequently a reflection of the cathode spot may be seen in the fused surface of the drop on the end of the positive electrode. All this has been the subject of much study, and when an arc between electrodes of similar area but carrying a welding current of 150 A is examined in the same way, some entirely other phenomena become apparent. For example, when a powerful current is used the drop grow to a certain diameter, usually somewhat larger than that of the rod and then there is a sharp explosion or crackling noise and the drop disappears, its motion being so fast that it is impossible to see it by means of movie cameras of ordinary speed.

In order to overcome some difficulties, the experiment is conducted at very short-arc time. The iron arcs are studied by means of an apparatus and designed based on the principle of the ballistic galvanometer. The experiments were conducted in a setup consisted of a brass rod about 8 mm in diameter and 20 mm in long terminated at either end by an iron electrode about 10 mm diameter and 30 mm long with a moving system which was suspended at the center of gravity by a bifilar suspension. The important part of the setup is a galvanometer mirror fixed to the suspended system and reflects a spot of light on a screen and a pin (moving electrode) projects from below, dipping into a mercury cup.

The current passes through the fixed electrode, the arc between it, and moving electrode, and then through the moving electrode and out through the mercury cup. In order to the arc ignition the electrodes must be momentarily touched together and then drawn apart by controlling a key. The method of operation consisted in ignition of the arc with a small current, 5 A, when the forces due to which were small and also small oscillation and then sharply passing to the high current for a short period varying from 0.1 to 0.6 s and observing the first deflection of the spot of light. From this and the moment of inertia, the impulse was calculated from data of the ballistic galvanometer. The impulse was defined as the force multiplied by time of application.

The results were obtained for wide range of arc currents (I = 38, 68, 90, 120, and 154 A). It was proved that the measured deflection obtained for electrode orientation used in experiment was entirely due to the arc forces and not by electromagnetic forces. A considerable number of observations were conducted to obtain the arc lengths L_{arc} dependences with constant current. Figure 11.9 shows the relation between forces and the arc length (with I as parameter) produced by the arc moving electrode relatively fixed electrode and spot light of the arc was projected on a screen (see detail of the experiment in [33]). As follows from theses dependences, the length increases with current. However, the force weakly depends on the L_{arc} for $I < 90$ A and the force significantly increases with L_{arc} decreasing depending on I.

The relation between the force and arc current was presented in Fig. 11.10. As can be seen that the force depends parabolic on the arc current significantly increasing after $I > 100$ A depending on L_{arc}.

Using the data from Figs. 11.9 and 11.10, the specific force in dyne per amperes as function on current can now be calculated. This dependence is illustrated in Fig. 11.11. In essence, this specific force increased with I and decreased with L_{arc}. The obtained results discussed in frame of that measurements were published previously. For low current, these results are close to the measured in published works [20–25].

Easton et al. 1934 [34] continued the experimental study of the force between electrodes of the arc in a vacuum. The relations between arc current and mass of condensed metal, force between electrodes, and velocity of the vapor stream were investigated. The experimental setup with a chamber in 43 cm diameter consisted

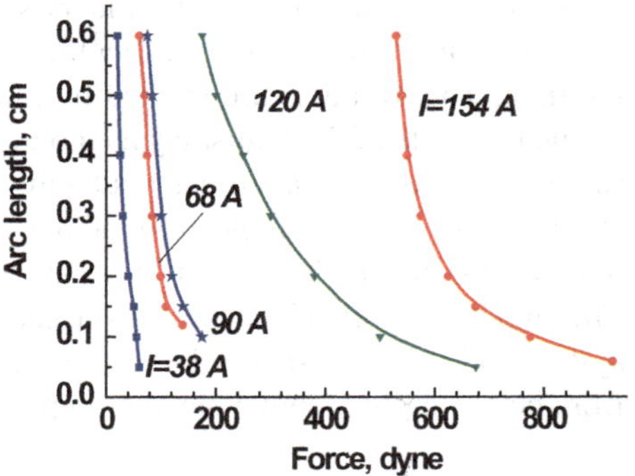

Fig. 11.9 Arc length versus force for arc current as parameter [33]

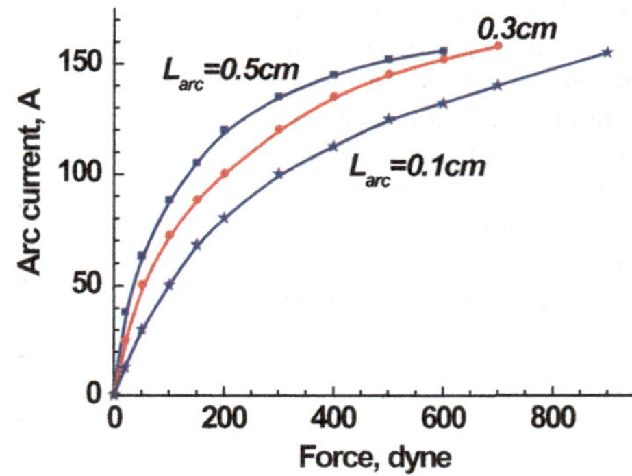

Fig. 11.10 Current versus force for arc length as parameter [33]

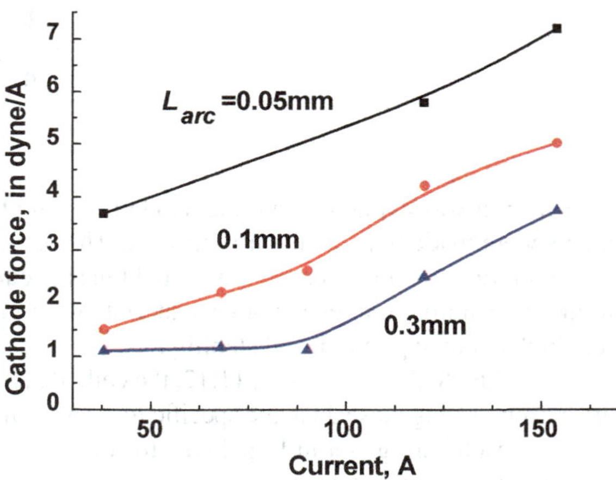

Fig. 11.11 Dependence of the force in dyne/A on arc current

of a fixed and moving electrodes connected with a measurement unit. The fixed electrodes consist of two steel or fiber disks. A mechanical measurement unit included an equipment with a vane, a cup filled with mercury, an aluminum float and this arrangement allowed to measure the forces produced by the arc onto the vane.

The authors noted that, owing to the construction of the instrument, it was not possible to obtain an extremely high vacuum; values of pressure varying from 20 to 50 μm (millitors) of mercury being the best obtainable. Careful studies showed that a further increase of vacuum would produce no effect other than slight increase of the forces measured. No fundamental changes in the phenomena due to higher vacuum, therefore, seem to us likely. The results of arc force measurements are presented in Fig. 11.12. It was observed that the force is produced whether the anode or the cathode is stationary. With the anode moving, the force is much smaller for low currents than with the cathode moving.

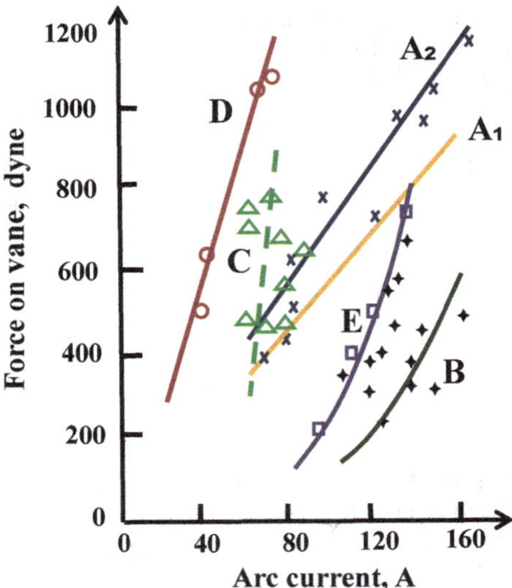

Fig. 11.12 Force on vane as function on arc current with the chamber pressure P (millitors) as parameter: A_1 & A_2-cooper cathode, $P = 50$; B-tungsten anode, $P = 40$. C-carbon cathode, $P = 42$. D-zinc cathode. $P = 56$. E-copper anode, $P = 50$ [34]

Hence, it was explained that the anode force might be overlooked when experiments were made with small currents only. The copper arc was found to be considerably more stable than certain others and thus the curves for copper show less scattering than similar curves for other materials where the arc could not be so readily manipulated owing to reduced stability.

According to the data of Fig. 11.12, the cathode force (in dyne) increases with the arc current. Using these data, the specific force per unit current can now be calculated and the results are given in Fig. 11.13 for different electrode materials. As can be seen, the force per unit current also increases with current but for copper cathode, it is constant and lower than the value measured by Tanberg [20] by about factor two in low range of currents.

Tonks 1936 [35] indicated that measurement of the force at a liquid mercury-arc cathode is problematic due to the force that can be directed not normal to the general

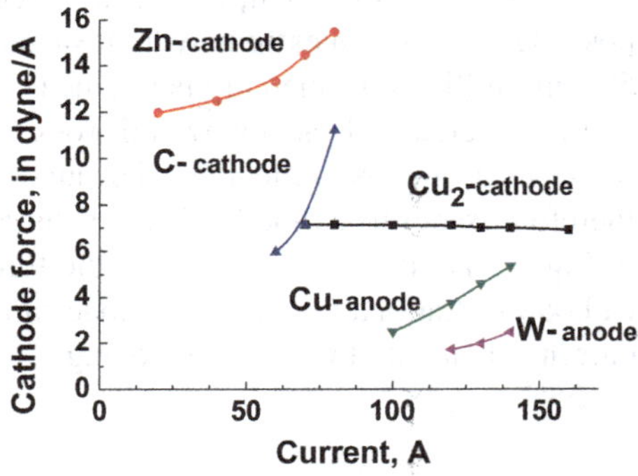

Fig. 11.13 Force per unit current as function on arc current for different materials of the cathodes and anodes

liquid level. It can be perpendicular to small portions of that surface, which, due to the violent inclination, make considerable angles with the horizontal direction. Confining the spot to a small cathode area of mercury surface [29] may increase the pressure through the constriction of the arc path above the spot.

Tonks studied the cathode force in mercury arc with anchored cathode spot by two methods. One was based on the depression of the meniscus edge carrying the cathode line, due to the pressure on it. Direct observation of the cathode line formed at the junction of mercury with a vertical Mo strip immersed in mercury showed that as the arc current was increased, the meniscus edge and cathode line were depressed. By measuring the lowering of the line, as the current was varied, Tonks determined the force per cm at the cathode line using theoretical model of a meniscus shape.

The second experiment used a torsion pendulum to measure the total force, which includes not only that exerted directly on the exposed anchor surface, but also the force of the first method. This was demonstrable either on the general ground of momentum balance or by an analysis of the hydrostatic pressures on the two sides of the anchor.

It has been noted that the measured force was a horizontal component. The direction was observed, in applying the magnetic field to the cathode line. When the field direction was turned from the horizontal through the current flow direction to parallelism with the anchor, the displacement of the cathode spot reversed. At some position of the field intermediate between these two extremes, the cathode line was not displaced. It is expected that the line of action of the mechanical force on the cathode coincide with the mean direction of current flow near the cathode. By deflecting this current, the magnetic field creates a force component parallel to the meniscus edge, which moves the cathode line to one side or the other.

The cathode line will be unaffected only when the magnetic field lies at angle in direction coincident with that of the current, and this neutral direction is the direction of the cathode force itself. This resultant force was obtained by measuring the neutral angle and using the horizontal component of the force. The results of the observations are shown in Fig. 11.14. The measured force rises slightly faster than linearly with current. According to these data the ratio of force to the arc current increases from 33 dynes/A at small currents to 42.5 dynes/A at 10 A (horizontal).

The resultant force vary from 35 dynes/A at low currents to 66 dynes/A at 10 A. The resultant force was explained by assuming that each emitting area of the cathode line exerts not only a pressure arising from it alone but also an additional force arising from the interaction of the currents from the neighboring areas and by more concentration of the electron emission at large current.

Robertson 1938 [36] experimentally demonstrated the variation of the force on the cathode as function on a gas pressure in an electric arc. The measurements were conducted as in previous works using a pendulum, which consisted of a duralumin rod supported on steel points resting in steel cups. The copper cathode rod was mounted in an aluminum block at the lower end of the pendulum. At the top was an adjustable counterweight, which controlled the sensitivity. Cathode was surrounded with a quartz sleeve to keep the cathode spot from wandering. The arc was initiated by moving the anode forward into contact with the cathode and withdrawing it via

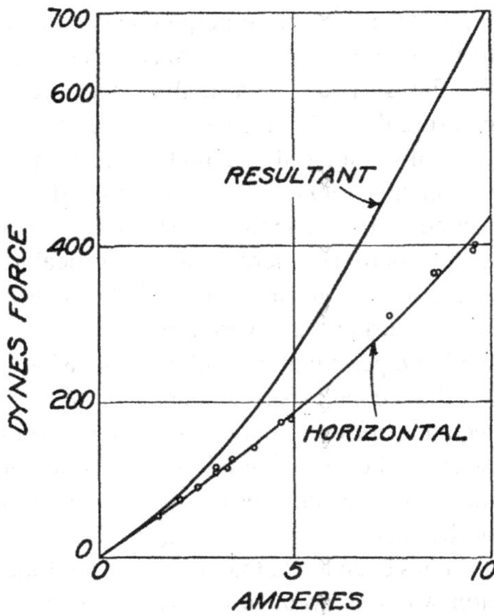

Fig. 11.14 Force dependence on arc current for anchored cathode spot of Hg arc. Figure taken from [35]. Permission number RNP/20/FEB/023162

a special joint. The pendulum deflection due to the arc action was measured with a mirror mounted at the axis of the pendulum and a scale 2.2 m away via an image projected onto a wall.

Measurements made of force as a function of current for different pressures of nitrogen and hydrogen down to 10 torr with the cathode diameter of 6.3 mm and the arc length 5 mm. Note, Duffield et al. [20] found the same order of magnitude as those observed the forces. In hydrogen, the force is greater than in nitrogen. Other measurements were made of force as a function of pressure for the 15 A copper arc in nitrogen with cathode diameter of 7.6 mm and the arc length 10 mm. The force was seen to rise from a value at 10 torr comparable to that observed by Duffield et al. [20] to a force at pressure of 0.3 torr of the order of magnitude of that found by Tanberg in the vacuum arc [27].

The force increased as the pressure in the chamber decreased with a sharp increase occurring as the pressure dropped below 5 torr. The largest force measured was ~15 dyne/A at a nitrogen pressure of 1 torr with currents in the range of 7–20 A. Some measurements were taken of the loss of material per second as a function of pressure. The cathode rod was weighed before and after a 30 s period of arcing at 15 A. The values found are much larger than the value observed by Tanberg [27] who observed a rate of loss at 15 A of about 0.23 mg/s. It was observed that when the arc has a tendency to strike in the crack between quartz and copper, then the force is greatly reduced. This fact indicates that the measured larger force characterizes the arc operation at the metallic cathode. There was no marked change in the appearance of the arc apart from a gradual broadening with decreasing pressure. The arc voltage remained constant at about 20 V throughout the range.

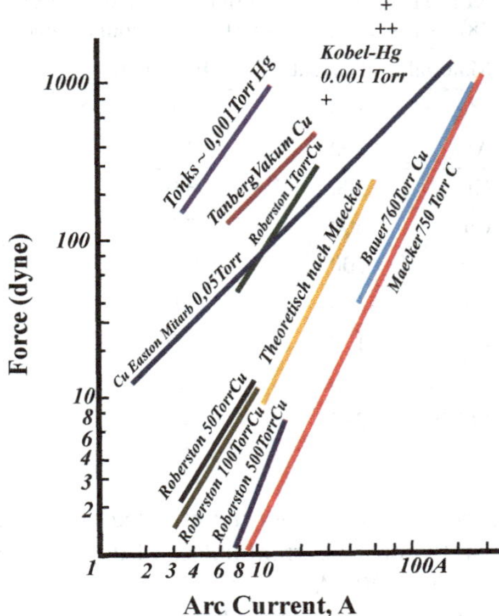

Fig. 11.15 Dependence of the force (dyne) on Hg, Cu, and C cathodes on arc current I in gases of different pressures [37]

Bauer in 1961 [37] reviewed the early studies of the cathodic jet and recoil on the cathode. He summarized the results of the force at the cathode early measured for different surrounding gas pressures and cathode materials. The results presented in form shown in Fig. 11.15, which indicate linear force dependence on arc current and significantly larger values measured in vacuum in comparison to that obtained in gas pressures.

11.4 Further Developed Measurements of the Force at Electrodes in Vacuum Arcs

Plyutto et al. 1965 [38] measured the cathode erosion rate G(g/C) by weighing method and the cathode force by pendulum method in vacuum arc in order to calculate the jet velocity V_j. The force F_c values were not presented, but it can be calculated using relation $F_c = G \times V_j$ (Table 11.10).

The calculated result shows relatively larger forces that were not observed in previously mentioned works that investigated the force of the arc in vacuum.

A sensitive device allowing the measurement of the forces exerted on an electrode by an electric arc is presented by Chabrerie et al. 1995 [39]. The force exerted on the electrode was measured with the help of a pre-stressed piezoelectric transducer. The help of ceramic piece and magnetic screening have insulated this quartz transducer thermally and electrically. This was achieved utilizing axisymmetric geometry, which allows very efficient protection of the quartz. A liquid metal (Hg) allows the arc

Table 11.10 Force for different cathode materials calculated using the measured G and V_j from [38]. Data given in brackets take in account oxidation of the metallic film condensed on the pendulum

Material	Current (A)	Erosion rate (10^{-5} g/C)	Velocity (10^6cm/s)	Force (dyne/A)
Mg	170	2.5	1.5 (0.88)	37.5
Al	300	2.3	0.65 (1.7)	39.1
Ni	300	3.9	0.7 (0.9)	35.1
Cu	300	5.2	7.8 (0.97)	50.44
Ag	300	7.2	0.84	60.48
Zn	300	13.0	0.23 (0.29)	37.7
Cd	170	31.0	0.18	55.8
LC-59	100	3.5	1.1	38.5
LC-59	285	6.3	0.76	47.88

current to flow from the electrode to the generator. The current generator permits one to deliver current pulses of duration from 5 to 200 ms and current peak values from 0 to 20 kA.

In order to insulate the quartz transducer from the vibrations that could be transmitted by the current leads themselves, it was used liquid metal contacts (Hg). Also, the base piece was fixed to a heavy cement block in order to minimize the influence of different vibrations. The used sensitive quartz was able to measure wide range of forces from 10 mN up to 500 N. The calibration of the device was conducted taking into account different force causes including the plasma and droplets flow, electrostatic and Lorentz forces as well as the forces exerted due to liquid metal and at electrode holder for uniform and constricted current distributions.

Force measurements have been realized for various gases (air and argon). The arc current, the arc voltage, and the force exerted on the anode or cathode were registrated. The arc root structure was observed by a high-speed framing camera. The arc attachment zone on the electrode surface is much larger in argon than in air. This fact was used to explain why the value of the force exerted on the surface in argon is weaker than in the case of air. The authors' previous study indicated that the force exerted on Cu electrodes was independent of the electrode gap d in the range 2–18 mm and for arc current in the range 2–10 kA.

The time dependent current pulse and measurement of the time dependent force at Ag electrodes [39] are re-presented in form of Fig. 11.16a, b, respectively. The force dependence is in accordance with the arc dependence on time, i.e., it is increases with current. The measured forces on the electrodes [39] have shown that the force on an anode is higher than the force on a cathode for Ag, in air at atmospheric pressure. Three time dependent current pulses with the same maximum values but different durations and corresponding measurement of the forces at Cu cathode are presented in Fig. 11.17a, b, respectively.

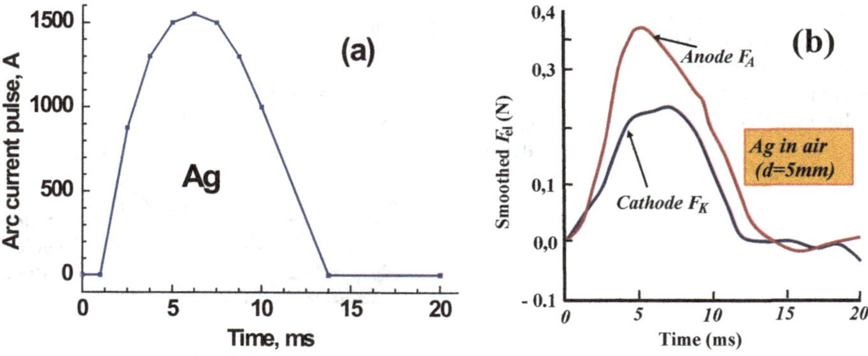

Fig. 11.16 Arc current (**a**) and force (**b**) as function on time for Ag electrode

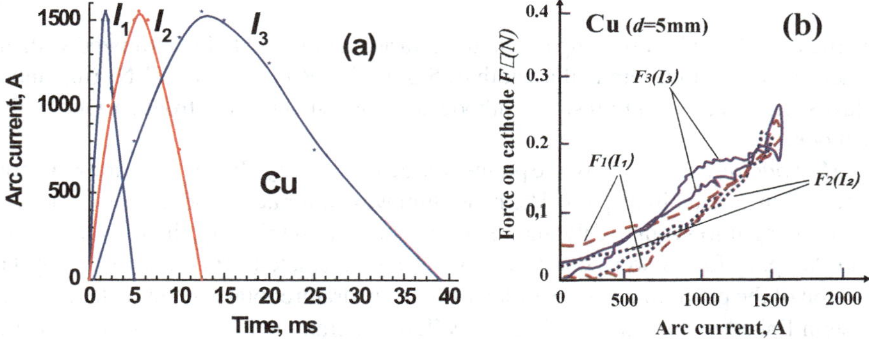

Fig. 11.17 Arc currents (**a**) as function on time for Cu cathode at three different arc pulse durations. Measured forces (**b**) on cathodes versus I^2 for corresponding Cu arc pulse durations

The results in Fig. 11.17 indicate the influence of the arc current pulse duration and of the value of dI/dt on the forces exerted by the arc on the cathode. As it can be seen this influence can be not monotonic and unique. The maximal force for Ag anode is about 0.37 N or 24.7 dyne/A, and Ag cathode is 23 N or 15.33 dyne/A at ~1.5 kA. The maximal force for Cu cathode is about 0.22 N or 14.7 dyne/A. It is significantly larger than that reported in above-described works for atmosphere arcs, but it can be understood due to substantially larger arc current.

Marks et al. 2009 [40] measure the force from the integral flux of the plasma jet from Cu and Al cathodes in a dc vacuum arc as a function of current by pendulum method. The arc was initiated by an "explosive fuse wire" technique. Images of each plasma arc, as well as the pendulum motion, were recorded using a digital camera. The vacuum chamber (530 mm length and 200 mm radius) was pumped down to a pressure of approximately (2–5) $\times$ 10^{-5} torr.

The electrode configuration that maximized the free expansion of the flux in the axial direction was determined and shown schematically in Fig. 11.18. The inner

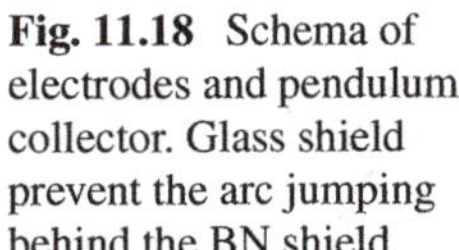

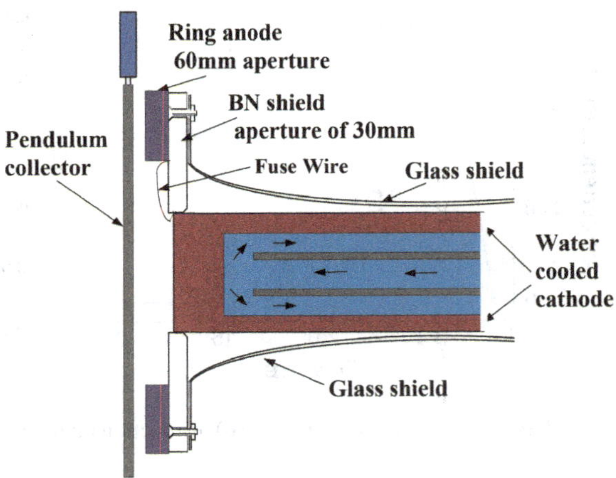

Fig. 11.18 Schema of electrodes and pendulum collector. Glass shield prevent the arc jumping behind the BN shield

diameter and width of the ring anode were varied and found to be optimized with an inner diameter of 60 mm and a width of 8.5 mm. A Boron Nitride (BN) ring and a glass shield were used to restrict cathode spot movement to the front surface of the cathode.

Methodology. The force of the plume was determined by observing its impact upon a pendulum placed in its path. The pendulum was calibrated in both air and vacuum by allowing it to swing freely and record its motion. In this way, the inertia, period, and damping forces were confirmed, verifying calculated properties. The angular motion of the pendulum was recorded using a quadrature rotary incremental encoder system linked to a computer via a LabVIEW program and a data acquisition card. The rotary encoder device was scavenged from a Logitech PS/2 computer mouse and connected to the axle of the pendulum via a coupling wheel, which served to amplify the rotation of the axle.

Teflon was chosen for the bearings as the sliding friction for metal on metal is drastically increased by placing it in vacuum due to the removal of the molecular contact film, which is present in air. The aluminum pendulum collector was at floating potential during arcing. The electrical output from the encoder consisted of two pulse trains, with ~90° phase difference between them. The number of pulses received indicated the angular displacement, and the sign of the phase difference between the pulse trains indicated whether motion was clockwise or anti-clockwise. The equation of motion of the freely swinging pendulum system can be expressed as below [41], additionally taking into account the damping term W due to the resistance of the encoder and contact wheel assembly:

$$I\frac{\partial^2\theta}{\partial t^2} + (W + T_j)\mathrm{sgn}\left(\frac{\mathrm{d}\theta}{\mathrm{d}t}\right) + MgL\theta = 0 \tag{11.5}$$

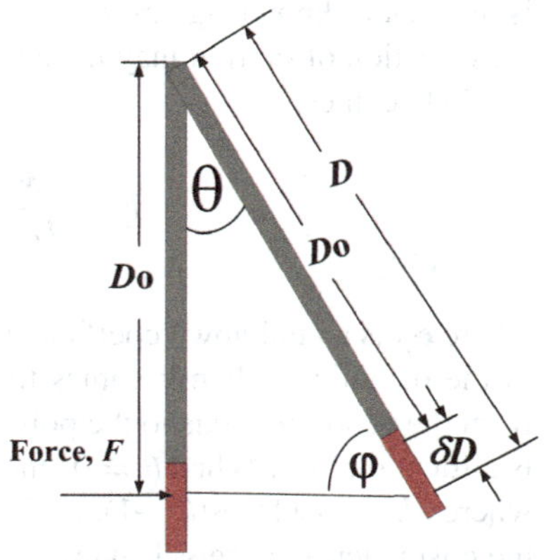

Fig. 11.19 Pendulum, applied force F, and lengths D, D_0 and δD (δD is exaggerated for clarity in the illustration). $D = D_0 + \delta D$

where $I = MR^2$ is the moment of inertia of the system, M is the total mass of the moving system, R is the radius of gyration of the pendulum, and θ is the angular displacement of the pendulum. The opposing torque exerted by the Teflon bearing on the pendulum axle is $T_j = \mu_f Mgr$, where μ_f is the coefficient of friction, and r is the radius of the axle. W is the opposing torque exerted on the axle by the contact wheel. sgn() is the sign function, whose value is -1 or 1, depending upon the direction of the pendulum swing, or 0 if it is stationary, and L is the distance to the center of mass (COM).

The solution of (1) without damping (when W and T_j are zero) is that of a simple harmonic oscillator with constant amplitude. Inclusion of the damping term produces a solution in which the amplitude $\theta(t)$ is reduced at a constant rate, that is, linear damping. The arc plasma exerts a force F that is approximately normal to the pendulum collector at a distance D from the pendulum axle. With deflection from normal at some angle; torque $F \times D$=FD sin(φ) is applied to the pendulum (see Fig. 11.19), and is equivalent to FD cos(θ). So, with a driving force, the situation described by (11.4) becomes:

$$FD\cos(\theta) = I\frac{\partial^2\theta}{\partial t^2} + \left(W + T_j\right)\mathrm{sgn}\left(\frac{\mathrm{d}\theta}{\mathrm{d}t}\right) + MgL\theta \tag{11.6}$$

Rearranging (11.6) we have the expression used to determine the force from the plasma, based on measured $\theta(t)$) data. The distance D from the pendulum axis where on average the plasma struck the collector was determined by measuring the distance from the axle to the center of the nearly circular deposition pattern on the collector.

Equations (11.5)–(11.6) only describe the pendulum in motion. In steady state, sgn(dθ/dt) = 0, which removes the resistive torque terms from the equations. The principal method used to determine the force was based on the data gathered from the initial deflection of the pendulum and (11.5). Steady-state measurements were more uncertain for two reasons. The first stems from the limits of the encoder wheel angular resolution (0.66°). The second cause of uncertainty for steady-state measurements is the lack of knowledge as to the amplitude and direction of the resistive torque, as the direction of motion may have reversed from the last reading. So in steady state, (11.5) becomes:

$$F = \frac{MgL\theta}{D\cos(\theta)} \pm \alpha_p \frac{W + T_j}{D\cos(\theta)} \tag{11.7}$$

where α_p is an unknown coefficient between 0 and 1, which depends upon the state of the pendulum when it comes to rest. In Fig. 11.19, the distance D of the force from the pendulum axis to the point at which it is applied to the pendulum collector is a function of θ. When $\theta = 0°$ the distance $D = D_0$. As θ changes $D = D_0 + \delta D$, where $\delta D = D_0(1/\cos(\theta) - 1)$. If $(D_0 + \delta D)$ is substituted for D in (11.6) and (11.7), the cosine term cancels, leaving a constant denominator of D_0. The value of θ was known at the instant when pulses arrived from the encoder each time the pendulum moved through the fixed angle $\Delta\theta = \theta_{n+1} - \theta_n$. The times at which the pulses arrived at the angular intervals $\Delta\theta = 0.66°$ were recorded by a LabVIEW program. The first derivative was calculated numerically using a central difference scheme

$$\left(\frac{d\theta}{dt}\right)_{n+0.5} = \frac{\theta_{n+1} - \theta_n}{(t_{n+1} - t_n)} \tag{11.8}$$

$\frac{d^2\theta}{dt^2}$ was calculated using the times at the intermediate points at which $\frac{d\theta}{dt}$ was calculated:

$$\left(\frac{d^2\theta}{dt^2}\right)_n = \frac{\frac{\Delta\theta}{t_{n+2} - t_{n+1}} - \frac{\Delta\theta}{t_{n+1} - t_n}}{(t_{n+1.5} - t_{n+0.5})} \tag{11.9}$$

The values of θ_n and the approximation to angular acceleration (11.6) were inserted into (11.6) to calculate the force at successive measured angles. From (11.9) it is clear that the times of three angles must be recorded for the acceleration to be calculated. The first six angles recorded and the respective times at which the pendulum moved through these angles after the arc began were used to calculate the average force on the pendulum.

The pendulum deflection was detected by the encoder and by visual inspection of photographs of the arm during arcing. As example, the photographic image of a Cu cathode, anode, and the pendulum collector during a 200 A arc is represented in Fig. 11.20. The pendulum position is shown through 423 ms after the arc ignition. So, the increasing deflection of the pendulum by the plasma is vividly demonstrated.

Though the arc was kept running for 1–3 s, the force measured initially, before the pendulum reached steady state, was determined in the period up to 0.4 s after the arcs had commenced. The force was measured as a function of current, using the average values of force calculated using (11.6) [40]. The results illustrated in form showed in Fig. 11.21. For both Cu and Al, the force increased linearly with the current. For

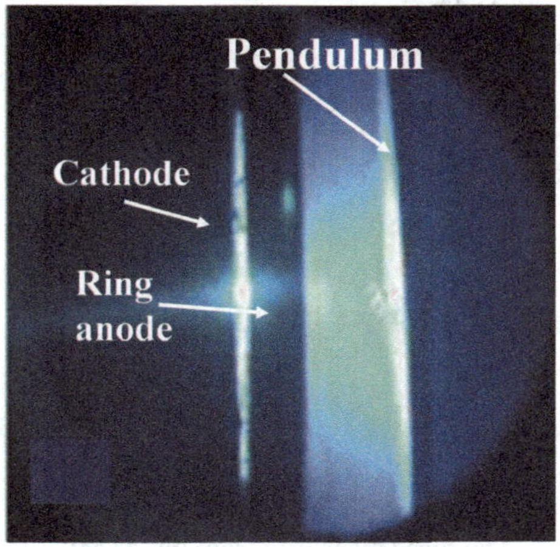

Fig. 11.20 Photographs of a Cu cathode, anode, and the pendulum with arc current 200 A

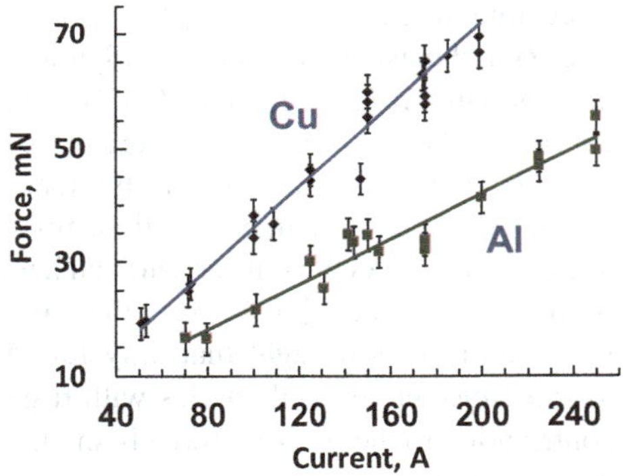

Fig. 11.21 Measured force as a function of arc current for Al and Cu cathodes during DC arcing

Cu, the slope was 0.36 mN/A, with an intercept of 1.37 mN, while for Al the slope was 0.21 mN/A and the intercept 2.2 mN.

Figure 11.22 is a comparison of the force per unit current measured at different times during the arc, initially (the first deflection of the pendulum) and after the pendulum reached steady state. The force per unit current is plotted against the current. It was useful to make the steady-state measurements as a rough verification of the veracity of the principal method, in which the force was measured before the pendulum reached steady-state deflection, despite the higher uncertainty of the steady-state measurements. In Fig. 11.22, the non-steady-state measurements are generally larger than the steady-state measurements for the same current, for currents above 100 A.

This can be partly attributed to the larger angular deflection at higher currents, that is, whereas the initial force was always measured using the same initial six angles, the deflection of the pendulum in the steady state was greater for higher currents, allowing less plasma to impinge upon the pendulum.

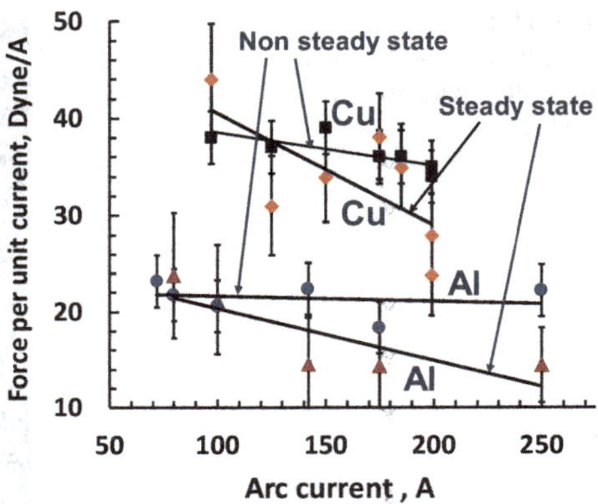

Fig. 11.22 Comparison of initial and steady-state measurements of the forces per unit current for Cu and Al cathodes

The force measured from the Al cathode was lower than from the Cu cathode. This result can be understood taking in account that the Cu mass density (9 g/cm^3) is about three times larger than Al (2.7 g/cm^3). The force is determined by the cathode erosion rates G and plasma jet velocities. Plyutto [38] measured these plasma parameters in dc arcs with Al (23 μg/C and 17×10^5 cm/s) and Cu (52 μg/C and 9.7×10^5 cm/s) cathodes. The force estimated using these values for a Cu cathode exceeds the force from an Al cathode, confirming the tendency measured in the present work. The measurements show a linear relationship between the force from the cathode plasma and arc current. As the ion current (and hence emission of plasma) is a linear function of the arc current [42], it is likely that the cathode force is mainly due to ion current momentum. It is thought that only few macroparticles hit the pendulum because they are ejected at small angles with respect to the cathode plane. Moreover, their contribution to the reaction force is small due to their significant lower velocity than that of the plasma.

The discrepancy between the force measured by Tanberg [27] and that presented in this work might be due to the following reasons: (i) the shape of the anode, (ii) the quartz fitting around the cathode, and (iii) the position of the pendulum. Tanberg's anode consisted of a rod-like contact held 15 mm from the cathode, while in this work a ring anode was used. The effect of the cathode spots concentrating in one region on Tanberg's cathode was much reduced by using a ring anode. Also, the quartz fitted around the cathode can influence the cathode spot location by its evaporation [30]. In the present work the collection of the plasma jet was improved as the pendulum was positioned closer (10 mm) than that used by Tanberg (20 mm). Finally, Tanberg's square pendulum vane had a width of 14.5 mm, while in the present work, the pendulum collector radius was 49.5 mm. The results presented did not take into account reflection of material from the pendulum collector, which would cause the measured force to be larger than the reaction on the cathode [43–45].

11.5 Early Mechanisms of Forces Arising at Electrodes in Electrical Arcs

The early works before 1929 indicated that existence of a repulsion of the electrodes was due to relatively large specific pressure generated near the electrodes. The relatively large force observed in vacuum arc was explained by Tanberg [27] assuming that large energy of the plasma flowed from the cathode was realized due to vaporization with extremely large cathode temperature (10^5 K). However, the later work of Tanberg and Berkey 1931 [28] experimentally showed that the cathode temperature was significantly low (~3000 K). Wellman [46] considered the reaction of atom flow leaving the cathode due to evaporation and gas molecule flow desorbed at cathode temperature of 3000 K using the relation between rate of metal evaporation and vapor pressure. It was obtained the same order of magnitude for both vapor stream and a gas stream and that cannot to be a mechanism explained the above mentioned experiment.

Compton 1931 [47] also indicate the physical difficulties of the interpretation based on metal evaporation. The inconsistency is between the possible temperature and the observed rate of evaporation, which supported a surface temperature of not more than 200 °C, as example for mercury arc. Therefore, not all evaporated cathode mass leaved the surface and a part of the flux returned to it. Compton suggested that a more reasonable interpretation of this pressure is to be found in the existence of an "accommodation coefficient" for ions, which accelerated in the cathode potential fall layer, strike and were neutralized at the surface. By means of an accommodation coefficient, the incoming copper ions still retain a fraction of their kinetic energy. These neutralized and energetic ions then rebound from the cathode with a random distribution of velocity forming the vapor stream suggesting a mechanism to account for the force of reaction upon the cathode. Compton's calculation showed agreement with measured result obtained by Kobel [29].

The approach of "accommodation coefficient" was discussed in [48, 49]. Experimentally, the momentum transfer to an auxiliary cathode has been studied in the positive column of a low-voltage helium arc [48]. The auxiliary cathode was a flat molybdenum plate insulated on one side by glass and suspended in such a way that its deflection gave a measure of the pressure against it. The measured pressure on the cathode is believed to be due to two phenomena. The first is the recoil of those ions, which retain some of their kinetic energy after neutralization. The second is a radiometer effect due to the heating of the cathode by positive ion bombardment. Based on these assumptions it was possible to calculate from the experimental data (pressure on the cathode) an accommodation coefficient for helium positive ions and the fraction of the measured current carried by electrons. It was noted that the accuracy of the experiment and obtained result was not high.

Slepian and Mason 1931 [50] indicated that Compton's suggestion that high-velocity neutral molecules leave the cathode as a result of the energetic positive ions neutralization cannot be accepted for explain the Tanberg's result, because the velocity of these ions cannot as great as the velocity obtained by Tanberg [27]. The maximum velocity of the neutralized ions cannot be expected to be much greater than that corresponding to the cathode potential drop, whereas Tanberg observed velocities of an order corresponding to 70 V. Also, the average energy of the all neutralized ions that participated in the reaction should be need to be less than 0.4 V to give the force measured by Tanberg [27].

Risch and Ludi [51] assumed that high-energy plasma flow can be realized due to potential energy of multiply ions by their neutralization at the cathode surface causing the observed force. Their assumption based on data of the ionizing potentials for the formation of Hg^+, Hg^{2+}, Hg^{3+}, H^{4+}, and Hg^{5+} by an electron impact at 10.4, 30, 71, 143, and 225 V, respectively, recently measured by Bleakner [52]. However in Bleakner's work (see here Fig. 11.4) even the fraction of Hg^{2+} is negligible at electron energy of 35 V that significantly lower than the cathode potential drop in a vacuum arc [2].

Tonks 1934 [53] developed a model where the electron pressure in a cathode spot plasma was considered to explain the cathode-force reaction. It was taken in account a momentum change due to the rebounding of plasma electrons from the positive ion sheath covering a negatively charged electrode. According to the kinetics of electrons with Boltzmann distribution the momentum must be appeared as a pressure $p = nkT_e$. Since the force cause the velocity reversal is the repulsion between the negative charge on the electrode and the electron, this pressure is exerted on the electrode. The plasma density n was obtained from a relation for ion current density j_i flowed to the cathode (later obtained by Bohm [54]):

$$j_i = K_{\text{Tonks}}\,\text{en}\left(\frac{2kT_e}{m}\right)^{0.5} \tag{11.10}$$

$$j_i = j(1-f) \tag{11.11}$$

where K_{Tonks} is the constant of Tonks, j is the current density, and f is the electron current fraction. By multiply of both sides of (11.10) and with (11.11), the following expression for the force was obtained [53]:

$$\frac{2nkT_e}{j} = \frac{(1-f)}{eK_{\text{Tonks}}}\left(\frac{kmT_e}{2}\right)^{0.5}\ \text{dyne/A} \tag{11.12}$$

Taking $T_e = 5 \times 10^4$ K in the plasma outside the cathode spot and $f = 0.9$ for mercury arc in (11.12), Tonks found the force of 61 dynes/A. This is to be compared with the average value 40 dynes/A calculated from Kobel's measurements [29] on the Hg arc. It should be noted that Tonks model [53] is actually up today, see Chap. 18.

Later, Robson and von Engel [55] have again attempt to understand the Tanberg's experiment. They indicated that the estimation of electrode temperature of 5×10^4 K has caused some concern. Their opinion is based on two questions including those who disbelieve in Newton law and those who question is accuracy of the experiment. A novel explanation was taken into account that some of evaporated atoms were ionized and returned to the cathode by electric field (like Compton [47]). Because of the dense vapor, the positive ions make many collisions and transfer the momentum to the neutrals. As result, the neutrals also returned to the cathode. So, the ions move through dense vapor a large fraction of this momentum was transferred by the returned vapor to the cathode. The balance between mass flux leaving the cathode surface and returned to it determines the force of reaction.

Using equation of particle conservation with returned particle flux and measured cathode force, Robson and von Engel estimated that the particle velocity should be 10^4 cm/s rather than 10^6 cm/s. This result seems not understandable because the measurement by Tanberg [27] cathode mass loss of course reflected all near-cathode phenomena including the returned particle fluxes. Loeb [56] also indicated about a difference between the number of atoms "evaporated" and those which actually escape into the vapor stream, since a large number of atoms is returned to the cathode as positive ions.

Reece [57] commented the work of Robson and von Engel [55] indicating that it was exceedingly difficult to reconcile this suggestion due to the results Easton et al. [34]. They [55] obtained velocities, which agreed fairly well with Tanberg's, by a method which did not directly involve the cathode, but measured the force on vane facing the cathode some two centimeters away from it. The authors noted, while the nature of the experiments is such that the results cannot be expected to be highly accurate there seems correct the order of magnitude and that vapor is in fact ejected from the cathode spots in vacuum.

Nevertheless, a special experiment was conducted by von Engel and Arnold [58] using a radioactive copper cathode and axial magnetic field or rotating drum method to stabilize the cathode spot. Using a Geiger–Muller counter, it was also detected particles with velocity not larger than 10^4 cm/s.

It was noted in [58] that Tanberg's results can be interpreted by taken in account a large aggregates (droplets) which transfer momentum to the vane but fail to stick on it. Hence, the measured rate of change of mass, found by weighing the vane, is much too small, which may give the apparent large average speeds. This explanation was based on a fact that the large aggregates passed through a collimating slit, which collected in the experimental system [58]. The authors also indicated that magnetic stabilization should not reduce the average velocity by two orders of magnitude. However, it seems unlikely that momentum of the droplets can be so much influence due to their very low velocities and comparable mass loss in the total erosion rate (see above). Thus, the jet velocity measured by Engel and Arnold remains unexplained even if their conclusion related only to the neutral atoms in [58].

Recently, Makrinich et al. [59, 60] studied the momentum delivered to a mixed ion-neutral flow by an electric force, which enhanced if ions collide with neutrals during the acceleration (similar to model of Robson & von Engel for vacuum arc [55]). This enhancement has been demonstrated for a configuration of a Radial Plasma Source. The momentum flux, carried by the mixed ion-neutral magnetized plasma flow after the flow has crossed the electric potential drop, was determined by measuring the force that the flow exerted on a balance force meter. The enhancement of the (assumed electric) force was shown to be proportional to the square root of the calculated number of collisions with neutrals an ion experiences as it crosses the potential drop.

The plasma density, electron temperature, and plasma potential were measured at various locations along the flow. These measurements were used to determine the local electric force on the ions. The total electric force on the plasma ions was then determined by integrating radially the local electric force. In parallel, the momentum flux of the mixed ion-neutral flow was determined by measuring the force exerted by the flow on a balance force meter. The maximal plasma density was between 6×10^{10} cm^{-3} and 5×10^{11} cm^{-3}, the maximal electron temperature was between 8 and 25 eV, and the deduced maximal electric field was between 2200 and 5800 V/m. The force exerted by the mixed ion-neutral flow on the balance force meter agreed with the total electric force on the plasma ions. This agreement showed that it is the electric force on the plasma ions, which is the source of the momentum acquired by the mixed ion-neutral flow in a discharge for configuration of a Radial Plasma Source [60].

11.6 Summary

The above review shows that the force in electrical arcs firstly observed in the end of nineteenth centurion as repulsion of the planar electrodes and generated by the vapor pressures of the electrode materials. The force on each separate electrode can be positive or negative depending on larger or lower vapor pressure of the opposite electrode. When the electrode gap was increased, the vapor flow produce, so named, vapor stream. It was observed cathode and anode streams. As the electrical arc was investigated in air at atmosphere pressure, the forces were measured relatively small (up to about 10 dyne) and the streams have moderate velocities, in range of 10^3–10^4 cm/s. The investigations of the electrode vapor flow in experiments with air arcs have been conducting relatively widely up to 1929. The important issue was establishing of an effect of accelerated vapor stream that in present, named as plasma jet (see Chap. 12), previously, unknown term). In essence Dewar (1879–1891) first who found this effect as "transit of the vapor material from electrode to electrode." The obtained data were very important to understand, not only qualitative but also quantitative, the role of electrode material losses in the formation of large pressures near the electrodes, generate forces and the streams (jets) with relatively not small velocities even in a surrounding gas of atmosphere pressure.

Distinctive feature results of these parameters were observed investigating the electrical arc in vacuum provided mainly in period of first decade after 1930. Tamberg in 1930 measured significantly large cathode force in vacuum arc. This work caused series new measurements mostly confirmed his data and some discussions of different interpretation of the results. Robson and Engel have been discussed this work titling it as "Tamberg effect" having in mind the effect of discrepancy between observed cathode force and velocity of the cathode jet calculated by Tanberg using his measured cathode mass loss. Afterwards, other authors used term "Tamberg effect" in their further publications understanding it as effect of cathode jet firstly found by Tanberg that is misperception in light of above description (studied by Dewar).

Robertson's experiment show wide variation of the force on the cathode from that, previously obtained in surrounding gas, to the values measured later in a vacuum. Namely, the force rise from a low value at in nitrogen pressure of 10 torr comparable with that observed by Duffield, Burnham, and Davis (1920) to a force at pressure of 0.3 torr of the order of magnitude of that found by Tanberg (1930) in the vacuum arc. The modern data are indicated on 40–60 dyne/A depending on arc conditions.

Discussed above, the early models developed to explain the force mechanisms indicate the complexity of the problem and that an unit point of view is absent. It is clear that the force F is a result of plasma jet reaction and therefore depends on the rate of cathode mass loss G (g/s) and the jet velocity V_j and expressed as $F = GV_j$. Therefore, an understanding of the force origin is related to the nature of the plasma jet, which will be considered in the next Chaps. 12 and 18.

References

1. Ecker, G. (1961). Electrode components of the arc discharge. *Ergebnisse der Exakten Naturwissenschaften, 33,* 1–104.
2. Kesaev, I. G. (1964). *Cathode processes in the mercury arc.* NY: Consultants Bureau.
3. Kesaev, I. G. (1968). *Cathode processes in electric arcs.* Moscow: NAUKA Publishers. (in Russian).
4. Rakhovsky, V. I. (1970). *Physical bases of the commutation of electric current in a vacuum.* Translation NTIS AD-773 868 (1973) of physical fundamentals of switching electric current in vacuum. Moscow: Nauka Press.
5. Lafferty, J. M. (Ed.). (1980). *Vacuum arcs. Theory and applications.* NY: Wiley.
6. Handbook of vacuum arc science and technology, edited by R. L. Boxman, P. J. Martin, and D. M. Sanders, Noyes Publ. Park Ridge, N.J. (1995).
7. Anders, A. (2008). *Cathodic arcs: From fractal spots to energetic condensation.* Springer.
8. Dewar, J. (1879). Formation of hydrocyanic acid in an electrical arc. *Proceedings of the Royal Society of London, 29,* 188–189.
9. Dewar, J., & Scott, A. (1879). On the vapor densities of potassium and sodium. *Proceedings of the Royal Society of London, 29*(196–199), 206–209.
10. Dewar, J., & Scott, A. (1879). Further experiment on the vapor densities of potassium and sodium. *Proceedings of the Royal Society of London, 29*(196–199), 490–493.
11. Dewar, J. (1880). Studies on the electric arc. *Proceedings of the Royal Society of London, 30*(1), 85–93.
12. Dewar, J. (1881). Manometric observations in the electric arc. *Proceedings of the Royal Society of London, 33*(1), 262–266.
13. Feddersen, W. (1861). Uber die electrische funkenentladung. *Annalen Der Physik Und Chemie, 113*(3), 437–467.
14. Schuster, A. (1887). Experiments on the discharge of electricity through gases. *Proceedings of the Royal Society of London, 42,* 371–379.
15. Schuster, A., & Lecture, B. (1890). The discharge of electricity through gases. *Proceedings of the Royal Society of London, 47*(286–291), 526–561.
16. Schuster, A., & Hemsalech, G. (1898). The constitution of the electric spark. *Proceedings of the Royal Society of London. Series A, Mathematical and Physical Sciences, 64*(1), 331–336.

17. Duddell, W. (1904). On the resistance and electromotive forces of the electric arc. *Philosophical Transactions of the Royal Society of London. Series A, Containing Papers of a Mathematical or Physical Character, 203*, 305–342.
18. Geoffrey Duffield, W. (1915). The consumption of carbon in the electric arc. I. Variation with current and arc-length. II. Influence upon the luminous radiation from the arc. *Proceedings of the Royal Society of London. Series A, Containing Papers of a Mathematical and Physical Character, 92*(636), 122–143.
19. Fleming, J. A. (1890). On electric discharge between electrodes at different temperatures in air and in high vacua. *Proceedings of the Royal Society of London, 47*(286–291), 118–126.
20. Duffield W. G., Burnham, T. H., & Davis, A. H. (1920). The pressure upon the poles of the electric arc. *Philosophical Transactions of the Royal Society of London. Series A. Mathematical or Physical Character, 220*, 109–136.
21. Tyndall, A. M. (1920). On the pressure on the poles of an electric arc. *Philosophical Magazine, 40*, 780–781.
22. Beer, H. E. G., & Tyndall, A. M. (1921). Manometric observations at the poles of the electric arc. *Philosophical Magazine, 42*(252), 956–971.
23. Tyndall, A. M. (1921). On the forces acting upon the poles of the electric arc. *Philosophical Magazine, 42*(252), 972–981.
24. Sellerio, A. (1916). Effetto di repulsione nell'arco electric. *Nuovo Cimento, 11*(1), 67–86.
25. Sellerio, A. (1922). Repulsive effect upon the poles of the electric arc. *Philosophical Magazine, 44*, 765–777.
26. Tanberg, R. (1929). Motion of an electric arc in a magnetic field under low gas pressure. *Nature, 7*(124 new volume), 371–372.
27. Tanberg, R. (1930). On the cathode of an arc drawn in a vacuum. *Physical Review, 35*(9), 1080–1090.
28. Tanberg, R., & Berkey, W. E. (1936). On the temperature of cathode in vacuum arc. Phys. Rev., 38, 296–304, (1931). L. Tonks. The Force at an Anchored Cathode Spot. *Physical Review, 50*, 226–233.
29. Kobel, E. (1930). Pressure and high velocity vapor jets at cathodes of a mercury vacuum arc. *Physical Review, 36*(11), 1636–1638.
30. Berkey, W. E., Mason, R. C. (1931). Measurements on the vapor stream of the cathode of a vacuum arc. *Physical Review, 37*(12), 1679.
31. Berkey, W. E., & Mason, R. C. (1931). Measurements on the vapor stream of the cathode of a vacuum arc. *Physical Review, 38*(5), 943–947.
32. Mason, R. C. (1933). High-velocity vapor stream in the vacuum arc. *Transactions of the American Institute of Electrical Engineers (A.I.E.E. Trans.), 52*(1), 245–248.
33. Creedy, F., Lerch, R. O., Seal, P. W., & Sordon, E. P. (1932). Forces of electric origin in the iron arc an explanation of overhead welding. *Transactions of the American Institute of Electrical Engineers (A.I.E.E. Trans.), 51*(2), 556–563.
34. Easton, E. C., Lucas, F. B., & Creedy, F. (1934). High velocity streams in the vacuum arc. *Electrical Engineering, 53*, 1454–1460.
35. Tonks, L. (1936). The force at an anchored cathode spot. *Physical Review, 50*, 226–233.
36. Robertson, R. M. (1938). The force on the cathode of a copper arc. *Physical Review, 53*(7), 578–582.
37. Bauer, A. (1961). Zur Feldbogentheorie bei kalten verdampfenden Kathoden I. *Zeitschrift für Physik, 164*(5), 563–573.
38. Plyutto, A. A., Ryzhkov, V. N., & Kapin, A. T. (1965). High speed plasma streams in vacuum arcs. *Soviet Physics—JETP, 20*, 328–337.
39. Chabrerie, J. P., Devautour, J., Gouega, A. M., & Teste, Ph. (1995). A sensitive device for the measurement of the force exterted by the arc on the electrodes. *IEEE Transactions on Components, Packaging and Manufacturing Technology. Part A, 18*(2), 322–328.
40. Marks, H. S., Beilis, I. I., & Boxman, R. L. (2009). Measurement of the vacuum arc plasma force. *IEEE Transactions on Plasma Science, 37*(7), 1332–1337.

41. Simbach, J. C., & Priest, J. (2005). Another look at the damped physical pendulum. *American Journal of Physics, 73*(11), 1079–1080.
42. Kimblin, C. W. (1973). Erosion and ionization in the cathode spot regions of vacuum arcs. *Journal of Applied Physics, 44*(7), 3074–3081.
43. Zalucki, Z., & Kutzner, J. (1976). Particles reflection from the anode surface in vacuum arc. In: *International Conference on Electrical Contact Phenomena* (pp. 178–182), Tokyo, Japan.
44. Daalder, J. E. (1977). Reflection of ions in metal vapour arcs? *Journal of Physics D: Applied Physics, 10*(7), 87–90.
45. Boxman, R. L., & Goldsmith, S. (1990). Characterization of a 1 kA vacuum arc plasma gun for use as metal vapour deposition source. *Surface & Coatings Technology, 43*(44), 1024–1034.
46. Wellman, R. (1931). Reactiondue to gas molecules leaving the cathode of an arc. *Physical Review. Letter to Editor, 38,* 1077.
47. Compton, K. T. (1931). On the theory of the mercury arc. *Physical Review, 37*(9), 1077–1090.
48. Lamar, E. S. (1933). Momentum transfer to cathode surfaces by impinging positive ions in a helium arc. *Physical Review, 43*(3), 169–176.
49. Compton, K. T., & Lamar, E. S. (1933). A test of the classical "Momentum Transfer" theory of accommodation coefficients of ions at cathodes. *Physical Review, 44*(5), 338–344.
50. Slepian, J., & Mason, R. C. (1931). Highvelocity vapor jet of cathode in vacuum arc. *Physics Review. Letter to Editor, 37*(6), 779–780.
51. Risch, R., & Ludi's, F. (1932). Die Entstehung des Strahles schneller Molekiile an der Kathode eines Lichtbogens. *Zeitschrift für Physik, 75*(11/12), 812–822.
52. Bleakner, W. (1930) Probability & critical potentials for the formation of multiply charged ions in Hg vapor by electron impact. *Physical Review, 35*(2), 139–148.
53. Tonks, L. (1934). The pressure of plasma electrons and the force on the cathode of an arc. *Physical Review, 46*(9), 278–279.
54. Bohm, D. (1949). *The characteristics of electrical discharges in magnetic field* (A. Guthry & R. K. Wakerling, Eds.). New York: McGraw-Hill.
55. Robson, A. E., & von Engel, A. (1957). An explanation of the Tanberg effect. *Nature, 179*(4560), 625.
56. Loeb, L. B. (1939). *Fundamental processes in electrical, discharge in gases* (p. 633). New York: Wiley.
57. Reece, M. P. (1957). Tanberg effect. *Letters to Nature, 100*(4598), 1347.
58. von Engel, A., & Arnold, K. W. (1962). Fast neutral particles from arc cathode. *Physical Review, 125*(3), 803–804.
59. Makrinich, G., & Fruchtman, A. (2014). The force exerted by a fireball. *Physics of Plasmas, 21*(2), 023505.
60. Makrinich, G., Fruchtman, A., Zoler, D., & Boxman, R. L. (2018). Electric force on plasma ions and the momentum of the ion-neutrals flow. *Journal of Applied Physics, 123*(17), 173302.

Chapter 12
Cathode Spot Jets. Velocity and Ion Current

Analysis of the experimental works published previously shows the presence of relatively large forces and as result generation of the high-velocity jet. These works mainly investigated the electrode material loss as vapor flow and some separate works considered also the presence of the ions. On other hand, the electrode material flow is a highly ionized complicated structure, which consists of a multi-charged and represented a supersonic plasma flux. The data obtained in period from 1930 to present time are important to understand the vacuum arc phenomena. Also, the plasma jet properties characterized by parameters are very useful for practical applications. In this chapter, the results of measurements illustrating mainly data of the plasma jet velocity, ion current, ion charge state, and energy of the plasma jet will be presented in different, respectively, sections as it will be possible to separate the topics (in chronological order), and then main characteristic points will be summarized. Some symbols for same parameter is remained, as it was denoted in the original papers, as example, the symbol Q or Z denoted as ion charge number.

12.1 Plasma Jet Velocity

Let us consider works that mainly studied the plasma expansion and its velocity as it is determined by different methodologies. Lawrence and Frank 1930 [1] determined the velocity of the plasma expansion studying the arc instability with the Kerr cell electro-optical shutter in the early stages (0.5 μs) of sparks between electrodes of Zn, Cd, and Mg. The method is based on phenomena that the metallic spark lines were emitted at a greater distance from the electrodes. This observation gives at once a measure of the average velocity of migration of the metallic ions away from the electrode surfaces. It was found from several such observations that the luminosity of the metallic vapors was spread from the electrodes with speeds of 2.1×10^5, 1.5×10^5, and 1.2×10^5 cm/s for Zn, Cd and Mg, respectively.

I. Beilis, *Plasma and Spot Phenomena in Electrical Arcs*, Springer Series on Atomic, Optical, and Plasma Physics 113, https://doi.org/10.1007/978-3-030-44747-2_12

Table 12.1 Jet velocity against arc current for different metals [2]

	Material	Current (A)	Velocity ($\times 10^6$ cm/s)
Cathode	Ag	166	0.97
	Al	110	0.695–1.84
	Au	53	0.406
	C	73	3.58–1.73
	Cd	43	0.164
	Cu	115	1.25
	Fe	142	0.88
	Mg	55	0.274
	Mo	135	2.72
	Sn	60	0.119
	Zn	57	0.531
Anode	Cu	120	1.27
	Fe	106	1.01
	W	145	1.19

Easton et al. 1934 [2] devoted to measurements of the velocity of the vapor jet from both anode and cathode for number of different materials using the force measurements described in Chap. 11. The anode jet was not noticeable until currents were of over 40–50 A.

The stream velocities, computed on the assumption of a Maxwellian velocity distribution and that all impinging particles adhere to the collector. The results are given in Table 12.1. The authors noted that the jet velocity increases with the melting temperatures of the cathode metals. It was indicated that the jet velocity for Zn, Cd, and Mg cathodes agrees with previous measurements of the work Lawrence and Frank [1].

In order to understand the influence of different materials on the vapor flow velocity in one discharge, Raisky 1940 [3] conducted experiments with cathodes from different alloy materials such as Mg-Zn, Zn-Li, and Li-K. A spectroscopic method was used to obtain the image of the discharge in light of spectral lines of the considered materials. Analysis of the image scan showed no difference in the vapor expansion. Both materials expand with same velocities at all distance. It was indicated that the vapor expansion cannot be as molecular, and it is a hydrodynamic flow.

Method photographs of microsecond spark duration in hydrogen was used by Haynes 1948 [4] to study mercury vapor flow and the jet velocities for different polarities. Spectroscopic examination was based on measuring the time required for the excited mercury atoms in the jet to go a known distance. A mercury pool was used as one electrode in a tube. The spark gaps consist simply of a rod of molybdenum, 1.5 mm in diameter, brought within 7 mm of a mercury pool in a tube containing purified hydrogen at a pressure of 90 cm of Hg. The sparks occurred between the tip of the molybdenum rod and the mercury pool at the rate of several hundred per second. The current in the sparks is a unidirectional pulse having a constant value of a few hundred amperes and a duration adjustable from 0.25 to 5 μs.

It was observed that during the time of discharge, the jets were ejected in a direction always normal to the electrode surface and the spectrograms show that the jets were largely composed of mercury vapor. At the cathode, all parts of the jet move with approximately the same velocity. The velocity decreases with distance (Fig. 12.1), but the initial velocity of the cathode jet was 1.9×10^5 cm/s, and that of the anode jet was 1.5×10^5 cm/s. The initial velocities were independent of current and gas pressure.

Hermoch 1959 [5] used also the optical methods considering both continuous in the time scanning and phase scan to obtain photographs of the vapor formation and to measure it expanding velocity. The validity of the results is shown for the case of a different experimental arrangement. A discharge between plane electrode (cathode or anode) was localized in the hole of the electrically insulated plate. On the plate, there were pasted strips of copper foil-probes to measure the electric field in the discharge, which passed through the hole in the plate. The opposite electrode was tungsten cylinder of 4 mm diameter.

The photographs showed that the ionized vapor flow was represented in form of separate jets (close one to other, apparently, generated by separate spots) expanding up to a front denoted a boundary between of the jets and the surrounding air. The front

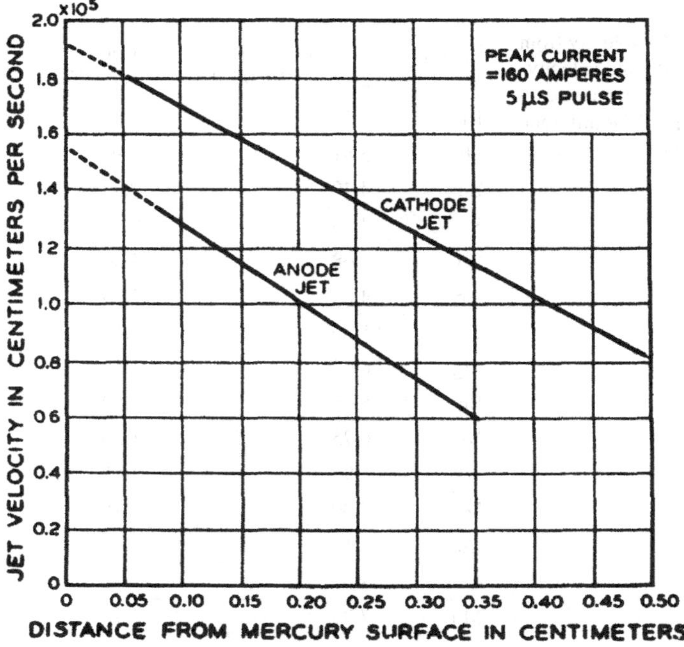

Fig. 12.1 Velocity of cathode and anode jet fronts as a function of distance from mercury surface. Figure taken from [4]. Permission number RNP/20/FEB/023163

velocity weakly increased with plate thickness at constant hole diameter. The current density j at the electrode surface was determined by hole diameter. As example, for Pd anode, the velocity increased from 3×10^5 to 4×10^5 cm/s when j increased from 1×10^4 to 8×10^4 A/cm^2, respectively. The electric field near the electrode depends on discharge time and current density. According to the probe measurements, as example, for Cu anode, the electric field increased from 200 to 700 V/cm when j increased from 2×10^4 to about 10×10^4 A/cm^2, respectively. The flow velocities are given in Table 12.2 (second column) for different electrode metals at maximum discharge current of 2 kA, plate thickness of 1 mm and hole diameter of 2 mm. The author indicate that all measured values for various cathode materials were larger by 10% than that for anodes. It was discussed that the vapor acceleration can be understood by considering the relation between the vapor velocities, estimated energy, and calculated mass losses from the electrodes. It was concluded that the velocity increased when the mass loss decreased. The intensity of electrode vaporization determines the vapor velocity, but the thermal phenomena determine the origin of the electrode material jets.

Optical investigation of the relation between the phenomena at the electrodes and the plasma formation and its development was provided in [6, 7]. Analysis of the photographs showed that the vapor formation occurred at each of the electrodes, the vapor expands axially, and then interacts between one and other producing a plasma cloud in form of disk expanding radially from the gap. The measured velocity

Table 12.2 Jet velocity from an electrode [5] and velocity of radial plasma expansion from the gap [6] for different electrode metals

Material	Velocity ($\times 10^5$ cm/s) 2kA N2, 1959 [5]	Velocity ($\times 10^5$ cm/s) $<$ 10kA N3, 1959 [6]
Bi	3.39	0.85
Pb	3.81	0.95
Zn	4.40	0.95
Sn	4.55	0.90
Cd	4.68	1.00
Ag	4.85	1.25
Co	5.80	1.15
Mg	6.13	1.35
Be	6.65	2.50
Ni	6.75	1.30
Al	6.96	1.65
Fe	7.00	1.15
Cu	7.1	1.55
C	7.6	2.50
Mo	7.7	1.70
W	8.6	1.40
Ta	8.9	1.45

for different numbers of materials is given in Table 12.2 (third column). In accordance with observation, the jet development has discrete character and therefore discrete interaction. It indicated that metallic vapor pushed the background gas and the discharge operated in the vapor of electrode material. However, the jet velocities were observed to be lower than that in a vacuum.

Approximately similar jet velocity was measured by Plyutto et al. [8] for different cathode material using pendulum method to determine the force and the amount of condensed cathode materials on the pendulum surface (see Chap. 11). Previously Plyutto [9] observed ion acceleration in a plasma expansion in vacuum spark of few kA. A very large plasma velocity of 10^7–10^8 cm/s was reported. The plasma velocity was measured directly using time-of-flight method at known distance of the expanding plasma front.

Handel et al. [10] studied the plasma propagation in pulsed discharges of 10 μs duration, with peak current of 70 A in vacuum (10^{-6} torr). The electrode arrangement consisted of silicon carbide cone anode of ~0.3 cm diameter at the base and 1 cm length. This pointed tip was in contact with a flat metal cathode of similar dimensions. Three sets of Langmuir probes were used to measure the drift ion and electron velocities, as well as the density and temperature of the electrons. The ion energy was derived from time-of-flight measurements between two probes. The electron temperature was measured of about 1 eV located at 10 cm from the electrodes, even less than 1 eV was derived from probes 10 and 15 cm distance from the discharge.

The observation shown that ions and electron were drifting at approximately the same velocity. The ion velocity dependence on atom mass was measured as $v \sim m^{-0.5}$. Using this relation, the ion velocity is shown in Fig. 12.2, the calculation (theoretical curve) and the measured data (circle with point). According to these results, the ion velocity decreases from about 3×10^6 cm/s for light Al to about 5×10^5 cm/s for heavy Pb. The density profile with distance r was determined as r^{-2}. The ion density of 10^{12} cm^{-3} was obtained at about 1 cm distance from the discharge.

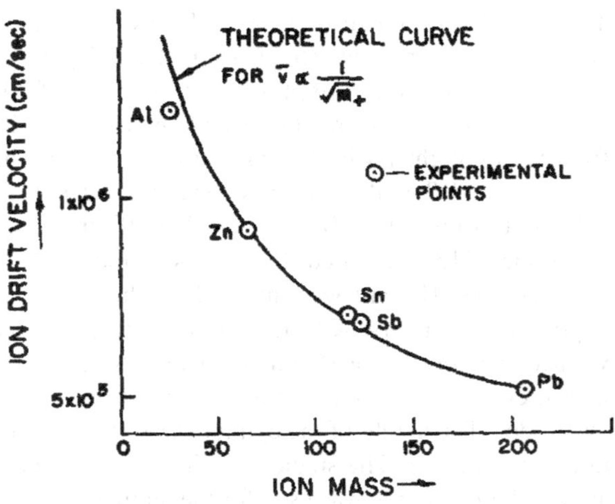

Fig. 12.2 Ion velocity as function on ion mass. Figure taken from [10]. Permission number 4777011054601

A probe method was used by Tyulina [11] to determine directly the velocity propagation of the plasma jet produced during current interruption by contacts in an experimental system similar to circuit breaker in a vacuum. The contacts made of were molybdenum or tungsten of cylindrical configuration. One contact was fixed, while the second could be displaced using electromagnetic control. The probes were placed at various distances from the circuit breaker axis. The jet velocity was determined by simultaneously registration of the oscillograms of the currents in two probes located on one side but at different distances from the axis. The arc voltage was 20 V, the current was in the range of 50–300 A, and the arc duration was varied between 5 and 25 μs.

It was measured the average velocity of ion propagation was of (1.3–2) $\times$ 10^6 cm/s at the front of a plasma at different interruptions of the current. The velocity of electrons can be equal to the ion velocity as well as the electron velocity can exceed the ion velocity by not greater than 1.5 times. The plasma velocity increased by 1.5 times when the interrupted current increases from 50 to 300 A. The ion current at discharge axis decreases inversely with square of distance at the axis during 400 μs. The ion density of 10^{12} cm^{-3} at distance of 1 cm was obtained using measured plasma propagation velocity and probe current. It was indicated that the energy of the measured high-speed flow several times larger than the gap voltage also was observed previously. The observed plasma velocity cannot be explained by thermal velocity of the ions and by magnetic field of the arc since the Lorenz force does not change the absolute value of the velocity.

The probe method was prolonged up to currents 1000 A investigating the plasma flow at current shut off [12]. According to the observations, when the contact was switched off by their opening in the vacuum, the velocity of the plasma propagations, measured at fixed distance, drops sharply downstream of the front of the expanding plasma cloud in the first 100–400 μs of the arc. The most abrupt drop in the rate downstream of the front was observed when the current was close to the value of 1000 A. The velocity of plasma propagation at the termination of the arc varied with time and current in the same manner as the velocity behind the plasma cloud front. These results are shown in Fig. 12.3. It was shown that the rate of plasma propagation across the front and downstream of the front depends on the level of current shut off as follows. The velocity of plasma ions across the front is only slightly dependent on the current in range from 100 to 1000 A (curve 1). The velocity behind the plasma front varies slightly with the current in case when the current shut off varies from 30 to 300 A. When the current stepped up to 1000 A, it was observed a drop in the velocity to values lying below the 1.3 $\times$ 10^5 cm/s (curve 2).

Rondeel [13] measured the average velocity of ions emitted from a copper arc in a vacuum. The arc was initiated by the contact separation with a velocity of approximately 1 m/s. All measurements were carried out with a DC current and with copper contacts of 15 mm diameter. The average ion energy was determined by measurements of the temperature rise of a thin cylindrical copper shield outside the electrodes. The height of the cylinder was 10 mm, the diameter was 25 mm, and the thickness 0.2 mm. The shield was thermally and electrically isolated. A chromel–alumel thermocouple was used for the temperature measurements. At an arc current

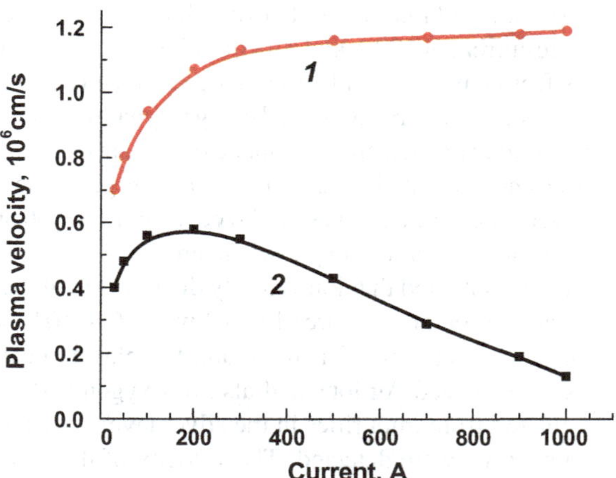

Fig. 12.3 Plasma propagation velocity as function of current shut off: 1-across plasma front; 2-arc mode and termination arc mode

of 450 A and an arcing time of 40 ms, the temperature rise was found to be about 40 K. The thermal time constant of the system was about loos, which was assumed sufficient. It was also assumed that the temperature rise of the shield was caused by the kinetic energy of the ions only which allowed determine the ion current to the shield.

The result shows that both the ion current and the temperature rise of the shield were dependent on the electrode separation, due to the shadowing effect of the anode. The positive ion current increased from 10 to 17 A when the electrode separation increased from 1 to 5 mm at an arc current of 450 A. A relation between the average ion velocity and the average degree of ionization of these ions has been found using the measurements of the temperature rise of a thin copper shield hit by the positive ions, which emitted from a copper–vapor arc, and the ion current to the shield. A second relation is found from the measurement of the impulse gain of a small copper plate hit by the ions. From these two relations, the average velocity was found to be 1.4×10^4 m/s. On the assumption of single ionization of the ions, it was found that 50% of the emitted vapor was ionized.

The velocity distribution function of vacuum arc ions was measured by a time-of-flight technique for wide cathode materials with a pulse length of 250 μs and current in range of 100–500 A [14]. A cylindrical cathode of 6.25 mm diameter with annular anode of inner diameter of 13 mm and the closest cathode-to-anode distance of about 5 mm were used. The distance from the cathode surface to the ion collector was 2.153 m. The measured ion current curves I(t) were used to derive the ion distribution functions using some analytical relation. A peak of ion velocity in region of (0.75–2) $\times$ 10^6 cm/s was found for different cathode materials (C, Mn, Mg, Ti, Cr, V, Ge, Sr, Y, Al, In, Ag, Co, Sm, Er, Ni, Cu, Zr, Mo, Al, Pt, Al, Au), while for Sb the peak is 4 $\times$ 10^6 cm/s. No significant change was found in the current range 100–400 A investigated for Ti cathode with the peak velocity of about 1.6 $\times$ 10^6 cm/s. This same was concluded for obtained ion velocity distributions for various cathode materials, measured at four different arc pulse repetition rates.

The study of ion charge state distribution and ion velocity distribution was reported for arc current of 100 A and 3 μs duration [15]. Two simultaneously operating vacuum arcs from a non-pure silver cathode produce two arc jets that contain Ag ions. The measurements were conducted using a dynamic time-of-flight diagnostics. The anode and cathode of each arc were made of two strips of silver layers pasted on an alumina wafer and separated by a narrow 100 μm gap. The ion velocity distribution of the merged plasma beam was analyzed 70 cm away from the arcs and compared with that of the ions emitted by a single arc.

It was obtained that ion velocity distribution of the plasma of each of the two arcs operating separately of arcs 1 and 2 was 0.6×10^6 and 0.8×10^6 cm/s, respectively. In case of operation of a single arc, the plasma beam consisted of more than 90% of single ionized Ag ions and also of oxygen and carbon ions (about 10%), which originated from impurities in the silver layer. When two arcs are serially connected, two maxima were detected. The velocity of the second maxima was always higher than the velocity observed when each arc was individually operated. Distribution of the plasma ion velocity depended on the connection order and had the following maximal points: velocities 0.67×10^6 and 1.1×10^6 cm/s, in case when the cathode of arc 2 was grounded, and velocities 0.57×10^6 and 1.2×10^6 cm/s, in the other case when the cathode of arc 1 was grounded (see [15]). Furthermore, the total ion charge measured on the ring electrode when the two arcs were serially connected was lower by 30% than the sum of the ion charge collected from each of the arcs separately.

12.2 Ion Energy

Plyutto 1960 [9] observed ion acceleration in an isothermal plasma expansion in vacuum of high-current spark at 7 kA by mass spectrometer raising a small fraction of the ions to energies several time greater than the applied voltage. Later, Plyutto et al. 1965 [8] used Retarding Field Analyzer (RFA) to measure the ion component and ion energy and analyzed of ion charge state in the extracted plasma jet from the cathode. The RFA was placed at 20 cm from the anode, and the arc duration was 1.5–2 s, which was repeated after about 0.5–1 min at 50–70 measurements. The arc current was mainly at 100 A. It was found that the average ion energy per unit ion charge was 5–10 eV for first group of metals Zn, Cd, Pb, and 18–25 eV for second group of metals Mg, Al, Ni, Cu, and Ag. The energetic specter was 0–70 eV. The better approach to Maxwellian velocity distribution was observed at 300 A arc. The voltage fluctuations were mostly detected at 20–100 A, which were significantly lower at 300 A arc.

The mass spectrometer experiment was conducted cathodes placed at distance of 15 cm from the hole of 1 mm diameter through what the plasma was penetrated. The ion current was 200–300 μA at arc current of 100 A. The degree of ionization was calculated using measuring total plasma flux and ion flux deposited on a plate. The observations found a number of multi-charged ions given in Table 12.3. According

Table 12.3 Degree of plasma ionization in percentage and the ion charge state indicated as from Me^{+} to Me^{n+} for different cathode materials [8]

Charge state	Mg^{+}–Mg^{2+}	Al^{+}–Al^{3+}	Ni^{+}–Ni^{3+}	Ag^{+}–Ag^{3+}	Zn^{+}–Zn^{2+}	Cd^{+}–Cd^{2+}	Pb^{+}–Pb^{2+}
Degree of ionization, %	80–100	50–60	60–70	50–60	15–20	12–15	18–25

to the measurements the one charged ions dominated for first group of metals, while two charged ions of about 50%, and three charged ions of about 2% were found for second group of metals, except copper for which was detected ions up to Cu^{4+} (no data were reported about degree of ionization). The absence of three charged ions for Mg was explained by relatively large potential of ionization (79.4 V). Also, the authors noted that the multi-charged ions could be interacted between one with other with changing their state during the plasma flow.

Similar but more detail investigation was provided by Davies and Miller 1969 [16] to measuring the relative flux of the various ions emitted by cathodes of a series of metals and determining energy distribution and charge-state distribution of each ion, all as a function of DC arc current. The used experimental setup consists of an arc tube section to produce the arc, an analyzer to determine the mass, and energy spectrum of the arc products (Fig. 12.4). The arc tube was constructed of 7.6 cm diameter glass tubing. The electrodes were attached to mating flanges by means of 1.27 cm diameter steel rods and used metals.

The arc was initiated mechanically by contacting the two electrodes and then immediately separating them about 0.5 cm. Short arc duration about few seconds limited the minimum current to about 25 A. Currents were limited to about 300 A, which is too low to observe the onset of anode spots. The molybdenum rod anode was used but after a few arcs, the molybdenum was coated with cathode material so that essentially the arc was between the same metals.

The mass and energy analysis of the neutral and ionic species emitted from the arc was accomplished by a double-focusing mass spectrometer with a narrow energy acceptance. Both the cylindrical electrostatic energy analyzer and the magnetic analyzer are 90° sectors with a 5 cm radius of curvature. With an entrance slit of 0.025 cm and an exit slit of 0.051 cm, the energy spectrum from the electrostatic analyzer for an incoming ion beam of uniform energy spread was a trapezoidal peak with a base $\Delta E/E$ of 1.5%. The neutral species from the arc were ionized by electron bombardment. The yield of Cu^{+} obtained by ionizing the neutral Cu from a 100 A arc was about 10^{-13} A at the detector. The yield of Cu^{+} directly from the arc was approximately 10^{4} times greater.

The data were obtained for the copper ions emitted axially by using a hollow cylinder of molybdenum as the anode. The arc tube was constructed so that, the analyzer, anode and cathode were aligned along the same axis. In this type of geometry, a larger proportion of the ions emitted normal to the cathode surface could able to enter the analyzer than in the original geometry. The stability of the signal obtained

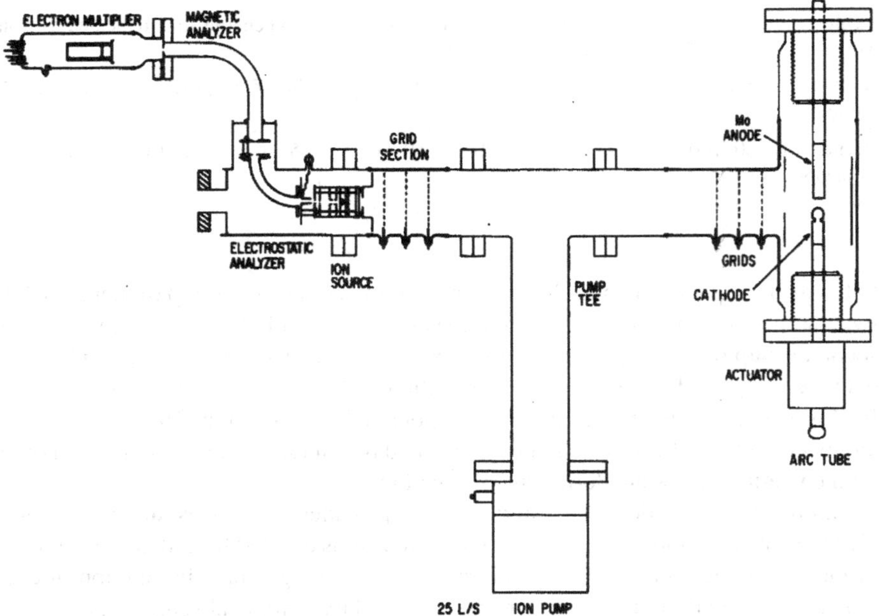

Fig. 12.4 Schematic diagram of arc tube and analyzer. Figure taken from [16]. Permission number 4776971374479

with the axial geometry was considerably poorer than that obtained with the radial geometry. It was caused by a greater tendency of the cathode spots to move off the end of the cathode in the axial case and to be shielded by the anode structure.

An ion energy distribution varied somewhat in shape, but in general was symmetric enough and characterized by three parameters: peak height, energy at which the peak occurred (Ep) and peak defined as full width of the ion-energy distribution at a height equal to one-half maximum height. The spectra were found to be fixed with respect to the electrode potentials. Normally, the cathode was grounded. The peaks of energy distributions per unit charge as dependence on arc current can be presented in the form shown in Fig. 12.5 for the elements investigated for the first three charge states separately. Most measurements were made on the radial flux, but ion-energy distributions of the axial (through anode) flux from a copper cathode are shown in Fig. 12.6 together with radial flux copper results.

In general, the obtained results show that significant quantities of multiply charged ions were measured for all elements examined. The energy distributions (normalizing energy as ion energy/ion charge) for the various ions are similar, peaking at potentials well above the arc voltage. The fraction of ions that are singly charged increases with increasing arc current. For a given element, as the degree of ionization increases the location of the ion-energy distribution peak shifts to lower energies. For a given degree of ionization, the location of the peak tends to lie at higher energies

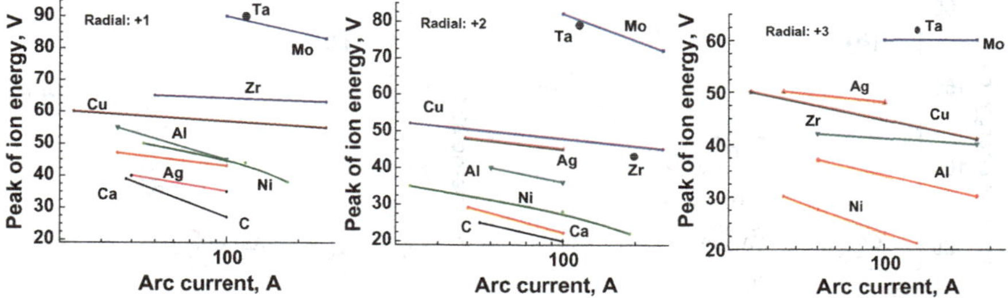

Fig. 12.5 Peak of energy of ion energy distribution as a function of arc current with radial flux for one charge state (Radial 1), two charges state (Radial 2), and three charges state (Radial 3). Energies are given per unit electronic charge in volts

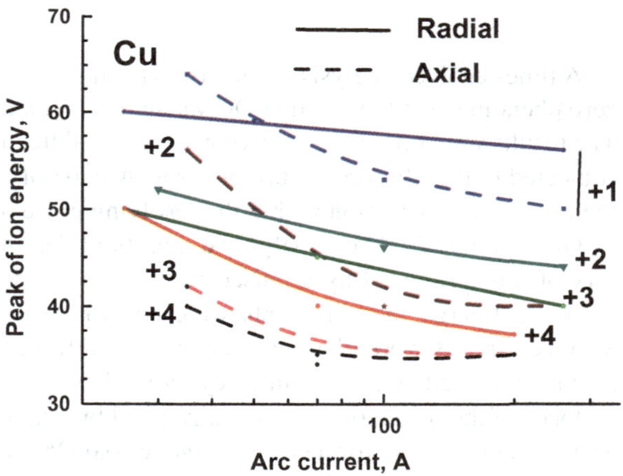

Fig. 12.6 Energy of peak of ion energy distribution as a function of arc current. Energies are given per unit electronic charge in volts. Solid and dotted lines indicate the radial and axial fluxes from copper cathode respectively

for elements with greater arc drops (Chap. 7). This dependence of the ion energy distribution and of the average ion charge, Kutzner and Miller showed in their later work in 1989 [17] as it is demonstrated in Fig. 12.7a, b.

The location of the peak of a particular ion-energy distribution shifts to lower energies as the arc current increases. A weakly difference was observed in the ion energy peak for radial and axial plasma expansion for different arc currents. It obtained a linear increase of the neutral atoms ejected from the arc for silver and copper cathodes with the increase of the current.

Some results of work [16] agree well with that measured by Plyutto et al. [8]. In particular, in this work, it was also shown that the fraction of singly charged ions tends to be larger for elements of lower arc drops, the multiply charged ions decreased noticeably with increasing arc current, and when the arc current increased to 300 A, the average and maximum ion energies decreased. However, the average ion energies are somewhat higher in work [16] than those found by Plyutto [8].

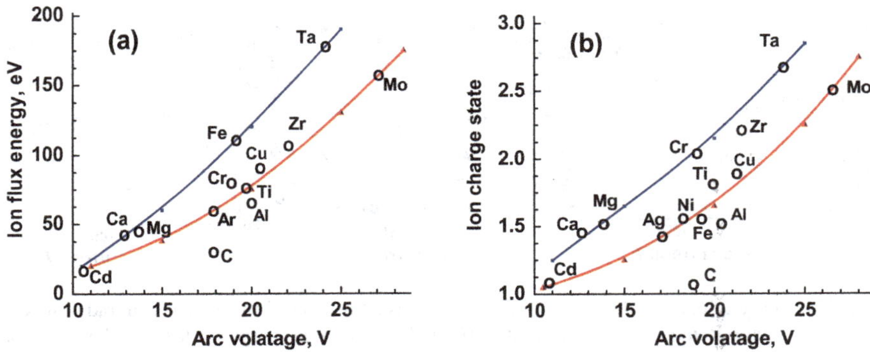

Fig. 12.7 Average ion flux energy (**a**) and average ion charge state Z (**b**) as function of arc voltage for different elements

A time-resolved analysis of the fluxes from vacuum arcs near and through current zero phenomena in low-current DC vacuum arcs ($dI/dt \sim 50$ A/μs) was studied [18]. The results were obtained in looking at the radial flux from 50 and 100 A copper arcs subjected to rapidly forced current zeros. A burst of low-energy ions is produced at the time of arc extinction within the resolving power of used apparatus <5 μs.

Grissom and McClure [19] have measured the energy distributions of the total flux of ions produced by a pulsed vacuum arc of 5 μs duration, with arc current of 60 A. Energy measurements were performed using a cylindrical electrostatic analyzer. In this work, the ions were extracted from the arc plasma over the anode. Al_2O_3 separated two concentric cylindrical electrodes of Kovar. The interfacing surfaces of the insulator were metalized and brazed directly to the metal electrodes. The center electrode formed as the anode (0.0635 cm diameter) and the outer shell was served as cathode. The electrode separation was 0.127 cm. Thus, the anode surface area was defined by a ceramic insulator limited the portion of the anode surface exposed to the arc. The anode was at positive potential, while the cathode was at grounded potential, and therefore, the ion energy measured with respect to the cathode potential. It was noted that in previous works the ions were extracted from the cathode plasma.

Spectroscopic analysis of the plasma indicated that the ion beam consisted of vanadium from film over the anode, Al and O ions (ceramic), and iron, cobalt and nickel from Kovar. The observations showed that the ion energy distribution consist of two types of ions: (i) single, broad peak with a low-energy cutoff and (ii) a high energy components that extends well beyond the total arc voltage applied to the discharge and that was already indicated in the previous works [8, 16]. The total voltage across the arc was 40–60 V. The ion energy corresponding to the low-energy cutoff distribution corresponds to the minimum potential of the plasma over the anode. It was reported that the ion peak energy was in range of 50–80 eV, the minimum energy in range of 13–24 eV, and the maximum energy was in range of 175–235 eV.

Table 12.4 Energy distributions of identified ions for 75 and 100 A vacuum arc [20]

Ion	Arc current (A)	Time (μs)	FWHM (eV/charge)	FW (eV/charge)
O^{2+}	100	4	147	277
	100	1	103	227
Al^{3+}	100	4	137	300
	100	1	76	225
Al^{2+}, O^{1+}	100	3.5	58	115
Cr^{3+}	100	0.5	62	135
Ni^{3+}	100	4	48	120
Co^{3+}	100	0.5	53	117
Cr^{2+}	75	3	47	108
	100	3	44	106
Al^{1+}	75	3	52	108
	100	3	56	129

Using the electrode configuration of the work [19], Grissom [20] performed a single-shot measurements of the complete energy distributions of mass selected ions from a 6 μs vacuum arc using a time-of-flight mass spectrometer. The energy distributions of ions mass separated according to flight times in a two meter drift tube were obtained by observing the spreading in time of individual 15 ns ion bursts produced by the spectrometer deflection plate assembly. The energy distributions of identified ions were measured, and the summarized data are given in Table 12.4. The results of full width of half maximum (FWHM) and full width (FW) of the measured energy distributions were indicated. It was concluded that the data indicate the presence of ions with energies per unit charge in excess of the applied voltage in general agrees with earlier measurements [8, 16, 19].

The measurements of ion flux, ion velocity v_i and ion energy from vacuum arc plasma were conducted by Yushkov et al. 2000 [21] and then by Anders and Yushkov 2002 [22]. The arc occurred between the front face of a cylindrical cathode of 6.25 mm diameter and an annular anode of 1.3 cm inner diameter. The closest cathode-to-anode distance was about 5 mm.

Nearly rectangular arc pulse shape of 250 μs duration and 100–500 A amplitude was used. The ion current modulation was detected by a Langmuir probe working in the ion saturation regime with a constant negative bias of −60 V. The drift time to the ion detector was determined to obtain kinetic ion energies. The ion velocity was treated as a constant for the length of the drift region between the cathode and the sheath edge of the detecting probe $s = 277$ mm. The ion velocity was determined as $v_i = s/t_{pl}$ and ion energy as $E_{\text{ien}} = m_i v_i^2/2$ where t_{pl} is the time of flight of ions in the plasma drift region as measured by the time delay between an arc current maximum (or minimum) and the corresponding ion current maximum (or minimum).

The method is based on the assumption that ion production is approximately proportional to the arc current and the plasma streaming velocity "far" from the spot (1 mm) is constant. The ion Mach numbers were defined as the ratio of the ion velocity to the ion sound speed that is in form:

$$v_{is} = \sqrt{\frac{\gamma_e kT_e + \gamma_i kTi}{m_i}} \tag{12.1}$$

where $\gamma_e = 1$ and $\gamma_i = 5/3$ and the ion sound speed was estimated using the temperature derived from ion charge state measurements. A wide range of cathode materials has been used. The result indicates that the kinetic ion energy is higher at the beginning of each discharge and approximately constant after 150 μs.

Table 12.5 presents the results of average ion velocity, kinetic energy, electron temperature T_e and approximate ion Mach number M_i for most conducting elements of the periodic table measured for arc currents 100–300 A at pressure 10^{-4} Pa. The result does not noticeably depend on arc current and is valid for pressures up to about 10^{-2} Pa. Data are valid for time t >150 μs after arc initiation. According to the obtained data, the Mach number is significantly large indicating that the plasma flow is supersonic. Taking into account that for 100 A arc the ion current is 8 A, the momentum of the total ions ejected from the cathode plasma, for example of Cu, is $1.39 \times 10^{-16} \times 8/1.6 \times 10^{-19} \approx 7 \times 10^2$ gcm/s^2. This ion momentum shows the possibility of energetic deposition of thin films and to use of vacuum arc for microthrusters for space applications.

12.3 Ion Velocity and Energy in an Arc with Large Rate of Current Rise *dI/dt*

In short-pulsed high-current vacuum discharges [9, 23, 24], a considerable increase of the ion energy in comparison to longer duration low-current arcs has been observed. The measured ion energy range in microsecond arcs with high current was 1–10 keV (*high energy*), while for 100 A millisecond arcs it was around or lower 100 eV (*low energy*) [8, 16]. The main parameter influenced the character of plasma jet expansion is the rate of current rise.

Astrakhantsev et al. [25] investigated a relation between the average energy of accelerated ions and the maximum rate of current rise in a vacuum discharge. The electrode arrangement mounted in a vacuum vessel consisted of a cylindrical copper cathode with a flat operating surface of 8 mm diameter and a parallel Ta anode with a 3 mm diameter hole at its center. The cathode–anode gap was 9 mm wide. A tungsten igniter was placed at the cathode axis. The pulse duration of the igniter was about 5 μs, and current magnitude was of about 10 A. The main discharge began about 0.5 μs after the onset of the ignition pulse. The initial capacitor voltage controlled

Table 12.5 Plasma jet velocity, ion kinetic energy, ion momentum, electron temperature T_e, and Mach number M_{ion} for different cathode metals [22]

Metal	Velocity (10^6 cm/s)	Energy (eV)	Momentum $m_i v_i$ (10^{-17} gcm/s)	T_e (eV)	M_{ion}
Li	2.31	19.3	2.67	2.0	3.1
C	1.73	18.7	3.45	2.0	3.0
Mg	1.98	49.4	7.98	2.1	4.8
Al	1.54	33.1	6.89	3.1	3.3
Si	1.54	34.5	7.18	2.0	4.1
Ca	1.39	39.9	10.2	2.2	4.2
Sc	1.46	49.6	10.9	2.4	4.5
Ti	1.54	58.9	12.2	3.2	4.3
V	1.63	70.2	13.8	3.4	4.5
Cr	1.63	71.6	14.1	3.4	4.6
Fe	1.26	45.9	11.7	3.4	3.7
Co	1.21	44.4	11.8	3.0	3.8
Ni	1.15	40.6	11.2	3.0	3.6
Cu	1.32	57.4	13.9	3.5	4.0
Zn	1.03	35.7	11.1	2.0	4.2
Ge	1.11	46.2	13.4	2.0	4.8
Sr	1.15	60.5	16.8	2.5	4.9
Y	1.32	80.3	19.5	2.4	5.8
Zr	1.54	112	23.3	3.7	5.5
Nb	1.63	128	25.1	4.0	5.6
Mo	1.73	149	27.6	4.5	5.8
Ru	1.39	139	23.3	4.5	4.8
Rh	1.46	142	24.9	4.5	5.1
Pd	1.21	80.1	21.4	2.0	6.3
Ag	1.11	68.7	19.9	4.0	4.1
Cd	0.68	26.6	12.6	2.1	3.6
In	0.60	21.6	11.5	2.1	3.2
Sn	0.70	29.5	13.6	2.1	3.7
Ba	0.79	44.6	18.0	2.3	4.4
La	0.69	34.6	16.0	1.4	4.9
Ce	0.79	45.5	18.4	1.7	5.1
Pr	0.84	51.5	19.6	2.5	4.5
Nd	0.81	49.7	19.5	1.6	5.6
Sm	0.81	51.8	20.3	2.2	4.9
Gd	0.81	54.1	21.3	1.7	5.6

(continued)

Table 12.5 (continued)

Metal	Velocity (10^6 cm/s)	Energy (eV)	Momentum $m_i v_i$ (10^{-17} gcm/s)	T_e (eV)	M_{ion}
Tb	0.84	58.1	22.1	2.1	5.3
Dy	0.84	59.4	22.6	2.4	5.0
Ho	0.86	64.1	23.7	2.4	5.2
Er	0.89	69.3	24.8	2.0	5.9
Hf	1.03	97.5	30.4	3.6	5.2
Ta	1.20	136	36.2	3.7	6.0
W	1.11	117	33.6	4.3	5.2
Ir	1.07	113	34.1	4.2	5.2
Pt	0.81	67.2	26.4	4.0	4.1
Au	0.69	49.0	22.6	4.0	3.5
Pb	0.58	35.8	19.8	2.0	4.2
Bi	0.47	23.9	16.3	1.8	3.6
Th	0.99	118		2.4	7.0
U	1.14	160		3.4	6.9

the current rise of the discharge and its parameters. Ion energy and mass distributions of the plasma jet were investigated by a time-of-flight method using a single channel electrostatic analyzer.

The ion energy distribution [25] throughout the discharge parameter range showed a relationship between the average plasma jet energy ε_i and the maximum rate of current rise $\dot{I}\rightarrow(\mathrm{d}I/\mathrm{d}t)$, which is illustrated in Fig. 12.8. It can be seen that for different ion species the particle energy linearly rises after 3×10^8 A/s over a wide range of discharge parameters. For smaller values of current rise, the corresponding value of energy was almost independent of the discharge parameters.

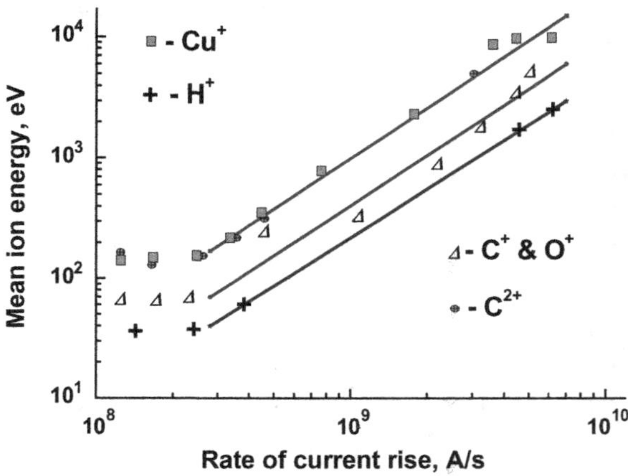

Fig. 12.8 The relationship between the mean ion energy ε_i and the maximum rate of current rise of different species of ion: □-Cu^+; •-C^{2+}; (**Δ**)-C^+ and O^+; and (+)-H^+

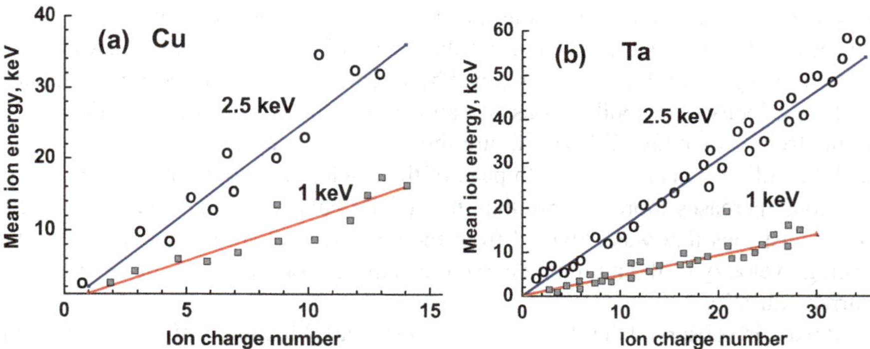

Fig. 12.9 Dependence of the mean energy E_Z on the charge multiplicity of Cu^{n+} (**a**) and Ta (**b**) ions on charge number Z for the voltage U_0= 1.0 (1) and 2.5 keV (2). The straight lines are linear approximations of the measured data

The parameters of Cu and Ta ions ejected from the plasma of a vacuum spark with a voltage up to 2.5 kV and a current rise rate up to 2×10^{10} A/s were studied [25, 26]. The experiments were carried out with a vacuum discharge with a capacitor voltage U_0 up to 2.5 kV. The electrode system consisted of a cathode of diameter 1 mm and a plane grounded grid-type anode separated by 9 mm from the end face of the cathode and was placed in a chamber with a vacuum not worse than (5–8) $\times$ 10^{-6} torr. The energy and charge state distributions of ions were measured by the time-of-flight method with an electrostatic analyzer of the "plane capacitor" type with the energy resolution $\Delta\varepsilon/\varepsilon \approx 2 \times 10^{-2}$ and a time resolution of the registering circuit of about 40 ns.

It can be seen from Fig. 12.9 that the average energy E_z of ions obtained by integrating their energy spectrum increases almost linearly with the charge number Z. The various components of Ta^{n+} ions are in considerably wider range of charge states as compared to copper ions. It was indicated that the multiply charged ions could be produced already at the initial stage of the discharge 400–600 ns prior (3 kA) to the attainment of the peak value of current.

An arc was studied between a copper wire cathode of 1 mm diameter with a cone-like working surface and a grounded grid anode, which was placed at 10 mm distance from the cathode [28]. The discharge was ignited by a high-voltage breakdown at the top surface of a dielectric insert. Two peaks of ion current were observed for the collector spaced on 10 and 35 cm from the anode. The first peak has a lower delay time with respect to the discharge onset and corresponds to the fast ions. The fast ions are the multiple charged accelerated ions. The second peak has a larger delay time and more amplitude than the first one. That peak corresponds to slow ions, which represent the bulk of ions of the cathode jet plasma.

According to Gorbunov et al. [28] for the given peak of the discharge current, the energy spectra of ion species at different charge states were produced from a set of signals of the ion energy analyzer. Then, the average velocities of individual ion species were obtained from the spectra. These velocities were averaged over all species to obtain the average velocity of the fast multiple charged ions. In a wide range of vacuum spark a relation was obtained between velocity of the fast ion component and the velocity of the bulk of ions (see above) of the cathode plasma jet (Fig. 12.10) found from data of [26, 27]. This figure shows that the velocities of both the fast ions and the bulk of ions increase with peak of the discharge current, but velocity of the fast ions increases more sharply than that of the bulk ions. The averaged velocity of the fast ions that was obtained from the collector measurements is close to the average velocity of the multiple charged ions throughout the range of the discharge current variation.

It was also observed the sharp and similar (linearly) increases of ion current at the collector at distance of 35 cm from the discharge gap versus peak of the discharge current up to 100 kA for the both bulk and fast ions of the cathode jet.

Ion acceleration at the initial stage of a pulsed arc was studied [29]. The electrode array comprised a cylindrical cathode of 6 mm diameter and an annular anode with an inner diameter of 13 mm. The anode was 9 mm away from the cathode surface. The vacuum chamber was evacuated to a residual pressure (4–6) $\times$ 10^6. The discharge was initiated on the cathode surface by a high-voltage breakdown via a dielectric insert between the cathode and the igniter. The measurements were conducted with discharge current pulse of 150 A amplitude at 30 μs rise time and of 200 μs duration. The plasma flux leaving the arc discharge region passed through the anode hole and expanded while streaming toward the anode grid. Ion energy distribution of the cathode plasma jet was measured by an electrostatic ion energy analyzer that was placed at a distance of 45 cm from the cathode.

The ion component throughout the discharge pulse was observed for two groups of ions, which were originated at different instants of the discharge pulse. The first instant t_1 corresponds to the initial transient stage of the pulse, approximately 25 μs after ignition, when the discharge current attains a maximum, and the second instant t_2, 100 μs later, denotes the stage when the parameters of the plasma relax to the

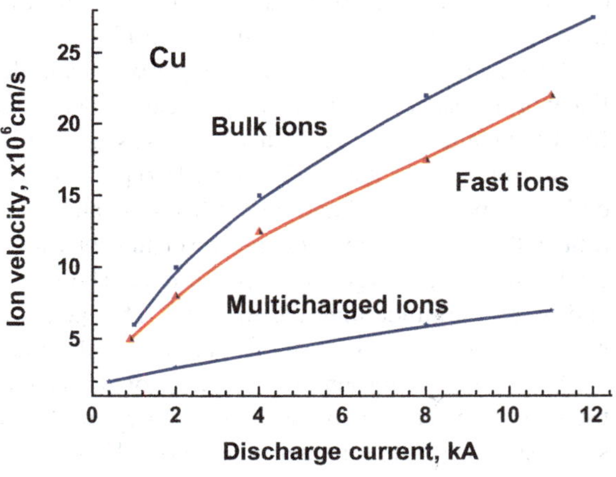

Fig. 12.10 Dependences of the cathode jet velocity on the discharge current for bulk ions, average velocity of the fast ion component, and for velocity of the multiple charged ions

Table 12.6 Directed energies E_{dir} of ion flux in cathode plasma jet obtained at different discharge stages [29]

Material	t_1	t_2
	E_{dir} (eV)	E_{dir} (eV)
Al	160	70
Ti	165	75
Mg	130	60
Zr	250	100

steady-state values. The respective data of directed energies E_{dir} of ion flux for each time interval are given in Table 12.6. While at early stage the ion energy E_{dir} is relatively large, the measured ion energies at the steady-state stage are close to those which have been measured earlier for a wide set of cathode materials.

It is found that at first stage the accelerated ions propagate within a narrow angle that is as much as, approximately, $\pm 15°$ in relation to the plasma flux axis. Also, it was show that at in the initial stage of the discharge <25 μs after ignition, when the discharge current rises rapidly, the macro-spot was located close to the igniter. In the last stage of the discharge, the cathode area occupied by the microspots enhanced due to their random motion.

12.4 Ion Current Fraction

Kimblin 1971 [30] measured the ion current emitted from DC vacuum arcs between copper electrodes in both a vacuum interrupter and a metal walled arc chamber. Data were obtained for arc currents of $\leq$3 kA. In the arc chamber, the arc plasma was generated with a 27 cm diameter metal wall, which biased to cathode potential. According to the experiment, the ion current was saturates for a bias of about 7 V. Therefore, to measure the ion current, the wall was connected to grounded cathode potential via a low impedance current-measuring shunt providing a useful bias of about 20 V (arc voltage). This electrical circuit used to study the ion current as function of the arc current, electrode spacing, electrode diameter, and wall diameter.

According to Kimblin [30], the ion current dependence on electrode spacing and electrode diameter for a 275 A arc can be shown in the form of Fig. 12.11, where the data points represent the mean ion current observed for several arcing sequences at the given electrode spacing. For 2.5 cm diameter electrodes, the ion current first increases linearly with electrode spacing and reaches a maximum saturated value for a spacing of 1.5 cm. For 5 cm diameter electrodes, the ion current approaches the same maximum value but at a longer electrode spacing of 3 cm.

A similar dependence of the ion current versus electrode separation and diameter is observed throughout the current range of 100–3000 A, although the ion currents attain maximum values at slightly smaller electrode spacing with increasing arc current. Figure 12.12 shows maximum of the total ion current as function on arc current. The data present average approximation of the measurements for electrode spacing

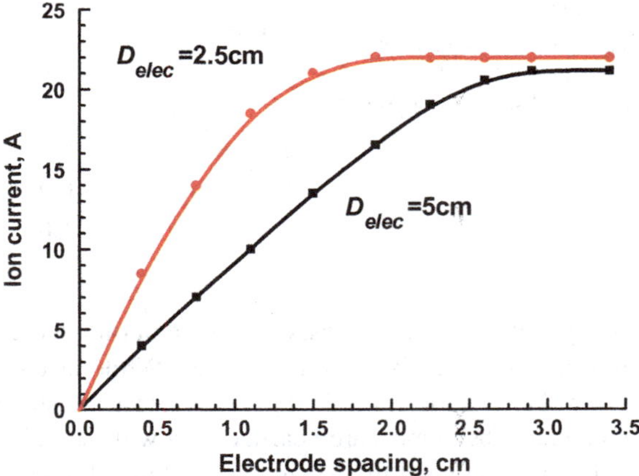

Fig. 12.11 Wall ion current as a function of electrode spacing for electrode diameters of $D = 2.5$ and 5 cm (arc current 275 A; chamber wall biased to cathode potential)

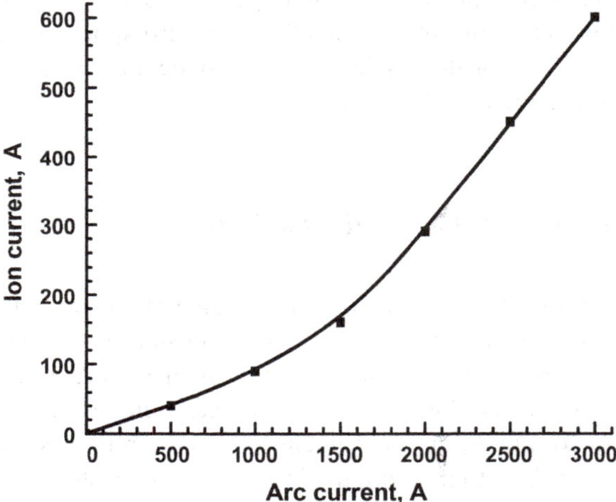

Fig. 12.12 Maximum ion current collected by a negatively biased wall versus total arc current (sum of ion "cathode current" and "wall current")

1.3 and 2.5 cm diameters electrodes at wall diameter 27 cm and electrode spacing 3 and 5 cm diameter electrodes at shield diameter 5.7 cm. The shield inserted into the chamber to measure the influence of the distance to the wall from the electrode axis. As follows from the measurements the maximum ion current values vary from 8% of the total arc current at 100 A–20% at 3000 A. This maximum ion current is a fundamental arc property independent of wall diameter, anode diameter, and electrode spacing.

However, the dependence of the wall ion current lower than the maximum indicates that the cathode regions adjacent to the cathode spots are the predominant sources of ionization for the plasma. Assuming single ionization, the author [30] concluded that 55% of the vapor leaving these regions is ionized. For the postarc current, the rate of arrival of ions at the wall decays exponentially with a time constant of about 3.5 μs. This characteristic time was used to obtain an average ion speed during arcing as 8×10^5 cm/s taking into account the known distance from the center of the cathode to the wall.

In further Kimblin's work, 1973 [31], the ion current emitted from vacuum arc was measured for various materials (Cd, Zn, Ag, Cu, Cr, Fe, Ti, C, Mo, W). The experiments were performed with pairs of 2.5 cm diameter electrodes, and both anode and cathode were manufactured from identical material. The saturated ion current escaping from the interelectrode region to the wall was conveniently detected for all experiments with the wall biased negative to cathode potential via a current-measuring shunt. Experiments were performed at DC arc current levels of 50–1000 A. The measurements with tungsten were performed using a pair of high-purity 1.9 cm diameter electrodes, and the ion current was observed over the current range of 50–600 A. For the used current range and for electrode spacing of 2 cm, the arcs were characterized solely by a multiplicity of cathode spots with no anode spot present. Thus, evaporation occurred solely from the cathode. Average arc duration was varied in range from 0.1 for W to 4.5 s for C.

It was observed that for each material, the dependences of the wall ion current on the electrode spacing and anode geometry were consistent with previous measurements [30]. By assumption of predominant vapor ionization in the cathode regions, it was concluded that the expansion was consistent with subsequent isotropic free flight motion from this region. Comparison of the net erosion rate with the wall ion current indicated that, for high-vapor-pressure materials such as Cd and Zn, about 15% of the vapor leaves the cathode regions ionized. For low vapor pressure materials such as C, Mo, and W, this fractional ionization is almost 100%. The ion current magnitudes observed at long electrode spacing were similar for each material, and it is about 8%, except 7% for carbon and 10% for W, of the arc current. Since the used materials span a wide range of thermal characteristics (from Cd with low boiling temperature T_b = 1038 K to the refractory materials C, Mo, and W up to T_b = 5973 K), it is concluded that ion currents of this magnitude can be expected from cathode materials in general. Furthermore, this comparison between experiment and theory indicates that the total probability for vapor ionization in the cathode region exceeds 50%.

As it is shown above, the ion current fraction significantly depends on electrode gap distance and anode size. Let us consider the works of Kutzner et al. [17, 32, 33] and Kutzner [34], which measured the ion current and angular ion current distribution dependence on the gap distance, and anode diameter. The arc current was varied from 50 to 340 A and arc duration was varied from 10 to 14 ms. To measure the current flux 14 ring, electrical probes of 108 mm in diameter, situated concentrically around the copper electrode system, were used. The cathode diameter of 30 mm and anode diameters were 11 and 20 mm. The total ion current to the probe shields I_{is} was obtained as sum of ion currents flowed to each probe. The electrode separation was conducted by the change in the anode position in range of 6–35 mm. A cone having an angle θ, with respect to the axis, which was normal to the cathode surface, subtended the solid angle ω in the form:

$$\omega = 2\pi(1 - \cos\theta) \tag{12.2}$$

The ion current I_{ia} in the anode shadow depends on the anodic solid angle ω_a that was determined by the anode diameter and gap distance. The total ion fraction emitted from the cathode is $I_{ip} = I_{is} + I_{ia}$. The measured angular ion fraction distribution $f_i(\omega)$ was determined as

$$f_i(\omega) = \frac{I_i}{I_a \Delta\omega} \tag{12.3}$$

where I_i is the ion current inflowing into the solid angle $\Delta\omega$ and I_a is the arc current. The measured angular ion fraction distribution is shown in Fig. 12.13.

According to the experiments, the angular distribution of ion current emitted from the cathode was approximated by a function defined as following equation:

$$f_i(\omega) = F_m \text{Exp}\left(\frac{\omega^2}{k_L^2}\right) \tag{12.4}$$

The shape factor k_L and maximum value F_m for Cu electrodes are given in Table 12.7.

As it follows, the ion current was obtained from vacuum arcs in which the anode subtends a minimum solid angle, i.e., by maximizing the electrode separation, while this frequently increases the arc voltage and its fluctuations, producing an unstable arc operation [35] (Tables 12. 8, 12.9, 12.10, 12.11 and 12.12).

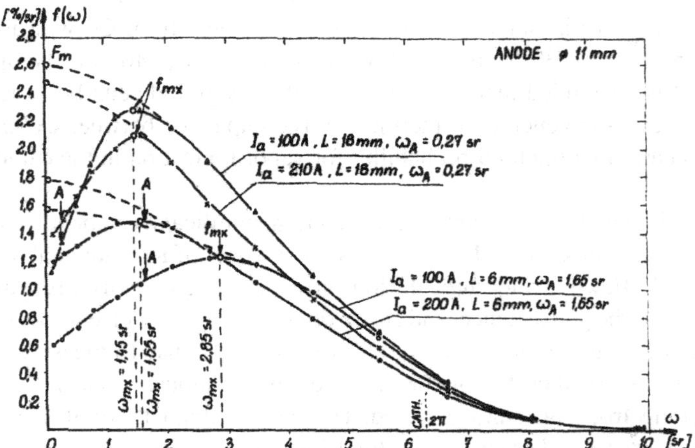

Fig. 12.13 Experimental data of angular distribution of probe ion current $f_i(\omega)$ (solid lines) and extrapolated distributions (dashed lines) at angles ω lower than maximal ω_{mx} for different currents, gap distances and anode diameter of 11 mm; f_{mx} is the maximal ion current fraction at ω_{mx} (F_m extrapolated value, Cu electrodes) [33]

Table 12.7 Experimental data for vacuum arc of 100 A and 11 mm anode diameter [17]

Gap distance (mm)	k_L (sr)	F_m (%/sr)	U_{arc} (V)	I_{ip} (%)	I_{is} (%)
6	4.77	2.17	16.3	8.9	6.3
9.5	4.69	2.37	19.0	9.6	7.7
14	4.50	2.63	21.5	10.2	8.8
18	4.46	2.83	24.0	10.9	9.5
20	4.43	3.05	25.0	11.3	10.9

Table 12.8 Ion fraction, in percentage, and energy per unit charge Z, in eV, for different ion charge states emitted from different metals [44]

Metal	Fraction,% energy	Ion charge state				
		1+	2+	3+	4+	5+
Fe	Fraction,%	54	46	0.5		
	E_i/Z, eV	92	61			
Cr	Fraction,%	16	68	14	2	
	E_i/Z, eV	73	37	34		
Ti	Fraction,%	27	67	6		
	E_i/Z, eV	65	39	34		
Cu	Fraction,%	38	55	7	0.5	
	E_i/Z, eV	59	45	44		
Mo	Fraction,%	3	33	42	19	3
	E_i/Z, eV	95	65	51	45	49

Table 12.9 Effect of cathode temperature T_sK on the ion flux parameters for Ti at arc current of 100 A [47]

T_sK	E_i/Z (eV)			Ion charge fraction (%)		
	Ti^{1+}	Ti^{2+}	Ti^{3+}	Ti^{1+}	Ti^{2+}	Ti^{3+}
390	57	35	21	17	81	2
620	33	22.5	9	29	69	2
770	32	22	7	34	65	1

Table 12.10 Effect of cathode temperature T_sK and arc current on the ion charge fraction for Mo. From [47]

T_sK	Arc current (A)	Ion charge fraction (%)					
		Mo^{1+}	Mo^{2+}	Mo^{3+}	Mo^{4+}	Mo^{5+}	Mo^{6+}
410	90	3	47	39	9	1.5	0.5
560	90	5	83	11	1	0.5	–
560	220	6	82	10	2	–	–

Table 12.11 Ion fraction with different charge state as dependences on angle for each cathode material [48]

Element	Angle (°)	Ion charge fraction (%)		
		1^+	2^+	3^+
Al	90	87	11	2
	60	78	20	2
	30	76	21	3
Cu	90	59	37	4
	60	47	49	4
	30	37	62	1
Ti	90	35	60	5
	60	19	73	8
	30	6	88	6

Cohen et al. 1989 [36] measured ion current distribution extracted from a vacuum arc between a Cu cathode and a conical ring anode by a set of five probes. The arc current pulse was generated using a 0.4 F capacitor bank, which was initially charged to 160 V. The current circuit also included a series inductor and resistor used to shape the current pulse, having an amplitude of 725 A and a 70 ms half-amplitude full-width. A drawn arc trigger to the cathode ignited the arc. The 19 mm diameter Cu rod cathode was mounted 3 mm from a 50 mm outer diameter by 12 mm thick anode, which placed at the center. The ion current was measured using a set of five (1 cm^2 area) Cu probes, which were supported by a holder having a 50 mm radius of curvature and mounted 50 mm from the cathode. The probes were mounted at angles of $\theta = 0°$, 25°, and 45° with respect to the electrode axis. Each probe was connected to the cathode potential via a 1 Ω resistor. Also, the mass flux distribution was determined by weighing the probes before and after a series of 20 arcs.

The results given in Fig. 12.14, in the case of 34 mm ID anode aperture, show a linear dependence [36]. Further tests with other probes resulted in the same linear dependence. The total ion current emerging through the aperture was estimated by numerically integrating hand-drawn curves from the data in the region $-45° \leq \theta \leq 45°$. The total ion current obtained was approximately 8.5% of the arc current. It was found that the measured ion distribution was a slightly flattened cosinusoidal function. This result is similar to that obtained in [30, 31].

Marks et al. 2009 [37] determined the ion current flux from a copper cathode using methods developed by Kimblin [31]. The current, arriving at a disc-shaped collecting surface, which was held at approximately cathode potential, was measured by reading the voltage dropped across a resistor (0.37 Ω) which was placed between the collecting surface and electrical ground (to which the cathode was also connected). The disc-shaped probe had dimensions equal to that of the pendulum collector (diameter 99 mm) and was placed about 2 mm from the anode plane in a position that was similar to the pendulum collector position during the force measurements (see

Table 12.12 Ion current fractions expressed at different average charge states [52]

Metal	Ion current fraction with different charges (%)					
	Charge ($Q = 1$)	2	3	4	5	Average charge (Q_{av})
C	100					1
Mg	23	77				1.77
Al	38	52	10			1.72
Si	38	58	4			1.66
Ti	3	80	17			2.14
Cr	14	73	13			1.99
Fe	18	74	8			1.90
Co	30	62	8			1.78
Ni	35	58	7			1.72
Cu	26	49	25			1.99
Zn	76	24				1.24
Zr	4	47	38	11		2.56
Nb	2	36	43	19		2.79
Mo	6	40	36	18		2.66
Rh	28	52	18	2		1.94
Pd	24	69	7			1.83
Ag	18	66	16			1.98
In	79	21				1.21
Sn	36	64				1.64
Gd	3	78	19			2.16
Ho	8	79	13			2.05
Ta	5	30	33	28	4	2.96
W	3	25	39	27	6	3.08
Pt	52	44	4			1.52
Au	28	69	3			1.75
Pb	47	53				1.53
Th	1	1	72	17		3.05
U	1	29	62	8		2.77

Chap. 11). The ion current was also measured at the chamber walls instead of at the probe. In this case, the plasma was free to expand into the chamber, the walls of which were approximately at cathode potential. Figure 12.15 shows the ion current fraction as a function of current for a Cu cathode measured at the probe and chamber walls. An average Cu ion current of 8.6% with a standard deviation of 0.41% was measured at the chamber walls, and an average of 9.5% with a standard deviation of 0.4% when using a probe positioned similarly to the pendulum collector.

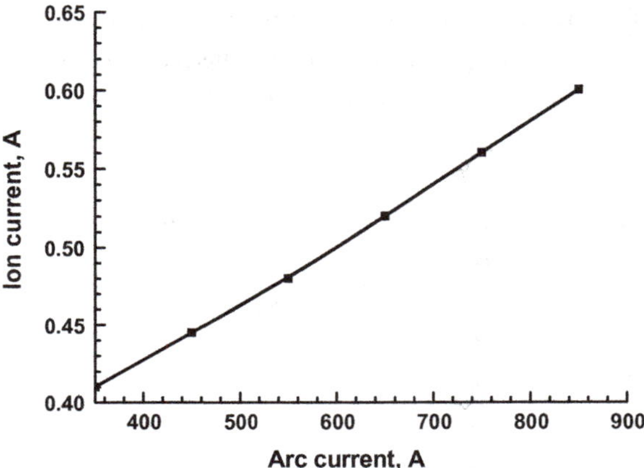

Fig. 12.14 Peak ion current ($\theta = 0°$) smoothed by a 10-Hz filter as a function of the arc current

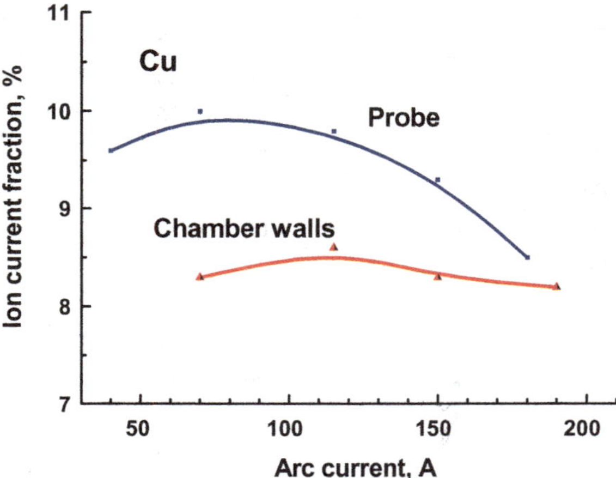

Fig. 12.15 Ion current versus arc current measured at the chamber walls (bottom curve) and at the probe (top curve)

Anders et al. [38, 39] measured the ion flux from vacuum arc cathode spots in two vacuum arc systems. The first was a vacuum arc ion source, which was modified allowing collecting ions from arc plasma streaming through an anode mesh. Cathode spots of the vacuum arc were ignited on the front surface of a rod cathode. The anode used as a semi-transparent mesh made from fine stainless steel. The mesh was spherical in shape with mesh openings of 0.8 mm × 0.8 mm and a geometric transparency of 60%. The anode mesh was electrically insulated from the original ion source anode and the extraction system. Two collector electrodes were negatively biased up to $U_b = -200$ V with respect to the mesh anode. They served as a large collector to all ions coming through the anode mesh. An eight-stage pulse-forming-network provided arc pulses of 250 μs duration. The arc current was 100 A for most experiments. The base pressure of the cryogenically pumped system was about 5×10^{-5} Pa.

The second discharge system essentially consisted of a cathode placed near the center of a spherically shaped stainless steel mesh anode, whose radius of curvature was 109 mm. Each opening of the mesh was 2 mm × 2 mm, and the geometric transmittance was 72%. Using this geometry, ions would "see" the same mesh regardless of their flow direction. The ion current from different cathode materials was measured for 50–500 A of arc current. It was found that the ion current fraction slightly depends on the material. The range of ion current fraction was varied from low value 5% for W to large value 19% for C. A linear relationship between ion flux and arc current was measured. Using the mean ion charge state data the ion erosion rates was determined in the range from 16 for Al up to about 173 μg/C for Pb or Bi for arc currents up to 500 A.

12.5 Ion Charge State

To obtain some insight into the processes occurring in vacuum discharges, Franzen and Schuy 1965 [40] investigated the time variation of the yield of individual ion species extracted from an arc. A method of time-resolved mass spectroscopy was developed [41]. It was studied the ion formation and energy mass spectra of ions extracted from the discharges between solid electrodes in vacuum with different durations. Fe ions produced by a low-voltage arc were examined. A relatively large amount of multiply charged ions as well as singly charged ions was found. Named from Fe^{+} to F^{6+} for short time in range 0–0.3 μs and from Fe^{+} to F^{4+} for low-voltage arc up to 3 μs. Mainly, all of the ions from the low-voltage-triggered condensed discharge were detected with energies less than 100 eV.

Davies Miller [16] obtained the ion yield results, which listed the fraction of the total radial ion flux from different cathode metals represented by each charge state of the ions. The data of ion charge energies were presented above in Figs. 12.5 and 12.6, and the fractions were presented below in Table 12.13.

The fraction of singly ionized ions was less mainly for metals with a higher arc voltage except for Ca and Ag, or Zr and Mo. So, it was observed a general tendency for the metals with higher arc voltages to have ion fluxes with a larger ion fraction and of higher degree. It was noted the influence of differences of the potential of ionization on the ion charge distribution. For example, the lack of Al^{+4} was explained due to the 92 V difference between the ionization energies of Al^{+3} and Al^{+4}. An analysis of the measured velocity and energy of the ions with different charge states considering mechanism of their origin was provided in [42].

Lunev et al. 1977 [43] used single and double Langmuir electrostatic probes and multi-grid analyzer to measure the plasma characteristics of vacuum arc with molybdenum cathode (chamber serve as anode) and arc currents in range of 120–250 A. A spark near the cathode initiated an arc between the cathode and chamber. The plasma flux moved along the axis of the cylindrical chamber. The ion and neutral atom fluxes were determined by weighing before and after film deposited at plates arranged along the arcs of circles centered at the center of the cathode plane. To

Table 12.13 Characteristic parameters of the ions including charge state, energy and velocity for arc current of 100 A except of 140 A for Ta [56]

Element	Charge number (Z)	Charge fraction	Peak energy E_{ip} (eV)	Arc voltage, (Cat. potential drop, u_c), V	Ion kinetic energy $E_{iz} = E_{ip} - Zeu_c$	Average ion velocity (10^6 cm/s)
Cu	+1	0.30	57	20.5 (20)	37	1.06
	+2	0.54	98		56	1.31
	+3	0.15	126		66	1.42
	+4	0.04	–		–	–
Ag	+1	0.65	45.5	16.5 (16)	29.5	0.73
	+2	0.34	88		56	1.0
	+3	0.01	141		93	1.29
Al	+1	0.49	47	20 (19)	28	1.41
	+2	0.44	73		35	1.58
	+3	0.07	93		36	1.60
Ni	+1	0.48	47	18.5 (18)	29	0.97
	+2	0.48	60		24	0.89
	+3	0.03	79		25	0.91
Zr	+1	0.14	66	21.5 (20)	46	0.98
	+2	0.60	95		55	1.08
	+3	0.21	129		69	1.21
	+4	0.05	146		68	1.20
Ti	+1	0.27	65	20.5 (19)	46	1.36
	+2	0.67	78		40	1.27
	+3	0.06	102		45	1.35
Mo	+1	0.09	90	25.5 (21)	69	1.18
	+2	0.48	164		122	1.56
	+3	0.32	186		123	1.57
	+4	0.11	230		146	1.71
Ta	+1	0.13	90	24 (23)	68	0.85
	+2	0.35	160		116	1.11
	+3	0.28	192		126	1.16
	+4	0.13	228		140	1.22
	+5	0.1	260		150	1.26
	+6	0.003	285		147	1.25
Ca	+1	0.53	36	13 (11)	25	1.10
	+2	0.47	46		24	1.07
C	+1	0.96	28	16 (14)	14	1.50
	+2	0.04	40		12	1.39

determine the contribution of the ion and neutral fluxes, it was compared the grows rates of the condensate on a substrate held at positive potential of +70 V and on a substrate at the floating potential with distance to the cathode of 25 cm and current of 180 A. The measurements show 80% of contribution of the ion flux. Electron temperature was measured of about 3 eV that independent of the probe position and the arc current. At this same distance and arc current, the average ion energy was determined by the retarding multi-grid analyzer that was found to be about 20–30 V by assumption that all ions were single charged [43].

The quantitative measurements of the ion composition in the plasma stream produced by DC vacuum arc were conducted using RF monopole mass spectrometer [44]. The multi-grid analyzer was also used to measure the energy of direct motion of the ion. The distance to the cathode was 50 mm. The measurements were begun 10 min after the arc ignition. The plasma potential measured by a single probe does not exceed a few volts. Therefore, the obtained ion energy spectra did not displaced due to the plasma potential. The plasma composition and the average ion energy weakly depend on the arc current. The composition of the ion charge state and the average ion energy for different cathode metals are given in Table 12.8.

As the arc current increased, the relative abundance of multiply charged ions in the plasma showed some very small reduction. The decrease in the average ion energy per unit charge the authors explained by the difference between the points at which the ions of different charge were produced in the acceleration region. The obtained data well agree with that for Cu cathode and different for molybdenum cathode, which were both measured by Davis and Miller [16].

The difference is explained by relatively low Mo thermal conductivity and shorter duration (0.1 s) of the arc. The authors noted that their data for average energies of the various charge ions was not measured in the previous works in the plasma of a vacuum arc. So, Plyutto et al. [8] measured the ion energy with multi-grid analyzer without charge separation and Miller [16] determined the energy at the maximum of the total ion distribution.

Aksenov et al. 1981 [45] and in 1983 [46, 47] investigated the plasma flux formation and an influence of the cathode temperature on the ion characteristics. It was used the apparatus and diagnostic equipment like to that described in [43, 44]. The charge and mass-energy analyzer was used to obtain the ionic component and the radial distribution of the ion current density. Also, the integral cathode temperature was measured by chromel–alumel thermocouple placed at 2–3 mm from the working surface. The measurements begin at 10–15 min after striking the vacuum arc, which indicated to the establishment of steady-state thermal and vacuum conditions [47].

The results show that when the integral cathode temperature increased, the ion energy spectra (dI/dU) were transformed in way that additional peaks occur and groups of particles were recorded with small energy not exceeding a few eV (see Fig. 12.16). The reduced average ion energy E_i/Z decreased when the cathode temperature rises and decreased the proportion of highly charged ions, while the proportion of singly charged ions increases (Tables 12.9 and 12.10).

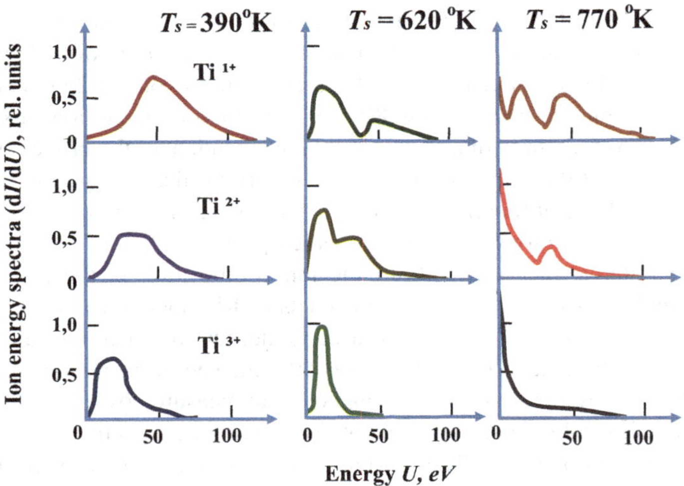

Fig. 12.16 Energy spectra of ions of different charges obtained at Ti cathode with temperature T_S as parameter

It can be seen from the tables that the average energy decreased with ion charge number and cathode temperature increasing. But the single charged ion fraction increased and larger ion charges decreased with cathode temperature increasing.

The radial distribution of the total ion current also decreased with cathode temperature rise (see Fig. 12.17) measured at the axis of 250 mm from the cathode plane with single probe biased at −60 V. While the ion current density lower than that by factor about 1.5 for cathode temperature 770 K this difference and absolute values of the ion current density decreased with radius and approximately equal at different temperatures for radius about 16 cm [47].

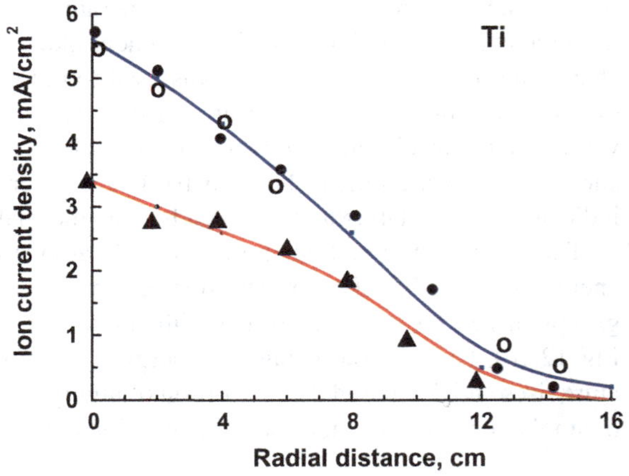

Fig. 12.17 Distribution of total ion current along the radius of the system at distance of 250 mm from Ti cathode plane at arc current of 100 A (pressure of 7 × 10^{-3} Pa) with cathode temperature as parameter: •-(1)-390 K; o-(2)-620 K; ·▲-(3)-770 K

It was concluded that the average energy and average ion charge as well as the individual ion charge slightly decreased with arc current due to increase of the integral cathode temperature with arc current and arc duration. This conclusion was applicable also for dependence of ion charge state on arc current for pulse vacuum arcs.

In general, the plasma jet is not space uniform due to the free plasma expansion. Khoroshikh et al. 1988 [48] studied the composition of the ionic component in the expanding plasma as a function of the angular coordinate and distance from the cathode surface. The diameter of water-cooled Al, Cu, and Ti cathodes were 64 mm, and arc current was 100 A. The measurements indicated that the angular distribution of singly charged ions for each of the cathode materials are substantially more uniform than that for double and triply charged ions. The difference in angular distribution of the ions with different charge is given in Table 12.11 for average data obtained at distance 30 cm from the cathode. The maximal ion charge fractions were observed in direction perpendicular to the cathode surface, i.e., at angle of 90°.

The divergence of the plasma jet is characterized by decreasing of the jet parameters with expanding distance [48]. Figure 12.18 demonstrates the dependence of charge components on distance L. It was obtained that the currents of titanium ions of different charges decreased with distance from plane cathode (Fig. 12.18). This decrease between 10 cm (from electrode axis) and 38 cm for Ti^+ falls off by about 3 times, Ti^{++} ions by 13 times and Ti^{+++} ions by 30 times. It was noted also that the change of ion charge state in the plasma jet at distance far from the cathode was due to charge exchange process between ions with different charge numbers.

Brown and co-authors [49, 50] developed an ion source in which the Metal Vapor Vacuum Arc (MEVVA) was used as the method of plasma production and from which high-quality, high current beams of metal ions can be extracted. The anode of the discharge is located on axis with respect to the cylindrical cathode and has a central hole through which a part of the plasma jets streams. The plasma jet drifts through the postanode region to a system of multi-aperture extraction grids in the accel–decel configuration, which was extract the ion component from the plasma

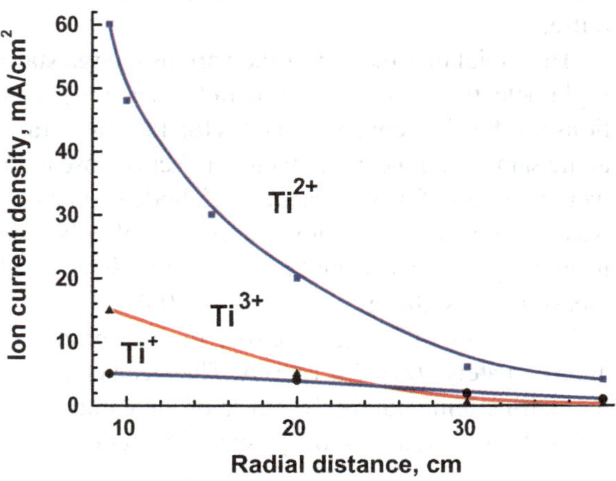

Fig. 12.18 Current density of ions of different charges as function on distance from plane cathode

plume. The ion source comprised of 16 separate cathodes mounted in a single cathode assembly, allowing the operational cathode to be changed simply by rotating a knob to position the desired cathode in line with the anode and extractor of the device. This construction allows to use many different cathode materials and compare in a relatively short experimental run and with confidence in maintaining the same experimental conditions. Pulse length was typically of 250 μs. The arc current was mostly in the range of 100–200 A, and voltage of the ion beam extraction was 20–60 kV. The source is operated on a test stand equipped with various diagnostics to monitor the source performance and the parameters of the extracted beam, including a time-of-flight (TOF) diagnostic for measurement of the ion charge-state spectrum.

Further, Brown et al. 1987 [51] used a time-of-flight ion charge-state diagnostic including a nested set of annular gating plates, greatly increasing the fraction of the beam that the diagnostic samples and allowing the radial structure of the beam charge-state distribution to be studied easily. The system has been used to survey the ion charge-state distribution of the beam produced by the MEVVA high-current metal ion source for a wide variety of cathode materials and operating conditions. Ions of mass number A and charge state denoted as Q are extracted from the ion source through a voltage drop U_{apl} and thus acquire a velocity

$$v_i = \sqrt{\frac{2eU_{\mathrm{apl}}Q}{m_i A}} \tag{12.5}$$

where e is the electronic charge and m the ion mass. The short sampled pulse is caused to drift through a distance L adequate to allow separation of the various Q/A components in the beam, and the flight time of these components is measured by a detector (Faraday cup) located at the end of the drift region. Since the beam was composed almost entirely of a single mass ion, it provided a charge-state analysis. The time in the main beam pulse at which the short time-of-flight pulse (typically a few tenths of a microsecond) is gated out was varied electronically, so allowing the charge-state spectrum to be sampled as a function of time throughout the beam pulse.

The radial distribution of the various charge states in the vicinity of the time-of-flight detectors was obtained by radial scanning with the Faraday cup. The distributions for the different charge states for Ta^{+}, Ta^{2+}, Ta^{3+}, Ta^{4+}, and Ta^{5+} were all peaked at the same radial position, which in fact is close to the axis. The charge-state spectra were measured for wide range of cathode metals (Table 12.12) [52]. The ion source was operated at a repetition rate of several pulses per second, up to a maximum of near 100 pulses per second for short-pulse length and low average power. For all the measurements, the arc current was 200 A.

The measured charge-state spectra were influenced weakly over the range covered approximately $100 < I < 1200$ A. The mean charge state is observed to increase with the boiling temperature T_{bol} in K of the cathode material [53]. A simple LC pulse line with pulse duration of typically 250 μs drives the arc. The average charge state

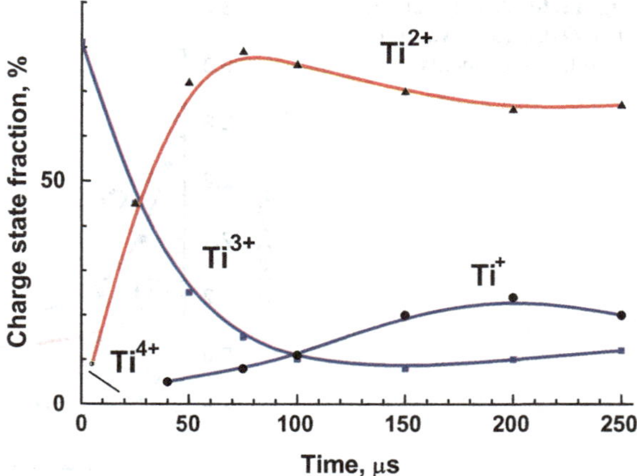

Fig. 12.19 Time dependence of the charge-state spectrum (normalized particle fractions) throughout the arc current pulse, for the case of the half-sinusoid arc current pulse shape for Ti [54]

Q_{av} was found fitted reasonably well by the following empirical approximation for arc current in the range of 100–200 A:

$$Q_{\mathrm{av}} = 0.38\left(\frac{T_{\mathrm{bol}}}{1000} + 0.6\right) \tag{12.2}$$

The charge-state data with particular emphasis on the time history of the distribution throughout the arc current pulse duration is given in [54]. Two types of pulse lines, with 250 μs duration, drove the arc. One was a low-loss line and having a very flat pulse shape (flat pulse), and the other was quite loss and having a pulse shape similar to a damped half-sinusoid (half sinusoid pulse). The flat pulse was of magnitude 100 A and the half-sinusoid pulse had a peak current of 400 A. It was found that the spectra remain quite constant throughout most of the beam pulse, as long as the arc current is constant. However, when the arc first initiated and the arc current is still raising, a transient behavior was observed that could be seen in Fig. 12.19. During this initial time, the ion charge states produced were obtained to be significantly higher than during the steady current region that follows.

Kutzner and Miller summarized their results in 1991 [55] and in 1992 [56]. The data for individual components of the cathode ion flux are given in Table 12.13 together with the kinetic energies and velocities. The ion kinetic energies measured with respect to the cathode potential were corrected by the cathode potential drop u_c.

Anders et al. 1993 [57, 58] verified this dependence. The arc current was in the range 50–400 A, and the arc duration was 250 or 630 μs with a rise time of 30 or 90 μs, respectively. It was found that the ion charge-state distribution changes over a timescale on the order of hundreds of microseconds. The average charge state decreasing is shown in Fig. 12.20 for different cathode materials and the time-dependent ion charge fraction in Fig. 12.21 (f is the fraction of the ion species).

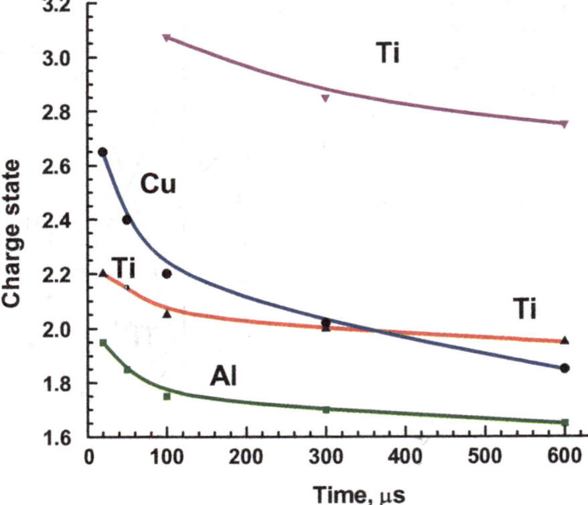

Fig. 12.20 Average charge state Q decrease over time for selected elements

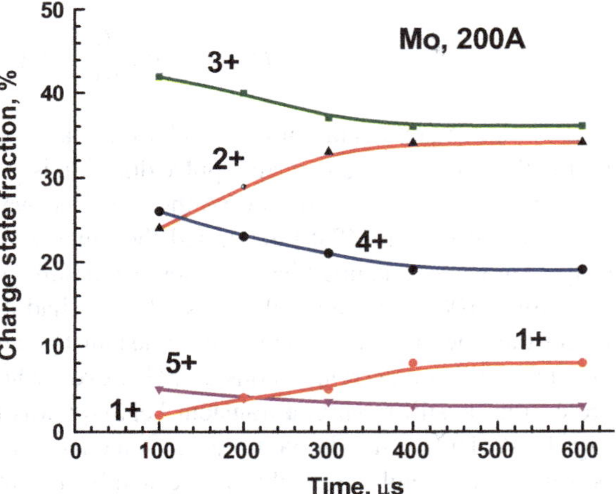

Fig. 12.21 Time evolution of the charge-state distribution for a 200 A molybdenum vacuum arc

The charge-state distribution was measured as a function of arc current in the range 50–400 A and found nearly no influence [59]. The total ion current detected by Faraday cup was electrical current I_{elec}. As the ions generated by the vacuum arc were multiply stripped with an average charge state, the particle current $I_{\text{part}} = I_{\text{elec}}/Q$. Indicated that it is important to keep this distinction, for example, in ion implantation where it is needed for estimating the implantation dose. The particle current fraction is summarized in Table 12.14. Lower boiling point metals tend to have lower mean charge state.

Tsuruta et al. used TOF method to study the ion charge state in period 1992–1997. An impulse voltage to generate metal ions applied a vacuum gap [60]. The rise time

Table 12.14 Vacuum arc ion charge-state fractions and average charge states, expressed in terms of particle current [59]

Metal	Particle current fraction (%)					
	Charge ($Q = 1$)	2	3	4	5	Average
Li	100					1.0
C	100					1.0
Mg 12	46	54				1.5
Al 13	38	51	11			1.7
Si 14	63	35	2			1.4
Ca 20	8	91	1			1.9
Sc 21	27	67	5			1.8
Ti 22	11	75	14			2.1
V 23	8	71	20	1		2.1
Cr 24	10	68	21	1		2.1
Mn 25	49	50	1			1.5
Fe 26	25	68	7			1.8
Co 27	34	59	7			1.7
Ni 28	30	64	6			1.8
Cu 29	16	63	20	1		2.0
Zn 30	80	20				1.2
Ge 32	60	40				1.4
Sr 38	2	98				2.0
Y 39	5	62	33			2.3
Zr 40	1	47	45	7		2.6
Nb 41	1	24	51	22	2	3.0
Mo 42	2	21	49	25	3	3.1
Pd 46	23	67	9	1		1.9
& 47Ag	13	61	25	1		2.1
Cd 48	68	32				1.3
In 49	66	34				1.4
Sn 50	47	53				1.5
Sb 51	100					1.0
Ba 56		100				2.0
La 57	1	76	23			2.2
Ce 58	3	83	14			2.1
Pr 59	3	69	28			2.2
Nd 60		83	17			2.2
Sm 62	2	83	15			2.1

(continued)

Table 12.14 (continued)

Gd 64	2	76	22			2.2
Dy 66	2	66	32			2.3
Ho 67	2	66	32			2.3
Er 68	1	63	35	1		2.4
Tm 69	13	78	9			2.0
Yb 70	3	88	8			2.1
Hf 72	3	24	51	21	1	2.9
Ta 73	2	33	38	24	3	2.9
W 74	2	23	43	26	5	3.1
Ir 77	5	37	46	11	1	2.7
Pt 78	12	69	18	1		2.1
Au 79	14	75	11			2.0
Pb 82	36	64				1.6
Bi 83	83	17				1.2
Th 90		24	64	12		2.9
U 92		12	58	30		3.2

and time constant for the decay of the arc current were 0.1 and 4.5 μs, respectively. The arc currents were of 320 and 420 A. The cathode of the arcing electrode was a rod of lead or copper with the diameter being 1 mm. The anode was a stainless steel rod with a diameter of 20 mm. The distance between the cathode and collected grid was 25 mm. The ion current was measured for varied ion extraction times after the arc ignition. At a lead cathode, Pb^+ and Pb^{++} ions were detected for the ion extraction times less than 45 μs. The average charge-state fractions of the Pb^+ and Pb^{++} ions were 91 and 9%, respectively. At a copper cathode, Cu^+, Cu^{++}, and Cu^{+++} ions were detected for the ion-extraction times less than 12.5 μs and the average charge-state fractions were 42, 41, and 17%, respectively.

Tsuruta et al. [61] applied this same materials and point-plane electrode (needle cathode) gap by a 13 μs duration sinusoidal arc or a 10 μs duration exponentially decaying arc. The ion measurements were made at variable times after the arc ignition. At the lead cathode, Pb^+ and Pb^{++} ions were generated and the upper limit times for the Pb^+ ion detection were 48 and 46 μs from the arc ignition, for the sinusoidal and exponential arcs, respectively. At the copper cathode, Cu^+, Cu^{++}, and Cu^{+++} ions only measured within 15 and 13 μs from the arc ignition, for the sinusoidal and exponential arcs, respectively.

The charge state for Ag and Zn was studied with needle cathode configuration by a 13 μs duration sinusoidal arc or a 9 μs duration exponentially decaying arc in the work [62]. It was detected Ag^+, Ag^{++}, and Ag^{+++} ions and Zn^+, Zn^{++} ions. The respective ion charge fractions are given in Table 12.15.

The ion motion from 13 μs pulse vacuum arc is investigated in further works by Tsuruta et al. [63, 64]. Pair of electrodes, a grid, an ion drift tube, and an ion collecting

Table 12.15 Ion charge fractions for different cathode materials. Ion velocities of measured different ion charge state. From [62]

Element	Arc current (A)	Duration (μs)	1^+	2^+	3^+	v_f	References
Cu	320–420	4.5	42	41	17		Tsuruta [60]
Pb		4.5	91	9			
Ag	Sin 240	13	51	39	10		Tsuruta [62]
	Exp 220	4.5	53	33	14		
Zn	Sin 240	13	89	11			
	Exp 220	4.5	80	20			
Ion velocities for each ion charge, 10^6 cm/s							
Cu	Sin 150	13	1.3	1.4	1.7	$v_f = 2.1$	Tsuruta [63, 64]
Ag	Sin 150	13	1.4	1.1	0.8	$v_f = 1.6$	

part were arranged in a metal vacuum chamber, the diameter being 100 mm. The cathode was circular cone and the anode was a tungsten rod of diameter 20 mm. The diameter of the ion drift tube is 50 mm and the length is 400 mm. The distance from the electrode axis to the grid is changed in the range from 20 to 140 mm. The waveform of the arc current was negative half-cycle sinusoidal with a peak current of 150 A. The ion charge was obtained measuring the ion time flight. It was detected that velocities of the ions generated at the ignition of the arc were fast and independent of the observed ion charge state. The measured average velocities were of 2.7 × 10^6 cm/s for Cu and 2.0 × 10^6 cm/s for Ag cathodes. However, the velocities of the ions generated at the extinction of the arc were relatively low and lower with charge number decreasing that is given in Table 12.15.

The influence of a vacuum arc current jump on the extracted plasma jet velocity and charge-state distribution was investigated in 1999–2000 by Bugaev et al. [65, 66]. The vacuum arc was powered from a LC line charged to 1 kV, which provided an arc current of up to 300 A at a pulse duration of 400 μs. A jump current rise was supported by an auxiliary power source, which was connected to the vacuum arc power source. The power source increases the vacuum arc jump current to 1 kA for 1–4 μs at any desired point of the discharge pulse. To measure the charge-state distributions, a specially designed TOF mass spectrometer was used which was similar to that in [51].

During the action of the jump current the additional voltage, applied to the gap, falls by the exponential law from several hundreds to one hundred volts and, within several microseconds after the completion of the current jump, the arc operating voltage takes its conventional value (20–40 V). The ion current step is delayed with respect to the beginning of the step current pulse, depending on the cathode material, by 10–40 μs. This time delay of the ion beam current pulse was significantly longer than the time required to the ions to be accelerated and drift to the collector.

Measurements were made for different cathode materials and the charge state as well the, respectively, directed velocities of ions are given in Table 12.16. It can be seen that the difference between the directed velocities for variously charged ions

Table 12.16 Charge state, velocity and energy of ions in the plasma jet for different cathode materials [65]

Element	Charge	Velocity (10^6 cm/s)	Energy (eV)
Mg	1^+	1.75	38.4
	2^+	1.68	35.6
Al	1^+	1.79	40.1
	2^+	1.61	36.2
	3^+	1.55	33.7
Ti	1^+	1.61	64.9
	2^+	1.53	58.3
	3^+	1.40	55.8
	4^+	1.47	53.4
Cu	1^+	1.43	67.2
	2^+	1.32	57.6
	3^+	1.28	51.3
	4^+	1.20	47.8
	5^+	1.24	50.5
Pb	1^+	0.54	31.6
	2^+	0.53	30.0
	3^+	0.50	26.6
	4^+	0.50	27.0
	5^+	0.52	28.8
Bi	1^+	0.39	16.6
	2^+	0.37	14.9
	3^+	0.36	14.2

is rather small, which allows the conclusion that the directed velocities of ions and the ion energies for variously charged ions of the same cathode material are almost equal. This result agrees with that obtained in [63, 64] at arc ignition and disagrees at the arc extinction.

In work of [66], the vacuum arc operated at the typical arc parameters of current 100–300 A, with pulse duration of 250 μs, at a repetition rate of about 1 pulse/s, the vacuum arc current was increased by several hundred amperes for about 5–15 μ s at any desired point of the main arc pulse. In most cases, the current spike was placed 150–200 μs after triggering the arc pulse. During the jump current, an increase in charge state is observed, and the fraction of Ti^{2+} and Ti^{3+} in the plasma is decreasing while Ti^{4+} fraction increases (Fig. 12.22).

The observed charge-state distributions were higher at the beginning of the vacuum arc [57, 58]. This has been attributed to changes of the cathode surface temperature and spot type. Anders 2005 [67] studied additional effect. It consists in that the charge-state distributions shifted, and it could be due to collisions with neutral atoms. It was noted that the neutral atoms can be produced due to (i) evaporating macroparticles; (ii) vapor from cooling, yet still-hot craters of previously active

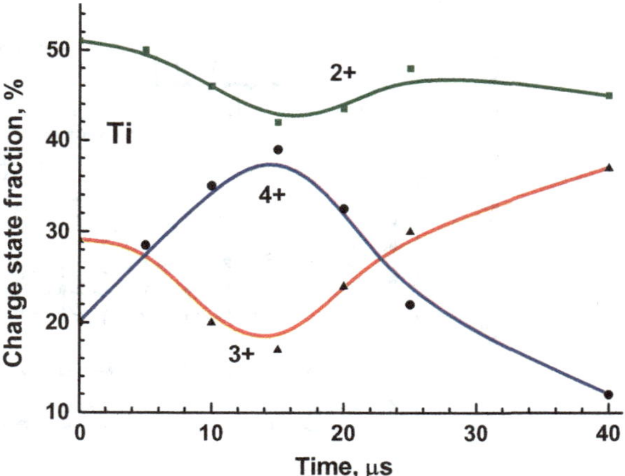

Fig. 12.22 Time dependence of the charge-state distribution spike for titanium (zero-time corresponds to the trigger instant of the arc jump current). The arc was operated with a quasistatic magnetic field

cathode spots; (iii) neutrals formed by recombining ions; and iv) the contribution of self-sputtering and in case of oblique angle condensation with low sticking probability of condensing ions. An analysis concentrated on estimation of characteristic times and mean free paths have shown that the experimentally observed time shift of the average charge state of cathodic arc plasma may at least in part due to charge exchange reactions of ions with neutrals that gradually fill the discharge volume.

Another work demonstrated Anders et al. [68] investigating an influence of neutrals presence on the ion charge state. A vacuum arc of 300 A and duration of 300 μs was generated by a pulse-forming network, at a repetition rate of up to 10 pulses per second (pps). Different metals (Al, Mg, Cu, Pb, Bi, and Mo) were used as cathode in the experiments. The charge-state distributions were measured at selected delay times after arc triggering using the TOF spectrometer. Each cathode was a metal rod of 6.25 mm diameter and about 25 mm length, inserted in an alumina tube. Cathode spots could only burn on the rod's front surface, with the metal plasma expanding through an annular anode of 12.5 mm inner diameter into a plasma expansion chamber of about 10 cm diameter and 10 cm length.

The effects of gas and metal vapor were studied by introducing nitrogen gas into the cathode region and by intentionally introducing surfaces with which the expanding metal plasma can interact. For study of the metal neutrals produced by the interaction of plasma with surfaces, different cases were considered including the original plasma expansion. A set of annular baffles made from 0.5 mm thick aluminum sheet metal, with an inner diameter of 40 mm and spaced 5 mm apart from each other, inserted in the expansion zone between anode and extraction grid; and an aluminum tube of the same length and inner diameter as the baffle system. Figure 12.23 shows that the reduction of charge states critically influenced by the density and nature of the neutrals. Neutrals are comprised of both metal species and gaseous species; each may have several sources and locations of origin. Experiments with additional surfaces in the expansion zone show that plasma–wall interactions can significantly contribute to the presence of neutrals in the discharge zone.

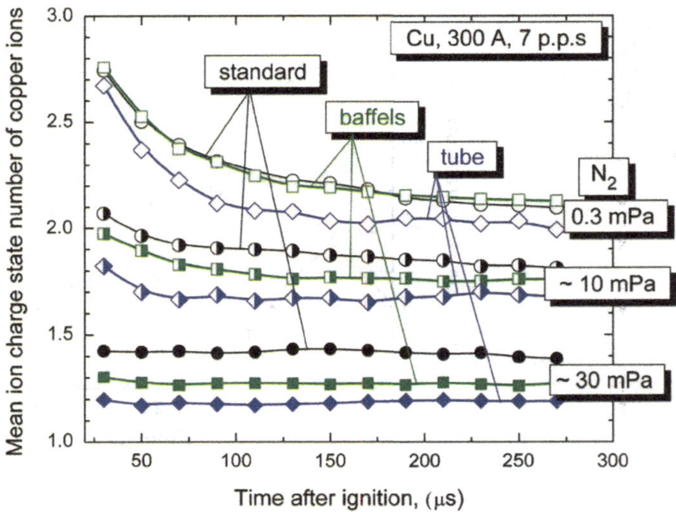

Fig. 12.23 Compilation of results of charge-state measurements for copper ions under various conditions: "standard" refers to the usual geometry of the "Mevva V" vacuum arc ion source; "baffles" and "tube" refer to measurements when additional baffles and a tube are inserted in the expansion zone, respectively. Additionally, three pressure conditions were used also. Figure taken from [68]. Permission number 4776960713721

The ion charge states were observed highest near the cathode spot. The charge reduced when the plasma is flowing toward a substrate, or ion extraction system. It was concluded that the data on ion charge states measured at some distance from the electrode assembly could be different from that produced at cathode spots due to the presence of neutrals.

An investigation of the ion charge state distributions from cathodic vacuum arcs using a modified TOF method was provided by Anders et al. work 2005 [69]. Experiments have been done in double gate and burst gate mode, allowing studying dynamic changes of ion charge-state distributions with a time resolution down to 100 ns. In the double gate method, two ion charge spectra are recorded with a well-defined time between measurements. A cathode rod 6.25 mm in diameter is placed in a ceramic tube allowing a cathode spot to burn only on the rod's front face. The plasma produced at cathode spots expanded through an annular anode and arrived after distance of 138 mm drift at a three-grid ion extraction system, which forms an ion beam that used for the TOF diagnostic of ion charge states. The mass of ions is characterized by the cathode material, and hence, charge states were interpreted from the TOF ion spectrum. The arc current was pulsed, and each pulse had duration of 250 μs with the amplitudes between 50 and 500 A. The vacuum base pressure was about 8×10^{-5} Pa. The time resolution was about of 150 ns. The ion detector was a Faraday cup with magnetic suppression of secondary electrons. A higher charge state arrives earlier at the detector, which allows detect the charge number using the relation between the ion kinetic energy, ion charge state, ion mass and measured ion velocity. The elements Mg, Bi and Cu were selected for tests, representing metals of very different properties.

The results of the study showed that fluctuations of ion charge states were always present, even at the shortest time difference between measurements. For all elements, a large stochastic change occurs even at the limit of resolution. Correlation of results for short times between measurements was found, but it was argued that this is due to velocity mixing rather than due to cathode processes. The burst mode of TOF measurements revealed the systematic time evolution of ion charge states within a single arc discharge, as opposed to previous measurements that relied on data averaged over many pulses. Above was reported a decrease in ion charge state throughout the pulse. This decrease can be observed up to a few milliseconds, until a "noisy" steady-state value that also was established. Since the extraction voltage is constant, a decrease in the ion charge state has a proportional impact on average energy of the ion beam.

The influence of arc parameters on the temporal development of the ion beam average charge state for a wide range of cathode materials was also investigated by Oks et al. 2008 [70]. Ion beam mass/charge-state distributions were measured by a TOF, TOF spectrometer. Arc discharge of 100–400 A current had duration of 300 μs. The evolution of the ion beam charge-state distribution was measured at ten different times after arc triggering from 25 to 250 μs with steps of 10 or 25 μs, and the results are shown in Figs. 12.24, 12.25 and 12.26.

As can be seen, the ion charge states are highest near the cathode spot and reduced when the plasma was flowing toward a substrate, or ion extraction system. It was noted that the data on ion charge states measured at some distance from the electrode assembly could be different from that produced at cathode spots due to the presence of neutrals.

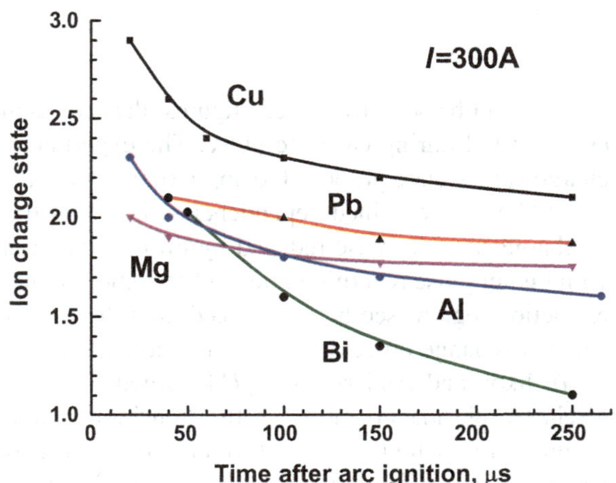

Fig. 12.24 Ion beam average charge-state number for different cathode materials as a function of time after arc ignition with the arc current of 300 A

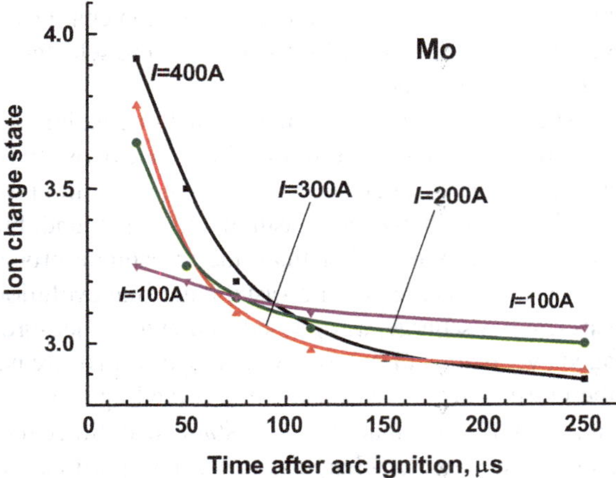

Fig. 12.25 Reduction of the ion beam average charge state for Mo cathode with the arc current as a parameter

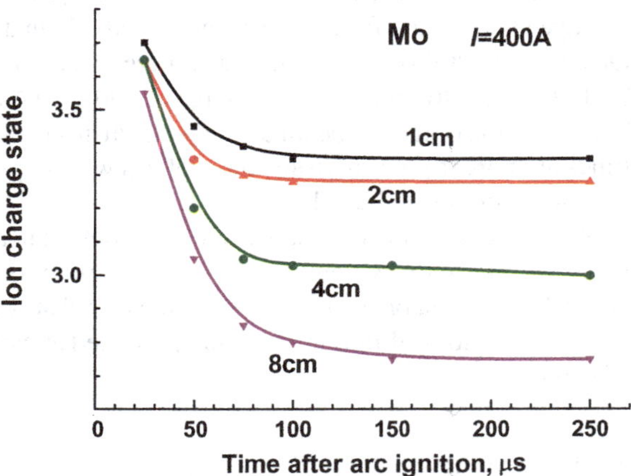

Fig. 12.26 Average ion charge-state number for molybdenum as a function of time for different distances between cathode surface (location of ion production) and extraction grid (location of ion analysis)

As it can be seen from these figures, the ion beam average charge state showed essential fall during each arc pulse. The experimental data of the decrease of the charge values were proposed to approximate with a first-order exponential function (see [70]). The obtained dependencies show that for fixed pulse duration, the rate of charge state decrease reduced by lower arc current. The charge state decreased with higher pulse repetition rate, and reduction of the distance between cathode and extraction region (see below). Noted, that the latter effect may be associated with charge exchange processes in the discharge plasma.

Yushkov and Anders 2008 [71] studied the ions which extracted from pulsed discharge plasma operating in the transition region between vacuum spark and vacuum arc. A high, transient voltage between anode and cathode characterizes vacuum sparks. The current was quite high, leading to short-lived "plasma points" containing highly charged ions. The experiments were carried out at the vacuum arc ion source "Mevva V" at Berkeley Laboratory with the pulse discharge of a full width at half-maximum duration of 8 μs and current of up to 10 kA.

The cathode was a 6.25 mm metal rod coaxially placed in a copper tube; anode was of 25 mm inner diameter. The following cathode materials has been used Ti, Pt, Sb, W, Au, Er, Al, and Bi. The observations showed principally similar behavior for all materials. Average ion charges states of about six for the used cathode materials was observed at a peak current of about 4 kA, and with a pulse duration of 8 s. The authors found for platinum the highest average charge state of 6.74 with charge states present as high as 10^{+}, and for gold traces of charge state of 11^{+}.

The ion charges state distribution showed a very distinct trend to lower charge stages, indicating the transition from vacuum spark to vacuum arc ultimately on the time scale of 1 ms, characterized the steady-state values of DC arcs. Figure 12.27 shows the decay of the ion charges state distribution (Pt cathode), with the highest charge state of 10^{+} present only at the beginning of the discharge pulse [71].

Sangines et al. 2010 [72] investigated the plasma near the cathode of a high-current pulsed arc using time-resolved optical emission spectroscopy. It was studied a coupling of the time evolution of the intensity of the emission from highly ionized species with the spot locations deduced from images obtained with a fast framing camera. The arc trigger pulse triggered the camera at intervals of 25 μs with an exposure time gate width of 2 μs. A rectangular arc current pulse of 600 μs duration was used with average values at the plateau region ranging from 800 to 2000 A. A 50 mm diameter disk Al cathode was used. The cathode has a hole in the center to accommodate the trigger electrode. The anode was a 60 mm diameter and 60 mm long copper cylinder concentric with the cathode.

It was observed a correlation between a reduction in the charge state and an increase in the spacing of cathode spots. As the cathode current was increased, the distances between spots were reduced, and these charge states were produced for longer times. Also, as the cathode current was increased, the peak of the emission

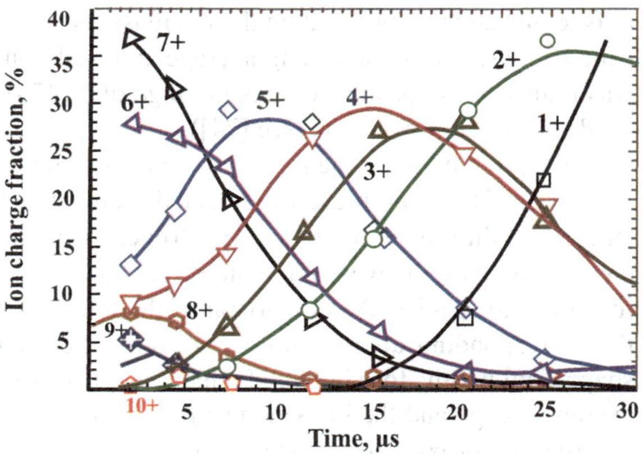

Fig. 12.27 Evolution of the ion charge-state distribution as measured with the TOF method. The time "zero" is defined here as the arrival time of the first fastest ions at the Faraday cup detector for Pt cathode according [71]

intensity from Al^{2+} shifts to later times. It was concluded that when the arc current increased at relatively high values (~kA) the dynamics of the cathode spots determine a mechanism for the production of high charge states.

The results of experimental studies of ion beams escaping from a plasma of laser-induced vacuum discharge were presented [73]. The experiment was conducted with discharge at current amplitude of about 1.7 kA, and current rise rate was about 7.5×10^9 A/s for Al cathode. The discharge is initiated with a laser pulse of 30 μs length, energy less than 10 mJ, and power density 5×10^{11} W/cm^2. It was observed highly charged ions of the cathode material in the discharge mode.

The production of Al ions was observed at the instant when the discharge current rise rate attains a peak value. Ion energy distributions were characterized by the presence of a significant non-Maxwellian tail of the accelerated ions. The maximum ion charge state and energy per charge unit were Al^{8+} and 13 keV/Z. The energy is comparable to the value observed under similar experimental conditions for the laser-produced plasma at pulse energy of 400 mJ and power density 2×10^{13} W/cm^2. The experimental dependence of average energy on the charge state for ions of both the discharge and laser plasmas fits the linear relation $E_a = k_a Z$, with the coefficient k_a as much as 2.5 keV/Z for the laser-produced plasma and 4.5 keV/Z for the discharge plasma.

Zhirkov et al. 2013 [74] used charge-state resolved mass spectrometry to characterize unfiltered DC arc plasma from 63 mm compound cathodes produced by powder metallurgy. Cathode based on Ti-C, Ti–Al, and Ti–Si compositions were selected to include large differences in ion mass. Such cathodes were used to study effects caused by change in relative cathode composition. The arc was operated at current of 65 A and a base pressure around 10^{-6} torr. The front orifice of the spectrometer was placed 33 cm from the cathode surface, and for each cathode, the plasma was characterized through mass-scans at fixed ion energy and energy-scans at fixed mass-to-charge ratio for all ions.

The measurements showed that both the energy peak and averaged energy increased with increasing of the ion charge state. The calculated peak velocities of different ion species in plasma generated from a compound cathode were found to be equal and independent on the ion mass. Increasing the C concentration in Ti-C cathodes resulted in increasing average and peak ion energies for all ion species. It was indicated the peak velocities in range of $(1.37\text{–}1.55) \times 10^6$ cm/s for ions from used cathode compositions (see [74]).

A coupling between the ion charge state, ionization energy, and ion energy distributions in DC vacuum arc for selected elements with different atomic mass was studied by Zhirkov et al. 2015 [75]. To determine the plasma composition, the ion energy distributions were integrated to obtain areas proportional to the number of ions of each species. A comparison of the total ion kinetic energy distribution and the corresponding distribution of the ion charge state showed close to equivalent shapes and widths for all cathode materials. It was found a correlation between the average charge and the ion kinetic energy in the energy region where most ions are recorded (approximately centered around the peak kinetic energies). Namely, the average charge state and related average ionization energy increase with increasing

Table 12.17 Peak kinetic energy, plasma composition, average charge state, and average ionization energy ($E_{ion.avg}$) for different cathode plasmas [75]

Element	E_{peak} (eV)	f_{+1} (%)	f_{+2} (%)	f_{+3} (%)	f_{+4} (%)	f_{+5} (%)	f_{+6} (%)	$E_{ion.avg}$ (eV)
Al	41.0	42	52.0	6			1.6	18.6
Ti	53.0	6.0	66.0	28.0			2.2	27.5
Cu	62.0	9.0	70.0	20.0	0.2		2.1	34.0
Mo	136.0	2.0	63.0	29.0	6.0		2.4	35.0
W	162.0	6.0	57.0	22.0	11	4	2.5	39.6

ion kinetic energy. Also, the average charge state tends to saturate for the highest energies. The average ionization energies in the plasmas were found to be linearly correlated with the kinetic ion energies. Table 12.17 presents the plasma composition (f_n,%), peak kinetic energy (E_{peak}), average charge state (Z_{avg}), and average ionization energy ($E_{ion.avg}$) of the studied plasmas from Al, Ti, Cu, Mo, and W cathodes.

The ion energy distribution showed that a higher peak kinetic energy accompanies presence of more highly charged ions. It can be seen that the maximum ion charge registered in the ion flux from the Al cathode is 3^+ and the corresponding peak kinetic energy is 41 eV. The maximum charge registered in the flux from the W cathode is 5+ , accompanied by peak energy of 162 eV.

Yiftah et al. 2017 [76] measured the charge state and velocity of ions extracted from the expanding plasma of 5 μs vacuum arc using a time-of-flight mass spectrometer. The arc was sustained between the cylindrical copper cathode of 1 mm diameter and Kovar cylindrical anode. The distance between the two electrodes was 0.1 mm. The ion charge state and velocity were detected after the arc pulse at different times (in range 20–45 μs) by opening an electronic shutter. The arc current was 35 A. Figure 12.28 shows the average source operation current from consecutive operations of 200 pulses (blue) followed by the ion current acquired around the shutter position located 75 cm away from the source (red) enclosed by the one standard deviation spread (green). The black circle illustrates a shutter operation time (20 μs) and its ions corresponding velocity of 3.75 cm/s.

Figure 12.29 shows the average signal (and the one standard deviation spread) acquired for ions travelling at the velocity corresponding to the black circle indicated in Fig. 12.28. An example of average (red) ion current collected from 25 consecutive pulses was presented. Arrows illustrate the expected arrival time of the different species.

Different copper ion species and ions velocity as dependence on the shutter time were observed. To compare the number-charged ion contributions, the charge from the different ion numbers was normalized to their maximal values. Figure 12.30 shows the normalized results obtained as a function of the used time points. This figure shows that the highest observed charge state Cu^{+3} is much more prominent

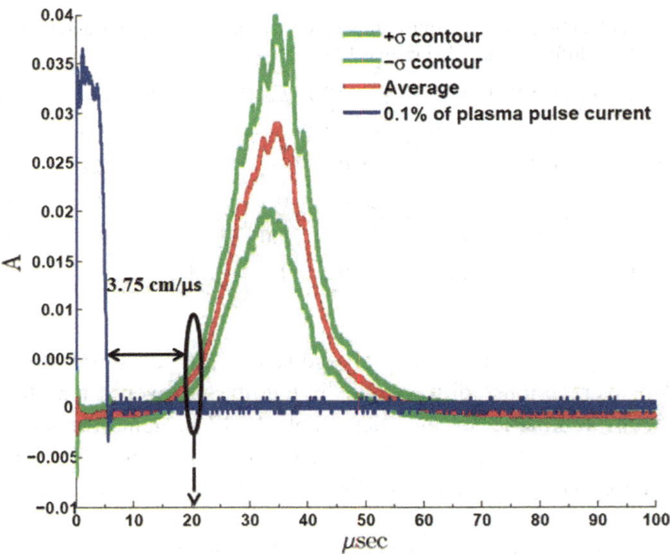

Fig. 12.28 Average source operation current from consecutive operations of 200 pulses (blue) followed by the ion current acquired around the shutter position. The pulse current in this graph is presented as one hundredth of its real magnitude in order for all traces to be presented on the same scale. Figure taken from [76]. Permission number 4777021394874

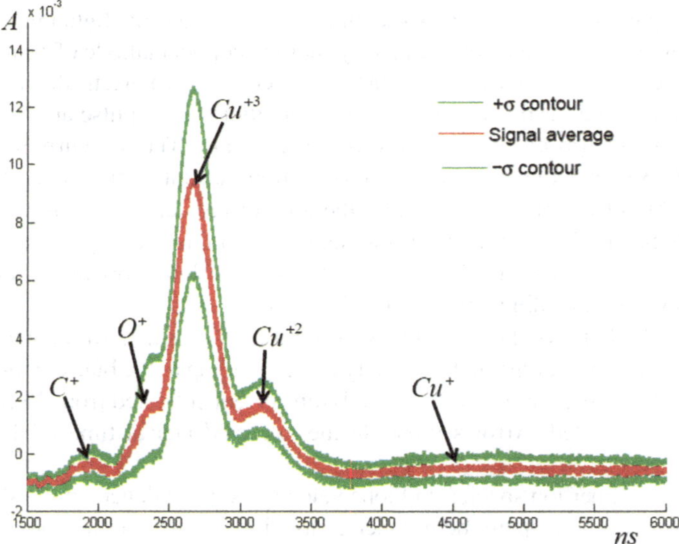

Fig. 12.29 Time dependence of ion current selected from the plasma plume moving through the shutter opening time of 20 μs. Figure taken from [76]. Permission number 4790880134297

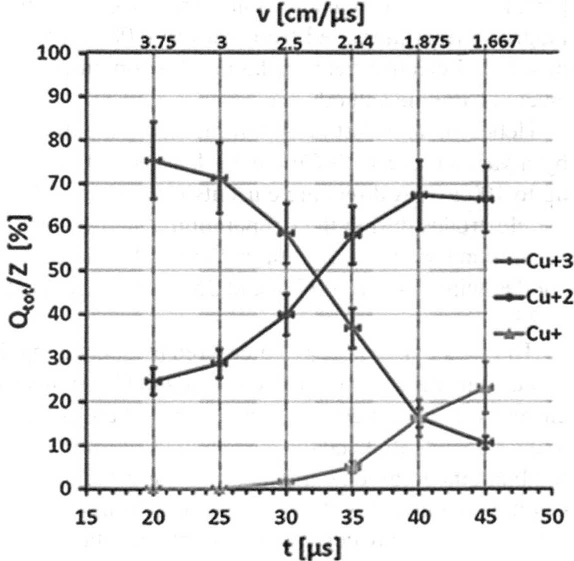

Fig. 12.30 Copper ion species (arbitrary units in percentage, vertical axis)) as dependence on the shutter time, horizontal axis) with and corresponding ion velocities. Figure taken from [76]. Permission number 4791410312031

at shorter shutter times, whereas at longer shutter times, the lower charge states, i.e., Cu^{+} and Cu^{+2} were dominated. This mean that the fastest ions were made of a large fraction of Cu^{+3} with some Cu^{+2} while the slowest ions were made of a larger fraction of Cu^{+2} with some Cu^{+}.The dependence of corresponding ion velocities was also obtained. It can be seen that the ion velocities are decreasing with detection times from 3.75 to 1.75 cm/s.

Recently, Shipilova et al. 2017 [77] measured an energy spectra of a metallic ion beam extracted from the plasma of a pulsed vacuum arc (pulse duration of 200 µs, arc current up to 100 A) by an electrostatic energy analyzer in a range of the extraction voltage U_{ext} of up to 10 keV. It was found that the most probable ion energy E_m/Z is significantly less than eU_{ext}, and the difference between these values as well as the width of the spectra decrease with increasing U_{ext} or/and decreasing with arc current. The spectra contain "tails" of ions with energies significantly exceeding E_m/Z, and the value of E_m/Z in the initial phase was considerably more than that in the quasistationary phase. The specifics of the ion optics of the extraction gap and the action of the space charge of the ion beam were considered in order to analyze of the observed effects.

12.6 Influence of the Magnetic Field

The dependence of the ion current collected by a shield biased to the cathode potential in a vacuum arc with current of 4.2 kA, 7 cm electrode diameters, and electrode spacing 19 mm of a vacuum interrupter was measured by Kimblin and Voshall, 1972

[78]. It was obtained that the ion current decreased from 280 to 120 A when the axial magnetic field increased from zero to 0.08 T. Plasma confinement and generation of a radial electric field explained this ion current decrease, which retard the ions ejected from the cathode spot.

Heberlein and Porto [79, 80] measured the distribution of ion currents generated by a vacuum arc with Cu electrodes without and with axial magnetic field varied up to 100 mT. A direct arc currents ranging from 70 to 2400 A was passed through the electrodes during their separation, and arc duration were 0.1 and 1 s. Electrode separation speed was 18 mm in 20 ms. The cathode diameter was 50.8 mm, and two anodes with diameters 12.7 and 25.4 mm were used. The arc gaps varied in range 7–25 mm.

The total ion current was measured by connecting the chamber wall to the cathode and also using a number of collector shields connected to the cathode. Also, the ion current was measured with one individual collector, while the other collectors were at floating potential and the measured ion current values to each collector were added to obtain the total ion current [80]. Figure 12.31 shows the ion current as a function of total arc current without magnetic field indicating that the ion current fraction is ~9% of the total arc current. The identical data were obtained for the total ion current for all mentioned measurement arrangements.

The measurements of angle ion distribution showed that the ion current distribution becomes strongly peaked in the forward direction with increasing axial magnetic field. The collector adjacent to the anode collects an amount of ion current, which is the same or even higher compared to the zero magnetic field case. The ion currents to the remaining collectors being strongly reduced with axial magnetic field increasing. The effect increases with increasing axial magnetic field strength up to value of 100 mT [80]. Figure 12.32 shows the reduction in the ion current fraction when an axial magnetic field of increasing strength is applied. After an initial drop to

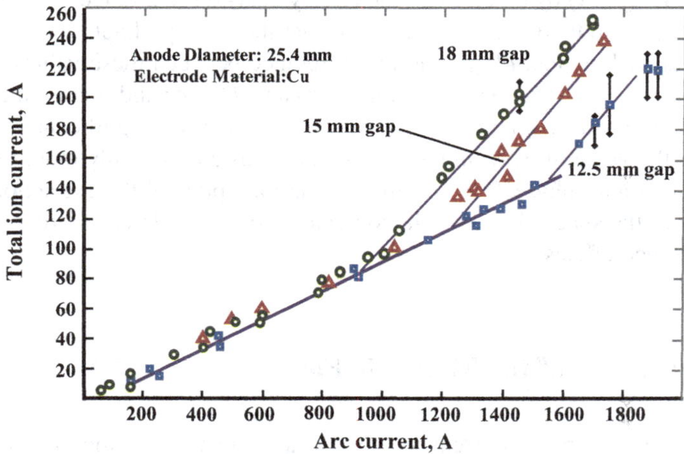

Fig. 12.31 Ion current as a function of total arc current without magnetic field

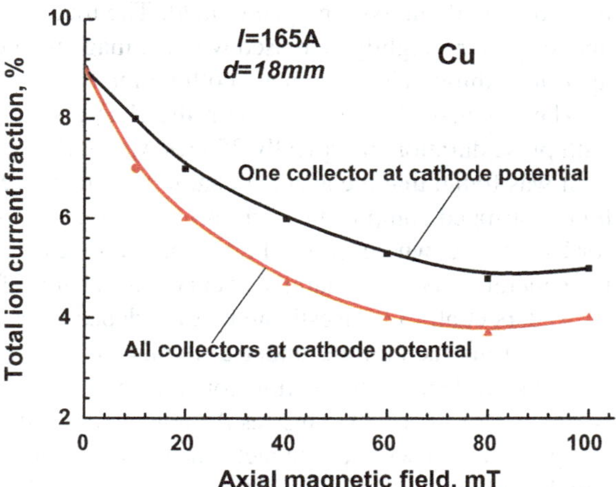

Fig. 12.32 Ion current fraction as a function of axial magnetic field

approximately half its original value, the collected ion current appears to become less dependent on the magnetic field. This reduction was explained by part of ion flux collected at the anode due to increase of the ion current in axial direction in comparison with that in the radial direction. Also, the observed effect was explained by assuming a curvature of the ion path due to the interaction with the magnetic field, and by assuming a radial electric field generated by the reduced mobility of the electrons as in work of [78].

Drouet and Meunier 1985 [81] studied the expansion of the metallic plasma cloud produced at copper cathode spot with an arc of 50 A, duration of 25–300 μs and in helium pressure from 0 to 50 torr without and with an applied magnetic field of 850 G (85 mT). Streak and frame high-speed photography and electrostatic probes were used. It was shown that at background pressures of up to 0.5 torr, the plasma expansion was similar to that observed in vacuum, i.e., the plasma expands at a constant velocity approximately equal to 0.7×10^6 cm/s. With further increasing the pressure, a background gas limits the expansion to a certain volume, which was smaller when a higher the gas pressure was presented. Over same range of pressures show slowing down progressively to zero velocity, with deceleration increasing as the pressure rises. A transverse magnetic field in vacuum confined the plasma close to the cathode plane and along the field lines; however, this effect of the magnetic field decreases and even disappears as a background gas was introduced and its pressure was raised in the experimental chamber. So, as the helium pressure was increased up to 10 torr, the luminous shape of the plasma cloud was found to be identical, irrespective of whether a magnetic field of used magnitude was applied.

Cohen 1989 [36] studied the ion current angle distribution by applying an axial magnetic field. It was showed that the plasma beam was collimated; i.e., the axial ion current substantially increased with the magnetic field to twice its 0-field value at 64 mT, while the off-axial current density decreased slightly with respect to its 0-field value. It was concluded that the ion current distribution became peaked along the z-axis with an axial magnetic field. The total ion current extracted through the anode aperture slightly increased with the magnetic field, and an anode with a larger aperture exhibited less magnetic collimation.

The magnetic field influence on the charge-state distributions in a vacuum arc with pulse duration of typically 250 μs was studied by Brown Galvin, 1989 [53].

It was found that the average charge state increased with field strength, and new highly stripped components were produced [53]. This effect is shown in Fig. 12.33 for Ur cathode, where shift to higher charge states with increasing field strength and the general shape of the family of curves for charges from 1^+ to 6^+ ions can be seen.

Anders et al. [57] investigated a time dependence of the charge state of the ions ejected from vacuum arc under a permanent coaxial magnetic field. The average charge-state dependence on time for Al cathode and arc current of 100 A in a vacuum is shown in Fig. 12.34. This result indicated that the large difference between ion charge with and without magnetic field was observed at the initial time of 100 μs and this difference decreases with time. At about 700 μs, the influence of the applied magnetic field on the average ion charge was found to be very small. A charge-state distribution of ions formed in a vacuum arc plasma in a magnetic field were measured using a magnetic spectrometer (GSI, Darmstadt) and a time-of-flight system at a time of 200 μs after triggering the pulse (Berkeley) by Oks et al. [82]. The results are illustrated in Table 12.18

As it follows from these data, the charge states of all of the metal species investigated significantly increased by a magnetic field. The high charge-state fractions increased and the low charge-state components decreased. For the highest charge states of a given spectrum, the increase in current fraction can be a very large factor, an order of magnitude or more. Some measurements were provided also at a field as high as 6 kg, and the degree of charge-state enhancement was further increased. Thus, for Ti the average charge state was 2 for $B = 0$, and 2.6 for $B = 3.75$ kg, while

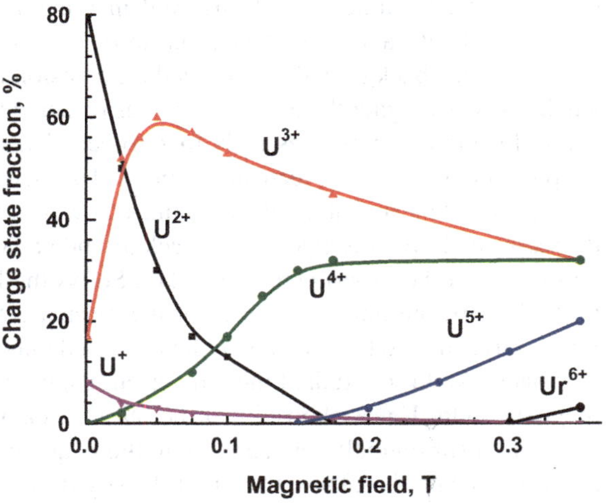

Fig. 12.33 Measured charge state fractions as a function of the applied magnetic field strength. Uranium cathode, $I = 100$ A

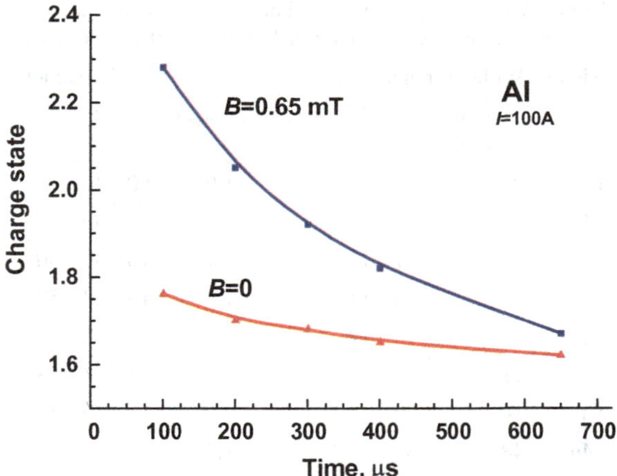

Fig. 12.34 Time evolution of the average ion charge state Q for Al cathode of 100 A arc without and with a permanent axial magnetic field of 65 mT

it was 3.0 for $B = 6$ kg. The particle fraction of Ti^{4+} in the beam is increased from 1% at $B = 0$, to 6% at $B = 3.75$ kg, and to 19% at $B = 6$ kg. Similar large increase was obtained for Sc, Ni, Co, and Hf.

Paoloni and Brown 1995 [83] investigated the dependence of charge-state distribution of ions on the arc current, in the presence of axial magnetic field from cathode materials Ti, Sn, and Pt. A pulsed vacuum arc was ignited between a water-cooled cylindrical cathode and an annular anode. An intense plume of highly ionized metal plasma is created at cathode spots for the 250 μs duration of the discharge. The charge-state distribution of ions in the beam was measured using a time-of-flight diagnostic, and the plasma beam passes through an annular electrostatic deflector. The charge-state distribution was determined at a time 50 μs after the initiation of the discharge. The measurements showed that in the absence of external magnetic field the charge-state distribution was independent of the arc current, while the ion charge state rises with an increasing current (in the range 20–100 A) under the 1 kG magnetic field [83]. The results are shown in Figs. 12.35 and 12.36

Oks et al. 1996 [84] have investigated the charge-state distributions of metal ions produced in a vacuum arc plasma at high current varied over the range from 200 to 4 kA and the magnetic field varied from zero up to 10 kG. The arc between cathode and anode was triggered by a short duration of 10 μs, high-voltage pulse (12 kV) with an arc pulse duration of 200–400 μs. The distributions of the ion charge state were measured by a TOF method after at least 100 μs after arc ignition. The experiments were conducted for low current (50–700 A), with a magnetic field of 10 G/A, and for high current (up to 4 kA), with a magnetic field of 1 G/A. According to Oks et al. [84] the influence of arc current on the charge-state distribution was weak at low current (Fig. 12.37), while for large current both the magnetic field and the arc current determined the charge-state distribution (Fig. 12.38 and Table 12.19).

Anders and Yushkov 2002 [22] showed that the kinetic energy of ions increased with increasing magnetic field, and the increase has a tendency to level off toward

Table 12.18 Charge-state distributions and mean charge states in particle current fractions with and without magnetic field applied. Arc current 220 A, $B_{max} = 3.75$ kG [82]

Metal	Without magnetic field						With magnetic field							
	Charge ($Q = 1$)	2	3	4	5	$Q_{aver}1$	1	2	3	4	5	6	$Q_{aver}2$	$Q_{aver}2/Q_{aver}1$
C	96	4				1.0	60	40					1.4	1.4
Mg	51	49				1.5	5	95					1.9	1.27
Al	38	51	11			1.7	10	40	50				2.4	1.4
Sc	23	66	11			1.9	16	23	59	2			2.5	1.31
Ti	11	76	12	1		2.0	5	35	54	6			2.6	1.3
V	11	72	15	2		2.1	13	31	48	8			2.5	1.2
Cr	14	70	15	1		2.0	11	26	55	8			2.6	1.3
Mn	48	52				1.5	26	47	25	2			2	1.33
Fe	28	68	6			1.8	7	58	35				2.3	1.28
Co	34	59	7			1.8	9	56	31	4			2.3	1.25
Ni	43	50	7			1.6	19	62	18	1			2.0	1.27
Cu	28	53	18	1		1.9	8	41	47	3	1		2.5	1.32
Y	7	63	29	1		2.2	6	9	77	8			2.9	1.32
Nb	3	40	39	16	2	2.7	1	9	23	52	13	2	3.7	1.37
Mo	7	30	40	20	3	2.8	5	11	26	48	10		3.5	1.25
Ba	3	97				2.0	2	41	53	3	1		2.6	1.30
La	4	65	31			2.3	3	16	61	20			3.0	1.30
Gd	8	81	11			2.0	1	43	41	15			2.7	1.35
Er	8	62	30			2.2	2	12	70	16			3.0	1.36
Hf	7	26	48	18	1	2.8	5	16	31	32	15	1	3.4	1.21
Ta	1	17	39	39	4	3.3	1	5	13	40	41	2	4.2	1.27
W	1	17	35	35	12	3.4	1	5	16	39	32	7	4.2	1.20
Pt	12	70	18			2.1	3	25	64	8			2.8	1.30
Pb	40	60				1.6	1	75	24				2.2	1.37
Bi	89	11				1.1	9	60	31				2.2	2.0

saturation (Fig. 12.39), similar to what has been previously observed with the average ion charge state as a function of magnetic field [84]. It was also indicated that if the plasma is produced in a region of low field strength and streaming into a region of higher field strength, the velocity might decrease due to the magnetic mirror effect.

A metal vapor vacuum arc-type ion source has been developed by Hollinger et al. [85] for the high-current injector at the GSI accelerator facility for the production of high charge-state ions. The production of high charge-state Bi ions using the mixed Bi-Cu cathodes (with 8–15% of Cu admixed) compared to pure Bi cathodes was reported by Adonin and Hollinger [86]. This ion source provided a high-intensity ion beam with a fraction of fourfold charged uranium ions up to 67% [87]. The ion

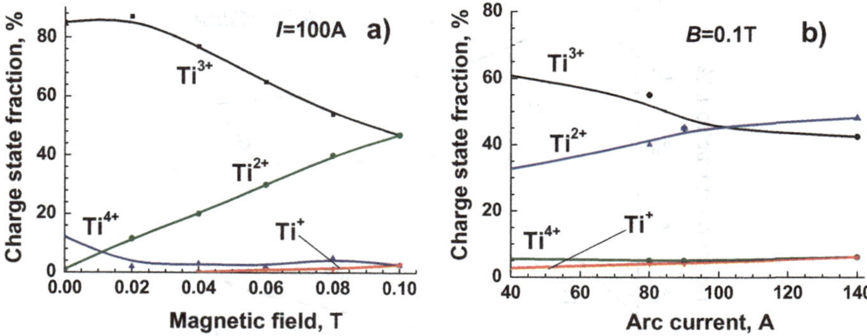

Fig. 12.35 Charge-state spectra for titanium versus magnetic field at an arc current of 100 A (**a**), and versus arc current at a magnetic field of 1 kG (**b**)

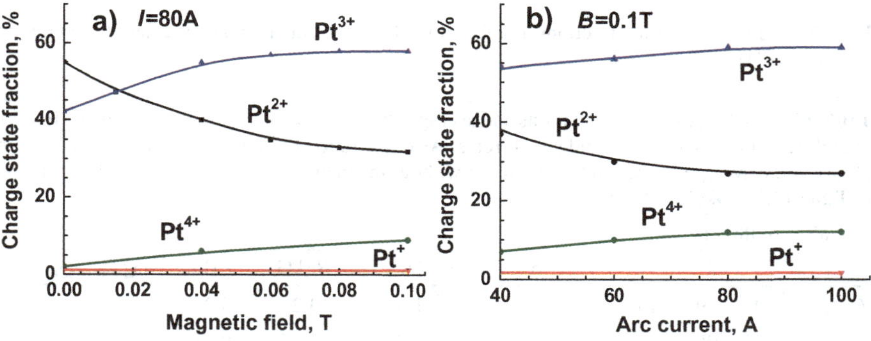

Fig. 12.36 Charge-state spectra for platinum versus magnetic field at an arc current of 80 A (**a**), and versus arc current at a magnetic field of 1 kG (**b**)

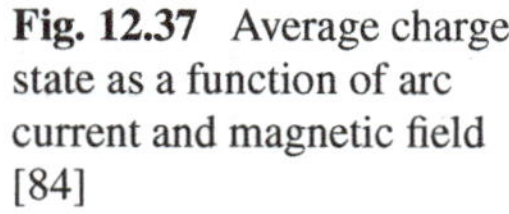

Fig. 12.37 Average charge state as a function of arc current and magnetic field [84]

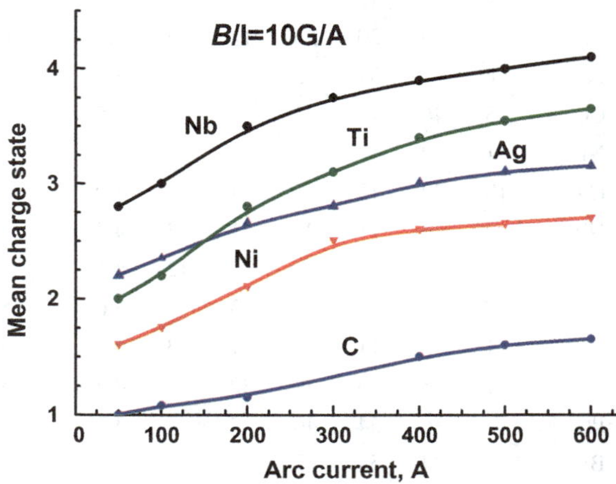

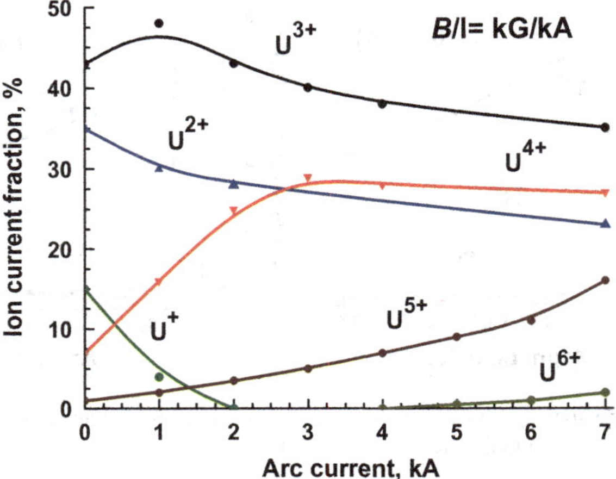

Fig. 12.38 Ion charge state fractions as a function of arc current and magnetic field for a uranium plasma jet [84]

Table 12.19 Charge-state distributions and average charge states $Q_{aver}2$ in particle current fraction with different magnetic fields and arc currents. Also, the ratio $Q_{aver}2/Q_{aver}1$ was presented, where $Q_{aver}1$ is the average charge states measured without magnetic field and for current 220 A (listed in Table 12.14. Oks1995) [84]

Metal	With magnetic field									
	1	2	3	4	5	6	$Q_{aver}2$	B (kG)	I (kA)	$Q_{aver}2/Q_{aver}1$
C	29	29	13				1.8	3.2	3.2	1.80
Mg	5	95					1.9	3.75	0.22	1.27
AI	5	11	85				2.8	1.2	1.2	1.65
Ti	1	6	15	58	20		3.9	10.0	1.3	1.95
V	13	31	48	8			2.5	3.75	0.22	1.20
Cr	4	9	20	53	12	2	3.7	6.2	0.8	1.85
Fe	6	20	34	38	2		3.1	2.2	2.2	1.72
Co	5	46	47	2			2.5	6.0	0.4	1.40
Ni	1	9	19	32	27	12	3.5	3.4	3.4	2.18
Cu	10	22	32	32	4		3.0	4.6	0.6	1.57
Nb		6	11	29	51	3	4.3	1.2	1.2	1.59
Mo		10	19	32	27	12	4.1	5.4	0.7	1.46
Ag	7	23	37	30	3		3.0	5.4	0.7	1.57
Hf	1	5	11	39	41	3	4.2	4.6	0.6	1.50
W	1	5	16	39	32	7	4.2	3.75	0.22	1.20
Pt	1	16	34	46	3		3.3	10.0	1.2	1.57
Bi	7	27	57	9			2.7	4.6	0.6	2.45
U	1	20	32	28	16	3	3.5	4.6	0.6	1.52

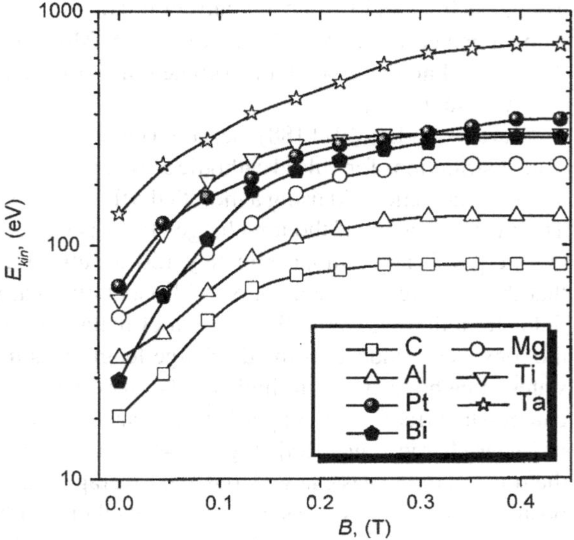

Fig. 12.39 Ion kinetic energy of indicated cathode materials as a function of magnetic field strength with arc current of 250 A and for arcing time $t > 150$ μs. Figure taken from [22]. Permission number 4776970475661

energy distributions were measured with an electrostatic cylinder spectrometer for C, Ti, and U with an arc current of 400 A as a function of the magnetic field *B*.

Data points have been obtained by averaging over 16 shots using a repetition rate of 2.3 pps with a pulse length of 500 μs. Data points are recorded 250 μs after triggering the arc [87]. Figure 12.40 illustrated the ion energy dependence on magnetic field. It can be seen that the ion energy has tendency to saturation with magnetic field.

Anders et al. 2006 [39] showed that applying an external magnetic field to the cathode region of the discharge is related solely to the appearance of ions with higher charge numbers in the plasma. At a magnetic field induction in the cathode region

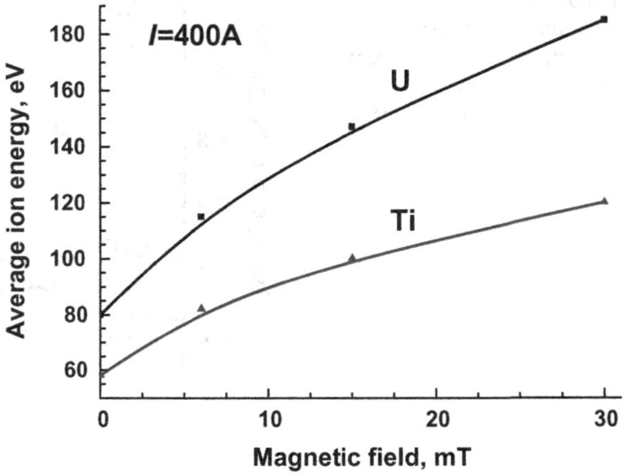

Fig. 12.40 Average ion energy <E_0> of titanium and uranium as function of the magnetic field, points are the experimental data, while the solid lines approximate the data for U and Ti, respectively [87]

as high as 0.45–0.50 T, the ion current fraction reached its maximum value for used cathode materials (C, Mg, Al, Ti, Co, Cu, Y, Mo, Cd, Sm, Ta, W, Pt), indicating that the effect of an increase in the average ion charge number with increasing magnetic field was saturated.

Oks and Anders 2011 [88] measured copper and gas ion species and their charge-state distributions for pulsed cathodic arcs in argon background gas in the presence of an axial magnetic field using a modified MEVVA-V source and a TOF spectrometer. The measurements of the ion charge-state distribution were made at the middle of the arc pulse, i.e., at approximately 150 μs after the arc has started. It was found that the burning voltage and power dissipation greatly increased with the magnetic field strength. In general, the fraction of metal ions and the average charge state of the ions were reduced as the discharge length was increased, i.e., the metal charge states at cathode spots are higher. Also was detected that the decrease of the high charge-state fraction of copper ions was accompanied by an increase of the fraction of lower charge states [88] (Fig. 12.41). Increasing the cathode position increases the fraction of the gas charged ions. The average charge state decreased with cathode position and with gas pressure. It was stayed that without magnetic field, only small amounts of gas ions can be found provided the cathode was positioned at a rather long distance to the extraction zone. The situation changes significantly when even a weak magnetic field was applied. The fractions of gaseous ions increased with magnetic field strength and cathode position. The observed dependencies were explained by the combination of charge exchange collisions and electron impact ionization.

Zhuang et al. [89] (similar as in [65, 66], named "jump current") studied the ion velocities in the plasma jet generated by the microcathode arc thruster by means of TOF method using enhanced ion detection system (EIDS). The EIDS triggers perturbations (spikes) on arc current waveform, and the larger current in the spike generates denser plasma bunches propagating along with the mainstream plasma.

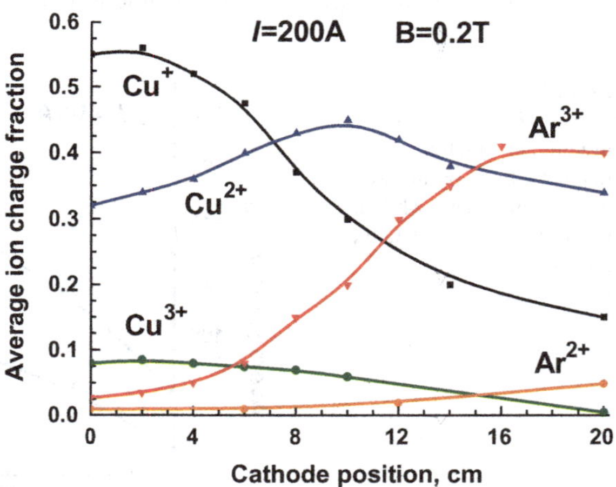

Fig. 12.41 Normalized composition of the extracted ion beam at different cathode positions with respect to the extraction area. Arc current 200 A, argon pressure 0.04 Pa, magnetic field 0.2 T

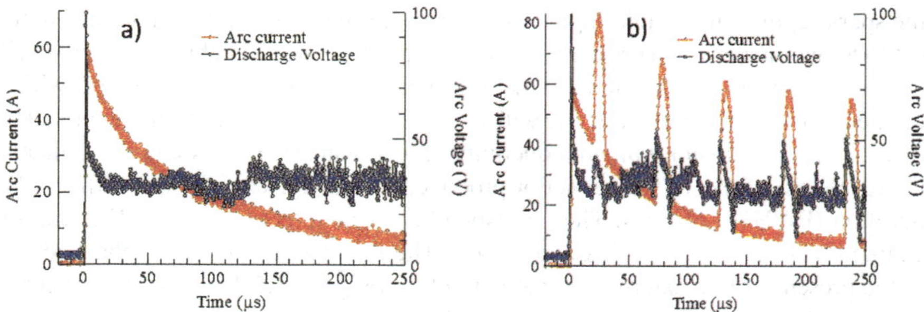

Fig. 12.42 **a** Arc current and arc voltage without superimposed spike. **b** Arc current and arc voltage with the superimposed spike (the spike position can be selected relative to the arc initiation). Figure taken from [89]. Permission number 4790880938696

The EIDS utilizes double electrostatic probes rather than single probes. Details of the method and results of the measurements (Fig. 12.42) were also summarized in [90]. The experiments were provided at residual gas pressure of about 10^{-4} torr using an annular titanium cathode with 4.85 mm inner diameter and 6.34 mm outside diameter, and same diameter annular copper anode with 1 mm width. The annular ceramic insulator tube having same inner and outer diameters and a width of about 1 mm was used as separator between the arc electrodes. The average Ti ion velocity was measured around 2×10^6 cm/s without a magnetic field.

It was found that the application of a magnetic field does not change ion velocities in the interelectrode region while it leads to ion acceleration in the free expanding plasma plume by a factor of two. Ion velocities of about 3.5×10^6 cm/s were detected for the magnetic field of about 300 mT at distance of about 100–200 mm from the cathode. The average thrust was calculated using the ion velocity measurements and the cathode mass consumption rate, and its increase with the magnetic field was demonstrated.

Anders 2012 [91] presented a review, and the works described the research at Lawrence Berkeley National Laboratory. The report was focused on clarifying the data of the observable ion charge-state distributions, which are known to involve multiple charge states and depend on the cathode material. Using a magnetic field it was discussed a possibility to control of such arc parameters as plasma density, the apparent spot motion, to guide plasma in a filter of the macroparticles and to control plasma impact on a substrate, and also to enhance the ion charge states. It was indicated that the strategies for obtaining higher charge states for applications include reducing the neutral density, use magnetic fields, or apply additional different ionization methods known from other types of ion sources.

Recently, Yushkov et al. [92] described different previously used techniques of metal ion charge state elevation including application of a short current pulse to the discharge as to transiently increase the discharge voltage and power, emulating the conditions of a high-current vacuum spark. New experimental results by utilizing the spark regime and combining it with a strong pulsed magnetic field applied to the cathode region were also presented. The cathodes from gold and bismuth were used.

The current of the vacuum spark between the end of the rod cathode with a diameter of 6.25 mm and the anode with an inner diameter of 30 mm was generated by discharging a capacitor with an inductivity of less than 20 nH. In some experiments, the discharge of the 10 μF capacitor produced a peak current of 4 kA with a pulse duration (FWHW) of 6 μs. The corresponding parameters for the 1 μF capacitor, charged to 15 kV, were 15 kA and 1.6 μs. The ion charge state was enhanced due to the presence of an external magnetic field (in range 1–2.5 T) at the arc current of 300 A and tends to saturate for Ti, Pt, and C when the magnetic field exceeds 1 T [93]. It was noted that the addition of a magnetic field to the spark plasma magnetizes the electrons and limits plasma expansion, which leads to an increase in the electron temperature relative to the free expansion case and to an increase in the likelihood of electrons to cause ionizing collisions.

In order to obtain ion with extra charged number, the experiment was conducted for bismuth cathode without magnetic field. This material has a high electrical resistance, leading to strong ohmic heating of the cathode, and a low thermal conductivity, reducing heat removal from the cathode surface. As a result, the energy dissipated in the arc spot region and the vapor flux produced in cathode spot is higher compared with those for other metals. The high vapor pressure of bismuth suggests a dominance of metal vapor, reducing the negative influence of gas desorption. The resulting TOF spectra for a bismuth cathode for pulses of 1.6 μs pulse duration the bismuth ion spectrum reached from Bi^{6+} to Bi^{13+}. The mean ion charge state of bismuth was $Bi^{10.4+}$. It was concluded that any gaseous impurity in a vacuum discharge plasma leads to a decrease of the metal ion charge state. Taking into account this point, the bismuth is particularity suited to produce high charge states.

12.7 Vacuum Arc with Refractory Anode. Ion Current

In last decades [94–97], a vacuum arc mode called the Hot Refractory Anode Vacuum Arc (HRAVA) was investigated, which has much less macroparticles (MP) contamination than conventional cathodic arcs. Refractory (Graphite, W, Nb, or Mo) HRAVA anode temperatures increased with time from room temperature in the initial stage to 2000–2500 K after arcing for 50–100 s in a 150–350 A vacuum arc. Metallic plasma of cathode material is re-evaporated by the hot anode. The HRAVA plasma flows and expands radially away from the electrode axis.

Beilis et al. [97] showed that the total ion current in the radially expanding plasma of a HRAVA was found to be larger than that measured in the conventional cathodic vacuum arc. The total ion current fraction f_i increased during the transition from the initial cathode spot stage to the HRAVA stage (Fig. 12.43a). While the f_i is independent of arc current in range of 150–350 A and other arcing conditions in

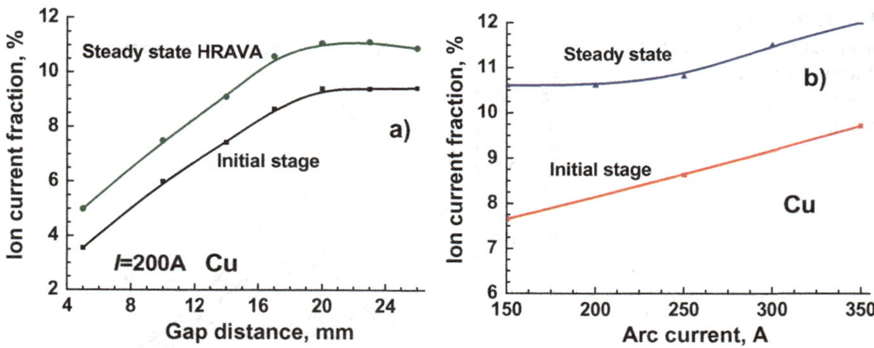

Fig. 12.43 Dependence of f_i on electrode gap h for $I = 200$ A during initial stage, and stage of steady state (**a**); dependence of f_i on arc current I for electrode gap $h = 17$ mm during initial stage and at steady state (**b**)

conventional arcs, the fraction increased with arc current in this current range in the HRAVA (Fig. 12.43b). The ion current fraction was found to be 50% larger for a 350 A HRAVA than maximal value obtained previously in conventional cathodic arcs.

The high value of f_i during steady state may in part be due to the increasing radial flow of the plasma as the anode is heated. The flowing re-evaporation of cathode material from the anode no longer allowed the axial plasma flow from the cathode to condense thereon. Another part may be due to evaporation and subsequent ionization of MPs striking the anode.

The total ion current and its angular distribution were measured in the radially expanding plasma of a HRAVA [98]. Arc currents from 175 to 340 A were supplied by a Miller XMT-400 welder (open circuit voltage of 70 V, current <400 A). In most of the experiments, a water-cooled copper cathode with diameter $D_c = 30$ mm and graphite anode (POCO DFP-1) with diameter $D_a = 32$ mm and length 30 mm were used. In some cases, other electrode materials were used. The electrodes were mounted on the same axis, and the axial gap between the electrodes, h, was varied from 3 to 26 mm. The ion current distribution along the z-axis (perpendicular to the electrode axis) was measured at two distances from the electrode axis, using ring probes placed coaxial to the electrode axis, with 1 cm width and radii $R = 5$ and 7 cm ($z = 0$ was chosen at the front cathode surface plane). The solid angle covered by the anode from the center of the cathode surface $\omega_a = 2\pi(1 - \cos\theta_a)$, where θ_a is the angle between the electrode axis and the line connected the cathode center with the anode face surface edge.

The angular distribution of the ion current at the initial arc stage ($t \sim 5$ s) for gap $h =$ 3, 6, 10, 16, and 22 mm at $R = 5$ cm and $h = 16$ mm at $R = 7$ cm is shown in Fig. 12.44 for 200 A). The angular distribution is given as a function of solid angle ω rather than angle θ in order to compare directly the results obtained in this work with those previously reported by Kutzner [34]. It is seen that all distributions had a maximum

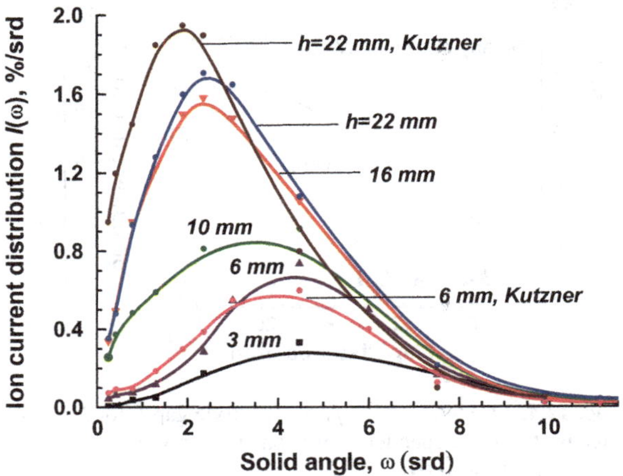

Fig. 12.44 Angular distribution of the ion current during the initial stage ($t \sim 5$ s). $I_{arc} = 200$ A, $R = 5$ cm, with cathode shields

(at $\omega = \omega_{max}$). For smaller gaps, the maximum decreased and ω_{max} increased. On the initial cathodic arc stage, the peak of the ion current angular distribution was shifted from the gap mid-plane to the anode direction due to the contribution of the cathode plasma jets (Fig. 12.44).

The two curves denote the results obtained by Kutzner [34] for $h = 6$ and 22 mm under similar conditions (30 mm copper cathode and $I = 200$ A). Kutzner [34] obtained slightly higher $f(\omega)$ for $\omega < \omega_{max}$ and vice versa for $\omega > \omega_{max}$. It is also seen that the distributions obtained at $R = 5$ and 7 cm (for $h = 16$ mm) were close.

The dependence of the maximum ion fraction (i.e., at large h) on arc current is illustrated in Fig. 12.45 for a graphite anode and Mo, W, and C cathodes in a few second duration arcs. It was observed that ion current fraction f_i was approximately

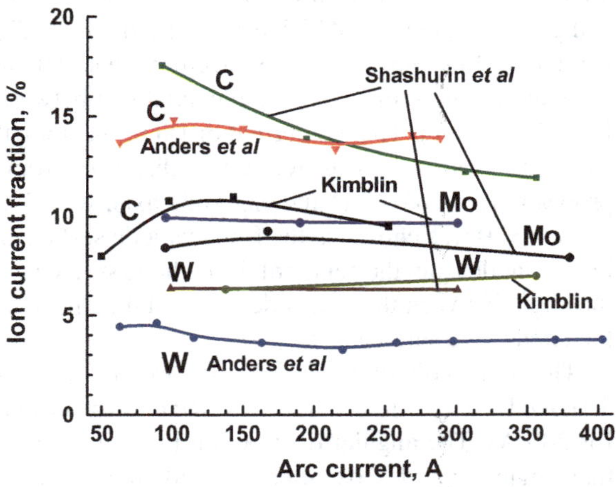

Fig. 12.45 Maximal ion fraction versus arc current for graphite, Mo, and W cathodes (solid lines) in [98] (initial stage) and reported in [31-Kimblin, 38-Anders et al.] for cathodic vacuum arcs

independent of the arc current for Mo and W cathodes (9.5 and 6.3%, respectively), while for graphite it decreased from 17.5 to 12% for $I = 150$ and 350 A, respectively. Kimblin's [31] and Anders et al. [38] data for cathodic vacuum arcs are shown also for comparison.

In the initial stage, the ion flux distribution was skewed in the anode direction due to the contribution of cathode plasma jets. In the developed HRAVA, the ion flux distribution is close to symmetric with respect to the mid-plane of the gap (for the gap sizes $h < 6$ mm and arc current of 200 A) suggesting that the directed motion of the cathode spot plasma jet ions is randomized by collisions in the dense plasma formed in the fully developed HRAVA mode.

Thus, in the developed HRAVA mode, the ion current distribution is approximately isotropic for short gaps ($h > 6$ mm) but skewed in the anode direction with larger gaps. HRAVA arc voltages and their fluctuations are weak and significantly lower than that in conventional cathodic vacuum arcs, resulting in higher arc stability, especially in large gaps ($h > 20$ mm).

An experiment to determine copper ion flux expanding from the HRAVA inter-electrode gap as a function of background gas pressure was conducted in work of [99]. The fraction of the ion flux in the radially expanding plasma flux was obtained by measuring the ion current and the film thickness. The measurements were conducted with arc currents of 145–250 A, a molybdenum anode and an electrode separation of about 10 mm. The saturation ion current was measured with a circular flat probe with 10 mm diameter biased at −30 V with respect to the anode. Distance L_{pr} from probe to the electrode axis was varied in the range from 30 to 170 mm.

The measurements showed that the collected ion current in vacuum was almost constant during the first 30 s of the arc duration ~2.5 mA/cm^2 at a distance of 110 mm from the arc axis, with an arc current of 200 A. In the developed HRAVA, at arc time >40 s the ion current increased to a steady-state value of ~5.5 mA/cm^2. The ion fraction in total deposition flux was 0.6 in vacuum and decreased with nitrogen pressure, except that a local maximum of ~0.8 was observed at ~13 Pa (Fig. 12.46).

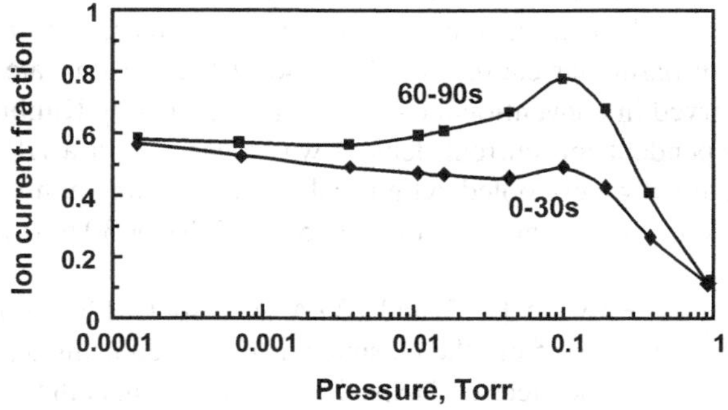

Fig. 12.46 Ion flux fraction in the depositing flux as a function of nitrogen pressure for the initial arc stage (averaged over 0–30 s) and the developed HRAVA (averaged over 60–90 s) at L_{pr} = 110 mm and arc current 200 A

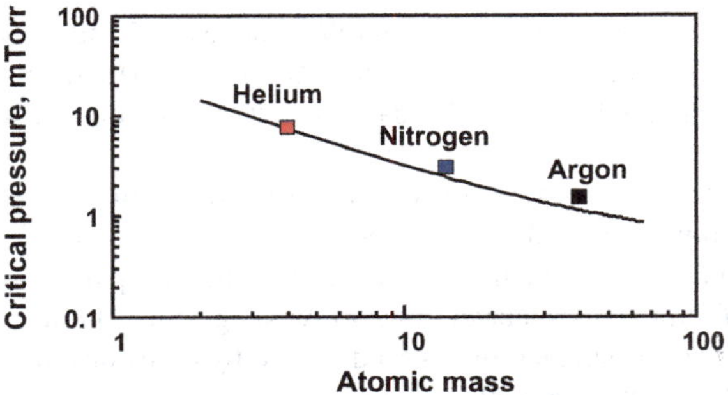

Fig. 12.47 Dependence of the critical pressure for ion flux on gas type (squares—experimental data, solid curve—approximation). Two points for the nitrogen correspond to one experiment by assumption of nitrogen presence only in molecular or in atomic state

The measured ion current in argon, nitrogen, and helium environments and the deposition rate in nitrogen remained approximately constant with background gas pressure up to some critical pressure, and then decreased with pressure eventually reaching zero.

As can be seen from Fig. 12.47, the critical pressures were 2, 4, and 10 Pa for argon, nitrogen, and helium, respectively. The critical nitrogen pressure for the deposition rate was 2 Pa in contrast with 4 Pa for the ion current. According to the observed temporal evolution, the ion current is relatively small during the initial stage of arcing ($t < 30$ s). This initial ion current is caused by the radial expansion of the cathodic plasma, while the most of Cu plasma condensed on the cold anode surface. The subsequent increase of the ion current, up to the maximum, was due to re-evaporation of this condensed Cu plasma (including MP's) from the heated anode. After this maximum, the ion current decreased with time to a steady-state value determined by the dynamic equilibrium between copper condensation and its re-evaporation from the hot anode surface in a vacuum.

The decreasing of the deposition flux with gas pressure for both arc stages was due to interaction of the expanding copper plasma plume with the background gas. As a result, part of the plasma flux was returned to the cathode, shields, etc., and therefore, the ion current and deposition rate were reduced with increasing background gas pressure. This in particular caused the decrease of the erosion rate with pressure, previously observed in conventional cathodic vacuum arcs by Kimblin [100].

The time-dependent ion current density was measured in a HRAVA sustained between a consumed water-cooled cylindrical Cu cathode and non-consumed cylindrical W anodes with different thickness $d = 5,10,15,20$, or 30 mm separated by a $h = 10$ mm gap [101].

Arc currents of $I = 130,150,175$, and 200 A were applied for period of 90 s. Ion current density J_i extracted from the plasma was measured using a probe located at varying distances from the electrode axis. The construction, cathode–anode configuration, probe position, and electrical source with the connections are illustrated in Fig. 12.48.

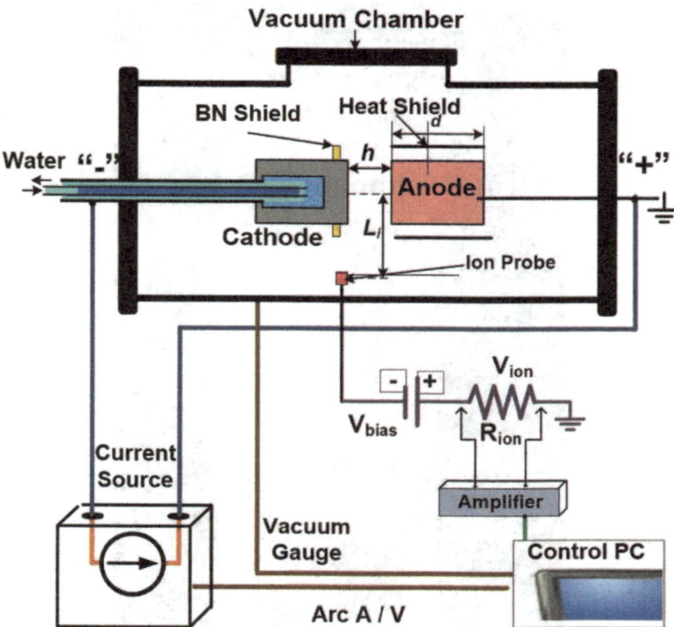

Fig. 12.48 Schematic diagram of the experimental setup

The active surface of the probe was oriented to be either perpendicular (⊥) or parallel (II) to the radially expanding plasma, to measure the directed or random component of J_i, respectively. The distances from the electrode axis probe surface or probe axis varied as $L_i = 50$, 80, 110, and 140 mm. The probe was negatively biased to $V_{bias} = -20$ V with respect to the grounded anode. A photograph of the developed at 60 s HRAVA is shown in Fig. 12.49. The brightness of the plasma is distributed almost uniformly in the gap. The arc voltage was measured to be ~21 V.

The ion current density started increase at arc ignition and grew slowly, passed through a peak and reached a final steady-state level. The total transient time consisted of three characteristic stages: (i) an initial constant current stage until time, t_{bp}, (ii) increasing current stage until the peak was reached at time t_p, (iii) and then a decreasing current stage until steady state was reached at time t_{ss}. The steady-state level increased with arc current (Fig. 12.50) and decreased with probe distance (Fig. 12.51).

The time to reach the steady state decreased when the anode thickness d was decreased from 30 to 5 mm: for $I = 200$ A from 48 to 12 s and for $I = 150$ A from 69 to 20 s, and weakly depended on probe orientation. The ion current density to the perpendicular probe significantly exceeded that for the parallel orientation.

The observed relatively low (weakly changed with time) ion current density during time $t < t_{bp}$ is the ion flux emitted from conventional vacuum arc in the initial stage, before the HRAVA mode is established, when most of plasma originating

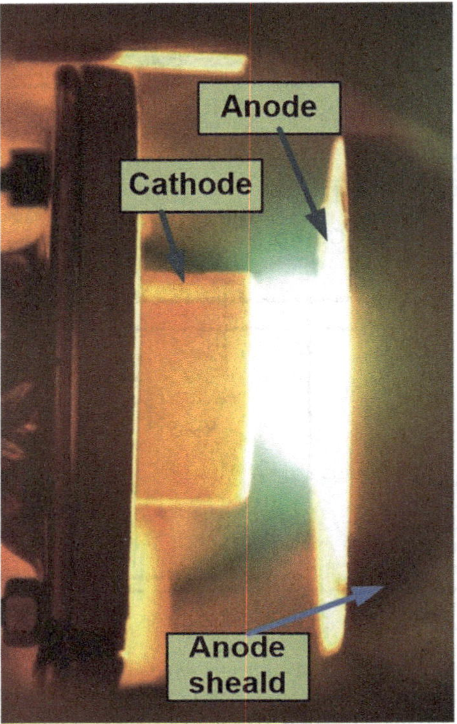

Fig. 12.49 Plasma plume photograph in HRAVA at $t = 60$ s ($I = 175$ A, $h = 10$ mm, $d = 15$ mm)

from cathode erosion was deposited on the cold anode surface. With time, the anode temperature increased due to anode heating by the plasma jet heat flux. The initial time t_{bp} of the rise to peak stage in essence marks the beginning of the HRAVA mode. When the anode was sufficiently hot ($t > t_{bp}$), the plasma jet was re-evaporated from the anode and also the previously deposited material was re-evaporated, generating an additional amount of ion current flux from the anode plasma plume.

The ion current density first increased to a maximal value ($t \geq t_p$), and then decreased, and finally, when all previously deposited cathode material was removed from the anode surface, a steady state was reached. During steady state, most of the ion flux emitted from the anodic plasma plume originated from plasma jet reflection. Thus, the ion current peaked when re-evaporation of cathode material was maximized, which occurred at time t_{bp} after arc ignition, needed to heat the anode. This time t_{bp} strongly depended on the anode thickness, because the time to heat the anode increased with d [102]. It was shown that the anode temperature transient time t_{ss} increased with d due to the increased thermal capacity of the larger anodes. The steady-state temperature decreased with d because radiation losses increased with larger anode surface area [103]. The observed larger ion current density in perpendicular direction in comparison to that in parallel direction at all arc stages was explained

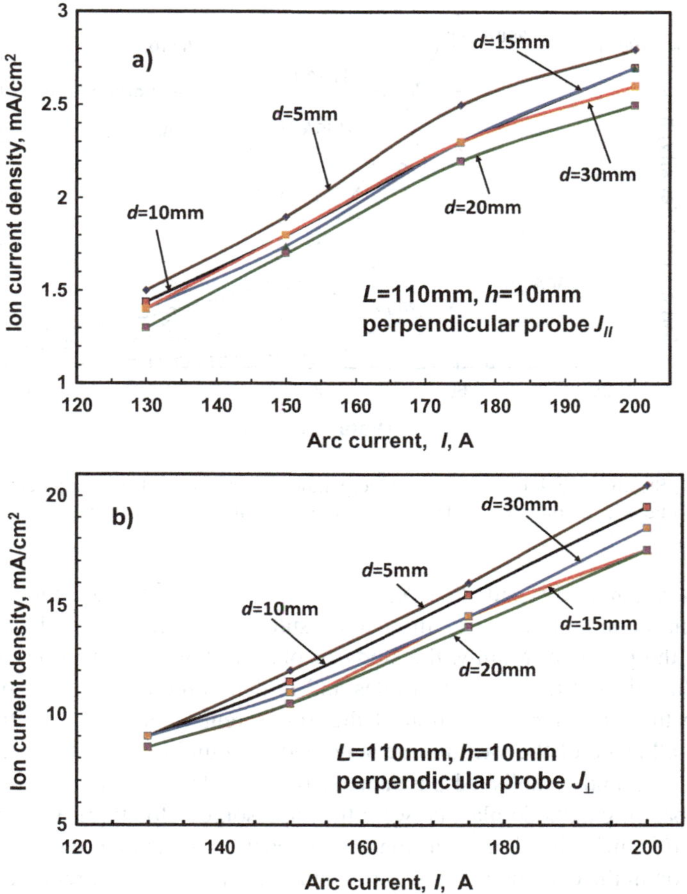

Fig. 12.50 Steady-state ion current with (**a**) parallel (II) and (**b**) perpendicular (⊥) probe orientation versus arc current ($L_i = 110$ mm, gap $h = 10$ mm) with anode thickness as parameter

taking into account the difference of ion velocity components of the plasma flow in parallel and perpendicular directions at the plasma–sheath probe boundary [101].

12.8 Summary

The present analysis of the plasma jets (in comparison to early found data of vapor streams, Chap. 11) shows that the further studies detected complicated structure of the expanding plasma generated in the electrode spots. The main results there discussed were obtained from different research group of Miller, Plyutto, Kutzner, Brown, Kimblin, Aksenov, Heberlein Anders, Oks, Yushkov, Paperny, Hollinger,

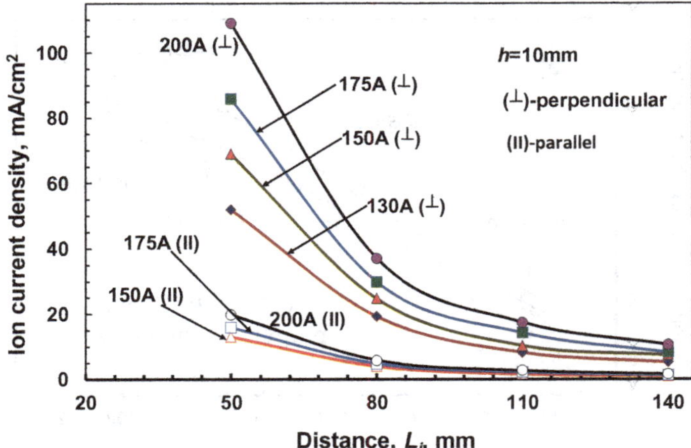

Fig. 12.51 Steady-state ion current with perpendicular (solid markers) and parallel (hollow markers) probe orientation as dependence on L_i ($d = 15$ mm, gap $h = 10$ mm)

Keidar, Boxman, Beilis, and other. Numerous works were published with wide range of different data, and here only the main results were able to consider as attempt to reflect the problem. Mainly, the study of plasma flow consist in observation of the ion flux ejected from the electrodes and in determine the physical parameters including the ion velocity, fraction of the ion current, charge state of the ions as well the influence of the vacuum arc state and surrounding condition. A common understanding and established parameters, obtained from numerous observations of the expanding cathode plasma, were the supersonic velocity of the ions at level of about 10^6 cm/s, the significant ion kinetic energy larger than the energy of ion acceleration in the cathode potential drop and total arc voltage. A relatively low ion current fraction observed, which was similar for different material, and it is about 8% like for Cu, except 7% for carbon and 10% for W, of the arc current. However, separate measurements indicated a range of ion current fraction varied from low value 5% for W to large value 19% for C.

In general, the mentioned parameters relatively weakly varied depending on current and cathode material for DC and moderate current of arcs with relatively low rate of current rise (dI/dt). The measured ion velocity can be varied by factor 2–3 and the energy as square of the velocity. From the above-published data, some uncertainty in the maximal ion current fraction in the vacuum arc as well as in ion current angular distribution can be noted. However, the results obtained for different electrode gap indicated the influence of the anode size when the radial ion flux was detected. Therefore, the authors of the previous works agree that maximal ion current may be obtained from vacuum arcs in which the anode subtends a minimum solid angle (i.e., by maximizing the electrode separation or the anode mesh transparencies). Thus, this result gives a maximal ion current fraction depending on cathode material that is a fundamental characteristic of the vacuum arc. It should be noted that sometime a relatively low ion fractions were obtained due to specifics of the measurement methods. These lower data can be connected with necessity to be corrected taking into account the geometric transmission factor of the mesh anode and therefore with possibility of an uncertainty appearing.

The general, the average ion energy as well as the average charge number Z tend to be greater for metals with larger arc voltages which was found mainly for refractory cathode materials (W, Mo, Ta). The ratio of average ion energy to the energy by arc voltage was determined in the range 1.5–3. The observation shown that ions and electron drifting was at approximately the same velocity. In some experiments for low melting cathode metals (Zn, Sn, and Pb), the ion velocity dependence on atom mass m was measured as $v \sim m^{-0.5}$, i.e., ion velocity decreased with m. Also for low-melting cathode metals, the total ion flux mainly consists of singly charged ions. The presence of multiply charged ions increased with melting point of the cathode metal.

At the same time the peak velocities of different ion species in plasma generated from a compound cathodes (based on Ti) in a DC vacuum arc for current of 65 A were found to be equal and independent on the ion mass at distance of 33 cm from the cathode surface. The peak velocities were in range of $(1.37–1.55) \times 10^6$ cm/s for ions from Ti-based cathode compositions.

The distribution of the **ion current** depends on angle with respect to the electrode axis. The maximal ion current is at axial direction (zero angle) and decreased with radius. This decreasing is larger at larger distance from the cathode. Therefore, the measured ion velocity at radial distance is lower than the velocity of plasma expansion in axial direction. The direct measurements of axial expansion with annular anode (250 μs duration and 100–500 A) indicated that the ion energy can be varied from 19.3 eV for Li to 120 eV for W and 160 eV for U; the Mach number is 3.1 for Li, 5 for W, and 6.9 for U.

In case of **pulse arcs**, the situation with plasma jet parameters changes quite significantly. The effect was appeared when the measurements were conducted for arcs with small pulse time, in the arc initiation or at high-current arcs with large $\mathrm{d}I/\mathrm{d}t > 10^8$ A/s. Also, the cathode temperature or axial magnetic field can affect the plasma structure. Thus, the directed ion energy and ion velocity are not constant throughout the arc operation time. The velocities of the ions generated at the ignition of the arc were fast and independent of the observed ion charge state. As example, the velocity of 3.75 cm/s was measured at distance 70 cm in the expanding plasma after 20 μs of 5 μs pulse vacuum arc with current of 35 A. In addition, the velocities of the ions generated at the extinction of the arc were relatively low and lower with charge number decreasing.

Thus, when the rate of current rise increases from 3×10^8 to 10^{10} A/s for copper cathode, the **ion energy** linearly rises from about $100–10^4$ eV. When the cathode temperature increased from 390 to 770 K in case of Ti cathode, the ion average energy decreased from 57 to 32 eV (Ti^+) and from 21 to 7 eV (Ti^{++}). The ion charge fraction changed in different ways. When the cathode temperature increased from 390 to 770 K, the single charged ion fraction increased (from 34 to 87%) and larger ion charges decreased (from 81 to 65%, Ti^{++}).

The **ion charge states** were widely studied. It was well established that ion CSDs, produced between the arc gap and in the expansion zone, depend on several factors including the cathode material. During the initial arc time, the ion charge states was significantly higher than at the steady current region. The average charge state decreases to a steady-state value with the vacuum arc operation time and the steady state was reached after few hundred μsec. The ion beam average charge state indicated essential fall during arc pulse in range from 25 to 250 μs.

In additional, the measurements of the evolution of the ion charge state distribution for very short pulsed discharge operating in the transition region between vacuum spark and vacuum arc showed grows of ion charges state. For example, when the pulse duration varied from lower than 1 μs to 30 μs for Pt cathode ion charge numbers varied from 1^+ to 10^+ and with gold cathode charge state of 11^+. Also, for selected bismuth cathode as a material of low electrical and heat conductivity, obtained record metal ion charge states, up to Bi^{13+}, extracted from a single discharge stage of relatively simple construction.

The highest ion charge states observed near the cathode spot reduces with distance toward a substrate, or ion extraction system. For example, Ti ions of different charges decreased with distance from the cathode and the difference of the current of the ion charges was relatively small at large distance (~40 cm). It was noted that the data of ion charge states measured at some distance from the electrode assembly could be different from that produced immediately near the cathode spots. The reduction of charge states is critically influenced by the density and nature of the neutrals and their production. This effect may be associated with charge exchange processes in the discharge plasma. The presence of low pressure ambient gas manifests as a decrease in the ions directed velocity. In this case, ions of different ionization level travel in different velocity.

The **magnetic field** influences the plasma jet parameters. In an Cu arc of 50 A, duration of 25–300 μs in an applied magnetic field of 85 mT, it was found the ions with constant **velocity** of about 0.7×10^6 cm/s up to helium pressure of 0.5 torr. The ion velocity of about 3.5×10^6 cm/s was detected from the Ti cathode in vacuum arc (peak current 60–80 A) during time increased up to 250 μs in the magnetic field of about 300 mT at distance of about 100–200 mm. For this arc and without a magnetic field the velocity was measured around 2×10^6 cm/s. The kinetic energy of ions increased with increasing magnetic field and the increase have a tendency to level off toward saturation when the magnetic field increased from zero to 0.4 T for different cathode materials at arc current of 250 A and arcing time >150 μs.

The **ion current** decreased from 280 to 120 A with magnetic field increasing from zero to 0.08 T when the ion current collected by a shield (at the cathode potential) in a vacuum arc of current 4.2 kA with electrode spacing 19 mm and 7 cm electrode diameters. The measurements of ion distribution on the angle in an Cu arc of duration 0.1 and 1 s and currents ranged from 70 to 2400 A showed that with increasing axial magnetic field up to 100 mT, the ion current distribution becomes strongly peaked in the forward direction. In these relatively small values (up to 100 mT), the ion current fraction was also decreased with axial magnetic field strength. After a drop of this fraction to approximately half its original value for small magnetic fields, in range up to 40 mT, the collected ion current fraction appears to become less dependent on the magnetic field up to 100 mT. This reduction was explained by part of ion flux collected at the anode due to increase of the ion current in axial direction.

In general, the **average charge state** increased with magnetic field strength, and new highly charged components were produced. A large difference between ion charge, measured without and with magnetic field of 65 mT, was observed at the initial arc time (100 μs, Al cathode arc current of 100 A) and this difference decreases with time (up to 700 μs). In addition, at about 700 μs, the influence of the applied magnetic field on the average ion charge was found to be very small. At a large magnetic field induction in the cathode region as high as 0.45–0.50 T, the ion current fraction reached its maximum value for used cathode materials (C, Mg, Al, Ti, Co, Cu, Y, Mo, Cd, Sm, Ta, W, Pt), indicating that the effect of an increase in the average ion charge number was saturated with increasing magnetic field.

According to DC Miller's 1969 experiment the fraction of single charged ions increased while the higher charged ions decreased noticeably with increasing the arc current measured at 50, 100, and 200 A. However, Brown and Galvin 1989 observed relatively weakly influence of the arc current on the ion charge state in the range between 100 and 1200 A with pulse length of 250 μs. This weakly influence was interpreted in accordance with experimental data for which an increasing of the arc current involved to simply the creation of more cathode spots, rather than a change in the plasma parameters within the spots. There is a shift to slightly higher charge states as the current is increased, but this trend often reverses for a sufficiently high arc current. Sangines 2010 also observed that, when the arc current was increased, the distances between spots were reduced, and therefore, the high charge states were produced for longer times. Also, as the current was increased, the peak of the emission intensity from Al^{2+} shifts to later times. It was concluded that when the arc current increased at relatively high values (~kA) **the dynamics of the cathode spots** determine a mechanism for the production of high charge states [104, 105].

References

1. Lawrence, E. O., & Dunnington, F. G. (1930). On the early stages of electric sparks. *Physical Review, 35,* 396–407.
2. Easton, E. C., Lucas, F. B., & Creedy, F. (1934). High velocity streams in the vacuum arc. *Electrical Engineering, 53,* 1454–1460.
3. Rayskiy, S. M. (1940). Propagation of the vapor of the electrode material at spark discharge. *Soviet Physics—JETP (ЖТЭФ), 10*(8), 908–909. (In Russian).
4. Haynes, J. R. (1948). The production of high velocity mercury vapor jets by spark discharge. *Physical Review, 73*(8), 891–903.
5. Hermoch, V. (1959). The vapor jets of electrode material of a short-time high-intensity electric arc. *Czechoslovak Journal of Physics, 9*(2), 221–228.
6. Hermoch, V. (1959). A short time channel in a high current electrical discharge. *Czechoslovak Journal of Physics, 9*(3), 377–387.
7. Hermoch, V. (1959). On processes in electrode spaces of a short-time high-intensity electric arc. *Czechoslovak Journal of Physics, 9*(4), 505–511.

8. Plyutto, A. A., Ryzhkov, V. N., & Kapin, A. T. (1965). High speed plasma streams in vacuum arcs. *Soviet Physics—JETP, 20*, 328–337.
9. Plyutto, A. A. (1936). Acceleration of the positive ions in expanding plasma of a vacuum spark. *Soviet Physics—JETP, 39*, 1589–1592.
10. Hendel, H. W., & Reboul, T. T. (1962). Adiabatic acceleration of ions by electrons. *Physics of Fluids, 5*(3), 360–363.
11. Tyulina, M. A. (1965). Acceleration of ions in a plasma formed by breaking a current in a vacuum. *Soviet Physics Technical Physics, 10*(3), 396–399.
12. Tyulina, M. A. (1967). Investigation of plasma propagation rate when current is switched off in a vacuum. *Soviet Physics Technical Physics, 11*(10), 1421–1423.
13. Rondeel, W. G. J. (1974). Investigation of the ions emitted from a copper-vapor arc in vacuum. *Journal of Physics. D. Applied Physics, 7*, 629–634.
14. Byon, E., & Anders, A. (2003). Ion energy distribution functions of vacuum arc plasmas. *Journal of Applied Physics, 93*, 1899–1906.
15. Shafir, G., & Goldsmith, S. (2007). Charge and velocity distribution of ions emitted from two simultaneously operating and serially connected vacuum arcs. *IEEE Transactions on Plasma Science, 35*(4) part 2, 885–890.
16. Davis, W. D., & Miller, H. C. (1969). Analysis of the electrode products emitted by dc arcs in a vacuum ambient. *Journal of Applied Physics, 40*(5), 2212–2221.
17. Kutzner, J., & Miller, H. C. (1989). Ion flux from the cathode region of a vacuum arc. *IEEE Transactions on Plasma Science, 17*(5), 688–694.
18. Miller, H. C. (1972). Measurements on particle fluxes from dc vacuum arcs subjected to artificial current zeroes. *Journal of Applied Physics, 43*(5), 2175–2181.
19. Grissom, J. T., & McClure, G. W. (1972). Energy distributions ions from the anode plasma of a pulsed vacuum arc. *International Journal of Mass Spectrometry Ion Physics, 9*, 81–93.
20. Grissom, J. T. (1974). Energy distributions of ions from a pulsed vacuum arc. In *Proceedings of VIth International Symposium on Discharges and Electrical Insulation in Vacuum* (pp. 253–258), Swansea, England.
21. Yushkov, G. Y., Anders, A., Oks, E. M., & Brown, I. G. (2000). Ion velocities in vacuum arc plasmas. *Journal of Applied Physics, 88*(10), 5618–5622.
22. Anders, A., & Yu Yushkov, G. (2002). Ion flux from vacuum arc cathode spots in the absence and presence of a magnetic field. *Journal of Applied Physics, 91*(8), 4824–4832.
23. Plyutto, A. A. (1961). Acceleration of positive ions in expansion of the plasma in vacuum spark. *Soviet Physics JETP, 12*, 1106–1108.
24. Korop, E. D., & Plutto, A. A. (1970). Acceleration of ions of cathode material in vacuum breakdown. *Soviet Physics Technical Physics, 15*, 1986–1989.
25. Astrakhantsev, N. V., Krasov, V. I., & Paperny, V. L. (1995). Ion acceleration in a pulse vacuum discharge. *Journal of Physics. D. Applied Physics, 28*, 2514–2518.
26. Artamonov, M. F., Krasov, V. I., & Paperny, V. L. (2001). Generation of multiply charged ions from a cathode jet of a low-energy vacuum spark. *Journal of Physics. D. Applied Physics, 34*, 3364–3367.
27. Artamonov, M. F., Krasov, V. I., & Paperny, V. L. (2001). Registration of accelerated multiply charged ions from the cathode jet of a vacuum discharge. *Soviet Physics JETP, 93*(6), 1216–1221.
28. Gorbunov, S. P., Krasov, V. P., Paperny, V. L., & Savyelov, A. S. (2006). Flow of multiple charged accelerated metal ions from low-inductance vacuum spark. *Journal of Physics. D. Applied Physics, 39*, 5002–5007.
29. Paperny, V. L., Chernich, A. A., Astrakchantsev, N. V., & Lebedev, N. V. (2009). Ion acceleration at different stages of a pulsed vacuum arc. *Journal of Physics. D. Applied Physics, 42*, 155201.
30. Kimblin, C. W. (1971). Vacuum arc ion currents and electrode phenomena. *Proceedings of the IEE, 59*(4), 546–555.
31. Kimblin, C. W. (1973). Erosion and ionization in the cathode spot regions of vacuum arc. *Journal of Applied Physics, 44*(7), 3074–3081.

32. Zalucki, Z., & Kutzner, J. (1976). Ion current in the vacuum arc. In *Proceedings of VII International Symposium Discharge Electrical Insulation in Vacuum* (pp. 297–302), Novosibirsk. USSR.
33. Kutzner, J., & Zalucki, Z. (1977). Ion current emitted from cathode region of a DC copper vacuum arc. In *Proceedings of 3rd International Conference Switching Arc Phenomena* (pp. 210–216), Lodz, Poland, part 1.
34. Kutzner, J. (1978). Angular distribution of ion current in a DC copper vacuum arc. In *Proceedings VIII International Symposium Discharge Electrical Insulation in Vacuum* (pp. A1–A15), Albuquerque, NM, USA.
35. Kutzner, J. (1981). Voltage-current characteristics of diffusion vacuum arc. *Physica, 104C,* 116–123.
36. Cohen, Y., Boxman, R. L., & Goldsmith, S. (1989). Angular distribution of ion current emerging from an aperture anode in a vacuum arc. *IEEE Transactions on Plasma Science, 17*(5), 713–716.
37. Marks, H. S., Beilis, I. I., & Boxman, R. L. (2009). Measurement of the vacuum arc plasma force. *IEEE Transaction on Plasma Science, 37*(7), 1332–1337.
38. Anders, A., Oks, E. M., Yushkov, G. Y., Savkin, K. P., Brown, I. G., & Nikolaev, A. G. (2005). Measurements of the total ion flux from vacuum arc cathode spots. *IEEE Transactions on Plasma Science, 33*(5), 1532–1536.
39. Anders, A., Oks, E. M., Yushkov, G. Y., Savkin, K. P., Brown, I. G., & Nikolaev, A. G. (2006). Determination of the specific ion erosion of the vacuum arc cathode by measuring the total ion current from the discharge plasma. *Technical Physics, 51*(10), 1311–1315.
40. Franzen, J., & Schuy, K. D. (1965). Time- and energy-resolved mass spectroscopy: The condensed vacuum discharge between solid electrodes. *Zs Naturfor. 20a,* 176–180.
41. Schuy, K. D., & Hintenbergeh, H. (1963). Massenspektroskopische Festkörperuntersuchungen verbesserter Reproduzierbarkeit mit dem Gleichstrom-Abreißfunken im Vakuum zur lonenerzeugung Z. *Naturforschg. 18a,* 926.
42. Miller, H. C. (1981). Constraints imposed upon theories of the vacuum arc cathode region by specific ion energy measurements. *Journal of Applied Physics, 52*(7), 4523–4530.
43. Lunev, V. M., Ovcharenko, V. D., & Khoroshikh, V. M. (1977). Plasma properties of a metal vacuum arc. I. *Soviet Physics Technical Physics, 22*(7), 855–858.
44. Lunev, V. M., Padalka, V. G., & Khoroshikh, V. M. (1977). Plasma properties of a metal vacuum arc. *II. Soviet Physics Technical Physics, 22*(7), 858–861.
45. Aksenov, I. I., Bren', V. G., Padalka, V. G., & Khoroshikh, V. M. (1981). Mechanism shaping the ion energy distribution in the plasma of vacuum arc. *Soviet Technical Physics Letter, 7*(19), 497–499.
46. Aksenov, I. I., Bren', V.G., Osipov, V. A., Padalka, V. G., & Khoroshikh, V. M. (1983). Plasma in a stationary vacuum arc discharge. Part1. Plasma flux formation. *High Temperature, 21,* 160–164, (1983).
47. Aksenov, I. I., Bren', V. G., Konovalov, I. I., Kudryavtsev, E. E., Padalka, V. G., Sysoev, Y. A., & Khoroshikh, V. M. (1983). Plasma in a stationary vacuum arc discharge. Part2. Affects of integral cathode temperature. *High Temperature, 21*(4), 484–488.
48. Khoroshikh, V. M., Aksenov, I. I., & Konovalov, I. I. (1988). Structure of plasma jets generated by cathode spot of a vacuum arc. *Soviet Physics—Technical Physics, 33*(6), 723–724.
49. Brown, I. G. (1985). The metal vapor vacuum arc (MEVVA) high current ion source. *IEEE Transactions on Nuclear Science, 32*(5), P1, 1723–1727.
50. Brown, I. G., Galvin, J. E., & MacGill, R. A. (1985). High current ion source. *Applied Physics Letter, 47*(4), 358–360.
51. Brown, I. G., Galvin, J. E., MacGill, R. A., & Wright, R. T. (1987). Improved time of flight ion charge state diagnostic. *Review of Scientific Instruments, 58*(9), 1589–1592.
52. Brown, I. G. (1988). Multiply stripped ion generation in the metal vapor vacuum arc. *Journal of Applied Physics, 63*(10), 4889–4898.
53. Brown, I. G., & Galvin, J. E. (1989). Measurements of vacuum arc ion charge-state distributions. *IEEE Transactions on Plasma Science, 17*(5), 679–682.

54. Galvin, J. E., Brown, I. G., & MacGill, R. A. (1990). Charge state distribution studies of the metal vapor vacuum arc ion source. *Review of Scientific Instruments, 61*(1), 583–585.
55. Miller, H. C., & Kutzner, J. (1991). Ion flux from the cathode region of a vacuum arc. *Contributions to Plasma Physics, 31*(3), 261–277.
56. Kutzner, J., & Miller, H. C. (1992). Integrated ion flux emitted from the cathode spot region of a diffuse vacuum arc. *Journal of Physics. D. Applied Physics, 25*(4), 686–693.
57. Anders, A., Anders, S., Juttner, B., & Brown, I. G. (1993). Time dependence of vacuum arc parameters. *IEEE Transactions on Plasma Science, 21*(3), 305–311.
58. Anders, A. (2001). A periodic table of ion charge-state distributions observed in the transition region between vacuum sparks and vacuum arcs. *IEEE Transactions on Plasma Science, 29*(2), 393–398.
59. Brown, I. G. (1994). Vacuum arc ion sources. *Review of Scientific Instruments, 65*(10), 3061–3081.
60. Tsuruta, K., Suzuki, K., & Kunitsu, K. (1992). Charge state and residence time of metal ions generated from a microsecond vacuum arc. *IEEE Transactions on Plasma Science, 20*(2), 99–103.
61. Tsuruta, K., & Yamazaki, N. (1993). Residence time of metal ions generated from microsecond vacuum arcs. *IEEE Transactions on Plasma Science, 21*(5), 426–430.
62. Tsuruta K, Yamazaki, N., & Watanabe, G. (1994). Residence time and charge state of silver and zinc ions generated from microsecond vacuum. *IEEE Transactions on Plasma Science, 22*(4), 486–490.
63. Tsuruta, K., Sekiya, K., Tan, O., & Watanabe, G. (1996). Velocities of copper and silver ions generated from impulse vacuum arc. In Proceeding of *17th International Symposium o Discharges Electrical Insulation Vacuum* (Vol. 1, pp. 181–184), Berkeley, CA.
64. Tsuruta, K., Sekiya, K., & Watanabe, G. (1997). Velocities of copper and silver ions generated from an impulse vacuum arc. *IEEE Transactions on Plasma Science, 22*(4), 603–608.
65. Bugaev, A. S., et al. (1999). Influence of a current jump on vacuum arc parameters. *IEEE Transactions on Plasma Science, 27*(4), 882–887.
66. Bugaev, A. S., Oks, E. M., Yushkov, G. Y., Anders, A., & Brown, I. G. (2000). Enhanced ion charge state in vacuum arc plasmas using a "current spike" method. *Review of Scientific Instruments, 71*(2), 701–703.
67. Anders, A. (2005). Time-dependence of ion charge state distributions of vacuum arcs: An interpretation involving atoms and charge exchange collisions. *IEEE Transactions on Plasma Science, 33*(1), 205–209.
68. Anders , A., Oks, E. M., Yushkov, G. Y. (2007). Production of neutrals and their effects on the ion charge states in cathodic vacuum arc plasmas. *Journal of Applied Physics, 102*(4), 043303.
69. Anders, A., Fukuda, K., & Yushkov, G. Y. (2005). Ion charge state fluctuations in vacuum Arcs. *Journal of Physics. D. Applied Physics, 38,* 1021–1028.
70. Oks, E., Youshkov, G. Y., & Anders, A. (2008). Temporal development of ion beam mean charge state in pulsed vacuum arc ion sources. *Review of Scientific Instruments, 79,* 02B301.
71. Yushkov, G. Y., & Anders, A. (2008). Extractable, elevated ion charge states in the transition regime from vacuum sparks to high current vacuum arcs. *Applied Physics Letter, 92*(4), 041502.
72. Sanginés, R., Israel, A. M., Falconer, I. S., McKenzie, D. R., & Bilek, M. M. M. (2010). Production of highly ionized species in high-current pulsed cathodic arcs. *Applied Physics Letter, 96,* 221501.
73. Romanov, I. V., Rupasov, A. A., Shikanov, A. S., Paperny, V. L., Moorti, A., Bhat, R. K., et al. (2010). Energy distributions of highly charged ions escaping from a plasma via a low-voltage laser-induced discharge. *Journal of Physics. D. Applied Physics, 43,* 465202.
74. Zhirkov, I., Eriksson, A. O., & Rosen, J. Ion velocities direct current arc plasma generated from compound cathodes. *Journal of Applied Physics, 114*(21), 213–302.
75. Zhirkov, I., Oks, E., & Rosen, J. (2015). Experimentally established correlation between ion charge state distributions and kinetic ion energy distributions in a direct current vacuum arc discharge *Journal of Applied Physics, 117*(9), 093301.

76. Silver, Y., Nachshon, I., Beilis, I. I., Leibovitch, G., & Shafir, G. (2017). Velocity and ion charge in a copper plasma plume ejected from 5 microsecond vacuum arcs. *Journal of Applied Physics, 121*(5), 053301.
77. Shipilova, O. I., Chernich, A. A., & Paperny, V. L. (2017). Characteristics of intense multi-species metallic ion beams extracted from plasma of a pulsed cathodic arc. *Physics of Plasmas, 24,* 103108.
78. Kimblin, C. W., & Voshall, R. E. (1972). Interruption ability of vacuum interrupters subjected to axial magnetic fields. *Proceedings of the Institution of Electrical Engineers, 119*(12), 1754–1758.
79. Heberlein, J. V. R., & Porto, D. R. (1982). The interaction of vacuum arc ion currents with axial magnetic fields. In *Proceedings of 10th International Symposium Dischargers and Electrical Insulation in Vacuum* (pp. 181–182), USA, Columbia.
80. Heberlein, J. V. R., & Porto, D. R. (1983). The interaction of vacuum arc ion currents with axial magnetic fields. *IEEE Transactions on Plasma Science, 11*(3), 152–159.
81. Drouet, M. G., & Meunier, J. -L. (1985). Influence of the background gas pressure on the expansion of the arc-cathode plasma. *IEEE Transactions on Plasma Science, PS-13*(5), 285–287.
82. Oks, E. M., Brown, I. G., Dickinson, M. R., MacGill, R. A., Emig, H., Spadtke, P., et al. (1995). Elevated ion charge states in vacuum arc plasmas in a magnetic field. *Applied Physics Letters, 67*(2), 200–202.
83. Paoloni, F. J., & Brown, I. G. (1995). Some observations of the effect of magnetic field and arc current on the vacuum arc ion charge state distribution. *Review of Scientific Instruments, 66*(7), 3855–3858.
84. Oks, E. M., Anders, A., Brown, I. G., Dickinson, M. R., & MacGill, R. A. (1996). Ion charge state distributions in high current vacuum arc plasmas in a magnetic field. *IEEE Transactions on Plasma Science, 24*(3), 1174–1183.
85. Hollinger, R., Galonska, M., Gutermuth, B., Heymach, F., Krichbaum, H., Leible, K.-D., et al. (2008). Status of high current ion source operation at the GSI accelerator facility. *Review of Scientific Instruments, 79,* 02C703.
86. Adonin, A., & Hollinger, R. (2012). Development of high current Bi and Au beams for the synchrotron operation at the GSI accelerator facility. *Review of Scientific Instruments, 83*(2), 02A505.
87. Galonska, M., Hollinger, R., & Spadtke, P. (2004). Charge sensitive evaluated ion and electron energy distributions of a vacuum arc plasma. *Review of Scientific Instruments, 75*(5), 1592–1594.
88. E.M. Oks, and A. Anders, Measurements of the Ion Species of Cathodic Arc Plasma in an Axial Magnetic Field, *IEEE Trans. Plasma Sci.* 39, N6, (2011)
89. Zhuang, T., Shashurin, A., Beilis, I. I., & Keidar, M. (2012). Ion velocities in a micro-cathode arc thruster. *Physics of Plasmas, 19*(6), 063501.
90. Keidar, M., & Beilis, I. I. (2016). *Plasma engineering*. Elsevier, London-NY: Acad Press.
91. Anders, A. (2012). The evolution of ion charge states in cathodic vacuum arc plasmas: a review. *Plasma Sources Science Technology, 21*(3), 035014.
92. Yushkov, G. Y., Anders, A., Frolova, V. P., Nikolaev, A. G., Oks, E. M., & Vodopyanov, A. V. (2015). Plasma of vacuum discharges: The pursuit of elevating metal ion charge states, including a recent record of producing Bi^{13+}. *IEEE Transactions on Plasma Science, 43*(8), Part I, 2310–2317.
93. Anders, A., Yushkov, G., Oks, E., Nikolaev, A., & Brown, I. (1998). Ion charge state distributions of pulsed vacuum arc plasmas in strong magnetic fields. *Review of Scientific Instruments, 69*(3), 1332–1335.
94. Beilis, I. I., Goldsmith, S., & Boxman, R. L. (2000). The hot refractory anode vacuum arc: a new plasma source for metallic film deposition. *Surface & Coatings Technology, 133–134*(1–3), 91–95.
95. Beilis, I. I., Koulik, Y., Yankelevich, Y., Arbilly, D., & Boxman, R. L. (2015). Thin-film deposition with refractory materials using a vacuum arc. *IEEE Transactions on Plasma Science, 43*(8), Part I, 2323–2328.

96. Beilis, I. I., Koulik, Y., & Boxman, R. L. (2017). Anode temperature evolution in a vacuum arc with a black body electrode configuration. *IEEE Transactions on Plasma Science, 44*(8), Part II, 2115–2118.
97. Beilis, I. I., Shashurin, A., Boxman, R. L., & Goldsmith, S. (2006). Total ion current fraction in a hot refractory anode vacuum arc. *Applied Physics Letters, 88,* 071501.
98. Shashurin, A., Beilis, I. I., & Boxman, R. L. (2008). Angular distribution of ion current in a vacuum arc with a refractory anode. *Plasma Sources Science and Technology, 17,* 015016.
99. Beilis, I. I., Shashurin, A., & Boxman, R. L. (2007). Measurement of ion flux as a function of background gas pressure in a hot refractory anode vacuum arc. *IEEE Transactions on Plasma Science, 35*(4), Part 2, 973–979.
100. Kimblin, C. W. (1974). Cathode spot erosion and ionization phenomena in the transition from vacuum to atmosphere arcs. *Journal of Applied Physics, 45*(12), 5235–5244.
101. Beilis, I. I., Koulik, Y., & Boxman, R. L. (2012). Ion current density measurements in a copper vacuum arc with different refractory anode thicknesses. *Journal of Applied Physics, 111*(4), 043302.
102. Beilis, I. I., Koulik, Y., & Boxman, R. L. (2011). Temperature distribution dependence on refractory anode thickness in a vacuum arc: Experiment. *IEEE Transactions on Plasma Science, 39*(6), 1303–1306.
103. Beilis, I. I., Koulik, Y., & Boxman, R. L. (2011). Temperature distribution dependence on refractory anode thickness in a vacuum arc: Theory. *IEEE Transactions on Plasma Science, 39*(6), 1307–1310.
104. Keidar, M., Beilis, I. I., & Brown, I. G. (1998). Multiply charged ion transport in free boundary vacuum arc plasma jet. *Journal of Applied Physics, 84,* 5956–5960.
105. Keidar, M., Brown, I. G., & Beilis, I. I. (2000). Axial ion charge state distribution in the vacuum arc plasma jet. *Review of Scientific Instrument, 71,* 698–700.

Chapter 13
Cathode Spot Motion in a Transverse and in an Oblique Magnetic Field

Different aspects of experimental investigation of cathode spot motion in a vacuum arc under transverse- and oblique-oriented magnetic fields will be reviewed and analyzed.

13.1 The General Problem

Cathode spot motion and dynamics, especially in a magnetic field, play an important role in the near-electrode phenomena and influence the arc performance in applications [1–3]. The spot motion and current per spot depend on the arc current, gap length, interelectrode pressure [1, 4–6] and determine the stability of the vacuum arc. The presence of a magnetic field changes the vacuum arc characteristics. While the transverse magnetic field changes the near-cathode plasma through the cathode spot behavior, the plasma jet expansion is influenced by an axial magnetic field. The action of the axial magnetic field on the interelectrode plasma (including experiment, models, and calculated results) was detailed in [3]. Under a transverse magnetic field, a specific spot motion named retrograde motion, spot grouping, and spot splitting was detected by measurements. This chapter summarizes the main experimental observations in order to understand the general features of the spot retrograde motion as well spot grouping and splitting.

The cathode spot generated conductive plasma jet expanding to the anode [7]. So, cathode–plasma–anode assembly supports a current carrying line in a closed electrical circuit. When a transverse magnetic field is applied, the arc column and spot move in a direction mutually perpendicular to the directions of the arc current and the magnetic field. Therefore, it was expected that such circuit conductive line in an external transverse magnetic field would move in accordance with Ampere law direction similar to usual conductive rod. However, the experiment showed two interesting phenomena when the vacuum arc is operated in a magnetic field. (1) While cathode spot motion is normally random without field, in the presence of

I. Beilis, *Plasma and Spot Phenomena in Electrical Arcs*, Springer Series on Atomic, Optical, and Plasma Physics 113, https://doi.org/10.1007/978-3-030-44747-2_13

a magnetic field the spots do not move in the direction of the force $\boldsymbol{j} \times \boldsymbol{B}$ ($\boldsymbol{j}$ and $\boldsymbol{B}$ being the vectors of the current density and the magnetic flux density, respectively). In contrary, they are move in the retrograde direction $-\boldsymbol{j} \times \boldsymbol{B}$, i.e., opposite to the Amperian direction [4]. (2) The cathode spot grouping in the vacuum arc [5, 8–10] and their expansion on the cathode surface in a ring away from their origin [5].

13.2 Effect of Spot Motion in a Magnetic Field

The effect of cathode spot retrograde motion was first observed by Stark in 1903 [11] and this motion he denoted as "Entgegengesetzt" effect. Let us discuss the details of this work that not considered in previous reviews. The work considered an arc with mercury cathode, and it is consisted of few sections reported about subjects that were investigated before the result of observation of the spot motion in a transverse magnetic field. In the beginning of the plasma expansion from the cathode, the formation of plasma jet and condensation of the mercury vapor on opposite wall of a glass chamber were studied. It was detected that when the opposite glass wall with condensed liquid mercury was heat locally the jet deflected into direction of cold part of the wall. The experiment showed that the mercury plasma expands from hot place in the direction to the cold place.

The next section related to the study of the plasma jet that expanded from the cathode between additional two Pt rods directed parallel to the cathode surface in the presence of a magnetic field also directed perpendicular to direction of the jet. As the jet expands with some plasma velocity V_j crossing the magnetic field lines, a voltage φ_{in} was induced between these two solid electrodes separated by distance $d = 1$ cm. Measuring this voltage, the plasma velocity was obtained from

$$V_j = \frac{H}{10^{-8}d} \tag{13.1}$$

For magnetic field strength, $H = 1850$ G the jet velocity in the mercury arc was calculated as 2.8×10^4 cm/s. It was commented that this value is lower limit value because the velocity depends on a mercury pressure [12], current registered between solid electrodes and jet properties in a point measured in the glass chamber. Also, a plasma jet deflection was observed even in the absence of a current between the solid electrodes.

The experimental results named paradoxical behavior of the mercury arc in a magnetic field were described in the following section. This description is characterized into different parts of the arc. The first part is the liquid mercury, which served as anode, was observed as covered uniformly by a glowing layer. From this anode part, a luminous plasma column is extended up to the cathode. The third, the cathode region was observed as not uniform luminescence region that consisted of certain number of plasma roots (spots). It was observed that in a transverse magnetic field the anode glow and the luminous plasma column were shifted in the direction coincident

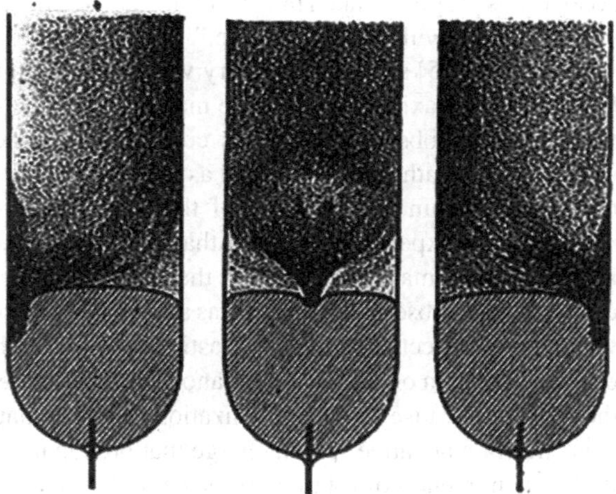

Fig. 13.1 Illustration of "Entgegengesetzt" spot motion under magnetic field (1850 Gauss) on mercury cathode showed by Stark 1903 [11]

with the direction of motion of usual current carrying conductor, while the cathode plasma roots (spots) moved in "entgegengesetzt richtung" (Fig. 13.1), i.e., in opposite direction. Stark in the last section of his work [12] suggested few mechanisms to explain the observed effect based on the experimental results obtained in the above sections. In this essence, it was suggested that the positive charged ions are shifted in the observed opposite direction under action of the voltage induced in parallel to the cathode surface due to plasma motion in a transverse magnetic field as well as due to Hall effect. In essence, this mechanism was used in later publications, but ion shifting of small density in quasineutral plasma cannot significantly influence the large space charge in the cathode sheath. The details of this mechanism will be discussed in the next theoretical Chap. 19. Later Stark [13] analyzed the cathode spot phenomena as basis for the support of the arc current.

After a relatively short period in 1904 Weintraub [14] also reported the observed retrograde phenomenon. The experiment was conducted with mercury cathode and graphite anode. In the absence of magnetic field, the spot on the Hg cathode usually wanders about the surface to give it a fixed position for iron wire protruding a little above the surface of the mercury. The experiment with magnetic field showed that "the arc is first deflected one way, and after a while the bright spot is seen to move away from the iron wire in the opposite direction." Weintraub described that "the action of the magnetic field on the arc presents, therefore, the features which are at first sight paradoxical, if the arc is vertical and the field horizontal, the deflection of the arc and that of the spot are in opposite directions." This result was explained as: "the action of the field on the cathode spot can be in most cases formally accounted by assuming that positive current elements leave the cathode surface in a direction perpendicular to that surface." It was also indicated that "the behavior of the arc in a magnetic field can be partly explained by assuming it to be a flexible conductor capable of changing its length, both ends being capable of sliding along with two surfaces, and that on the anode being much less mobile than that on the cathode."

Also, it was proposed taken into account "the fact that the conductivity of the mercury vapor changes with its temperature."

Minorsky [15] studied a mercury vapor arc which produced in the annular space between two coaxial cylinders, the magnetic field being radial with respect to these cylinders. The observation shows complex phenomena that characterized by the rotation of the cathode spot as well as by the deformation of the arc. It was observed that, for a certain low pressure of the argon, the rotation of the cathode spot is reversed. The experiment showed that the phenomenon apparently depends on the intensity of the magnetic field and the argon pressure (average mean free path). To understand the observed result, it was assumed that the free electrons propagate along with certain trajectories and the elastic collisions with the atoms cause the passage of electrons from one trajectory to another. Among these passages, it was shown that the collisions cause the atom ionization (change shape of the arc) and bring to an existence of a negative spatial charge that propagates in the annular space, resulting in the phenomenon of rotation of the cathode spot in one or other direction.

Minorsky's work was commented by Tanberg [16] by the following report "In the *Journal de Physique et le Radium* for April 1928, Prof. N. Minorsky describes the observations he has made on the behavior of a mercury arc in a magnetic field under low gas pressure. Under certain experimental conditions, he found that the arc moved in a direction opposite to that which should be expected from the electrodynamic laws governing the movement of a conductor carrying current in a magnetic field and proposed a space charge theory to explain this phenomenon. His theory is based upon the assumption that by the combined effects of collisions and the magnetic field, the electrons in the arc stream are made to travel in the electrodynamic sense around the annular space in which his arc is drawn and will thereby establish a negative space charge on the opposite side of the arc. The action of this space charge on the positive ions of the arc is assumed to cause the arc to move in the direction observed. It will be realized that this theory requires a circular or at least a closed path for the arc to move along." The last Tanberg's conclusion indicates some doubt regarding the mentioned theory because the requested arc configuration was not indicated in Minorsky's work. Instead, Tanberg [16] proposed another model assuming a stream of positive ions moving away from the cathode that can be deflected in reverse direction.

13.3 Investigations of the Retrograde Spot Motion

The future experimental studies of retrograde spot motion have been presented extensively in the literature [1, 4, 5]. Retrograde motion has been studied in self-magnetic fields and in external magnetic fields. The effect a transverse magnetic field (TMF) can be different depending on it orientation with respect to the cathode surface. When the direction of TMF is parallel to the cathode surface, the spot moves strongly perpendicular (in the $-\boldsymbol{j} \times \boldsymbol{B}$, direction) to the magnetic strength lines (named as "normal motion"). When the direction of TMF oriented at an acute angle (oblique magnetic fields) with respect to the cathode surface, the spot motion was declined at some angle with respect to the direct motion, i.e., to the $-\boldsymbol{j} \times \boldsymbol{B}$, direction. Let us consider these phenomena.

13.3.1 Magnetic Field Parallel to the Cathode Surface. Direct Cathode Spot Motion

A direct cathode spot motion will be considered for electrical arcs burning at different conditions in the electrode gap and at the cathode.

13.3.1.1 Cathode Spot Velocity Moved in Transverse Magnetic Field

In this section, the results of investigations of the spot motion are presented for *vacuum or for low-pressure electrical arcs*, for which the retrograde effect occurs. Froome [17] has observed a maximal velocity at around 100 m/s for spots moving in retrograde direction on mercury in both self-generated and externally applied magnetic fields (0.17 and 0.3 T). The cathode spot of transient mercury arcs has been studied by means of a Kerr cell camera. Mainly, Froome reported about linear dependence of the spot length on arc current. It was noted that the total length of the emitting line is not affected by the strength of the field, but the field affects the spot velocity only.

Smith presented a number of works investigating the phenomena of reverse cathode spot motion in a transverse magnetic field. In one of his first experiments 1942 [18], Smith reported that a magnetic field transverse to the arc current drives the spot in a direction transverse to the field but in a sense opposite to the ponderomotive force. The experiment conducted in an arc tube with annular mercury cathode was used, and a radial magnetic field was employed. With a south pole at the center, the arc spot raced around clockwise. The mercury circulated slowly in the opposite direction. The arc stream curled backward as the spot rushed forward. Velocity was measured focusing light from the spot upon a photo-cell arrangement and using the amplified current to deflect the beam of a cathode-ray oscillography.

Definite patterns observed upon the screen of the oscillograph from which the number of excursions per second made by the arc spot around its track was deduced. An arc track with very smooth edges was observed for speeds above about 3×10^3 cm/s while for lower speeds showed a ragged arc path. The arc track arises in the region where the horizontal component of the magnetic field was strongest, i.e., the path where it could go fastest.

The spot arise so that by it motion the disposition of the magnetic field varied to give the strongest horizontal component in the desired region. As result, the arc track arises in the region where the horizontal component of the magnetic field was strongest, i.e., the path where it could go fastest [18]. A plot of velocity of spot as a function of magnetic field strength is shown in Fig. 13.2. The result for a constant

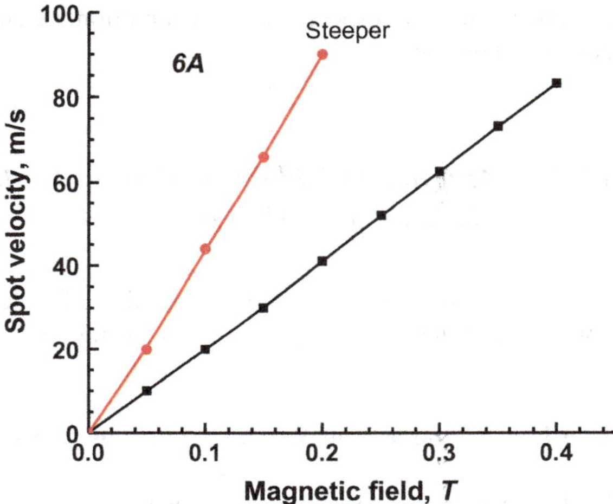

Fig. 13.2 Velocity of arc spot verses horizontal component of magnetic field. Arc current—6 A. Steeper line observed with cooler mercury

temperature of' the mercury is a straight line through the origin. The steeper straight line was observed with cooler mercury. The magnet was turned on and the speed observed in less than a second. Then the magnet was shut off, and after several minutes, another reading at higher field strength was made.

The mercury temperature was at about the same low value for each observation. The lower curve was taken by allowing the arc to race around till it reached a constant lower limit of speed, the mercury presumably reaching a constant temperature. The measured velocities were too high for mechanical effects to cause turbulence of the liquid surface. The large condenser for the vapor insured an inconsiderable pressure everywhere except at the arc spot. Thus, the observed arc spot moved over a smooth mercury surface and going in the direction opposite to that determined by the effect of the field on the arc stream with the speed was accurately proportional to the magnetic field.

The next Smith's investigations were published in a series of letters. Thus, the work 1943 [19] described observation with an arc between equal parallel rings of copper is subjected to a transverse radial magnetic field. It was observed that with lower ring as cathode and south pole at center, the arc rotates counterclockwise at atmospheric pressure, but below a critical pressure, it rotates clockwise. The pressure for reversal varies approximately inversely as the distance between rings. Smith suggests that the peculiar clockwise rotation results from the thermomotive force or Righi–Leduc factor operative in the solid cathode (Chap. 19). According to Righi–Leduc phenomenon, if a magnetic field is applied at right angles to the direction of a temperature gradient in a conductor, a new temperature gradient is produced perpendicular to both the direction of the original temperature gradient and to the magnetic field [20]. Mass movement of ionized gas caused by electromagnetic force tends to drive the arc around counterclockwise. This tendency becomes less at reduced pressure since viscosity factors increase due to spreading of the arc column. Below the critical pressure, phenomena in the solid cathode predominate. Some additional details of the above experiment were presented in [21] in order to discuss published different mechanisms of retrograde spot motion.

In a letter published in 1951 [22], the arcs were realized in mercury vapor between clean massive polished terminals of both Ta and Mo. The lower end of the cathode was down into a pool of mercury, which serves as a coolant. A transverse magnetic field drives the arc at abnormally high maximum retrograde speed. A field parallel to arc stream between plane parallel terminals raises the cathode voltage causing changes of the spectrum. For a field of 1 T, the HgII spectrum is brilliant, like HgI, and even HgIII is prominent. The maximum voltage between anode and cathode was about 16.5 V. The observed random fluctuations of voltage not to exceed 0.15 V and hence cannot serve to explain the spark spectrum. For all studied cases, the cathode current density was below 100 A/cm^2. The high retrograde mobility, low-current density, peculiar spark spectrum, variation of cathode voltage in a longitudinal field are all characteristic for the studied arc.

The work 1951 [23] used a cylinder 2 cm in diameter of polished molybdenum with axis vertical, projects well above a surface of mercury in an evacuated tube. The anode is above and radially larger than the cylinder. An arc anchors along the circle where the mercury wets the molybdenum. An axial magnetic field of 0.1 to 1 T is applied along with the arc stream except for a few thousandths of a cm where the arc current is radial to the molybdenum. The cathode spot races around on the cylinder in the retrograde direction. The spot motion observed through a rotating toothed wheel and with photocell and oscillography as well as with radial probe and oscillography. Not observed any change of the velocity from retrograde to proper for any field strength. The velocity approaches asymptotically to about 120 m/s with magnetic field. Current density is greatest at the leading edge of the spot. Spot may be single or of two or more equal segments separated by darker regions. A mode of one, two, or more, once established is stable.

In another work, Smith [24] studied a mercury arc raced around a carboloy cylinder projecting about three mm above the mercury in a vertical magnetic field. Arc current and field were parallel except at the carboloy anchor. Revolutions per second were observed by probe methods. The motion was retrograde, i.e., contrary to amperes law for all values of fields. The measurements show that the speed curve rose rapidly between 0.1 and 0.3 T and then leveled at 120 m/s. At field of about 0.9 T the curve rose rapidly to about twice the plateau value (250 m/s) and continued rose with a less rate of rise (to 800 m/s) to the highest field, 1.65 T. When an arc running around in a shallow groove cut in the carboloy near its top, the arc was above the junction of liquid and carboloy. In this case, the speed dependence showed a short plateau but its value about three times those noted above. Ultimate speeds about six times sonic were found. These results, as Smith indicated, seem to leave the experiment without a satisfactory theory of the retrograde motion.

In a short letter, Gallagher and Cobine [25] indicated that in a mercury arc the cathode spot under the influence of a magnetic field perpendicular to the arc axis may be made to move contrary to the direction predicted by electromagnetic theory. It was found also that "this retrograde effect is exist not only in the mercury arc, but also in arcs between solid electrodes with different gases, including helium, argon, nitrogen, hydrogen, oxygen, and carbon dioxide. Electrode materials studied include carbon, tungsten, molybdenum, nickel, copper, aluminum, and cadmium. For retrograde motion, the arc column and anode spot tend to move in the conventional direction, but are "dragged" in the reverse direction by the cathode spot." The reversal motion was determined by gas pressure, magnetic field, and current. At constant current, the critical reversal pressure is the proportional to the magnetic field *B*. Increasing current at constant *B* decreases the value of critical pressure. Currents were used in range from 2 to 10 A, magnetic fields from 10 to 450 G and reversal pressures from 60 to 0.8 cm Hg. It was noted that oxidize of the electrodes maintain a stable arc.

Gallagher continued such investigation presenting the results in 1950 [26]. A series of experiments was conducted with a low-pressure mercury arc, using a mercury pool cathode. The cathode was ring-shaped. The magnetic field was radial, so that the current entering the cathode spot was always perpendicular to the field. Both tube and magnet were immersed in oil, which was cooled by circulating through a heat exchanger. The experimental glass tube was operated at a line voltage of 250 V DC with a variable series resistance to regulate the current. The velocity of the spot was obtained from the frequency of rotation.

The velocities of the spot as a function of magnetic field, arc current, and pressure were obtained. As shown in Fig. 13.3, the velocity varies almost linearly with magnetic field (increased from about 500 up to 3300 G) over the middle range. The slopes of the linear portions of the curves appear to be independent of current in range from 0.75 to 5 A.

It was observed that the voltage across the arc at high magnetic fields is high and may cause instability. However, in the described experiment the arc voltage did not show any unusual fluctuations in this region such as might be expected if the arc were unstable. In Fig. 13.4, the spot velocity is presented as dependence on arc current.

There is a definite leveling-off point for each value of field, with the constant velocity region setting in at smaller values of current as the field is increased. St.

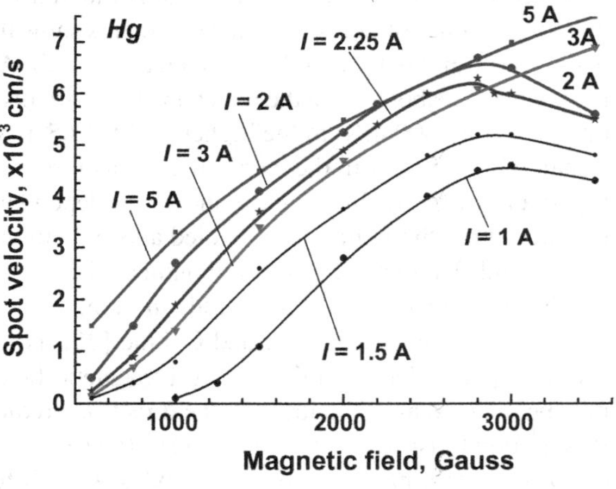

Fig. 13.3 Velocity of cathode spot on Hg pool *vs.* magnetic field for various arc currents. Pressure = 0.7 μ [26]

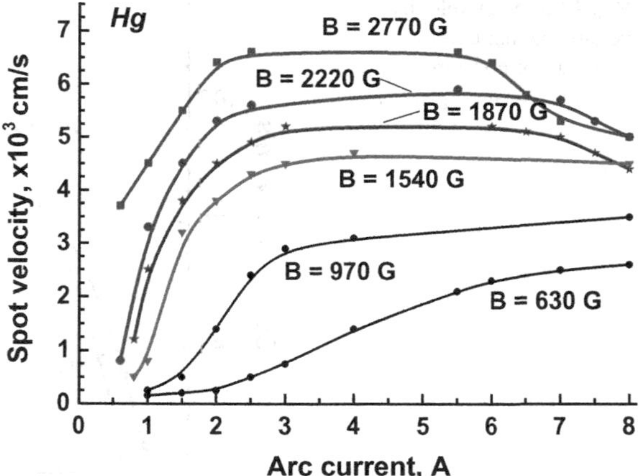

Fig. 13.4 Velocity of cathode spot on Hg pool versus arc current for various magnetic fields. Pressure = 0.7 μ [26]

John and Winans extended Galagher's measurements on a mercury cathode up to magnetic fields of 1.2 T [27] and 2.07 T [28] and also observed the spectrum of the arc. The discharge was ignited in a tube mounted in the magnet. The cathode was composed of a molybdenum stump standing in a pool of mercury. The anode was a circular metal disk. When the molybdenum has a very clean surface, it is wet by the mercury, which then rises around the stump. The arc spot is formed at the junction of the mercury and the molybdenum stump. The cathode spot moves around the stump at the junction between the mercury and the molybdenum stump. The spot velocity is determined from the frequency and the diameter of the stump.

The use of two photoelectric cells, permitted the determination of the direction of the spot motion and showed that there was one spot. The spectrum of the cathode spot region was obtained by focusing an image of the spot on the slit of a spectrograph. Figure 13.5 shows a plot of the retrograde spot velocity as a function of magnetic field strength for constant arc currents of 5.5 and 2.0 A and no inert gas in the tube. An increase in magnetic held strength is accompanied by an increase in retrograde velocity.

With the higher arc current (5.5 A), the spot velocity is changed discontinuously to a value nearly twice as great, at a certain magnetic field strength. The change was from about 112 m/s to about 193 m/s (not presented in Fig. 13.5) at field strength about 1.05 T.

With an inert gas (argon) in the tube at a pressure of 150 mm Hg, the spot velocity depends on the magnetic field strength as shown in Fig. 13.6 for the arc current as parameter. These dependencies are similar to those for no inert gas present. However, the curves are shifted downward along the velocity axis toward the velocities indicating forward motion of the spot, and the velocities are about one-hundredth of the velocities with no inert gas.

It was indicated that the tube temperature influence the spot velocity due to mercury evaporation and flowing increase of the Hg pressure depending on the dissipated power with the arc current. At lower tube temperature, the sharp rises

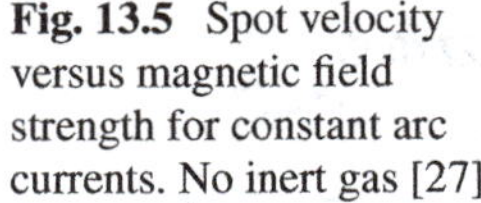

Fig. 13.5 Spot velocity versus magnetic field strength for constant arc currents. No inert gas [27]

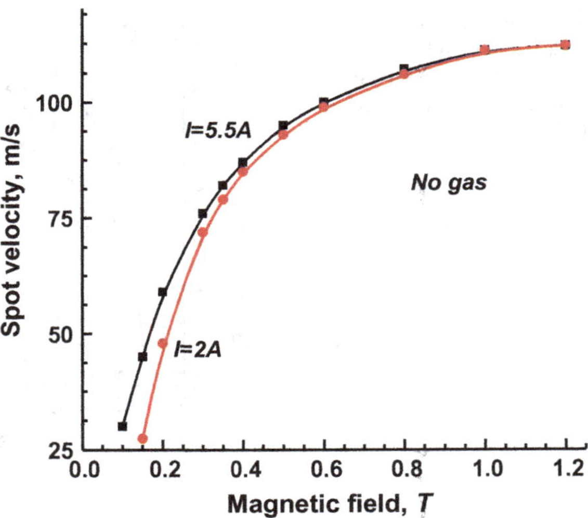

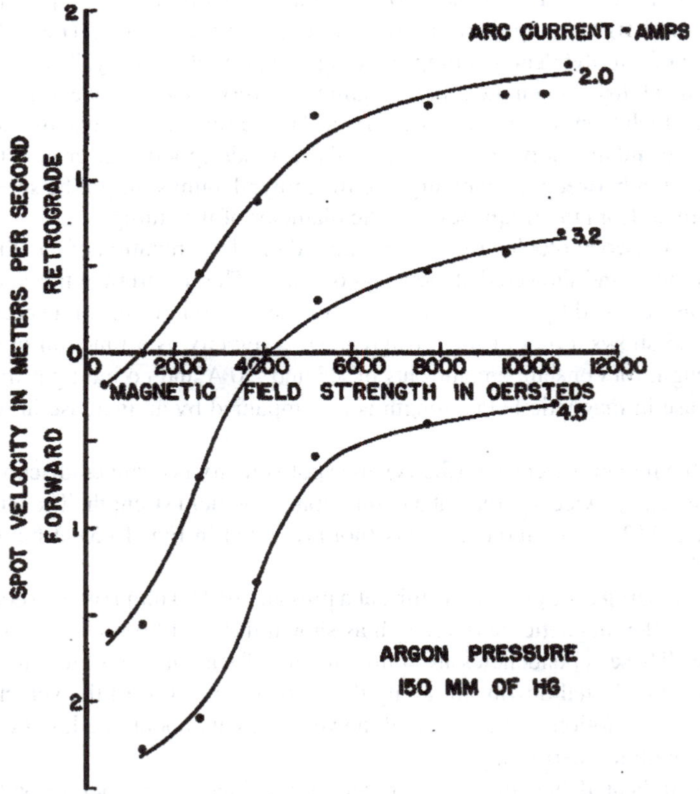

Fig. 13.6 Spot velocity as dependence on B (in oersteds) in argon arc at constant current as parameter. Figure taken from [27]. Used with permission

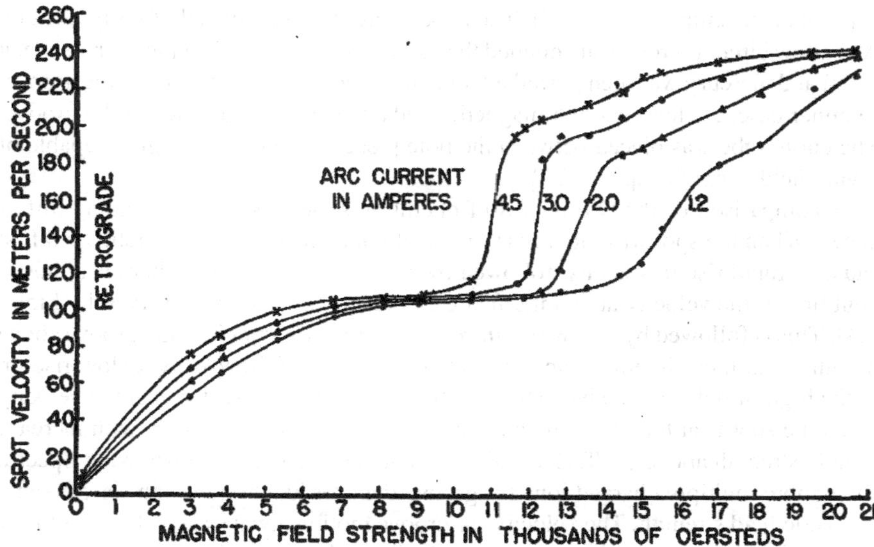

Fig. 13.7 Spot velocity as a function of magnetic field strength, for constant arc current, with no background gas. Figure taken from [28]. Used with permission

in velocity occurred at lower magnetic field. Figure 13.7 shows that with increasing field strength, and constant arc current, the velocity first increase rather linearly and then approaches a saturation value. With further increasing field, the approximate doubling of retrograde velocity was observed from 1.1 to 1.5 T, which followed by an additional rapid rise of velocity beginning at about 1.5 T and more. The highest velocity of the spot occurred at a 2.07 T, and it was 2.3 times the value at the first plateau.

An analysis of the spot velocity as a function of arc current for constant magnetic field strengths show that the velocity increases in retrograde direction as the current was increased when no gas is present. Over a limited range of magnetic field strength from 1.16 T to about 1.54 T, a small change in arc current with constant field causes the retrograde velocity to nearly double.

John and Winans [28] studied the spectra of mercury arc with the spot racing about the beveled edge of the top of the stump for magnetic field strengths from 0.015 to 2.2 T with no inert gas. The lines were enhanced and new lines appeared with field strength such as HgI, HgII from single charged ions, HgIII from doubly charged ions and a continuous spectrum, which is especially intense at the lines. The Hg lines are broadened symmetrically and also asymmetrically. Assuming that the broadening was due to a Stark effect, the electric field strength in the cathode spot region was obtained greater than 6×10^6 V/cm.

A retrograde motion of mercury arc when the spots move in a groove in the cathode above the level of mercury was compared by Zei and Winans [29] with motion of spots at the cathode junction between mercury and the molybdenum. It was used the experimental setup as in [27, 28]. The anode was also constructed of molybdenum. A shallow circular groove surrounded the cathode near its top. The junction between metal and mercury was then placed either at the groove, or slightly below the groove. In either case, for high enough magnetic fields, the spot was formed at the groove. The entire tube was placed between the pole pieces of an electromagnet capable of giving field strengths up to 2.2 T.

A comparison of the two modes of operation showed some outstanding differences. When the spot was moving at the junction, an increase in magnetic field first causes a rapid rise in velocity, followed by a virtual plateau. Then there is almost a doubling of the velocity at an intermediate field strength as first observed by Smith [24]. This is followed by a slow rise in velocity. When the spot is in the groove above the junction, there is first a rapid rise in spot velocity, followed by a slow rise up to the highest fields attainable. The spot velocity is always greater than the velocity when the spot is at the junction, and there is no rapid rise of velocity with increase of field strength above 0.5 T. Zei et al. [30] first observed this spot behavior. Spectra show more multiply charged ions for groove than for junction motion for the same magnetic field strength. The velocity for groove motion always exceeds the velocity for junction motion, and it increases markedly with a reduction in mercury vapor pressure. The retrograde velocity passed through a maximum with arc current for groove spot motion.

Hernqvist and Johnson presented an interesting work [31]. They reported that a similar effect of retrograde spot motion on cold cathode was observed in externally heated hot cathode gas discharges of the ball-of-fire mode. The experimental construction consists of a cylindrical diode tube with an oxide-coated and indirectly heated cathode C of 0.050 inch diameter and a nichrome (an alloy of nickel with chromium (10–20%) and sometimes iron up to 25%) anode of 1 inch diameter (Fig. 13.8). The tube structure is 1 inch long and has mica insulators covering the ends. A 0.010-inch-wide slit is cut along the entire anode length parallel to the cathode.

Behind this slit is placed a collector. The anode is grounded. The collector is biased slightly positive relative to ground. An adjustable magnetic field was applied parallel to the cathode surface by a coil surrounding the tube. Experimental study of externally heated hot cathode discharges in noble gases has shown that this discharge can exist in a four different modes depending on the current and characterized by different glow patterns and potential distributions schematically shown in Fig. 13.8. These modes have been termed as anode glow, ball-of-fire, Langmuir, and temperature-limited mode.

When operated in the ball-of-fire mode, the glow region appeared at some place in the cathode–anode region, presumably determined by irregularities in the tube structure. If a magnetic field is applied parallel to the cathode, the glow region is found to rotate around the cathode in a direction opposite to the electromagnetic forces on the charges flowing between cathode and anode. The shape of the glow region was practically undisturbed during the rotation.

The frequency of rotation of the glow region was measured. A frequency of rotation of 100 cycles per second corresponds to an average velocity of about 6 m/s.

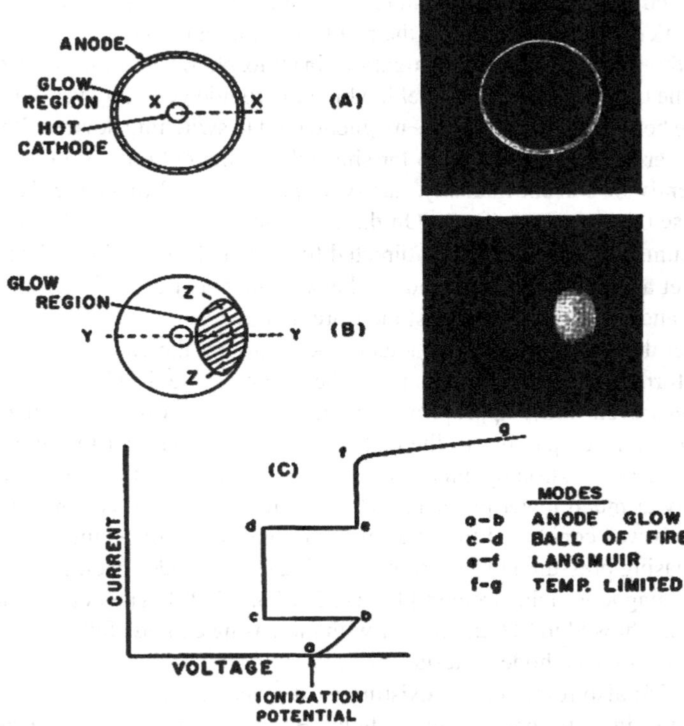

Fig. 13.8 Schematic drawing and photograph of cylindrical diode operating in (**a**) anode glow mode and (**b**) ball-of-fire mode, together with idealized volt-ampere characteristic (**c**). Figure taken from [31]. Used with permission

It was found that the frequency of rotation is practically independent of the anode current except for high currents under which conditions the glow region occupies almost half of the cathode–anode space. Comparison between the retrograde motion of the ball-of-fire mode of discharge in different inert gases and that of the mercury pool arc showed that the dependences of the two phenomena on gas pressure, current, and magnetic field were quite similar.

Eidinger and Rieder [32] reviewed the behavior of the arc in transverse magnetic field describing the results of publications before 1957 as well their own extensive investigations. The description includes not only the physical conditions of the researches, but also the electrotechnical of the essential information about the switching device construction. They analyzed the minimum field intensity to respond its influence, conditions for motionless persistence of an arc determining by the gap pressure and the magnetic field strength between parallel conductors. Studied the influence of different arc lengths and arc currents on the speed dependence, at which an arc moved through a magnetic field. The effect of the magnetic blast on the arc characteristics was compared with the results of other authors.

The authors' experimental investigations in atmospheric air are described the light waves in the transverse magnetic field, which to provide generally valid quantitative data on the behavior of the arc in the magnetic field in switching devices. It was noted that all experiments carried out so far showed that the column of an arc is moved independently of current intensity, gas type, pressure, and magnetic field strength in the sense of the Lorentz force. On the cathodic discharge parts, however, it acts under circumstances an oppositely directed force, which slows down the movement of the sheet at falling pressure (under otherwise identical conditions), then let him stand still and finally moves against the Lorentz force.

The fact that this counterforce actually acts only on the cathode spot is already apparent form of a low-pressure arc in the magnetic field [11, 15, 25, 26]. The column runs even under appropriate conditions a lesser distance in the direction of the Lorentz force parallel to the cathode and then turns first to the anode [33]. Furthermore, when strongly inhomogeneous fields are used, the motion reversal occurs only at much lower pressures when the magnetic field is concentrated near the anode. It was concluded that the shifts in favor of the opposing force occurred with decreasing pressure [11], current [15, 25–27], electrode spacing [19], and with increasing magnetic field strength [15, 18, 22, 23, 25–28]. The dependence on the gas type was showed in [27, 28, 31], or with increasing cathode fall [25, 26], and the dependence on the cathode material.

Farrall [34] also reviewed the existing works since the early 1900s on arc retrograde motion in transverse magnetic field. He analyzed the dependences of spot velocity on magnetic field strength, on arc current, electrode gap length, and on electrode surface state. The published experimental data were discussed in frame theoretical mechanisms of the phenomena presented in the literature.

Gundlach [35] studied the cathode spot motion caused by self-magnetic field in a high-current (500–4000 A) vacuum arc on Mo, Cu, C, and Ti electrodes. The electrodes with diameter 7.2 cm for Cu and 4.4 cm for other materials were placed in Pyrex glass vacuum (10^{-6} torr) chamber. The electrodes were cleaned by heating and by discharging before the measurements. The arc started by trigger voltage pulse. The rectangular current pulse was 2 ms. Cathode spot motion was recorded by image converter. A number of photographs of the spot motion from, at which the retrograde spot velocities were obtained as the spot size shift ratio the shifting time, illustrated the results. The velocities were expressed by following approximations: $V = 0.65(I)^{0.5}$ for Ti, $V = 0.48(I)^{0.5}$ for Mo and $V = 0.33(I)^{0.5}$ m/s for Cu. The results were discussed assuming asymmetry of the heat flux due to returned ions. The increase of the velocity of the spot ring expansion on gap distance d was obtained for Cu, Mo, and Ti electrodes as increase of ring radius with d in form characterized by relation $R_r \sim d^{\beta}$, where $\beta = 0.1$–0.2 for $d < 0.5$ cm and $\beta = 0$ for $d > 0.5$ cm.

The influence of an external magnetic field (up to 0.5 T) and dependence of the retrograde spot velocity for DC vacuum (10^{-5} torr) arc for range of currents $I =$ 20–150 A were investigated by Seidel and Stefanik [36]. The electrode materials were Cu, Al, and steel with spacing of 2 and 5 mm. The spot velocity was measured

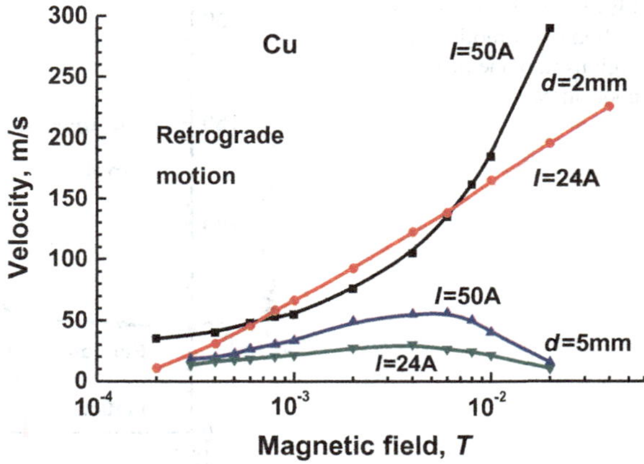

Fig. 13.9 Spot velocity as function on magnetic field for Cu cathode with arc current and electrode spacing as parameters [36]

using high-speed camera. Anode and cathode were made as two concentric rings, and the anode was the external one. Cathode outer diameter remained unaltered. The external magnetic induction B varied up to 0.5 T. The self-magnetic field of the electrodes was eliminated by their specific connection.

Dependence of spot velocity on magnetic field for Cu cathode for different arc current and electrode spacing can be seen in Fig. 13.9. Similar dependencies were obtained also for Al cathode, but the spot velocity is larger at Cu cathode than at Al and also Fe cathodes.

The velocity dependencies, as well their values, are strongly depending on electrode spacing for this same arc current. The measurements showed that the spot velocity increases linearly with magnetic induction up to certain value of B depending on and I and spacing. The future increase of B causes at first decrease in the velocity and at last with still greater values the reversing of the motion direction occurs, which is then in accordance with law of Ampere for all materials. According to Seidel and Stefanik [36], the possible retrograde or Amperian spot motion the region in I-B coordinates can be indicated in Fig. 13.10.

The results shows that similarly to arc in surrounding gas pressure in the short vacuum arc also a minimum value of magnetic field is necessary above which the spot can move in the retrograde direction. This experimental fact was explained by a high vapor density produced in the electrode gap due to plasma jets generated from the high dense spot plasma. The spot plasma density was estimated by the interpretation of the measured results using data for cathode spot current density and cathode thermal regime.

The direction of erosion tracks generated by an arc moved by a transverse magnetic field was studied by Daalder [37] on polished copper cathodes. Coils placed outside the vacuum chamber generated the field. The arc was initiated by separating disk electrodes. To this end, a molybdenum pin of one mm diameter protruded 2 mm at the anode edge, touching the cathode surface. Increasing of the field caused a rise of the direct spot velocity and an increase of the arc voltage. The arc moves in retrograde direction with velocity determined by the arc current and magnetic field.

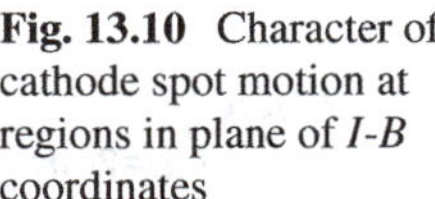

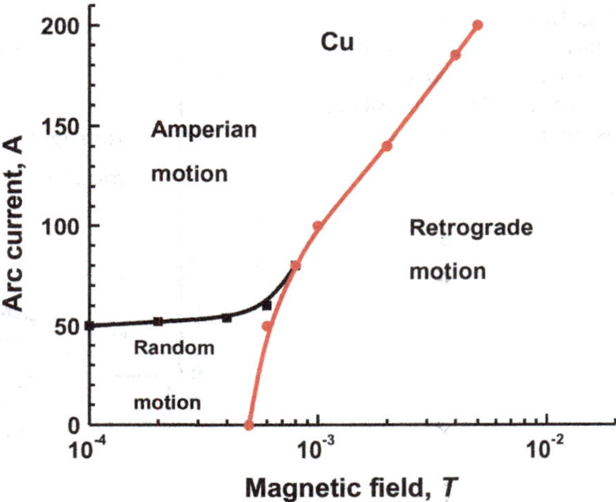

Fig. 13.10 Character of cathode spot motion at regions in plane of *I-B* coordinates

For a magnetic field of 0.081 T and an average arc current of 75 A, the erosion trace was oriented perpendicular to the field direction. An average spot velocity of 5 m/s was estimated using the trace length. The spot velocity significantly increased with the field of 0.16 and 0.2 T. The velocities have been around 42 and 95 m/s for arc currents 80 and 180 A, respectively. The traces were mainly straight. They consist of large numbers of craters mostly superimposed. The crater sizes were of several micrometers in diameter similar to that observed for a randomly moved arc on Cu for 50–150 A. It was noted that magnetic field primarily promoted a transition from chaotic to directed movement of the spot, which indicated previously by Kesaev [4].

The interaction between diffuse vacuum arcs and magnetic fields applied transverse to the electrode axis has been investigated by Emtage et al. [38]. The authors indicated that in general the transverse magnetic fields tend to confine the cathode plasma, forcing it out of contact with the anode. As result, the arc voltage then rises and current can extinction. This action is different from those associated with axial fields, which stabilize the arc and reduce the arc voltage [39]. It was noted that the transverse magnetic field bows the arc plasma in the forward direction, because the principal force on the ions is the Hall electric field rather than the Lorentz force. Experiment showed that at low magnetic fields, the arc voltage remains low, but at an initial field (which depends on the electrode configuration), the cathode plasma is bent out of contact with the anode and the arc voltage increases. The arc current can then be extinguished if a parallel circuit of low impedance is provided; the observed current interruption times bear a predicted relationship to the rate of rise of magnetic field. It was obtained that for currents exceeding 6 kA, the arc current does not fall completely to zero. Under these circumstances, arc extinction can be achieved by applying an oscillating magnetic field transverse to the electrode axis. Arc extinction via magnetic field–vacuum arc interaction could have applications to AC current limiters and DC breakers.

An investigation of the motion of low-current vacuum arcs in a high transverse magnetic field was conducted with regard to arcing impurities by Sethuraman and Barrault [40, 41]. Cathode spot velocity was measured using streak, and the arc tracks were analyzed using optical and scanning electron microscopes. The electrodes (titanium, stainless steel, and molybdenum) are in the form of discs with about 60 mm diameter, and the average electrode spacing is about 6 mm. The cathode, which is normally well polished, is mounted directly on to the copper rod using a thin collar. The cathode disk can be gently rotated from outside the vacuum system. The anode disk is slightly tapered in order to photographic study of the cathode spot movement. The arc is initiated by applying a high-voltage positive pulse. Arc currents were in the range of 1–100 A. The total lifetime of the arc was more than 600 μs. In the absence of the magnetic field, the cathode spot movement was random, while in the presence of the magnetic field the cathode spot follows retrograde motion.

Two kinds of spots were detected from the streak pictures of the cathode spot motion and from subsequent track analysis. One type was called "fast spots" that moved quickly over the cathode producing a relatively weak erosion and small crater sizes. The second type called "slow spots," moves relatively slower with strong erosion and crater sizes of 10-50 μm. By cleaning the disk using glow discharge, it was possible to obliterate all the self-initiated tracks except those in the close vicinity of the main track. Therefore, the glow discharge cleaning can reduce the arcing tendency of the electrode.

The slow spots, which were the true representation of the material, give a more consistent result with the experimental scatter comparatively small. The rate of the velocity increases with the magnetic field is much lower at higher fields. The measured large scatter of fast spots velocity is understandable since the velocity depends on the amount of contamination available on the surface at any time. In contrary, the dependence for slow spot shows monotonic velocity increase with the arc current corresponding to a magnetic field of 0.125 T and the rate of this increase is larger with lower values of arc current, but rather slowly at higher currents (in range of 10–70 A). Under the influence of the magnetic field, the linear tracks were predominant. However, at arc currents above 50 A the linear tracks tend to change to the fern type with the characteristic branching observed by SEM analysis. The track is about 100 μm wide and was formed by a series of overlapping craters with an average crater dimension of about 25 μm. Discontinuity in the overlapping craters shows the evidence of the cathode spot jumping.

According to Sethuraman and Barrault [41], the linear relationship between the velocity and B is kept until $B = 0.5$ T shown in form of Fig. 13.11. At fields greater than 0.5 T, the velocity seems to saturate to more or less a constant value which we call the "saturation velocity." All the three materials investigated show very similar characteristics and reach a saturation velocity at about 1 T. Titanium and molybdenum have velocities within 20% of each other with the molybdenum velocity lower. Stainless steel shows a higher velocity than the other two materials.

Nurnberg et al. [42] measured the velocity of the arc is as a function of the applied transverse magnetic field in the vacuum arcs. Cathodes were made of 0.2–0.5 mm thick sheet metal attached to the copper cathode support with clips. The

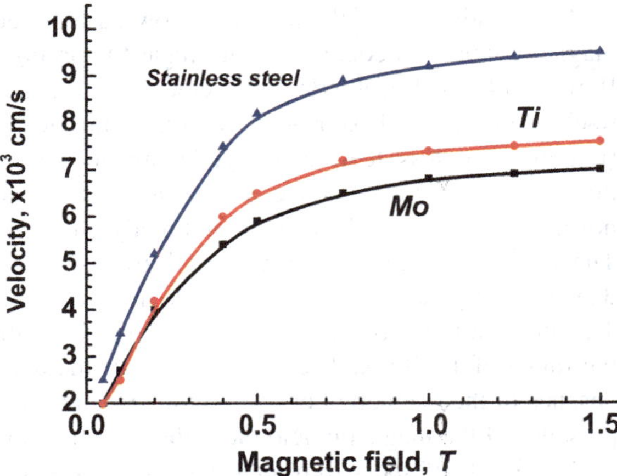

Fig. 13.11 Dependence of cathode spot velocity on magnetic field. Arc current 35 A. The experimental scatter is ±10 m/s

vacuum arc was ignited by contacting and separating the electrodes. The interelectrode distance used in these experiments was from 2 to 4 mm. A magnetic field of 0.01 to 0.1 T drives the arc in the retrograde direction. The velocity was determined from the known characteristic sizes and the operating time that was obtained from the current/voltage oscillogram. Arc velocities were measured on unconditioned surfaces and surfaces exposed to more than 50 previous arc discharges. In both cases, the measurements were conducted in a high background vacuum. The effect of the previous arc discharge operation, i.e., the conditioning consists of an elimination of the oxide layers and an increase of the surface roughness. The arc gap was always 3 mm. Arc currents of 16 A and 40 A have been used. The results of velocity measurement of the arc, using SS 316 as cathode material [42], are represented in Fig. 13.12. In all cases, the arc velocity increases with increasing magnetic

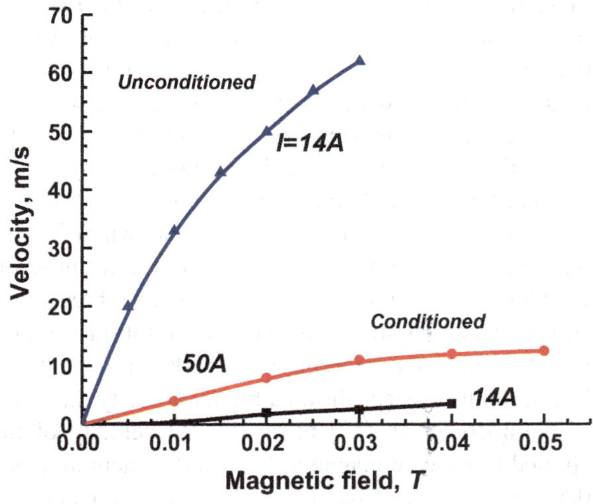

Fig. 13.12 Velocities of vacuum arcs on unconditioned (current 14 A) and conditioned (currents 14 and 50 A) stainless steel 316 cathode surfaces as function on magnetic field

induction B. The highest arc velocities were found for the unconditioned stainless steel 316 surfaces. The lower data have been obtained for conditioned stainless steel 316 surfaces.

Fang [43] measured the arc velocity of the vacuum arc spot motion under a magnetic field. The cathode sample with dimensions 34 × 24 × 1 mm was attached to a molybdenum cathode support, which could be heated to 1200 K by two heating elements. The anode was pure graphite (EK506-Ringsdoff-Werk). The electrode spacing was varied between 0.5 and 10 mm. The arc ignited by applying a positive voltage pulse to the trigger pin at the side of the cathode. The trigger assemblies and the electrodes were installed in a high vacuum system, which was evacuated by a turbomolecular pump to 10^{-7} mbar. The magnetic field was generated by an electromagnet placed outside the vacuum chamber, the field direction parallel with the cathode surface. The retrograde velocities arcs were measured using two opto-transistors. The distance between opto-transistors was 20 mm. The arc velocity time determined using the time between the two optical signals. The retrograde velocities were measured as function of the magnetic induction, the arc current, the electrode spacing, and the cathode material (copper, titanium, and 316 stainless steel).

Prior the arcing the cathode samples are degreased using acetone. Then they are degassed at high temperature (800–1200 K) in a high vacuum. From scanning electron microscope, analysis of the arc tracks and from direct visual observation two kinds of spots was detected. On an unconditioned cathode surface, the spot moves quickly and randomly in a somewhat jumpy manner producing weak erosion. Therefore, at the first ten arc operations very high average velocities have been observed (for stainless steel ~30–40 m/s). After some arc operations (~20 arcs), the velocity of the arc gradually decreases (up to 5–3 m/s). At this time, spots of the second type occur. They move slower and produce much erosion; the craters left by the arcs here are much greater (about 10–20 μm). According to Fang [43], the arc velocity as a function of the magnetic induction B can be shown in Fig. 13.13 for different oxygen-free cathode materials and arc currents I, the electrode spacing d being 1.5 mm.

It can be seen that the arc velocities considerably depend on cathode materials properties. The velocity increases linearly (except for Ti at high field) as the magnetic field raise. The slope of each of the plots depends on the cathode material. With increasing arc current I at constant magnetic field, the arc velocity increases.

The effect of electrode spacing at low arc current and magnetic field on the cathode spot velocity was studied [43]. For a copper cathode, with a magnetic field of 0.08 T and arc current of 30 A, it was demonstrated that as the spacing increased, the spot velocity increased as well up to a saturation value. The spot velocity sharply increased from 0.9 to 1.5 m/s when the electrode spacing was increased from about 0.5 to 1.7 mm and the velocity saturated when electrode spacing was increased beyond 5 mm ($B = 0.08$ T) as shown in Fig. 13.14.

It was found that in a transverse magnetic field the arc voltage depends significantly on the magnetic field B and the electrode spacing, and slightly on the arc current. The arc voltage rises from 19 to 25 V when B increased from 0.02 to 0.1 T

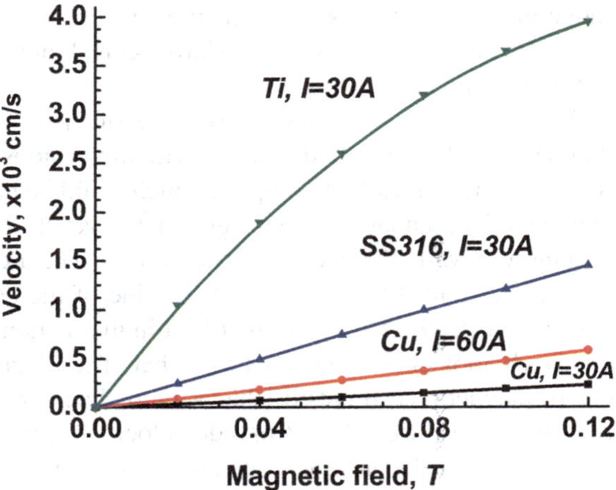

Fig. 13.13 Arc velocities as a function of magnetic field for different cathode materials and arc currents. The electrode spacing d being 1.5 mm

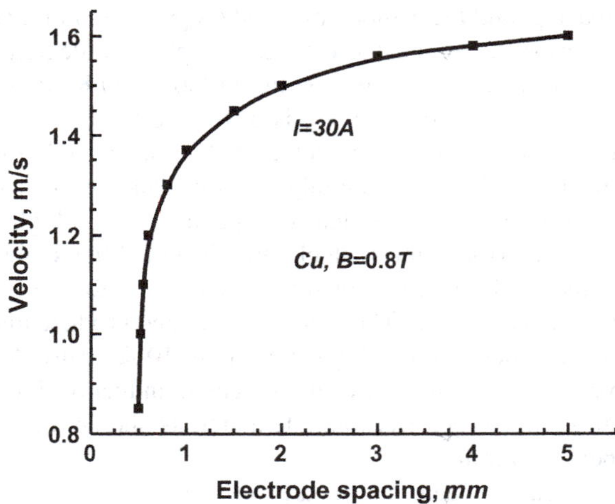

Fig. 13.14 Arc velocities as a function of electrode spacing. The experimental scatter is ±0.1 m/s

at $I = 60$ A and from 19 to about 60 V when electrode spacing increased from 1 to 6 mm at $I = 30$ A and $B = 0.08$ T.

A research group in University of Liverpool studied the retrograde spot motion. Djakov and Holmes [44, 45] reported some simple experiment in order to base their theoretical model to explain the mechanism of retrograde spot motion. High-speed photography with an image converter camera in the streak mode operation was used to registration the spot motion. The streak photographs with duration of 200 μs and speed of 0.25 mm/μs were studied with microphotometer. It was observed that rapid motion of the spot produced a pronounced deflection of the track in the streak photograph. All densitometer traces showed asymmetrical distribution of luminescence in the spot region indicating on some curvature of the plasma column in the direction of the spot motion.

Sherman et al. [46] investigated the retrograde spot motion under self-magnetic field in a high-current vacuum arc (typically pressure 5×10^{-8} torr). The electrodes had a circular configuration, with anode and cathode of equal size. The electrode materials were hard-drawn, high-conductivity (HDHC) copper of 99.9% purity, and vacuum-cast (VC) copper of more than 99.95% purity. The HDHC electrodes were 100 mm in diameter with radiused edges, giving a plane area of 75 mm diameter, and the VC electrodes were 75 mm in diameter with 1 mm radiused edges. The arcs were initiated by a surface-breakdown trigger mounted in a hole at the center of the cathode. The current rise time was 170 μs, and then a constant current was of 1.65 ms with the current amplitude varied between 2.7 and 6.9 kA. A magnetic probe was used to measure the self-generated magnetic field of the arc. The cathode spots were photographed with a high-speed framing camera at 15,000–20,000 frames per second. A magnetic probe was used to measure the self-generated magnetic field of the arc. For VC electrodes, the magnetic flux density was calculated by assumption that all the current I enters the cathode through the spots that were positioned on a ring of radius r the average azimuthal magnetic flux density at a spot is given by $B = \mu I_0/4\pi$.

The experiment showed that during the current pulse, the cathode spots moved radially outwards from the cathode center and they expanded across the cathode in a ring-shaped pattern (Fig. 13.15) [47, 48]. Some spots initially were observed inside the ring, and these appeared less bright than those on the ring. The number of spots observed later, when the spots within each cluster could be resolved, was consistent with a mean current per spot of approximately 100 A.

This basic behavior was reproducible and independent of current from the lowest current (2.7 kA) up to approximately 7 kA. A larger fraction of spots occurred inside the ring on the HDHC electrodes than on the VC electrodes, and there was a tendency on both materials for the fraction inside the ring to be larger and to be non-uniformly distributed at higher currents. A different type of behavior was observed for currents above approximately 7 kA. The spots started to expand away from the ignition hole, but within 300 μs a second expansion center formed spontaneously on the cathode surface. This resulted in two expanding rings of spots each behaving similarly to those previously depicted and, in addition, moving away from each other. This behavior was always observed for currents above 8 kA with a 15 mm electrode separation.

Fig. 13.15 Expanding ring of cathodic spots: current, 5.3 kA; ring diameter, 50 mm [47]

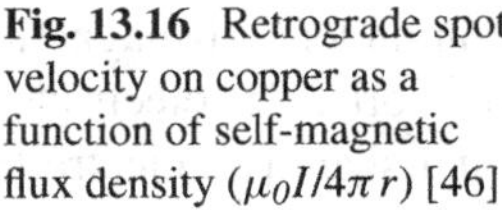

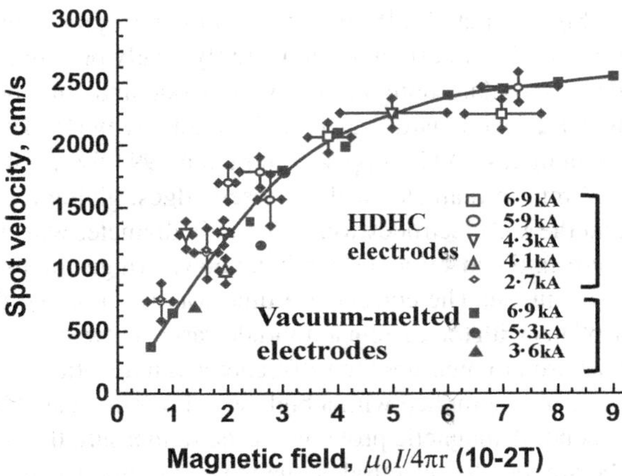

Fig. 13.16 Retrograde spot velocity on copper as a function of self-magnetic flux density ($\mu_0 I/4\pi r$) [46]

The dependence of the spot velocity obtained with a gap of 15 mm [46] is shown in Fig. 13.16. The spot velocities are proportional to the flux density up to about 0.02 T, with a proportionality constant of 700 ± 140 m/s/T. Above 20 m/s and 4×10^{-2} T, the velocity increases relatively slowly with increasing flux density. No significant dependence of the spot velocity on the gap was found for the HDHC electrodes at gaps of 5, 10, and 20 mm.

The motion of vacuum arc cathode spots under the influence of self-generated azimuthal and externally applied axial magnetic fields has been investigated by Agrawal and Holmes 1984 [49]. The experiments were conducted with circular electrodes, having 75 mm diameter and fixed gap length of 15 mm. The anode was of copper, and the cathode was constructed from Cu, Al, Mg, Bi, Zn, and Cd of high purity (99.985–99.999%) with low gas content. A surface-breakdown trigger assembly was used to the arcs initiation. The pressure was of about 10^{-8} Torr before each arc. A low-current arc (200–400 A) was initiated and run for 100–200 μs before the main current pulse was applied. The current rise time was 100–250 μs with amplitude up to 10 kA and the pulse duration of 1.5–5 ms. The spot motion was recorded by means of a high-speed framing camera, and the self-generated magnetic field was measured with an inductive magnetic probe.

Agrawal and Holmes [49] observed (as in work of [46]) the ring pattern of spot expansion on the cathode surface and studying the spot velocities. The spontaneous spot formation with high rates of current rise was also detected. Depending on the presence of an axial magnetic field, two different modes of spots expansion were observed. In the absence of an axial field, the spots remained on the ring giving a "curvilinear" density distribution. The rings with diameters up to 10–15 mm appeared to consist of a number of spots distributed close one to other in line segments. At larger diameters, individual spots were distinguished. At high currents, the ring was more nearly perfect, although the spots still not uniformly were distributed on the ring.

The distribution of cathode spots was quite different in the presence of an axial magnetic field. During the ring expansion, more spots tended to appear inside the ring. At high arc currents and in weak axial fields, many spots remained on the ring even at large ring diameter. In strong fields and at large diameter, the spots appeared to be fairly uniformly spread over the whole area inside the ring. The above two modes of ring expansion were reproducible up to certain currents which was dependent on the cathode material, and it was for copper 9.6 kA. At higher currents, the rate of current rise can influence the shape of spot distribution.

In particular, a spontaneous second ring can be appeared. In this case, an anode spot located opposite to the second ring and bowing of the plasma column from the first group of spots in the Amperian direction by the magnetic field of the second parallel discharge was also be detected. According to the observations, the initial rate of current rise up to 3×10^7 A/s had no effect on the initial symmetrical ring expansion of cathode spots on all the metals except zinc and cadmium, for which the limits were approximately 1.5×10^7 A/s and 6×10^6 A/s, respectively.

A measured relationship [49] between the retrograde spot velocity and the self and axial magnetic flux densities on a copper cathode is represented in form of Fig. 13.17. It can be seen that at low self-generated fields, the velocities directly proportional to the flux density but at high field, the velocity saturation was observed. The approximate upper limits of the linear range and the slope of the linear portion of the curves for different cathode materials are given in Table 13.1. The axial magnetic field decreases the retrograde velocities. On a bismuth cathode, only linear dependence of retrograde spot velocity on self-magnetic field up to 0.04 T was observed for axial magnetic field varied from 0 to 0.05 T as parameter. The general dependencies of cathode spot velocity versus self-generated magnetic field for Al, Sn, and Mg were similar to that dependence for copper. The velocity for Zn and Cd as function on both magnetic fields was similar to the dependence for Bi.

It has been found that the relationship between the velocity and self-generated field was independent of the arc current, and dependent only on self-generated field characteristic and cathode material. This is due to the fact that a discharge at high

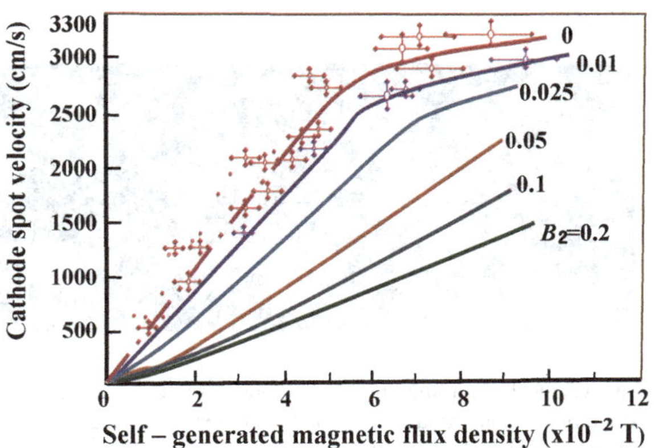

Fig. 13.17 Retrograde spot velocity on a copper cathode at various axial magnetic fields (in T) as a function of self-generated magnetic flux density

Table 13.1 Velocity and self-generated field limits and the slope of the linear range of spot velocity characteristics of the metals investigated [49]

Cathode material	Velocity m/s	Self-magnetic field, 0.01 T	Slope, m/s/T
Cu	30	4.8	520 ± 100
Al	35	2.1	1670 ± 140
Sn	35	3.5	1000 ± 110
Mg	30	2.8	1100 ± 120
Bi	22	3.9	560 ± 80
Zn	15	5.4	280 ± 50
Cd	7	3.1	230 ± 80

currents consists of a large number of spots each carrying approximately the same current. It was noted that in spite of widely differing physical properties of the investigated cathode materials (Table 13.1), the measured maximum spot velocities are within the same order of magnitude (10–65 m/s).

Some similar observed characteristics of cathode spot retrograde motion were summarized by Drouet in 1981 [50] and in general repeated by Drouet [51] in order to use the published data for theoretical description of the spot motion.

Also, recently Song et al. [52] demonstrate the magnetic field influence on the cathode spot distribution and spot motion trajectory. In the experiment, electromagnetic coils and permanent magnets were used and the vacuum chamber is evacuated to 10^{-3} Pa. The water-cooled cathode was fabricated from titanium (99.99% purity) of 95 mm in diameter and 50 mm in thickness.

The authors have shown that increasing the number of magnets and decreasing the distance between magnets and cathode, both lead to enhancing cathode spots motion velocity. The shape of spot distribution is illustrated in Fig. 13.18a. It can be seen ring distribution at zero and small coil current. However, the radii of cathode spots trajectory decrease gradually as well the distribution pattern changed with the increasing of electromagnetic coil's current. The spot trajectory begins to contract after electromagnetic coil's current exceed 1.5 A and reaches to the cathode center at

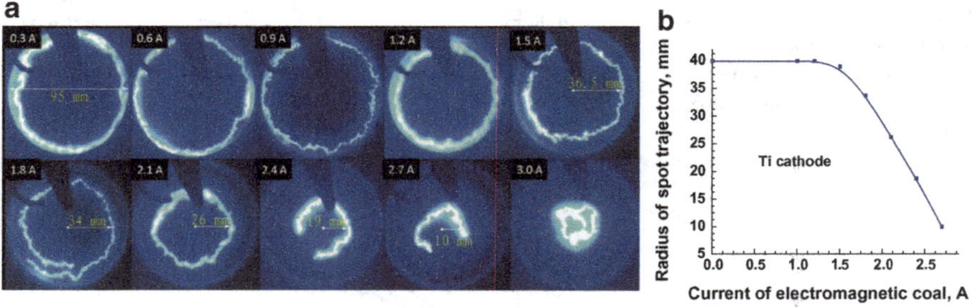

Fig. 13.18 Expanding ring of cathode spots dependence on a magnetic field (**a**, left) and spots trajectory's radii reduction with the increase of electromagnetic coil's current. The spot image taken from [52]. Used with permission

3 A shown in Fig. 13.18b. Parallel magnetic field component intensity influences the speed of cathode spots rotate motion, and perpendicular magnetic field component drives spots drift in the radial direction.

Swift et al. [53] studied the motion of cathode spots on a titanium cathode, in a vacuum, in a closed magnetic field configuration was studied by. The tangential and normal components of the magnetic field were measured using a Hall probe with an active area of 1 mm^2. The cathode was a water-cooled titanium disk of 100 mm diameter, and the anode was the cylindrical stainless-steel vacuum chamber of 600 mm diameter connected to ground. The cathode–anode separation was 150 mm. The chamber was evacuated by an oil diffusion pump and has a base pressure of 10^{-4} Pa. The arc was initiated by momentarily touched tungsten-tipped trigger. The arc is sustained by a high-current welding-type power supply and normally arc current used in the range of 80–100 A. A conventional camera with a lens shutter and an image converter camera (Imacon-700) operated as framing camera with an exposure time of 20 μs were used to record the motion of the cathode spot.

The cathode spot trajectory was observed as a stable one in the investigated geometry. Measurement of the position of the deepest part of the erosion track at a constant current and magnetic field showed that the cathode spot tended to move in a circle of the same diameter where the normal component (with respect to the plane of the cathode) of the magnetic field was zero. It was shown that the measured velocity increased nonlinearly with magnetic field passing to asymptotic behavior and to a value of about 25 m/s at large values of magnetic field of about 0.01 T. The high-speed photograph study showed that at zero magnetic field the trajectory of the cathode spot appears chaotic. For the nonzero magnetic field, the spot still executes a chaotic motion, but has the retrograde drift velocity superimposed on it. In the presence of an oxygen at low partial pressures, the spot shows an increased velocity and a tendency to branch. At higher partial pressures, the spot splits into a number of separate spots and there is a large increase in the spot velocity.

Also Swift [54] conducted experiment in a system where anode was the 400 mm diameter vacuum chamber or a stainless steel disk 250 mm in diameter located 40 mm from the cathode. The experiment showed that at low pressures as 3 kPa an anode spots formation occurred, and in all instances of their formation, the motion of the arc was observed to be non-retrograde. This pressure is much lower than the pressure at which cathode spots begin to move in the non-retrograde direction with no anode spot formation. This fact allows concluding that the anode spot formation determined by the arc conditions and on the cathode spot motion.

Klajn [55] provided an experimental analysis of the vacuum arc behavior in a pulsed or oscillatory transverse magnetic field with amplitudes up to 120 mT and frequency of 900 Hz. The arc was produced between Cu flat contacts of 20 mm diameter, opened mechanically with the speed of 0.4 m/s during the switching-off operation of the current half-wave of amplitude 500 A and frequency of 30 Hz. The contacts were separated of about 3.5–4 mm, and the pressure was of 5×10^{-3} Pa. A high-speed photography (2000 frames/s) and a shielded Langmuir probe were used. The main results of high-speed observation were related to a behavior of the cathode spots, and any data related to a dependence of the spot velocity were not presented.

Each frame showed about 0.5 ms of the arcing time were studied before and after the magnetic field action. An analysis of the photographs showed that the character of plasma movement and the displacement of the arc in the oscillatory magnetic field were observed in accordance with reported previously the retrograde spot motion. The value of plasma potential during the arc operation was obtained between 13 and 17 V, before experiments with the magnetic field. Arc voltage increase up to about 80 V was observed during the magnetic field action.

Juttner and Kleberg 2000 [56–58] conducted experiments that reported on the retrograde arc spot motion on copper and tantalum cathodes in vacuum in the presence of a magnetic field. Later Juttner [59] reviewed the data. In those experiments, a base pressure was less than 10^{-7} Pa. Three electrode systems were studied. One was with small needle cathode (Cu) was of 750 μm diameter facing with its end a copper anode cylinder of 10 mm diameter, thus forming a point-to-plane gap of approximately 200 μm electrode separation. The magnetic flux density increased from 0 to 0.4 T during 500 μs. On the side faces of the cathode, the external magnetic field was parallel to the surface, thus yielding a situation suitable for retrograde motion. An electrode system constructed for study the phenomenon of hot tantalum cathode (see below). Another used geometry was similar to system for hot cathode with copper electrodes. The U-shaped anode was placed 1–3 mm above a strip cathode. The cathode had a width of 12 mm and an effective length of 5 cm made of a 0.5-mm-thick metal sheet. Two permanent magnets with field-oriented parallel to the cathode surface of −0.4 and +0.36 T (the + and − signs denoting orientation up and down the system) were used. Rectangular high-voltage pulses with amplitudes of 5–13 kV were applied from a pulse-forming network. The arc was initiated by a gap breakdown. For sufficiently small electrode distances (<100 μm for needle cathode and <1 mm for strip cathode), the breakdown stage changed to a low-voltage arc after times <100 ns. The spots are imaged with time and space resolutions of <100 ns and <10 μm, respectively. The arc currents varied in range of 2–100 A.

Generally, the experiment provided at small electrode configuration showed that luminous plasma of a single cathode spot has a diameter of 50–80 μm on Cu electrodes. An inner structure consisting of fragments with diameters of 5–30 μm with 10–30 A per fragment was obtained. The authors indicated that no obvious difference in fragment dynamics has been found with and without a magnetic field. The plasma structure near the spots was studied with U-shaped anode and Cu strip cathode. A plasma glow near the spot expanding in the retrograde direction is shown in Fig. 13.19. This part of plasma is interpreted by the authors as a plasma jet. It can be seen that the spot luminous is significantly intense than the plasma glow. In general, the plasma glow takes place around the spot but elongated in the retrograde direction on length equal about of the spot size.

The dynamics of near spot plasma is presented in Fig. 13.20 where time regime is shown in four frames with the exposure increasing. At 500 ns in the first frame, the plasma edge near the spot denoted by 1. In the following frames, the plasma region furthermore expands at exposures of 1, 2, and 5 μs. The glow ends at edge 2, as indicated in the third fame, but it is also reached in second frame, i.e., through one microsecond. According to the scale indicated in Fig. 13.20, the plasma region

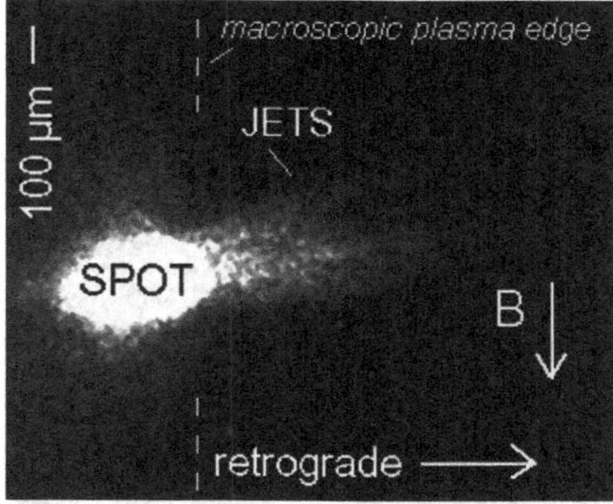

Fig. 13.19 Plasma structure around the spot in a magnetic field. From [57, 58]

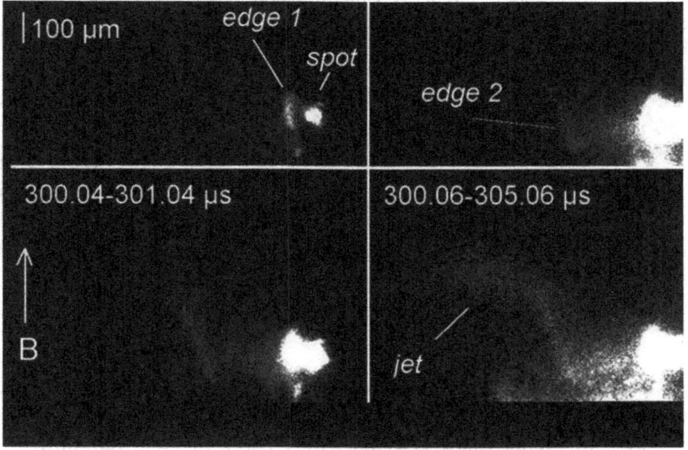

Fig. 13.20 Sequence of frames down the columns with increasing exposure: 0.5, 1.0, 2.0, 5.0 μs. The fourth frame shows a jet emanating from the upper corner of edge 2, $I = 25$ A. From [57]

between edges 1 and 2 is about 2.5×10^{-4} cm, i.e., the velocity of expansion is about 250 cm/s. However, Juttner and Kleberg noted that "changes in the plasma structure between edges 1 and 2 occurred within 50 ns" and the velocity may be significantly larger, as 3 km/s, estimating from intensity profiles along the streak trace.

A spot exposed four times with an interval of 10 μs is shown in Fig. 13.21. It can be seen that the spot size is about 50 μm. The spot displaced together with expanding plasma. The authors indicated that the new spot ignited in the direction of the expanding plasma.

A spot displacement is shown in Fig. 13.22, which is not exactly straight. It can be seen that a new spot was ignited in most cases on length equal about of the spot size.

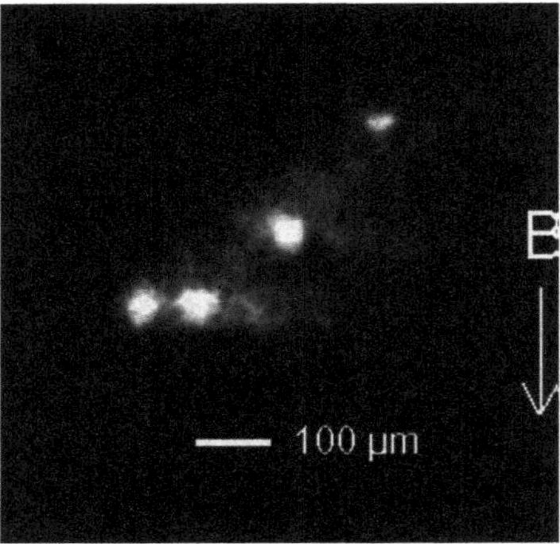

Fig. 13.21 Fourfold exposure of 200 ns of a single spot with 10 μs separation of the exposures. *I* = 20 A. From [57, 58]

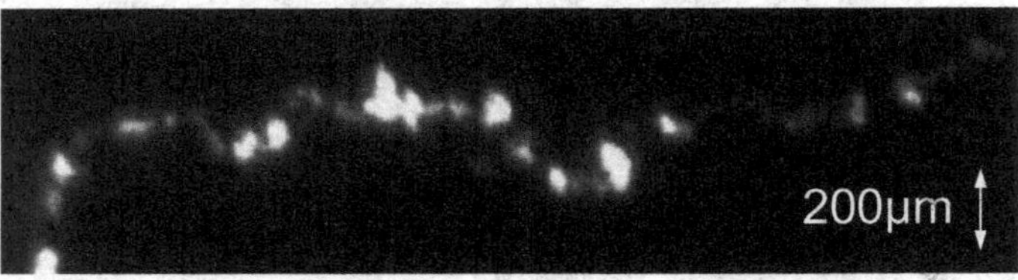

Fig. 13.22 , Integral picture of long exposure (200 μs) of a single spot. From [57]

Dukhopel'nikov et al. [60] demonstrated a high-speed video filming of the process of combustion of a vacuum arc for both with and without the presence of an external magnetic field. The experiment provided in a vacuum chamber with diameter of 700 mm and length of 700 mm. The residual pressure was $5{\cdot}10^{-3}$ Pa. A cylindrical cathode made of a titanium pipe with diameter of 56 mm and length of working part 450 mm mounted on the axis of the chamber. At one end, the cathode attached to the cathode assembly on the flange of the vacuum chamber. The other end of the cathode is freely positioned in the vacuum chamber. Cooling water and supply voltage supported to from the direction of the cathode assembly. Ignition of the discharge realized by means of a spark-over of a thin conducting film on the surface of an insulator located on the free end of the cathode. The arc was powered by a welding rectifier with no-load voltage 75 V and maximum current 315 A. The discharge current was of 110 A with voltage 23 V. The magnetic system creates a magnetic field of induction B = 0.005 T on the surface of the cathode. The discharge burns at a pressure in the chamber 1×10^{-2} Pa.

In the absence of a magnetic field, the discharge was ignited at the free end of the cathode, traveled to the cathode assembly, and was extinguished in the arc suppressing device. Video filming demonstrated that some of cathode spots burn

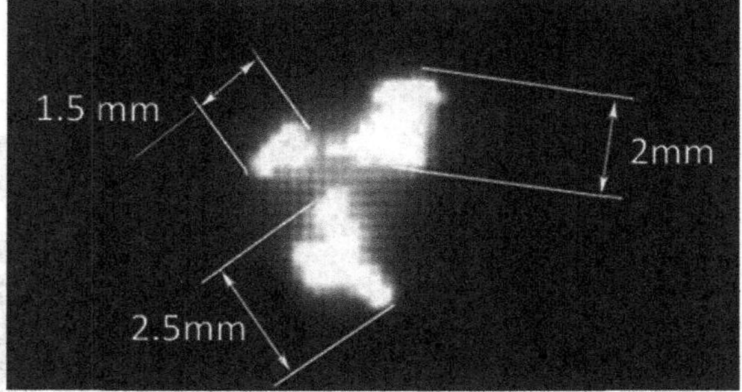

Fig. 13.23 A glow of cathode spots in vacuum arc in the absence of a magnetic field (arc current 110 A, voltage 23 V, cathode of titanium, speed of filming 5000 frames per second). Spots travel slowly along the cathode. Figure taken from [60]. Used with permission

simultaneously for 200 μs. The group of spots moved along the cathode in random fashion. Two modes of burning that replace each other were detected.

It was established, in the absence of the external magnetic field, the spots move intermittently. The experiment showed that one of spot group slowly travel into region of 5–10 mm in dimension and the spots were observed with size of the illuminated region of amounts to 1–3 mm (Fig. 13.23). The spots have significant brightness. In this mode, the arc burns for 1.8–4.2 ms.

In the second mode, a rapid jump of the group of spots to a new location over a distance of 14–30 mm during a period lasting 2×10^{-4}–1×10^{-3}s was observed. The dimension of the illuminated region of the spot here amounts to 0.2–0.8 mm, which is significantly less than in the first mode. The brightness of the spot is low, and the speed of the group of spots in this mode was in the range 15–60 m/s. Another mode of group of spots ignited at the free end of the cathode traveled along the cathode in random fashion. This group of cathode spots was found at the same location at 1.8–4.2 ms and jump to another location in the same time; the mean speed of a cathode spot was 5–6 m/s. Nevertheless, it was summarized that in the absence of a magnetic field, the mean speed of the group of cathode spots along with the cathode amounts to 5–6 m/s. The dimension of the pit left on the cathode by a single spot is in the range of 0.22–0.38 mm.

Figure 13.24 shows the spot motion in the presence of a magnetic field. The group of cathode spots moves uniformly at a speed of 15–20 m/s. The dimensions of the illuminated region of the cathode spot of the field amount to 0.45–0.65 mm. The dimension of the pit left on the cathode by a single spot is 0.14–0.26 mm. A variation in the direction of magnetic field leads to a variation in the direction of the cathode spot travel. It was established that in the presence of an external magnetic field the direction of a cathode spot is opposite to Ampere's force. In general, the results show that the size, behavior of motion and velocity of group of spots significantly depends on magnetic field presence.

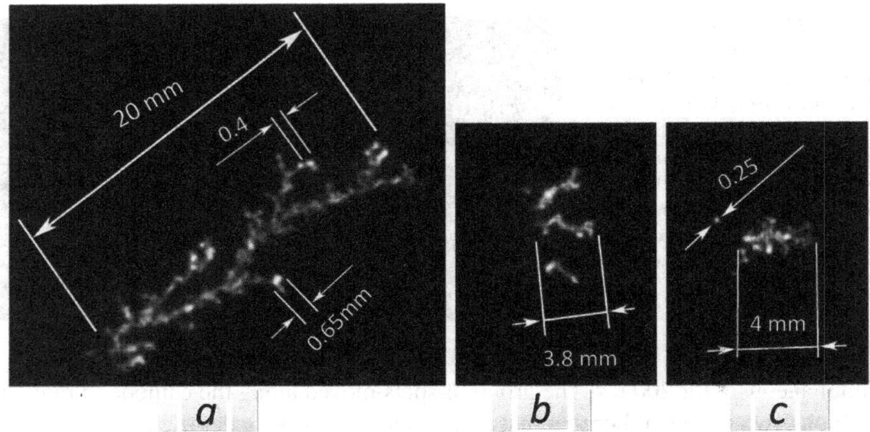

Fig. 13.24 A glow of cathode spots in vacuum arc where there is a magnetic field with induction $B = 0.005$ T present (arc current 110 A, voltage 23 V, cathode of titanium). Speed of filming 1000 frames per second (**a**) and 5000 frames per second (**b** and **c**). Figure taken from [60]. Used with permission

13.3.1.2 Cathode Heating and Retrograde Cathode Spot Motion

Smith [61] first reported on retrograde spot motion along a hot cathode. He observed an arc between concentric tantalum electrodes in argon and mercury vapor that was driven around by an axial magnetic field. Argon pressure was from 5 to 30 torr. Estimated Hg vapor pressure was above 0.1 mm. Arc current was approximately 13 A for a cathode 2.5 cm in diameter. The cathode was so constructed that the arc spot operated upon its outer surface and could in a few seconds bring it to a white heat. When the cathode was only at a red (approximately 900–1300 K) heat or less, the arc raced rapidly in the retrograde direction. When the cathode in the spot was reached white (1700–2300 K) glow, the arc would turn suddenly and then travels very slowly in the proper direction (the way the electromagnetic forces were acting). Arc drop was 16 V when the cathode was relatively cool and 14 V when white hot. No conditions of pressure, field intensity, or arc current were detected that would make the arc go in the retrograde direction. Thus, Smith observed that for Ta cathode the retrograde motion was only characteristic of arcs at stage of not very hot cathode. Gallagher and Cobine [62] also observed independently that the reverse motion does not take place on tungsten cathodes which are at sufficiently high temperatures. However, Hernqvist and Johnson [31] reported experiment in which the retrograde motion observed on an externally heated thermionic cathode.

So, the experimental results of the early investigations different and furthermore were studied later. Nurnberg et al. [63] measured the arc velocity for Al cathode in a magnetic field 0.03-0.1 T for heated cathode conditioned by a number of arcs. All arcs moved in the retrograde direction. The spot velocity decreased from about 3.4 m/s at room temperature to about 1.7 m/s for 770 K.

Fang et al. [64] and Fang [65] investigated the velocity of vacuum arcs in a magnetic field at the cathode for Cu, Al+3% Mg, and 316 stainless steel. The sample used as the cathode and attached to a Mo cathode support, which can be heated to 1200 K by two heating elements. The cathode temperature was measured with a thermocouple. The anode below is made of graphite and is wedge-shaped so that any material coming from the cathode in solid or droplet form was deflected to the side. The electrode spacing is about 2 mm. The cathode samples were degassed at a temperature of 770 K in a vacuum for 8 h. The working vacuum was 10^{-8} mbar. The vacuum arc discharge is triggered by injecting plasma into the electrode gap, and after the arc initiation, the arc voltage drops from 500 V to about 20 V. Arc currents were varied from 15 to 70 A, and homogeneous magnetic inductions were varied in range of 0.02–0.1 T. The velocity of the moving discharge was measured optically by projecting a magnified image of the electrode gap onto a screen (Fig. 13.25).

The measurements at room temperature show the arc velocity consistent with the other recent measurement in which the cathode surface was free of a contamination. According to Fang [65], the arc velocity decreases with increasing the cathode temperature for all investigated cathode materials at constant arc current and magnetic induction. The higher value of arc current leads to higher velocities that can be seen for the case of Cu cathode. The higher value of magnetic induction leads to higher velocities that were obtained [65] in case of 316 stainless steel (Fig. 13.26). The decrease course of retrograde velocities with the increase of the cathode temperature mostly keeps unchanged. It can be seen also, when the cathode temperature not high, about 0.5 of the material melting temperature T_m, the retrograde velocities decrease slightly with increasing the cathode temperature. When the cathode temperature was above 0.6–0.8 T_m, the retrograde velocities decrease steeply.

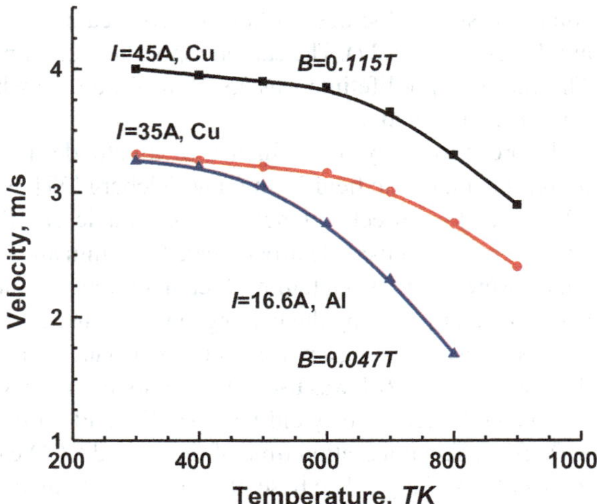

Fig. 13.25 Arc velocity as a function of the cathode temperature with Cu and Al as cathode material

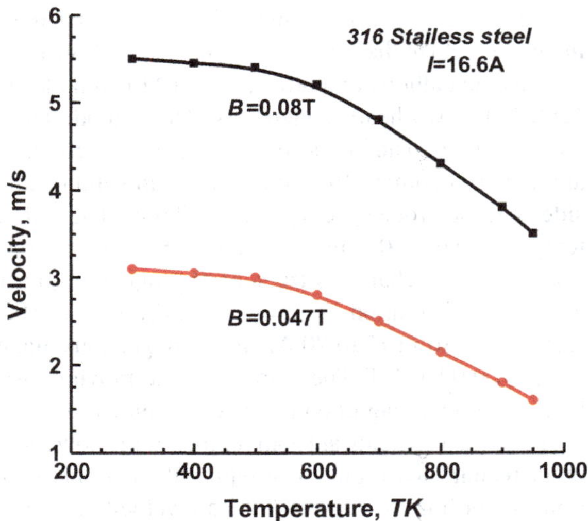

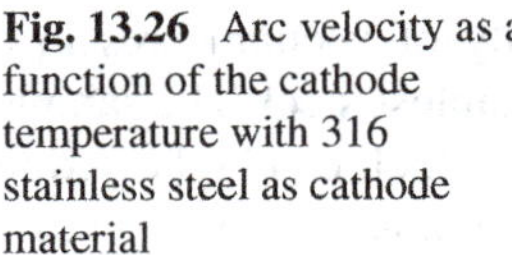
Fig. 13.26 Arc velocity as a function of the cathode temperature with 316 stainless steel as cathode material

Puchkarev and Murzakayev [66] used autograph method to investigate the arc spot motion over a cleaned tungsten (wires 0.1–0.2 mm in diameter) cathode in oil-free vacuum (10^{-6} Pa) for current in range of 2–50 A. The duration of a rectangular arc pulse was 1.2 μs. To identify the track width more accurately, the cathode spot motion was directed by the magnetic field *B* created with the pulse current passed through the wire cathode. The magnetic field varied in the range of 0.1–0.7 T. The cathode cleaned prior to the experiment by heating at 2000 K.

The experiment was provided also with heated cathode at 1800 K. The cathode–anode distance was small as 0.1–0.2 mm. The measurements [66] showed mostly equal the spot velocities on hot and cold cathodes, which depend on the current amplitude (Fig. 13.27). The cathode spot starts dividing at a current of a few amperes. The cathode spot lifetime was 25–50 ns for a cold while it increased to 150–200 ns for a heated cathode.

In order to study the influence of a cathode, heating on the spot motion in a transverse magnetic field Juttner and Kleberg [57] used strip cathode with width of 12 mm, and an effective length of 5 cm made of a 0.5-mm-thick metal sheet. The anode was a U-shaped electrode placed 1–3 mm above the cathode. The cathode and anode were made of tantalum. A cathode temperature up to 2100 K (white glow) was reached by varying the heating current through the cathode. The temperature of Ta was measured using a pyrometer. A permanent magnet with an absolute magnetic flux density of 0.36 T was used. The cathode surface was placed at an angle of 45° to the magnetic field, thus yielding a parallel and normal field component of $B_p = B_n = 0.255$ T. The integration time of the cooled CCD camera was 8 ms, thus grabbing the whole discharge. The base pressure was about 10^{-5} Pa.

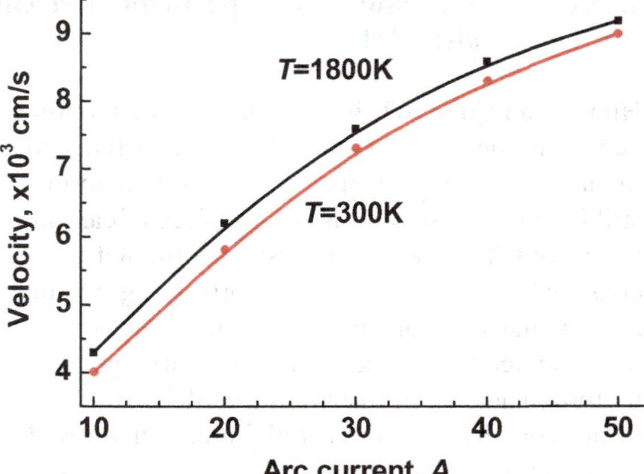

Fig. 13.27 The velocity of the CS motion as a function of current at cold and heated cathodes

Note that with increasing temperature, it becomes more difficult to ignite an arc in the presence of the magnetic field. Therefore, the arc was ignited outside the magnetic field on cold parts of the electrode assembly if the gap distance was >5 mm, or inside the magnetic field with the gap <1 mm. Observed, every discharge ignited on the hot cathode showed spot motion and spot traces under the same angle as in the case when no heating occurred. Also, no change in the spot direction occurred; the motion of the cathode spots and therefore the arc remained retrograde up to 2100 K (white glow). Above 2100 K tantalum becomes brittle, so this was the limiting temperature in conducted experiments with hot cathode in the described experiment.

Juttner and Kleberg [57] concluded, the retrograde motion is not a surface effect, since drastic changes in the cathode heating have no impact on the motion. This statement appears to be in contradiction to results obtained by Smith [60]. The contradiction was explained indicating that Smith observed retrograde motion as long as the cathode showed a red glow (approximately 900–1300 K) and a motion reversal of the whole arc at white glow (1700–2300 K). However, at this cathode heating the pressure inside Smith's [61] vacuum vessel was 13 Pa for Hg and 1–4 kPa for Ar. As the heating probably increased the gas pressure, the motion reversal could be a gas effect. Also, it was noted that possible a transition of the discharge to a spotless thermionic arc at a temperature of 2200 K and therefore a transition to a diffuse mode. In this case no is the retrograde motion.

While the Juttner et al. [57–59] conclusion concedes with the result of Hernqvist and Johnson [31], Puchkarev and Murzakayev [66], it is different from the result of Gallagher and Cobime [62], Nurnberg et al. [63] and Fang [65] (showed spot velocity change with cathode temperature) indicating that the problem remains to be investigated.

13.3.1.3 Gas Pressure and Gap Distance Influence on the Spot Motion Under TMF

Himler and Cohn [67] investigated the arc motion using few experimental systems depending on gas pressure. The first apparatus consisted of a pair of brass plates mounted in such a manner that a gap was formed in a horizontal plane. The arc striking mechanism consisted of a piece of carbon, which moved across the gap by a solenoid. The use of this system is limited by plate melting. The next apparatus consisted of a pair of concentric carbon rings mounted in such a manner to constitute a horizontal, circular gap, one-quarter of an inch wide. A circular coil around the bell jar generated the magnetic field. This arrangement of electrodes insured a perfectly uniform field over the path of travel of the arc. The next apparatus consisted of a pair of horizontally mounted parallel tungsten wires. The use of this system is limited by oxidation of the tungsten. The final apparatus consisted of two copper electrodes are mounted in a horizontal position. One is fixed rigidly to the baseboard, while the other one is attached to a backing plate, which can be moved back and forth with the aid of a small motor.

It was obtained that when the pressure was lowered, the velocity of travel of the arc decreased rapidly until, at the critical pressure, the arc stops moving altogether. At that pressure, the arc changed in appearance. The cross section of the arc decreases greatly, and its brilliance decreases markedly. The anode and cathode spots no longer are well defined, and the arc appeared unstable. If still the pressure lowered, the arc will travel in the opposite direction as it was traveled at atmospheric pressure. Extensive attempts were made to obtain accurate data of this critical pressure for conditions at different experimental systems. This information was presented in form of critical pressure variation by change in gap width and field current. In particular for an arc 7.5 A, the critical pressure decreased in the range of 50–700 torr, when the electrode gap increased from 1/16 to 1/4 inches. When field current increased from 2 to 3 A, the critical pressure decreased.

Yamamura 1950 [68] showed a phenomena related to some difficulties of driven of an arc by the magnetic field at very small gaps between the electrodes. These phenomena called as immobility phenomena. The immobility time and the driving velocity of the arc were measured under various air pressures. Under a certain low air pressure, the arc motion was unstable under the magnetic field. At the air pressure lower than a critical value, the arc can drive again by the magnetic field, and the arc moves in a direction opposite to the direction of the electromagnetic force.

It was obtained that the electrode gap for Cu, at which an electric arc begins to be driven, decreased parabolically from 0.4 to 0.2 mm when the magnetic field intensity increased from 0.005 to 0.04 T. The critical air pressure (the pressure at which permanent immobility, i.e., the reversal arc motion occurred) increased linearly with the magnetic field intensity. Figure 13.28 demonstrates this measured critical pressure from [68] depending on magnetic field intensity. It was informed that the

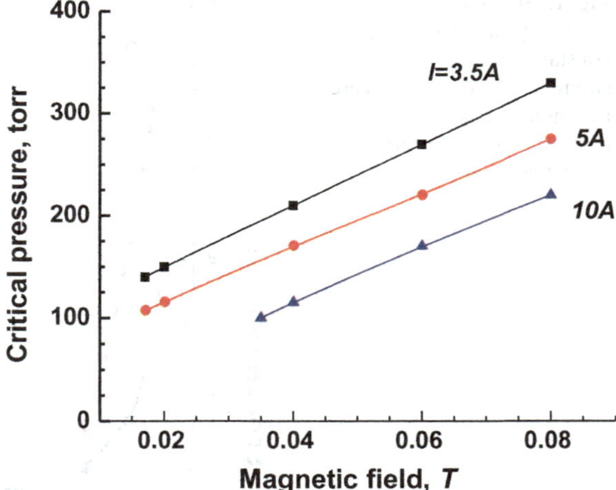

Fig. 13.28 Relations between driving magnetic field intensity and critical air pressure at which the permanent immobility occurred for Cu and gap = 2.5 mm

permanent immobility was also detected for Fe, Ag, Sn, and liquid Hg and that the aluminum electrodes have a wider pressure range where permanent immobility occurs.

In vacuum arcs, the retrograde velocity depends on the electrode gap distance and pressure. Gallagher and Cobine 1949 [62] and Gallaher 1950 [26] showed that the velocity of the spot depends on the pressure, decreasing as the pressure increases for a mercury arc using a mercury pool cathode. When the pressure was sufficiently increased, the spot can be made to slow down to zero velocity. The pressure at which reversal occurs linearly depends on the magnetic field and the current, as shown in Fig. 13.29 for mercury arc. As it can be seen, the critical pressure is significantly

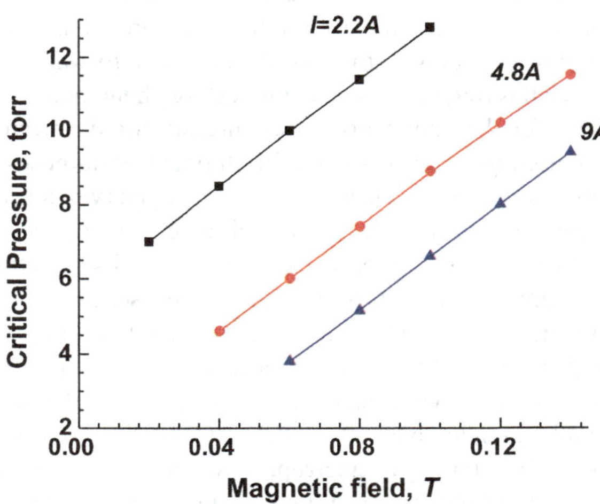

Fig. 13.29 Reversing pressure versus magnetic field in mercury pool tube from [26]

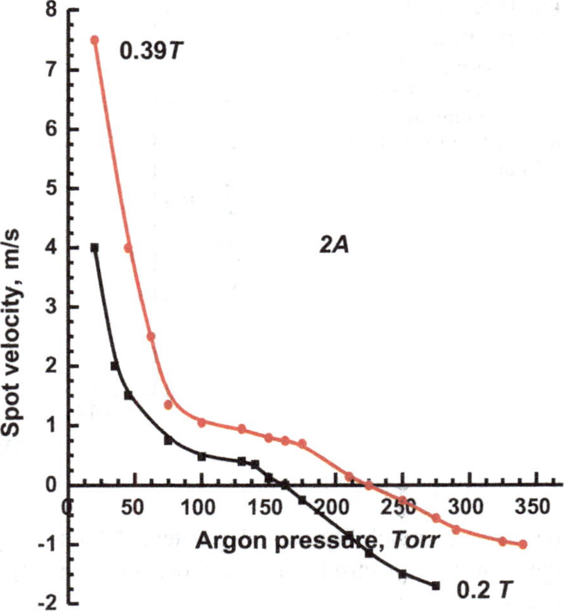

Fig. 13.30 Spot velocity versus argon pressure for constant magnetic field strengths. Positive values are retrograde velocities, and negative values are forward velocities from [27]

lower than for solid metallic cathode like Cu and similar materials. In general, fewer fields were required to reverse the spot at high currents than at low. Gallagher [26] suggested that this fact represents a condition of balance between the forces acting on the spot and the forces acting on the column, since it has been observed that the column always tries to move in the "correct" direction.

John and Winas [27] observed similar dependence. They obtained the spot velocity dependence on argon pressure for mercury arc shown in Fig. 13.30. With a low gas pressure, the spot moves in retrograde direction. Increasing the pressure causes the spot to slow down, stop, and then move in forward direction.

Dallas [69] discussed an interesting phenomenon of the arc motion in a magnetic field for the application in an aircraft circuit interruption. It was discovered that the used interrupter is valid for all practical purposes of circuit breaking equipment operating at sea level, but something happens when Ampere's law is applied to similar equipment at altitude. As the altitude is increased, an electric arc in a magnetic field may falter, stop, and finally reverse its direction of motion. Arc immobility and arc reversal phenomena limit the use, affect the design, and dictate the test procedure required for aircraft circuit interrupting equipment using magnetic arc suppression. This dilemma was illustrated by the comparison of a low-travel snap-action switch with magnetic arc suppression that shown to open consistently and stably an inductive 120 V DC circuit of 9 A from sea level to an altitude of 25,000 feet while failing to interrupt 1 A. This means that a switch, which will interrupt a large current successfully, will not interrupt 10% of that value. It was noted that this disconcerting discontinuity in switch performance was caused by a phenomenon known as "arc immobility" which is a precondition to "arc reversal" motion. A design of an interrupter was proposed for which the arc immobility zones in range of current from 1 to 9 A and altitude from 10,000 to 50,000 feet were determined by arc motion observation when the contact gaps were 0.035 and 0.07 inch with permanent magnet field of 0.03 T.

Dunkerley and Schaefer [70] presented experiments that carried out in air at atmospheric pressure, and only the forward spot motion was observed on oxide tungsten cathode for 30 A and 20 ms duration. The tracks were obtained by using parallel anode and cathode surfaces between which the arc was initiated in the presence of a transverse magnetic field of 150 gauss. The observation shown an arc track indicating the cathode spot was momentarily fixed while the plasma column continues to move under the influence of the transverse magnetic field (Fig. 13.31a).

According to Dunkerley and Schaefer [70], schematically the spot motion is illustrated in Fig. 13.31b. The arc has been anchored at some point due to the surface condition of the cathode [Fig. 13.31(1)]. However, the plasma continues to move due to the transverse magnetic field [Fig. 13.31(2)]. Then positive ion bombardment the cathode surface preparing it for the formation of a cathode spot at new location due to distorted plasma column [Fig. 13.31(3)]. So, the cathode spot transfers to the "preconditioned" next area (y) and extinguishes at the previous location (x) [Fig.13.31(4)].

Also, a track was observed for a 1 A DC arc traveling along a surface of lightly oxidized tantalum. The track gives the appearance of a discontinuous cathode spot although the arc current was continuous. Discontinuous tracks of this type were produced when the surface of the cathode is very lightly oxidized or oxidized non-uniformly.

Robson and Engel 1956 [71] observed retrograde motion in a short arc in air burning between copper electrodes with a 0.5 mm separation, for $I < 3$ A and $B > 0.3$ T. They also indicated a reversal spot motion, which occurred for arc current increased to 5A even for $B = 0.4$ T. Robson's experiment 1978 [72] showed that in arcs with currents of 5 and 10 A, gap 1.9 mm and air+argon pressure $P_{sur} = 30$ Torr, the spot moves in the retrograde direction when the external magnetic field $B_{em} <1$ T. For stronger magnetic field B_{em} varied from about 1.2 to 5 T, the spot moved in Amperian direction.

A series of works investigating the spot motion in a transverse magnetic field and for different gas pressure were published by Guile and co-authors. In one of their first works [73], the spot movement studied in the normal direction for the air arc at atmospheric pressure, which burned between parallel electrodes fed at one end with current. No external magnetic field was applied (only self-magnetic field was influenced). The high-speed photography showed that arc movement was partly regular and partly random. The regular movement was controlled by the cathode spot, which gave unbroken straight tracks up to 12 inches long. Random jumps occur when the arc column touches an electrode.

The arc characterized by two groups according to whether the material was magnetic or non-magnetic. The velocity was measured as dependence on arc current for mild steel and extruded brass. It was found that the arc velocity increased with

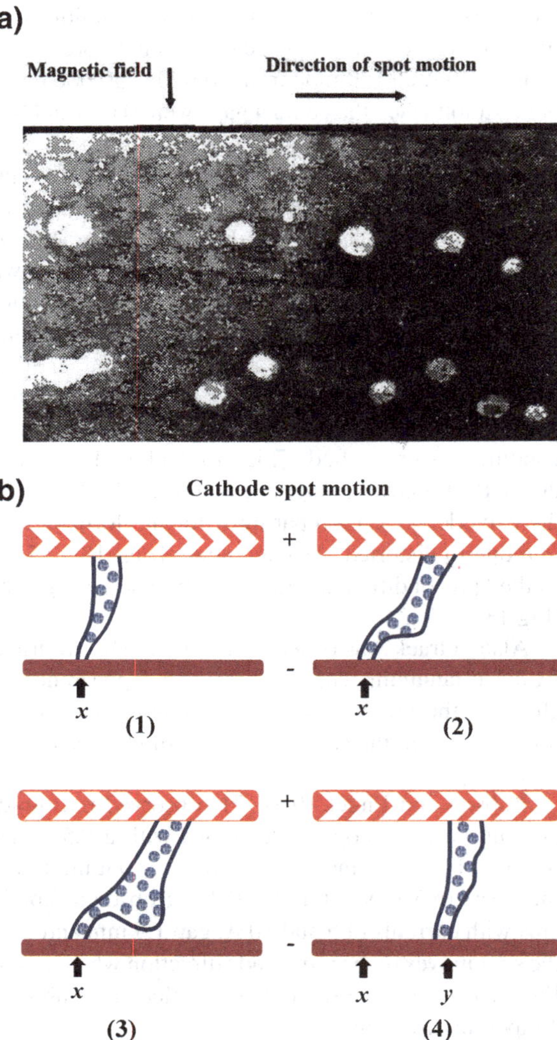

Fig. 13.31 **a** Intermittent cathode spots formed by continuous-current arc. Figure taken from [70]. Used with permission. **b** Schematic illustration of phenomenon of intermittent cathode spots motion

current and dependent on the nature of the electrode materials. The greater velocity observed for the magnetic steel. These results were discussed by Secker [74].

Guile et al. [75] used drum camera operating at 960 frames to photograph arcs in free air. The arc burned between two horizontal and parallel cylindrical electrodes of approximately 1 cm diameter so that anode situated above the cathode (Figs. 13.32 and 13.33). The spacing was 3.2 cm, and current was fed into and out of the electrode system at one end. The electrodes were smooth but not highly polished. The arc initiated by an exploded wire at the current-feed end of the electrodes.

After initiation, the arc travelled along the electrodes either under the action of the self-magnetic field, or else under the combined action of this field and an axial

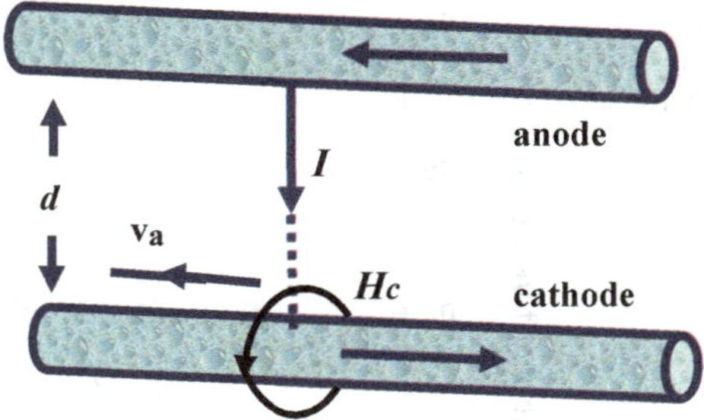

Fig. 13.32 Axial motion due to circumferential flux. From [75]

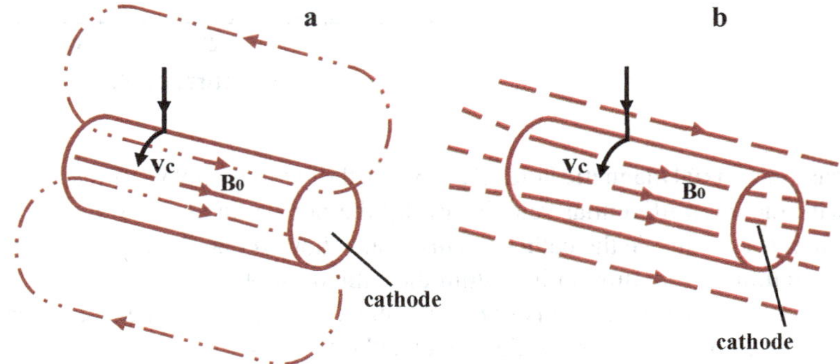

Fig. 13.33 Circumferential motion due to axial flux: a permanently magnetized cathode. b electromagnetized cathode from [75]

field set up in the electrodes by external means. If an electrode is ferromagnetic, then the magnetic flux density within it is increased and there was a discontinuity in the flux density at the surface. Interaction between arc current I and circumferential flux density B causes the arc to move in an axial direction with a velocity v_a (Fig. 13.32). For ferromagnetic electrodes, it was also produced appreciable axial flux density B, either by means of an external electromagnet or else by magnetizing them permanently as shown in Fig. 13.33. Interaction between I and B will produce a circumferential force and a circumferential motion with velocity v_c.

The first tests were made with copper, aluminum, brass, mild steel, and magnetic stainless steel electrodes without axial magnetic field to determine the relationship between axial velocity v_a and current I. In the case of mild steel and at lower currents, the track was regular and continuous and melting had occurred as the arc moved forward. The track became discontinuous when the current was increased. At currents of about 500 A, a third type of continuous high velocity track was detected with surface discoloration and slight pitting, but no evidence of extensive melting. At still higher currents, the cathode movement became random. These various tracks are interesting and indicate that the arc motion determined by cathode phenomena.

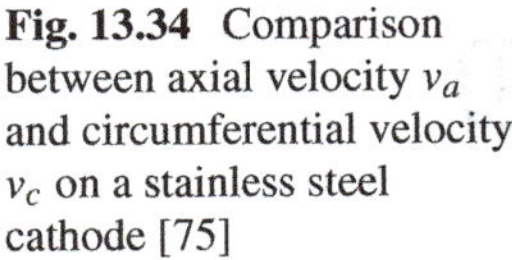

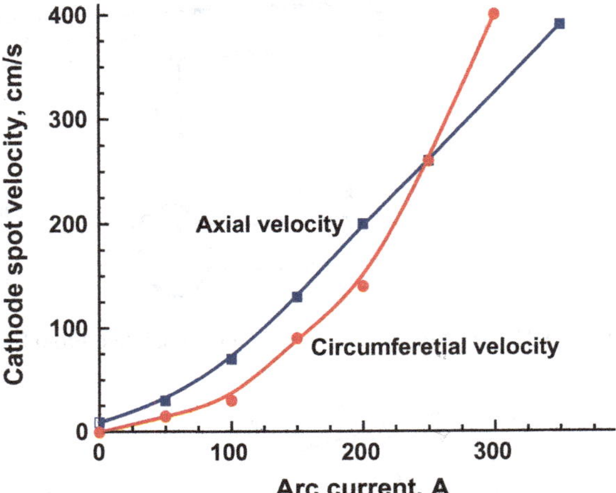

Fig. 13.34 Comparison between axial velocity v_a and circumferential velocity v_c on a stainless steel cathode [75]

In case of an axial magnetic field (Fig. 13.33), the cathode spot was moved a helical path with the circumferential velocity v_c, by the normal laws for a magnetic field in the direction of inside the cathode. This result showed that the forces responsible for the circumferential motion lie within the cathode spot itself. Linear relationship between axial magnetic flux density B_a (in range of 0–1.2 T) and the circumferential velocity v_c was observed for stainless steel cathode spots.

The comparison between axial and circumferential velocities is presented in Fig. 13.34, which demonstrated the spot velocities increase with arc current. It can be seen that these dependencies are not linear and v_a larger up to about 250 A and then it is lower than v_c. The axial was very much greater in magnitude for the magnetic steel electrodes than that of non-magnetic materials.

Guile and Secker [76] studied an arc cathode spot in a transverse magnetic field showing the spot moving either in the Amperian (forward) direction, or in the opposite (retrograde) direction.

As in previous works [73, 75], the arcs were photographed at 960 frames per second as they moved in air at atmospheric pressure between two horizontal and parallel cylindrical electrodes of about 1 cm diameter, situated one above the other in the same vertical plane with the anode uppermost as shown in Fig. 13.32. The distance between electrodes was 3.2 cm, and current was fed into and out of the electrode system at one end. The arc initiated by explosion of a fine wire near the current feed end, and it then moved parallel with the axis of the electrode.

The self-magnetic field was eliminated by feeding equal currents into the anode at each end, and taking equal currents from each end of the cathode. It was do it to determine the dependence of the cathode spot velocity upon arc current and magnetic field. A cylindrical glass chamber enclosed the electrodes, and the gas was evacuated to pressure of about 10^{-3} torr. The electrode system was mounted inside a solenoid. Tests on mild steel electrodes 3 mm apart were made at arc currents between 40 and 670 A and at fields up to 0.055 T. Up to this field, the cathode motion was wholly in the continuous mode but at higher fields the motion became discontinuous. When the gap was increased, less of the motion was continuous. The cathode velocity found in the continuous mode, which independent of arc current in the used range.

A tendency for continuous motion was observed when the cathode spot moved in the retrograde direction than when it moves forward, since in both cases there is a forward force on the arc column as evidenced by the air blast observed by Smith [19]. In certain cases found that the cathode root after moving some distance in the retrograde direction jumped randomly in the forward direction and retrograde motion then started again. It has been found that the velocity of retrograde movement was dependent both on the material of the cathode and on its surface condition, as has been shown for forward motion. For example, at an air pressure of 12 torr, an interelectrode gap of 1.5 mm, a magnetic field of 0.06 T, and an arc current of 1.0 A, the retrograde velocities on polished brass, tarnished brass, and aluminum were 1.0 m/s, 5.1 m/s, and 2.4 m/s, respectively.

Lewis and Secker [77] studied the effect of cathode oxide layers on arc velocity. The experimental setup was used as in the mentioned above series of experiments. The electrodes of brass were prepared by mechanical polishing, followed by etching for 2 min in a 10% solution of ammonium persulfate. They then thoroughly washed in water, dried in warm air, and allowed to oxidize in a dry atmosphere for various times. This method of treatment produced uniform, smooth surfaces with only a very thin oxide layer immediately after etching, which could then be increased by the subsequent oxidation. A new electrode was used for each experiment to enable satisfactory control of the surface conditions to be maintained and to allow direct correlation between the arc tracks left on the cathode and the film of the arc motion. A series of measurements was made with electrode oxidation times covering the range 0–180 min. The transverse magnetic flux density (0.024 T) and the arc current (70 A) were constant, and measurements were made at a pressure of 25 torr, with an interelectrode spacing of 6.4 mm. Under these conditions, movement always occurred in the forward or Amperian direction.

In practically every case after initiation, the arc root on the cathode moved part way along the electrode with a uniform velocity, the magnitude of which depended on the degree of the oxidation of the surface. A uniform velocity as function on oxidation time was obtained for two different brasses with oxidation times up to 45 and 80 min. The arc velocities vary inversely with time, passing through a peak value at 30 and 50 min, respectively, for both brasses. The retrograde arc velocity was measured as a function of time from arc initiation for current 2.1 A, gas pressure 25 torr, magnetic flux density 0,015 T. It was obtained that in air the arc velocity decreased continuously with arcing time. In hydrogen and nitrogen, the velocity increased initially to a peak value. A peak velocities were found in air (~2 m/s), nitrogen (~6 m/s), and hydrogen (~12 m/s). The difference between the characteristics for air and either hydrogen or nitrogen suggests that in air, continuous oxidation was occurring and in the other gases reduction was taking place. A decrease of the interelectrode spacing has in all cases led to an increase of the arc velocity. Guile *et al* 1961 [78] discussed the retrograde spot motion.

Later Guile and Secker [79] have demonstrated experimentally that retrograde running does in fact occur on cathodes of both lead and zinc. The arc velocity was presented as a function of pressure for lead and zinc cathodes and both at currents of 2.3 and 6 A. Transverse magnetic flux density was of 0.08 T. The interelectrode gap was 1.6 mm, and ambient gas was 99.995% argon. The retrograde spot motion was observed with velocity that reduced from 1 m/s to about zero when the pressure increased from zero to 160 torr and then a forward motion of the spot was appeared. It was noted that the similar experimental results as well as the retrograde spot motion observed by Yamamura [80] for magnetically induced arc motion on lead and zinc electrodes. Yamamura indicated that the Righi–Leduc effect is not the primary phenomenon, which recently has been suggested by Kingdon [81].

Guile et al. [82] presented the results of measurements of the velocity of high-current electric arcs driven by transverse magnetic fields in air at atmospheric pressure. Details of some experiments, which have been presented previously, were also summarized. Ranges of parameters covered are: electrode spacing up to 10 cm, magnetic field up to 1.75 T, and current up to 20 kA. The experimental data for different electrode configurations, sizes, and various electrode materials are shown in Table 13.2. Conditions have been found where arc movement is virtually independent of the electrode material and spacing. For different materials and electrode configurations, Fig. 13.35 shows the arc velocity as dependence of arc current with magnetic field as parameter, and Fig. 13.36 shows arc velocity as dependence on transverse magnetic field, B in Wb/m^2 (= in T) with arc current as parameter.

It is seen from the curve fields in the region of 0.12 Wb/m^2 that the velocity and its rate of change with current are higher for rectangular rails than for cylindrical rails and are higher still for the racetrack arc configuration. The curve for 6 kA at 0.32 cm spacing is below that for 3.6 kA at 1.27 cm spacing (Fig. 13.36), and again it appears that the narrow rectangular electrodes give higher velocities than the cylindrical electrodes, probably mainly owing to the arc running at the edge adjacent to the insulating surfaces of the rectangular electrodes. Even though the velocity exceeds Mach 1.0 for the ambient gas conditions, there are no discontinuities in the velocity curves up to the highest magnetic fields (1 or even 1.75 T), and at highest currents 20 kA. According to the measurements shown in Fig. 13.37 shows that the arc velocity is decreased as the electrode spacing, and consequently, the arc length is increased. This holds for the certain conditions for spacing greater than 0.16 cm. For a spacing of a less than a few millimeters increase in spacing can, under some conditions, cause the arc velocity to increase to a peak value, beyond which a fall occurs. The spacing at which the peak velocity occurs depends upon a number of factors, which include the gas pressure and the arc current.

Reviewing the electrode phenomena in arcs, Guile [83], indicated that the almost all arcs at atmospheric pressure and above move in the same direction (forward, or Amperian) as a normal current carrying conductor.

But at reduced pressures, some arcs move in the reverse direction. The column of the arc mostly hasn't any tendency to move in the retrograde direction, even when the arc is in fact moving retrogressively, and the seat of this reverse movement appears to lie in the cathode spot or fall region.

Table 13.2 Experimental conditions and results according to the data from [82]

Electrode configuration	Material	Electrode size, cm	Space, cm	Measuring distance or time	Magnetic field, B, T	Arc current, A	Velocity, m/s
Straight, horizontal, cylindrical, DC arcs	Carbon Al	15 × 0.95 diameter	0.32	5–13, cm	0.006–0.05 0.01–0.1	40–670 40–670	3.5–18 5–13
Straight, horizontal, cylindrical, AC arcs, 0.5 cycle	Polished brass	30 × 0.96 diameter	10.2 0.32	30, cm	0.032–0.16 0.02–0.128 0.021 0. 128 0.032 0.128	2500–6000 400–6000 100–2000 100–800 1200–20,000 800–6000	45–135 60–215 26–140 26–140 80–215 80–215
Straight, horizontal, open-ended rails, single-end connection, DC arcs	Carbon Brass	120 × 5.5 × 2.5 120 × 3.8 × 0.35	3.1 1.27	50, cm	0.01–0.09 0.026–0.037 0.04–0.49 0.012–1	350 100–1200 100-1250 200–3700	10–50 30–100 30–275 70–580
Straight, horizontal, open-ended rails, double-end connections, DC arcs	Brass Copper	120 × 3.8 × 0.35 20 × 3.8 × 0.33	1.27	50, cm	0.03–0.108 0.155 0.06–0.108 0.02–0.108	100–1000 100–1000 200–1000 200	30–124 20–46 27–125 27–80
Horizontal, long, closed-path length, racetrack DC arc	Brass	Total path length = 219, Thickness = 0.33	1.27	Several cycles of arc revolution	0.02–0.12 0.04–0.12 0.06–0.12 0.02–0.12	250–500 230–1200 230–2700 230–1200	40–70 55–205 65–340 40–200
Circular electrodes with annular gap, DC arcs	Carbon	Inner radius = 0.65, Outer thickness = 1.9	0.65	Several cycles of arc revolution	0.006–0.094	100–750	16–190
	Brass	Inner radius = 12.7 or 17.8, Outer thickness = 0.33	1.27		0.02–0.13	200–450	50–130

(continued)

Table 13.2 (continued)

Electrode configuration	Material	Electrode size, cm	Space, cm	Measuring distance or time	Magnetic field, B, T	Arc current, A	Velocity, m/s
Straight, vertical, cylindrical, arc motion upwards, DC arcs	Copper	–	Few tenths	–	0.032 max	60 max	50 max
Ring electrodes with circular path, DC arcs	Brass	Mean radius = 8.1 Thickness = 0.6	0.2	Many cycles of arc revolution	0.034–0.106	20	25–250
					0.034–0.106	87	50–100
					0.034-0.106	10-80	18-95
Ring electrodes with annular gap, AC arcs, 0.5 cycle	Copper	Inner radius = 2.0 Thickness = 0.3	0.1	1st cycle of arc	0.046	10–1000	20–107
				Several cycles of arc	0.03–0.13	200–2000	80–390
Ring electrodes with annular gap, DC arcs	Copper	Inner radius = 0.48	0.8	Many cycles of arc	0.2–1.75	200, 500	70–180 150–470

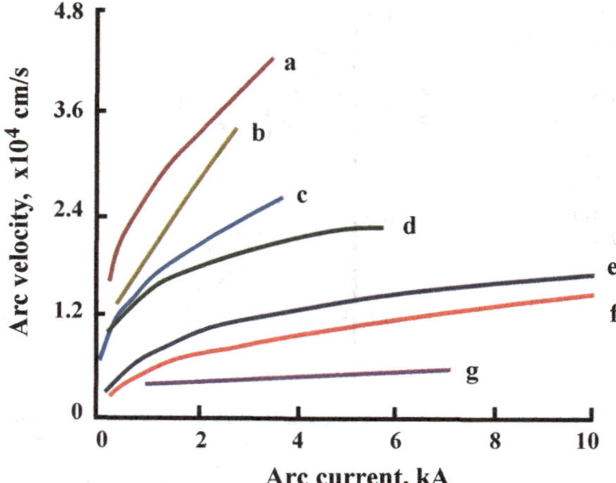

Fig. 13.35 Arc velocity as function on arc current [82]. *a*-Rectangular brass rails, $B = 0.44$–0.49 Wb/m^2, $d = 1.27$ cm; *b*-Brass electrodes, racetrack arc, $B = 0.12$ Wb/m^2, $d = 1.27$ cm *c*-Rectangular brass rails, $B = 0.12$–0.15 Wb/m^2, $d = 1.27$ cm; *d*-Cylindrical brass rails, $B = 0.128$ Wb/m^2, $d = 0.32$ cm *e*-Cylindrical brass rails, $B = 0.032$ Wb/m^2, $d = 0.32$ cm; *f*-Cylindrical brass rails, $B = 0.11$ Wb/m^2, $d = 0.32$ cm *g*-Cylindrical brass rails, $B = 0{,}032$ Wb/m^2, $d = 6.4$ and 10.2 cm

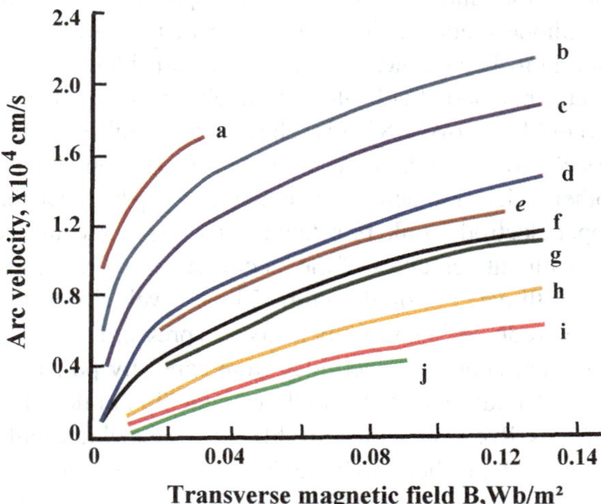

Fig. 13.36 Arc velocity as function on transverse magnetic field, $B < 0.15$ Wb/m^2 [82]. *a*-Cylindrical brass rails, $I = 10$ kA, $d = 0.32$ cm; *b*-Cylindrical brass rails, $I = 6$ kA, $d = 0.32$ cm *c*-Cylindrical brass rails, $I = 3$ kA, $d = 0.32$ cm; *d*-Cylindrical brass rails, $I = 1$ kA, $d = 0.32$ cm *e*-Brass electrodes, racetrack arc, $I = 510$ A, $d = 1.27$ cm; *f*-Cylindrical brass rails, $I = 500$ A, $d = 0.32$ cm *g*-Brass electrodes, $I = 350$ A, annular gap $d = 1.27$ cm, inner radius between 2.5 and 17.8 cm. *h*-Cylindrical brass rails, $I = 200$ A, $d = 0.32$ cm; *i*-Cylindrical brass rails, $I = 100$ A, $d = 0.32$ cm *j*-Rectangular carbon rails, $I = 300 + 50$ A, $d = 3$ cm

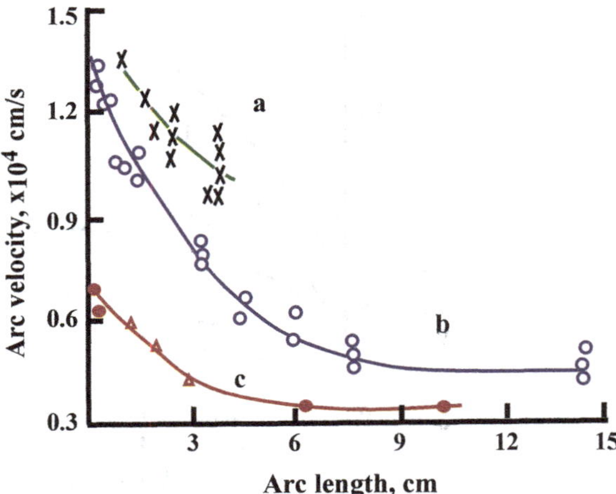

Fig. 13.37 Arc velocity as function on arc length. From [82]. *a*-Rectangular brass rails, $B = 0.125$ Wb/m^2, $I =$ 800 A *b*-Concentric carbon electrodes, rotating arc, $B =$ 0.047 Wb/m^2, $I = 360$ A *c*-Cylindrical brass rails • | B $= 0.032$ Wb/m^2, $I = 600$ A Rectangular brass rails Δ |

The retrograde movement is more likely to occur at low currents, high magnetic fields, small electrode spacing, and low gas pressures. It has been observed in atmospheric air, but only at currents below 5 A, magnetic fields above 0.3 T and spacing below 2 mm. At reduced pressures, it can occur at much higher currents, e.g., 800 A at pressures below 1 torr, and it is believed to be responsible for the mutual repulsion of the separate cathode spots which occur in high vacuum arcs.

Another work published by these authors group considered the motion of the arc under the transverse magnetic field with annular electrodes where either the anode or cathode surface could be moved [84, 85]. The experimental system consisting of two cylindrical brass electrodes each of mean radius 2.5 cm, mounted on a common axis one above the other inside an evacuable chamber. The gap between the electrodes was 1.5 mm. The upper electrode could be rotated on its axis by means of a variable speed motor, and the rotational velocities of the electrode and the arc could be monitored separately using collimated photodetectors. The arc velocity was determined as a function of the electrode surface velocity in argon a pressure of either 160 or 660 torr. Prior to each measurement, the electrodes were carefully polished, left to oxidize overnight and then "conditioned." A radial magnetic flux of density was maintained normal to the arc axis. Rotation of the movable electrode, which could be made either the cathode or the anode, is then modified the movement of the arc.

Figure 13.38a shows that, with the rotating electrode either as the anode turning in the Amperian (forward) sense or as the cathode moving retrograde, the column and anode jet processes both favor new electron emission sites being set up on the Amperian side of the existing cathode spot [85]. If the electrodes moved in the opposite sense, however, ions from the positive column continue to arrive at the Amperian side of the spot but the anode jet acts now to stimulate emission sites on the retrograde side [85] (Fig. 13.38b).

At 160 torr with the cathode electrode stationary, the arc moved in the retrograde direction under the action of the applied magnetic field. The essential difference

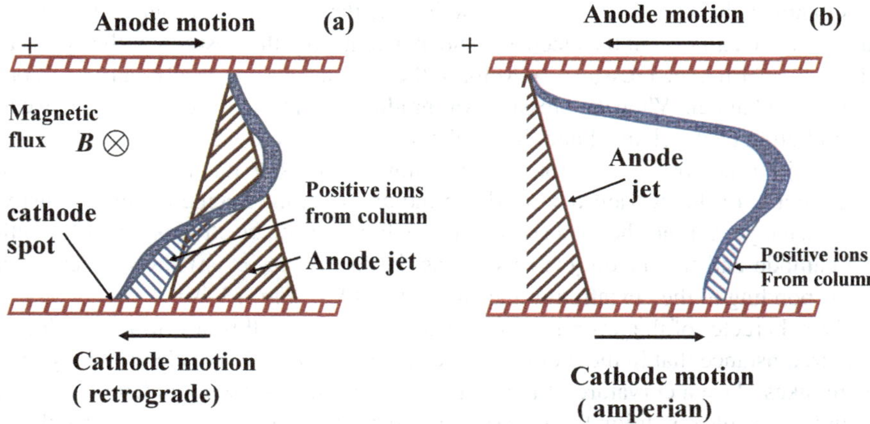

Fig. 13.38 Schematic presentation of electrode root and arc column dispositions with an applied transverse magnetic field and one moving electrode. **a** cathode moving retrograde or anode moving in Amperian sense. **b** cathode moving in Amperian direction or anode moving retrograde

caused by rotating the cathode electrode in opposite directions for what the anode jet interacts with the cathode surface on different sides of the existing spot (Fig. 13.38). In this case, the measured arc velocity follows the cathode surface velocity quite closely, indicating that cathode spot processes have a dominant effect in the selection of new spot locations. The role of the anode jet was detected in the characteristics of the arc when the cathode was rotating in the retrograde direction. At argon pressure of 660 torr, the arc moved in the Amperian direction.

Experiments, in which the anode rotated, indicate somewhat ambiguous evidence for the importance of the anode jet in establishing new emission sites on the cathode surface. From the measurements at both 160 and 660 torr, the authors [85] conclude that anode jets do not exercise a greater control over arc motion than the processes in the cathode spot except at very low currents (5 A). Noted, ions from the positive column do not appear to play a significant role at all.

Hermoch and Teichmann [86] investigated the influence of a transverse magnetic field on the jets of an impulse discharge. The discharge chamber was placed in the field of an air-type solenoid fed with direct current. The intensity of the magnetic field varied continuously up to 1 T. The jets coming from the electrodes and the entire discharge were then photographed by both a high-speed framing camera (the time of exposure most frequently adopted was 50 μs) and by a smear camera. A current impulse of the discharge of an approximately rectangular form, amplitude up to 400 A and a duration of 200 μs was obtained by a discharge of a time-delay line. The discharges were produced in an air environment at a reduced pressure. A jet image depends primarily on the polarity and the material of the electrodes, on the pressure and on the current of the discharge. Let us consider the main results. It was assumed that any gas motion induced by rotating the upper electrode does not significantly affect the arc motion.

Obtained, that in absence of the magnetic field, the majority of the jets issue at right angles to the end face of the electrode, independent from the position of the opposite electrode. At higher pressures (<20 torr), the jets form an intensely radiating core inside the channel. When the pressure was gradually reduced, the difference between the radiation of the channel and that of the jets tends to diminish.

The jet trajectories were found a curvature in transfer magnetic field (TMF). The amount of the curvature depends on the pressure, the magnetic field intensity, the discharge current, the material and the polarity of the electrodes. In TMF, both the cathode and the anode jets (insofar as the latter ones exist) show a deviation corresponding to the deviation of the channel in TMF (Fig. 13.39).

The character of the jets remains unchanged even after this deviation was due to the circumstance that in the vicinity of the electrode spots (in the electrode regions) there takes place a curvature of the jet trajectories. As has been pointed out earlier, an indispensable condition for a curved trajectory is a passage of the jets through the discharge channel. A detail data were obtained by the measurements of the deviation of the jets, issuing from the cathode region, in dependence on TMF B, pressure p, and arc current I of the discharge.

Influence of the pressure was observed from 200 to 1 torr at I = const for two values of TMF (Fig. 13.40). It can be seen that the jet trajectory at lower pressures an instantaneous curvature gradually tends to diminish and, at certain values of p (and evidently also of B), the trajectory becomes a straight one.

The curvature of a jet trajectory disappears at a more distant point, either where the jet leaves behind the current channel, or where, due to losses, the jet temperature decreased. For low B values, a small deviation of the jet occurred all the time within the channel and there ensues a permanent, but insignificant curvature of the trajectory (Fig. 13.40a). When the pressure lowered, the radial distribution of the intensity of the jet radiation is changed. Up to a pressure of about 25 torr, the jets are nearly rotationally symmetrical. When the pressure decreased significantly, the intensity of

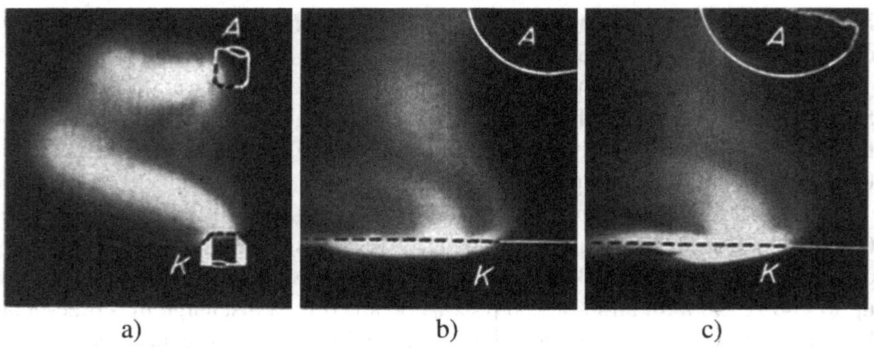

Fig. 13.39 Hermoch **a**, **c**. Framing picture of the jets in a transverse magnetic field, **a** Cu cathode, $p = 200$ torr, $B = 0.106$ T, $I = 400$ A. **b** AI cathode, $p = 3$ torr, $B = 0.63$ T, I–400 A. **c** Cu, $p = 3$ torr, $B = 0.63$ T, $I = 40$ A. Figures taken from [86]. Used with permission

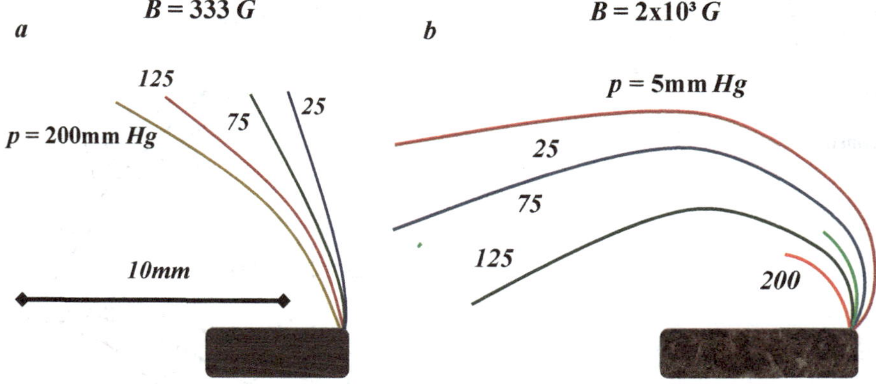

Fig. 13.40 Course of the outward jet trajectory in dependence on the pressure (B = const., I = const). **a** $B = 333$ G; **b** $B = 2 \times 10^3$ G [86]

the radiation and also the density of the jets are changed in such a way that they tend to increase in a transverse section in a direction opposite to that of the Lorentz forces and form a sharp plasma boundary.

Deviation of the jets in dependence on the intensity of the magnetic field with p = const, I = const was presented in Fig. 13.41. An increase of the lateral magnetic field results, in principle, in an increase of the deviation of the jets. The jet curvature increased with the magnetic field strength and for lower ambient gas pressure at constant arc current.

Deviation of the jets in dependence on the discharge current is indicated in Fig. 13.42 for $p = 50$ torr and for two intensities of the magnetic field. The overall deviation of the jets is inversely proportional to the discharge current.

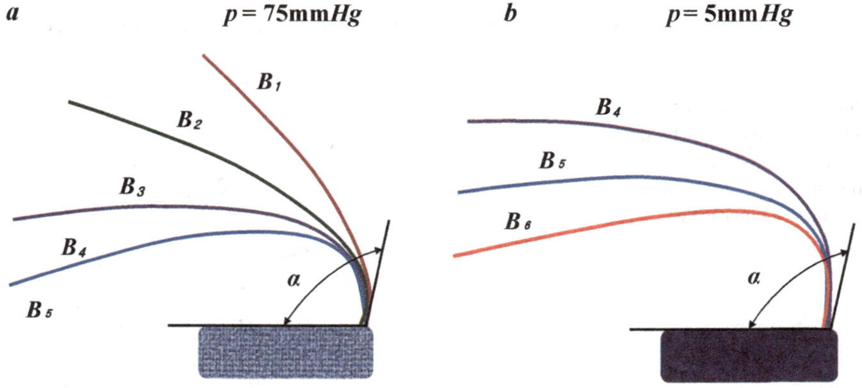

Fig. 13.41 Course of an outward jet trajectory in dependence on B ($B_1 = 0.0333$ T, $B_2 = 0.066$ T, $B_3 = .0106$ T, $B_4 = 0.14$ T, $B_5 = 0.2$ T, $B_6 = 0.245$ T, (p = const., I = const.). **a** $p = 75$ torr; **b** $p = 5$ torr [86]

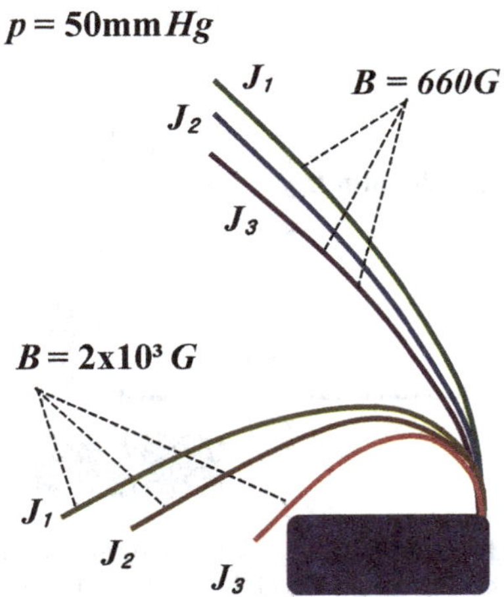

Fig. 13.42 Course of an external jet trajectory in dependence on the discharge current. $I_1 = 400$ A, $I_2 = 300$ A, $I_3 = 200$ A, ($p =$ const, $B =$ const) [86]

Dependence of the jets deviation on the material of the cathode was presented in Fig. 13.43. As electrode materials, there were used Mg, A1, Cd, Zn, Sn, Pb, Ni, Cu, and C. With the parameters *p*, *B* and *I* constant, the jet deviation depends on the electrode material (Fig. 13.43). The measurements of the velocity of the jets extracted from a Cu cathode show that the velocity increase ($I =$ const) when the pressure is reduced, and it changed from approximate of 5×10^5 cm/s at $p = 760$ torr up to a velocity of (8–9) $\times$ 10^5 cm/s at $p = 1$ torr. The curvature of the jets was explained because of the action of forces arising in a plasma flow through a magnetic and electric field.

Later Hermoch [87] studied the influence of the cathode material on the retrograde velocity of individual spot for different cathode materials (Fe, Cu, Pb, and Zn) for currents from 10 to 40 A and magnetic field 0.6 T and pressure of 3 torr during 200 μs. A correlation between the spot velocity (order of 10^4 cm/s) and velocity of the cathode plasma flow (order of 10^5 cm/s) was observed. Both values decreased

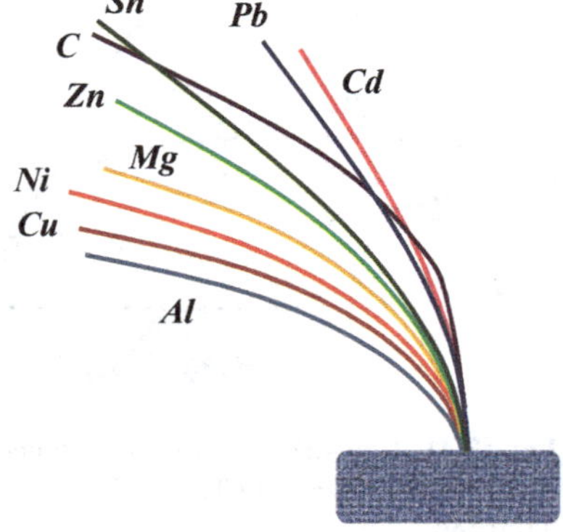

Fig. 13.43 Course of a central trajectory of the jets for different cathode materials ($p =$ const. $B =$ const., $I =$ const) [86]

when the cathode properties changed from refractory to low-melting materials. Also on cathode with low-melting materials, the splitting rate of spots and the motion of individual spots were small, whereas the spots on refractory materials were highly mobile, and the splitting rate was large and the lifetime was short.

Murphree and Carter [88, 89] presented photographic observation of the retrograde motion of an arc in argon pressures 5–760 torr. The experiment used the cathode and anode from oxygen free copper in form of two concentric cylinders with gap of 1.65 cm. The arc current varied from 110 to 220 A, magnetic field from 0.05 to 0.15 T. A 16 mm Hycam K1001 camera used to obtain the pictures of cathode, anode, and column regions in the arc. It was observed that the cathodic arc region moved in retrograde direction with reduction in pressure (to 40–45 torr), while the positive column and anodic arc region move in the Lorentz direction for all pressures. Two different cases of retrograde motion were observed: (i) the cathode spots form an apparently continuous cathodic region; (ii) a groups of discrete cathode spot were detected. This spot behavior was determined by strength of applied magnetic field and arc currents.

13.3.1.4 Magnetic Field and Group Spot Dynamics

As it was described in Chap. 7, the low velocity "group spot" consisting of a number of sub-spots [90] occurred at low pressure in relatively high-current arc (up to 1 kA) and when a vacuum arc operates with a sufficiently hot cathode by its high current [91].

In arcs burning in a background gas, the sub-spots are not grouped together, staying discreet on the cathode surface. For copper cathodes, a velocity less than 10 cm/s, and group spot currents of 200 A [5], 75 A [8], 100 A [46], and 100 $\pm$ 30 A [9] were observed. The spot fixing at the molybdenum-mercury boundary in form line group spot was investigated by Khromoi and co-authors [92–94].

Khromoi and co-authors [95–97] studied the group spot dynamics. In these experiments parallel-plate electrodes with gaps 3–5 mm and a replaceable disk cathode with diameters between 25 and 50 mm were used. A high-voltage pulse applied at a trigger rod with diameter 1.2 mm placed in the disk center. Two types of arc pulses were used with linearly current rise 10^6–10^9 A/s and rectangular pulse with amplitude 0.025–10 kA and duration up to 4 ms.

The dynamics of Ti cathode spot in a self-magnetic field and the expanding ring configuration of the cathode spots are shown in Fig. 13.44. The observed number of spots was used to obtain the current per spot I_s. The measured shifting between two following frames was used to calculated velocity of the spot. The spot velocity as function B is presented in Fig. 13.45. The spot velocity is proportional to the magnetic field when two or more spot can be occurred at given arc current.

When the arc current is relatively large ($I > 100$ A, i.e., $I > 2I_s$) and multiple spots are observed on the cathode surface, the directed retrograde spot velocity increased linearly with the magnetic field. Without an external magnetic field, the group spots expand in a ring form away from the ignition point due to symmetrical distribution of self-magnetic field of spots.

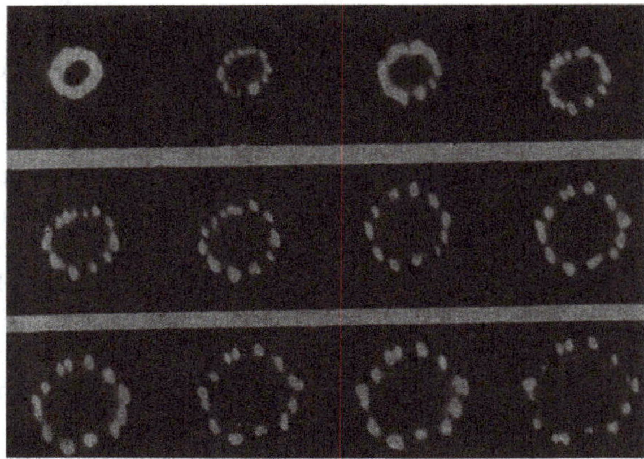

Fig. 13.44 High-speed photographs of titanium cathode region for high current arc in a self-magnetic field with exposition time of 10 μs/frame. The discharge development should be considered from left to right and from upper to down. The last ring is 6.5 cm in diameter. From [97]. Used with permission

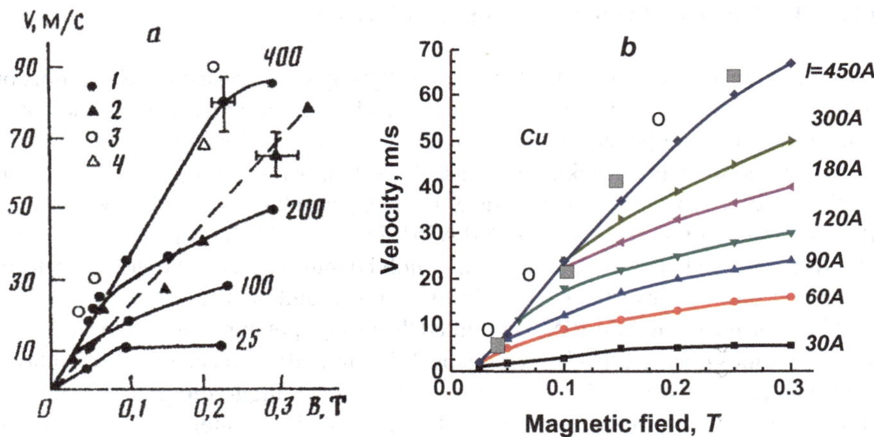

Fig. 13.45 Velocity of the group spot as function on magnetic field *B*. **a**—molybdenum cathode; • 1-low current arc (25–450 A) in an applied external magnetic field; ▲ 2 and ○ 3—high current arc (2–10 kA; 3–10^6–10^9 A/s) in a self-magnetic field. Figure taken from [97]. Used with permission. **b**—copper cathode in an applied magnetic field indicated currents from 30-to 450 A. High current arc in a self-magnetic field indicated by ▩—rectangular current pulse and ○—linear current rise [97]

The proportional coefficient is this same for applied and self-magnetic field and is about 220, 310, and 400 m/(sT) for Cu, Mo, and Ti, respectively. For lower arc currents when the arc current is small ($I < 2I_s$) just one group spot is observed. The proportional coefficient decreased and the velocity declined from linear dependence and then the velocity tends to saturation to a value determined by the arc current. The

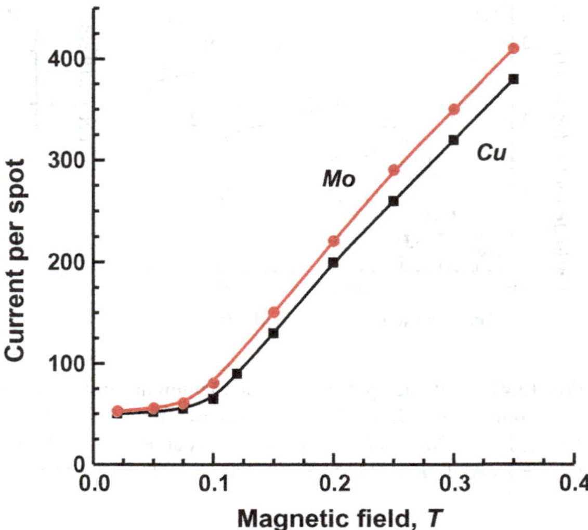

Fig. 13.46 Dependence of current per group spot on magnetic field for Mo and Cu cathodes. The curves are representing averaged values of experimental data obtained for low current in an applied magnetic field and for high current with only self-magnetic field [97]

spot velocity found proportional to the arc current $V = K_i I$, and the average coefficient K_i can be obtained as 0.2–0.25 cm/(sA) from the data presented in Fig. 13.45.

Figure 13.46 shows the average current per spot as dependence on magnetic field. The group spot increased with both external and self-magnetic fields B. It can be seen that spot current weakly depends up to about B=0.1 T, and then this current linearly increases with magnetic field. The group spot current increased from 60 to 300 A when the transverse magnetic field increased from 0.05 to about 0.4 T. So, the spot current increases and the number of spots decrease with increasing magnetic field for fixed arc current.

Recently, Chaly et al. [98] provided similar investigation of the cathode spot velocity in a vacuum on refractory metal electrodes in a transverse magnetic field. Molybdenum and tungsten-butt electrodes of 30 mm each in diameter were used. The rectangular pulses of current with durations of several milliseconds powered the arc. The arc current can be varied up to $I = 300$ A. The arc was ignited on the cathode surface near its edge by using a molybdenum needle.

The interelectrode gaps were $h = 6$ and 2 mm. The experiments were performed under pressure of 10^{-4} Pa. The surface of electrodes was cleaned prior to starting the measurements. A high-speed photography device VFU-1 photographed the arc. This device provides 60 consecutive frames on the film. The exposure time was varied from 25 μ/frame up to 3 μs/frame. The start of the filming could be delayed relative to the arc initiation. The filming was done at the angle ~10° with respect to the cathode surface.

Figure 13.47a shows the dependence of the cathode group spot velocity (V) on the induction of the transverse magnetic field B_t at different arc currents I and gap lengths h on Mo and W cathodes. Figure 13.47b demonstrates a comparison of these dependencies between data for W (curves 1, 2) and Mo (curves 3, 4, 5) cathodes.

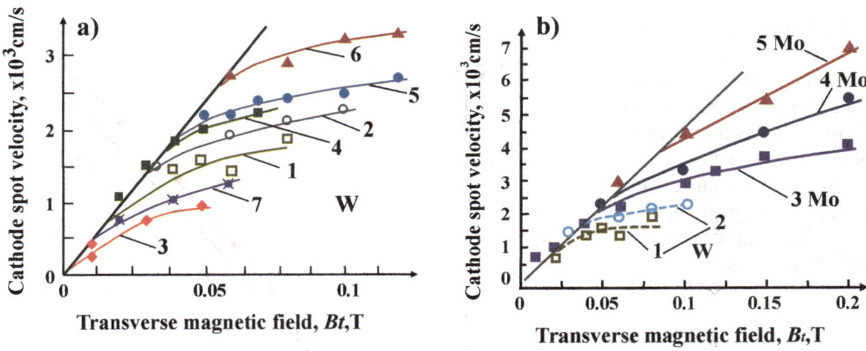

Fig. 13.47 Cathode spot velocity of vacuum arc in transverse magnetic field [98]. **a** W cathode. $h = 2$ mm, 1–$I = 75$ A; 2–I=150 A. $h = 6$ mm, 3 $I = 30$ A; 4–$I = 75$ A; 5–$I = 150$ A; 6–$I =$ 300 A; 7–$I = 70$ A. **b** Comparison of W (curves 1, 2) and Mo (curves 3, 4, 5) cathodes. $h = 2$ mm. 1–$I = 75$ A; 2–$I = 150$ A; 3–$I = 90$ A; 4–$I = 180$ A; 5–$I = 300$ A

Those results were obtained in arcs with a single cathode spot with current I_s, i.e., at such values of B_t at which $I_s \geq I$. Similar result was obtained in the arc with Cu electrodes.

It can be seen that the spot velocity increases linearly with magnetic field, and then at some critical value B_t (depended on current I), this function provided to be weaker. The group spot I_s on molybdenum is somewhat lower than on tungsten and is $I_s \approx 65$ A for $h = 2$ and decreases down to $I_s \approx 55$ A for larger arc length of $h =$ 6 mm. As in the case of tungsten, this result for spot current I_s is close to the result in [97], which states that $I_s \approx 50$ A on molybdenum in the arc with the length $h =$ 3–5 mm. The group spot velocity V increases at moderate B_t values according to the relation $V = KB_t$, i.e., linearly, on tungsten, like on other metals, and, in particular, on copper, which has been carefully studied [43, 97]. In contrast to copper, the spot velocity on tungsten at $I \geq I_s$ not depends on current and on the length of the gap on the linear part of the $V(B_t)$ dependence. However, the magnetic field induction B_t, at which saturation of the dependence $V(B_t)$ starts, depends on I and h, and, accordingly, the level of V at which the saturation takes place. Similar results were obtained for molybdenum at $h = 3$–5 mm [96].

The proportional coefficient K in the relationship between cathode spot velocity and the magnetic field induction was almost the same for both materials (Mo and W) and equal $K \approx 460$ m/(sT). This K value is different from the value $K = 370$–380 m/(s T) for molybdenum according to the dependence V (B) measured in [97]. As indicated in [98], this difference is possible due to different technology of molybdenum production. According to the measurements (Fig. 13.47) with increasing B_t, the dependence $V(B_t)$ for cathode spots on tungsten starts to be saturated earlier than that on molybdenum. As a result, spots on W at comparable currents are moving slower than those on Mo at $B_t > 0.05$ T.

13.3.2 Phenomena in an Oblique Magnetic Fields

Two main phenomena were observed when the magnetic field intersects the cathode surface by an angle: (i) change of the direction of the cathode spot motion and (ii) the cathode spots division. Below the main details of the experimental study will be presented.

13.3.2.1 Cathode Spot Motion in Oblique Magnetic Fields

For the case of mercury cathode Smith [99] first observed that "besides the rapid motion due to the transverse component of the field, the arc moves slowly away from the obtuse angle between field and mercury surface and toward the acute angle." Smith obtained this effect for an experiment, where in addition to the transverse component a normal component of magnetic field occurred, known now as the "acute angle effect." Kesaev in 1957 [100, 101] observed the spot drift angle in a non-uniform magnetic field.

The drift angle was also found to occur by Wroe [102], and he developed a method utilizing this effect to control the motion of the cathode spot. In one case, Wroe used the magnet pole-piece located at the pointed anode. This geometry provides a shaped field, which intersects the cathode surface at an angle, the arrangement being symmetrical with respect to the axis of the system. Thus, the cathode spot always has a tendency to move toward the axis, and in practice the motion is confined to a small region.

In another case, a cathode with a pointed end was used in a uniform axial magnetic field which is thus at an angle to the conical surface and the cathode spot motion was confined to a small region near the tip as it is shown in Fig. 13.48. It was showed that

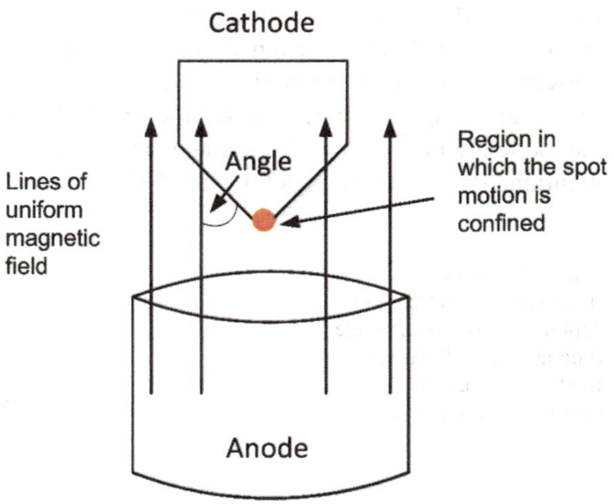

Fig. 13.48 Schematic presentation of method of stabilizing a cathode spot on the end of a cylindrical cathode by means of a uniform magnetic field [102]

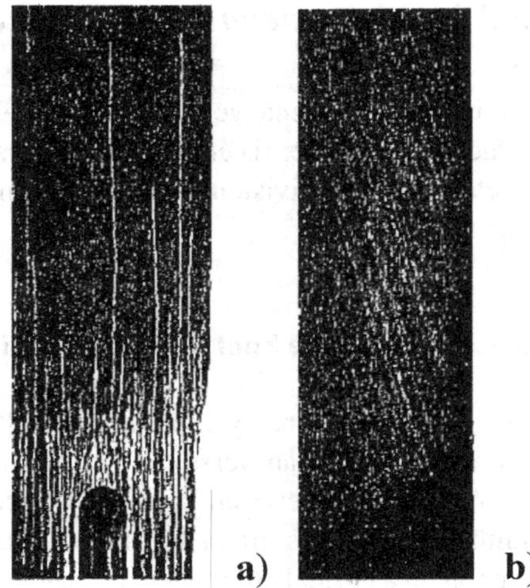

Fig. 13.49 Traces on cathode left by spot motion **a** without magnetic field; **b** with magnetic field [103]

field strengths of at least 0.05 T were required at the cathode for reliable stabilization. The inclination of the field does not seem to be very critical, and an angle of about 45° was normally used. It was found that field strength of 0.1 T would stabilize an arc on a cathode 0.25 inches in diameter.

Later, Robson [103] reported quantitative experimental data characterized the spot drift motion in an oblique magnetic field and using the data from above-mentioned references. His arc was operated with currents in range of 2.8–10 A between a flat aluminum cathode and a parallel copper anode placed in an evacuated chamber in uniform field (0.05–0.2 T) generated by an electromagnet. The direction of the subsequent cathode spot motion was determined by examining the continuous traces left on the cathode surface, which can be seen in Fig. 13.49.

Robson's contribution was measuring the drift angle θ between the retrograde direction and direction of the cathode spot motion direction (Fig. 13.50) as a function of the acute angle of the magnetic field with respect to the cathode surface, φ, the current and the magnetic field. As example, Fig. 13.51 demonstrates dependence of the drift angle on magnetic field strength. It can be seen that the drift angle parabolically increases with the magnetic field strength. Sometimes the drift angle mentioned as "Robson angle" [2, 59, 68]. It is now obvious that this is erroneous

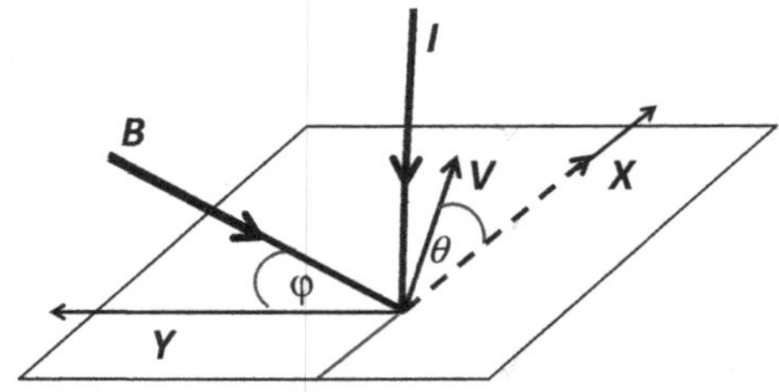

Fig. 13.50 Schematic presentation of retrograde spot motion with drift angle θ under inclined magnetic field with respect to the cathode surface by angle φ

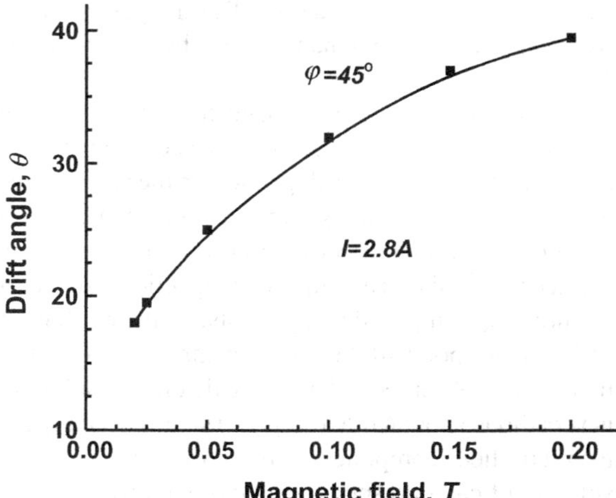

Fig. 13.51 Drift angle as function on magnetic field

name as authors in published works of Refs [99, 100, 102] detected this spot drift angle previously. It is sometimes occurred when somebody use not exact knowledge about first author naming a measured data, and then it was repeated in a further publications.

Below for a case when the magnetic field lines obliquely intersected the cathode surface, the drift component of cathode motion in the direction of the opening of the acute angle between the magnetic field lines and the cathode surface will be named *as the acute angle effect*. It should be noted here about additional data obtained by Smith [99]. He has photographed a region for a cathode spot moving in a magnetic field by a microscope in direction parallel to the mercury surface. Smith indicated that for 10 μm out from the mercury surface a "cathode dark space" emitting a continuous spectrum was found, and then for a further 40 μm a very much more intense region of excited vapor was taken place.

Relatively recently Lang et al. [104] used a magnetic field having axial and transverse components applied to the cathode surface to investigate the influence on the cathode spot motion and the MPs reduction on TiN films. A water-cooled titanium cathode (99.99% purity) of 64 mm in diameter and 30 mm in thickness with an inclined wall shoulder was mounted inside a vacuum chamber, which acted as the anode. A circular-shaped magnetic flux guide placed around the cathode surface to increase the transverse magnetic component. The transverse magnetic flux density, which varies in the range of 0–35 G, was chosen as the parameter and the arc current was at 60 A.

The distributions of the transverse and the normal components of different magnetic flux density on the cathodic target surface were simulated by using the finite element model [104]. The distribution of the intensity of normal magnetic flux decreased from a large value near the cathode to a small value far away from the cathode, and the magnetic flux intersected with the target surface at acute angles directed to the edge of the cathode. The transverse component of the magnetic flux density increased gradually from zero at the center of the target to a relatively larger value at the edge of the target. The images of cathode spot motion (macroscopic) were captured from a dynamic video by a high-speed camera with the exposure time 1 ms.

The author [104] summarized their results in the following. The cathode spot moved at random on the cathode surface when magnetic field was relatively weak. In this case, the spot motion depended on the surface conditions, including the presence of a dielectric layer and geometric enhancement, which had a random distribution; therefore, the probability of initiation of a new cathode spot would occur on the target surface randomly. In an imposed magnetic field, the cathode spot motion was defined by three superimposed components: (i) the random walk, (ii) the retrograde motion, and (iii) the spot drift due to acute angle effect. With increasing transverse magnetic field, it was observed a deterministic effect leading to a preferential initiation of the new cathode spot on the retrograde side. The axisymmetric distribution of transverse magnetic field component caused a rotational cathode spot motion. The rotational velocity of cathode spot was determined by the transverse magnetic field intensity. With a relatively strong magnetic field, the cathode spot rotates near the edge of the cathode surface and is confined to a circular trajectory. The acute angles formed by magnetic field intersected with the target surface directed to the edge of the cathode, increasing the tendency of spot drift by acute angle effect toward the cathode target edge. With a relatively strong magnetic field (30 G), the tendency of rotational cathode spot motion and acute angle drift was strong enough that the random walk can be neglected.

The stabilizing effect of the axial magnetic field was studied by Chaly in 1999 [105], which showed that mobility of the cathode spot decreased when the axial magnetic field component increased. Song in 2012 [106] also observed the decreased mobility of the cathode spot with axial magnetic field. When an axial magnetic field was applied, a more uniform spot distribution was obtained over the cathode surface. Yanabu in 1979 [107] showed how this characteristic was used to improve the performance of vacuum circuit breakers.

13.3.2.2 Cathode spot motion with a long roof-shaped cathode under magnetic field

Controlled motion of the cathode spot using a magnetic field is commonly used to influence erosion uniformity on the cathode surface and plasma beam uniformity [108, 109]. The cathode design is a critical element in deposition apparatus. Controlled cathode spot motion on long cathodes will be required for filtered vacuum arc deposition of uniform coatings on large areas, needed for applications such as transparent conductive coatings for large flat displays and energy conserving coatings on architectural glass. The spot motion can be controlled by the cathode geometry and the applied magnetic field configuration.

A circular transverse magnetic field to control the spot motion of a cathodic arc on a large rectangular target was designed by Li et al. [110]. The targets made of pure graphite (99.99%) and red copper (99.95%) were used. The graphite target was

450 mm long and 120 mm in width, whereas the copper target was 450 mm long and 60 mm wide. The vacuum chamber was evacuated to 1×10^{-4} Pa. Argon was inputted at a rate precisely controlled by a mass flow controller into the vacuum chamber to pressure lower than 1 Pa. The magnetic field was produced by permanent magnets (0.3 T) below the target. A camera with a 0.25 ms exposure time recorded the cathode spot motion. The average velocity of spot is calculated by s/t, where s represents trajectory perimeter and t represents the, respectively, time per the trajectory.

It was shown that the retrograde spot velocity increased linearly with the arc current at fixed magnetic field. For graphite cathode, the spot velocity was relatively small and increased from 4.25 to 12.44 mm/s while for Cu the larger velocity increased from 0.8 to 1.8 m/s when the current increased from 60 to 120 A. Both the gas pressure and arc current affect the stability of the cathode arc. A higher gas pressure stabilizes the arc but reduces the spot velocity. The difference in the velocity between graphite and copper was explained by the graphite negative temperature coefficient of resistance at low temperature while the resistivity of copper increased linearly with temperature.

The authors [110] indicated that during the erosive action of the spot, the target surface uneven. In this case, at inclined area the cathode spot tends to drift toward the opening of the acute angle formed between the field line and the surface. The spot moves faster on an oblique surface because of the smaller resident time.

One was an elongated "roof-top" configuration with an external applied magnetic field investigated experimentally [111, 112]. A roof-shaped aluminum cathode was manufactured and placed inside a copper anode frame. This plasma source was connected to a vacuum chamber and was surrounded by a rectangular magnetic coil. The arc burned on a cathode that included a flat roof-top and four sloped sides (Fig. 13.52), inclined by an angle α and the cathode spot moved in transverse magnetic field in the retrograde direction.

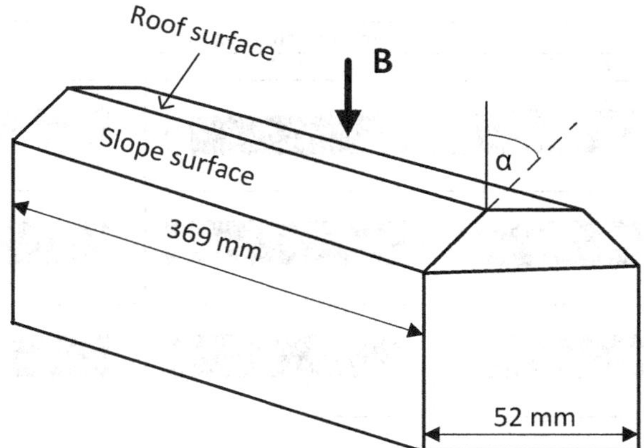

Fig. 13.52 Roof-shaped Al cathode geometry with height 74 mm for all angles

Also, in the inclined magnetic field on the cathode slopes, the cathode spot motion was deflected from the direct retrograde motion toward the direction of an acute angle. A high-speed camera was used to observe cathode spot motion. These videos were analyzed in order to calculate the spot velocity and surface distribution. The spot velocity was calculated with the aid of an image analysis software tool (P.C.C). It was determined the cathode spot velocity on the "roof top" and slopes, and the distribution of cathode spots between the roof and slope surfaces with different slope angles α, magnetic fields, and arc currents.

The transverse component of magnetic field at slope surfaces produced retrograde spot motion, resulting in the spot motion V_x and a spot distribution. Manually sampling and analyzing selected frames used to calculate the distribution of the cathode spots on the surface. It was obtained that under a magnetic field, the spot motion on the roof was slow (<1 m/s) and mainly random, while on the slopes fast retrograde motion was observed.

The circumferential spot distribution due to retrograde motion of the spots between the 4 slopes of the cathode and spot appearing at cathode top surface is presented in Fig. 13.53 showing the effect of different magnetic fields and slope angles, for $I = 200$ A. It can be seen that when $\alpha = 17°$ and $B = 5$ mT, the spots were located mostly on the slopes and were distributed fairly uniformly along a race-track shaped path.

However with all other values of B and α, either an increasing fraction of the spots was on the roof (e.g., for α = 40° and $B = 5$ mT, and for α = 17° and $B = 9$ mT), and/or the spots were concentrated at one end of the cathode.

The spot velocity V_x increased with the magnetic field approximately linearly, as seen in Fig. 13.54 for an arc current of 200 A. When the magnetic field was increased from 5 mT to 14 mT, the velocity increased from about 5 m/s up to roughly 17 m/s. The rise in average velocity only weakly depended on the slope angle and V_x slightly decreased with α.

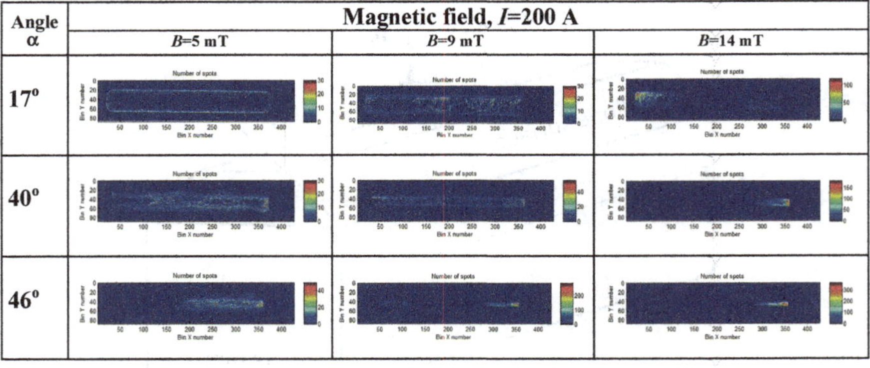

Fig. 13.53 Cathode spot distribution maps with I = 200 A for different slope angles and magnetic field strengths. The figure was taken from [112]. Used with permission

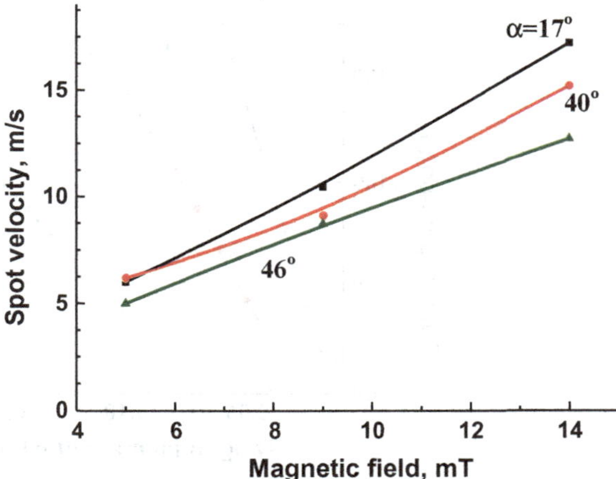

Fig. 13.54 Average velocity V_x on the cathode slope as a function of B, with α as a parameter, $I = 200$ A

13.3.2.3 Cathode Spot Splitting in an Oblique Magnetic Field

In the last decade, the current per spot was investigated also in fields, which obliquely intercept the cathode surface [113, 114]. The arc was operated with constant current rectangular pulses with amplitude up to I = 300 A and durations up to 4 ms. Two pairs of coils produced a uniform magnetic field with components normal and tangential to the cathode surface, whose strengths could be varied independently in the ranges of $B_n \leq 0.35$ T and $B_t \leq 0.25$ T, respectively.

The arc was photographed by a framing camera to analyze the structure of the cathode attachment of the arc with an exposure 25 μs/frame. Cathode spots could be spatially resolved if the distance between them was at least 0.25 mm. The current per spot I_s was found as the mean over many frames of I/N, where N is the number of observed cathode spots.

It was found [113, 114] that for an arc on a Cu cathode without a magnetic field, $I_s \approx 65$ A. When only a transverse magnetic field was imposed, I_s increased linearly with B_t (see above). However, when a fixed value of B_n was used imposed, a linear relation was again found for I_s vs B_t, but shi*f*ted to larger ranges of B_t with increasing B_n (see Fig. 13.55). This means that the spot current decreases with B_n. These results were obtained with an arc current I sufficiently large to produce the observed dependences, i.e., the needed number of spots.

The investigation of cathode spot current per spot under magnetic field having transverse and normal components was extended for molybdenum and tungsten cathodes by Chaly et al. [98]. The results are presented in Fig. 13.56.

It can be seen that I_s depends not only on the magnetic field and its inclination, but also on the length of electrode gap h. The current per cathode spot in an arc without magnetic field for molybdenum cathode at $h = 2$ mm is about 65 A and decreases down to 55 A at $h = 6$ mm. In the case of tungsten cathode, the current per spot is about 90 A and 75 A for $h = 2$ and 6 mm, respectively. According to the results presented in Fig. 13.56 at B_t/B_n « 1, the current per spot $I_s \rightarrow I_s(B = 0)$ and weakly

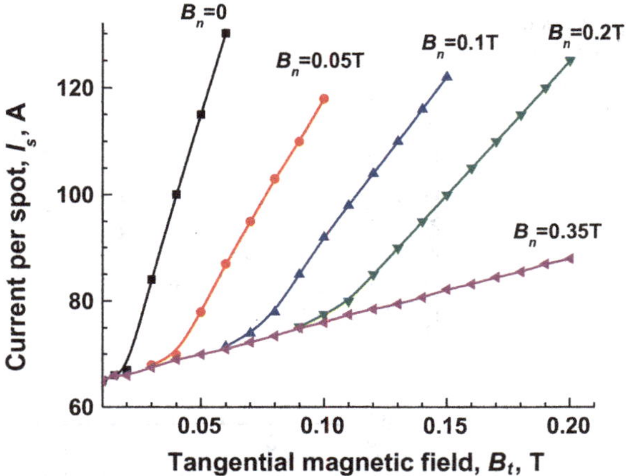

Fig. 13.55 Dependence of mean current per spot I_s for Cu cathode on the magnetic field induction and its inclination to the arc axis. Gap distance was 4 mm [113]

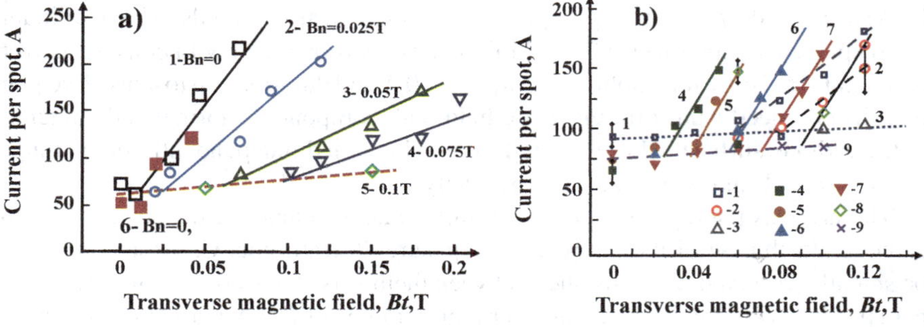

Fig. 13.56 Current per cathode spot [98]. **a** Mo cathode. $h = 2$ mm, $1{-}B_n = 0$; $2{-}B_n = 0.025$ T; $3{-}B_n = 0.05$ T; $4{-}B_n = 0.075$ T; $5{-}B_n = 0.1$ T. $h = 6$ mm, $6{-}B_n = 0$. **b** W cathode. $h = 2$ mm, $1{-}B_n = 0$; $2{-}B_n = 0.025$ T; $3{-}B_n = 0.05$ T. $h = 6$ mm, $4{-}B_n = 0$; $5{-}B_n = 0.025$ T; $6{-}B_n = 0.05$ T; $7{-}B_n = 0.75$ T; $8{-}B_n = 0.1$ T; $9{-}B_n = 0.12$ T

depends on the magnetic field amplitude $|B|$. For a case when $B_t/B_n \gg 1$, the field amplitude has a strong effect on the current per the cathode spot. Figure 13.57 shows the dependence of arc voltage U on the normal field induction B_n for arcs with Cu cathode of different length at $Bt = 0$ with h as parameter. The arc voltage increases with normal magnetic field and the rate of such rise is larger for larger electrode gap distance.

Relatively recently the effects of an axial magnetic field on the vacuum arc characteristics between transverse magnetic field (TMF) contacts were studied by Ma et al. [115]. In the experiments, an external AMF was applied to a pair of TMF contacts. The external axial field flux density varied from 0 to 110 mT. The arc current in the tests varied over a range from 0 to 20 kA rms at 45 Hz. The diameters of both

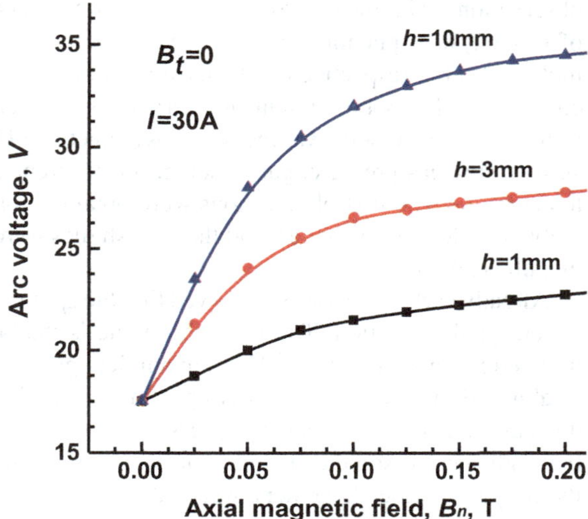

Fig. 13.57 Dependence of average arc voltage on normal component of magnetic field [113]

the anode and the cathode are 60 mm. The contact material was CuCr25 (25% Cr). The full contact gap was set at 20 mm. The contacts were then installed inside a demountable vacuum chamber, which was subsequently evacuated and maintained at a pressure of 10^{-3}–10^{-4} Pa. A High-speed charge-coupled device (CCD) video camera was used to record the vacuum arc evolution. The CCD video camera was set at a recording speed of 10^4 frames/s.

The experimental results were concentrated on the vacuum arc characteristics included the time-dependent vacuum arc behavior illustrated by high-speed photographs and the arc voltage waveform. The measurements show that the application of the axial magnetic field effectively reduces the TMF voltage noise component of the arc and reduces the formation of liquid metal drops between the contacts. The diffuse arc duration increases linearly with increasing axial component of the magnetic field flux density; but in this case, it also decreases linearly with increasing arc current. The results also indicate that the diffuse arc duration before the current zero is usually more than 1 ms under the condition that the axial field per kA was more than 2.0 mT/kA. Finally, under the application of the AMF, the arc column of the TMF contacts may constrict and remain in the center region without transverse rotation. No data were reported about spot current development in the combined magnetic field.

13.4 Summary

When a transverse magnetic field, i.e., magnetic field parallel to the surface, is applied in an arc, the cathode spot moves in the "retrograde" $-\mathbf{j} \times \mathbf{B}$ direction. Since the effect of retrograde cathode spot motion was studied more than one century from its first observation, a significant progress can be distinguished regarding to the overall state of this physical phenomenon. The experimental investigations of retrograde spot motion and the respective results have been presented extensively in the literature, considering the spot motion in an external magnetic and in self-magnetic fields. The retrograde spot velocity was measured as a function of the magnetic field, arc current, surrounding gas pressure, gap spacing, for different cathode material, and cathode temperature. The first observations were obtained for arc with mercury cathode in ambient of low-pressure gas, and then the studies were continued for arcs with solid metallic cathodes.

Already early observations showed that the spot motion reversed from Amperian to retrograde direction. The important issue is the behavior of spot motion under transference magnetic field of arcs in ambient gases with different pressures. It was established that the velocity of Amperian spot motion decreased with gas pressure decrease. At some ‘critical’ gas pressure, a phenomenon of arc motion was found that indicates of some difficulties of driven of an arc by the magnetic field up to its stopping. These phenomena were called immobility phenomena. The immobility time and the critical pressure at which the reverse spot motion occurred and driving velocity of the arc were widely measured showing their dependence on magnetic field, arc current, electrode gap distance, low gas pressure, and cathode materials. For example, the critical pressure for Cu exceeds that for Hg cathode by two orders of magnitude, both materials indicated it an increase with arc current (2–10 A) and a linear dependencies on magnetic fields in range 0.01–0.12 T. The interesting experiment showed the retrograde motion in air but observed in a short arc burning between Cu electrodes with a 0.5 mm separation under relatively large magnetic field of $B > 0.3$ T.

In number of future studies, the retrograde velocity was obtained as increased linearly (proportional) with the magnetic field and then saturates at the maximal velocity determined during the random spot motion observed when a magnetic field is not applied. Thus, the phenomena can be detailed as follows. When the transverse magnetic field is imposed, the trajectory of the spot motion becomes partially straight and a direct component of the motion was appeared. The future increase of the magnetic field increased the direct component of the spot motion up to point when the spot velocity reaches the maximal value and then not changed (saturated) with magnetic field for constant arc current and for single spot. Sometime the saturation velocity dependence was not reached in experiments where the imposed magnet field was used not enough large (<0.1 T).

With arc current increasing the spot velocity continue increase linearly up to maximal saturation velocity with magnetic field. The velocity saturation begins in the case when the arc current was lower than twice of threshold current per spot. This experimental result demonstrated an increase of current per spot I_s with magnetic field B and a single cathode spot was produced at certain relation between B, I_s and arc current I. The arc current determines the maximal spot velocity, and this velocity increases linearly with the arc current. In general, the spot at relatively large current (about 100 A and larger) is a group spot that consists of a number of sub-spots. The current per group spot linearly increased with magnetic field (after 0.05–0.1 T). The current per group spot increased due to increase of the number of sub-spots in the group. The spot motion can as continuity trajectory at relatively low magnetic field and in the form of discontinuous jumps between consecutive spot locations for large (>0.1 T) of used magnetic field.

Analysis of the experimental results showed that the relationship between velocity and magnetic field is not depending on the nature of magnetic field. It is the same in case of imposed magnetic field (at low current arcs) as well in case of self-magnetic field at high-current arcs. The proportional coefficient K in behavior of the velocity on magnetic field mainly depends on cathode materials. The coefficient K does not depend on the nature of how this field produced and determines the spot mobility. Although the velocity in low-current arc (1 kA) depends on current, it has been found that in high-current arc the relationship between the velocity and self-generated field (coefficient K) was independent of the arc current and dependent only on self-generated field characteristic and cathode material. This is due to the fact that a discharge at high currents consists of a large number of spots each carrying approximately the same current. When the external magnetic field is absent, the spots expand in a ring form due to self-magnetic field influence and for sufficient value of arc current supported sufficient number of spots. With the rates of current rise higher than critical value (10^7–10^8 A/s), spontaneous spot formation has been observed on Cu, Al, Zn, Sn, Mg, and Cd cathodes. In spite of widely differing physical properties of the cathode materials, the measured maximum spot velocities were within the one order of magnitude, i.e., between 10^3 and 10^4 cm/s.

In some cases, it has been evidenced an influence of a transverse magnetic field on the jets born in the electrode regions, at a low pressure, particularly initiated in the cathode region of high-current arc. The importance of the plasma column in determining Amperian arc motion was observed for high-current and high-pressure arcs. In low-pressure arcs, the plasma jet action becomes most pronounced at high arc velocities and for large interelectrode spacing. At a high arc velocity, the cathode spot may move by a stepping process rather than in a continuous manner. A large interelectrode gap allows the arc jet to decline out ahead of the electrode spots, and jet shorting to the cathode support then gives rise to stepwise initiation of cathode spot motion. A similar stepping motion was also detected at plasma jet domination in short arcs acted upon by high transverse magnetic fields. The experiment showed that the jet curvature increased with the magnetic field strength and for lower ambient gas pressure at constant arc current. When the pressure and magnetic field were constant, the jet curvature was larger for lower arc current.

Nevertheless, different conclusions were published about some data for retrograde spot motion. It was related to influence of cathode heating. Most of published results indicated the spot velocity of such motion was decreased with cathode temperature. Nevertheless, a contrary data were published also, and respective discussion was considered in Sect. 13.3.2.1 of this chapter. Although it was observed that, there is discrepancy in data related to proportional coefficient K, which characterizes the linear dependence between retrograde velocity of the cathode spot and magnetic field, a significant increase of this coefficient with arc current which was obtained for different cathode materials. While the spot velocity increased with electrode spacing for retrograde motion in vacuum arcs, the velocity of such motion decreased for arcs burned in air at atmospheric pressure. The rate of the velocity change was larger for small gap distances (<5 mm) in comparison with that for further increase of the distance.

Once more phenomenon described above is related to the applied magnetic field that inclined to the cathode surface. Two effects of spot behavior in an oblique magnetic field were described above. The first, the cathode spot motion is deflected from the retrograde direction by a spot drifting angle toward the direction of the acute angle formed by the field line and its projection on the cathode surface. This is spot drifting due to "acute angle effect" first observed by Smith [96], Kesaev [98] and Wroe [100] but significantly later studied by Robson. A specific behavior of spot velocity and spot splitting in an oblique magnetic field is the other effect. The experiment showed that the drift angle parabolically increased with the magnetic field strength. The spot current in an oblique magnetic field increased linearly with its transverse component B_t, but at fixed value of normal component B_n to the cathode surface the linear dependence shifted to larger ranges of B_t with increasing B_n indicating that the spot current decreases with B_n.

Finally, the above state of experimental results show that understanding retrograde spot motion requires understanding the condition of new spot development after it ignition at nearest new locations in the retrograde direction from a spot that extinguishes. The mechanism of this effect is considered in Chap. 19.

References

1. Boxman, R. L., Sanders, D. M., & Martin, P. J. (Eds.). (1996). *Handbook of vacuum arc science & technology: Fundamentals and applications*. Park Ridge New Jersey: Noyes Publications
2. Anders, A. (2009). Cathodic arcs: From fractal spots to energetic condensation (Vol. 50). Springer Science & Business Media.
3. Keidar, M., & Beilis, I. I. (2018). *Plasma engineering*. London, New York: Academic Press, Elsevier.
4. Kesaev, I. G. (1968). *Cathode processes in electric arcs*. Moscow: Nauka Publishers. (in Russian).
5. Rakhovskii, V. I. (1976). Experimental study of the dynamics of cathode spots development. *IEEE Transactions on Plasma Science*, *4*(2), 81–102.
6. Juttner, B. (1997). The dynamics of arc cathode spots in vacuum: new measurements. *Journal of Physics D: Applied Physics, 30,* 221–229.
7. Juttner, B. (1987). Characterization of the cathode spot. *IEEE Transactions on Plasma Science, PS-15*(5), 474–480.
8. Djakov, B. E., & Holmes, R. (1974). Cathode spot structure and dynamics in low current vacuum arcs. *Journal of Physics D: Applied Physics, 7,* 569–580.
9. Beilis, I. I., Djakov, B. E., Juttner, B., & Pursch, H. (1997). Structure and dynamics of high-current arc cathode spots in vacuum. *Journal of Physics D: Applied Physics, 30,* 119–130.
10. Beilis, I. I. (2002). Vacuum arc cathode spot grouping and motion in magnetic fields. *IEEE Transactions on Plasma Science, 30*(6), 2124–2132.
11. Stark, J. (1903). Induktionserscheinungen am Quecksilber-lichtungen im Magnetfield. *Physikalische Zeitschrift, 4,* 440–443.

12. Stark, J., & Reich, M. (1903). Druckbeobachtungen am Quecksilberlichtbogen. *Physikalische Zeitschrift, 4,* 321–324.
13. Stark, J. (1904). Quecksilber als kathodische basis des lichtbogens. *Physikalische Zeitschrift, 5,* 750–751.
14. Weintraub, E. (1904). Investigation of the arc in metallic vapours in an exhausted space. *Philosophical Magazine, 7*(of Series 6), 95–124.
15. Minorsky, N. (1928). Rotation of the electric arc in a radial magnetic field. *Journal of Physical Radium, 9*(4), 127–136.
16. Tanberg, R. (1929). Motion of an electric arc in a magnetic field under low gas pressure. *Nature, 124*(3123), 371–372.
17. Froome, K. D. (1949). The behaviour of the cathode on an undisturbed mercury surface. *Proceedings of the Physical Society. Section B, 62*(12), 805–812.
18. Smith, C. G. (1942, July). The mercury arc cathode. *Physical Review, 62*(1–2), 48.
19. Smith, C. G. (1943). Motion of the copper arc in transverse magnetic field.*Physical Review, 63*(5–6), 217.
20. Smith, C. G. (1943). Erratum: The mercury arc cathode. *Physical Review, 64,* 40.
21. Smith, C. G. (1957). Motion of an arc in a magnetic field. *Journal of Applied Physics, 28*(11), 1328–1331.
22. Smith, C. G. (1951, January). A new cold electric arc. *Physical Review, 83*(1), 194–194.
23. Smith, C. G. (1951, January). Motion of an anchored arc impelled by a magnetic field. *Physical Review, 82*(4), 570–570.
24. Smith, C. G. (1951, January). Retrograde arc motion of supersonic speed. *Physical Review, 84*(5), 1075–1075.
25. Gallagher, C. J., & Cobine, J. D. (1947). Retrograde motion of an ac cathode spot in a magnetic field. *Physical Review, 71*(7), 481.
26. Gallagher, C. J. (1950). The retrograde motion of the arc cathode spot. *Journal of Applied Physics, 21*(8), 768–771.
27. John R. M. St., & Winans, J. G. (1954). Motion of arc cathode spot in a magnetic field. *Physical Review, 94*(5), 1097.
28. John, R. M. St., & Winans, J. G. (1955). Motion and spectrum of arc cathode spot in a magnetic field. *Physical Review, 98*(6), 1664.
29. Zei, D., & Winans, J. G. (1959). Motion of high speed arc spots in magnetic fields. *Journal of Applied Physics, 30*(11), 1813–1819.
30. Zei, D., John, R. M. St., & Winans, J. G. (1955, January). Some properties of arc cathode spots in magnetic fields. In *Physical Review* (Vol. 100, No. 4, pp. 1232–1232). One Physics Ellipse, College Pk, Md 20740-3844 USA: American Physical Soc.
31. Hernqvist, K. G., & Johnson, E. O. (1955). Retrograde motion in gas discharge plasmas. *Physical Review, 98*(5), 1576–1583.
32. Eidinger, A., & Rieder, W. (1957). Das Verhalten des Lichtbogens im transversalen Magnetfeld Magnetische Blasung. *Archiv für Elektrotechnik, 43*(2), 94–114.
33. Robson, A. E., & Engel, A. (1954). Origin of retrograde motion of arc cathode spots. *Physical Review, 93*(6), 1121–1122.
34. Farrall, G. A. (1962). *A review of reverse motion in magnetically driven arcs. Seminar on electrical contacts* (pp. 1–34). USA: University of Maine.
35. Gundlach, H. C. W. (1972). Experimental study of retrograde motion of the arc cathode spot in high vacuum.
36. Seidel, S., & Stefaniak, K. (1972). Retrograde motion of the electric arc in vacuum and its mechanism on the solid electrodes. *Proc. Vth ISDEIV*, 237–247.
37. Daalder, J. E. (1978). *Cathode erosion of metal vapor arcs in vacuum.* Thesis: Eindhoven Univ. Tech.
38. Emtage, P. R., Kimblin, C. W., Gorman, J. G., Holmes, F. A., Heberlein, J. V. R., Voshal, R. E., et al. (1980). Interaction between vacuum arcs and transverse magnetic fields with application to current limitation. *IEEE Transactions on Plasma Sciences, 8*(4), 314–319.

39. Kimblin, C. W., & Voshall, R. E. (1972). Interruption ability of vacuum interrupters subjected to axial magnetic fields. *Proceeding of Institution of Electrical Engineers, 119*(12), 1754–1758.
40. Sethuraman, S. K., & Barrault, M. R. (1980). Study of the motion of vacuum arcs in high magnetic field. *Journal of Nuclear Materials, 93*, 791–798.
41. Sethuraman, S. K., Chatterton, P. A., & Barrault, M. R. (1982). A study of the erosion rate of vacuum arcs in a transverse magnetic field. *Journal of Nuclear Materials, 111*, 510–516.
42. Nurnberg, A. W., Bauder, U. H., Mooser, C., & Behrisch, R. (1981). Cathode erosion in vacuum arcs and unipolar arcs. *Contributions to Plasma Physics, 21*(2), 127–134.
43. Fang, D. Y. (1982). Cathode spot velocity of vacuum arcs. *Journal of Physics D: Applied Physics, 15*(4), 833–844.
44. Djakov, B. E., & Holmes, R. (1970, September). Cathode spot motion in a vacuum arc under the influence of the inherent magnetic field. In *Proceedings of international conference on gas discharges* (pp. 468–472).
45. Djakov, B. E., & Holmes, R. (1972, September). Retrograde motion of a cathode spot and conduction of heat in the cathode. In *Proceedings of 2nd International Conference on Gas Discharges* (pp. 183-185).
46. Sherman, J. C., Webster, R., Jenkins, J. E., & Holmes, R. (1975). Cathode spot motion in high-current vacuum arcs on copper electrodes. *Journal of Physics D: Applied Physics, 8*, 696–702.
47. Webster, R., Holmes, R., Jenkins, J. E., & Sherman, J. C. (1975). *The characteristic behaviour of a high current vacuum arc (No. ULAP-T-41)*. Liverpool University.
48. Bushik, A. I., Juttner, B., & Pusch, H. (1979). On the nature and the motion of arc cathode spots in UHV. *Beitrage Plasma Physical, 19*(3), 177–188.
49. Agarwal, M. S., & Holmes, R. (1984). Cathode spot motion in high-current vacuum arcs under self-generated azimuthal and applied axial magnetic fields. *Journal of Physics D: Applied Physics, 17*(4), 743–756.
50. Drouet, M. G. (1981). The physics of the retrograde motion of the electric arc. *Japanese Journal Applied Physics, 20*(6), 1027–1036.
51. Drouet, M. G. (1985). The physics of the retograde motion of the electric arc. *IEEE transactions on plasma science, 13*(5), 235–241.
52. Song, X., Wang, Q., Lin, Z., Zhang, P., & Wang, S. (2018). Control of vacuum arc source cathode spots contraction motion by changing electromagnetic field. *Plasma Science and Technology, 20*(2), 025402.
53. Swift, P. D., McKenzie, D. R., Falconer, I. S., & Martin, P. J. (1989). Cathode spot phenomena in titanium vacuum arcs. *Journal of Applied Physics, 66*(2), 505–512.
54. Swift, P. D. (1990). Cathode- & anode-spot tracks in a closed magnetic field. *Journal of Applied Physics, 67*(4), 1720–1724.
55. Klajn, A. (1999). Switching vacuum arc in a pulsed transverse magnetic field. *IEEE Transactions on Plasma Sciences, 27*(4), 977–983.
56. Jüttner, B., & Kleberg, I. (2000). The retrograde motion of arc cathode spots in vacuum. *Journal of Physics D: Applied Physics, 33*, 2025–2036.
57. Kleberg, I. (2001). *Dynamic of cathode spot in external magnetic field*. Berlin, Germany: Humboldt University.
58. Jüttner, B., & Kleberg, I. (2000). *Retrograde arc spot motion in vacuum* (pp. 188–191). Xian, Cina p: Proceedings XIXth International Symposium on Discharges and Electrical Insulation in Vacuum.
59. Juttner, B. (2001). Cathode spots of electric arcs. *Journal of Physics D: Applied Physics, 34*(17), R103–R123.
60. Dukhopel'nikov, D. V., Zhukov, A. V., Kirillov, D. V., & Marakhtanov, M. K. (2005). Structure and features of the motion of a cathode spot on a continuous titanium cathode. *Measurement Techniques, 48*(10), 995–999.
61. Smith, C. G. (1948, January). Arc motion reversal in transverse magnetic field by heating cathode. *Physical Review, 73*(5), 543–543)

62. Gallagher, C., & Cobine, J. D. (1949). Reverse blowout effect. *Electrical Engineering., 68,* 469.
63. Nürnberg, A. W., Fang, D. Y., Bauder, U. H., Behrisch, R., & Brossa, F. (1981). Temperature dependence of the erosion of A1 and TiC by vacuum arcs in a magnetic field. *Journal of Nuclear Materials, 103,* 305–308.
64. Fang, D. Y., Nurnberg, A., & Bauder, U. H. (1982). Arc velocity and erosion for stainless steel and aluminum cathodes. *Journal of Nuclear Materials, 111&112,* 517–521.
65. Fang, D. Y. (1983). Temperature dependence of retrograde velocity of vacuum arcs in magnetic fields. *IEEE Transactions on Plasma Science, 11*(3), 110–114.
66. Puchkarev, V. F., & Murzakayev, A. M. (1990). Current density and the cathode spot lifetime in a vacuum arc at threshold currents. *Journal of Physics D: Applied Physics, 23*(1), 26.
67. Himler, G. J., & Cohn, G. I. (1948). The reverse blowout effect. *Electrical Engineering, 67*(12), 1148–1152.
68. Yamamura, S. (1950). Immobility phenomena and reverse driving phenomena of the electric arc. *Journal of Applied Physics, 21*(3), 193–196.
69. Dallas, J. P. (1953). Arc interruption phenomena in a magnetic field at altitude. *American Institute of Electrical Engineers Part II Applications & Industry., 71*(6), 419–422.
70. Dunkerley, H. C., & Schaefer, D. L. (1955). Observations of cathode arc tracks. *Journal of Applied Physics, 26*(11), 1384–1385.
71. Robson, A. E., & Engel, A. (1956). Motion of a short arc in a magnetic field. *Physicl Review, 104*(1), 15–16.
72. Robson, A. E. (1978). The motion of low-pressure arc in a strong magnetic field. *Journal of Physics D: Applied Physics, 11,* 1917–1923.
73. Guile, A. E., & Mehta, S. F. (1957). Arc movement due to the magnetic field of current flowing in the electrodes. *Proceedings of the IEE-Part A: Power Engineering, 104*(18), 533–540.
74. Secker, P. E. (1960). Explanation of the enhanced arc velocity on magnetic electrodes. *British Journal of Applied Physics, 11*(8), 385–388.
75. Guile, A. E., Lewis, T. J., & Menta, S. F. (1957). Arc motion with magnetized electrodes. *British Journal of Applied Physics, 8*(11), 444–448.
76. Guile, A. E., & Secker, P. E. (1958). Arc cathode movement in a magnetic field. *JJournal of Applied Physics, 29*(12), 1662–1667.
77. Lewis, T. J., & Secker, P. E. (1961). Influence of the cathode surface on arc velocity. *Journal of Applied Physics, 32*(1), 54–63.
78. Guile, A. E., Lewis, T. J., & Secker, P. E. (1961). The motion of cold-cathode arcs in magnetic fields. *Proceedings of the IEE-Part C Monographs, 108*(14), 463–470.
79. Guile, A. E., & Secker, P. E. (1965). Retrograde running of the arc cathode spot. *British Journal of Applied Physics, 16*(130), 1595–1597.
80. Yamamura, S. (1957). *Journal* of the *Faculty* of *Engineering, Tokyo, 25,* 57–145.
81. Kigdon, K. H. (1965). The arc cathode spot and its relation to the diffusion of ions within the Cathode Metal. *Journal of Applied Physics, 36*(4), 1351–1360.
82. Guile, A. E., Adams, V. W., Lord, W. T., & Naylor, K. A. (1969). High current arcs in transverse magnetic fields in air at atmospheric pressure. *Proceedings of the Institution of Electrical Engineers, 116*(4), 645–652.
83. Guile, A. E. (1971, September). Arc-electrode phenomena. In *Proceedings of the Institution of Electrical Engineers* (Vol. 118, No. 9R, pp. 1131-1154). IET Digital Library.
84. Chaudhry, N. R., Lewis, T. J., Newton, R. H., & Secker, P. E. (1968). Gas-space and electrode effects in the motion of low-current arcs. *Journal of Physics D: Applied Physics, 1,* 1163–1169.
85. Secker, P. E., Sanger, C. C., & Lewis, T. J. (1972). Behaviour of low current arcs on moving electrodes. *JJournal of Physics D: Applied Physics, 5,* 580–588.
86. Hermoch, V., & Teichmann, J. (1966). Cathode jets and the retrograde motion of arcs in magnetic fields. *Zeitschrift f. Phys., 195,* 125–145.
87. Hermoch, V. (1973). On the retrograde motion of arcs in magnetic fields. *IEEE Transactions on Plasma Science, 1*(3), 62–64.

88. Murphree, R. P., & Carter, D. L. (1969). Low-pressure arc discharge motion between concentric cylindrical electrodes in a transverse magnetic field. *AIAA, 7*(8), 1430–1437.
89. Murphree, D. L., & Carter, R. P. (1970). Photographic observations of the retrograde rotation of an arc discharge. *Physical Fluids, 13*(7), 1747–1750.
90. Zykova, N. M., Kantsel, V. V., Rakhovsky, V. I., Seliverstova, I. F., & Ustimets, A. P. (1971). The dynamics of the development of cathode and anode regions of electric arcs I. *Soviet Physics Technical Physics, 15*(11), 1844–1849.
91. Bushik, A. I., Shilov, V. A., Juttner, B., & Pursh, H. (1986). Spot behaviour and cathode surface local heating. *High Temperature, 24*(3), 445–452.
92. Khromoi, Y. D., Zemskova, L. K., & Korchagina, Y. (1978). Cathode spot anchoring in a pulsed vacuum discharge current I. *Soviet Physics Technical Physics, 23,* N8.
93. Khromoi, Y. D., & Sysun, V. I. (1984). Cathode spot anchoring in a pulsed vacuum discharge current II. *Soviet Physics Technical Physics, 29*(7), 774–776.
94. Gura, P. S., Sysun, V. I., & Khromoi, Yu D. (1984). Motion of the channel of a pulsed high current vacuum discharge in a magnetic field. *High Temperature, 22*(2), 200–204.
95. Sysun, V. I., & Khromoi, Yu D. (1984). Plasma parameters of a pulsed vacuum discharge. *High Temperature, 22*(3), 366–371.
96. Perskii, N. E., Sysun, V. I., & Khromoi, Yu D. (1985). Magnetic field dependence of the current through multiple spots on a molybdenum cathode. *Soviet Physics Technical Physics, 30*(11), 1358–1359.
97. Perskii, N. E., Sysun, V. I., & Khromoi, Y. D. (1989). Dynamics of vacuum discharge cathode spots. *High Temperature, 27*(6), 832–839.
98. Chaly, A. M., Barinov, Y. A., Minaev, V. S., Myatovich, S. U., Zabello, K. K., & Shkol'nik, S. M. (2013). Characteristics of vacuum-arc cathode spots on the refractory metal electrodes. *IEEE Transactions on Plasma Science, 41*(8), 1917–1922.
99. Smith, C. G. (1946). Cathode dark space and negative glow of a mercury arc. *Physical Review, 69*(3–4), 96–100.
100. Kesaev, I. G. (1957, January). On the Causes of Retrograde Arc Cathode Spot Motion in a Magnetic Field. In *Soviet Physics Doklady* (Vol. 2, p. 60).
101. Kesaev, I. G., & Pashkova, V. V. (1959). The electromagnetic anchoring of the cathode spot. *Soviet Physics Technical Physics, 4*(3), 254–264.
102. Wroe, H. (1958). The magnetic stabilization of low pressure dc arcs. *British Journal of Applied Physics, 9*(12), 488.
103. Robson, A. E. (1960). The motion of an arc in a magnetic field. In *Ionization Phenomena in Gases, Volume I* (p. 346).
104. Lang, W. C., Xiao, J. Q., Gong, J., Sun, C., Huang, R. F., & Wen, L. S. (2010). Study on cathode spot motion and macroparticles reduction in axisymmetric magnetic field-enhanced vacuum arc deposition. *Vacuum, 84,* 1111–1117.
105. Chaly, A. M., Logatchev, A. A., & Shkol'nik, S. M. (1999). Cathode processes in free burning and stabilized by axial magnetic field vacuum arcs. *IEEE transactions on plasma science, 27*(4), 827–835.
106. Song, X., & Shi, Z. (2012). *Experimental investigation on the expansion speed of cathode spots in high-current triggered vacuum arc* (pp. 301–304). Tomsk: Proceedings XXV International Symposium on Discharges and Electrical Insulation in Vacuum.
107. Yanabu, S., & Souma, S. (1979). Vacuum arc under an axial magnetic field and its interrupting ability. *Proceedings of IEEE, 126*(4), 313–320.
108. Chaly, A. M. (2005). Magnetic control of high current vacuum arcs with the aid of an axial magnetic field: A review. *IEEE transactions on plasma science, 33*(5), 1497–1503.
109. Zabello, K. K., Barinov, Yu A, Chaly, A. M., Logatchev, A. A., & Shkol'nik, S. M. (2005). Experimental study of cathode spot motion and burning voltage of low current vacuum arc in magnetic field. *IEEE Transactions on Plasma Science, 33*(5), 1553–1559.
110. Li, L., Zhu, Y., He, F., Dun, D., Li, F., Chu, P. K., et al. (2013). Control of cathodic arc spot motion under external magnetic field. *Vacuum, 91,* 20–23.

111. Beilis, I. I., Sagi, B., Zhitomirsky, V., & Boxman, R. L. (2015). Cathode spot motion in a vacuum arc with a long roof-shaped cathode under magnetic field. In *XXVIth Intranational Symposium on Discharges and Electrical Insulation in Vacuum* (pp. 213–216). India, Mumbai.
112. Beilis, I. I., Sagi, B., Zhitomirsky, V., & Boxman, R. L. (2015). Cathode spot motion in a vacuum arc with a long roof-shaped cathode under magnetic field. *Journal of Applied Physics, 117*(23), 233303.
113. Chaly, A. M., Logatchev, A., Zabello, K., & Shkol'nik, S. M. (2007). Effect of amplitude and inclination of magnetic field on low current vacuum arc. *IEEE Transactions on Plasma Science, 35*(4), 946–952.
114. Chaly, A. M., & Shkol'nik, S. M. (2011). Low current vacuum arc with short arc length in magnetic fields of different orientations: A review. *IEEE Transactions on Plasma Science, 39*(6), 1311–1318.
115. Ma, H., Wang, J., Liu, Z., Geng, Y., Wang, Z., & Yan, J. (2016). Vacuum arcing behavior between transverse magnetic field contacts subjected to variable axial magnetic field. *Physical Plasmas, 23,* 063517.

Chapter 14
Anode Phenomena in Electrical Arcs

For long time, traditionally, the anode mechanism of an electrical arc commonly was accepted to be less complicated than the processes in cathode region. Nevertheless, in the second half of the twentieth century and, especially, in the last few decades, the experimental investigations showed different important forms of existence of vacuum arcs at the anode. A discussion of the importance for anode phenomena study is presented in this chapter.

14.1 General Consideration of the Anode Phenomena

In general, the anode is a passive collector of electrons closing the current of an electrical circuit of the arc discharge. In vacuum, the anode collects the electrons incoming to the surface from the cathode plasma jet. When the electron current to the anode is limited and it is lower than the arc current, an anode spot can appear. In the presence of an external gas of low pressure, the cathode plasma jet can be shielded from the anode surface; as result, the electron flux impinge upon the anode decreases due to their scattering by gas atoms. In this case, at the anode surface, an electrical sheath arises in which the electrical field accelerates the electrons to the anode supporting the requested arc current. When the pressure of the external gas is enough large, the near-anode electron density significantly decreased (so-called starvation state). Acceleration of the electron is substantially enhanced. As result, a relatively large energy flux to the anode is formed causing local surface overheating and intense anode vaporization. The evaporated from hot anode the metallic atoms are ionized increasing the plasma density and producing conditions for arise an anode spot. Similar process occurs in high-current vacuum arcs when the anode plasma can be generated due to gap filling by the intense cathode plasma jets and by a large high-current energy dissipation. Another case for anode spot appearing is in vacuum arcs with relatively small anode diameter when anode surface area is not enough to collect the requested arc current.

I. Beilis, *Plasma and Spot Phenomena in Electrical Arcs*, Springer Series on Atomic, Optical, and Plasma Physics 113, https://doi.org/10.1007/978-3-030-44747-2_14

The anode as passive and active electrode current collector was widely reviewed. Cobine [1] reviewed the behavior of the anode region in vacuum arcs. A comprehensive review of anode phenomena and anode modes in vacuum arcs was provided by Miller [2–6]. Boxman et al. [7] conducted another review of different anode structures and their modeling. Recently, Heberlein et al. provided a review of anode arc in gases [8]. Shkol'nik [9] reviewed the anodic arc behavior in a wide pressure range, as well in vacuum arcs and also considering the conditions of transitions between the different anode modes. It was noted that although there are numerous studies of the anode region of electric arcs, the understanding of the mechanisms of anode phenomena is still limited. The reason for this lack is that numerous discharge parameters influence on the anode arc modes that dependent on the complicated plasma–solid interaction and difficult for measure.

Let us consider the main experimental results, which indicate appearance of the anode modes and the specifics of anode role in electrical arcs. The primary observations were published for discharges mainly appeared in the presence of a surrounding gas.

14.2 Anode Modes in the Presence of a Gas Pressure

In an external gas, the anode spot appearance occurred at relatively low current (<100 A) in contrast to that in high-current vacuum arcs.

14.2.1 Atmospheric Gas Pressure

Short duration arcs between various metals in air have been studied by Somerville et al. [10] by means of a Kerr cell camera. It was observed a transient arc with durations ranging from one microsecond to one millisecond, in air at atmospheric pressure, between solid metallic electrodes. Arc durations of 1, 5, 20, 200 and 1200 μs were used with currents ranging up 200 A. Different metals were used for electrodes. The material properties varied from a low melting like thin (232 °C) to a relatively high boiling point (2270 °C). The electrode marks left by these arcs were examined by microscopy. Photographs of arcs between a fine wire cathode and the plane surface of a tin anode show a bright glow near the anode, which defines the anode " spot." It was observed that after the initiation of the arc, the diameter of the anode spot remains approximately constant with the arc duration. In the case of an 80 A arc, the diameter was 0.03 cm. Assuming that all the current passes through the luminous spot, the average current density was of the order of 10^4 A/cm^2. The area of the anode spot increases with the current. Microscopic examination of marks left on the anode by arcs with a moderate rate of rise of current shows that the arc affected a single circular region. For most metals, this region appears to have been molten, but for Cu and W, the surface was only a discoloration.

The melted area has a regular circular boundary and increases both with arc current and with arc duration. Marks left on tin anodes by 80 A and 50 μs arcs have radii a little over twice those of the corresponding spots. It appears that the increase of the molten area with arc duration is due to melting around a central active spot. It was noted that in accordance with the observations, the anode marks were consistent with the belief that they are caused by the heating of the anode through electron bombardment, which is concentrated over an area a few tenths of a millimeter in diameter. On polished copper anodes, the anode mark is a discolored area, probably consisting of oxides, which increases with arc duration. The area and general appearance of the anode mark are independent of the material and shape of the cathode and also of the interelectrode distance, unless this is less than a few tenths of a millimeter. At these short distances, the change in appearance is due to ejection of liquid metal from the molten area by the expanding air confined between the electrodes.

Blevin [11] used the techniques that have been employed by Somerville et al. [10] to study the behavior of arcs at the anode with constant current pulse. The examined arc had durations varying from 1 μs to 1 ms. The arc arises between a Cu wire cathode and a polished plane Al anode in air at atmospheric pressure. The arc is initiated by moving the electrodes one to other till sparking occurred. The instantaneous potential across the gap was of the order of 30 V.

A photograph of the anode region was conducted for arc with 50 A and 200 μs duration. The Kerr cell shutter was open throughout the duration of the arc. Considerable radial contraction was observed near the anode in form of a highly luminous and circular "anode spot". The spot appeared within the first few microseconds of the arc duration and does not change significantly with time. Similar results were obtained for the low melting point metals like tin and lead, while was not detected in this short time at the Ag anode with higher melting point. Considering the anode spot area, current density was estimated of 5×10^4 A/cm^2 for different metals.

The influence of electrode evaporation on the anode region formation and the anode energy dissipations of short time arcs has been widely studied and significantly contributed by Hermoch. The obtained results relate to the spots with a high current density, to the spot splitting, and to the intense evaporation of the electrodes. It can be regarded as the main characteristics of the anode region of short time (impulse) discharges. This is true of discharges not only with currents of 1–10 kA but also with much lower currents about 10 A, having a duration of 10–10^{3}μs. Let us consider the details.

Hermoch in 1954 [12] and in 1959 [13] used limitation of the anode surface to control the collected current density in order to investigate influence of the current contraction on the anode phenomena in a high-current and short-duration arcs. The surface area was limited by mask from a celluloid film or by specific capillary. The experiment showed that in contrast to discharge without artificial contraction, in which the arc voltage decreased with electrode distance, in the case of limited anode area, the voltage rise is observed with arc length. Such dependence was measured for electrodes covered by mask with cross Sect. 5×10 mm and a hole of 2 mm in the mask for different materials Cu, AI, Zn, Sn, and Pb [12]. The sinusoidal current with

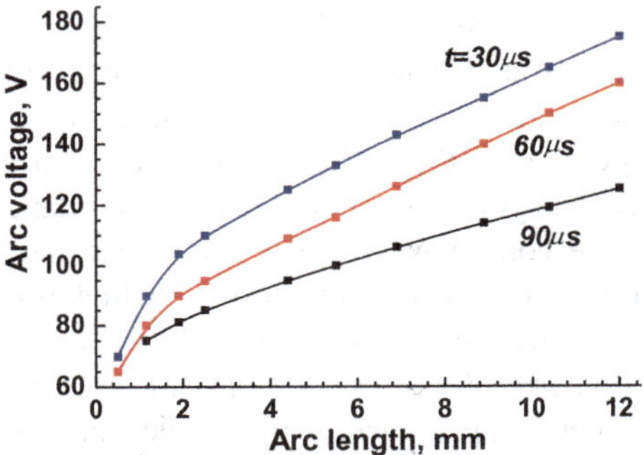

Fig. 14.1 Arc voltage as dependence on the arc length with arc time t as parameter [12]

maximal value of $I_{max} = 2500$ A and arc duration of 120 μs was used. Figure 14.1 shows the behavior of arc voltage with its length for different arc durations t. While the dependences indicate strongly voltage decreasing at very short arcs (<2 mm), the voltage increases linearly with larger arc length.

In [13], the maximal current was $I_{max} = 1250$ A and arc duration was 150 μs. A graphite electrode was used to determine the energy measuring the rate of the electrode evaporation. Only qualitative dependence on an average current density during the current pulse was discussed.

The voltage dependences were obtained at the moment of maximal arc current. The arc voltage dependence on arc length is presented in Fig. 14.2 with radius R_{con} as parameter for copper anode. Different anode areas (contraction) were characterized by different radii indicated in the figure by values R_{con}. The voltage increased with arc length and decreased with R_{con}, i.e., when the arc contraction reduced. The dependence of the arc voltage on R_{con} was measured for different pairs of cathode and anode materials from Cu, Cd, Zn, Pb, Sn, Fe, Al, Ni, and W at arc length of 6 mm. The arc voltage significantly decreased from 200,180, 160, and 140 V depending on the pairs when R_{con} was increased from 1 to 3 mm and then practically not changed at about 25 V. The voltage for pair Al–Al decreased from about 100 V. The voltage increasing with arc length was explained by arc spot localization on the limiting anode

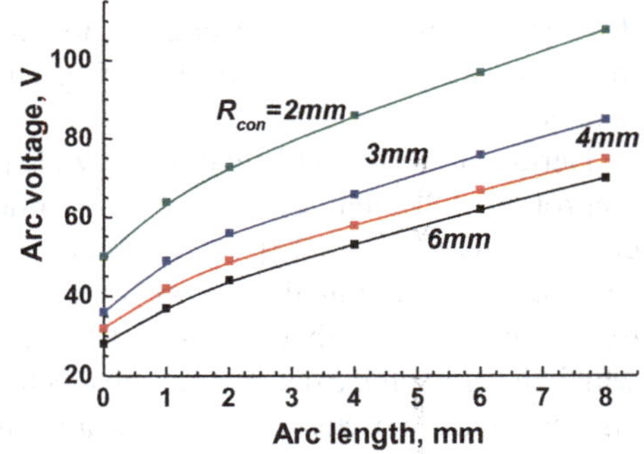

Fig. 14.2 Arc voltage as dependence on the arc length with Cu anode active area (indicated by radius R_{con}) as parameter. The anode area indicated in the figure was characterized by different radii [13]

surface. This location causes the strong anode evaporation and filling the electrode gap with an erosion material reaching conditions like as intense high-current arcs.

Hermoch [14] studied the spot phenomena at the anode by using stratified electrodes instead of the usual anodes made from one material. Stratified electrodes permitted the magnitude of the thermal flow to the electrodes to be changed while keeping the anode material the same. A layer of tin with 0.01–0.1 mm thick was formed on a massive copper basis. In order to make a comparison of the results, measurements were also performed on massive electrodes from tin and copper. A framing camera with an exposure time of 5 μs allows photographically to observe the form and size of the anode spots. The polished anode was with an active surface area of 10 $\times$ 10 mm. The discharge always burnt on a part of the surface not yet disturbed by it. A cylindrical W rod was used as cathode. The current impulse was aperiodic and lasted 300 μs. The maximum value of the current was varied from 600 to 4000 A. The current density in the anode spots was determined from the area of the spots and the instantaneous value of the current. The anode spot current density for arc current 1 kA and arc time of 100 μs is presented Table 14.1 reported in [15]. In this work, the role of strong anode vaporization in the anode spot formation is discussed in detail and the spot splitting was studied.

Summarizing, the following experimental results were obtained [14]:

(1) For Sn anode (in form of prism), the spot was observed with current density 2.4 $\times$ 10^4 A/cm ^{2}, and its magnitude does not change with a change in current. The discharge gave rise to a relatively deep crater with the signs of strong melting.
(2) The current density on Cu anode increases with increasing current. For maximum current 600 and 1.5 kA, the current density was 2.17 $\times$ 10^4 and 2.8 $\times$ 10^4 A/cm^2, respectively. The anode was colored with oxide films and without melting marks. For currents above 3 kA, the current density was 6 $\times$ 10^4 A/cm^2, and the spot splits into a series of partial spots. The craters on the electrode surface exhibited signs of strong melting of the material.
(3) For stratified anode, the current density depends both on the thickness of the layer and on the current. For anode with Sn layer of the order 0.01 mm thick, the layer disappeared in a very short time at current larger than 1 kA. At lower current, the current density corresponds to the density of the spot on the copper and the layer remains undisturbed. At maximal current of 1.5 kA, the anode Sn layer of the order 0.1 mm thick was weakened but not reached the copper surface, and the current density was 3.26 $\times$ 10^4 A/cm^2 showing greater contraction than a Cu or Sn electrode.

Hermoch [16] presented results of intense arc mode in a high-current (<10 kA) short-time (10–100 μs) arcs accompanied by intense electrode vaporization. Cylindrical electrodes of 6 mm diameter produced from different materials were used. The photograph of the interelectrode region showed a symmetrical expansion of luminous gas after arc initiation. At the next time step, a discharge channel was formed. The plasma jets were ejected from the electrode spots. The impinging jets produced a plasma plume in a disk form that expands radially between the electrodes. Figure 14.3 shows the photographs of the expanding plasma for different materials. It was also

Table 14.1 Current density in the anode spot for different anode materials [15]

Material	Al	Ag	Be	Bi	Cd	Co	Cu	Mg	Ni	Pb	Sb	Sn	Zn	Fe	C	Mo	Ta	W
Current density, (10^4 A/cm^2)	4.83	5.8	7	2.25	2.5	5.82	7.32	3	5.7	2.93	3	5.5	2.8	5.5	2.8	6.13	4.82	6

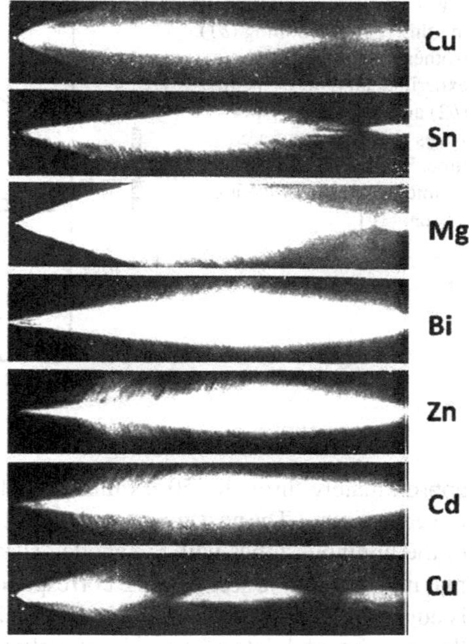

Fig. 14.3 Intense mode in form of expanding plasma channel of short-time high-current electrical discharge. Figure taken from [16]. Used with permission

observed that the jets could be produced discontinuously, resulting in corresponding discrete plasma expansion during development of the discharge (see the last picture in Fig. 14.3).

While the radial expansion between refractory electrodes was relatively minor, this expansion is significant in case of volatile materials like Pb, Bi, Cd, and Zn. The experiments show that during development of the discharge, the electrode vapor was ionized, pushed the previous original gas, and the intense arc operated mainly in the plasma of electrode material. For Cu electrodes with gap of 3 mm, the current density was near constant value of 4.5×10^4 A/cm^2 up to peak current of 1 kA and increased by factor 2.4 when the arc current increased to 4 kA. It was observed that the current density increased in direction to the electrodes, and at the cathode, it was larger than at the anode.

Beilis and Zykova [17] investigated the anode spot development optically by high-speed camera with exposition time varied in range from 2 to 10 μs in air at atmospheric pressure. Cu electrode of 10 mm diameter was used with inclined surface that was perpendicular to the optical axis of the camera. The cathode of pointed shape was placed at 2.5 mm from the anode surface. The current was varied from 5 to 40 A. The arc duration was lower 12 ms. In contrast to Somerville et al. experiments [10], in the present work, a fixed anode spot was observed that changed during the arc time in the considered interval of the arcs. Few spots were produced at 40 A.

Three concentric zones in the anode spot were detected. The central zone, the most luminous, appeared during the first 30 μs. The second zone was observed

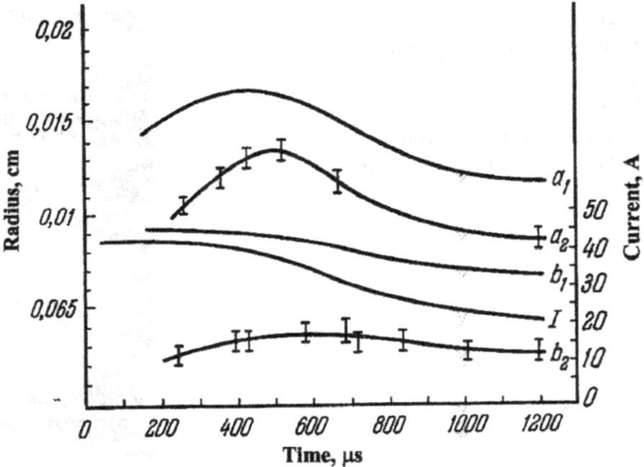

Fig. 14.4 Calculated radii of melting ($a1$) and boiling ($b1$) isotherms as well the experimental radii of the first ($b2$) and second ($a2$) spot zones as functions of arc time. The line denoted by I' is time dependent of the arc current [17]

approximately through 150 μs that placed around the central zone with relatively lower luminous. The next zone not glowed, but weakly reflected the light. Comparison of the luminous zones with the anode mask brings to the following conclusion. The maximal size of the second zone corresponds to the melting area, while the last zone is equal to the size of the oxide film at the anode surface. In order to interpretation, the experimental data of the radii of melting and boiling isotherms were calculated. Since the observed spot areas have circular form, all experimental and calculated results as radius dependences are presented in Fig. 14.4 (see caption). The calculated model assumed the heat flux from the spot was approximated by a continued point heat source acted on a semi-infinite bulk body. The boiling temperature for Cu was 2595 °C at which the vapor is at atmosphere pressure.

Since the arc current was changed during the arc, the calculation was conducted for current, which was taken for all period from the arc beginning up to each calculated time. The results show that the experimental and theoretical curves are similarly dependent on the arc time. However, the calculated dependences exceed the measured data. This difference appears due to neglecting the energy losses by the anode vaporization and others in order to simplify the calculated model for principal description of the observations. It was indicated that the current density in the first bright spot area was 10^6 A/cm^2, while for two other areas, the current density was decreased from 10^4 to 10^3 A/cm^2. These values were obtained by assumption that all current of the arc was flowed through the considered areas.

Shih [18] reported about a method using a split anode to measure the anode current density distribution in high-current pulsed arcs. A copper anode in air at one atmosphere was studied. A nearly rectangular arc current pulse was 750, 1200, 1750, and 2250 A. All data were obtained using a pulse length of 40 μs. The split copper anode consisted of two anode plates, separated electrically by an insulating material. The data were obtained using the split anode with a 0.001 cm gap. Rogowski coils used to detect the current to each half of the split anode as a function of arc position (cathode can shift) relative to the splitting plane. The measured radial distribution showed that anode current density decreased from a peak value more than an order of magnitude at radius about 0.1 cm depending on arc current. The peak current densities ranged from 3.4×10^5 to 5.5×10^5 A/cm^2. The value of peak current first increased and then decreased with increasing total current. This trend agrees with the observations of the size of the erosion spot at different currents.

Pfender [19] reported the results of the anode phenomena investigated in argon arc at 1 atm for currents between 100 and 300 A. A specific apparatus was used in the experiment. The arc burned in constricted tube between a tip-shaped thoriated tungsten cathode and plane water-cooled anode (Cu flat disk of 44 mm diameter), which is perpendicular to the axis of a symmetric arc. The total length of the arc can reach up to 120 mm. Another electrode configuration was used in which the cathode is parallel to the anode surface.

Two types of anode arc roots were studied. In relatively short arcs, the action of the cathode jet attached the anode surface in a diffuse form. The cathode plasma jet characterized by a bell shape is impinged on the anode surface. This is cathode jet-dominated arc mode. In contrast, at long arcs (arc length much larger than the arc diameter), a plasma constriction at the anode and an anode jet were detected and named as anode jet-dominated arc mode. For parallel electrode configuration, anode plasma constriction and an anode plasma jet were observed. It was noted that experiments with different arc currents reveal a transition from anode jet-dominated (I >150 A) to the cathode jet-dominated arc mode (I <150 A).

14.2.2 *Low-Pressure Gas*

The anode spot behavior significantly changed when the surrounding pressure decreased in comparison with atmospheric state. Zykova [20] and Zykova et al. 1970–1971 [21] investigated the influence of low-pressure gas on development of the anode spot. The cathode of pointed shape was placed at 2.5 mm from a polished anode of 20–24 mm diameter. The photographic high-speed camera with about 10^5 frames/s was used. The pressures of He, Xe, and Ar were varied from 400 to 10^{-4} torr with current of 600 and 1200 A. The anode materials were Zn, Cd, Sn, Cu, and W. The experiment showed that for 400 torr, a few spots were detected and placed near one to other so that only one trace was observed. The dynamics showed that in the arc beginning, only one spot was observed, and then, a splitting to few spots was appeared during the arc development. The splitting was occurred earlier for materials with lower heat conductivity and lower boiling temperature like Cd and Zn. For example, the splitting occurred early for Sn than for Cu. It was noted that the spot behavior was similar to that observed for atmospheric arcs at relatively moderate currents.

The anode region was significantly changed when the gas pressure decreased from 400 to 100 torr. In this case, a few spots were observed from the arc beginning. The spots placed separately so that also separate traces were detected. Although the spots are fixed, the spot group can be moved due to dead of old and appearing of new spots. The group spot mobility increased in gases with lower atom number and for

Table 14.2 Anode spot parameters for different anode materials [20]

Pressure, torr	Material	Spot current, A	Spot diameter, cm
100	W	20	0.002–0.009
	Zn	50	0.015–0.09
10	W	5–10	
10^{-4}	W	10–60	
	Zn	30	

materials with larger melting temperature. For example, the group spot at Zn anode in Xe for pressure 100 and 10 torr was fixed, while in He and Ar, this group moved with velocity 1–2 m/s. A correlation between mobility of the cathode and anode spots was observed. The anode spot lifetime was varied from 30 to 400 μs. At pressure of 10^{-4} torr, this time life was 30–60 μs at Zn anode. The anode spot current for different He pressure materials is presented in Table 14.2. The spot numbers increased with arc current and also changed with gas pressure and anode materials. The anode area occupied by the spots increased with arc current and spot distribution at the anode surface can be not uniform. For example, at W anode, the distance between spots can be increased from 0.05 to 0.5 mm.

Dyuzhev et al. [22, 23] considered the condition for anode spot initiation. The plasma region at the anode was investigated in experiments arcs with thermionic cathode at atmospheric pressure and in different pressures of argon and xenon using alkaline dopant vapors of Cs and Rb. The cathode was produced from Ni and Mo tubes of 1–3 mm diameter and 5–15 mm length dense inserted in another tube of 10 mm. The anode was produced from Cu, Mo, and stainless still with working area changed from 0.8 to 0.01 cm^2 to reach anode current density up to few kA/cm^2. The cathode was grounded, and a pulse voltage was applied to the anode that can support arc current up to 1.5 kA. The plasma parameters (density and temperature) were measured by probe method.

The experimental results showed that at low and moderate pressures of inert gases in wide ranges of discharge conditions and plasma parameters with high and weakly ionized plasma, a negative anode potential drop was formed. In this case, only a distributed anode discharge was observed, and anode spots were absent. It was noted that although occurring of some triggering phenomena (local heating, vaporization, and ionization), the plasma density in the anode region cannot be developed at the negative anode potential. The anode potential drop formation and condition of the anode spot initiation were studied using the measured plasma parameters and the volt–current characteristics, and the following results were obtained.

When the anode area is enough large, an increase of the plasma density and temperature with the applied voltage u_{ap} was observed. Therefore, the random current I_r from the plasma to the anode also increased. As a result, at some point, a negative anode potential drop formed in order to decrease the current from the plasma up to value equal to arc current I_a. Another case appears when the anode area is relatively small. The anode current I_a approaches to the random current I_r from the anode plasma with voltage u_{ap}. In this case, a volt–current characteristic with a saturation region is formed, and with further increase of the voltage, the plasma parameters are not changed. So, the discharge current is saturated due to current limiting by the random current from the plasma. As a result, the plasma potential near the anode can be smaller than the anode potential, i.e., a positive anode potential drop is produced.

It was noted that the saturation of the volt–current characteristics also depends on an electron emission possibility of the cathode to support the current $I_a \approx I_r$. When the anode plasma is weakly ionized, the increase of u_{ap} also does not change the anode polarity sign due to rise of the anode current I_a because of an increase of the plasma density. The measurements showed that the anode polarity sign could be changed from negative to positive only in case when the anode plasma is highly ionized and when the condition of $I_a \approx I_r$ was fulfilled. Further increase of u_{ap} leads to increase of the positive anode potential drop and consequently to increase of the energy flux to the anode surface. The anode temperature increased, and anode vaporization and vapor ionization were intensified. As a result, a significant rise of the arc current and an anode spot ignition was realized.

The condition of change of the anode potential drop to a positive value to formation of the anode spot was indicated in [24]. It was observed that anode spot arises for tungsten at positive anode potential drop between 10 and 20 V. The anode current density reached a value that was about of the random electron current density flowing to the anode. At this condition, the thermal and electrical parameters were investigated. The results showed that the anode temperature and anode vapor increased with the arc current, which finally leads to anode spot appearing.

Thus, the above experiments and also the analysis [25] of the experiments showed that in low- and moderate-pressure arcs, the anode spot formation was possible by transition of the discharge to a mode with positive anode potential drop.

Miyano et al. [26] investigated the different anode modes in an arc as function of ambient gas pressure ranging from 0.01 to 300 Pa. The stainless steel chamber acted as the anode. The low current was constant of 50 A. The cathode (64 mm in diameter) materials were Al, Ti, Fe, Ni, and Cu. The ambient gases were He, Ne, Ar, H_2, N_2, O_2, and CH_4. A transition of the anode mode was observed from diffuse arc to footpoint to plane luminous to anode spot mode as the pressure was increased. The anode voltage depends strongly on gas species (influence of the potential of atom ionization) and weakly on the cathode material. The pressure of the arc mode transition were lower for diatomic than that for monoatomic gases. It was assumed that this result is due to gas dissociation and further changes in the potential of ionization of the species.

14.3 Anode Modes in High-Current Vacuum Arcs

The transition of the arc into the different anode modes and the boundary arc parameters between the different regions depend strongly upon the electrode material, the electrode and gap geometry, and the current waveform [27, 28]. The modes are appeared on form of diffuse arc mostly at low current, so named footpoint mode characterized by intermediate arc current, anode spot, and intense arc appeared at high current.

Schellekens [29] conducted an experiment, and he concluded that the previously named diffuse arc based on high-speed photography is ambiguous. It was noted that to describe more exactly of the diffuse arcing state, the luminous from background plasma and also a luminous from multi-sport jets should be considered. An optical probe method was used, which enables one to quantify the arcing state by a new measurable quantity named by Schellekens [29] as *diffusity*. The photographic observations indicate that the vacuum arc at moderate axial magnetic field strengths consists of arc channels and background plasma. The plasma density and therefore the light emissivity are thus not uniform. It was shown that in accordance with the observation, the plasma intensity I_ε could be described as the following sum

$$I_\varepsilon = \int_{-L}^{L} \varepsilon_d \mathrm{d}x + N \int_{-R}^{R} \varepsilon_0 \mathrm{d}x \tag{14.1}$$

where N is the number of arc channels in the observational volume, L is the radius of background luminous, R is the radius of the channel luminous, and ε_d, ε_o are, respectively, the emissivity of the background plasma and arc channels determined in [29]. Using (14.1), the diffusity was defined as

$$\text{diffusity} = \frac{\int_{-L}^{L} \varepsilon_d \mathrm{d}x}{\int_{-L}^{L} \varepsilon_d \mathrm{d}x + N \int_{-R}^{R} \varepsilon_0 \mathrm{d}x} \tag{14.2}$$

Diffusity defined by relation (14.2) attains 100% when the plasma is absolutely diffuse. This is achieved either when no arc channels are produced ($N = 0$) or when the arc channels have very low emissivity. When plasma consists of many independent arc channels (cathode multi-spot arc) and the emissivity of the background plasma is zero, the relation (14.2) attains 0% In the intermediate case, when both the background plasma and the arc channels are present, the relation (14.2) indicated some percentage between the mentioned extreme cases. It was noted that the expression (14.2) is independent of the observational volume and can hence be used to compare the results for different channel conditions. However, the experiment in [29] showed that the above diffusity definition is mostly important *in the presence of a magnetic field* promoting the channel's production.

The experimentally studied diffuse or spotless mode as well as the anode spot mode forming for different conditions of arc burning is reported below.

14.3.1 Anode Spotless Mode. Low-Current Arcs

Drouet et al. [30, 31] measured a current distribution in vacuum arc with anode spotless mode using a multi-ring anode. The experiments were conducted for arc currents of 58, 137, 204, and 316 A for arc duration of 200 μs, copper cathode of 2 mm diameter, and anode of 90 mm outer diameter. Photographs of the arc detected the symmetric plasma luminous with respect to the cathode–anode axis. The average density of the current collected by each anode element was determined taking into account the current and area of the elements. The results are presented in Fig. 14.5. Observed that the shape of the current distribution varies with the electrode gap distance and arc current. As it can be seen, the anode current density significantly decreases with anode radius.

Rosenthal et al. [32] conducted a spectroscopic study of the temporal evolution of a vacuum arc with graphite anode. This arc begins as a multi-cathode spot vacuum arc operating for 150–200 s duration at relatively low current of 175 and 340 A in a DC arc. The intensity of spectral lines of Cu I, Cu II, and Cu III was investigated as a function of time near the cathode, near the anode, and in the middle of the 10 mm electrode gap. The 30-mm-diameter copper cylinder cathode was water cooled and had a cylindrical protrusion of 10 mm diameter and 5 mm height at its center, to facilitate a contact near its axis with the mechanical trigger, thus ensuring that the discharge was initiated near the axis of the electrodes. The anode was made from a 32 mm diameter and 30-mm-long graphite cylinder. The optical arrangement allowed the imaging of the interelectrode plasma region onto the entrance slit of a Czerny–Turner monochromator. A charge-coupled device (CCD) black-and-white TV camera was used to record the radiation intensity distribution across the two-dimensional image of the interelectrode plasma. Line intensity measurement was performed with 40 ms time resolution and 0.4 mm spatial resolution.

The experiment demonstrated that the different anode modes observed in a high-current vacuum could be appeared also at relatively low current in arcs with refractory non-consumable anode. The transition of the anode modes for 175 A with arc time development at, respectively, video frames is shown in Fig. 2 in paper of [32]. The

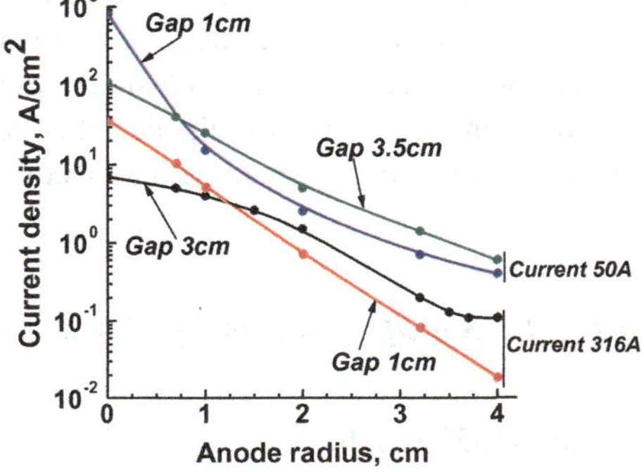

Fig. 14.5 Radial dependence of the current density distribution on the anode surface for arc currents 50 and 316 A. The current densities measured for interelectrode gap distances of 1.5, 2.0, and 2.5 cm were distributed accordingly and between the curves denoted by gap 1.5, 3, and 3.5 cm

advanced illustration of the time development arc modes (anode plasma plume generation and its expansion in the gap) can be seen in Fig 10.9 (Chap. 10). As the anode was heated by the long-duration arc, the anode temperature increased with time. As a result in arc beginning, the anode is cold and only diffuse mode was observed without any luminosity. At 15 s of arcing, the Cu plasma partially reflected from the heated surface, and a footpoint anode mode is occurred. Finally, at 50 s of arcing and significant cathode plasma jet reflection and plasma plume filling in the gap, the intense anode mode is produced. However, during the arc development, no anode spot mode was observed due to relatively large anode surface, which is enough for supporting the arc current.

The spectroscopic data indicated that the anode acts as a collector for ions emitted from the cathode. These data indicated also the existence of Cu III near the anode, whereas no spectral lines of copper ions were detected in the middle of the interelectrode gap.

14.3.2 Anode Spotless Mode. High-Current Arcs

Mitchel [33] conducted a series of experiments to measure the distribution of current at the anode by multi-element probes during various stages of the vacuum arc development. A 50-mm-diameter copper cathode was separated at 10 mm from seven probe elements at the anode. The probe elements were placed at the anode in circular form. The current in each element was recorded from the potential difference of the element. The current is recorded in various elements during discharges in which the total arcing voltage was less than 40 V, i.e., when current peak was less than the starvation current. The measurement for arc with diffuse anode mode was conducted at the anode of 45 mm diameter at current of 3.5 kA. In this case, about 30% of the total current passes through the center elements, while the rest of the current is distributed between the peripheral elements at the start of each current loop.

Schellekens [34] measured the axial current density distribution in front of the anode. This distribution was obtained using multi-probe anode. Six 1.5-mm-thick copper disks were mounted on equidistant separation of 5 mm in radial direction. The anode diameter was 6 cm. The current density was deduced from the potential distribution on the anode. The measurements were conducted prior to the anode spot formation for arc-constricted plasma at the anode region with current of 5 kA. As it can be seen from Fig. 14.6, the current decreases parabolically from the anode axis at maximal value to zero at the anode edge.

14.3.3 Anode Spot Mode for Moderate Arc Current

Kimblin [35] was the first to study the arc transition of the anode modes and showed experimentally an appearing of the diffuse and spot states. This study was further

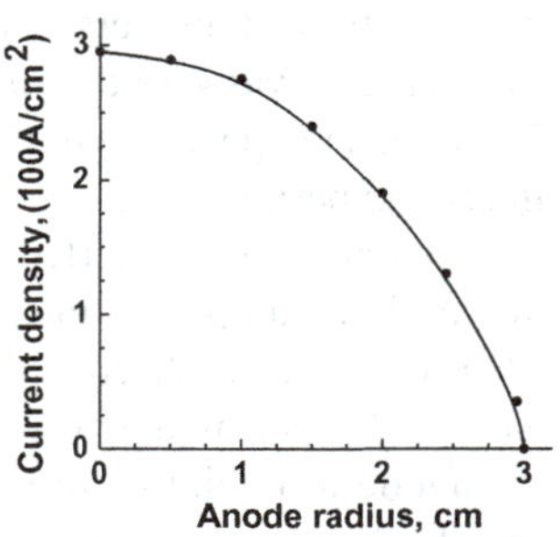

Fig. 14.6 Radial dependence of the current density distribution on the anode surface for arc currents 5 kA prior to the anode spot formation [34]

developed including analysis of a technological importance of anode spots in vacuum 1974 [36]. The specifics of anode geometry used to investigate the transition and arc voltage for copper electrodes over the DC current range of 0.1 to 2.1 kA. The cathode diameter was 5 cm, the anode diameter was either 1.3 or 5 cm, and the electrodes were separated to 2.5 cm in a time of 150 ms.

For both anodes, the average arc voltage increased with electrode gap distance to values significantly higher than the voltage of 20 V observed for short electrode gap (see Fig. 14.7). The anode mode remained diffuse up to voltage about 75 V with the 1.3-cm-diameter anode at electrode separation of 2.5 cm and for arc currents of <0.4 kA. At 0.4 kA (threshold current), the anode plasma configuration changed from the diffuse to the contracted spot mode. With increasing arc current, the transition to the evaporating spot mode occurred at shorter electrode gaps. However, for the 5-cm-diameter anode, spot was observed at the 2.5-cm gap for an arc current of 2.1 kA.

Kong et al. [37] used method of photography of the anode plasma region in order to understand the influence of the opening electrode velocity on the anode phenomena. The high-speed video camera Phantom V10 was used to record the vacuum arc mode evolution in the experiments. The recording velocity was set to be 4000 frames/s with the exposure time of 2 μs. The experiments were conducted in a vacuum interrupter

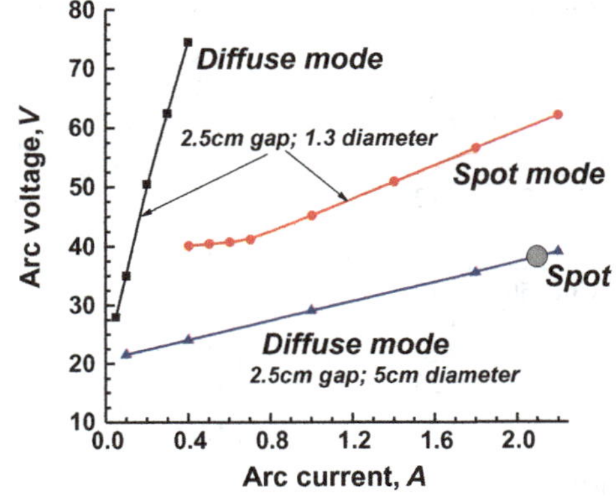

Fig. 14.7 Dependence of the arc voltage versus current at 2.5-cm electrode gap, which is obtained by approximation of the data from Kimblin [36]. Spot forms at 0.4 kA. Lower curve is for 5-cm-anode diameter where denoted that spot forms at 2.1 kA (by circle)

with electrode diameters of 12 and 25 mm for materials of Cu, CuCr25, and CuCr50. The velocity was varied from 1.3 to 1.8 m/s. The thickness of the electrode was 4 mm. The contact gap was 18 mm, and the upper electrode was the anode. The arcing time in each experiment was about 9 ms with the 50 Hz frequency current.

The experiments started from a low current level (the peak current was several hundreds of amperes) in a vacuum interrupter, at which the anode was passive. Therefore, the arc discharge was a low-current anode mode. However, the footpoint and anode spot modes were studied preferable at high current. The photographs of the anode footpoint or spot appearance at Cu and CuCr25 anodes are presented in Fig. 14.8. It can be seen that the footpoint appears in different forms depending on the average velocity and current. Table 14.3 demonstrates the first threshold current at which the footpoint or spot appears.

The experimental results showed that the high-current anode mode that first appeared was a footpoint at the opening velocity of 1.8 m/s, regardless of the type of the electrode material, being Cu, CuCr25, and CuCr50, and the size of the contact diameter, being 12 and 25 mm (Table 14.3). A small size luminous footpoint on the anode and a diffuse arc between the electrode gaps in the footpoint mode were

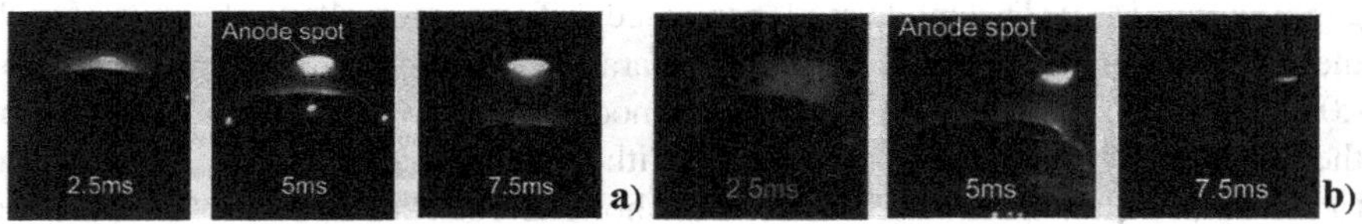

Fig. 14.8 Photographs of the high-current anode modes (nanocrystalline CuCr25) with opening electrode velocity 180 cm/s and peak arc currents indicated in the pictures for **a** electrode diameter of 12 mm, current 2.55 kA and **b** electrode diameter of 25 mm, current 2.6 kA [38]

Table 14.3 First threshold current at which arc mode transfers from a low-current to high-current mode

Materials	Diameter, mm	Velocity, m/s	Footpoint threshold current, kA	Spot threshold current, kA
Cu	12	1.3	–	4.4
		1.8	3.42	
	25	1.3	–	5.92
		1.8	4.72	
CuCr25	12	1.3	–	2.91
		1.8	2.29	
CuCr50	12	1.3	–	2.64
		1.8	2.48	
	25	1.3	–	3.68
		1.8	2.88	

Kong et al. [37]

observed. At the velocity of 130 m/s, the first appeared anode mode was a large bright spot with the arc current increasing. Also, there was a columnar or jet arc with a distinguishable boundary between the electrodes in the anode spot mode. The first threshold current is higher at the average opening velocity of 130 m/s than that at 180 m/s independent of what electrode material or diameter is used in the experiment. Also, this current is larger for Cu than for CuCr25. It was indicated that at the high velocity of electrode opening, a high electrode gap is produced with only footnote anode mode. In this case the density of the plasma adjacent to the anode was too low and the ion starvation current incident on the anode was smaller.

14.3.4 Anode Spot Mode for High-Current Vacuum Arc

Reece [38] qualitatively described the phenomena at the anode in a cathode spot mode when the anode is only a current collector. He indicated that the cathode plasma jet condensed at the anode surface transferring the kinetic and potential energy of the plasma and as result heat the anode. It was noted that no necessity for an electron sheath because positive ions impinge on the anode and condense at its surface. The anode becomes coated with a film of metal coming from cathode jet.

Mitchel [33] has studied the vacuum arc with sinusoidal current up to 100 kA peak experimentally. Oxygen-free copper electrodes from 6.5 to 75 mm diameter, commercial-grade tough-pitch copper, 25-mm-diameter cadmium, and lead-tipped electrodes were tested at separations in range of 0–40 mm. All electrodes were from 80 to 140 mm long and machined to a smooth finish. The experiment showed that during sinusoidal current loops up to about 10 kA peak, between copper electrodes of a few centimeter diameter, the arc plasma in the anode region is generally diffuse. When a high vapor pressure material, such as lead, was used as a cathode with a copper anode, a diffuse arc develops at currents in excess of 30 kA. If the peak current during a 50 Hz exceeds 10 kA, an anode spot of 100 mm^2 may develop. This area is clearly liquid, and a bright constricted column, which extends from the anode toward the cathode, generally obscures the luminosity of the surface. It was also indicated that anode spot formation and a constricted column normally occur at arc current of about 6–8 kA peak.

Voltage dependence across a pair of 2.5-cm-diameter copper butt contacts at 5-mm separation was measured. The voltage is about 20 V, with oscillations of 1–2 V, and it is largely independent of current, separation, and electrode geometry up to 1 kA and is determined by the cathode material. From 1 to 6.5 kA, the voltage rises fairly linearly with current from 20 to about 40 V at 6.5 kA. For currents in excess of 6.5 kA, the voltage is almost proportional to the current and may give arcing voltages in excess of 120 V.

Rich et al. [39] concerned with investigations associated with the development of an anode spot for vacuum arcs. The plane-parallel electrode geometry was kept fixed in which the gap spacing was 0.95 cm and the electrode diameter was 5.72 cm. The threshold current density for anode spot formation was determined for electrode

Table 14.4 Threshold current at anode spot formation against the electrode materials and with accordance to correlation parameter of their thermal characteristic

Material	Threshold current, A	Current density, 10^2 A/cm^2	$T_m(k\rho c)^{1/2}$
W	13.8	5.4	1693
Mo	13.6	5.3	1204
Cu	10.3	4.0	954
Ag	9.7	3.8	723
Al	6.8	2.7	383
Sn	2.3	0.9	57

Rich et al. [39]

materials chosen in wide range of thermal and electrical properties, named of Sn, Al, Ag, Cu, Mo, and W. Arcing was over one half cycle of a 60 Hz current wave. The anode spot formation was determined from high-speed streak photographs of the arc. An anode spot formation was associated with the sharp changes in arc voltage and noise voltage. The noise voltage and arc drop decrease as the spot develops. The measured threshold current for anode spot formation for different electrode materials, characterized by their thermal constants, is presented in Table 14.4. This current decreases with reduction of melting temperature of the metals. The current density was obtained under the assumption that is prior to the anode spot initiation; the discharge was uniform over the anode surface.

It can be seen that the anode spot is formed at larger arc current for material with larger thermophysical constants, where T_m, k, ρ, and c are the melting point (°C), thermal conductivity, mass density, and specific heat, respectively. The thermal constants are in CGS units. Figure 14.9 shows the maximum arc voltage appearing

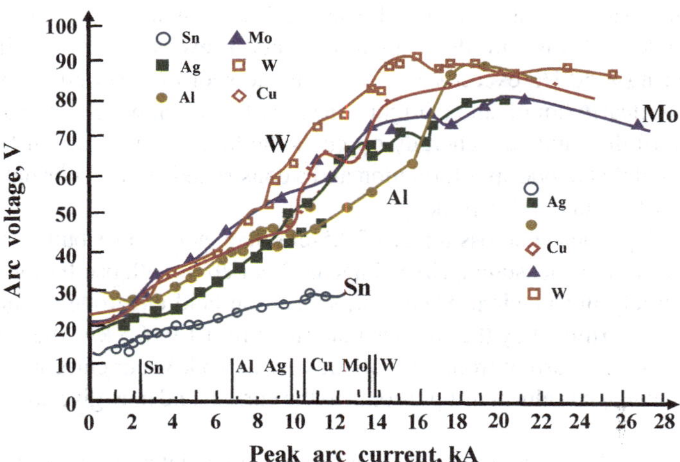

Fig. 14.9 Maximal arc voltage of the data obtained for the six electrode materials plotted as dependence on peak arc current [40]

during the arcing cycle for all studied electrode materials as dependence on peak of the current wave. The vertical bars indicate the positions of the threshold currents for anode spot formation or, as in the case of tin, the current at which traces of surface melting first occurred. In general, the arc voltage for the high vapor pressure materials is lower than that for the low vapor pressure materials, while the threshold currents increase with this vapor pressure. Some saturation of these characteristics can be seen at currents larger than 15 kA.

Boxman [41] measured the threshold current for the transition between the low-current vacuum arc mode and the high-voltage noisy mode associated with anode spot formation as a function of peak current I_p, current waveform frequency, and electrode separation. The arc was sustained between 25-mm-diameter Cu and Ni butt electrodes. Ambient pressure was less than 10^{-6} torr. Anode spot were formed only near the center of the anode. Tests were conducted at current waveform frequencies at 347 Hz and 66 Hz on Cu electrodes and 543 Hz, 81 Hz, and 58 Hz on Ni electrodes. The frequency was defined as 1/4 T_p, where T_p is the time from current initiation to current peak I_p.

Low-amplitude noise was often observed on the voltage waveform prior to transition, as observed between 1.8 and 2.4 ms. The transition voltage jump was located accurately in time. In particular, the transition onset was defined by the jump in voltage that occurred at 2.4 ms. Figure 14.10 shows the transition current I_t as a function of peak current that is plotted for various electrode separations. Data at 66 Hz on Cu electrodes show only a slight tendency for I_t to increase with I_p. However, at high frequencies, I_t is found to be nearly proportional to I_p for a range of peak currents from the lowest current at which transition is observed, I_1, to some higher current I_2, at which a constant I_t is reached in Fig. 14.11. Low-amplitude voltage noise on Cu electrodes was observed prior to transition for currents exceeding I_1. When the peak current just exceeded I_1, transition would occur after the peak current.

Heberlein and Gorman [42] observed vacuum arcs modes between separating butt-type electrodes of 100 mm diameter. Sinusoidal current wave with a frequency of approximately 50 Hz at peak values of the AC arc current half-wave was ranged from

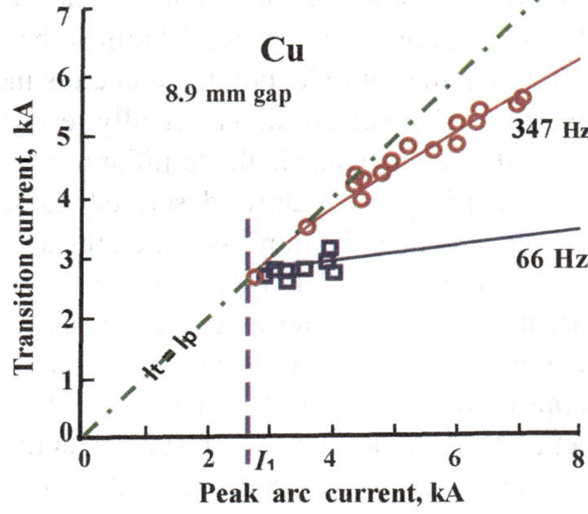

Fig. 14.10 Transition current as a function of peak current for Cu arcs with electrode separations 8.9 mm for 66 Hz and 347 Hz. According to the probabilistic explanation of transition current variation (see [40]), transition occurs after current maximum for values of I_p located to the left of the tangent point with $I_t = I_p$ line [41]

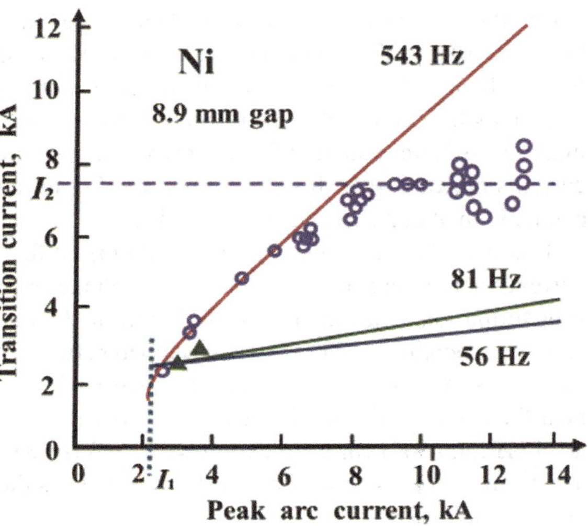

Fig. 14.11 Transition current as a function of peak current for Ni arc with electrode separation of 8.9 mm. Data points for 543, 81, and 58 Hz are represented by symbols **o**, ▲, and • respectively. The solid curves are result of some treatment of the observed data described in the work of [41]

5 to 67 kA. Electrode separation speed was 2.4–5 m/s, and the final electrode gaps varied in range of 12–25 mm. The arc was observed with a high-speed movie camera (Hycam) running at a speed of 5–8 frames per millisecond. Exposure time of the individual frames was 50 μs. The authors noted that additional to study the traditional parameters as instantaneous current gap, electrode diameter, also dependence on the current I_{sep} at the instant of contact separation and I was investigated as well.

The results were summarized as follows. At current levels below 7 kA, a transient arc column forms (***bridge column***) were appeared. The diameter of the column increased continuously until the arc is diffuse. The time between electrode separations and the time of diffuse arc appearance increased with the current I_{sep} at the instant of electrode separation, but the time was always less than 1 ms. When the electrodes are separated at currents between 7 and 15 kA, a ***diffuse column*** was appeared. The column diameter increased up to a certain value, at which the column becomes stable with diameter about 10 mm. For relatively small gaps, anode spot was formed, and this leads to the appearance of another arc column, the ***jet column***. Typical dimensions for the column diameters are 12 mm at the anode, 10 mm at the narrowest point, and 20 mm at the cathode, but the diameters increased at currents above 36 kA. Further increase of the electrode gap finally results in separation of the ***anode jet*** from the cathode jet and hence in the familiar mode of the anode spot-dominated appearance.

According to the above described results, Heberlein and Gorman [42] presented an "appearance diagram," in which the arc appearance is correlated to values of gap and current (Fig. 14.12). The "arcing" line in the diagram shows one representative arcing sequence. After electrode separation at approximately 29 kA, a constricted column forms, which changes into a jet column when the gap reaches a value of approximately 4 mm. At a gap of 9.5 mm, the jet column breaks up into an anode jet and a cathode root, which subsequently disappears. The anode spot dies away at current values below 18 kA and a diffuse arc appeared. The boundaries between the

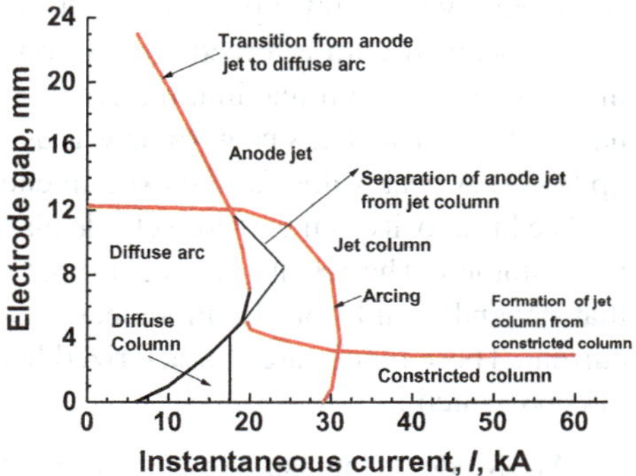

Fig. 14.12 . Physical arc appearance as a function of current and electrode gap, for one half cycle of arcing, 50–60 Hz, electrode diameter 100 mm, $I_{sep} > 7$ kA

regions of different appearances are the averages of a large number of points. It is mainly true for the boundary between the jet column and anode jet regions. In the triangular central area, all types of arc appearances were observed, depending on the previous appearance of one specific arcing sequence. The investigated [42] anode modes with drawn arcs, depending of the gap length, were discussed in detail by Miller [28].

Kaltenecker [43] studied the transition of vacuum arcs modes by an optoelectronic equipment and streak photography with time resolution in the range of microseconds. The arc was ignited between separating electrodes with fixed anode and grounded moved cathode. Butt-type electrodes with 30 and 40 mm diameter of copper, Cu/W, and stainless steel were used. During contact, opening a low-current arc was sustained by a 10-Hz resonant circuit. A 50-Hz high-current loop up to 20 kA peak was superimposed during or after contact opening. Fifteen photodiodes were used to measure the light emission along an arc diameter near the anode surface.

The total transition time from the diffuse arc mode to a fully established anode spot was determined as 760 μs. At fully open gaps of 10–15 mm, the transition times were shorter of 50–300 μs. The transition time (in range 50–800 μs) varies with the gap lengths when the threshold current is reached and the rate of current rise is determined by the 50 Hz. No influence of the contact material on the transition phenomenon was detected. The threshold current was about 10 kA for both copper and stainless steel electrodes of 40 mm diameter. It was of 1–2 kA higher for Cu/W electrodes and decreases with decreasing anode diameter. At the threshold current, locally instable anode spots were formed, which disappeared after 50–70 μs. After that, the anode spots were reappeared at some microseconds later in another site. Simultaneous formation of two anode spots can increase in diameter and finally merge. The spot diameter increases with increasing arc current. Spot diameters up to about 15 mm were observed corresponding to the diameter of various molten anode areas.

In further work, Kaltenecker and Schussek [44] also were used optoelectronic measurements, streak photographs, high-speed movies, and correlated arc voltage/current records to study the power frequency vacuum arcs. The average velocity of the electrode opening was between 0.2 and 1.2 m/s. Experiments were conducted with a butt-type contacts 20–40 mm in diameter at gap distances up to 15 mm as well with a forced arc motion using CuCr ring electrodes with 80 mm in diameter and with maximum electrode distance about 10 mm. The high-current half-wave up to a 20 kA peak value was used for butt-type electrodes and arc currents up to 40 kA peak value for the experiments on ring electrodes.

The luminosity of the interelectrode plasma, anode spot appearing, and arc voltage were studied. The results indicated three different high-current vacuum arc modes that depend mainly on the momentary electrode distance and a certain threshold current. These modes are characterized below in form as Kaltenecker and Schussek [44] presented it.

1. Arc Mode 1 (Intense Arc): Corresponds to the classical constricted arc mode. It appears at an electrode distance, which is not greater than about 4 mm. During the diffuse mode at threshold, a small luminous point at the anode begins to glow and forms a bright anode spot within typical 500 μs. The final anode spot diameter varies between 10 and 20 mm. The anode spot was steady, and it was eroded the anode surface heavily leaving a crater. The arc voltage increases from 20 to about 40 V and remains smooth even when the anode spot was formed.
2. Arc Mode 2 (Unsteady Arc): appeared at intermediate gap lengths of 4–8 mm, and it was characterized by one or more anode spots and a luminous arc. If this arc mode emerges from the diffuse arc mode, several flashing anode spots appear with a short lifetime of a few tens of microseconds. The transition time was 150 μs, and the anode spots tend to fast motion immediately after formation. The mean arc voltage jumps to values about 60 V, but high-frequency peaks up to 80 V were superimposed.
3. Arc Mode 3 (Footpoint Arc): Appears at gap lengths greater than 8 mm. Many anode spots, always moving with speeds up to 2500 m/s, characterize this mode. This arc type may develop directly from the diffuse mode or from the arc mode 2. The mean arc voltage is around 90 V, with high-frequency oscillations up to 150 V. Anode spots are established within 50–100 μs, i.e., much quicker than steady anode spots of the intense arc. The spots tend to run toward the electrode edges and to stop movement temporarily at molten sites deriving from the previous tests. The existence of footpoint was associated with high and noisy arc voltage indicating the relative unstable character of the anode activities.

At high-current vacuum arcs on ring electrodes, it was observed that the modes were the same as those observed on butt-type electrodes. The intense arc was forced by the generated magnetic blast field to move in Amperian direction, which avoided severe electrode melting. Different characters of high-current arc motion were detected. The first step was immediately after formation of a bright anode spot and arc column constriction, and the arc remains motionless spot. It was characterized by an almost constant and smooth arc voltage of about 25 V and by severe electrode erosion. The second step was a beginning slow direct arc motion at a gap length about 1 mm with mean voltage of about 40 V. The next was a fast arc motion step at a gap length of about 3 mm which was characterized by a fast increase in the arc voltage with significant high frequency of oscillations and an unsteady arc. The arc velocities were about 10–20 m/s in second step and 200–3000 m/s in accelerated step. Sometimes the arc speed possibly increases up to 6000 m/s or more.

Toya et al. [45] studied the vacuum arc that sustained between a 60-mm-diameter anode and a 30-mm-diameter cathode spaced 4 mm apart in the current range of 5–20 kA. The electrodes were made of oxygen-free, high-conductivity copper. The arc was photographed with a high-speed camera (HYCAM K2001), which was operated at a speed of 5000 frames/s, corresponding to a time resolution of 200 μs. An exposure time of 4 μs was obtained with the aid of a rotating shutter. Table 14.5 summarizes the results of observations of the anode mode with connection of the cathode spots state and with the observation of arc voltage waveforms.

Janiszewski and Zalucki [46] experimentally tested a high-current vacuum arc between butt-type copper electrodes at fixed gap of 10 mm. Current pulses of up to 30 kA peak amplitude at an initial value of rate of current rise $(\mathrm{d}I/\mathrm{d}t)_o$ that varied from 1 to 10 kA/ms were applied. Arc duration was approximately 14 ms.

The butt-type electrodes of the diameters 30, 55, and 80 mm were made of OFHC copper. Arcs were photographed with a high-speed framing camera, mostly at 10^4 frames/s. A detailed study of discharge modes in phase transition from a high-current diffuse arc to a constricted arc with an anode spot was conducted. Most of the measurements were obtained at a peak current slightly in excess of 10 kA for electrodes of 55 mm diameter. Figure 14.13 shows that the most interesting result

Table 14.5 Experimental results of different anode modes developed with arc current

Current, kA	Trace of cathode spot	Trace of anode surface	Arc voltage, V	Anode mode at current crest
5	Numerous single spots, uniformly spread	Neither melting nor erosion is observed	Less than 40 V. no high-frequency oscillation	Diffuse arc mode before the transition begins
10	Numerous single spots, uniformly spread	Slight melting with approx. 4-mm-diameter erosion trace	Approx 40 V with superimposed peaks up to 50 V, no sudden jump	Diffuse arc mode when the transition begins (footpoint mode)
12	Some spots bunched at the edge, while most of the spots spread in the region of one-third of the cathode surface	Medium melting with approx 10 mm diameter erosion trace no crater	Approx 40 V with superimposed peaks up to 60 V, sudden jump from 40–60 V near the current crest	Anode spot mode transition has occurred near the current crest
15 and 20	Many spots bunched at the edge, while most of the spots spread in the region of one- fourth of the cathode surface	Gross melting, severe erosion, approx 10-20-mm diameter crater trace. Molten-metal ejection	Sudden jump from 40 to 80 V at I = 7–8 kA, while decreasing to approx 40-50 V at the current crest.	Anode spot mode transition is completed before the current crest constricted arc column appears

Toya et al. [45]

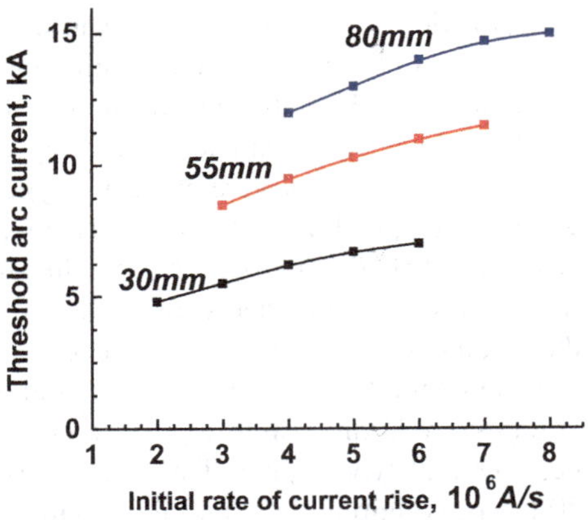

Fig. 14.13 Threshold current at the moment of initial thermal activity at the Cu anode as a function of the initial value of $(\mathrm{d}i/\mathrm{d}t)_o$. Electrode diameters of 30, 55 and 80 mm

obtained in this research [46] is the dependence of the anode transition threshold current on initial rate of current rise $(\mathrm{d}I/\mathrm{d}t)_o$.

It can be seen that higher values of threshold current are found for larger electrode diameters and for higher values of $(\mathrm{d}I/\mathrm{d}t)_o$. It was noted that lower threshold currents for smaller diameters were the result of greater concentration of spots per unit surface of the cathode as well as an earlier attainment of the contact edge by expanding cathode spots.

Zalucki [47] measured the vacuum arc voltage for sinusoidal semi-wave currents of the amplitude (peak) I_{am} up to 13 kA and frequencies 0.9 and 6.0 kHz for electrodes with small gap length. The electrodes were made from Cu60Cr40 and spaced 2 mm apart in plane butt-contact geometry. The anode was 30 mm in diameter and the cathode 55 mm.

The cylindrical vapor shield with an inner diameter of 100 mm was electrically insulated. The pressure in the vacuum chamber was 10^{-5} Pa. Measurements showed that in the case of alternating current, the main parameters, which affect the shape and values of arc voltage, are current, frequency, and amplitude I_{am}. At a given instantaneous value of current I, the arc voltage was higher during current increase (i.e., for $\mathrm{d}I/\mathrm{d}t > 0$) than during its decrease ($\mathrm{d}I/\mathrm{d}t < 0$). A qualitative correlation between the measured arc voltage and an expected known arc mode in the interelectrode gap was found. Figure 14.14 presents a maximal arc voltage u_{am}, at the maximal amplitude (peak) I_{am} of the current semi-wave and an arc voltage u_{aim} at time t_m as a function of current amplitude. t_m is the time from the onset of the current half cycle to the crest value of arc voltage. As the amplitude and current increase, both voltages u_{am} and u_{aim} change considerably.

Using the traces presented for $I_{am} \leq 9$ kA, the arc voltage was obtained, decreasing in the time range from t_f to half of a semi-wave $T/2$ in spite of the further increase of the current instantaneous value. At a certain current instantaneous value during its increase (about 10.5 kA), large voltage fluctuations start, henceforth called high amplitude fluctuations (HAF).

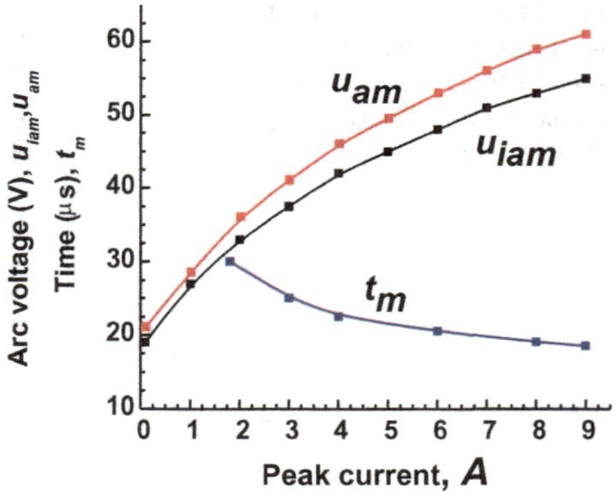

Fig. 14.14 Mean arc voltage at peak current (u_{aim}), a maximum arc voltage (u_{am}), and time (t_m) from the onset of the current half cycle to the crest value of arc voltage versus peak current I_{am}. Current frequency is of 6.0 kHz [47]

Table 14.6 The characteristic values of current I_{am}, arc voltage u_{af}, and time t_f at HAF onset I_{af}, and Δt_f at frequency f. From [47]

I_{am}	f	t_f	I_{af}	u_{af}	Δt_f
8	900	150	6.2	49	–
9	900	140	6.9	51	177
11	900	119	7.0	58	260
11	6000	35	10.6	78	20
13	900	111	8.1	60	383
13	6000	27	11.2	82	26

Table 14.6 presents the characteristic arc values, which describe the onset and duration of HAF. There, I_{af} is the current instantaneous value, u_{af} is the arc voltage just before the onset of HAF, t_f is the period from the high current initiation to the onset of HAF, and Δt_f is the oscillation duration. The data indicate that both the mean voltage and current value I_{af} at the moment of HAF onset rise with the increase of current amplitude (peak) I_{am} and frequency f.

The obtained data indicated that mean voltage u_a increases with increasing of $(\mathrm{d}I/\mathrm{d}t)_o$ and that u_a is a complex function $(\mathrm{d}I/\mathrm{d}t)_o$ (see Fig. 14.15). At $(\mathrm{d}I/\mathrm{d}t)_o \approx 50$ A/µs, there is a fast slope change of the function $u_a = f(\mathrm{d}I/\mathrm{d}t)_o$. Within the range from about 50 to 400 A/µs, the arc voltage increases approximately linearly with the $(\mathrm{d}I/\mathrm{d}t)_o$ increase. It was noted that the slope change of the function $u_a = f(\mathrm{d}I/\mathrm{d}t)_o$ in the vicinity of threshold value $(\mathrm{d}I/\mathrm{d}t)_o \approx 50$ A/µs can be due to the current conduction change.

Series of optical and electrical investigation of anode phenomena in high-current vacuum arc was provided by Batrakov et al. [48] and Khakpour et al. [49–51]. An arc was formed between two identical electrodes made from Cu or CuCr25 of 2 cm diameter in a vacuum circuit breaker [48]. The electrode separation was about 1 m/s. A sine half-wave current pulse of up to $I = 15$ kA in peak and 10 ms duration was used. The atomic and ionic distribution of electrode material during the formation of

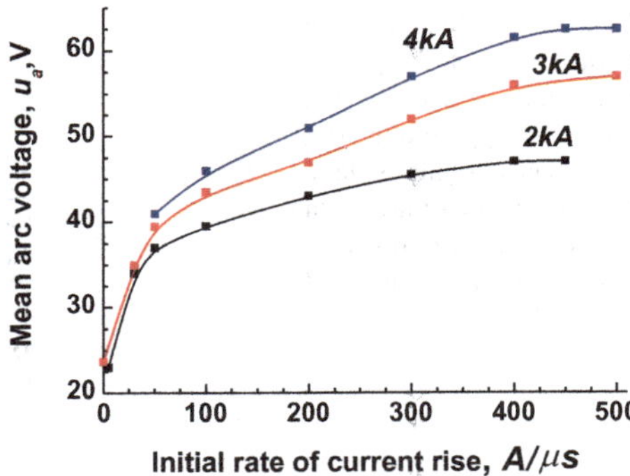

Fig. 14.15 Mean arc voltage u_a at instantaneous current I_{af} of 2.0, 3.0, and 4.0 kA as a function of $(dI/dt)_0$ during the current increase [47]

high-current modes or the transition between them was investigated. The results for Cu electrodes showed that when the arc started, the gap voltage was almost equal to the cathode voltage drop, and the arc was in the diffuse mode in spite of a relatively short gap. At a certain time, the discharge mode transforms into the anode spot, which accompanied by an increase in arc voltage and the onset of voltage noise (Fig. 14.16a) [52].

In case of CuCr25 electrodes, an anode plume at high current with a shell was observed (Fig. 14.16b) [52]. Batrakov et al. [48] assumed that the balance between pressures in both the anode and cathode jets controlled the plume shell. Plumes were observed in experiments with copper electrodes also. However, unlike the case of CuCr25 electrodes, their shape was not regular and also because it follows heat-insulated liquid metal protrusions or around flying droplets (Fig. 14.16b, right image). Noted that the plume appearance was sensitive to a spectral range under observation. It was indicated that anode plume was supported by electrode evaporation. The neutrals emitted from the light anode shell. The plume was covered by halo from which the light was emitted by ions. The evaporation rate of the CuCr25 was estimated comparable with the evaporation rate of hot liquid Cu protrusions and droplets.

Khakpour et al. [49] studied the anode phenomena in a vacuum arc between cylindrical CuCr7525 electrodes with a diameter of 10 mm. Experiments have been

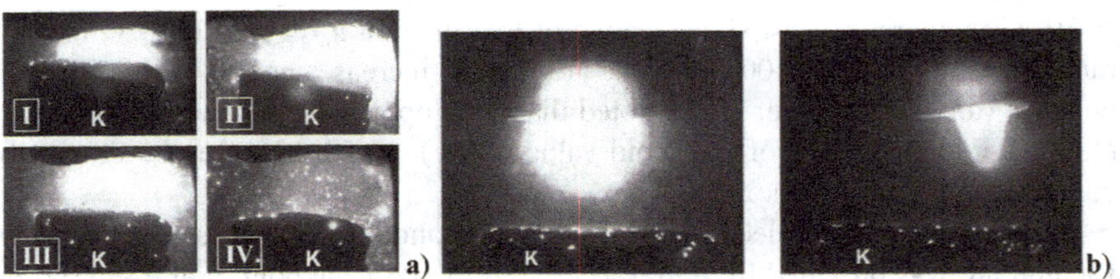

Fig. 14.16 **a** Set of images taken at arc in a gap with copper electrodes. The numbers I-IV correspond to different time in ms range; **b** image of arcing formed by copper–chromium electrodes at different shot and time gates. The beginning of the time gate is 950 μs before current zero [52]

performed applying AC 50 Hz waveforms and peak currents between 2 and 6 kA as well as 10-ms pulsed DC with peak currents between 1.5 and 3 kA.

Electrode separation speed has been varied between 1 and 2 m/s, and the final gap distance was 20 mm. Arc dynamics was controlled by two high-speed cameras with recording speed of 10^4 frames/s and an exposure time depending on the current amplitude to avoid saturation. Time- and space-resolved optical emission spectroscopy was used to examine the temporal and spatial distribution of different atomic and ionic copper lines. The transition from low-current mode to different high-current modes was examined observing the intensity of Cu I, Cu II, and Cu III line radiation near the anode, the cathode, and in the interelectrode gap for different discharge current waveforms. High-speed camera images were produced which indicate the formation of footpoint, anode spot, and intense modes for AC 50 Hz arc (the details see in Fig. 3 in [49]).

At the beginning of the discharge, a diffuse anode mode was observed. The anode spot mode continues up to about 5 ms. It was noticed that the detected bright area near the anode during the anode spot first increases and then decreases, with time following the current waveform. Then, the intense arc mode with high emission starts on both anode and cathode. In the intense mode, both anode and cathode are brighter than anode spot mode. The results show that during the formation of anode spot and intense mode, the distribution and the intensity of all lines change noticeably in the different spectral regions. The high ionization states indicated the arc dynamics behavior during transition to high-current anode modes. According to the author's conclusion [49], the axial intensity distribution of atomic and ionic copper lines was determined during different mode transitions:

(i) From diffuse mode to footpoint mode for the AC case. In both modes, the intensity of Cu I changes only slightly near the cathode but is lower in case of footpoint than diffuse mode in the gap. Near to the anode, the intensity is also lower in case of footpoint mode. The relative intensity of the Cu II line behaves very similar to the Cu I line. The relative intensity of the Cu III line shows different behaviors than Cu I and Cu II with a much broader spatial profile along the discharge axis that changes with the discharge mode.
(ii) Before and after transition from footpoint to anode spot mode. The intensity of the Cu I line is almost similar in both footpoint and anode spot mode with a pronounced increase toward the cathode, but the maximum intensity near the anode is higher in the footpoint mode. The intensities of the Cu II line are very low near the anode and show a broad profile near the cathode. In contrast to that, there is an abrupt change in the distribution of the Cu III line during transition from footpoint to anode spot near the anode, cathode, and interelectrode gap. The intensity of Cu III shows local maxima near the anode and the cathode during anode spot but one maximum in the gap in the footpoint mode.
(iii) From anode spot to intense mode in the AC 50 Hz half-wave. This transition typically occurs at larger gap distances and accompanied with larger dark areas (i.e., low intensities of Cu I, Cu II, and Cu III lines) in the interelectrode gap. However, all line intensities increase considerably near the anode and

more moderate near the cathode during the transition to the intense mode. The intensity of atoms and ions shows intensity maxima near the anode, which are considerably higher than the intensities near the cathode in the intense mode. The transition from anode spot to intense mode occurs very fast, which is indicated as jump in voltage.

Transition between different high-current modes was investigated in [53] by means of video spectroscopy in case of AC 50 Hz and DC pulse. The current, voltage, and electrode distance as dependences on time and time transition to high-currentanode modes of the vacuum arc for 10 ms are presented in Fig. 14.17.

It can be seen that during footpoint and anode modes, the arc voltage increases. The formation of intense mode is accompanied by an abrupt increase by approximately 20 V in the voltage. High-speed camera images (Fig. 14.18) show the formation of footpoint, anode spot, and intense modes. At the arc beginning, a diffuse mode is observed at the anode (Fig. 14.18a). A dark region appears in the interelectrode gap and near the anode. After about 7.1 ms, the anode spot appears in form of a high brightness area (Fig. 14.18c). The anode spot mode continues up to ~10.5 ms. Then, the intense arc mode starts as shown in Fig. 14.18d.

Khakpour [54] and Khakpour et al. [50] investigated the high-current anode modes with different waveforms including the alternative current (AC) pulses of 50, 180, and 260 Hz and the direct current (DC) pulses of 5 and 10 ms. Three contact materials (Cu, CuCr50, and CuCr7525) with electrode diameters of 25, 20, and 10 mm were applied, respectively. Two opening velocities of 1 and 2 m/s are also examined. To study the high-current modes, both electrical signals and high-speed camera images are evaluated.

The transition to different anode modes is illustrated in Fig. 14.19 [54], for which the conditions are not mentioned. It can be seen influence of the gap distances. However, it was reported that the electrical parameters are important in the formation of anode modes. The results for different AC frequencies show that the first is the anode spot mode, which starts around maximum current of 2.7 kA for AC 50 Hz,

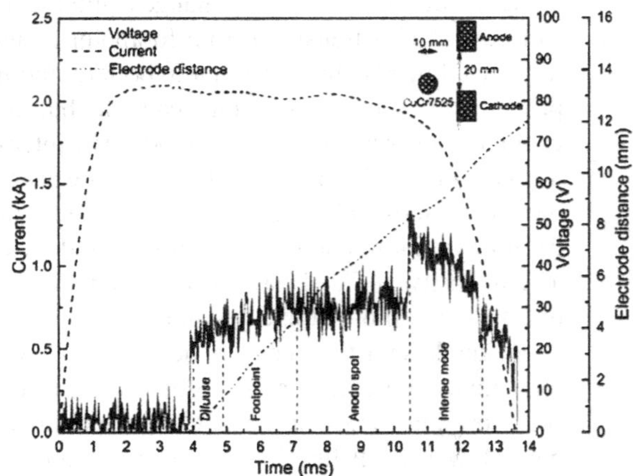

Fig. 14.17 Time dependent current, voltage and electrode distance as well the time transitions of the different arc modes for dc 10 ms arc [53]

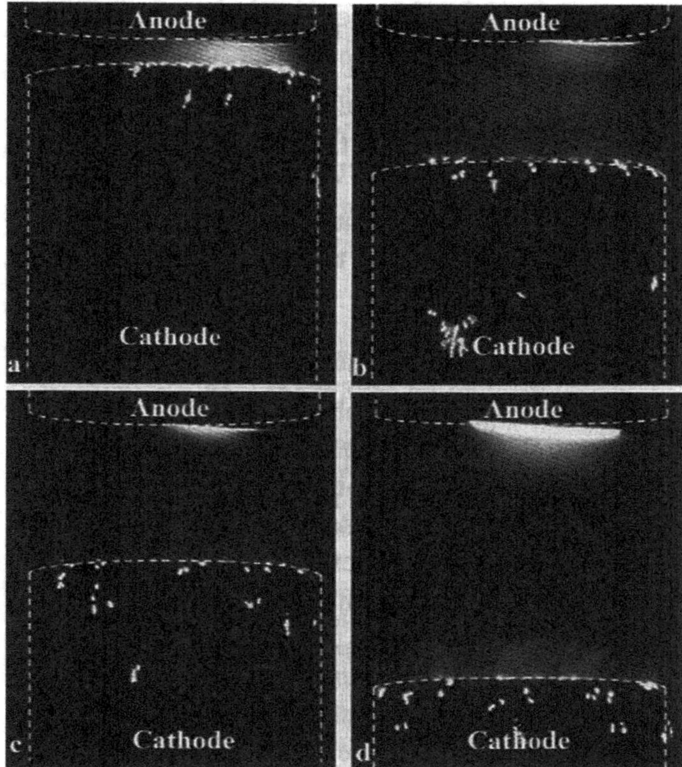

Fig. 14.18 Appearance of diffuse mode (**a**), footpoint (**b**), anode spot (**c**) and intense mode (**d**) at corresponding time and gap distances indicated in Fig. 14.17 [53]

while in the case of 180 Hz, the threshold current) is about 2.85 kA [49]. When the frequency increases to 260 Hz, the threshold current) of the first high-current mode is increased to 3.7 kA, and it appears as intense mode. In the case of AC 50 Hz operation, the anode spot mode appears at a contact distance of 6.7 mm. However, in the cases of AC 180 and 260 Hz, it is started at 1.8 and 1 mm, respectively. In the intense arc mode, very bright luminosity is observed that covers both anode and cathode as well the interelectrode gap. The arc voltage in the intense arc mode is always low and with low fluctuation, though higher than in the low-current diffuse arc mode.

Two different types of high-current anode spot modes have been identified in the anode spot mode regarding the arc voltage behavior [50]. It was observed that the first type of anode spot appears at a current of 3.5 kA x, and the arc voltage is increased by 10–15%. By increasing the current and gap length, a second type of high-current anode mode is recorded at a current of 4.3 kA and a gap length of 5.8 mm. An abrupt change of the voltage by about 40–50% was detected in the case of the second type. The radiation intensity in the case of the second type was quite

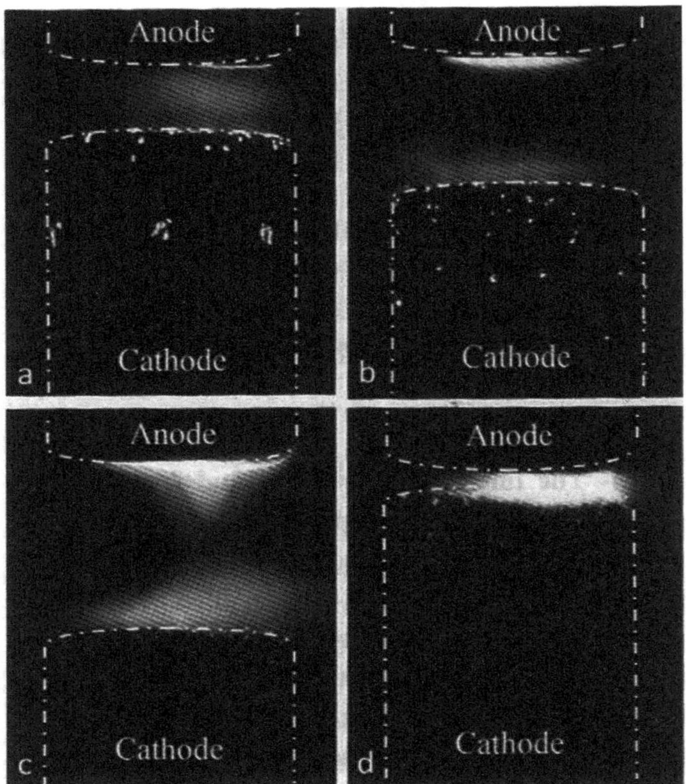

Fig. 14.19 Typical high-speed camera images of high-current anode modes: **a**—footpoint, **b** and **c**—Anode spot, **d**—intense mode [54]

higher. During the formation of the second mode, strong anode jets are formed, which probably interact with the cathode jets. Spectroscopic analysis showed the increase of Cu atomic lines near the anode during the second mode [49]. The high radiation intensity in the vicinity of both the electrodes observed in the second type of anode spot mode corresponds to the characteristics of the intense mode.

Khakpour et al. [51] investigated the differences between anode spot type 1 and type 2 and the discharge conditions for a transition from anode spot type 1 to type 2 by means of optical and electrical measurements. Experiments were performed applying pulsed DC waveforms over 10 ms with peak currents between 1.2 and 3 kA. Two cases (illustrated by images in Figs. 2 and 6 from [50]) were examined: (i) case 1—only diffuse, footpoint, and anode spot type 1 appear; (ii) case 2—anode spot type 2 occurs after the formation of anode spottype 1.

For first case, the maximum current is about 2.7 kA. The average arc voltage during high-current modes is about 35 V. *No abrupt change in arc voltage* (spot type 1) was observed during high-current modes.

In second case, anode spot type 2 appeared at about 10 ms and gap length of 8.5 mm with considerable larger and brighter illumination areas at both anode and cathode compared to anode spottype 1. The maximum current was 2.7 kA. With the transition to anode spot type 2, the *arc voltage changes abruptly* (spot type 2) by about 20 V at a time instant of 10 ms and a gap length of 8.25 mm (e). It was observed that by transition to the anode spot type 2, the distribution patterns of Cu I, Cu II, and Cu III lines are similar to anode spot type 1; however, the relative intensities of all species increased near the anode at transition to anode spot type 2. The relative intensity close to the cathode was almost similar to that of anode spot type 1. Near to the cathode, the intensity of atomic lines is almost the same during different arc modes.

Data presented by Khakpour et al. [51] illustrated the differences between anode spot type 1 and type 2 with respect to current and voltage waveforms, transferred charges, aspect ratios, temporal and spatial distributions of Cu I, Cu II, and Cu III line intensities. The main different characteristics of the anode spot types 1 and 2 were summarized as follows.

(1) An abrupt change in the spatial distribution of Cu III lines is observed at transition to anode spot types 1 and 2.
(2) A jump in the arc voltage appears at the transition from anode spot type 1 to type 2.
(3) The abrupt changes of optical emission and arc voltage hint at a fast transition from one mode to another.
(4) The transition from anode spot type 1 to type 2 is dominated by the ratio of electrode diameter to gap length. However, increasing the current can decrease the value necessary for the formation of anode spot type 2.
(5) The formation of an anode plume always observed after extinction of anode spot type 2 and some hundred microseconds before current zero crossing.

Popov et al. [55] further studied the transition from diffuse arc to anode spot modes in high-current vacuum arc, and recently, Batrakov [56] provided the further analysis of conditions for appearance of both anode spot types 1 and 2, which images shown in Fig. 14.20. It was concluded that all experiments up to now were performed for CuCr25. So, the further investigations appearance of the modes should conducted for other contact material and for cases, when the cathode and the anode are made of different materials. Also, it is important to consider experiments in the presence of a magnetic field. Recently Miller [57] reviewed the works presented results of investigations of the anode phenomena considering the publications up to 2017.

14.4 Measurements of Anode and Plasma Parameters

The anode and anode plasma are characterized by anode surface temperature, by plasma electron density and electron temperature. The corresponding experimental data are analyzed depending on the arc burning conditions.

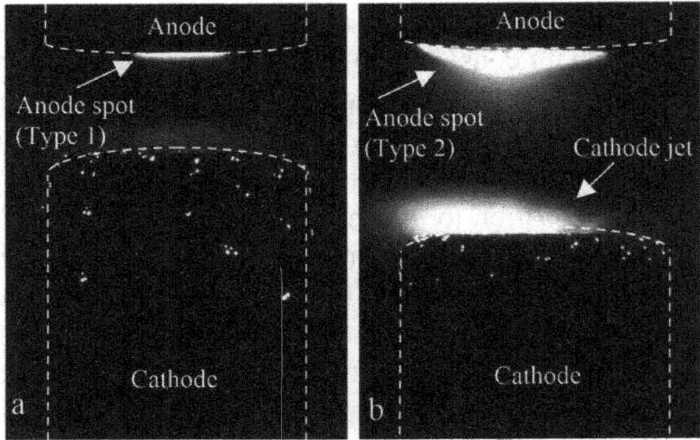

Fig. 14.20 Appearance of the Anode spot type 1 (**a**) and Anode spot type 2 (**b**) in high-current vacuum arc with CuCr25 butt-type electrodes of 10 mm diameter [56]

14.4.1 Anode Temperature Measurements

Mitchel [33] has measured the equilibrium anode spot temperature of Cu electrodes by measuring material evaporation rates and obtained values ranging from 2970 to 3080 °C for 10 μs pulse widths and peak current densities of approximately 2×10^4 A/cm^2,

Grissom and Newton 1972–74 [58, 59] measure the anode surface radiance as a function of time for microsecond vacuum arcs using infrared radiometric microscope. Anode surface temperatures were derived from the measured radiance values using the surface spectral emittance. The anode spot temperatures as a function of time were determined for AI, Cu, and Kovar anodes as well as the temperatures due to plasma heating of Al_2O_3 (alumina) and BeO (beryllia) surrounding the active anode area. The arc was operated with rectangular-shaped current pulses of 1–12 μs duration at current levels of 20–180 A. It was indicated that the anode spots tended to repeat in the same location on several consecutive operations of the arc.

The anode spot temperature s were obtained taking the normal spectral emittance values between 0.2 and 0.3, in accordance with the published data. The time-dependent anode spot temperatures are given in Fig. 14.21. The temperature rise to a stationary condition for Al and also for Cu anodes is approximately 2 μs. The anode spot temperatures at the end of the arc pulse are between 2000 and 2450 °C and the latter value corresponding to the boiling point of Al. For Cu, the temperature at the end of the pulse is between 2750 and 3450 °C. The boiling point of Cu is approximately 2600 °C. The measurements showed that the maximum surface temperatures of the ceramic insulators were between 2900 and 3500 °C for AI_2O_3 and between 2000 and 2300 °C for BeO. An estimation from anode conductivities using the measured temperatures gives the anode spot current density for Al and Cu which fall between 10^5 and 10^6 A/cm^2.

Si photovoltaic detectors at wavelengths of 0.6, 0.9, and 1.0 μm selected by interference filters were used by Boxman [60] to direct time-resolved (of about 10 μs) measurements of anode temperature. The measured voltages were converted to anode temperatures by using two limited values of emissivity. For the minimum

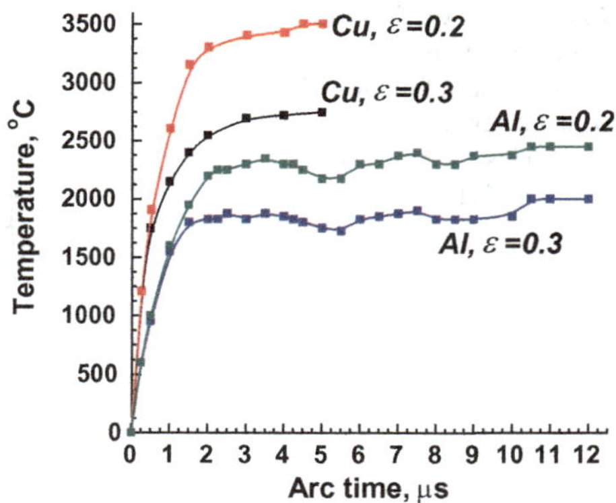

Fig. 14.21 Anode spot temperature as a function of time after arc initiation for Cu and Al anodes with emittances of 0.2 and 0.3 [59]

emissivity, the maximum and for unity (black body), the minimum temperatures were obtained. The overlap between the three measurements gave the overall range of the equilibrium temperature of the anode spot. Transverse measurements of the gap plasma alone showed that it contributed less than 10% of the total signal. For nickel electrode 2.5 cm diameter, 0.9 cm gap, 7.2 ms sinusoidal current pulse, the anode spot formation occurred at a current of 2.7 kA, about 1.65 ms into the arc. The temperature of the anode before transition to the spot mode was less than 1370–1500 K, that is, below the melting point of nickel. The equilibrium anode spot temperature was 2790–2960 K, which is slightly less than the atmospheric boiling point of nickel. Also, a high-speed movie of the arc was observed. It was noted that the arc had a diffuse appearance when the arc voltage was quiet (before the anode spot formed) and had a constricted appearance near the anode when the voltage was noisy. Measuring the total power input to the arc apparatus at the initial time, it was showed that this power was insufficient to produce the observed initial rate of rise of anode temperature (10^7 K/s) even if all the power were distributed uniformly over the entire anode (area 3.9 cm^2). Therefore, it was concluded that the arc had to constrict to area at least of 2.4 cm^2 or lower (near 1 cm^2) before the anode spot developed.

In another work Boxman [41] measured the temperature near the center of the Ni anode surface also by the filtered Si photovoltaic detector apparatus previously described [60], but at higher frequency of the current pulse. The temperature at the center of the Ni anode surface before transition to the spot mode is plotted as function of peak 543 Hz current pulse in Fig. 14.22. The results show a generally increasing anode surface temperature prior to transition with increasing peak current, with values ranging from around 1550 to 2250 K. After transition of the spot mode the anode surface temperature rose very rapidly, as previously reported for lower frequencies [60].

It should be noted that however, at low frequencies, the anode spot would form consistently in the center of the anode. At high frequencies, the location of the anode spot was more random, though generally, it would be located in the central portions of the anode surface, rather than on the edges. Thus, much of the scatter in the

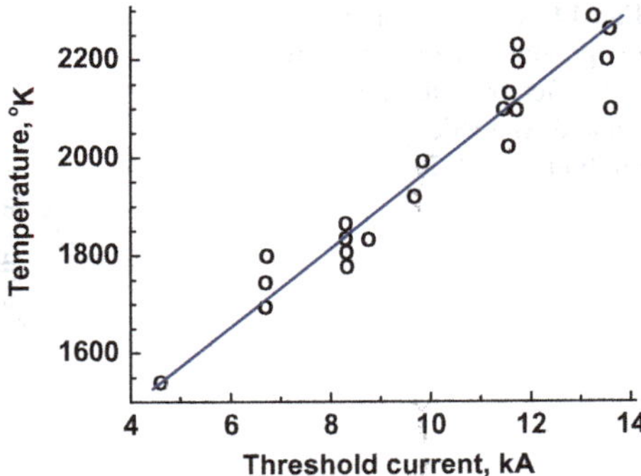

Fig. 14.22 Anode surface temperature just prior to transition as a function of peak current for arc on Ni electrodes separated by 8.9 mm [41]

temperature data is related to the location of the subsequent anode spot, which may form on the fringes or outside of the field of view of the measurement system.

Dullni et al. [61] determined the surface temperature of the anode area melted during the anode spot mode for vacuum arcs by pyrometer technique and also using measurements of thermionic currents. The electrodes used were of 25 mm diameter which is made of CuCr 75/25 and spaced by 8–13 mm apart. The used geometry allowed the stabilization of the anode spot in the center of the gap. Sinusoidal current pulses of up to a 20 kA peak and of 45 Hz half-wave were applied. To use the thermionic current method, the streak photography of the anode spot was obtained. It was indicated that it is restricted and has a diameter of 10 mm for a sinusoidal current pulse of 7.5 kA peak. The observed melted region was almost constant during the measured period.

To measure the temperature in the anode spot mode, the current was switched off within 40 μs and after 7.7 ms arcing time, when the anode spot was present. The temperature decay after a 7 kA current peak measured by pyrometer technique and using thermionic currents [61] is shown in Fig. 14.23. The pyrometrical data show the temperatures more than 2500 K observed at beginning of the measurements, and it decreases then below 1500 K within 3 ms. Noted that according to the thermionic

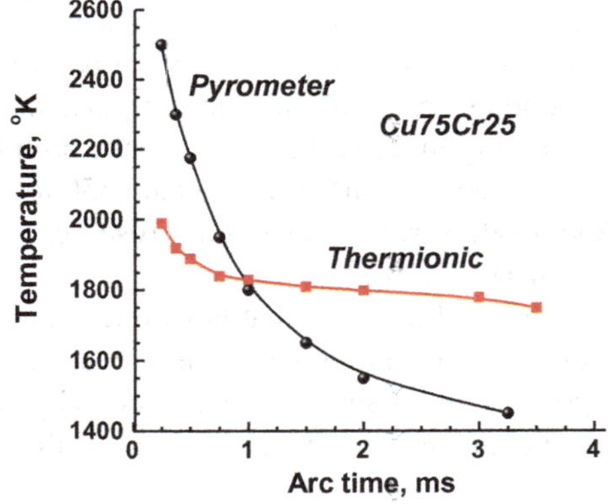

Fig. 14.23 Time-dependent temperature decay in melted area on the anode after a fully developed Anode spot. The dependence is obtained by an approximation of data measured by pyrometer and thermionic current [61]

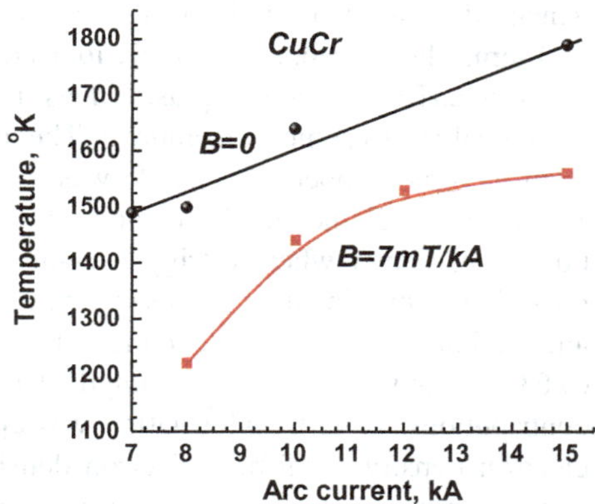

Fig. 14.24 Maximum anode surface temperature at the current zero as function on the arc current amplitude

current measurements, the temperature at the beginning of the measurement was 100 K below those obtained by pyrometer for the forced extinction in the anode spot mode. The difference was due to more sensitive to the maximum surface temperature using the method of thermionic current measurements. Later, the measured thermionic current gives larger temperatures than that by pyrometer.

The anode temperature immediately after the current zero was determined using the optical pyrometer method by Schneider et al. [62]. This synthetic circuit generated 10 ms arcs with a peak current of 8–15 kA. Two identical copper–chromium electrodes of diameter 2 cm formed the contact gap. The electrode separation speed was about 1 m/s. The anode surface temperature depending on the arc current amplitude both with and without AMF [62] is shown in Fig. 14.24.

The temperature obtained in pyrometric measurements by Dullni [61] was higher than the temperature obtained by Schneider et al. [62]. However, the temperature computed from thermionic currents is very close to temperature measured here. The maximum anode surface temperature, in Schneider et al. [62] case, can reach the values measured by Dullni [61]. It was noted that the difference in the temperature obtained could be due to the difference in thermal history.

14.4.2 Anode Plasma Density and Temperature

Kudryavtsev [63] studied the anode plasma jet temperature in a high-power pulsed short vacuum discharge with 2-10-mm gaps and of 4 μs current durations. The arc was ignited between steel disk cathode and wire anodes (Mo, Fe, Ni, Ti, and Al) of 1.5 mm diameter. The current was varied from 0.7 to 3.3 kA. The plasma temperature was obtained from the relative intensities of spectral lines. The radial temperature distribution was measured at distance of 1.5 mm from the anode in arc gap of 5 mm. The temperature was about 19,000 K at the arc axis and parabolically decreased to zero at radius of 4 mm. Using the measured temperature, the calculated plasma density decreased similarly from about 8×10^{15} cm^{-3} at the axis to zero at radius of 4 mm. The Al vapor was highly ionized and was mainly composed of Al^{++} ions.

Boxman [64] studied the plasma densities in a 0.4–4.0 kA high-current vacuum arc by optical interferometry technique. The arc was sustained between butt electrodes from gas-free copper. The anode was a flat disk with rounded edges, either 1 or 1.5 inches in diameter. The cathode was a 1-inches-diameter disk with diameter hole in its center where a trigger electrode was placed. The arc was initiated by breakdown applying a 4 kV pulse to the trigger electrode. Gap lengths between 0.27 and 0.35 inches were used. The measurements were made with a visible wavelength (0.63 μm) and an infrared wavelength (10.6 μm) to separate the electron and neutral contributions. Abel transformation was applied to obtain the radial profile of the electron density. A diffuse electron density profile was observed, with the axial densities ranging from 1×10^{14} at 0.4 kA to 2×10^{15} cm^{-3} at 4 kA. At the transition from the diffuse mode to the high-current constricted mode, which occurred at about 3 kA, a sudden decrease in electron density from 1.7×10^{15} cm^{-3} to 0.7×10^{15} cm^{-3} was observed in the center of the discharge. Figure 14.25 shows the change in the electron density distribution that occurred as a result of transition to the high-current mode [64]. Before the transition, a diffuse density distribution with maximum value occurred near the center.

Immediately after transition, the electron density in the center of the distribution drops by more than a factor of 2. However, as the current increased from 3 kA at transition to 4 kA at current maximum, the electron density increases approximately to the electron density distribution that was before the transition. An upper limit for the neutral vapor density was detected to be 2.8 times the electron density.

Boxman [53] explained the decrease in electron density at high-current onset due to an increase of anode surface heating by the arc. As result begins, a cold vapor was emitted in comparison with the plasma temperature. The interaction of the cold vapor with the hot plasma would cool the plasma and might result in a decrease in the average degree of ionization of the plasma and hence of the electron density. It

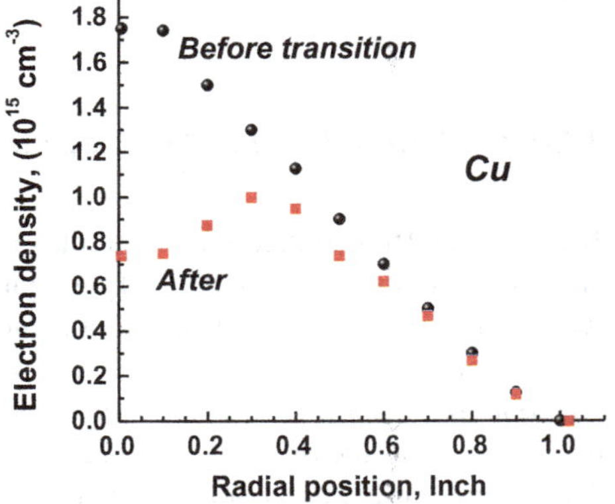

Fig. 14.25 Electron density profile before and after transition from diffuse to high-current arc in a 4 kA peak current discharge between 1-inch-diameter butt electrodes spaced 0.35 inches apart. Transition occurred at approximately 3 kA

was noted that this explanation consisted with the observed jump in arc voltage, as a decrease in plasma temperature that causes a decrease in plasma conductivity.

A spectroscopic investigation of the anode plasma of a pulsed vacuum was performed by Bacon [65], and a discussion of the measured data by collisional–radiative model was given by Bacon and Watts [66]. The Al anode was 0.89 mm o.d. and was press fitted into a 10-mm-diameter Al_2O_3 or BeO ceramic sleeve. The 0.38-mm o.d. Mo cathode was brazed into the center of an Al_2O_3 insulator, which was brazed into a 4.06-mm o.d. Kovar sleeve. The cathode assembly was grounded. The arc currents with approximately rectangular pulse of 10 μs duration and of about 50 A or larger were confined to less than a 1-mm-diameter region. The electrodes were spaced 3 mm. Radial distributions of AlI, AlII, and AlIII excited-state densities in the plasma produced by anode spots were tested. It was shown that Peak AlIII 4D excited-state densities were determined to be about 5×10^{17} m^{-3} within 0.05 mm of the anode. This peak occurred as the arc was extinguished. The maximum AlIII 4D excited-state density during the steady-state portion of the arc discharge was approximately 3×10^{17} m^{-3} within 0.05 mm of the anode. From ratios of AlIII excited-state densities, the Boltzmann distribution temperature was determined to be 2.0 ± 0.5 eV. The excited-state densities of AlIII have cylindrically symmetrical distribution relative arc axis dependence. An one-dimensional variation was observed within about 0.3 mm of the anode surface.

Harris [67] used laser interferometry to determine electron density distribution vacuum arcs before and after the transition from the low-voltage mode to the high-voltage mode associated with anode spot formation. The arc was sustained between 25-mm-diameter copper butt electrodes. The current wave frequencies of 66 and 347 Hz with a peak of 2 and 2.8 kA were used. The electrodes were spaced at 7.6 and 11 mm.

It was found that 347-Hz current waves yielded much the same electron density data as 66-Hz current waves. The transition to the anode spot mode occurred at threshold current of 2.6 kA when the arc peak was 2.8 kA and electrode spacing 11 mm. Before transition, the electron density profiles are peaked on axis, with maxima of 9×10^{14} and 6×10^{14} cm^3 in the cathode and anode regions, respectively. At about 0.1 ms after transition, the electron density profile in the cathode region remains unchanged, while the central electron density near the anode drops by a factor of about 2. The plasma density decreased to about (2–4) $\times$ 10^{14} cm^3 The decrease in electron density at threshold current occurs in a 3-mm-thick region in front of the anode. As the current was increased to the threshold for transition, the electron density profile near the cathode remains its central peak, while the density profile near the anode flattens.

An interesting work was presented by Seliverstova and Zygankov [68] which investigated the influence of the anode thermophysical properties and anode temperature on the anode spot dynamics. A current pulse with an amplitude of 70 A with rate of current rise of 10^6, 5×10^6, and 10^8 A/s was used. The anode plasma was photographed with SFR-1 camera of 1 μs resolution. Anode materials Al, Sn, Zn, Fe, and Pt presented different anode properties. Typical results were obtained for 10^6 A/s and pulse duration of 800 μs.

Table 14.7 Dependence of initial anode spot appearance on the anode vapor pressure

	Sn	As	Pt	Fe	Zn
Saturated pressure at melting temperature, torr	4.4×10^{-23}	2×10^{-5}	6.5×10^{-3}	5.3×10^{-3}	1.5×10^{-1}
Time of the first spot appearance, μs	0	4–5	7–9	12–20	25–40

Data from [68]

It was indicated that initially, one or few spots were observed which accompanied by the anode melting. A plasma constriction occurred at earlier at anode material with lower saturated vapor pressure like tin (Table 14.7). Initially, only one spot was appeared at Al anode, and then, the spot diameter increased with arc current. At Fe anode, the splitting of the spot occurred after 450 μs. At Sn anode, the first spot appeared with radius of 0.1 mm, and the spot splitting was observed after 200 μs. The time-dependent Al surface area covered by anode spots for cooled and heated anode surface is presented in Fig. 14.26. The significant anode cooling (−196 °C) was reached by insert the anode in a liquid nitrogen. The anode heating was performed up to melting temperature.

It can be seen that the anode spot area for both anode cooled and heated is different from such area appeared in the arc with the anode at room temperature. The spots are constricted when the anode is cold, while it is significantly increased for hot anode in comparison with room temperature anode. The dependencies are not one-valued functions and were possible due to effect of spot splitting. It was observed that for larger anode temperature, the splitting of the spot appeared earlier. The anode spot splitting was occurred better near the current maximum, at which a larger heat flux to the surface was caused. The authors assumed that intense local anode evaporation and further reducing of the plasma electro-conductivity could explain this phenomenon. It was noted that this explanation agreed with the observed intense spot splitting when the rate of energy rise at the anode surface was produced due to larger rate of current rise, of 10^8 A/s. In this case, an intensive vaporization produces local arc flames appeared in the first microseconds of the arc operation. The cold flames prevent the

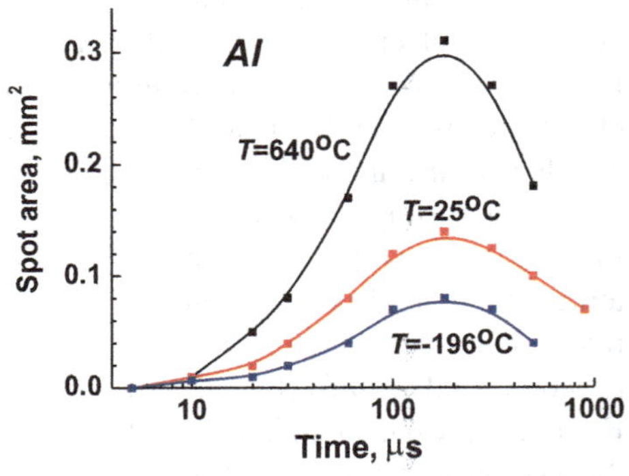

Fig. 14.26 Variation of Al anode area occupied by the Anode spot with time for different anode surface temperatures T. Arc current pulse 800 μs [68]

current continuity, and as result, the anode region is splitted for a number of spots on all tested metals. The present explanation is in accordance with the interpretation reported by Boxman [64].

Lyubimov et al. [69] studied the anode area dynamics, anode erosion, and anode plasma parameters of an arc carried out in air at atmospheric pressure. The cathode had a pointed shape, and the anode was a cylinder which is 10 mm in diameter and 20 mm in height. The separation between electrodes was 2, 5 mm. Both electrodes were made of 4DA-grade aluminum and the anode surface being polished. The duration of the discharge pulse was of 0.8 ms. The current was increased up to 100 A during the first 0.1 ms and then decreased to 30 A at the end of the pulse.

An SFR-1 camera with a supplementary optical device with time resolution of about 1 μs and spatial resolution of about 10^{-3} cm was used. The optical axis of the camera was perpendicular to the anode plane. The erosion region on the electrode surface was investigated with the help of optical and electron microscopes. An STE-1 spectrograph with a single-lens optical system was utilized to determine the electron density n_e and temperature T_e in the anode plasma. In this case, the anode surface was arranged at an angle of 45° to the optical axis of the spectrograph, and the interelectrode distance was reduced to 0.15 cm. The electron density was determined studying the broadening of spectral lines by secondary Stark effect. The electron temperature was determined in terms of relative intensity of spectral lines of iron impurity excited atoms with the impurity content of 0.052%.

Experimental results showed that a distributed discharge was realized at the anode during the initial period of 4–5 μs after which a single spot of circular form and practically stationary arises. For used time-dependent arc current (see above), the spot radius increased with the pulse time and then passed through a maximum as shown in Fig. 14.27. The current density also increased up to 2×10^5 A/cm^2 in the first 100 μs and then decreased to about 10^4 A/cm^2 in the end of the arc pulse.

It was also found that microspots appear within the anode spot area at nearly maximum current, and their location and glow intensity varied with time. The size of such spots was several times smaller than that of the main spot. The erosion at the anode surface was a regular circular hole surrounded by a small circular raised

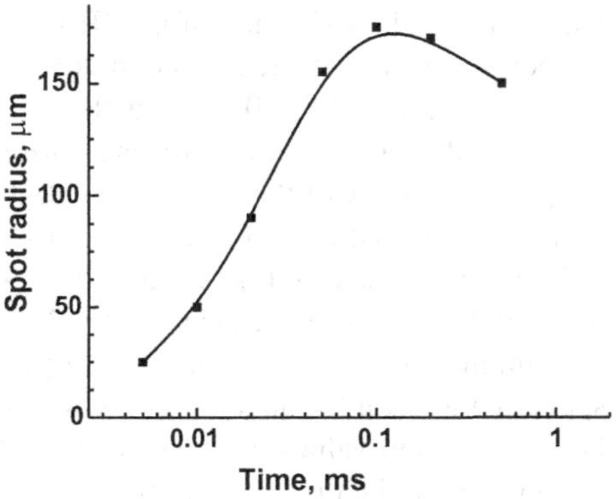

Fig. 14.27 Variation of the radius of the Anode spot with time[69]

rim. The maximum depth of the damaged layer (including rim) was $(4.7 \pm 0.4) \times 10^{-3}$ cm, and the hole depth (without rim) is $(2.1 \pm 0.7) \times 10^{-3}$ cm. The erosion region area (including rim) was $(0.23 \pm 0.03) \times 10^{-2}$ cm^2. Traces of action of the microspots were observed on the side surface and bottom of the hole. The erosion averaged over the discharge period was estimated on the basis of the measured hole sizes to give 0.6 μg without and 4.7 μg with rims taken into account.

The average electron temperature $\langle T_e \rangle$ for chosen three line pairs in the experiment was reported as the following data: (i) $Te = 0.073$ eV, (ii) $Te = 0.46$ eV, (iii) $Te = 0.67$ eV.

Several lines determined the density n_e from the relation:

$$\Delta = 2 \times 10^{-17} \gamma n_e$$

where **Δ** is the line broadening and ***y*** is the broadening constant. The $\langle n_e \rangle$ value calculated from this relation with $n_e = n_e(t)$ is the time average of the electron concentration because of the linearity of the above equation. The order of magnitude of $\langle n_e \rangle$ averaged over all the lines is obtained to be 10^{17} cm^{-3}.

Rosenthal et al. [32] conducted a spectroscopic study of the temporal evolution of a refractory anode vacuum arc. The arc begins as a multi-cathode spot mode, and the anode was heated by the current with time reaching a Hot Anode Vacuum Arc (HAVA). The arc duration was of 150–200 s at 175 or 340 A DC current. The 30-mm-diameter copper cylinder cathode was water-cooled and had a cylindrical protrusion of 10 mm diameter and 5 mm height at its center, to ensure that the arc was initiated near the axis of the electrodes. The anode was of 32 mm diameter and 30-mm-long graphite cylinder. Two Mo radiation shields surrounded the anode to reduce radiation heat losses. Spectral lines of Cu I, Cu II, and Cu III were investigated as a function of time near the cathode, near the anode, and in the middle of the electrode gap.

The spectroscopic study showed the development of the plasma plume and the changes in the radiation from different regions in the arc. It was observed that in the cathode spot mode, the cold anode acts as a collector for ions emitted from the cathode. A transition of the arc plasma from the cathode spot (cold anode) mode to the HAVA mode was detected. The transition was characterized by the formation of a radiating plume near the anode that expands toward the cathode. The spectroscopic data indicate the existence of Cu III near the anode, whereas no spectral lines of copper ions were detected in the middle of the interelectrode gap.

Spectral lines of Cu II and Cu III were observed near the anode only at $t \geq 35$ and 15 s after arc initiation, for arc current values of 175 and 340 A, respectively [32]. The intensity of three Cu III spectral lines at the anode region as a function of time, each normalized with respect to its peak intensity, is shown in Fig. 14.28 at $t = 80$ for 175 A and in Fig. 14.29 at $t = 20$ s for 340 A arcs. It can be seen that the measured intensity increased from $t = 30$ and 10 s (transition beginning) to nearly its maximum value at $t = 60$ and 20 s, respectively. The absolute intensity of the ionic spectral lines near the anode was always smaller than that of neutral species lines, for both arc current values. During steady-state HAVA mode operation, neutral copper is, most probably, the dominating atomic species in the interelectrode plasma.

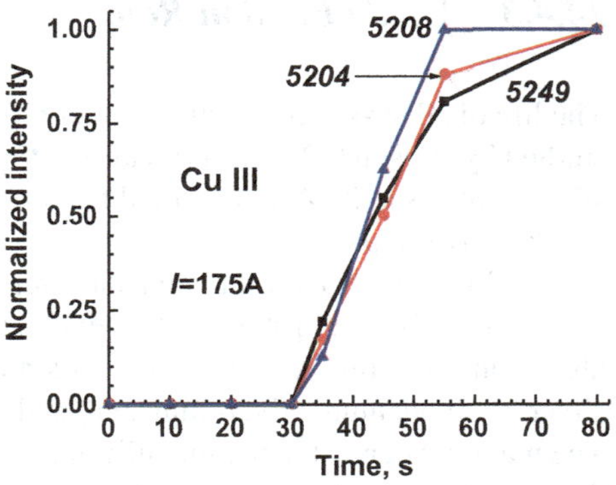

Fig. 14.28 The intensities of Cu III 5204, 5208 and 5249°A spectral lines near the anode as a function of time, each normalized with respect to its peak intensity at 80 s, in a 175 A arc [32]

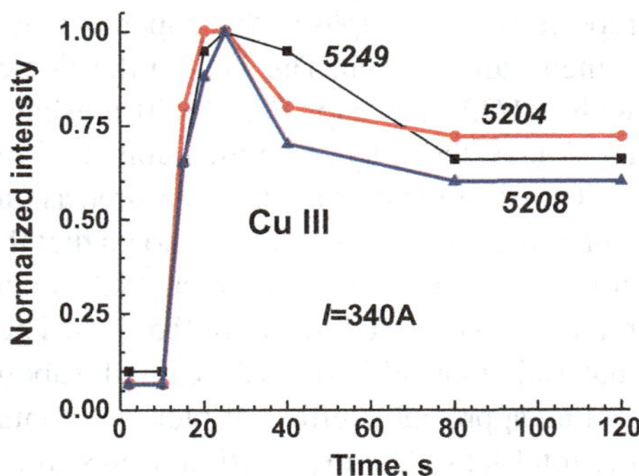

Fig. 14.29 Intensities of Cu III 5204, 5208, and 5249 °A spectral lines near the anode as a function of time, each normalized with respect to its peak intensity, in a 340 A arc [32]

Assuming that the collisions with electrons are the main mechanism for atomic excitation Cu I, the electron temperature of the plasma was determined from Planck spectral line intensity flux. It was also considered that the dominant mechanism of line broadening is Doppler broadening. The values of the electron temperature were determined in range 1.2–1.3 eV.

Recently, Khakpour et al. [70] and Khakpour et al. [71] studied the electron and heavy particle densities in a high-current vacuum arc considering the transition of anode modes including footpoint, spot of type 1 and type 2, and intense arc formation. The excited state densities for Cu I, Cu II, and Cu III lines are determined during anode spot types 1 and 2 and anode plume at a position of 0.1 mm from the anode surface. According to this study, the excited state densities of the Cu I and Cu II lines are observed significantly higher (up to a factor of 10) in the case of anode spot type 2 compared to anode spot type 1. Cu neutral gas densities of Cu I are found to increase by a factor of 6 during the transition from anode spot type 2 compared to anode spot type 1. The results indicate an increase of Cu vapor density in front of the anode during the transition between these two anode modes. Also, chromium densities observed during anode spot type 1 and type 2 indicated two times higher metal vapor during anode spot type 2 compared with anode spot type 1. The density difference almost remains up to the extinction of anode spot modes.

14.4.3 Anode Erosion Rate

The life of silver-surface contact under repetitive arcing at low currents 2–25 A was studied by Wilson [72]. It was measured that the loss of weight and thickness of solid silver contacts at 20 A were varied from 6×10^{-4} to 6.6×10^{-3} g depending on the number of arcing.

The high-current arc erosion data reported by Wilson [73] for the following elements: carbon, tungsten, molybdenum, nickel, iron, titanium, copper, silver, zinc, aluminum, and tin. Also four typical sintered alloy materials as copper–tungsten, silver–molybdenum, silver–tungsten, and silver–tungsten carbide were considered. The maximal current in the metallic arcs varied from about 20 to 25 kA. For carbon, the current was low reaching maximum of 12 kA. The erosion rate was measured in cm^3/kAs and in grams as a function of number of arc pulses. All arc erosion data reported on the graphs of this paper were taken with ½-inch diameter contacts with a 2-inch-radius crown. The data for the elements and the sintered alloys were obtained with a 1/32-inch gap. Fig. 14.30 presents typical measurements of the electrode erosion as dependence on the number of arcs that had been applied in succession.

The erosion rate was also measured as function of spacing for silver–molybdenum sintered alloy material. It was shown that the erosion was reduced from 0.3 cm^3/kAs at minimal gap to about 0.18 cm^3/kAs at distance of 0.05 inch, and then, this value remains constant up to gap of 0.5 inch. The anode of the carbon arc lost much more material (about 80%) than the cathode (about 20%). However, with silver, the polarity had no appreciable effect. Whichever contact was on top experienced about 60% of the total loss. The explanation seems to be simply that melted material falls from the top contact and lands on the bottom one. With copper, the anode losses average about 65% of the total.

A comparison of the results at low arc current (10–100 A) with that at high-current arcs was provided using an estimation of the electrode temperature. Wilson [73] indicated that when the arc current starts around 1 μs, the electrode surfaces heated to sufficiently high temperature at which the metal significantly evaporated.

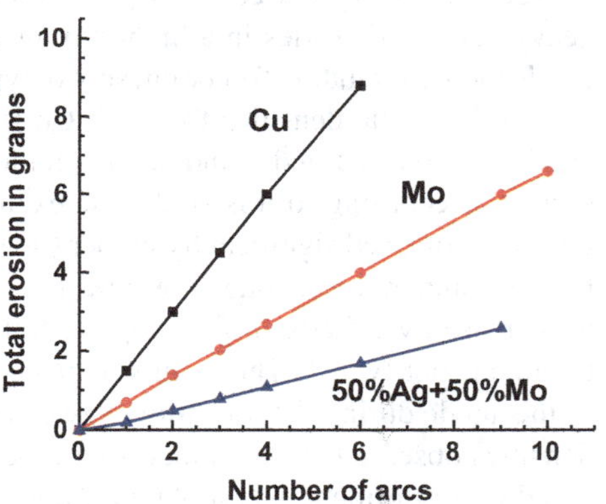

Fig. 14.30 Arc erosion loss as a function of the number of half-cycle arcs [73]

Therefore, the metal vapor emitted from the electrode surfaces by the end of a high-current half-cycle arc is surprisingly large. A vaporization rate equation was based on the assumption that all of the electrical energy released at the anode and cathode goes into vaporizing contact material and a predicted order yields the same as that determined in the experiment. According to the experiment with copper contacts carrying the 12 kA arc current, the weight of metal vaporized per operation is about 1 gram (see Fig. 14.30). The amount of metal vapor is relatively large. It appears certain that the original gas between the contacts must be swept out very early in the half cycle by this metallic vapor.

Thus, the arc, under the conditions of test, must flow through a vapor consisting almost solely of the vapor from the contacts. This is explanation of the fact that contacts tested at high current showed the same erosion rates in air and in oil. On base of this estimation, it was concluded that the result reached from experience at low current could not be applied at high current without specific verification. As example, the relative life of silver contacts in air 100 A will have many times larger than the arcing life of copper contacts, while at 12 kA, the test data show erosion of silver and copper to be practically identical. For copper and similar elements (except W and C), the total erosion loss is quite closely represented by the sum of the estimated loss by vaporization, and the loss of liquid phase in the form of drops blown out as this vapor escapes. Furthermore, the liquid loss constitutes a fairly high portion of the total loss.

Mitchel [33] indicated the erosion rate increasing from 10 μg/C to about 0.01 g/C when the arc current increases from 10 to about 100 kA. For low-current vacuum arc, Kimblin [74] reported that the anode erosion rate increased from 25 μg/C at 500 A to 120 μg/C at current of 3 kA for 15-mm-anode diameter. The comparison of erosion rate with the measured ion current showed that the anode vapor is highly ionized.

There should be mentioned the work of Hermoch [15] where the rate of anode evaporation was determined from the power required for evaporation of the anode. This power was determined from the anode energy balance using experimental values of the anode current density, electron temperature for an arc with current of 1 kA at a time 100 μs on anodes of different elements. The power loss by the anode heat conduction was determined by giving the boiling temperature of the anode in the spot and the specific thermal coefficients. In essence, the used approach is arbitrary because it is very difficult to obtain the measured values with the requested accuracy and also due to arbitrary giving of the boiling temperature. Therefore, Table 1 of [15] presented also arbitrary results. In addition, no any experimental data of anode erosion rate were indicated.

14.5 Summary

Considering the above experimental results, it can be emphases that the anode phenomena could be able to support the arc current. This role of the anode was controlled by different forms of anode region occurred with diffuse form or by appearing of the spot depending on condition of the cathode plasma jet propagation. The experimental investigations showed that for arcs with gas-filled gaps, the anode spot mode produced at relatively low arc current <100 A. It is because the cathode jet meets an obstacle (gas) at the anode region and the cathode jet disappeared before reaching the anode surface. The cathode jet relaxed, as a result, heated, and ionized the gas and anode vapor increasing local density of the charge particles. The plasma heated the anode, possible locally, and contracted in the anode region. Finally an anode spot can be arise.

In atmospheric arcs with current 1–2 kA, the arc voltage drop increases with current and can reach relatively large values (up to 160 V) for electrode gap of 12 mm. The pressure of the arc mode transition was lower for diatomic than that for monoatomic gases. The anode spot dynamics significantly changed when the gas pressure decreased from 400 to 100 torr. For lower pressure, a few spots were observed at beginning of the arc. The spots placed separately in a spot group that can weakly move due to dead of old and appearing of new spots. The group mobility depends on the atomic number of the filled gas. Few spots were observed even at 40 A in atmospheric arcs. In low-current vacuum arc, the cathode spot jet freely reached the anode.

The necessary arc current is provided by electron conductivity of the plasma jet and by the electron flux controlled in the anode sheath. The anode mode remained diffuse increasing the arc voltage. An appearing of the spot mode is determined by the gap distance and by anode size with respect to the cathode size. As example, at anode of 1.3 cm diameter, cathode of 5 cm diameter and electrode separation of 2.5-cm anode spot were produced at 0.4 kA (threshold current) and voltage of about 75 V. However, for the 5-cm-diameter anode, spot was observed at the 2.5-cm gap for an arc current of 2.1 kA. In vacuum, the arc voltage decreased to about 20 V with the spot mode.

In high-current vacuum arc, a large amount of ionized cathode material filled the electrode gap. The particle density in the gap can be so large that the further cathode jet expanding meets the obstacle similar to that produced in the arc with surrounding gas pressure. In this case, experimentally, it was detected that the arc operation passes different modes.

Three different high-current arc modes were distinguished. The above analysis of the experimental works indicates a transition process from the diffuse low-current mode to the high-current mode characterized by anode spot. At short electrode gap, an active mode was established. The transition of one anode mode to another strongly depends on arc current, surrounding gas pressure, gap length, geometry, and the material of electrodes. The transition to the spot mode in high-current vacuum arc characterizes by a ***threshold current***. Small electrode diameters cause anode spot formation at lower threshold current values. The higher values of threshold current was found for larger electrode diameters and for higher values of rate of current rise $(\mathrm{d}I/\mathrm{d}t)_0$ Forced arc motion does not change arc characteristics but avoids gross anode melting. Anode spot formation does not necessarily cause high-frequency arc voltage oscillations.

The threshold current in kA decreases with melting temperature reduce of the metals as follows: 13.8(**W**), 13.6(**Mo**), 10.3(**Cu**), 9.7(**Ag**), 6.8(**Al**), 2.3 kA(**Sn**). This kind of mode transition during change from anode spot type 1 to type 2 was interpreted from an energetic point of view. The increased evaporation of electrode material due to increase of the energy and local heating during the anode spot mode leads to a cloud of high-density vapor in front of the anode. Such spot differs from that produced at cold surface location. Table 14.8 summarizes the main characteristics of the anode modes in qualitative manner for different conditions of the arc and electrode geometry.

Current density in the anode spot varies significantly with one or two orders of magnitude. It is dependent on gas presence, its pressure, electrode geometry and material, arc current, and arc pulse duration. In the most studied case of copper electrodes, the anode current density varied in range of 10^3–10^6 A/cm^2. The results of measurements of the ***anode spot temperature*** are presented in Table 14.9. The data show some interval of the measured values. It can be understood taking in account different conditions of arc operation which determined by the arc current, arc pulse, and electrode configuration. However, the present data are informed at least some presentation about the anode temperature in contrast to that information in the cathode spot in which the data practically absent.

Table 14.8 Anode discharge modes according [28]

Anode mode	Characteristics	Conditions
Diffuse-1	Relatively low current (few kA). The anode is non-luminous. Act as collector of cathode plasma	At small and large gap sizes. Arc voltage increased with arc current. Anode erosion is zero or anode gain of cathode material is possible
Diffuse-2	Relatively low current. The anode is non-luminous. For anode sputterable materials. The sputtering particle density is lower than cathode plasma at anode	
Footpoint	Intermediate currents. Bright diffuse glow with small luminous spots associated with anode melting	Appearance of anode material. Increase of arc voltage and voltage noise
Spot	High current. The anode is active as a source of copious highly ionized material from one or several spots. Arc column appears in the interelectrode gap, and an anode plasma jet can be formed. Preferable appears at smaller anode radius	The anode surface in the spot area is at about boiling temperature. The arc voltage is frequently low, but may be also high with noise with arc current. Severe erosion of the anode is present
Intense	High current with very active anode with very bright luminous to cover the anode and fill the electrode gap. Large spots occur at both electrodes, which are brighter than those in the anode spot mode	It is appeared at shorter gap than in the spot mode Arc voltage is always low and quiet. Severe erosion from both electrodes with liquid droplets (from anode the droplets are about of mm)

Table 14.9 Anode temperature in spot and other modes

Author, References	Method	Type of spot	Current form	Temperature	Material
Mitchell [33]	Erosion rate	AS	50 Hz $I_{peak} = 17.5$; 24.6; 28 kA	2230–3350 K	Cu
	Optical	AS	50 Hz $I_{peak} = 35$; 38 kA	2730–2800 K	
Gunderson [5, 75]	High speed	AS	Dc	2850 K	
	Photograph	FP	Dc	1370 K	
Grissom and Newton [58, 59]	Detector	AS	Pulse, 1–2 μs 20–180 A	2000–2450 °C	Al
				2750–3450 °C	Cu
				2900–3500 °C	Al_2O_3
				2000–2300 °C	BeO
Clapas and Holmes [5, 76]	Detector	FP	Dc	1300–1570	Cu
Boxman [60]	Si photovoltaic detectors	AS	70 Hz 2.7 kA	2790–2960 K	Ni
Boxman [41]	Si photovoltaic detectors	At transition to AS	543 Hz $I_{peak} =$ 4–12 kA	1550–2250 K	Ni
Agarwal and Katre [6, 77]	Erosion rate	AS	Dc	3100–3150 K	Al, Cu
Dullni et al. [61]	Pyrometer technique;	AS	45 Hz $I_{peak} = 7$ kA	2500 K during 3 ms to 1500 K	Cu75Cr25
	By thermionic currents	AS	45 Hz $I_{peak} = 7$ kA	2000 K during 3 ms to 1800 K	
Rosenthal et al. [78]	Thermocouple	Diffuse	DC arc 150 s 175–340 A	1900–2300 K	Graphite
Beilis et al. [79]	Thermocouple	Diffuse	DC, arc 150 s 130–175 A	2300–2550 K	W
Schneider et al. [62]	Optical pyrometer. At CZ	AS	10 ms arc $I_{peak} =$ 8–15 kA	1500–1800 K without B	CuCr
				1200–1550 K B = 7mT/kA	

The electron density for copper butt electrodes was observed with the axial densities ranging (diffuse mode) from 1×10^{14} at 0.4 kA to 2×10^{15} cm^{-3} at 4 kA. At the transition from the diffuse mode to the high-current constricted mode, which occurred at about 3 kA, a sudden decrease in electron density from 1.7×10^{15} cm^{-3} to 0.7×10^{15} cm^{-3} was observed in the center of the discharge. Another data for copper butt electrodes showed that the transition to the anode spot mode occurred at threshold current of 2.6 kA when the arc peak was 2.8 kA. Before transition, the electron density profiles are peaked on axis, with maxima of 9×10^{14} and 6×10^{14} cm^3 in the cathode and anode regions, respectively. At about 0.1 ms after transition, the electron density profile in the cathode region remains unchanged, while the central electron density near the anode drops by a factor of about 2. The density decreased to about (2–4) $\times 10^{14}$ cm^3. Thus, the electron density was obtained in range of 1×10^{14}–2×10^{14} cm^3.

The ***erosion rate*** for high-current vacuum arc increased from 10 μg/C to about 0.01 g/C when the arc current increases from 10 A to about 100 kA, respectively. For low-current vacuum arc, the anode erosion rate increased from 25 μg/C at 500 A to 120 μg/C at current of 3 kA for 15-mm-anode diameter

References

1. Cobine, J. D. (1980). Vacuum arc anode phenomena. In J. M. Lafferty (Ed.), *Vacuum arcs*. New York: Wiley.
2. Craig Miller, H (1977). Vacuum arc anode phenomena. *IEEE Transactions on Plasma Science*, *5*(3), 181–196.
3. Craig Miller, H. (1983). Vacuum arc anode phenomena. *IEEE Transactions on Plasma Science*, *11*(2), 76–89.
4. Craig Miller, H. (1983) Discharge modes at the anode of a vacuum arc.*IEEE Transactions on Plasma Science*, *11*(3), 181–125.
5. Craig Miller, H. (1989). A review of anode phenomena in vacuum arcs. *Contributions to Plasma Physics 29*(3), 223–249.
6. Craig Miller, H. (1990). A review of anode phenomena in vacuum arcs. In Proceedings of the IEEE *International Symposium* on *Electrical Insulation in Vacuum* (pp. 140–149). USA: Santa Fe.
7. Boxman, R. L., Goldsmith, S., & Greenwood, A. (1997). Twenty-five years of progress in vacuum arc research and utilization. *IEEE Transactions on Plasma Science*, *25*(6), 1174–1186.
8. Heberlein, J., Mentel, J., & Pfender, E. (2009). The anode region of electric arcs: a survey. *Journal of Physics D: Applied Physics*, *43*(2), 023001.
9. Shkol'nik, S. M. (2011). Anode phenomena in arc discharges: a review. *Plasma Sources Science and Technology*, *20*, 013001.
10. Somerville, J. M., Blevin, W. R., & Fletcher, N. H. (1952). Electrode phenomena transient arcs. *Proceedings of the Physical Society. Section B*, *B65*(12), 963–970.
11. Blevin, W. R. (1953). Further studies of electrode phenomena in transient arcs. *Australian Journal of Physics*, *6*(2), 203–208.
12. Hermoch, V. (1954). The influence of a non-conducting layer on the surface of the electrode on the condensed arc. *Cechoslovackij fiziceskij zurnal*, *4*(2), 161–166.
13. Hermoch, V. (1959). A contribution to the study of electrode spaces of high current short duration electrical discharge. *Cechoslovackij fiziceskij zurnal*, *9*(1), 84–90.
14. Hermoch, V. (1963). On the formation of the anode space of short time high intensity electrical discharge. *Cechoslovackij fiziceskij zurnal*, *B13*(5), 321–326.
15. Hermoch, V. (1963). The anode space of short time high intensity electrical discharge. *Cechoslovackij fiziceskij zurnal*, *B13*(5), 327–334.
16. Hermoch, V. (1959). Channel of short time high intensity electrical discharge. *Cechoslovackij fiziceskij zurnal*, *9*(3), 377–387.
17. Beilis, I. I., & Zykova, N. M. (1968). Dynamics of the anode spot. *Soviet Physics Technical Physics-USSR*, *13*(2), 235.

18. Shih, K. T. (1972). Anode current density in high current pulsed arcs. *Journal of Applied Physics, 43*(12), 5002–5005.
19. Pfender, E. (1980). Energy transport in thermal plasmas. *Pure and Applied Chemistry, 52*(7), 1773–1800.
20. Zykova, N. M. (1968). *Investigations of the cathode and anode spots development of an electrical arc*. Krasnoyarsk: Thesis. Institute of Physics of Siberian Academy of Science SSSR.
21. Zykova, N. M., Kantsel, V. V., Rakhovsky, V. I., Seliverstova, I. F., & Ustimets, A. P. (1971). The dynamics of the development of cathode and anode regions of electric arcs I. *Soviet Physics-Technical Physics, 15*(11), 1844–1849.
22. Dyuzhev, G. A., Shkol'nik, S. M., & Yur'ev, V. G. (1978a). Anode phenomena in the high current density arc. *Soviet Physics-Technical Physics, 23,* 667–671.
23. Dyuzhev, G. A., Shkol'nik, S. M., & Yur'ev, V. G. (1978b). Anode phenomena in the high current density arc. *Soviet Physics-Technical Physics, 23,* 667–671.
24. Vainberg, L. I., Lyubimov, G. A., & Smolin, G. G. (1978). About change of a discharge form and anode destruction of an end accelerator at limiting regimes. *Soviet Physics-Technical Physics, 23*(4), 746–753. in Russian.
25. Dyuzhev, G. A., Lyubimov, G. A., & Shkol'nik. S. M. (1983). Conditions of the anode spot formation in vacuum arc. *IEEE Transactions on Plasma Science, 11*(1), 36–45.
26. Miyano, R., Saito, T., Kimura, K., Ikeda, M., Takikawa, H., & Sakakibara, T. (2001). Anode mode in cathodic arc deposition apparatus with various cathodes and ambient gases. *Thin Sold Films., 390,* 192–196.
27. Craig Miller, H. (1995) Anode phenomena. In R. L. Boxman, P. J. Martin, & D. M. Sanders (Eds.), *Handbook of vacuum arc science and technology* (pp. 308–364). Park Ridge, New Jersey: Noyes Publications.
28. Craig Miller, H. (1997). Anode modes in vacuum arcs. *IEEE Transactions on Plasma Science, 4*(4), 382–388.
29. Schellekens, H. (1985). The diffuse vacuum arc: a definition. *IEEE Transactions on Plasma Science, 13*(5), 291–295.
30. Drouet, M. G., Poissaed, P., & Meunier, J.-L. (1986). *Current distribution at the anode and current flow in the interelectrode region of the vacuum arc* (pp. 120–124). Proceedings of XII International Symposium on Discharge and Electrical Insulation in Vacuum. Tel Aviv-Yafo: Israel.
31. Drouet, M. G., Poissaed, P., Meunier, J.-L. (1987). Measurements of the current distribution at the anode in a low current vacuum arc. *IEEE Transactions on Plasma Science, 15*(5), 506–509.
32. Rosenthall, H., Beilis, I. I., Goldsmith, S., & Boxman, R. L. (1996). Spectroscopic investigation of the development of hot anode vacuum arc. *Journal of Physics. D. Applied Physics, 29*(5), 1245–1259.
33. Mitchell, G. R. (1970). High-current vacuum arcs. Part 1: An experimental study. In *Proceedings of the Institution of Electrical Engineers* (Vol. 117, No. 12, pp. 2315–2326). IET Digital Library.
34. Schellekens, H. (1981). A current constriction phenomenon in high current vacuum arcs. *Physica, 104C,* 130–136.
35. Kimblin, C. W. (1969). Anode voltage drop and anode spot formation in dc vacuum arcs. *Journal of Applied Physics, 40,* 1744–1752.
36. Kimblin, C. W. (1974). Vacuum arc ion currents and electrode phenomena. *IEEE Transactions on Plasma Science, 59*(4), 546–555.
37. Kong, G., Liu, Z., Wang, D., & Rong, M. (2011). High-current vacuum arc: The relationship between anode phenomena and the average opening velocity of vacuum interrupters. *IEEE Transactions on Plasma Science, 39*(6), 1370–1378.
38. Yu, L., Wang, J., Geng, Y., Kong, G., & Liu, Z. (2011). High-current vacuum arc phenomena of nanocrystalline CuCr25 contact material. In *Proceedings of* XXIVth *International Symposium on Discharges, Electrical Insulation in Vacuum,* (pp. 229–232) Braunschweig, Germany.
39. Reece, M. P. (1963). The vacuum switch, Part 1. Properties of the vacuum arc. *Proceedings of the Institution of Electrical Engineers, 110*(4), 793–811.

40. Rich, J. A., Prescott, L. E., & Cobine, J. D. (1971). Anode phenomena in metal-vapor arcs at high currents. *Journal of Applied Physics, 42*(2), 587–601.
41. Boxman, R. L. (1978). Time dependence of anode spot formation threshold current in vacuum arcs. *IEEE Transactions on Plasma Science, 6*(2), 233–237.
42. Heberlein, J. V. R., & Gorman, J. G. (1980). The high current metal vapor arc column between separating electrodes. *IEEE Transactions on Plasma Science, 8*(4), 283–288.
43. Kaltenecker, A. (1981). Anode spot formation in vacuum arcs. *IEEE Transactions on Plasma Science, 9*(4), 290–291.
44. Kaltenecker, A., & Schussek, M. (1985). Anode-spot formation and motion of vacuum arcs. *IEEE transactions on plasma science, 13*(5), 269–276.
45. Toya, H., Uchida, Y., Hayashi, T., & Murai, Y. (1986). Anode discharge mode and cathodic plasma state in high-current vacuum arcs. *Journal of applied physics, 60*(12), 4127–4132.
46. Janiszewski, J., & Załucki, Z. (1996). Photographic appearance of high-current vacuum arcs prior to and during anode spot formation. *Czechoslovak Journal of Physics, 46*(10), 961–971.
47. Załucki, Z. (1996). Voltage/current characteristics of high-current vacuum arc at a small gap length. *Czechoslovak Journal of Physics, 46*(10), 981–994.
48. Batrakov, A. V., Popov, S. A., Schneider, A. V., Sandolache, G., & Rowe, S. W. (2011). Observation of the plasma plume at the anode of high-current vacuum arc. *IEEE Transactions on Plasma Science, 39*(6), 1291–1295.
49. Khakpour, A., Gortschakow, S., Uhrlandt, D., Methling, R., Franke, S., Popov, S., … & Weltmann, K. D. (2016). Video spectroscopy of vacuum arcs during transition between different high-current anode modes. *IEEE Transactions on Plasma Science, 44*(10), 2462–2469.
50. Khakpour, A., Uhrlandt, D., Methling, R., Franke, S., Gortschakow, S., Popov, S., … & Weltmann, K. D. (2016). Impact of different vacuum interrupter properties on high-current anode phenomena. *IEEE Transactions on Plasma Science, 44*(12), 3337–3345.
51. Khakpour, A., Franke, S., Methling, R., Uhrlandt, D., Gortschakow, S., Popov, S., … & Weltmann, K. D. (2017). Optical and electrical investigation of transition from anode spot type 1 to anode spot type 2. *IEEE Transactions on Plasma Science, 45*(8), 2126–2134.
52. Batrakov, A., Schneider, A., Rowe, S., Sandolache, G., Markov, A., & Zjulkova, L. (2010, August). Observation of an anode spot shell at the high-current vacuum arc. In *Proceedings of* XXIVth *International Symposium on Discharges, Electrical Insulation in Vacuum* (pp. 351–354). Braunschweig, Germany.
53. Khakpour, A., Gortschakow, S., Popov, S., Methling, R., Uhrlandt, D., Batrakov, A., & Weltmann, K. D. (2016, September). Time and space resolved video spectroscopy of the vacuum arc during the formation of high-current anode modes. In *2016 27th International Symposium on Discharges and Electrical Insulation in Vacuum (ISDEIV)* (Vol. 1, pp. 1-4). IEEE.
54. Khakpour, A. (2016, September). Impact of AC and pulsed DC interrupting currents on the formation of high-current anode modes in vacuum. In *2016 27th International Symposium on Discharges and Electrical Insulation in Vacuum (ISDEIV)* (Vol. 1, pp. 1-4). IEEE.
55. Schneider, A. V., Popov, S. A., Lavrinovich, V. A., & Yushkov, A. Y. (2018, September). High Speed Registration of the Anode Spot Evolution of High Current Vacuum Arc Combined With Spectrally Selective Images. In *2018 28th International Symposium on Discharges and Electrical Insulation in Vacuum (ISDEIV)* (Vol. 1, pp. 213-216). IEEE.
56. Batrakov, A. V. (2018, September). Vacuum-Arc Anode Phenomena: New Findings and New Applications. In *2018 28th International Symposium on Discharges and Electrical Insulation in Vacuum (ISDEIV)* (Vol. 1, pp. 163-168). IEEE.
57. Miller H.C. (2017) Anode Modes in Vacuum Arcs: Update, IEEE Transactions on Plasma Science, 45(8), 2366-2374
58. Grissom J. T., & Newton, J. C. (1972). Anode temperatures and ion energy distributions during a vacuum arc discharges. In *Proceedings of V International Symposia on Discharges and Electrical Insulation in Vacuum* (p. 236). Poznan.
59. Grissom, J. T., & Newton, J. C. (1974). Anode surface radiance from microsecond vacuum arcs. *JJournal of Applied Physics, 45*(7), 2885–2894.

60. Boxman, R. L. (1975). Measurement of anode surface temperature during high current vacuum arcs. *Journal of Applied Physics, 46*(11), 4701–4704.
61. Dullni, E., Gellert, B., & Schade, E. (1989). Electrical and pyrometric measurements of the decay of the anode temperature after interruption of high-current vacuum arcs and comparison with computations. *IEEE Transactions on Plasma Science, 17*(5), 644–648.
62. Schneider, A. V., Popov, S. A., Batrakov, A. V., Sandolache, G., & Schellekens, H. (2013). Anode temperature and plasma sheath dynamics of high current vacuum arc after current zero. *IEEE Transactions on Plasma Science, 41*(8), 2022–2028.
63. Kudryavtsev, V. S. (1969). A spectroscopic and photoelectric investigation of pulsed discharges in vacuum. *High Temperature, 7*(3), 380–383.
64. Boxman, R. L. (1974). Interferometric measurement of electron and vapor densities in high current vacuum arcs. *Journal of Applied Physics, 45*(11), 4835–4846.
65. Bacon, F. M. (1975). Vacuum arc anode plasma-I-Spectroscopic. *Journal of Applied Physics, 46*(11), 4750–4757.
66. Bacon, F. M., & Watts, H. A. (1975). Vacuum arc anode plasma. II. Collisional-radiative model and comparison with experiment. *Journal of Applied Physics, 46*(11), 4758–4766.
67. Harris, J. H. (1979). Electron density measurements in vacuum arcs at anode spot formation threshold. *Journal of Applied Physics, 50*(2), 753–757.
68. Seliverstova, I. F., & Zygankov, N. F. (1975). Heat conditions of the electrodes and structure of the anode region. *Izvestiya Academii Nauk Sibirian Branch of SSSR, Ser. Tech. Nauk, 3*(1), 47–53. (in Russian).
69. Lyubimov, G. A., Rakhovsky, V. I., Seliverstova, I. F., & Zektser, M. P. (1980). A study of parameters of anode spot on aluminium. *Journal of Physics. D. Applied Physics, 13*(9), 1655–1664.
70. Batrakov, A. V. (2018, September). Vacuum-Arc Anode Phenomena: New Findings and New Applications. In *2018 28th International Symposium on Discharges and Electrical Insulation in Vacuum (ISDEIV)* (Vol. 1, pp. 163–168). IEEE.
71. Khakpour, A., Gortschakow, S., Franke, S., Methling, R., Popov, S., & Uhrlandt, D. (2019). Determination of Cr density during high-current anode modes in vacuum arc. *Journal of Applied Physics, 125*(13), 133301.
72. Wilson, W. R. (1953). Life of Silver-Surfaced Contacts on Repetitive Arcing Duty [includes discussion]. *Transactions of the American Institute of Electrical Engineers. Part III: Power Apparatus and Systems, 72*(6), 1236–1243.
73. Wilson, W. R. (1955). High-Current Arc Erosion of Electric Contact Materials [includes discussion]. *Transactions of the American Institute of Electrical Engineers. Part III: Power Apparatus and Systems, 74*(3), 657–664.
74. Kimblin, C. W. (1974). Cathode spot erosion and ionization phenomena in the transition from vacuum to atmospheric pressure arc. *Journal of Applied Physics, 45*(12), 5235–5244.
75. Gunderson, G. (1971). *Anode spot formation in vacuum arcs*. Trondheim: Thesis Norwegian University of Science and Technology.
76. Clapas, D., & Holmes, R. (1973). Anode spot temperatures in vacuum arcs. In *Proceedings of XI International. Conference on Phen in Gases*. Prague.
77. Agarwal, M. S., & Katre, M. M. (1982). Study of high-current vacuum arcs in axial magnetic fields. In *2014 International Symposium on Discharges and Electrical Insulation in Vacuum (ISDEIV)* (pp. 157–160). IEEE.
78. Rosenthall, H., Beilis, I. I., Goldsmith, S., & Boxman, R. L. (1995). Heat fluxes during the development of hot anode vacuum arc. *Journal of Physics. D. Applied Physics, 28*(1), 353–363.
79. Beilis, I. I., Koulik, Y., & Boxman, R. L. (2011). Temperature distribution dependence on refractory anode thickness in a vacuum arc: Theory. *IEEE Transactions on Plasma Science, 39*(6), 1307–1310.